The Mechanical and Physical Properties of the British Standard En Steels

(B.S. 970-1955)

Volume 3

En 40 to En 363

COMPILED BY

J. WOOLMAN, M. Sc. AND R. A. MOTTRAM, A. I. M.

STEEL USER SERVICE
BRITISH IRON AND STEEL RESEARCH ASSOCIATION

PERGAMON PRESS
OXFORD · LONDON · EDINBURGH · NEW YORK
TORONTO · SYDNEY · PARIS · BRAUNSCHWEIG

Pergamon Press Ltd., Headington Hill Hall, Oxford
4 & 5 Fitzroy Square, London W. 1
Pergamon Press (Scotland) Ltd., 2 & 3 Teviot Place, Edinburgh 1
Pergamon Press Inc., 44–01 21st Street, Long Island City, New York 11101
Pergamon of Canada Ltd., 207 Queen's Quay West, Toronto 1
Pergamon Press (Aust.) Pty. Ltd., 19a Boundary Street, Rushcutters Bay, N.S.W. 2011, Australia
Pergamon Press S.A.R.L., 24 rue des Écoles, Paris 5e
Vieweg & Sohn GmbH, Burgplatz 1, Braunschweig

First edition 1968

Library of Congress Catalog Card No. 63-21774

PRINTED IN HUNGARY

08 011167 X

CONTENTS

FOREWORD

The main object of these three volumes is to have available in one source of reference data on the most commonly used range of steels in the United Kingdom — B.S.970 En Steels. Some of the information has been published previously, some properties have been determined but not published, whilst the remaining data had not, until this work started, been investigated.

These volumes have been compiled by the Steel User Service of the British Iron and Steel Research Association which has financed the project jointly with the Department of Scientific and Industrial Research, to satisfy the many enquiries received from industry on the properties of steels and equivalent foreign specifications which could only be answered by access to the wide variety of sources in which this information was previously contained.

Today's exacting engineering designs, however, create a new demand for more detailed technical data and information in addition to mechanical properties which, in the main, are not readily available.

At all stages industry has been consulted and an advisory panel was formed to guide the work. All steelmakers consulted during the preparation have welcomed the opportunity of providing such a wealth of information, and in doing so I believe have made a very valuable contribution towards the more efficient use of steel.

E. W. SENIOR, C.M.G. J.P.
Director of the
British Iron and Steel Federation

INTRODUCTION

It is very necessary that engineers, designers, and all users of steel should have access to adequate information relating to the mechanical and physical properties of various steels, in order that they may ascertain the most suitable steel to use for a particular purpose. Hitherto the required data have not always been readily available, and in many cases extensive searching, and even special investigations, have been necessary in order to obtain them. A reference book which would provide this information in relation to all available steels is, therefore, very desirable. A start on this task has been made by the authors, in compiling data on the steels included in the En Series of British Standard 970. Volume 1 includes the most important properties of En 1 to En 20 inclusive, volume 2 is concerned with En 21 to En 39, and the present work, volume 3, with En 40 to En 363.

In carrying out their task the authors have collected information from a wide variety of sources, including published literature, works brochures, and test records, both published and unpublished, from many research laboratories. In addition, several laboratories have carried out special tests to obtain properties not previously determined, and various government-sponsored organisations have granted permission to extract data from hitherto unpublished reports. A list of references to sources is given at the end of each section.

Special attention has been given to showing the nearest foreign specifications for each steel, and the most recently recognised International Standard Symbols have been used for each property in addition to the British designation.

It is recommended that all users of these volumes should read the "Notes on the Use of Tables" which follow the Introduction. These notes not only indicate the care needed when making use of the data provided, but also give some useful hints for estimation of certain properties when data are lacking.

It must be emphasised that while the source of all data is indicated, the properties shown should not be used to compare steels from different suppliers. The information has been taken from a vast amount of data collected by the respective steel suppliers and the values quoted might be influenced not only by the manufacturing process and the deoxidation technique used but also by such factors as whether or not the material is treated in bulk, the degree of agitation in the quenching bath and the amount of oxide on the surface. Similarly, individual values should not be used for specification purposes since they do not necessarily indicate what can readily be obtained in all cases from materials which come within a particular composition specification.

While every care has been taken to ensure that the information is correct, the Steel User Service of the Association would welcome information relating to any errors or omissions.

ACKNOWLEDGMENTS

The task of compiling the information contained in this work has been aided by a grant from the former Department of Scientific and Industrial Research, under the Special Assistance to Industry Scheme.

The Association and the authors are indebted to the members of the Advisory Group and the Sheffield Steelworks Group for their help, encouragement and criticisms throughout the preparation of this publication. Members of one or both of these two Groups were:-

H. Allsop, Brown Bayley Steels Ltd.; W. H. Bailey, Jessop-Saville Ltd.; P. Bennison, Rolls-Royce Ltd.; J. Cameron, Colvilles Ltd.; P. E. Clary, Ford Motor Co. Ltd.; A.J. Fenner, National Engineering Laboratory; P. G. Forrest, National Physical Laboratory; W. H. Goodrich, Edgar Allen & Co. Ltd.; F. Henshaw, Kayser Ellison & Co. Ltd.; R. F. Johnson, The United Steel Cos. Ltd.; P. Jubb, The Brown Firth Research Laboratories; R. Lamb, The Alloy Steel Association Technical Committee and Hadfields Ltd.; R. L. Long, Park Gate Iron and Steel Co. Ltd.; R. A. McKinstry, British Standards Institution; J. R. Russell, English Steel Corporation (Chairman of Sheffield Steelworks Group); G. Weston, British Standards Institution (Chairman of the Advisory Group).

The authors gratefully acknowledge the help and encouragement given by the Director and staff of the Association during the preparation of this compilation.

J. Woolman
R. A. Mottram

NOTES ON THE USE OF THE TABLES

For easy reference, the compilation is divided into sections, each devoted to one En number, and within the sections the various items of information are arranged in the following order:-

(1) Specification
  Chemical Composition
  Mechanical Properties

(2) Related Specifications

(3) Applications

(4) Welding

(5) Machinability

(6) Hot Working and Heat Treatment Temperatures

(7) Physical Properties
  Specific Gravity
  Specific Heat
  Co-efficient of Thermal Expansion
  Electrical Resistivity
  Thermal Conductivity
  Young's Modulus, Shear Modulus and Poisson's Ratio
  Magnetic Properties

(8) Isothermal and Continuous Cooling Diagrams

(9) Hardenability

(10) Mechanical Properties at Room Temperatures

(11) Mechanical Properties at Low Temperatures

(12) Mechanical Properties at High Temperatures (including Creep Properties)

(13) Torsional Properties

(14) Fatigue Properties

Some of the En specifications are sub-divided into steels of slightly different composition which, in many cases, overlap and, on account of this and other reasons, it has been found impracticable to separate them. Where plentiful data are available, as is the case for the mechanical properties at room temperature, the steels have accordingly been arranged in order of increasing carbon content.

Apart from the tables of data, curves have been reproduced where information warrants in order to show graphically the effects of tempering temperature and of ruling section as heat treated and also to indicate the range of properties which might be expected from steels conforming to a particular En number.

The following notes may be of interest to users of the data.

Related Specifications

A list of the specification numbers for the nearest equivalent British, American and certain European Standard specifications is included in each section. This list should not only simplify the work required in dealing with enquiries concerning such specifications but it will no doubt be of value to other than British users who are more familiar with the standard steels of their own particular country.

Applications

The general characteristics and main uses of each steel are listed. Typical applications have been obtained from various works catalogues and from information supplied by a number of steel users.

Weldability

Weldability depends on a number of factors such as cooling rate of the heat-affected zone, the type of electrode used in the case of metallic arc welding as well as the composition of the steel. The cooling rate depends on the amount of metal which can conduct heat away from the weld junction (called the Thermal Severity Number) and on the degree of preheat. The effect of composition may be judged by use of the carbon equivalent (CE) formula developed by the British Welding Research Association for metal arc welding.

$$\text{C.E.} = \text{C} + \frac{\text{Mn}}{20} + \frac{\text{Ni}}{15} + \frac{\text{Cr} + \text{Mo} + \text{V}}{10}$$

The higher the carbon equivalent and the higher the Thermal Severity Number (TSN) the greater is the necessity to preheat and the higher must be the pre-heating temperature. It is not possible to state the necessary pre-heating temperature in simple terms of TSN and CE since it also depends on the type of electrode used. We would, however, refer readers to the pamphlet "Arc-Welding Low-alloy Steels" published by the British Welding Research Association, 29, Park Crescent, London, W.1.

The welding properties tabulated for each steel have been taken from a paper published by H.M. Stationery Office to whom we are grateful for permission to extract the details given. (Ref. 1)

The meaning of the various symbols used in the tables for describing the welding characteristics is given in the list of abbreviations following the notes on the use of the tables.

Machinability

It is not possible to give any definite quantitative measurement to machinability although many efforts have been made to find a machinability constant. This is due to the fact that the response to cutting tools depends on so many factors, e.g.

The type of machining operation (turning, drilling, sawing, screwing, etc.)
The rigidity of the machine
The tool (shape and rigidity of mounting)
The cutting condition (speed, feed, depth of cut)
The material being cut

The machinability of steel is appreciably affected by its composition (especially the sulphur content or presence of other free cutting additions), by the presence of certain hard constituents (both metallic and non-metallic), by the grain size and by the microstructure. Generally speaking, the annealed condition is best for machinability although cases are known where other microstructures give better results for certain machining operations. In the softer steels improved machinability may be obtained by giving the steel a small degree of cold work. Machinability, therefore, can only be expressed in very approximate relative terms. An attempt has been made to give some such rough relative assessment of the various En steels which it is hoped may be of some help to the machinist and production engineer. Apart from this Wickman Wimet Ltd. have carried out comparative machinability tests using carbide tools on many of the En steels and these results have been published by the company.

Hot Working and Heat Treatment Temperatures

These are affected to some extent by the carbon content. A low carbon permits somewhat higher hot working annealing and hardening temperatures than a steel with carbon at the high end of the specification range. Generally speaking, however, there is a fair degree of tolerance in the ranges of temperature which may be used for these heating operations and those quoted will be found to be reasonably satisfactory for all carbon contents within the particular specification.

The temperatures quoted for hot working are, in general, conservative. In many cases higher temperatures may be used, but in order to ensure that the overheating temperature is not exceeded the upper re-heating temperature has been limited to that which may be considered safe for all steels within the specification. The use of such lower re-heating temperatures for hot working has the added advantages of reducing the amount of decarburization and the extent of the grain growth.

Carburizing Data

The tables giving the approximate times to produce a given case depth are intended as a guide only. They are based on information provided by Wild Barfield Limited, (Ref.12).

The times for "solid" carburizing are based on work carried out using "Eternite" pack carburizing compound. The times for "salt" are based on the use of I.C.I. "Rapideep" salt. The values given under "gas" are based on the use of the Wild Barfield patented P. T. G. process, the Wild Barfield "Carbo-drip" liquid method, or Endogas as a carrier gas with suitable additions of hydrocarbon gas (usually propane). These carburizing processes give identical results when carburizing at the maximum rate and given a diffusion period to arrive in every example at eutectoid composition at the surface.

Physical Properties

Specific Gravity (d)

The specific gravity of a steel is only slightly affected by composition and heat-treatment. Values have been quoted only for a temperature of 20°C, but if it is desired to know the specific gravity d_T at any other temperature T(°C) this may easily be calculated from the formula

$$d_T = d_{20}[1-3\alpha(T-20)]$$

where α is the mean linear coefficient of thermal expansion between 20°C and T°C.

The specific gravity at room temperature of pure iron is close to 7·880. The specific gravity of steels is affected by the state of the carbon. In the case of fully annealed steels the specific gravity is diminished by 0·026 for each 1% of carbon up to 1·6%. In a fully quenched steel (with no free carbide or austenite) the decrease is 0·122 for each 1% of carbon up to 1·2%. The following table shows the approximate effect of 1% by weight of various alloying elements.

Alloying Element		*Effect on S.G. of 1% of element*
Silicon	(up to 3%)	—0·061
Manganese	(" 2%)	—0·012
Sulphur		—0·167
Phosphorus		—0·037
Nickel	(" 6%)	+0·006
Chromium	(" 4%)	—0·009
Aluminium	(" 6%)	—0·142
Molybdenum	(" 1%)	+0·020
Copper	(" 4%)	+0·006
Cobalt	(" 7%)	+0·006
Tungsten	(" 20%)	+0·065

These figures may help to indicate by how much the actual specific gravity of a steel may differ from the quoted data. With the amount of elements normally present in the low alloy steels the actual value will not differ from the quoted value by more than 0·02.

The density of steels in pounds per cubic inch is obtained by multiplying the specific gravity by 0·0361 and in pounds per cubic foot by 62·35.

Specific Heat (c)

The specific heat of low alloy steels is not very sensitive to changes in composition and heat treatment. For example, of the 16 carbon and low alloy steels tested by the N.P.L. (Ref. 2) the values of the mean specific heat between 50° and 100°C varied only between 0·114 and 0·119. Even the 13% chromium steels had a corresponding value of 0·113. The specific heats of the austenitic steels are slightly higher, namely, between 0·120 and 0·124.

The values of the mean specific heats are naturally affected by heats of transformation of changes occurring within the temperature range considered. Thus one may expect appreciable differences between the steels in the critical regions.

The values for the mean specific heat have been adjusted to a temperature range commencing at 20°C where the published data give values for ranges beginning with a different temperature.

Coefficient of Thermal Expansion (k)

The values quoted are mean coefficients of linear expansion per °C from 20°C to the stated temperature. This physical property is again not very sensitive to composition in low alloy steels provided no transformation occurs within the temperature range in question. The coefficient of expansion in the hardened condition differs only very slightly from that in the annealed condition up to 100°C, but beyond this temperature transformations occur which alter the properties of the steel and cause contraction of the material with consequent reduction of the mean coefficient of expansion.

Electrical Resistivity (ϱ)

At room temperatures the electrical resistivity is very sensitive to composition and structure especially as regards the state of the carbon. At temperatures above about 800°C this sensitivity is much reduced. The resistivity of pure iron at 20°C is close to 9·8 microhm-cm, whilst in fully annealed pure iron-carbon alloys carbon appears to raise this value by 3·5 for each 1% of carbon. In commercially annealed steels, however, the effect of carbon appears to be somewhat greater than this, namely 4·2 for each 1% of carbon. Radcliffe and Rollason (Ref. 3) give the following formula for annealed pure iron-carbon alloys at room temperature:-

Log $\varrho = 0{\cdot}90\ C_m + 0{\cdot}996$ where C_m = percentage of Fe_3C by volume. Their results agree very closely, up to 1·3 per cent of carbon, with the formula

$\varrho = 9{\cdot}94 + 3{\cdot}73C$ (where C = per cent of carbon by weight). This corresponds to a value of 9·94 for pure iron, a value which they considered rather high due possibly to a retention of 0·005% of carbon in solid solution. It could, however, be due to the combined effect of the small amounts of impurities in the iron. Thus only 0·001% of silicon could account for a rise in the resistivity of 0·013.

In hardened steels, carbon increases the resistivity appreciably more than in annealed steels. The amount of increase is not linear, the following being the resistivities at room temperature of pure iron-carbon alloys in the fully quenched condition:-

Carbon %	0	0·2	0·4	0·6	0·8	1·0	1·2	1·4
Resistivity, microhm cm	9·8	12·7	16·2	20·5	26·5	37·0	47·0	47·5

The influence on the electrical resistivity of the alloying elements in low alloy steels is approximately as follows:

Element	Effect on Electrical Resistivity of 1% by weight (microhm cm)	
Silicon	(up to 1·8%)	+13·6
Silicon	(1·8 to 6%)	+ 8·5
Manganese	(up to 6%)	+ 5·5
Phosphorus	(up to 1·2%)	+11·0
Nickel	(up to 5%)	+ 2·5
Chromium*	(up to 5×carbon%)	+ 0·6
Molybdenum	(up to 3%)	+ 1·0
Titanium	(up to 5%)	+ 0·5
Aluminium	(up to 4%)	+11·0

* In hardened low alloy steel the effect of 1% of chromium is to increase the resistivity by about 5·5 microhm cm.

Thermal Conductivity (k)

Data on thermal conductivity are relatively scarce due, no doubt, to the difficulty of measuring this property with a high degree of accuracy.

The thermal conductivity of steels at room temperature is sensitive to composition and heat-treatment but it is much less sensitive at temperatures above 700°C.

When assessing the influence of composition on the thermal conductivity at room temperature it seems to be better to deal with the reciprocal of this property, namely with the thermal resisitivity (λ), which tends to give more of a straight line relationship.

The thermal conductivity of pure iron at 20°C appears from the best results to be close to 0·175 cal/cm s deg C giving a value of the thermal resistivity of 5·72.

Information concerning the influence of composition is very scare but we have examined what data are available and have derived the following formula:-

$$\lambda = \frac{1}{k} = 5{\cdot}80 + 1{\cdot}6C + 4{\cdot}1S + 1{\cdot}4Mn + 5{\cdot}0P + 1{\cdot}0Ni + 0{\cdot}6Cr + 0{\cdot}6Mo$$

where λ = Thermal Resistivity at 0°C and k = Thermal Conductivity at 0°C; C, Si, Mn etc. = per cent by weight of carbon, silicon, manganese etc. This formula has been checked against the 18 low alloy steels for which thermal conductivity data have been determined by the National Physical Laboratory (Ref. 2). The values calculated for the low alloy steels agree with the determined values to within 5% for all the steels and to within 3% for 14 out of the 18 steels. The formula gives 5·8 for the thermal resistivity of pure iron (5·72) and the difference may be accounted for by the presence of trace impurities not allowed for in the formula.

The above formula does not apply to steels in the austenitic condition since the effect of carbon in solution is probably different from that in the form of cementite, and it is doubtful whether the coefficients of the formula can be applied to the large quantities of the alloying elements present in the austenitic steels. It is also probable that the thermal resistivity of face centred iron is different from that of body centred iron.

In view of the difficulty of obtaining accurate data for the thermal conductivities of steels, it may be useful to indicate two methods for obtaining reasonably reliable values where no data exist, especially for the values at elevated temperatures. The first of these methods can be used where the thermal conductivity at or near room temperature is known or can be reasonably estimated by the use of the above formula, and depends on the fact that the thermal conductivity of all steels tends to a constant value of about 0·065 at temperatures of the order of 800°C. The N.P.L. data mentioned previously give the thermal conductivity of 22 steels of a wide variety of composition. If, for any steel, the variation of thermal conductivity with temperature is required, the curve can be drawn with a reasonable degree of certainty, provided the value at or near room temperature is known or can be estimated, by reference to the curve for a similar steel drawn from the N.P.L. data.

The second method makes use of the Lorenz function $L = \frac{\varrho k}{T}$ (T = absolute temperature = 273+°C) and applies in the case where the electrical resistivity is known for a range of temperatures. Where ϱ is measured in microhm-cm and k in Cal/cm s deg. C, L is of the order of 6 to11 $\times 10^{-3}$. Theoretically the Lorenz function would be expected to be constant for all steels at all temperatures, but for the 22 steels listed by the N.P.L. the value varies between 6·8 and $11{\cdot}5\times10^{-3}$ at room temperature and between 6·1 and $7{\cdot}3\times10^{-3}$ at 800°C. The value of the Lorenz function can be deduced without a great deal of error from the known values of steels of near composition and treatment, and from the curve giving the variation of electrical resistivity with temperature, the thermal conductivities at different temperatures can be readily calculated, using this assumed value for L. Values for the electrical resistivity, thermal conductivity and Lorenz function for the 22 steels covered by the N.P.L. report are given in Appendix I.

Young's Modulus (E), *Shear Modulus* (G) *and Poisson's Ratio* (σ)

Young's Modulus (or Modulus in Tension) and Shear Modulus (or Modulus in Torsion) are readily determined by use of extensometers and torsionmeters respectively, but accurate values are not obtained unless very great care is taken and those obtained in normal routine testing are frequently in error by $\pm 5\%$. More accurate values appear to be obtained by vibration or pulse methods but the adiabatic values so obtained are not necessarily identical with the static values obtained with extensometers or torsionmeters on specimens under load. We have, therefore, indicated the method used for the determination where such information is known.

Most values quoted in the literature for Poisson's Ratio (σ) have been derived from separate determinations of E and G by use of the relationship $\sigma = \frac{E}{2G} - 1$. The values so derived are subject to considerable errors, since slight errors in either E or G are magnified in the subsequent estimation of σ. Thus the values of E and G of plain carbon steel are of the order of 13,500 and 5,250 tons/sq. in. respectively, giving a value of $\sigma = 0{\cdot}285$. If E were measured 2% high and G 2% low the calculated value of σ would be 0·335, giving an error of 17%. The values of σ determined by this method must, therefore, be regarded with considerable suspicion. Better values of σ from static tensile tests are obtained by simultaneous measurements of the lateral and longitudinal strains, the ratio of which gives σ directly. The maximum error by this method would be merely the sum of the errors in determining the two strains.

The values of the three elastic constants are not sensitive to structure or composition. Thus the values of E for the low alloy steels vary only between about 13,000 and 13,500 tons/sq.in., and of G from 5,000 to 5,300 tons/sq. in. The best estimates of σ are all between 0·27 and 0·30 with a general average of about 0·285. If E is known then G may be reasonably accurately estimated by using the value of 0·285 for σ since a small error in the latter produces only a negligible error in G. As an example for a steel having E = 13,500 tons/sq.in. then assuming σ to be 0·300 and 0·270 (i.e. errors of $\pm 5\%$) the calculated values for G are respectively 5,190 and 5,310 tons/sq.in. values which differ from the mean 5,250 by only 1·16%, which is an amount smaller than normal errors of measurement.

Jones and Nortcliffe (Ref. 4) suggest that for ferritic steels the ratio $\frac{E_t}{E_{20}}$ is a constant (f) at any elevated temperature. Values for f at various temperatures are:

Temperature	20°C	200°C	400°C	600°C
f	1·000	0·948	0·875	0·775

The same authors also show that steels in the hardened condition have slightly lower values of E than in the softened condition as illustrated by the following values on a sample of En 31.

Treatment °C	*Hardness* (H.V.)	*Young's Modulus* tons/sq.in. (*Vibration Method*)
O.Q. 830°C	800	13,100
" " T.400°C	600	13,600
" " T.720°C	260	13,800

Magnetic Properties

The magnetic properties of steels are highly sensitive to changes of composition and structure. This applies particularly to the permeability in low magnetic fields and to the coercivity, but not so much to the value of the saturation induction which depends almost entirely on the amount of magnetic phase present. Unfortunately, the amount of information available appears to be relatively small so that it is not possible to evaluate the effect of the alloying elements, but only to give an indication of the various trends. Silicon and aluminium and possibly nickel tend to improve the permeability in low magnetic fields and to reduce the coercivity. Carbon and chromium on the other hand reduce the permeability in low fields and increase the coercivity. The coercivity of any steel reaches its highest value when the steel is in the hardened condition and has the lowest value in the fully annealed condition.

Transformation Characteristics

Isothermal and continuous cooling diagrams have been included for each steel for which they are available. Continuous cooling diagrams may be drawn on either a time basis or a bar diameter basis, each having its own sphere of usefulness, a diagram with a time basis when considering controlled heat-treatments, e.g. for large forgings and the other for estimating the effect of different quenching rates on bars of varying diameters. Both types of diagram must be used with caution, however, since they are influenced by many factors such as melting procedure, deoxidation treatment and composition.

In some of the continuous cooling diagrams, namely those published by l'Institut de Recherches de la Sidérurgie (IRSID), lines are superimposed showing the different cooling cycles adopted to establish the different zones produced by the transformations which occurred. On each of these cooling curves is indicated the percentage of transformation of the austenite which has occurred during each transformation zone. The amount of transformation does not always add up to 100% and the deficiency gives the amount of austenite available for transformation to martensite. These curves also give the hardness at room temperature resulting from each particular cooling procedure adopted.

In these diagrams we have included, where possible, the temperatures for the AC_1 and AC_3 transformations as well as for the temperatures corresponding to the start of martensite formation on cooling (M_s) and the completion of the transformation (M_f) as well as temperatures corresponding to the formation of a stated percentage of the transformation product (M_{10}, M_{50}, M_{90}). All these transformation temperatures are affected by composition. Some of the diagrams give Ae_3 and Ae_1 instead of Ac_3 and Ac_1. The former are temperatures corresponding to true equilibrium conditions. Many attempts have been made to find formulae for calculating the transformation temperatures, but whilst the formulae agree tolerably well with the data from which they were derived, they do not agree so well with other published data. In an attempt to find a formula more universally applicable Dr. K.W. Andrews (Ref. 5) has examined data from British and foreign sources on some 150 steels. These have been dealt with statistically using the electronic computer at the United Steel Companies Ltd. and he has put forward the following formula for calculating the Ac_3 temperature:-

$Ac_3(°C) = 910 - 203\sqrt{C} - 15{\cdot}2\ Ni + 44{\cdot}7\ Si + 104\ V + 31{\cdot}5\ Mo + 13{\cdot}1\ W$ where the composition is quoted in weight per cent of the alloying element.

Other elements were discarded by the computer, as their variations were not such as to give a significant correlation and when an attempt was made to bring them into the formula by giving values for the elements, e.g. (—30Mn —110Cr—20Cu+700P+400Al+120As+400Ti), values which were derived from Dr. Andrews' previous estimates for the effect of such elements on the true equilibrium temperature (Ae_3), the calculated results were not, in general, so good as when these elements were neglected. The above formula was based on steels containing up to 0·6% C and less than 5% of other alloying elements. The formula gave calculated values which agreed with the observed values to within ±17°C in 67% of the steels and to within ±33°C in 95% of the steels. The greatest errors occurred in steel with the higher alloy contents but the differences were not systematic. It is possible that some of the errors were due to errors of determination of the observed temperature and possibly to the effect of interaction between the various elements. It was considered that further analysis at this stage would not lead to much improvement.

Dr. Andrews has made a similar analysis for estimating the Ac_1 and M_s temperatures and has given the following relationships:- $Ac_1(°C) = 723 - 10{\cdot}7Mn + 29{\cdot}1Si - 16{\cdot}9Ni + 16{\cdot}9Cr + 290As + 6{\cdot}38W$

$M_s(°C) = 512 - 410C - 14Mn - 18Ni$.

The latter formula for M_s might be compared with the following published formulae:-

$M_s = 538 - 361C - 39Mn - 19{\cdot}5Ni - 39Cr - 28Mo$ (Grange and Stewart) (Ref. 6)

$M_s = 561 - 474C - 33Mn - 17Ni - 17Cr - 21Mo$ (Stevens and Haynes) (Ref. 7)

$M_s = 500 - 317C - 33Mn - 17Ni - 28Cr - 11Si - 11Mo$ (Payson and Savage) (Ref. 8)

See Addendum on page xviii.

Formulae for calculating the temperatures for varying degrees of transformation have been proposed by Grange and Stewart and by Stevens and Haynes. These merely alter the constant in their formula for M_s as follows:-

	M_s	M_{10}	M_{50}	M_{90}	M_{99}	M_f	
Constant Term (°C)	538	513	488	452	416	—	(Ref. 6)
" "	561	551	514	458	—	346	(Ref. 7)

In the discussion of Grange and Stewart's paper, Jaffe proposed a modification to the formula as follows

$M_x = 538 - b(361C + 39Mn + 19{\cdot}5\ Ni + 39Cr + 28Mo)$

where x = percentage of martensite formed

b = 1·0 for M_s, 1·084 for M_{10}, 1·18 for M_{50}, 1·29 for M_{90} and 1·45 for M_{99}.

Hardenability

Hardenability is ascertained either from the Jominy test or from quenching tests on bars of different diameters, and both types of information are included.

The relative hardenability on quenched bars was obtained from hardness determinations across a number of diameters, and curves through the average values have been drawn neglecting any peaks of hardness resulting from segregation effects.

The Jominy test provides a ready means of assessing the relative hardenability of a particular cast of steel and, therefore, is very suitable for illustrating the range of hardenabilities to be expected from steels to a particular specification. Where they are available such ranges have been given for each specification.

It is not possible to relate accurately the Jominy hardenability with that obtained from quenched bars, but to a first approximation the relationship shown in Fig. 1. (Ref. 9) may be used.

Mechanical Properties

Proof Stress, Yield Strength and Tensile Strength values are given to the nearest 0·1 ton/sq. in. (these are long tons of 2240 lbs.). Conversion Tables of tons per square inch to pounds per square inch and to kilograms per square millimetre are given in Appendix IV to Volume I.

Values of percentage elongation and percentage reduction of area have been reported to the nearest ½ unit since the usual methods of measurement rarely improve on this accuracy. Elongation values are mainly those obtained on the former British Standard Test Pieces for which the gauge length is $4\sqrt{A}(=3{\cdot}54d)$ where A and d are respectively the area of cross section and the diameter of the parallel portion of the test piece. The recently published revision of BS.18 "Method for Tensile Testing of Metals", following recommendations of the International Standards Organisation, has introduced the gauge length of $5{\cdot}65\sqrt{A}\ (=5\times d)$ commonly employed on the Continent as that for British Standard test pieces. We have given wherever possible the elongation values corresponding to both gauge lengths. The amount of comparative data is, however, very small and existing conversion charts or curves appear to be far from accurate. The conversion from one gauge length to the other will obviously depend on the reduction of area and few conversion charts take account of this factor. From tests carried out by the National Engineering Laboratory, however, it would appear that for ferritic and martensitic steels the elongation per cent on $l = 5d$. is approximately $0{\cdot}83 \times$ Elongation on $l = 4\sqrt{A}$ and for austenitic steels the corresponding factor is 0·90. More accurate conversions can be obtained from the table we reproduce in Appendix II. This table was initially based on a paper by Kuntze (Ref. 10), and was adjusted to give the best fit for a large number of comparisons we had available. The table has been checked against the results of careful tests carried out by the National Engineering Laboratory on 52 specimens of low alloy steels of widely differing tensile properties. All the estimations of percent elongation on $l = 5{\cdot}65\sqrt{A}$ based on the table agreed within 1 unit and in 44 cases within 0·5 units with the determined values. This table cannot be used for austenitic steels. No distinction has been made between the elongation values of British and American test pieces. These rarely differ by more than 1 unit and, in any case, the values obtained on the American test piece will be lower than those obtained

Distance along jominy end quench bar corresponding to the centre of hardened round bars

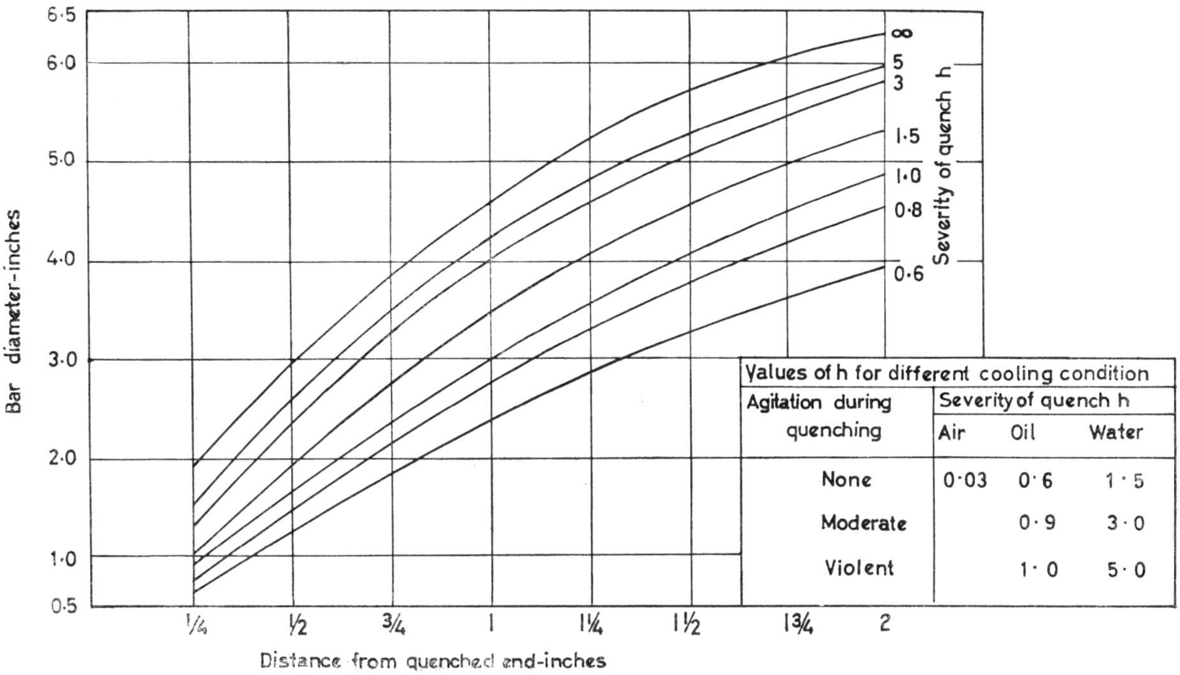

Fig. 1.

on a British test piece and consequently conservative in nature so far as concerns British usage.

We have not included, except in a few cases, values of the limit of proportionality. This value is very difficult to determine with a high degree of precision and can be very much affected by the presence of internal stresses such as are produced by cold straightening operations.

Izod Impact values and Charpy type tests with an Izod V notch have been reported as foot-pounds to fracture, to the nearest unit. Unless otherwise indicated Charpy key-hole and Charpy U-notched tests (Mesnager or DVM) are reported in the manner typical of Continental practice, namely in terms of kilogram-metres per square centimetre of section behind the notch. There is no definite relationship between the various notch Impact tests. Approximate relationships are shown in the Appendix VI.

The mechanical properties of steels are affected not only by composition but also by the steelmaking process and whether grain refining additions were used. Thus within any particular specification there is, in general, a fairly wide spread of properties; we have accordingly, where possible, drawn curves showing the range of values that have been recorded for each steel. We have also included curves showing the effect of section size (the so-called mass effect) for the most frequently used hardening and tempering treatments.

Mechanical Tests at Low Temperatures and Impact Transition Temperature Data

These properties are influenced by composition and, especially in the case of notch impact value, by the presence of trace elements such as phosphorus, tin etc. and by the ferritic grain size of the material. Unfortunately for some of the data, the fullest information regarding the material is not available and such important information as steel-making process, degree of de-oxidation or amount of aluminium added or the grain-size of the material tested cannot be stated. Such data must, therefore, be treated with extreme caution, and are only indicative of what is obtainable under certain (not necessarily stated) conditions.

Various methods have been proposed for determining the transition temperature, e.g. the temperature at which the impact value is a given percentage of the value at the lowest temperature when the fracture is 100% fibrous; the temperature at which the impact value is the average of the maximum and minimum values; the temperature for a specified impact value, or the temperature at which the fracture surface shows 50% fibrous and 50% brittle fracture. Where possible curves showing both the energy to fracture and the percentage amount of fibrous fracture have been included so that any of these methods of assessment may be determined. Where curves were not given in the original report the method of determining the transition temperature is stated.

Mechanical Properties at Elevated Temperatures

Short time tensile properties at temperatures above room temperature may be affected to a certain extent by the rate of pulling, information for which, however, is not always stated in the reports from which the data have been abstracted.

At the time of writing there is no British Standard Specification for the Short Time Tensile Test. The B.S.

Specification 3082 Pt. 1 1959 has now been withdrawn and a revised version is at the moment in course of preparation. In B.S.3082 Pt. 1 the rate of testing was specified to be not greater than 2·5 tons/sq. in. min., the recommended rate for arbitration purposes being between 1·25 and 2·5 tons/sq. in. min. In the revised version* the rate of pulling will be specified in terms of strain rate which it is proposed is to be within the range 0·001 to 0·003 in./in. min. when determining the Proof Stress and Lower Yield Stress values. No limitation is envisaged on the strain rate during the major part of the elastic range but the suggestion is that it should be reduced so that the above conditions can be met before the elastic limit or yield point is reached. There are two tentative specifications issued by the American Society for Testing Materials. E21-58T recommends 0·005 or 0·05 in./in. min. up to the yield point and 0·05 to 0·10 in./in. min. beyond this point. On the other hand E151-61T specifies three rates of pulling:-

(a) Conventional 0·005 $\pm$0·002 in./in. min. up to 0·6% strain offset and 0·10 $\pm$0·002 beyond this point

(b) Rapid 0·5 $\pm$0·02 in./in. min. in the elastic range, and

(c) 5·0 in./in. min. in the elastic range.

The present International Standards Organisation recommended rate (I.S.O./R. 205–1961) specifies a rate not greater than 5 tons/sq.in. min. and a rate of between 1·25 and 2·5 tons/sq.in. min. for arbitration purposes. We understand, however, that this specification is also under review and that rates similar to those proposed for the revised British specification are being put forward for consideration.

We have stated the rate of loading or the speed of pulling where this information was given.

* B.S. 3688, pt. 1, 1963.

Creep and Stress Rupture Properties

Many of the En steels are not suitable for High Temperature service and for such steels there are very few data on Creep and Stress Rupture Properties. Indeed the data on steels which are used for such service are not as full as we would have liked.

In cases where creep is likely to be of importance, the usual method of obtaining the design stress limits is to plot, on a stress/temperature diagram the following:-

(a) A line representing some fraction of the tensile strength of the material at the temperature concerned.

(b) A line representing the 0·2% proof stress or 1% proof stress at temperature.

(c) A line representing the stress to cause either rupture or 1% creep strain in 100,000 hours.

Safe design stresses are considered to be those below all of these lines. There appears to be appreciable scatter (of the order of $\pm$20%) in the stress values for times to reach specific creep strains or rupture at given temperatures for any one type of steel. Special heat treatments or fabrication may exploit the short term potentialities of the material without improving, and possibly even diminishing, the long term properties. Such treatments should be taken into account when considering the creep and rupture data quoted for any one steel.

The selection of carbon steels for high temperature service demands great care since the features which make for ease of manufacture, such as free machinability or ease of forming are those most likely to be associated with poor creep resistance.

Figure 2 shows a series of curves tentatively prepared by the National Engineering Laboratory for different steels used in steam plant.

In using creep data for a particular steel, one should take into account the characteristic shape of the creep curves for the material and the normally expected elongation at rupture. Some of the more creep resistant materials have curves with a flat secondary stage which continues for the majority of the creep test, but which accelerates rapidly into the tertiary stage to final rupture with a fairly small elongation. Such behaviour would not be satisfactory for some applications, where the normal behaviour of materials with greater creep ductility would be preferable. In these the tertiary stage is often prolonged throughout much of the test, but because of the considerable ductility, stress

TYPICAL PROPERTIES OF SOME STEELS USED IN STEAM POWER PLANT
ESTIMATED STRESS TEMPERATURE RELATIONSHIPS FOR RUPTURE IN 100 000 HOURS

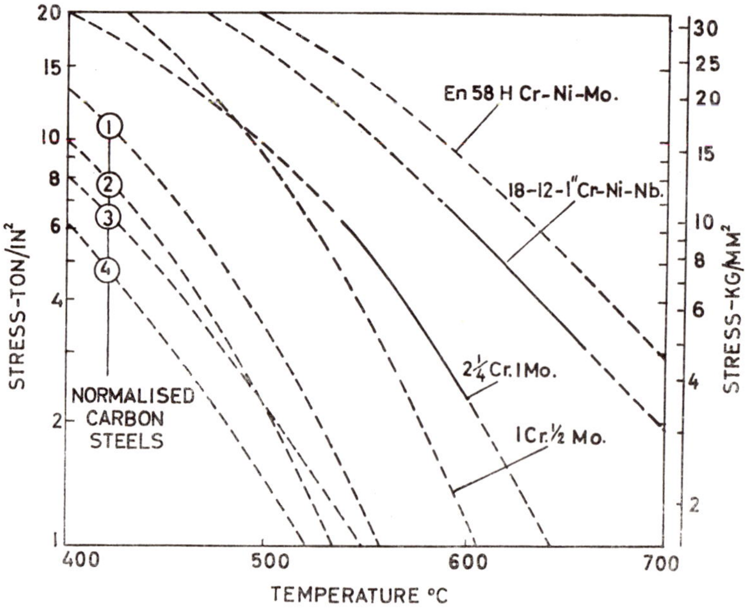

Fig. 2.

(1) about 0·25% carbon steel with about 1% Manganese, deoxidized by Silicon.
(2) ditto, but deoxidized by Aluminium or roughly group (1) steels subjected to prolonged stress relieving heat treatment.
(3) about 0·25% carbon steel with about 0·5% Manganese, deoxidized by Silicon.
(4) ditto, but deoxidized by Aluminium or roughly group (3) steels subjected to prolonged stress relieving heat treatment.

gradients will be reduced within the component and there is unlikely to be a sudden unexpected failure of the component.

It is strongly recommended, especially when there is any doubt in the use or interpretation of creep data, that the experience of the material manufacturer or one of the research laboratories specializing in creep testing should be invoked, so as to ensure that the design is as safe as can be achieved with the data available.

Torsion Tests

As in the case of tensile test data the strength values are reported to the nearest 0·1 tons/sq.in.

Torsional properties are usually determined on solid cylindrical specimens and may be calculated from the twisting moment in accordance with two different assumptions. According to the first of these, it is assumed that the material behaves elastically throughout the test; in other words that the shear stress at any point in the cross section of the test piece is proportional to its distance from the axis.

This gives the following formula for calculating the shear stress from the observed twisting moment.

(1) Shear Stress $= \frac{2}{\pi r^3} \times$ twisting moment (where r is the radius of the test portion)

The second method assumes that the stress is uniform from the axis to the surface of the specimen and gives the formula:-

(2) Shear Stress $= \frac{3}{2\pi r^3} \times$ twisting moment.

The formula (1) applies to all stresses up to the yield point since up to this point only a very small portion near the outer surface of the test piece suffers plastic deformation. It also applies to the maximum shear strength of materials which are very low in ductility. The formula (2) corresponds more closely to the conditions near breaking point in the case of ductile materials.

The various proof stress and yield point values have been calculated according to formula (1). As regards the torsional strength (or maximum shear strength) we have quoted the values using both these formulae. That obtained from formula (1) is sometimes called the modulus of rupture or apparent shear strength. That obtained from formula (2) is called probable shear strength for ductile materials. It will be noticed that the probable shear strength is equal to three-quarters of the apparent shear strength when testing cylindrical solid specimens.

In a few cases tests have been carried out on tubular specimens for which the above formulae do not apply. For thin walled tubes the stress can usually be considered to be uniform across the section, but we have estimated both the apparent and probable shear stress for tubular specimens used, taking into account the wall thickness and diameter in each case. The amount of data on the torsional properties of steels is comparatively small compared with that for the tensile properties. However, there appears to be a purely empirical relationship between the probable shear strength and the tensile strength of materials at or near normal atmospheric temperatures. One formula deduced by the late G. Stanfield of the Brown-Firth Research Laboratories, Sheffield, is:-

Shear Strength $= \frac{1}{2} \times$ Tensile Strength + 10 (tons/sq. in.) This formula appears to give values within about 2 tons/sq.in. for all steels above about 25 tons/sq.in. tenslie strength. A somewhat similar formula is quoted by the International Nickel Company (Mond) Ltd. in their publication *The Mechanical Properties of Nickel Alloy Steels* viz.: Shear Strength = 0·53 × Tensile Strength + 6·5 (tons/sq.in.) These two formulae agree at about 120 tons/sq.in. tensile strength whilst at a tensile strength of 25 tons/sq.in. the Brown-Firth formula indicated a shear strength of 22·5 tons/sq.in. compared with 19·7 by the International Co. (Mond) Ltd. formula. The Brown-Firth formula has the advantage of being the easier to use and to remember.

There does not appear to be such a simple relationship between the yield points in torsion and tension but the ratios of the two values appear to range themselves about a mean value of 0·58 which is the theoretical value according to the Huber-Mises-Henky criterion which deduces that the ratio should be $\frac{1}{\sqrt{3}} = 0{\cdot}577$.

Fatigue Properties

The fatigue properties of metals are usually determined on mechanically polished specimens. This method of preparation, unfortunately, not only work-hardens the surface layers but also induces compressive stresses in the surface layers, both of which tend to raise the fatigue limit of the material, and Cina has shown that the difference between the fatigue limits determined under conditions of reversed axial loading (Haigh test) and of reversed rotary bending (Wohler test) is largely removed when electrolytically polished specimens were used (Ref. 11). Since this cold work effect can be influenced appreciably by the different procedures adopted in preparing the test pieces it is clear that fatigue test data must always be treated with a good deal of caution.

Where possible we have included data on both notched and unnotched test pieces, but values for the former have only been given when either the theoretical stress concentration factor (K_t) is known, or the dimensions of the test piece and of the notch are stated from which K_t may be calculated. In all cases of notch-fatigue data we quote either the relevant details of the notch and the specimen or K_t or both.

It is, however, considered necessary to offer a word of warning concerning the possible extrapolation of these data to other cases, particularly where differences in specimen size and notch geometry are involved. Two factors may operate to render the extrapolation of data obtained on test pieces with shallow notches (such as would be used in a small rotating bending fatigue test) an unsafe procedure by which to assess the fatigue strength of a component with a deeper notch, even where different notches have the same K_t value. The first is that, for a certain notch-geometry, unbroken test pieces in a notched specimen fatigue determination may contain non-propagating cracks and their presence or absence will not, in general, have been established. The notched fatigue strength, estimated from the S—N curve using complete fracture as the criterion of failure, will, therefore, depend on the stress required to propagate a crack through the material rather than on the stress needed to initiate a fatigue crack. Recent researches have shown that the alternating stress required to propagate such cracks depends on the depth of the notch. For a given material, non-propagating cracks are more likely to occur in shallow than in deep notches having the same K_t value, and extrapolation from the case of the shallower notch would tend to be optimistic.

The second feature which can influence such extrapolations is the unknown difference in residual surface stresses in the roots of notches of different geometry and formed by different machining processes.

The practice of deriving a "notch-sensitivity" index for a material from a series of notched fatigue test pieces, and using the index directly to deduce the probable fatigue strength of a differently notched component may, therefore, lead to serious errors.

We are indebted to Mr. D.J. Armstrong and Mr. A.J. Fenner of the National Engineering Laboratory for assistance in preparing the notes on High Temperature properties and notched fatigue tests.

REFERENCES

1. *Welding Notes on the Combined Services Steel Specifications* (Spec. D.G.6 Part II), H.M.S.O.
2. *Physical Constants of Some Commercial Steels at Elevated Temperatures*, Butterworth Scientific Publications, 1953.
3. Radcliffe & Rollason, *J. Iron and Steel Inst.* Vol. 189, 1958, II, page 45.
4. Jones and Nortcliffe, *J. Iron and Steel Inst.* Vol. 157, 1947, page 535.
5. Dr. K.W. Andrews, The United Steel Companies Research and Development Department, Private Communication.
6. Grange and Stewart, *Trans. A.I.M.E.*, Vol. 167, 1946, page 467.
7. Stevens and Haynes, *J. Iron and Steel Inst.* Vol. 183, 1956, II. page 349.
8. Payson and Savage, *Trans. A.S.M.* Vol. 33, 1944, page 261.
9. *Steel Specification Handbook*, English Steel Corporation Ltd.
10. Kuntze, *Arch. für das Eisenhüttenwesen* Heft 10, April 1936, page 510.
11. Cina, *Metallurgia*, Vol. 55, 1957, No. 327, page 11.
12. Wild Barfield Heat Treater's Pocket Book.

ADDENDUM

Since the publication of Volume 1 work by Dr. Andrews based on an examination of 184 steels has led him to a change in the formula given on page xiv for calculation of M_s, i.e.

$$M_s = 539—423C—30{\cdot}4\,Mn—17{\cdot}7Ni—12{\cdot}1Cr—7{\cdot}5Mo$$

A formula has also been derived to include product terms for the interaction of carbon and chromium and carbon and iron:-

$$M_s = 512—453C—16{\cdot}9Ni+15Cr—9{\cdot}5Mo—71{\cdot}4C\times Mn—67{\cdot}6C\times Cr+217(C)^2$$

This formula is most satisfactory when the chromium content lies between 2·0% and 5·0%

LIST OF ABBREVIATIONS

General

AOH	Acid Open Hearth
BEA	Basic Electric Arc
Bess	Bessemer
BOH	Basic Open Hearth
C	Celsius (Celsius Centigrade)
cm	centimetre
deg,°	degree, degrees (temperature or angle)
dia., D	diameter
E	Electric
ft.	foot, feet
hex.	hexagon
hr	hour, hours
H.F.	high frequency
in.	inch, inches
mm	millimetre
NOH	Neutral Open Hearth
RT	Room Temperature
sq.	square
°C	degrees Celsius (Centigrade)
μ	microns (10^{-4} cm)

Chemical Composition

Al	Aluminium
As	Arsenic
C	Carbon
Cr	Chromium
Mn	Manganese
Mo	Molybdenum
N	Nitrogen
Nb	Niobium (Columbium)
Ni	Nickel
O	Oxygen
P	Phosphorus
Pb	Lead
S	Sulphur
Se	Selenium
Si	Silicon
Ti	Titanium
V	Vanadium
W	Tungsten

Welding

B	Brazing and Bronze Welding by Gas
F	Flash and Resistance Welding
G	Gas Welding
IG	Inert Gas Welding
MA	Metal Arc Welding
S	Spot, Seam and Projection Welding
a	Readily welded; no special precautions are required provided the correct filler rod or wire is used when fusion welding.
b	Readily welded, but post-welding heat treatment is necessary to restore the mechanical properties of the parent metal.
c	May be welded, but special precautions may be required, e.g. pre-heating, post heating or both, or with metal arc welding the use of particular electrodes.
d	May be welded, but special precautions may be required, e.g. pre-heating and the use of particular electrodes. Post welding heat-treatment is necessary to restore the mechanical properties of the parent metal.
e	Advice should be sought before attempting to weld.
f	Welding is not recommended.
g	Sufficient information on the welding of this steel is not yet available.
h	Brazing and Bronze Welding satisfactory.
j	Brazing and Bronze Welding is possible but the mechanical properties of the parent metal are liable to modification.
k	Brazing and Bronze Welding are not recommended.
m	The process is not applicable to steel in this form.
x	With currently available filler rods, wires and electrodes, the strength of the weld metal may be less than that of the parent metal. (This symbol is always used with one of the above.)

Heat Treatment

A	Annealed
AC	Air Cooled
AH	Air Hardened
BC	Blank Carburized
CD	Cold Drawn
EQ	End Quenched
FC	Cooled in Furnace
N	Normalized
OH	Oil Hardened
OQ	Oil Quenched
SC	Slow Cooled
T	Tempered
WH	Water Hardened
WQ	Water Quenched

Physical Properties

B	Magnetic Induction
B_{rem}	Magnetic Remanence
(c)	Specific Heat
(d),S.G.	Specific Gravity
E	Young's Modulus
G	Shear Modulus
Cal	Gram calories
H	Magnetizing Field
H_c	Magnetic Coercive Force
k	Thermal Conductivity
α	Coefficient of Thermal Expansion
λ	Thermal Resistivity
ϱ	Electrical Resistivity
σ	Poisson's Ratio

Transformation Characteristics

A	Austenite
Ac	Transformation Temperature on Heating
Ae	Equilibrium Transformation Temperature
B	Bainite
C	Cementite
F	Ferrite
M	Martensite
M_s	Start of Austenite to Martensite Transformation
M_f	Temperature for 100% Austenite to Martensite Transformation
M_x	Temperature for x% Austenite to Martensite Transformation
P	Pearlite

Mechanical Properties

A	Area
(A)	Elongation per cent (International Symbol)
(C)	Centre of test piece
El	Elongation per cent
ft.lb.	Foot-pounds
HB	Brinell Hardness Number
HRC	Rockwell C Hardness Number
HV	Vickers Diamond Hardness Number
KgM	Kilogramme-force metres
K_t	Neuber's theoretical stress concentration factor
l	Gauge length
Long, (L)	Longitudinal Test
(M)	Midway between centre and outside of test piece
(O)	Outside position of test piece
PS	Proof Stress
r	Radius of bend in a bend test
RA	Reduction of Area per cent
(R_m)	Tensile Strength (International Symbol)
(xR_p)	Proof Stress in Tension for x per cent departure from the Proportionality line
RPM	Revolutions per min, Reversals per min
(R_s)	Yield Point in Tension (International Symbol)
t	Thickness of Bend test piece
Trans, (T)	Transverse Test
TS	Tensile Strength
YP	Yield Point
(Z)	Reduction of Area per cent (International Symbol)

EN 40A, EN 40B: 3 PER CENT CHROMIUM-MOLYBDENUM NITRIDING STEEL EN 40C: 3 PER CENT CHROMIUM-MOLYBDENUM-VANADIUM NITRIDING STEEL (HIGHER TENSILE)

SPECIFICATION

CHEMICAL COMPOSITION (%)

	C	Si	Mn	S	P	Ni	Cr	Mo	V
En 40A	0·10—0·20	0·10—0·35	0·40—0·65	0·050 max.	0·050 max.	0·40 max.	2·90—3·50	0·40—0·70	—
En 40B	0·20—0·30	0·10—0·35	0·40—0·65	0·050 max.	0·050 max.	0·40 max.	2·90—3·50	0·40—0·70	—
En 40C	0·30—0·50	0·10—0·35	0·40—0·80	0·050 max.	0·050 max.	0·40 max.	2·50—3·50	0·70—1·20	0·10—0·30

MECHANICAL PROPERTIES

	Hardened and Tempered Condition				
	En 40A and En 40B				En 40C
	R	S	T	U	—
Limiting ruling section, in.	6	6	6	6	$2\frac{1}{2}$
Tensile strength, tons/sq.in., min. (Rm)	45	50	55	60	85
Yield strength [1], tons/sq.in., min. (Rs)	34	38	44	48	72
0·2% proof stress [2], tons/sq. in., min. (0·2Rp)	32	36	41	46	68
Elongation, %, min. (A)	22	20	18	17	10
Izod impact value, ft.lb., min.	40	40	40	35	15
Brinell hardness No. [3], (HB)	201/255	223/277	248/302	269/321	375/444

(1) Not to be used as acceptance tests except by special arrangement between the purchaser and manufacturer.
(2) When specifically requested in the enquiry or order.
(3) For information purposes only. Not contractual part of the specification.

RELATED SPECIFICATIONS

United Kingdom			France
En No.	BS Air	DTD	AFNOR
40B	S106		30CD12
40C		730 5012	

APPLICATIONS

The En 40 steels are used for parts requiring a high resistance to wear, En 40A and En 40B in those cases where a medium core strength and high resistance to shock are desired, and En 40C for parts requiring a core strength in excess of 85 tons per sq. in. Nitriding, besides increasing the wear resistance, also enhances the resistance to fatigue failure and reduces the notch sensitivity. These steels are also used in the unnitrided condition for applications requiring high tensile strength and high creep strength at temperatures up to 600°C.

Typical applications are:

En 40A: Aircraft and motor car engine cylinders, lightly stressed wear resisting parts for automatic packing machinery, textile machines and boot and shoe manufacture, elevators and conveyors, and plastic moulds.

En 40B: Aircraft, motor car and diesel engine crankshafts, components and gears, gudgeon pins, various types of spindles, machine tool parts, plastic moulds, compressor discs for gas turbines, and bolts for steam service.

En 40C: Heavy duty aircraft and motor car crankshafts, airscrew shafts, compressor discs, gears, thrust rings and aircraft components subjected to high stresses.

WELDING

(See Introduction (List of Abbreviations) for key to symbols used in this table.)

	Welding Process					
	MA	G	IG	S	F	B
En 40A, En 40B	d	cx	g	e	b	j
En 40C	f	cx	g	e	b	j

Reproduced by kind permission of H.M.S.O.

Note: (1) Sufficient information is not available to enable a recommendation to be made on welding before nitriding, in view of the unknown effects of the nitriding treatment on weld-metal toughness.

(2) Welding or brazing on the nitrided case is not recommended.

(3) None of the steels should be furnace-brazed after nitriding.

MACHINABILITY

In the softened condition the machinability of En 40A and En 40B is about 35% of that for mild steel, and of En 40C, about 30% of that for mild steel (En 3).

HOT WORKING AND HEAT TREATMENT TEMPERATURES

Forging, rolling and stamping	1200°C. Finish above 850°C, retarded cooling is advisable for large or intricate sections.
Softening	700°C—750°C, air or furnace cool.
Hardening (En 40A, En 40B)	880°C—910°C, oil quench.
(En 40C)	900°C—940°C, oil quench.
Tempering (En 40A, En 40B)	570°C—750°C, air cool.
(En 40C)	570°C—650°C, air cool.
Stabilizing	520°C—550°C, air cool.

NITRIDING DATA

Nitriding is carried out at a temperature of 490°C to 510°C. Fig. 40.1 shows the relation between case depth and nitriding period.

(Ref. 1)

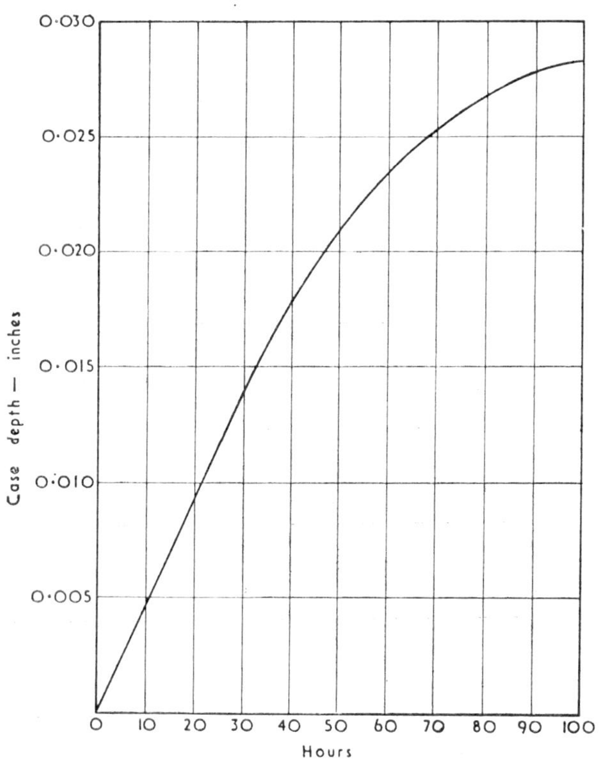

Fig. 40.1

HARDNESS DEPTH CURVES AFTER NITRIDING

Figs. 40.2, 40.3 and 40.4 show typical case hardness/depth curves for En 40A, En 40B and En 40C.

En 40A
Treated size 1¼ in. dia.
Oil quenched 900°C. Tempered 650°C.
Nitrided 500°C, 80 hrs.
20% dissociation.

(Ref. 2)

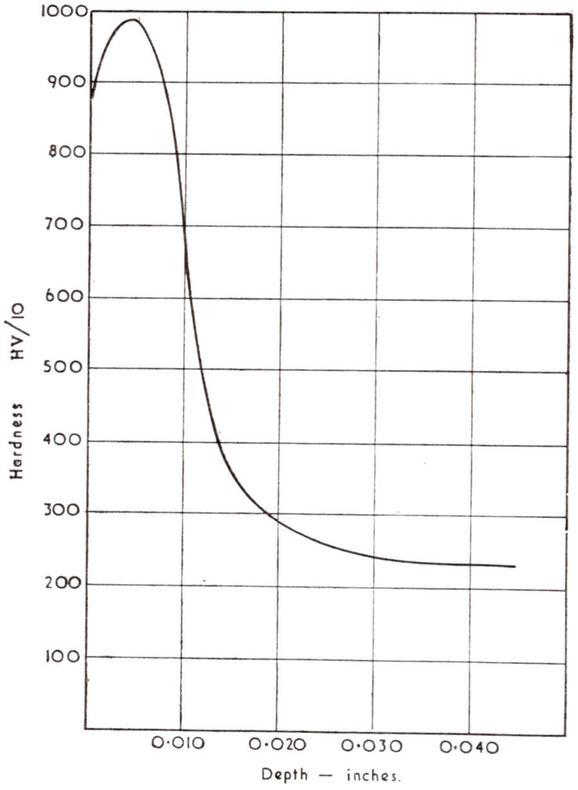

Fig. 40.2

En 40B
Treated size 1⅛ in. dia.
Oil quenched 900°C. Tempered 600°C.
Nitrided 500°C, 80 hrs.
20% dissociation.

(Ref. 2)

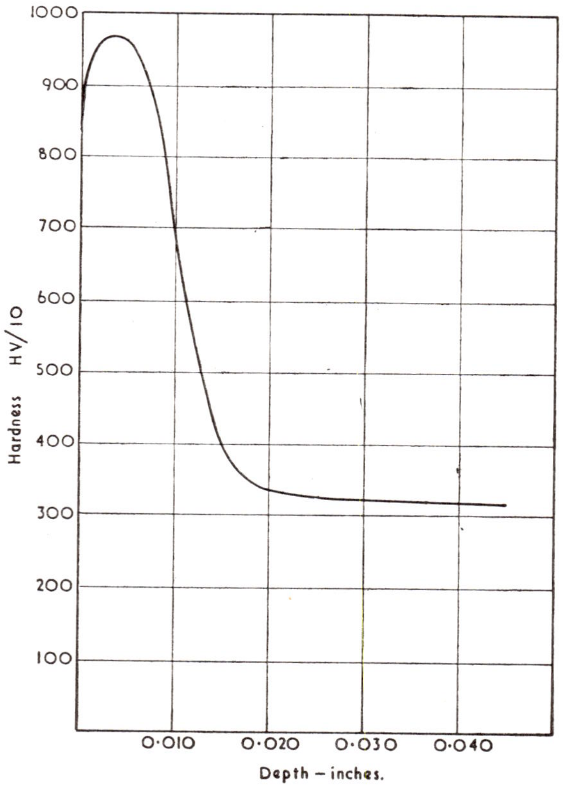

Fig. 40.3

En 40C
Treated size 1⅛ in. dia.
Oil quenched 970°C. Tempered 625°C.
Nitrided 500°C, 80 hrs.
20% dissociation.

(Ref. 2)

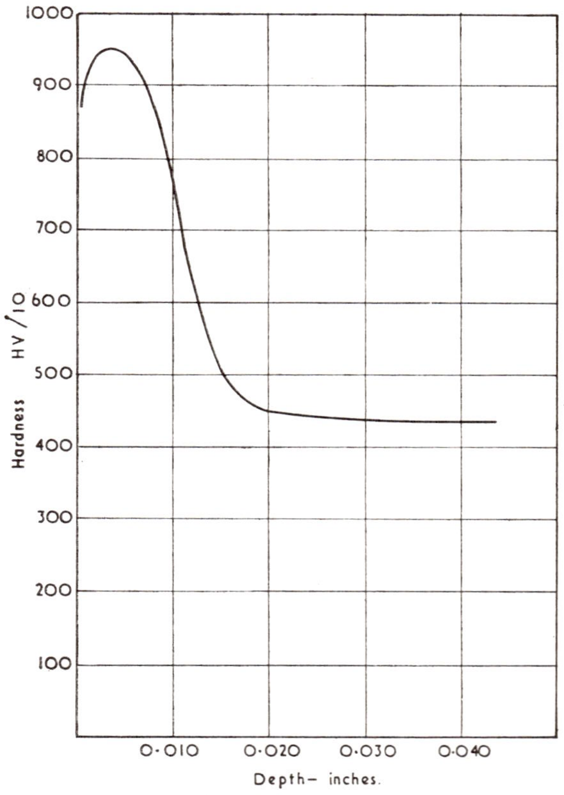

Fig. 40.4

PHYSICAL PROPERTIES

(See also En 29)

SPECIFIC GRAVITY (d)

Chemical Composition (%)								Heat Treatment °C	Specific Gravity at 20°C	Ref.
C	Si	Mn	S	P	Ni	Cr	Mo			
En 40A composition								Softened	7·86	1
En 40B composition								"	7·86	"
0·21	0·22	0·49	0·008	0·011	0·13	3·17	0·39	OQ900 T700	7·84	"
En 40C composition								Softened	7·86	"
" "								—	7·84	2
" "								—	7·85	3
" "								OQ960 T670	7·83	4

SPECIFIC HEAT (c)

	Heat Treatment	Mean Specific Heat cal/g deg. C			Ref.
		20 to 100°C	20 to 300°C	20 to 500°C	
En 40B composition	Softened	—	0·124	0·136	
En 40C composition	—	0·116			2

MEAN COEFFICIENT OF THERMAL EXPANSION (α)

Chemical Composition (%)						Heat Treatment °C	$10^6 \times$ Mean Coefficient of Expansion per deg. C								Ref.
C	Si	Mn	Ni	Cr	Mo		20 to 100°C	20 to 200°C	20 to 300°C	20 to 400°C	20 to 500°C	20 to 600°C	20 to 700°C	20 to 800°C	
0·21	0·22	0·49	0·13	3·17	0·39	OQ900 T700	12·5	13·3	13·7	14·1	14·4	14·7	14·9	14·9	1
En 40A and En 40B composition						Softened	12·2	—	12·9	—	13·6	—	14·0	—	2
En 40B composition						Softened	12·0	—	12·8	—	13·6	—	—	—	5
En 40C composition						—	11·6	—	12·8	—	13·6	—	14·0	—	2
"						—	—	12·8	—	13·5	13·8	14·0	—	—	3
"						OQ960 T670	12·5	12·9	13·2	13·5	13·8	14·0	14·2	—	4

ELECTRICAL RESISTIVITY (ρ)

Chemical Composition (%)						Heat Treatment °C	Electrical Resistivity microhm-cm			Ref.
C	Si	Mn	Ni	Cr	Mo		20°C	100°C	200°C	
0·21	0·22	0·049	0·13	3·17	0·39	OQ900 T700	27·8	—	—	1
En 40B composition						Softened	30	—	—	2
" "						—	34	38·9	46·2	5
En 40C composition						—	35·5			6

En 40

THERMAL CONDUCTIVITY (k)

	Heat Treatment	Thermal Conductivity cal/cm s deg. C								Ref.
		20°C	100°C	200°C	300°C	400°C	500°C	600°C	700°C	
En 40B composition	—	—	0·080	—	0·082	—	0·077	—	—	5
En 40C composition	—	0·092	0·085	—	0·075	—	0·071	—	0·069	6
" "	—	0·097	—	0·100	—	0·108	—	—	—	3
" "	Softened	—	0·090	0·090	0·088	0·083	0·079	0·074	0·070	4

YOUNG'S MODULUS (E), SHEAR MODULUS (G)

Chemical Composition (%)							Heat Treatment °C	Test Temperature °C	Young's Modulus tons/sq.in.		Shear Modulus tons/sq.in. Static	Ref.
C	Si	Mn	Ni	Cr	Mo	V			Static	Dynamic		
0·28	0·29	0·47	0·14	2·93	0·45	—	—	RT	13,500			1
0·38	0·20	0·50	0·23	3·00	0·80	0·18	—	RT	13,250		5,190	"
0·37		0·50	0·30	3·50	0·70	0·30	H and T	RT		13,500		6
"		"	"	"	"	"	"	100		13,500		
"		"	"	"	"	"	"	200		13,100		
"		"	"	"	"	"	"	400		12,200		
"		"	"	"	"	"	"	600		10,700		
"		"	"	"	"	"	"	700		9,500		
En 40C composition							OQ900 T700	RT	13,500			4
							" "	100	12,900			
							" "	200	12,200			
							" "	300	11,400			
							" "	400	10,800			
							" "	500	9,700			
							" "	600	8,300			
							" "	700	5,700			

MAGNETIC PROPERTIES

Chemical Composition (%)								Heat Treatment °C	Magnetic Induction B (Gauss) for magnetizing field H (Oersted) =								Brem (Gauss)	Hc (Oersted)	Ref.
C	Si	Mn	S	P	Ni	Cr	Mo		H = 10	20	30	40	50	100	150	200			
0·21	0·22	0·49	0·008	0·011	0·13	3·17	0·39	OQ900 T700	2,000	13,500	15,000	16,000	16,500	17,500	18,000	18,500	14,800	11·0	1

TRANSFORMATION CHARACTERISTICS

ISOTHERMAL TRANSFORMATION

(For transformation curves for En 40B see En 29.)

En 40C

Chemical Composition (%)								
C	Si	Mn	S	P	Ni	Cr	Mo	V
0·40	0·29	0·62	0·009	0·012	0·21	3·49	0·97	0·26

Basic Electric Arc.

Austenitizing Temperature 900°C, 30 min. (Ref. 1)

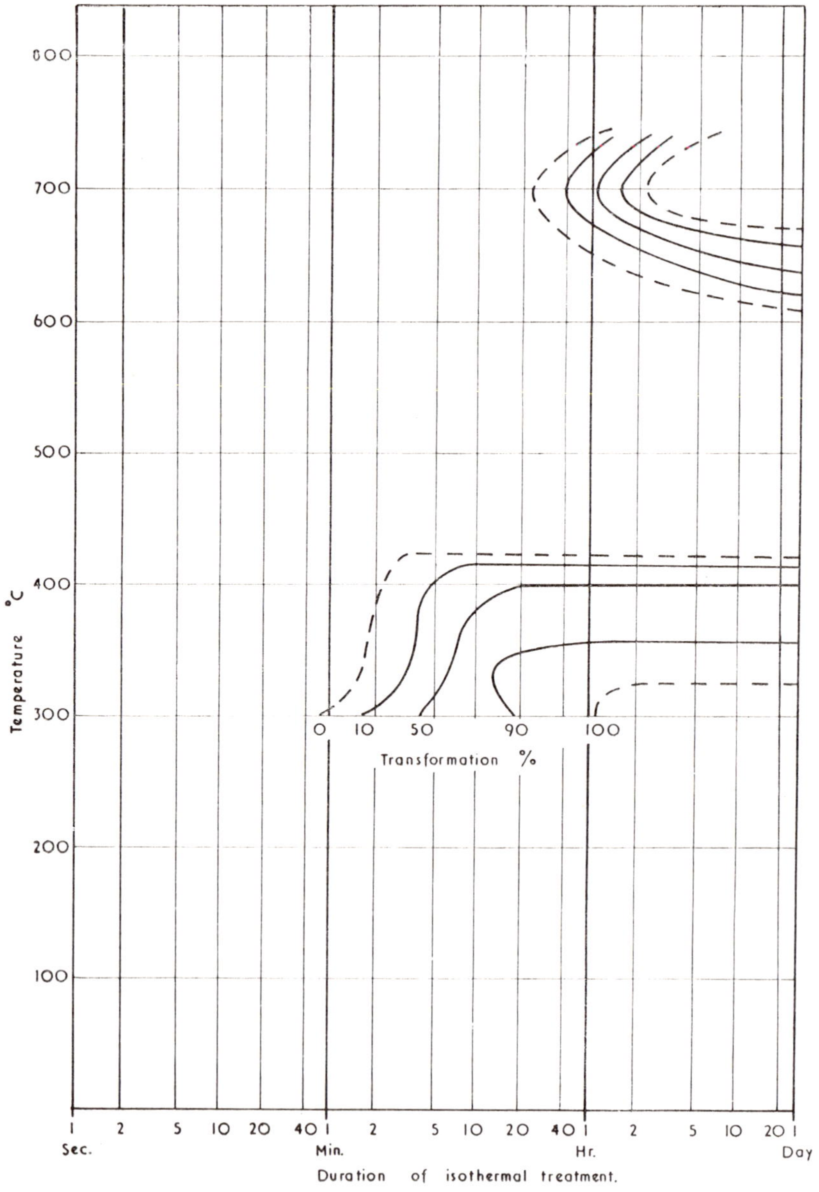

Fig. 40.5

HARDENABILITY

(See data given for En 29.)

END QUENCH HARDENABILITY DATA

Typical curve for En 40C.

(Ref. 2)

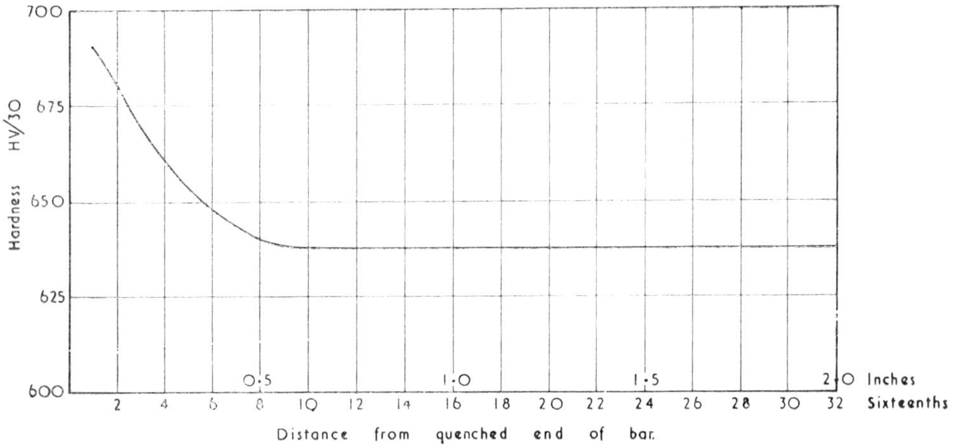

Fig. 40.6

MECHANICAL PROPERTIES AT ROOM TEMPERATURE

ASTM Grain Size	Chemical Composition (%) C	Si	Mn	S	P	Ni	Cr	Mo	V	Section Size dia. (in.)	Heat Treatment °C	Tensile 0·1% P.S. (0·1Rp)	0·2% P.S. (0·2Rp)	0·5% P.S. (0·5Rp)	Y.P. (Rs)	T.S. (Rm)	El % (A)	R.A. % (Z)	Izod ft.lb.	Ref.
												tons/sq.in.								
	0·16		0·50			0·29	3·02	0·41		0·564	OQ900 T150				80·0	92·0	16	55	34	5
	0·18	0·22	0·52			0·25	3·19	0·63		2	OQ900 T600				54·0	61·5	25	68	91	2
	0·19	0·17	0·44			0·07	3·14	0·55		1	OQ900 T630				49·0	56·0	25	70½	110	2
8	0·19		0·50				3·04	0·56		1⅛	OQ900 T600				60·6	64·0	20½	66	76	"
"	"		"				"	"		"	" T650				44·9	52·2	25	72	89	"
"	"		"				"	"		"	" T700				38·0	44·0	28½	76	115	"
"	"		"				"	"		3	" T600				66·8	71·4	18½	62	29	"
"	"		"				"	"		"	" T650				41·9	50·0	25	72	98	"
"	"		"				"	"		"	" T700				37·4	47·1	27	74	106	"
	0·19		0·50				3·14	0·54		⅝	BC920 OQ900 OQ860				88·1	95·8	15	47	22	"
	"		"				"	"		1⅛	" " "				48·6	67·4	16	50	25	"
	0·19						3·08	0·45		1	OQ900 T650				44·0	49·2	25		108	1
	"						"	"		1 7/16	" "				44·4	48·8	27		108	"
	0·19	0·19	0·50			0·14	3·20	0·63		3¼	OQ900 T610				49·0	56·0	25	71½	105	2
	0·20						3·23	0·43		1⅛	OQ900 T650				42·0	49·8	27		110	1
	0·20						3·20	0·43		1½	" T650				41·4	48·6	27		110	"
	"						"	"		2	" T630				43·0	50·2	25½		106	"
	0·20	0·23	0·56	0·024	0·014	0·12	2·91	0·42		1⅝	OQ900 T580 OQ	53·7	55·5			61·8	18½	68	76	7
	"	"	"	"	"	"	"	"		1½	OQ900 T600	50·6	52·6			61·2	21	72	104	"
	0·20	0·30	0·44	0·023	0·020	0·15	3·04	0·43		⅝	OQ900 T620	51·8				60·4	23	66½	67	4
	0·21	0·22	0·49	0·008	0·011	0·13	3·17	0·39		1⅛	OQ900 T650				44·2	52·3	25½	75	120	1
	"	"	"	"	"	"	"	"		"	" T700				35·8	44·4	29	77	120	"
	"	"	"	"	"	"	"	"		"	" T750				29·2	39·0	33	79	120	"
	0·21						3·16	0·43		¾	OQ900 T650				49·4	54·2	22½		101	"
	"						"	"		1⅛	" "				46·4	51·0	26		110	"
	"						"	"		2½	" "				42·6	49·2	25½		110	"
	0·21		0·50			0·24	3·09	0·41		2	OQ900 T650				42·0	50·6	24		109	8
	0·22		0·53				3·28	0·51		4¼	OQ900 T700				36.5	49.8	29		118	8
	0·22		0·55			0·27	2·96	0·50		4	" T630				53·8	60·1	21½		93	"

MECHANICAL PROPERTIES AT ROOM TEMPERATURE

(Continued)

ASTM Grain Size	Chemical Composition (%) C	Si	Mn	S	P	Ni	Cr	Mo	V	Section Size dia. (in.)	Heat Treatment °C	Tensile 0·1% P.S. (0·1Rp) tons/sq.in.	0·2% P.S. (0·2Rp)	0·5% P.S. (0·5Rp)	Y.P. (Rs)	T.S. (Rm)	El % (A)	R.A. % (Z)	Izod ft.lb.	Ref.
8	0·22		0·50			0·15	3·16	0·55		0·564	AC900 T200				73·6	84·5	19	50	46	5
"	"		"			"	"	"		"	OQ900 T200				92·8	99·1	15	47	36	"
"	"		"			"	"	"		"	" T300				93·5	98·6	15	54	23	"
"	"		"			"	"	"		"	" T400				88·8	94·4	15	53	22	"
"	"		"			"	"	"		"	" T450				86·4	91·0	16	55	23	"
"	"		"			"	"	"		1 1/16	" T500				81·5	90·0	16	55	22	"
"	"		"			"	"	"		"	" T550				76·4	84·4	19	59	36	"
"	"		"			"	"	"		"	" T600				64·7	69·8	20	62	66	"
"	"		"			"	"	"		"	" T650				56·0	60·6	21 1/2	68	81	"
"	"		"			"	"	"		"	" T675				50·1	56·6	23	68	82	"
"	"		"			"	"	"		"	" T690				37·7	48·3	20	72	—	"
"	"		"			"	"	"		3 (C)	" T550				77·0	81·1	17	59	24	"
"	"		"			"	"	"		" (O)	" "				82·4	84·9	17 1/2	59	24	"
"	"		"			"	"	"		" (C)	" T600				57·4	61·5	20	68	75	"
"	"		"			"	"	"		" (O)	" "				59·1	64·6	20	68	72	"
"	"		"			"	"	"		" (C)	" T650				47·8	55·8	22	70	87	"
"	"		"			"	"	"		" (O)	" "				48·2	56·1	21 1/2	70	83	"
"	"		"			"	"	"		" (C)	" T675				44·4	53·0	23 1/2	72	93	"
"	"		"			"	"	"		" (O)	" "				46·0	52·5	23 1/2	72	88	"
8	0·22		0·53				3·28	0·51		3 1/4 (C)	OQ900 T700				38·0	46·2	28 1/2	76	118	"
"	"		"				"	"		" (O)	" "				37·2	46·6	28	76	119	"
"	"		"				"	"		" (C)	" T725				33·9	44·9	32	78	119	"
"	"		"				"	"		" (O)	" "				35·1	44·6	30	76	119	"
"	"		"				"	"		4 1/4 (C)	" T700				35·5	45·4	28	74	119	"
"	"		"				"	"		" (O)	" "				36·5	49·8	29	76	118	"
"	"		"				"	"		" (C)	" T725				33·7	43·1	30	76	119	"
"	"		"				"	"		" (O)	" "				34·5	44·2	30	76	119	"
	0·23						3·05	0·51		1 1/8	OQ900 T600				59·6	65·0	22		83	1
	"						"	"		2 1/8	" T580				60·6	65·2	20 1/2		72	"
	0·23						2·99	0·44		1 1/8	OQ900 T600				48·0	56·0	22	70	90	"
	"						"	"		"	" T650				43·0	51·0	25	72 1/2	100	"
	"						"	"		"	" T700				34·0	43·0	29	74	110	"
	"						"	"		6	" T650				44·4	52·8	24		95	"
	0·23		0·53			0·19	2·93	0·56		1 1/8	OQ900 T630				54·9	62·8	20 1/2		85	8

MECHANICAL PROPERTIES AT ROOM TEMPERATURE
(Continued)

Process of Manufacture	ASTM Grain Size	Chemical Composition (%) C	Si	Mn	S	P	Ni	Cr	Mo	V	Section Size dia. (in.)	Heat Treatment °C	Tensile 0·1% P.S. (0·1Rp) tons/sq.in.	0·2% P.S. (0·2Rp) tons/sq.in.	0·5% P.S. (0·5Rp) tons/sq.in.	Y.P. (Rs) tons/sq.in.	T.S. (Rm) tons/sq.in.	El % (A)	R.A. % (Z)	Izod ft.lb.	Ref.
		0·24		0·50			0·24	3·10	0·46		4	OQ900 T630				55·1	62·3	23		82	8
		0·24	0·22	0·51			0·14	3·40	0·51		4	OQ900 T640				50·0	56·8	23	71	115	2
		0·25		0·50			0·18	2·92	0·54		4	OQ900 T630				55·1	60·2	22 1/2		93	8
		0·25		0·56			0·37	3·11	0·51		1 1/8	" T650				54·0	62·4	22		81	"
		0·26		0·59			0·27	3·11	0·56		4	OQ900 T630				56·8	63·7	21 1/2		84	8
B.E.A.		0·27	0·39	0·57	0·017	0·015	0·01	3·43	0·51		1 1/8	OQ900 T620	51·0				64·0	22	68	78	4
B.E.A.		0·27	0·25	0·53	0·013	0·011	0·29	3·08	0·55		1 1/8	" "	54·9				67·8	22	68	76	"
		0·27	0·24	0·52			0·20	3·32	0·52		3 3/4	OQ900 T610	52·0				64·5	22		84	2
		0·28						2·99	0·42		2	OQ900 T570				64·4	69·2	20		71	1
		"						"	"		2 1/8	" T590				59·4	64·4	20		82	"
		"						"	"		"	" T625				50·6	57·6	23		93	"
		"						"	"		"	" T650				47·0	55·0	26		94	"
		0·28						3·43	0·44		1 3/4	" T610				56·6	63·2	22 1/2		91	"
		"						"	"		2 7/8	" T600				57·4	64·4	22		84	"
		0·28						3·14	0·45		3 5/8	" T590				57·0	63·4	22		85	"
		"						"	"		"	" T620				56·0	63·2	22		85	"
		0·28		0·55			0·23	3·02	0·56		1 1/8	OQ900 T650				56·5	63·8	21		81	8
		0·28		0·53				3·10	0·57		2 1/2	" T620				58·9	66·8	20		76	"
		0·28	0·16	0·54			0·23	3·23	0·53		7/8	OQ900 T610	52·7				67·0	23		91	2
		0·29						3·26	0·44		1 1/8	OQ900 T600				58·8	65·9	23		82	1
		0·29						3·11	0·42		3	" T610				57·2	63·8	20 1/2		84	"
		0·29		0·55				3·14	0·62		6	OQ900 T650				51·2	59·0	21 1/2		69	8
		0·29		0·58			0·16	3·00	0·55		4	" T630				59·1	66·5	21 1/2		76	"
	8	0·29		0·51				3·22	0·58		0·564	OQ900 T200 OQ				117·7	118·8	14	40	25	5
	"	"		"				"	"		"	" T300 "				104·3	108·5	12	43	18	"
	"	"		"				"	"		"	" T400 "				98·9	103·3	15 1/2	53	19	"
	"	"		"				"	"		"	" T450 "				94·9	99·7	16	53	20	"
	"	"		"				"	"		1 1/8	OQ900 T500 OQ				92·4	98·1	16 1/2	52	20	"
	"	"		"				"	"		"	" T550 "				89·6	92·2	15 1/2	52	34	"
	"	"		"				"	"		"	" T600 "				68·0	73·2	19	64	70	"
	"	"		"				"	"		"	" T650 "				52·6	63·3	23	68	87	"
	"	"		"				"	"		"	" T650 FC				57·1	62·9	21 1/2	66	85	"
	"	"		"				"	"		"	" T675 OQ				50·1	57·3	23	68	94	"
	"	"		"				"	"		"	" T700 "				47·6	54·0	23 1/2	72	94	"
	"	"		"				"	"		"	" T725 "				41·7	49·0	26 1/2	74	110	"

MECHANICAL PROPERTIES AT ROOM TEMPERATURE (Continued)

Process of Manu-facture	ASTM Grain Size	Chemical Composition (%) C	Si	Mn	S	P	Ni	Cr	Mo	V	Section Size dia. (in.)	Heat Treatment °C	Tensile 0·1% P.S. (0·1Rp) tons/sq.in.	0·2% P.S. (0·2Rp)	0·5% P.S. (0·5Rp)	Y.P. (Rs)	T.S. (Rm)	El % (A)	R.A. % (Z)	Izod ft.lb.	Ref.
	8	0·29		0·51				3·22	0·58		3½ (C)	OQ900 T550				91·0	92·6	15	50	28	5
	,,	,,		,,				,,	,,		,, (O)	,, ,,				86·2	93·0	16	55	26	,,
	,,	,,		,,				,,	,,		,, (C)	,, T600				66·3	72·1	19	64	66	,,
	,,	,,		,,				,,	,,		,, (O)	,, ,,				65·6	72·4	20	66	64	,,
	,,	,,		,,				,,	,,		,, (C)	,, T650				51·1	59·2	23	70	92	,,
	,,	,,		,,				,,	,,		,, (O)	,, ,,				51·9	59·6	23	70	92	,,
	,,	,,		,,				,,	,,		,, (C)	,, T675				52·0	58·6	22	59	87	,,
	,,	,,		,,				,,	,,		,, (O)	,, ,,				53·2	58·0	23½	62	88	,,
	,,	,,		,,				,,	,,		6¼ sq. (C)	,, T600				59·6	66·2	17½	52	62	,,
	,,	,,		,,				,,	,,		,, (M)	,, ,,				60·0	67·6	20	59	59	,,
	,,	,,		,,				,,	,,		,, (O)	,, ,,				62·6	68·0	18	62	52	,,
		0·30						3·12	0·46		1 5/16	OQ900 T625				55·4	64·8	21½		82	1
		,,						,,	,,		3	,, T600				58·4	67·2	23		86	,,
B.E.A.		0·36	0·28	0·55	0·007	0·026	0·26	3·33	0·91	0·15	¾ × 1	OQ940 T550 AC	83·6				112·5	13	45	13	9
		0·36	0·30	0·61				3·01	1·05	0·23	2½	OQ950 T610	70·4				85·0	19		33	2
		0·37	0·23	0·61			0·22	3·02	0·90	0·21	TP	OQ950 T200	85				132	15	36	17	10
		,,	,,	,,			,,	,,	,,	,,	,,	,, T300	93				125	10½	34	16	,,
		,,	,,	,,			,,	,,	,,	,,	,,	,, T400	89				121	12	35	16	,,
		,,	,,	,,			,,	,,	,,	,,	,,	,, T500	79				118	12	35½	17	,,
		,,	,,	,,			,,	,,	,,	,,	,,	,, T575	75				103	15	53	24	,,
		,,	,,	,,			,,	,,	,,	,,	,,	,, T630	65				86	20	61	61	,,
		,,	,,	,,			,,	,,	,,	,,	,,	,, T660	60				65	22	63	82	,,
		,,	,,	,,			,,	,,	,,	,,	,,	,, T775	45				55	25½	66	103	,,
B.E.A.		0·37	0·27	0·48	0·012	0·010	0·24	2·94	0·82	0·15	0·424	OQ950	64		98		131	10	23	18	2
,,		,,	,,	,,	,,	,,	,,	,,	,,	,,	,,	,, T200	84		106		121	11	35	30	,,
,,		,,	,,	,,	,,	,,	,,	,,	,,	,,	,,	,, T400	80		95		107	13	40	20	,,
,,		,,	,,	,,	,,	,,	,,	,,	,,	,,	,,	,, T550	79		94		102	15	50	25	,,
,,		,,	,,	,,	,,	,,	,,	,,	,,	,,	,,	,, T600	76		86		94	16	57	31	,,
,,		,,	,,	,,	,,	,,	,,	,,	,,	,,	,,	AH950	54		91		124	11	33	27	,,
,,		,,	,,	,,	,,	,,	,,	,,	,,	,,	,,	,, T200	76		101		120	12	36	31	,,
,,		,,	,,	,,	,,	,,	,,	,,	,,	,,	,,	,, T400	81		97		111	14	41	22	,,
,,		,,	,,	,,	,,	,,	,,	,,	,,	,,	,,	,, T550	79		94		104	15	46	26	,,
		,,	,,	,,				,,	,,	,,	,,	,, T600	81		87		94	16	51	32	,,
		0·37	0·27	0·51				2·95	0·91	0·15	19/32 × ¼	OQ950 T600				76·3	87·6	20			,,
		0·37	0·23	0·50				3·09	0·94	0·23	TP	AH970 T540				85·0	116·0	17		27	,,
		,,	,,	,,				,,	,,	,,	2½	OQ950 T600	74·9				88·5	16		30	,,
		,,	,,	,,				,,	,,	,,	,,	,, T620	68·3				82·5	18		40	,,

MECHANICAL PROPERTIES AT ROOM TEMPERATURE (Continued)

Process of Manu-facture	ASTM Grain Size	Chemical Composition (%) C	Si	Mn	S	P	Ni	Cr	Mo	V	Section Size dia. (in.)	Heat Treatment °C	Tensile (tons/sq.in.) 0·1% P.S. (0·1Rp)	0·2% P.S. (0·2Rp)	0·5% P.S. (0·5Rp)	Y.P. (Rs)	T.S. (Rm)	El % (A)	R.A. % (Z)	Izod ft.lb.	Ref.
B.E.A.		0·38	0·20	0·50	0·010	0·010	0·23	3·00	0·80	0·18	TP	OQ960 T200	86·5	93·6	105·0		124·8	11			11
"		"	"	"	"	"	"	"	"	"	"	" T300	86·5	91·2	98·0		110·2	13			"
"		"	"	"	"	"	"	"	"	"	"	" T400	85·0	89·3	95·5		107·0	13½			"
"		"	"	"	"	"	"	"	"	"	"	" T500	77·5	84·0	92·0		103·0	15			"
"		"	"	"	"	"	"	"	"	"	"	" T600	69·4	73·4	76·5		84·5	18			"
"		"	"	"	"	"	"	"	"	"	"	" T650	54·8	56·9	59·2		63·9	20			"
"		"	"	"	"	"	"	"	"	"	"	" T700	44·9	46·0	47·6		52·9	26			"
B.E.A.		0·38	0·31	0·57	0·010	0·010	0·18	3·12	0·96	0·20	0·564	AH920 T550	86·0		102·0		113·8	15½	47	22	1
"		"	"	"	"	"	"	"	"	"	"	" T575	86·5		100·5		110·0	16	54	24	"
"		"	"	"	"	"	"	"	"	"	"	" T600	83·6		93·2		100·8	15	51	31	"
"		"	"	"	"	"	"	"	"	"	"	AH950 T550	81·5		100·8		112·6	15	47½	23	"
"		"	"	"	"	"	"	"	"	"	"	AH980 "	78·0		100·0		112·0	16	47½	22	"
"		"	"	"	"	"	"	"	"	"	"	AH1000 "	82·0		99·6		113·0	15	45	22	"
"		"	"	"	"	"	"	"	"	"	1⅛	AH920 T550	82·8		98·0		112·8	15	42	17	"
"		"	"	"	"	"	"	"	"	"	"	" T575	87·0		100·2		109·6	16	51	19	"
"		"	"	"	"	"	"	"	"	"	"	" T600	82·8		93·0		100·0	16	53	20	"
"		"	"	"	"	"	"	"	"	"	"	" T625	79·8		89·0		97·0	18	53	18	"
"		"	"	"	"	"	"	"	"	"	"	" T650	76·4		86·2		94·7	19	56	20	"
"		"	"	"	"	"	"	"	"	"	"	AH950 T550	81·0		98·0		112·8	15	43	15	"
"		"	"	"	"	"	"	"	"	"	"	AH980 "	80·0		100·4		113·2	16	45	15	"
"		"	"	"	"	"	"	"	"	"	"	AH1000 "	81·0		104·4		113·2	14	40	12	"
"		"	"	"	"	"	"	"	"	"	"	OQ920 T550	77·0		92·6		109·3	16	46	15	"
"		"	"	"	"	"	"	"	"	"	"	" T600	79·0		90·4		100·0	17	56	28	"
"		"	"	"	"	"	"	"	"	"	"	" T650	73·5		84·5		93·0	20	59	37	"
		0·39	0·25	0·51				3·20	0·94	0·21	1⅛	OQ950 T600				88·0	95·0	18	54	23	2
		0·39						2·96	0·92	0·20	1⅛	OQ950	86		106		123	9	29	17	"
		"						"	"	"	"	" T200	89		109		121	10	32	17	"
		"						"	"	"	"	" T400	86		100		112	12	35	15	"
		"						"	"	"	"	" T550	81		95		105	15	47	18	"
		"						"	"	"	"	" T600	74		85		93	16	54	25	"
		"						"	"	"	"	AH950	58		87		117	13	37	23	"
		"						"	"	"	"	" T200	65		92		106	14	41	26	"
		"						"	"	"	"	" T400	71		89		103	15	42	16	"
		"						"	"	"	"	" T550	75		90		101	15	47	13	"
		"						"	"	"	"	" T600	73		83		90	16	51	17	"
B.E.A.		0·39	0·30	0·66	0·013	0·013	0·15	3·05	0·95	0·20	0·564	OQ900 T590	76·5		86·0		94·0	16½	53	33	1
B.E.A.		0·39	0·24	0·64	0·010	0·011	0·16	3·12	0·91	0·19	0·564	AH920 T550	80·0		92·4		108·4	16	52	26	"
"		"	"	"	"	"	"	"	"	"	"	" T575	95·0		98·0		106·4	16	53	25	"
"		"	"	"	"	"	"	"	"	"	"	" T600	80·8		89·6		97·6	17	56	37	"

MECHANICAL PROPERTIES AT ROOM TEMPERATURE
(Continued)

Process of Manu-facture	ASTM Grain Size	Chemical Composition (%) C	Si	Mn	S	P	Ni	Cr	Mo	V	Section Size dia. (in.)	Heat Treatment °C	Tensile 0·1% P.S. (0·1Rp) tons/sq.in.	0·2% P.S. (0·2Rp)	0·5% P.S. (0·5Rp)	Y.P. (Rs)	T.S. (Rm)	El % (A)	R.A. % (Z)	Izod ft.lb.	Ref.
B.E.A.		0·39	0·24	0·64	0·010	0·011	0·16	3·12	0·91	0·19	$\frac{3}{4}$	AH920 T250, 1 hr.	96·8	103·2	112·2		126·3	14	42	14	1
"		"	"	"	"	"	"	"	"	"	"	" T300, 1 hr.	96·0	102·0	111·8		124·5	14	42	15	"
"		"	"	"	"	"	"	"	"	"	"	" " 2 hrs.	96·8	103·0	110·0		122·5	14	39	"	"
"		"	"	"	"	"	"	"	"	"	"	" " 6 hrs.	98·6	104·0	110·6		125·7	14	42$\frac{1}{2}$	"	"
"		"	"	"	"	"	"	"	"	"	"	" " 12 hrs.	96·8	103·0	111·0		123·5	14$\frac{1}{2}$	42	"	"
"		"	"	"	"	"	"	"	"	"	"	" T350, 1 hr.	93·0	99·0	106·5		120·6	14$\frac{1}{2}$	41	15	"
"		"	"	"	"	"	"	"	"	"	"	" T400, 1 hr.	92·5	97·5	104·3		120·2	14	42	13	"
"		"	"	"	"	"	"	"	"	"	"	" " 6 hrs.	85·2	90·4	97·0		117·2	15	42$\frac{1}{2}$	—	"
"		"	"	"	"	"	"	"	"	"	"	" T500, 1 hr.	77·0	84·8	93·0		115·5	15$\frac{1}{2}$	42$\frac{1}{2}$	16	"
"		"	"	"	"	"	"	"	"	"	"	OH920 T200, 1 hr.	88·3	95·4	104·5		125·3	16	42	19	"
"		"	"	"	"	"	"	"	"	"	"	" T300, 1 hr.	92·5	96·7	102·0		117·4	14	44		"
"		"	"	"	"	"	"	"	"	"	"	" T300, 6 hrs.	94·4	98·0	103·0		119·6	14	43		"
"		"	"	"	"	"	"	"	"	"	"	" T400, 1 hr.	90·5	94·6	101·3		113·0	14	43		"
"		"	"	"	"	"	"	"	"	"	"	" T400, 6 hrs.	86·5	91·0	98·0		111·5	16	45		"
"		"	"	"	"	"	"	"	"	"	"	" T500, 1 hr.	83·7	88·0	93·6		109·0	17	47	16	"
"		"	"	"	"	"	"	"	"	"	1$\frac{1}{8}$	AH920 T200	67·2		92·4		125·4	14	32	18	"
"		"	"	"	"	"	"	"	"	"	"	" T500	82·0		97·5		114·0	17	49	16	"
"		"	"	"	"	"	"	"	"	"	"	" T550	86·0		100·0		110·0	17	50$\frac{1}{2}$	20	"
"		"	"	"	"	"	"	"	"	"	"	" T575	85·0		96·8		106·8	15	49	20	"
"		"	"	"	"	"	"	"	"	"	"	" T600	81·6		90·0		97·6	17	54	19	"
"		"	"	"	"	"	"	"	"	"	"	OQ920 T200	88·3		104·5		125·3	16	42	19	"
"		"	"	"	"	"	"	"	"	"	"	" T500	75·5		92·2		104·0	18	50	18	"
	8	0·39		0·55			0·17	3·12	0·88	0·17	0·564	OQ930 T575				95·2	100·4	17	55	34	"
	"	"		"			"	"	"	"	"	" T600				93·1	97·4	17$\frac{1}{2}$	55	44	"
	"	"		"			"	"	"	"	1$\frac{1}{8}$	" T625				82·0	92·2	18$\frac{1}{2}$	57	42	"
	"	"		"			"	"	"	"	"	" T650				72·2	84·5	18	61	54	"
	"	"		"			"	"	"	"	"	" T675				66·4	74·0	20$\frac{1}{2}$	61$\frac{1}{2}$	73	"
	"	"		"			"	"	"	"	"	" T700				56·0	62·3	21$\frac{1}{2}$	66	94	"
	"	"		"			"	"	"	"	3 sq. (C)	" T600				90·1	95·4	14$\frac{1}{2}$	47	28	"
	"	"		"			"	"	"	"	" " (M)	" "				83·8	94·2	16	49$\frac{1}{2}$	28	"
	"	"		"			"	"	"	"	" " (C)	" T650				72·2	80·8	14	42	45	"
	"	"		"			"	"	"	"	" " (M)	" "				73·0	81·7	15	43$\frac{1}{2}$	47	"
	"	"		"			"	"	"	"	" " (C)	" T700				51·9	60·6	23$\frac{1}{2}$	63$\frac{1}{2}$	89	"
	"	"		"			"	"	"	"	" " (M)	" "				55·4	62·3	21	63$\frac{1}{2}$	82	"
		0·40	0·23	0·54				3·10	0·95	0·20	TP	AH970 T540	83·1				114·0	15		24	2
		"	"	"				"	"	"	2$\frac{1}{2}$	OQ950 T620	71·1				87·5	17		33	"
		0·43	0·28	0·59	0·021	0·012	0·21	3·20	0·79	0·27	1$\frac{1}{8}$	OQ960 T670	52·5	53·3	54·5		64·0	25$\frac{1}{2}$	65$\frac{1}{2}$	60	4

EFFECT OF TEMPERING ON CASE HARDNESS

Chemical Composition (%)								Nitriding Treatment	Hardness, HV, after tempering at stated temperature											Ref.
C	Si	Mn	S	P	Ni	Cr	Mo		20°C	100°C	200°C	300°C	400°C	500°C	550°C	600°C	650°C	750°C	800°C	
0·21	0·22	0·49	0·008	0·011	0·13	3·17	0·39	50 hrs. at 500°C	780	828	848	844	828	820	812	716	560	300	280	1
"	"	"	"	"	"	"	"	80 hrs. at 500°C	820	848	860	860	820	820	812	740	560	288	—	"

EFFECT OF TEMPERING ON THE MECHANICAL PROPERTIES (See also data given for En 29.)

Chemical Composition (%)			
C	Mn	Cr	Mo
0·29	0·51	3·22	0·58

McQuaid Ehn grain size 8 (ASTM).
1⅛ in. dia. bars. Oil quenched 900°C. Tempered as indicated.

En 40B (Ref.5)

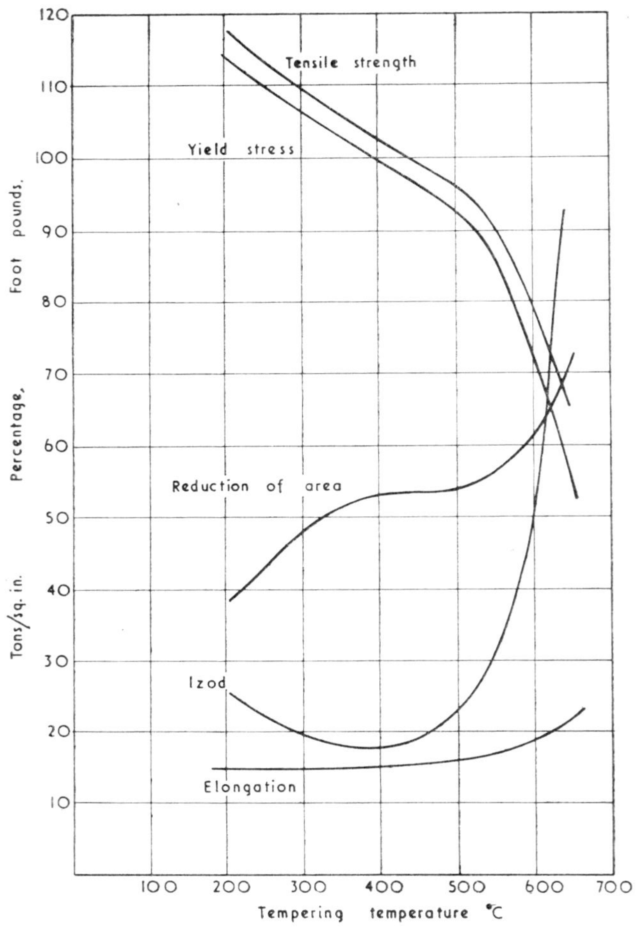

Fig. 40.7

Chemical Composition (%)			
C	Cr	Mo	V
0·39	2·96	0·92	0·20

1⅛ in. dia. bars. Oil quenched 950°C. Tempered as indicated.

En 40C (Ref. 2)

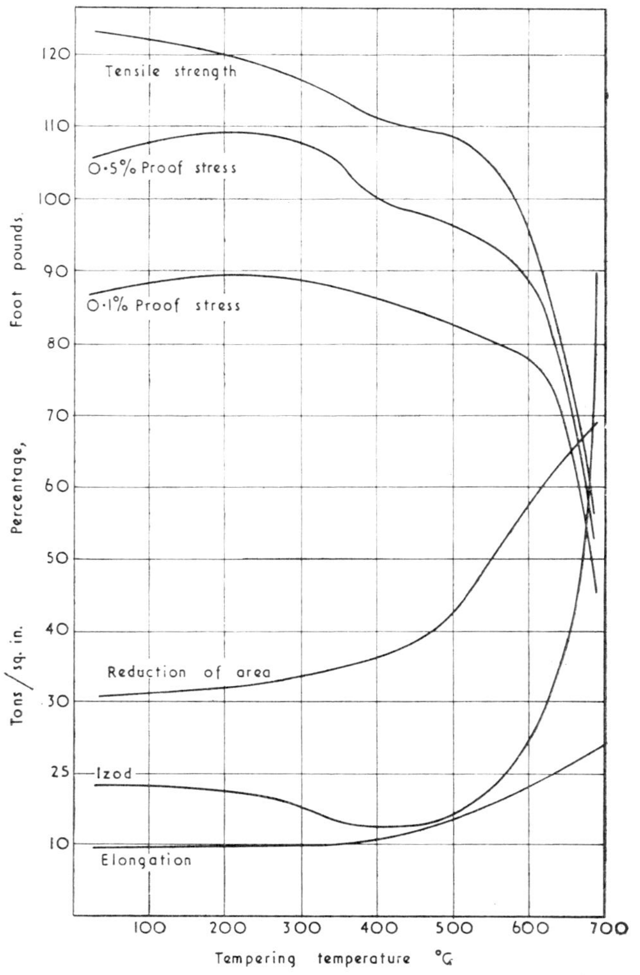

Fig. 40.8

EFFECT OF TEMPERING ON CASE HARDNESS

Typical curve for test pieces heated for 1 hour at temperature indicated.

En 40B (Ref. 1)

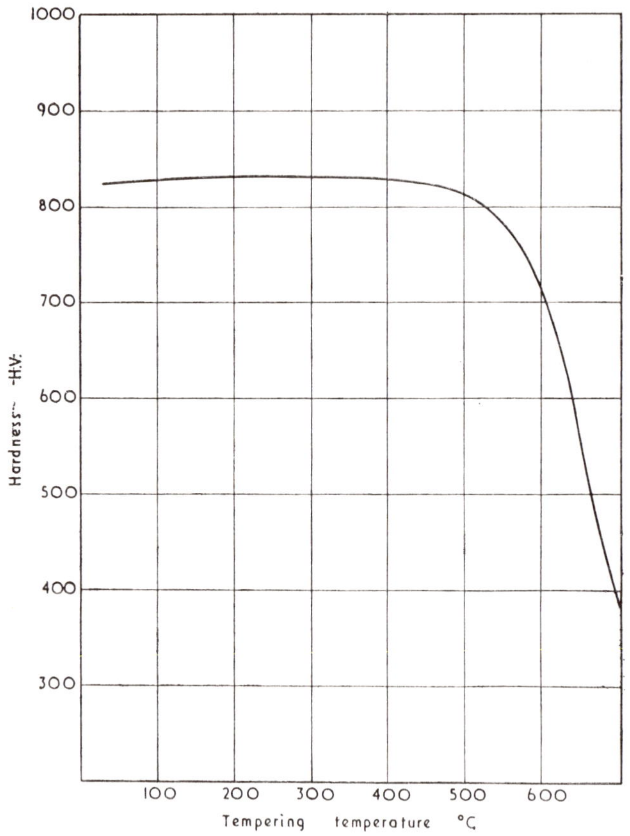

Fig. 40.9

EFFECT OF SECTION SIZE ON THE MECHANICAL PROPERTIES AT ROOM TEMPERATURE

(See En 29 for properties obtainable from En 40A and En 40B.)

TYPICAL CURVES FOR EN 40C

Oil quenched 970°C.
Tempered 600°C.

(Ref. 2)

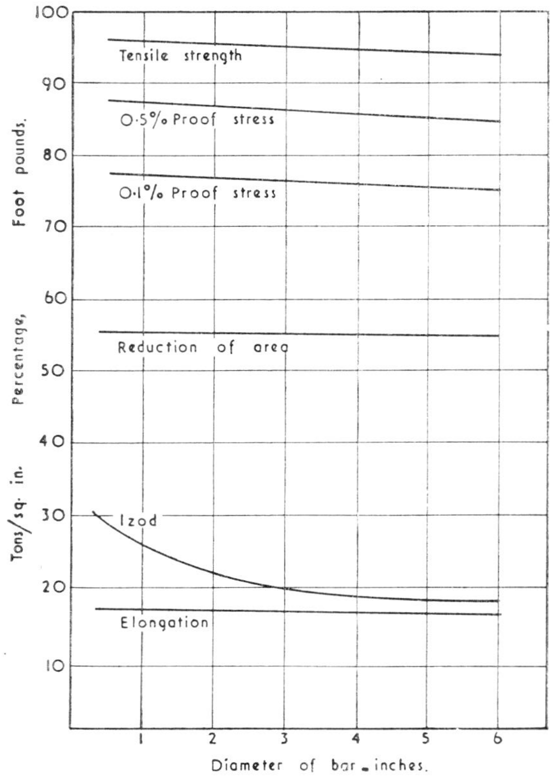

Fig. 40.10

Oil quenched 970°C.
Tempered 625°C.

(Ref. 2)

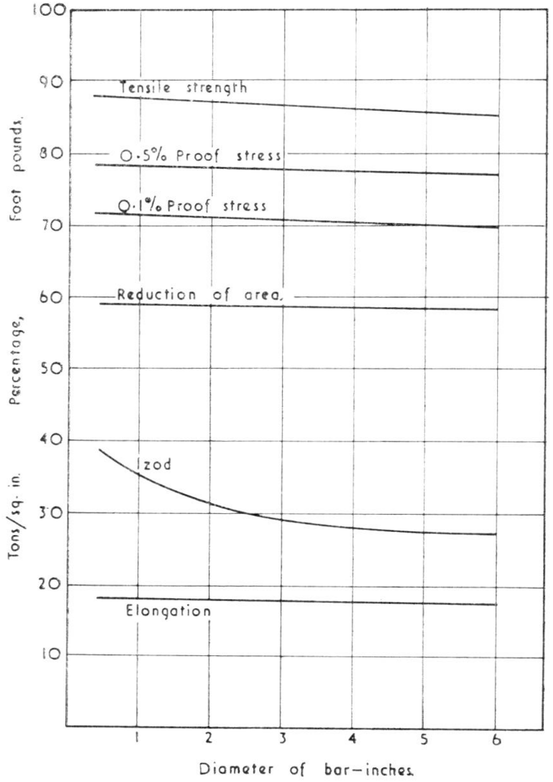

Fig. 40.11

Oil quenched 970°C.
Tempered 650°C.

(Ref 2.)

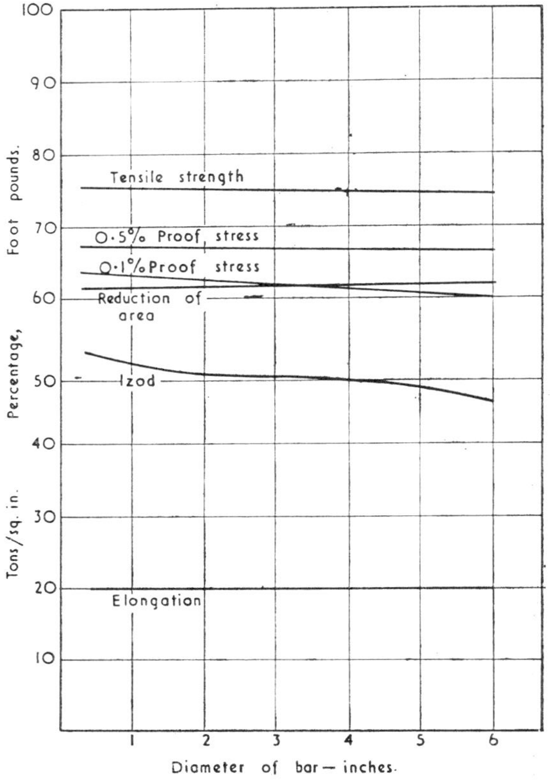

Fig. 40.12

MECHANICAL PROPERTIES AT LOW TEMPERATURE

Chemical Composition (%)								Heat Treatment °C	Test Temperature °C	Tensile				Ref.
C	Si	Mn	S	P	Ni	Cr	Mo			0·2% P.S. (0·2Rp) tons/sq.in.	T.S. (Rm) tons/sq.in.	El % (A)	R.A. % (Z)	
0·24	0·12	0·58	0·024	0·019	0·38	3·32	0·47	OQ900 T600	25	53·0	61·1	22	69½	12
"	"	"	"	"	"	"	"	" "	— 50	56·5	66·6	24	65	"
"	"	"	"	"	"	"	"	" "	—120	62·4	73·0	23½	59½	"
"	"	"	"	"	"	"	"	" "	—196	84·4	90·7	21½	50	"

IMPACT TRANSITION TEMPERATURE DATA

(See also En 29)

Chemical Composition (%)								Heat Treatment °C	Izod impact strength (ft.lb.) at stated temperature							Ref.
C	Si	Mn	S	P	Ni	Cr	Mo		20°C	0°C	—20°C	—40°C	—60°C	—80°C	—100°C	
0·24	0·12	0·58	0·024	0·019	0·38	3·32	0·47	OQ900 T600	80	76	42	30	24	18	15	12

IMPACT TRANSITION TEMPERATURE DATA
(Continued)

Process of Manufacture	Chemical Composition (%) C	Si	Mn	S	P	Ni	Cr	Mo	Section Size dia. (in.)	Heat Treatment °C	Hardness HB	Tensile Y.P. (Rs) tons/sq.in.	T.S. (Rm) tons/sq.in.	El % (A)	R.A. % (Z)	Charpy V notch Transition Temperature °C 50% max. energy	50% cleavage fracture	Ref.
B.E.A.	0·26	0·25	0·52	0·006	0·012	0·18	3·25	0·56	5 sq. (L)	OQ900 T590		55·0	63·8	23	70	48	50	1
"	"	"	"	"	"	"	"	"	" (T)	" "		55·6	64·0	17	43	48	60	"
"	"	"	"	"	"	"	"	"	" (L)	" T700		36·6	47·4	30	76	—45	—37	"
"	"	"	"	"	"	"	"	"	" (T)	" "		37·5	48·3	24	60	—17	—15	"
"	"	"	"	"	"	"	"	"	T.P.S.* (L)	" " T620		56·7	66·7	24	71	—45	—45	"
"	"	"	"	"	"	"	"	"	" (T)	" " "		58·4	66·6	20	53	—25	—15	"
"	"	"	"	"	"	"	"	"	" (L)	" T590 T550 24 hrs.	299					32	32	"
"	"	"	"	"	"	"	"	"	" (T)	" " " "	"					32	45	"
"	"	"	"	"	"	"	"	"	" (L)	" " T570 WQ	298					37	31	"
"	"	"	"	"	"	"	"	"	" (L)	AC900 T590	302					—20	—17	"

* 5 in. sq. treated at T.P.S.

MECHANICAL PROPERTIES AT HIGH TEMPERATURE

Process of Manufacture	Chemical Composition (%)									Section Size dia. (in.)	Heat Treatment °C	Test Temperature °C	Tensile						Izod ft.lb.	Ref.
	C	Si	Mn	S	P	Ni	Cr	Mo	V				0·1% P.S. (0·1Rp)	0·2% P.S. (0·2Rp)	0.5% P.S. (0·5Rp)	T.S. (Rm)	El % (A)	R.A. % (Z)		
													tons/sq.in.							
	0·25		0·55				3·20	0·50		6¾	OQ900 T610	20		52·0		59·4	17½	48	92	2
	"		"				"	"		"	" "	300		49·6		56·7	15½	53		"
	"		"				"	"		"	" "	400		47·2		54·5	16	52½		"
	"		"				"	"		"	" "	500		43·8		50·2	20½	67		"
	"		"				"	"		"	" "	550		36·6		43·4	22½	71½		"
B.E.A.	0·38	0·20	0·50	0·010	0·010	0·23	3·00	0·80	0·18		OQ920 T550 (1 hr.)	RT	77·2	84·4	91·2	99·5	17			11
"	"	"	"	"	"	"	"	"	"		" " "	100	71·7	78·6	87·0	96·6	18			"
"	"	"	"	"	"	"	"	"	"		" " "	150	71·0	78·0	86·2	95·7	17½			"
"	"	"	"	"	"	"	"	"	"		" " "	200	68·5	75·3	84·1	93·7	17½			"
"	"	"	"	"	"	"	"	"	"		" " "	250	67·1	74·2	83·0	93·2	16½			"
"	"	"	"	"	"	"	"	"	"		" " "	300	66·0	73·2	82·0	93·0	15½			"
"	"	"	"	"	"	"	"	"	"		" " "	350	63·7	70·8	79·4	89·6	15½			"
"	"	"	"	"	"	"	"	"	"		" T625 (1 hr.)	RT	73·7	77·7	81·2	88·0	18½			"
"	"	"	"	"	"	"	"	"	"		" " "	100	71·6	76·1	81·4	88·5	18½			"
"	"	"	"	"	"	"	"	"	"		" " "	150	67·8	72·6	77·6	85·0	19½			"
"	"	"	"	"	"	"	"	"	"		" " "	200	63·6	68·6	73·6	82·0	18½			"
"	"	"	"	"	"	"	"	"	"		" " "	250	63·7	69·7	76·5	84·9	17½			"
"	"	"	"	"	"	"	"	"	"		" " "	300	61·2	66·4	72·1	79·6	15			"
"	"	"	"	"	"	"	"	"	"		" " "	350	58·9	63·9	69·4	76·9	17			"
"	"	"	"	"	"	"	"	"	"		" T675	RT	51·2	51·8	52·5	58·6	25			"
"	"	"	"	"	"	"	"	"	"		" "	100	49·8	50·2	50·7	57·6	22½			"
"	"	"	"	"	"	"	"	"	"		" "	150	45·4	47·4	49·8	57·2	21			"
"	"	"	"	"	"	"	"	"	"		" "	200	43·4	45·1	46·4	54·8	22			"
"	"	"	"	"	"	"	"	"	"		" "	250	40·7	43·8	47·1	55·5	21			"
"	"	"	"	"	"	"	"	"	"		" "	300	42·3	44·1	47·5	54·9	21			"
"	"	"	"	"	"	"	"	"	"		" "	350	41·9	43·8	46·2	55·5	21½			"
"	"	"	"	"	"	"	"	"	"		A750	RT	28·2	28·5	29·1	40·0	30			"
"	"	"	"	"	"	"	"	"	"		"	100	28·2	28·7	29·7	39·2	28			"
"	"	"	"	"	"	"	"	"	"		"	150	27·2	27·7	29·0	39·7	26			"
"	"	"	"	"	"	"	"	"	"		"	200	26·2	26·3	26·5	35·9	27			"
"	"	"	"	"	"	"	"	"	"		"	250	24·9	25·5	26·7	35·5	26½			"
"	"	"	"	"	"	"	"	"	"		"	300	24·1	24·7	26·3	35·4	23½			"
"	"	"	"	"	"	"	"	"	"		"	350	23·0	24·2	25·5	35·6	23			"
"	"	"	"	"	"	"	"	"	"		OQ960 T200	200	54·3	66·7	87·6	122·0	18½			"
"	"	"	"	"	"	"	"	"	"		" T250	250	57·7	71·2	89·0	115·0	19½			"
"	"	"	"	"	"	"	"	"	"		" T300	300	58·3	70·3	86·2	107·7	19½			"
"	"	"	"	"	"	"	"	"	"		OQ960 held 1 hr.—70	200	62·6	78·0	98·7	126·3	17			"
"	"	"	"	"	"	"	"	"	"		" "	250	57·5	71·1	90·0	116·7	16½			"
"	"	"	"	"	"	"	"	"	"		" "	300	59·0	71·2	86·6	107·2	18			"

MECHANICAL PROPERTIES AT HIGH TEMPERATURE

(Continued)

Process of Manufacture	Chemical Composition (%) C	Si	Mn	S	P	Ni	Cr	Mo	V	Section Size dia. (in.)	Heat Treatment °C	Test Temperature °C	Tensile 0·1% P.S. (0·1Rp) tons/sq.in.	0·2% P.S. (0·2Rp) tons/sq.in.	0·5% P.S. (0·5Rp) tons/sq.in.	T.S. (Rm) tons/sq.in.	El % (A)	R.A. % (Z)	Ref.
B.E.A.	0·40	0·24	0·65	0·011	0·014	0·21	3·28	1·10	0·21	T.P.	AH920 T550	RT	91·0	96·7	102·3	110·0	17	43	1
"	"	"	"	"	"	"	"	"	"	"	" "	400	63·6	74·4	85·7	95·6	20	57	"
"	"	"	"	"	"	"	"	"	"	"	" "	450	63·6	73·5	84·6	92·8	20	59	"
"	"	"	"	"	"	"	"	"	"	"	" "	500	61·5	68·4	79·0	88·4	19	61	"
"	"	"	"	"	"	"	"	"	"	"	" "	550	58·1	64·5	74·0	81·2	20	63	"
	0·43	0·28	0·59	0·021	0·012	0·21	3·20	0·79	0·27	1⅛	OQ960 T670	20	52·5	53·3	54·5	64·0	25½	65½	4
	"	"	"	"	"	"	"	"	"	"	" "	100	49·2	50·7	52·5	61·8	24½	63½	"
	"	"	"	"	"	"	"	"	"	"	" "	200	46·0	48·0	50·2	59·0	23½	63	"
	"	"	"	"	"	"	"	"	"	"	" "	300	42·1	45·5	48·0	56·1	22½	63	"
	"	"	"	"	"	"	"	"	"	"	" "	400	38·5	42·0	45·1	52·6	21½	64	"
	"	"	"	"	"	"	"	"	"	"	" "	500	32·4	34·6	37·5	43·3	24½	71	"
	"	"	"	"	"	"	"	"	"	"	" "	600	19·1	21·7	24·7	32·6	33½	85	"
	"	"	"	"	"	"	"	"	"	"	" "	700	6·2	7·2	8·4	17·8	55½	92	"
	En 40C composition									1¼	OQ & T	20	54·5	55·4	56·4	65·0	19		3
	"									"	"	300	50·0			58·4	16		"
	"									"	"	400	45·0			55·2	18		"
	"									"	"	450	40·5			51·0	22		"
	"									"	"	500	29·5	34·0	38·5	48·0	26		"
	"									"	"	600	15·0			37·0	33		"

CREEP PROPERTIES

Chemical Composition (%)									Section Size dia. (in.)	Heat Treatment °C	Time hrs.	Creep Strain %	Stress tons/sq.in. give stated creep at						Ref.
C	Si	Mn	S	P	Ni	Cr	Mo	V					350°C	400°C	450°C	500°C	550°C	600°C	
En 40A and En 40B composition									1	OQ900 T650	1,000	0·1				5·6	2·7		2
" " "									"	" "	"	1·0				14·0	8·0		
" " "									"	" "	"	Rupture				19·0	12·0		
" " "									"	" "	3,000	0·1				4·2	2·0		
" " "									"	" "	"	1·0				12·0	6·0		
" " "									"	" "	"	Rupture				17·0	9·0		
" " "									"	" "	10,000	0·1				(3·1)	(1·5)		
" " "									"	" "	"	1·0				(10·0)	(4·0)		
" " "									"	" "	"	Rupture				(14·5)	(7·0)		
" " "									"	" "	100,000	0·1				(1·8)	—		
" " "									"	" "	"	1·0				(5·0)	—		
Steel to En 40C composition									1¼	OQ and T	100	0·1	46·0	(31·0)	(20·0)	(13·5)	(7·0)		3
" " "									"	"	"	0·2			(22·5)	(16·5)	(9·0)		
" " "									"	"	"	0·5				(19·5)	(11·0)		
" " "									"	"	"	1·0				(22·5)	(13·0)		
" " "									"	"	300	0·1	(40·0)	(27·5)	(17·5)	(11·0)	(5·5)		
" " "									"	"	"	0·2				(15·0)	(7·5)		
" " "									"	"	"	0·5				(18·0)	(9·0)		
" " "									"	"	"	1·0				(19·0)	(11·0)		
0·39	0·25	0·60	0·016	0·016	0·35	3·08	0·89	0·23	1⅛	OQ950 T640	100	0·1			24·8	11·9	—	—	4
"	"	"	"	"	"	"	"	"	"	" "	"	0·5			32·8	22·0	12·0	5·9	
"	"	"	"	"	"	"	"	"	"	" "	"	1·0			36·2	24·2	13·0	7·1	
"	"	"	"	"	"	"	"	"	"	" "	300	0·1			21·0	—	—	—	
"	"	"	"	"	"	"	"	"	"	" "	"	0·5			30·5	18·0	10·0	3·8	
"	"	"	"	"	"	"	"	"	"	" "	"	1·0			33·5	22·5	12·3	6·0	
"	"	"	"	"	"	"	"	"	"	" "	1,000	0·5			28·0	13·5	6·5	—	
"	"	"	"	"	"	"	"	"	"	" "	"	1·0			(31·0)	19·5	7·0	(3·7)	
"	"	"	"	"	"	"	"	"	"	" "	3,000	0·5				(10·0)			

Values in parenthesis are extrapolated or approximate values.

CREEP PROPERTIES (Continued)

Chemical Composition (%)									Section Size dia. (in.)	Heat Treatment °C	Test Temperature °C	Stress for a minimum creep rate of			Ref.
C	Si	Mn	S	P	Ni	Cr	Mo	V				10^{-5}/hr.	10^{-6}/hr.	10^{-7}/hr.	
0·39	0·25	0·60	0·016	0·016	0·35	3·08	0·89	0·23	1⅛	OQ950 T640	450	30·0	24·0	20·0	4
"	"	"	"	"	"	"	"	"		" "	500	(19·0)	12·0	(8·0)	"
"	"	"	"	"	"	"	"	"		" "	550	8·0	5·0	(3.0)	"
"	"	"	"	"	"	"	"	"		" "	600	4·2	(2·5)		"

CREEP RUPTURE DATA

Chemical Composition (%)								
C	Si	Mn	S	P	Ni	Cr	Mo	V
0·43	0·28	0·59	0·021	0·012	0·22	3·20	0·79	0·27

Basic Electric Arc.
1⅛ in. dia. bars. Oil quenched 950°C. Tempered 640°C. (Ref. 4)

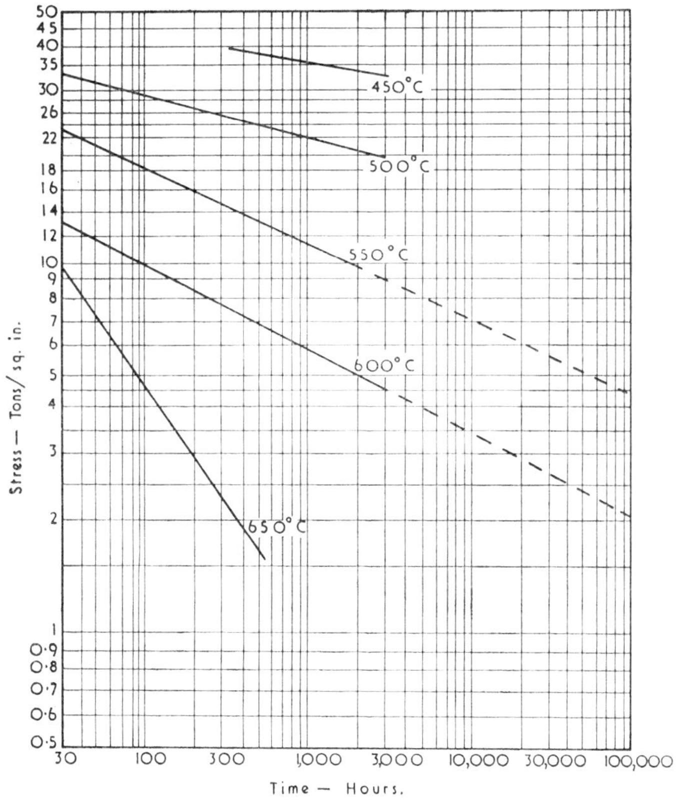

Fig. 40.13

Broken curves have been extrapolated by the Larson and Miller method (typical values).

MECHANICAL PROPERTIES IN TORSION

Process of Manufacture	Chemical Composition (%)									Section Size dia. (in.)	Heat Treatment °C	Tensile						Izod ft.lb.	Torsion					Ref.
	C	Si	Mn	S	P	Ni	Cr	Mo	V			0·1% P.S. (0·1Rp)	0·2% P.S. (0·2Rp)	0·5% P.S. (0·5Rp)	T.S. (Rm)	El % (A)	R.A. % (Z)		0·1% P.S.	0·2% P.S.	0·5% P.S.	Ap-parent max. shear stress	Prob-able max. shear stress	
												tons/sq.in.							tons/sq.in.					
B.E.A.	0·38	0·20	0·50	0·010	0·010	0·23	3·00	0·80	0·18	T.P.	OQ940 T500	78·3	84·1	92·3	106·0	15½	47	29	48·2	53·1	60·4	89·8	67·3	11
B.E.A.	0·40	0·28	0·55	0·013	0·010		3·24	0·89	0·23	—	—	74·4			94·0	17	54		57·1			80·7	60·5	13

FATIGUE PROPERTIES

ASTM Grain Size	Chemical Composition (%)									Section Size dia. (in.)	Heat Treatment °C	Tensile					Fatigue Limit tons/sq.in.								Ref.
	C	Si	Mn	S	P	Ni	Cr	Mo	V			0·1% P.S. (0·1Rp)	0·2% P.S. (0·2Rp)	T.S. (Rm)	El % (A)	R.A %. (Z)	Rotating Bending		Reversed Bending		Reversed Axial		Reversed Torsional		
												tons/sq.in.					Un-notched	Notched	Un-notched	Notched	Un-notched	Notched	Un-notched	Notched	
	0·20	0·23	0·56	0·024	0·014	0·12	2·91	0.42		1⅝	OQ900 T580, OQ	53·7	55·5	61·8	18½	68			33·0				19·9		7
	0·25	0·26	0·55	0·014	0·018	0·17	3·09	0·51		1½	OQ900 T600, OQ	50·6	52·6	61·2	21	72				14·5*				10·5*	"
	"	"	"	"	"	"	"	"		"	OQ900 T600, OQ ntd. 485, 10 hrs.								38·2	27·5*			23·8	19·8*	"
	"	"	"	"	"	"	"	"		"	OQ900 T600, OQ ntd. 485, 22 hrs.								38·4	27·4*			22·7	19·5*	"
	"	"	"	"	"	"	"	"		"	OQ900 T600, OQ ntd. 485, 72 hrs.									27·4*				18·7*	"
	0·20	0·23	0·56	0·024	0·014	0·12	2·91	0·42		1⅝	OQ900 T580 OQ ntd. 485, 72 hrs.								40·7				23·8		"
	0·25	0·26	0·55	0·014	0·018	0·17	3·07	0·51		1½	OQ900 T600 OQ ntd. 485, 72 hrs.								38·6†				23·5†		"
												* (Notch) hollow specimen 0·60 in. dia., 0·232 in. I.D., containing transverse hole 0·04 in. dia. † Plain specimen 0·004 in. ground from nitrided surface prior to testing.													
	0·24	0·16	0·55	0·014	0·016	0·12	3·18	0·53		1½	OQ900 T600	50·1	51·9	61·3	1½	74½	±33·7		±33·0						7
	0·29	0·24	0·55	0·022	0·018		3·10	0·62		¾	OQ900 T550			84·2			±47·5	±17·7*							14
	"	"	"	"	"		"	"		"	OQ900 T550 ntd. 480, 72 hrs.			86·1			±57·2	±51·2†							"
												* Notch ⅜ in. Whitworth form on 0·355 in. dia.													
7—8	0·29		0·53				3·28	0·51		1¼	OQ900 T620			63·5			±32·8	±14·0*			±32·0				5
												† Notch 55°V, 0·008 in. root radius, 0·02 in. deep, O.D. 0·335 in., root dia, 0·295 in. Neuber kt=2·85													
	0·31						3·26	0·58		—	OH and T			66			±35·4 (3″ fillet radius)								14
	"						"	"			" " "						±31·2 (¼″ fillet radius)								
	"						"	"			" " "						±26·8 (⅛″ fillet radius)								
	"						"	"			" " "						±11·0 (sharp corner)								
	"						"	"			OH and T ntd. 72 hrs.						±45·2 (3″ fillet radius)								
	"						"	"			OH and T ntd. 72 hrs.						±45·2 (¼″ fillet radius)								
	"						"	"			OH and T ntd. 72 hrs.						±43·0 (⅛″ fillet radius)								
	"						"	"			OH and T ntd. 72 hrs.						±34.5 (sharp corner)								
	"						"	"			OQ900 T610			67·9			±38·2	±12·5*							
	0·31	0·30	0·62	0·022	0·020	—	3·16	0·60		¾	OQ900 T610														
	"	"	"	"	"		"	"		"	" ntd. 480, 72 hrs.			70·0			±52·0	±50·2*							
	"	"	"	"	"		"	"		"	OQ900 T620			63·9			±35·8								
	"	"	"	"	"		"	"		"	" ntd. 590, 7 hr.			66·9			±51·0								
	"	"	"	"	"		"	"		"	" " " 16 hr.			62·6			±50·0								
	"	"	"	"	"		"	"		"	" " 500, 72 hr.			65·8			±46·2								
												* Notch ⅜ in. Whitworth form on 0·355 in. dia.													

FATIGUE PROPERTIES (Continued)

ASTM Grain Size	Chemical Composition (%)									Section Size dia. (in.)	Heat Treatment °C	Tensile					Fatigue Limit tons/sq. in.				Ref.
	C	Si	Mn	S	P	Ni	Cr	Mo	V			0·1% P.S. (0·1Rp)	0·2% P.S. (0·2Rp)	T.S. (Rm)	El % (A)	R.A. % (Z)	Rotating Bending		Reversed Axial		
												tons/sq.in.					Plain	Notched	Plain	Notched	
	0·36	0·28	0·55	0·007	0·026	0·26	3·33	0·91	0·15	¾	OQ940 T550 AC	83·6		112·5	13	45	±53·7	±27·3*			9
	0·38	0·25	0·59	0·017	0·013		3·33	0·86	0·18	¾	OQ930 T580			97·8			±41·5	±18·7†	±40·0	±10·5†	
	"	"	"	"	"		"	"	"	"	OQ930 T580 ntd. 510, 72 hrs.			96·8			±51·5	±50·5†	±38.0	±25.5 †	14

* Notch 55°V, 0·013 in. root radius, 0·030 in. deep, 0·300 in. root dia. Neuber Kt=2·63.

† Notch corresponding with ⅜ in. Whitworth thread on 0·355 in. dia. specimens

CORROSION FATIGUE

Chemical Composition (%)							Section Size dia. (in.)	Heat Treatment °C	T.S. (Rm)	El % (A)	R.A. % (Z)	Rotating bending fatigue strength (5×10^7 rev.) in sea water	Ref.
C	Si	Mn	S	P	Cr	Mo			tons/sq. (in.)			tons/sq.in.	
0·31	0·30	0·62	0·022	0·020	3·16	0·60	¾	OQ900 T620	64·0	25½	69	± 6·4	14
"	"	"	"	"	"	"	"	" ntd.	66·5			±37·2	"

FATIGUE AT HIGH TEMPERATURES

Chemical Composition (%)	Section Size dia. (in.)	Heat Treatment	Test Tempera-ture °C	Tensile					Rotating Bending Fatigue strength (10^7 rev.)	Ref.
				0·1% P.S. (0·1Rp)	0·2% P.S. (0·2Rp)	0·5% P.S. (0·5Rp)	T.S. (Rm)	El % (A)		
				tons/sq.in.					tons/sq.in.	
within En 40C specification	1¼	OH and T	20	54·5	55·4	56·4	65·0	19	±35·7	3
	"	"	300	50·0			58·4	16	±30·0	"
	"	"	500	29·5	34·0	38·5	48·0	26	±26·0	"

LIST OF REFERENCES

1. Private correspondence with Firth–Brown Ltd.
2. Private correspondence with the English Steel Corporation Ltd.
3. Private correspondence with Rolls-Royce Ltd.
4. Private correspondence with Jessop-Saville Ltd.
5. Private correspondence with The United Steel Companies Ltd.
6. C. J. Smithells. *Metals Reference Book*, Second Edition. Butterworths Scientific Publications, London, 1953.
7. P. H. Frith. Fatigue Tests on Crankshaft Steels. *JISI* 1948, Vol. 159, Aug., No. II, pp. 385—409.
8. B.S. 971 : Commentary on the British Standard Wrought Steels En Series. British Standards Institution, London, 1950.
9. Private correspondence with Westland Aircraft Ltd.
10. Private correspondence with De Havilland Aircraft Co. Ltd.
11. Ministry of Supply, unpublished report.
12. War Office, unpublished report.
13. Private correspondence with Dowty–Rotol Ltd.
14. W. E. Bardgett. Factors affecting the Influence of Nitriding on the Fatigue Strength. *Metal Treatment*, 1943, Vol. 10, p. 87.

1½ PER CENT CHROMIUM-ALUMINIUM-MOLYBDENUM NITRIDING STEEL

SPECIFICATION

CHEMICAL COMPOSITION (%)

	C	Si	Mn	S	P	Ni	Cr	Mo	Al
En 41A	0·25—0·35	0·10—0·45	0·65 max.	0·050 max.	0·050 max.	0·40 max.	1·40—1·80	0·10—0·25	0·90—1·30
En 41B	0·35—0·45	0·10—0·45	0·65 max.	0·050 max.	0·050 max.	0·40 max.	1·40—1·80	0·10—0·25	0·90—1·30

MECHANICAL PROPERTIES

	Har ened and Tempere Condition (4)		
	R	S	T
Limiting ruling section, in.	6	4	2½
Tensile strength tons/sq.in., min. (Rm)	45	50	55
Yield stress (1) tons/sq.in., min. (Rs)	34	38	44
0·2% proof stress (2) tons/sq.in., min. (0·2Rp)	32	36	41
Elongation, %, min. (A)	20	19	17
Izod impact value, ft.lb., min.	40	40	35
Brinell hardness No. (3) (HB)	201/255	223/277	248/302

(1) Not to be used as acceptance tests except by special arrangement between the purchaser and manufacturer.
(2) When specifically requested in the enquiry or order.
(3) For information purposes only. Not contractual part of the specification.
(4) The selection of En 41A or En 41B will be governed by the tensile range required.

RELATED SPECIFICATIONS

United Kingdom		America	France	Germany		Sweden	U.S.S.R.	
En	DTD	AMS	AFNOR	Werkstoff No.	Name	SIS	GOST	Mark
41A	87B	6470E	30CAD 6—12 40CAD 6—12	1.8504 1.8507 1.8550	34CrAl6 34CrAlMo5 34CrAlNi7	14.2940	4543—61	38KhMYuA

APPLICATIONS

En41 is used in the nitrided condition for parts requiring a very high degree of resistance to wear. In this condition the steel has a relatively high fatigue resistance and low notch sensitivity and is frequently used where resistance to fatigue is the predominant requirement. The steel when nitrided has a good resistance to corrosion in fresh and sea water, and steam and moist atmospheres, although its corrosion resistance is inferior to that of stainless steel.

En 41A is used for applications where a core strength of 45—55 tons/sq.in. is required. Typical uses are die casting plungers, gears, pins, and bolts, and for many applications in the automobile, railway, mining, textile and machine tool industries, where the maximum surface hardness of a nitrided case, combined with medium tensile core strength, is required.

En 41B is used for applications requiring a core strength of 55—65 tons/sq.in. Typical uses are brick press plates, gudgeon pins, clutch plates, collets, rollers, pneumatic hammer parts, ball joints, and track and shackle pins.

WELDING

See Introduction (List of Abbreviations) for key to symbols used in this table

	Welding Process					
	MA	G	IG	S	F	B
En 41A	g	g	g	g	b	j
En 41B	g	g	g	g	b	j

Reproduced by kind permission of H.M.S.O.

(1) Sufficient information is not available to enable a recommendation to be made on welding before nitriding, in view of the unknown effects of the nitriding treatment on weld metal toughness.
(2) Welding or brazing on the nitrided case is not recommended.
(3) The steels should not be furnace-brazed after nitriding.

MACHINABILITY

In the softened condition the machinability is approximately 45% of that for mild steel (En 3). When hardened and tempered to 45—55 tons/sq.in. the machinability is about 40%, and when hardened to 55—65 tons/sq.in. about 35%, of that for mild steel.

HOT WORKING AND HEAT TREATMENT TEMPERATURES

Forging, rolling and stamping	1200°C, finish above 850°C. Retarded cooling is advisable for large or intricate sections.
Softening	700°C—750°C, air or furnace cool.
Hardening	880°C—910°C, oil quench.
Tempering	550°C—720°C, air cool.
Stabilizing	530°C—550°C, air cool.

NITRIDING DATA

Nitriding is carried out at a temperature of 500°C to 520°C. Fig 41.1 shows the relation between case depth and nitriding period.

En 41A, En 41B

(Ref. 1)

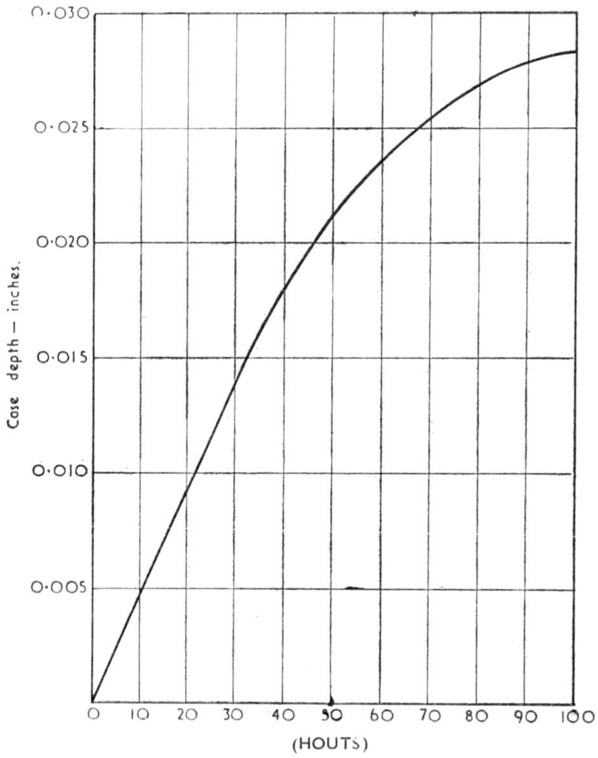

Fig. 41.1

TYPICAL CASE HARDNESS—DEPTH CURVES

Treated size $1\frac{1}{8}$ in. dia.
Oil quenched 900°C. Tempered 650°C.
Nitrided 515°C, 80 hrs.
25% dissociation.

En 41A, En 41B (Ref. 2)

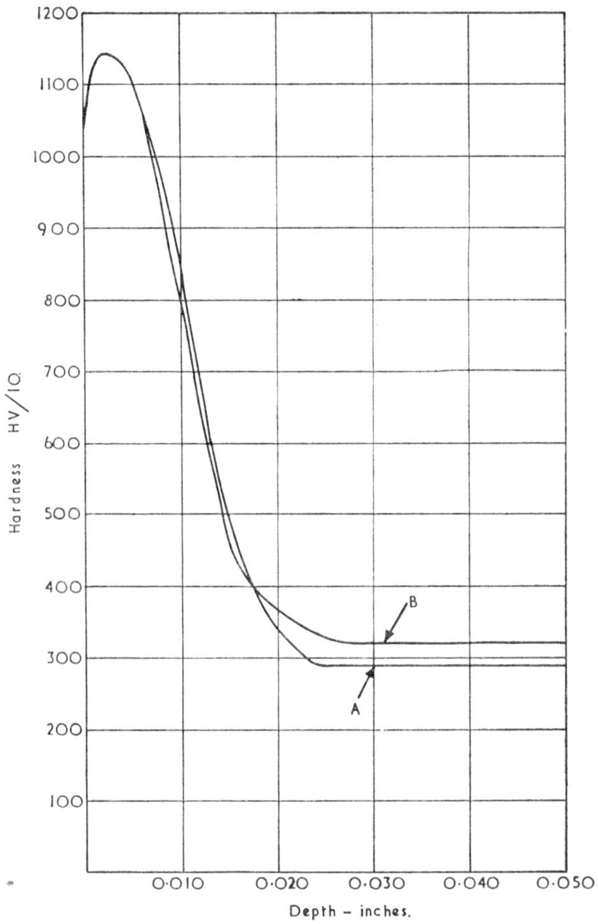

Fig. 41.2

En 41A (Ref. 1)

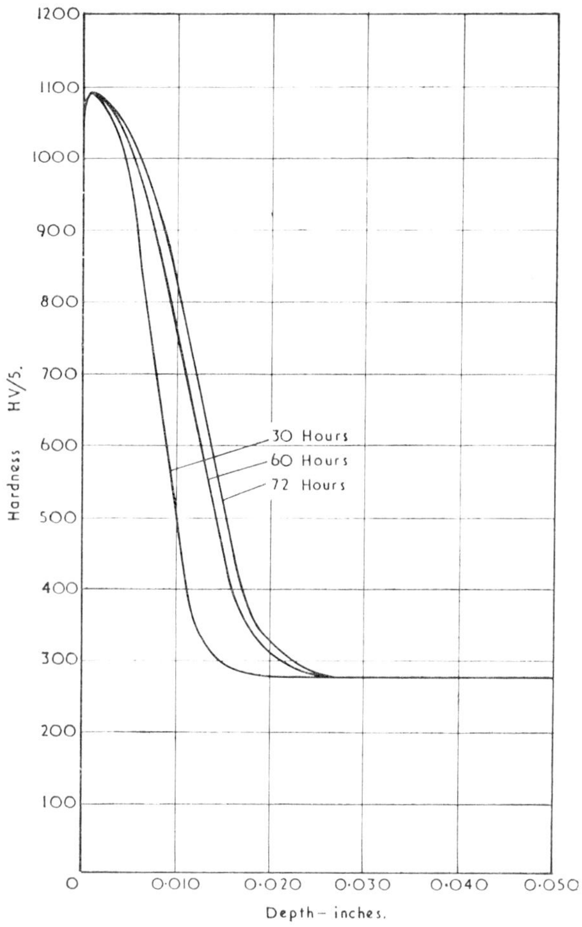

Fig. 41.3

PHYSICAL PROPERTIES

SPECIFIC GRAVITY (d)

Chemical Composition (%)	Heat Treatment	Specific Gravity at 20°C	Ref.
En 41A composition	Softened	7·74	1
" " "	Hardened and tempered	7·71	2
En 41B composition	Softened	7·73	1
" " "	Hardened and tempered	7·69	2

SPECIFIC HEAT (c)

Typical value of Mean Specific Heat (20—100°C) is 0·114 cal/g deg. C.

MEAN COEFFICIENT OF THERMAL EXPANSION (α)

Chemical Composition (%)	Heat Treatment	$10^6 \times$ Mean Coefficient of Expansion per deg. C							Ref.
		20 to 100°C	20 to 200°C	20 to 300°C	20 to 400°C	20 to 500°C	20 to 600°C	20 to 680°C	
En 41A composition (core)	Softened	12·5	12·7	13·0	13·3	13·5	13·7		1
" " " (case)	Nitrided	11·0		12·0		13·0			"
" " "	Hardened & tempered	12·5		13·3		14·0		14·4	2
En 41B composition (core)	Softened	12·5		13·0		14·0			1
" " " (case)	Nitrided	11·0		12·0		13·0			"
" " "	Hardened & tempered	12·7		13·4		14·0		14·4	2

ELECTRICAL RESISTIVITY (ρ)

Chemical Composition (%)						Heat Treatment	Electrical Resistivity microhm-cm 20°C	Ref.
C	Si	Mn	Cr	Mo	Al			
En 41A composition						Softened	37	1
En 41B composition						Hardened & tempered	37	2
0·39	0·23	0·61	1·63	0·18	1·10	Softened	39	1

THERMAL CONDUCTIVITY (k)

Typical value for En 41B hardened and tempered 0·07 cal/cm s deg. C at 20°C.

YOUNG'S MODULUS (E)

Chemical Composition (%)						Heat Treatment	Young's Modulus at 20°C tons/sq.in.	Ref.
C	Si	Mn	Cr	Mo	Al			
0·26	0·32	0·59	1·67	0·22	1·01	Softened	13,550	1
0·39	0·23	0·61	1·63	0·18	1·10	Softened	12,800	"

MAGNETIC PROPERTIES

Chemical Composition (%)	Magnetic induction B (Gauss) for magnetizing field H (Oersted) =								Brem (Gauss)	Hc (Oersted)	Ref.
	H = 10	20	30	40	50	100	150	200			
En 41 composition	1,000	4,000	10,000	13,000	14,000	16,500	17,500	17,800	12,850	17·0	1

TRANSFORMATION CHARACTERISTICS

ISOTHERMAL TRANSFORMATION

Chemical Composition (%)								
C	Si	Mn	S	P	Ni	Cr	Mo	Al
0·40	0·53	0·56	0·001	0·012	0·21	1·65	0·23	1·10

Basic Electric Arc.
Condition-annealed.
Austenitizing Temperature 900°C, 30 min.

(Ref. 3)

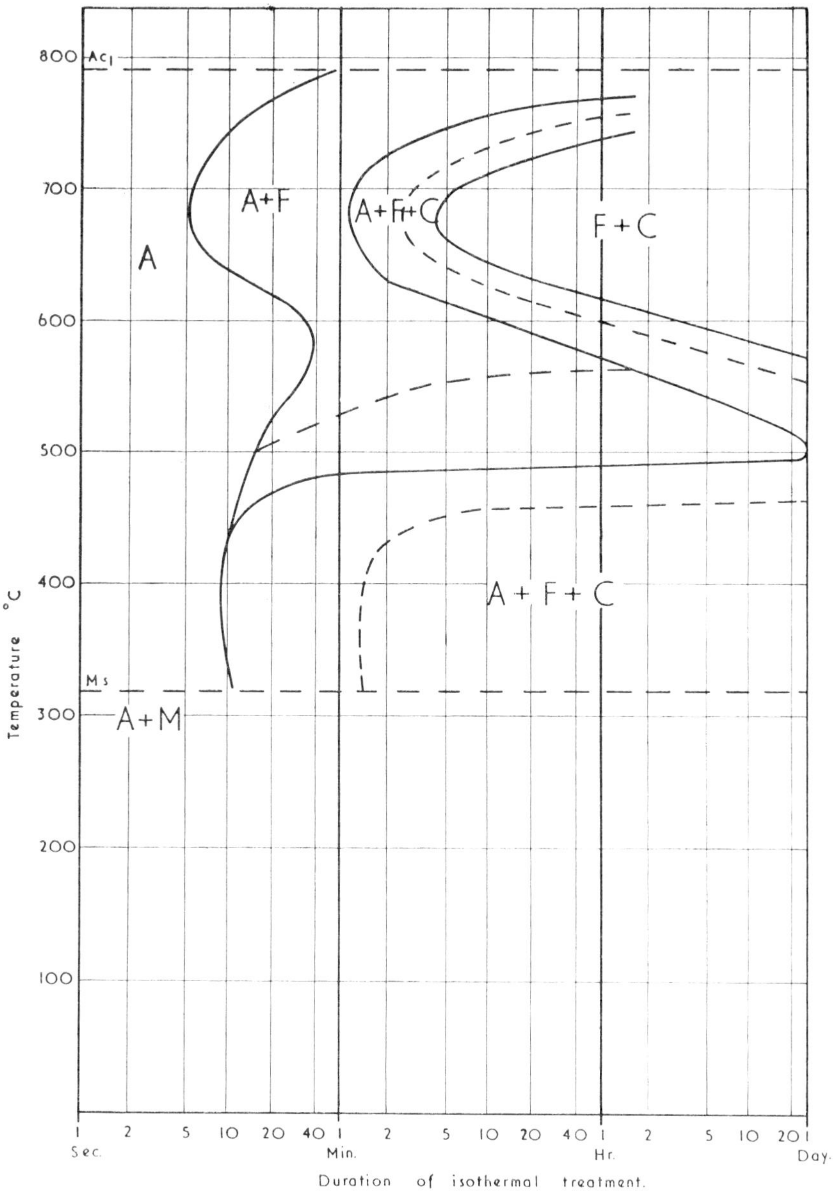

Fig. 41.4

CONTINUOUS COOLING TRANSFORMATION

Chemical Composition (%)								
C	Si	Mn	S	P	Ni	Cr	Mo	Al
0·40	0·53	0·56	0·001	0·012	0·21	1·65	0·23	1·10

Basic Electric Arc.
Austenitizing Temperature 900°C, 30 min.

(Ref. 3)

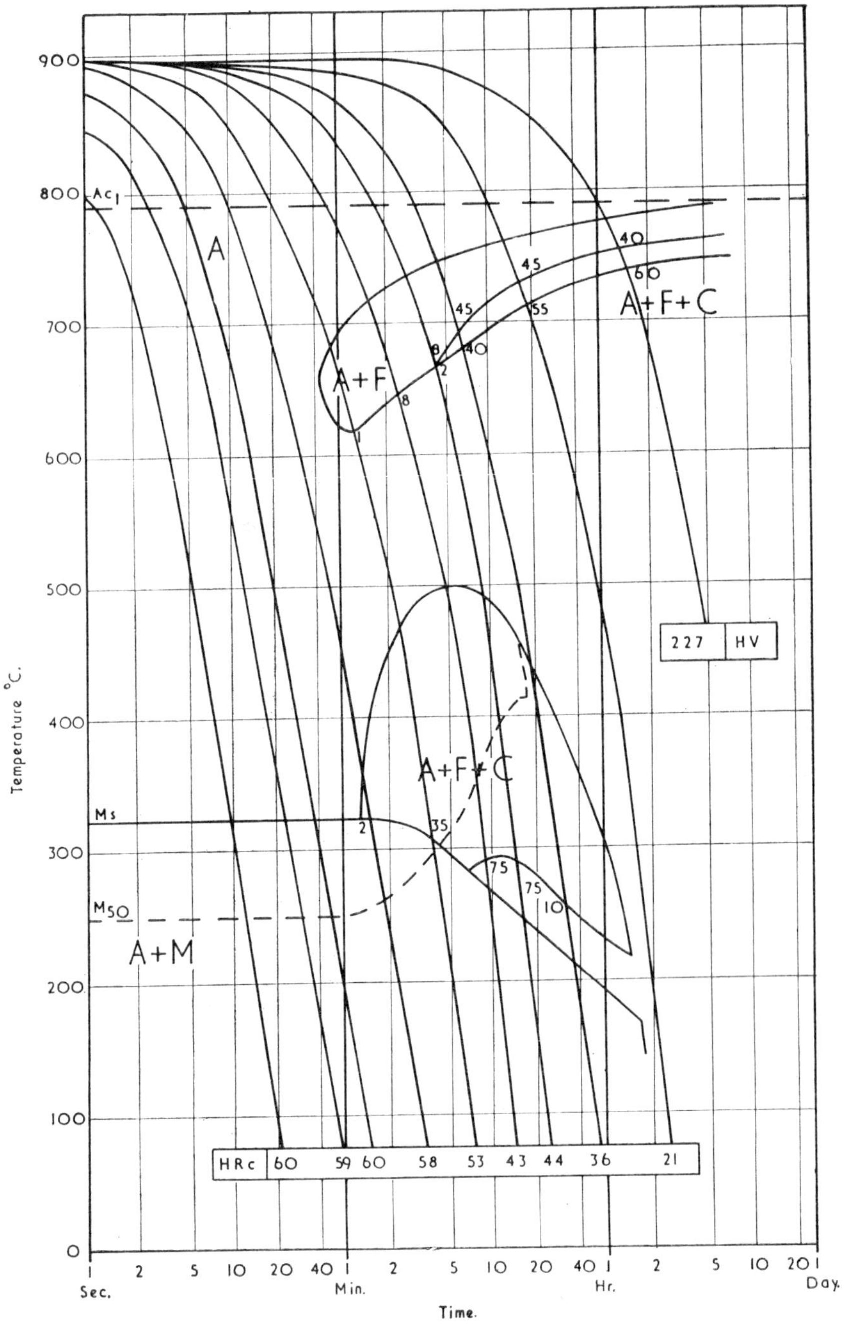

Fig. 41.5

HARDENABILITY

END QUENCH HARDENABILITY DATA

Curve No.	Chemical Composition (%)									Section Size dia. (in.)	ASTM Grain Size Actual	Ref.
	C	Si	Mn	S	P	Ni	Cr	Mo	Al			
1	0·40	0·27	0·46	0·011	0·016	0·20	1·58	0·25	1·01	$1\frac{1}{16}$	7	4
2	0·30	0·18	0·39	0·027	0·024	0·06	1·56	0·17	1·32	$1\frac{1}{4}$	6—8	"
3	0·29	0·19	0·39	0·030	0·023	0·08	1·57	0·19	1·32	$2\frac{1}{4}$	6—8	"

Basic Electric Arc.
Bars normalized 925°C.
End quenched from 900°C (curve 1)
925°C (curve 2 and 3).

As quenched.

(Ref. 4)

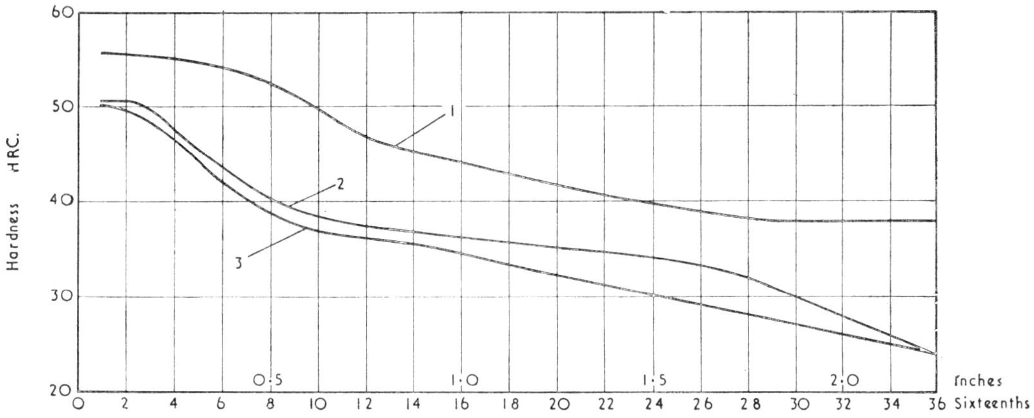

Fig 41.6

Quenched and tempered 650°C.

(Ref. 4)

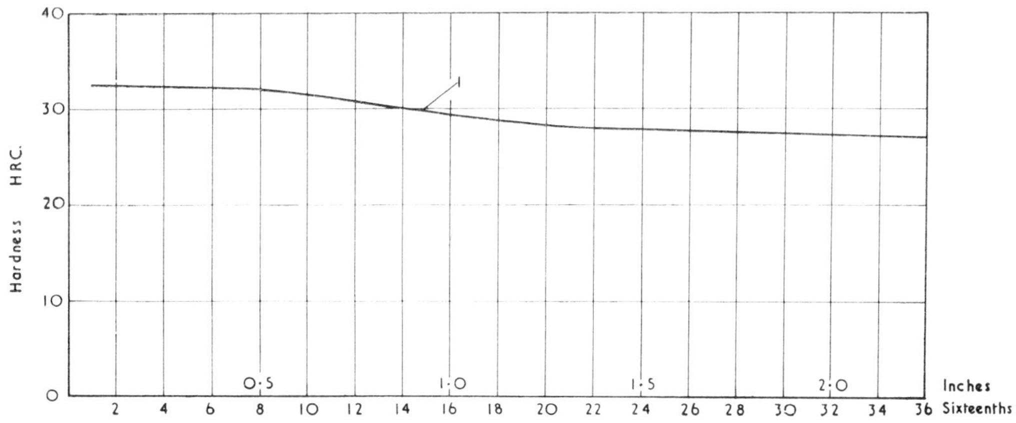

Fig. 41.7

En 41

END QUENCH HARDENABILITY DATA (Continued)

Typical curves for En 41A, En 41B

(Ref. 2)

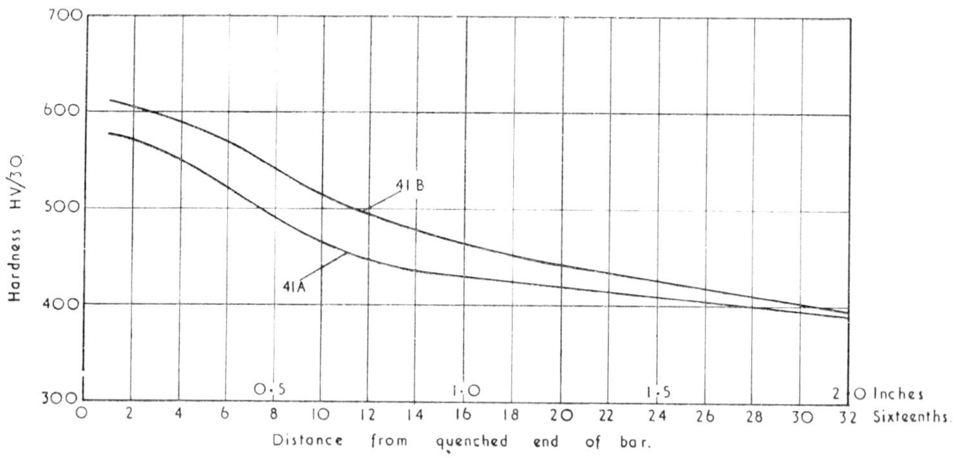

Fig. 41.8

MECHANICAL PROPERTIES AT ROOM TEMPERATURE

Process of Manu-facture	ASTM Grain Size	Chemical Composition (%)									Section Size dia. (in.)	Heat Treatment °C	Tensile									Izod ft.lb.	Ref.
		C	Si	Mn	S	P	Ni	Cr	Mo	Al			0·05% P.S. (0·05Rp)	0·1% P.S. (0·1Rp)	0·2% P.S. (0·2Rp)	0·5% P.S. (0·5Rp)	Y.P. (Rs)	T.S. (Rm)	El % (A)	R.A. % (Z) 5D	$4\sqrt{A}$		
													tons/sq.in.										
		0·25		0·58			0·21	1·58	0·22	1·01	$1\frac{1}{8}$	OQ900 T650					33·1	39·2	29			68	5
		0·27						1·51	0·20	0·98	$\frac{7}{16}$	OQ900 T650					50·2	57·4	25				1
		0·27						1·53	0·25	1·03	$2\frac{5}{8}$	" "					38·6	48·2	24			60	"
		0·28	0·34	0·49				1·46	0·18	0·97	$1\frac{1}{16}$	OQ900 T670					41·0	53·0	28		66	92	2
		0·29	0·23	0·59			0·16	1·72	0·27	1·40	$1\frac{1}{8}$	OQ875 T550					55·4	61·8	19		51	29	1
		"	"	"			"	"	"	"	"	" T600					46·0	55·0	$22\frac{1}{2}$		$52\frac{1}{2}$	58	"
		"	"	"			"	"	"	"	"	" T650					42·4	51·6	$24\frac{1}{2}$		$60\frac{1}{2}$	66	"
		"	"	"			"	"	"	"	"	" T700					33·2	45·4	27		66	79	"
		0·29						1·54	0·24	1·06	$2\frac{1}{4}$	OQ900 T650					39·2	48·6	25			75	"
		0·29						1·62	0·24	1·02	$3\frac{1}{4}$	" "					33·2	48·0	$26\frac{1}{2}$			43	"
		0·29						1·60	0·20	1·01	$3\frac{3}{8}$	" T600					41·0	50·2	21			35	"
B.E.A.	Actual 6—8	0·29	0·19	0·39	0·030	0·024	0·08	1·57	0·19	1·32	1	N925 OQ925 T650					39·8	49·9	24		$57\frac{1}{2}$	54	4
		0·29	0·32	0·58				1·59	0·21	1·10	$3\frac{1}{2}$	OQ900 T650					33·4	48·0	25		60	54	2
		0·30						1·67	0·24	0·99	$2\frac{1}{4}$	OQ900 T650					41·2	50·8	22			71	1
		0·30						1·58	0·19	1·08	$3\frac{3}{8}$	" "					37·6	47·0	26			65	"
B.E.A.	Actual 6—8	0·30	0·18	0·39	0·027	0·024	0·06	1·56	0·17	1·32	1	N925 OQ925 T650					41·0	51·4	24		$59\frac{1}{2}$	54	4
		0·31		0·63			0·23	1·66	0·24	0·97	$1\frac{1}{8}$	OQ900 T700					38·3	51·2	28			65	5
		"		"			"	"	"	"	"	" T750					36·1	46·9	31			69	"
		0·31						1·64	0·24	0·91	$3\frac{1}{2}$	OQ900 T650					40·0	49·0	25			41	1
		0·31						1·70	0·21	1·07	$4\frac{1}{2}$	" "					38·8	48·4	21			45	"
		0·31	0·42	0·54	0·010	0·017	0·23	1·73	0·28	0·93	$3\frac{1}{2}$	A720 OQ900 T550				47·2		56·0	22	18		44	"
		0·31	0·35	0·55				1·64	0·22	0·99	$1\frac{1}{8}$	OQ900 T690					38·3	50·2	27		$64\frac{1}{2}$	85	2
		0·31	0·38	0·60				1·65	0·19	1·11	$6\frac{3}{4}\times 2$	" T650					42·6	51·0	$22\frac{1}{2}$		57	78	"
		0·32	0·33	0·55				1·59	0·22	1·21	$2\frac{1}{2}$	OQ900 T650					42·0	52·0	26		58	49	2
B.E.A.		0·32	0·45	0·57	0·009	0·018	0·24	1·65	0·20	1·04	$\frac{3}{4}$	AH900	35·1	40·3	45·9	53·8		70·5	24		47	23	"
"		"	"	"	"	"	"	"	"	"	"	" T500 2 hrs. AC	38·9	43·3	49·2	55·8		74·9	18		$38\frac{1}{2}$	12	"
"		"	"	"	"	"	"	"	"	"	"	" T550 "	43·9	48·1	52·1	57·3		72·8	$17\frac{1}{2}$		$42\frac{1}{2}$	12	"
"		"	"	"	"	"	"	"	"	"	"	" T600 "	44·7	46·8	48·1	50·1		60·1	19		51	18	"
"		"	"	"	"	"	"	"	"	"	"	" T650 "	40·0	40·3	40·4	41·1		54·1	$23\frac{1}{2}$		$61\frac{1}{2}$	37	"
"		"	"	"	"	"	"	"	"	"	"	" T700 "				37·5		51·4	$25\frac{1}{2}$		$60\frac{1}{2}$	67	"
"		"	"	"	"	"	"	"	"	"	$1\frac{1}{8}$	AH900	34·7	39·1	45·3	52·6		69·0	22		44	17	"
"		"	"	"	"	"	"	"	"	"	"	" T2hr 500 AC	38·7	43·4	48·6	54·8		72·9	$17\frac{1}{2}$		37	8	"
"		"	"	"	"	"	"	"	"	"	"	" " 550 "	42·1	46·9	50·5	55·9		71·2	$17\frac{1}{2}$		$40\frac{1}{2}$	10	"
"		"	"	"	"	"	"	"	"	"	"	" " 600 "	43·5	45·0	46·3	48·2		58·8	19		$50\frac{1}{2}$	16	"

MECHANICAL PROPERTIES AT ROOM TEMPERATURE (Continued)

Process of Manufacture	Chemical Composition (%) C	Si	Mn	S	P	Ni	Cr	Mo	Al	Section Size dia. (in.)	Heat Treatment °C	Tensile 0·05% P.S. (0·05Rp) tons/sq.in.	0·1% P.S. (0·1Rp)	0·2% P.S. (0·2Rp)	0·5% P.S. (0·5Rp)	T.S. (Rm)	El % (A)	R.A. % (Z)	Izod ft.lb.	Ref.
B.E.A.	0·32	0·45	0·57	0·009	0·018	0·24	1·65	0·20	1·04	1⅛	AH900 T2hrs.650AC				40·6	53·6	24	61	36	2
”	”	”	”	”	”	”	”	”	”	”	” ” 700 ”				35·0	50·0	24	62½	53	”
”	”	”	”	”	”	”	”	”	”	2½	AH900 T2hrs.500AC	27·9	33·8	37·6	43·2	62·6	17	38	16	”
”	”	”	”	”	”	”	”	”	”	”	” ” 600 ”	33·6	35·5	37·2	40·1	54·8	23	53½	22	”
”	”	”	”	”	”	”	”	”	”	”	” ” 650 ”				27·5	49·1	28	59	42	”
”	”	”	”	”	”	”	”	”	”	”	” ” 700 ”	24·5	26·5	27·5	29·5	46·5	28	61½	46	”
”	”	”	”	”	”	”	”	”	”	4¼	AH900 T2hrs.500AC	25·4	27·3	29·5	34·0	49·0	26½	58	28	”
”	”	”	”	”	”	”	”	”	”	”	” ” 600 ”	24·8	26·3	27·4	31·3	48·5	26½	54½	25	”
”	”	”	”	”	”	”	”	”	”	”	” ” 650 ”	25·5	25·9	26·3	28·1	47·9	27	56½	28	”
”	”	”	”	”	”	”	”	”	”	”	” ” 700 ”	20·8	23·1	24·2	26·0	45·6	29	59½	30	”
”	”	”	”	”	”	”	”	”	”	6 (C)	AH900 T2hrs.600AC	25·1	26·1	27·0	29·4	50·0	26½	55	22	”
”	”	”	”	”	”	”	”	”	”	” (M)	” ” ”	24·8	26·3	27·4	30·1	50·9	24	41	21	”
”	”	”	”	”	”	”	”	”	”	” (C)	” ” 650 ”	25·0	26·0	26·5	28·2	49·5	26	54	26	”
”	”	”	”	”	”	”	”	”	”	” (M)	” ” ” ”	24·6	25·8	26·4	28·9	49·6	24	52	26	”
”	”	”	”	”	”	”	”	”	”	” (C)	” ” 700 ”	24·2	24·9	25·0	26·0	47·1	27	54½	32	”
”	”	”	”	”	”	”	”	”	”	” (M)	” ” ” ”	23·9	24·7	25·0	26·5	46·8	27½	53½	28	”
”	”	”	”	”	”	”	”	”	”	¾	OQ900	60·0	66·9	75·8	88·3	102·6	13	46½	35	”
”	”	”	”	”	”	”	”	”	”	”	” T2hrs.500AC	68·8	72·4	75·5	78·9	85·5	15½	46	17	”
”	”	”	”	”	”	”	”	”	”	”	” ” 550 ”	63·8	66·2	68·3	70·3	75·9	17½	51½	28	”
”	”	”	”	”	”	”	”	”	”	”	” ” 600 ”	53·9	54·8	55·8	56·7	64·6	20½	59	61	”
”	”	”	”	”	”	”	”	”	”	”	” ” 650 ”	48·5	48·8	49·0	49·6	59·5	23	63½	74	”
”	”	”	”	”	”	”	”	”	”	”	” ” 670 ”				44·5	55·7	25½	64½	85	”
”	”	”	”	”	”	”	”	”	”	”	” ” 700 ”				41·0	52·9	27	67	88	”
”	”	”	”	”	”	”	”	”	”	1⅛	OQ900	43·8	50·7	59·4	71·3	84·7	16	40½	53	”
”	”	”	”	”	”	”	”	”	”	”	” T2hrs.100AC	50·2	57·0	70·2	75·2	85·3	17	50½	47	”
”	”	”	”	”	”	”	”	”	”	”	” ” 200 ”	53·0	60·0	67·0	75·7	84·5	16½	52½	51	”
”	”	”	”	”	”	”	”	”	”	”	” ” 300 ”	56·6	62·0	68·6	77·3	85·6	16½	55½	42	”
”	”	”	”	”	”	”	”	”	”	”	” ” 400 ”	59·0	64·4	70·1	76·0	83·8	17	50½	28	”
”	”	”	”	”	”	”	”	”	”	1⅛	OQ900 T2hrs.500AC	62·8	66·5	70·0	74·0	82·1	15½	45½	17	”
”	”	”	”	”	”	”	”	”	”	”	” ” 550 ”	58·5	61·7	64·3	65·9	74·3	19	50	22	”
”	”	”	”	”	”	”	”	”	”	”	” ” 600 ”	49·5	50·5	51·3	52·3	61·8	20	57	68	”
”	”	”	”	”	”	”	”	”	”	”	” ” 625 ”	47·0	47·8	48·5	49·3	59·6	21	60½	72	”
”	”	”	”	”	”	”	”	”	”	”	” ” 650 ”	43·8	44·1	44·3	44·8	56·8	24	62	77	”
”	”	”	”	”	”	”	”	”	”	”	” ” 675 ”	41·9	42·2	42·3	43·1	54·7	25½	63	90	”
”	”	”	”	”	”	”	”	”	”	”	” ” 700 ”				39·1	51·6	27½	66	91	”
”	”	”	”	”	”	”	”	”	”	”	” ” 720 ”				35·9	48·9	29	67½	93	”
”	”	”	”	”	”	”	”	”	”	”	” ” 740 ”				35·0	48·3	29½	68	98	”
”	”	”	”	”	”	”	”	”	”	”	” ” 760 ”				33·2	46·1	31	67	98	”
”	”	”	”	”	”	”	”	”	”	2½	OQ900	40·4	47·0	54·5	64·1	76·2	19	50	32	”

MECHANICAL PROPERTIES AT ROOM TEMPERATURE (Continued)

Process of Manufacture	Chemical Composition (%)									Section Size dia. (in.)	Heat Treatment °C	Tensile							Izod ft.lb.	Ref.
	C	Si	Mn	S	P	Ni	Cr	Mo	Al			0·05% P.S. (0·05Rp) tons/sq.in.	0·1% P.S. (0·1Rp) tons/sq.in.	0·2% P.S. (0·2Rp) tons/sq.in.	0·5% P.S. (0·5Rp) tons/sq.in.	T.S. (Rm) tons/sq.in.	El % (A)	R.A. % (Z)		
B.E.A.	0·32	0·45	0·57	0·009	0·018	0·24	1·65	0·20	1·04	2½	OQ900 T2hrs.400AC	51·6	56·6	62·4	69·7	76·3	19	56½	31	2
"	"	"	"	"	"	"	"	"	"	"	" " 500 "	41·5	53·0	58·1	63·8	76·7	18	42	10	"
"	"	"	"	"	"	"	"	"	"	"	" " 550 "	53·5	57·6	60·4	64·3	74·4	18½	45½	15	"
"	"	"	"	"	"	"	"	"	"	"	" " 600 "	50·5	52·4	53·8	55·5	64·1	16	49½	22	"
"	"	"	"	"	"	"	"	"	"	"	" " 650 "	41·5	43·0	43·3	44·2	56·3	22½	60½	66	"
"	"	"	"	"	"	"	"	"	"	"	" " 675 "	40·1	40·3	40·5	41·0	54·4	24	62	75	"
"	"	"	"	"	"	"	"	"	"	"	" " 700 "				35·9	50·1	27	64	89	"
"	"	"	"	"	"	"	"	"	"	4½	OQ900	32·1	39·2	46·6	57·0	74·8	19½	41	23	"
"	"	"	"	"	"	"	"	"	"	"	" T2hrs.400AC	42·5	47·6	53·6	62·5	71·6	18	45	19	"
"	"	"	"	"	"	"	"	"	"	"	" " 500	43·1	49·0	54·5	63·6	72·3	17	40	12	"
"	"	"	"	"	"	"	"	"	"	"	" " 550	39·0	45·0	51·4	60·2	70·2	16½	42	12	"
"	"	"	"	"	"	"	"	"	"	"	" " 600	42·6	47·3	50·6	54·0	63·4	17½	47	17	"
"	"	"	"	"	"	"	"	"	"	"	" " 650	39·7	40·3	40·6	41·6	54·4	23	60½	51	"
"	"	"	"	"	"	"	"	"	"	"	" " 675	38·7	39·1	40·3	41·4	52·7	25	62½	57	"
"	"	"	"	"	"	"	"	"	"	"	" " 700	37·0	36·8	37·5	40·0	51·1	26	62	68	"
"	"	"	"	"	"	"	"	"	"	6 (O)	OQ900	37·6	42·7	48·3	55·1	67·9	23	48½	20	"
"	"	"	"	"	"	"	"	"	"	" (M)	"	40·2	44·9	49·9	56·6	72·1	18	43½	23	"
"	"	"	"	"	"	"	"	"	"	" (O)	" T2hrs.400AC	44·3	48·3	53·9	60·6	66·9	22	57½	12	"
"	"	"	"	"	"	"	"	"	"	" (M)	" " "	44·5	49·1	54·5	61·8	69·5	18	50½	8	"
"	"	"	"	"	"	"	"	"	"	" (O)	" " 500 "	41·9	46·6	51·5	57·6	71·1	19	40	9	"
"	"	"	"	"	"	"	"	"	"	" (M)	" " " "	42·0	46·9	52·0	57·7	72·1	17	36½	7	"
"	"	"	"	"	"	"	"	"	"	" (O)	OQ900 T2hrs.550AC	43·8	48·4	52·4	57·2	70·7	17½	39½	11	"
"	"	"	"	"	"	"	"	"	"	" (M)	" " "	41·8	47·2	52·3	57·6	71·1	16½	37½	5	"
"	"	"	"	"	"	"	"	"	"	" (O)	" " 600 "	43·1	44·9	46·4	47·6	57·8	21½	52	18	"
"	"	"	"	"	"	"	"	"	"	" (M)	" " "	42·6	44·5	45·9	47·5	58·7	19	47½	16	"
"	"	"	"	"	"	"	"	"	"	" (O)	" " 650 "	41·1	41·9	42·6	44·5	55·2	21	55	28	"
"	"	"	"	"	"	"	"	"	"	" (M)	" " " "	40·3	41·3	42·5	44·6	55·0	20	53	22	"
"	"	"	"	"	"	"	"	"	"	" (O)	" " 675 "	36·0	36·6	37·2	39·0	51·9	24½	59	47	"
"	"	"	"	"	"	"	"	"	"	" (M)	" " " "	36·3	37·0	37·8	39·9	52·3	25	59	30	"
"	"	"	"	"	"	"	"	"	"	" (O)	" " 700 "	37·3	37·5	37·8	38·6	52·5	24	59½	45	"
"	"	"	"	"	"	"	"	"	"	" (M)	" " " "	37·0	37·6	38·0	39·1	52·5	23	59	33	"
"	"	"	"	"	"	"	"	"	"	¾	WQ900	81·5	87·5	95·6		113·2	12	46	27	"
"	"	"	"	"	"	"	"	"	"	"	" T2hrs.500AC	70·3	72·0	74·0	77·8	83·2	15	50	25	"
"	"	"	"	"	"	"	"	"	"	"	" " 550 "	63·8	64·5	65·4	67·5	73·6	18	55½	39	"
"	"	"	"	"	"	"	"	"	"	"	" " 600 "	55·5	56·0	56·4	57·7	66·0	21½	59½	63	"
"	"	"	"	"	"	"	"	"	"	"	" " 650 "	48·3	48·5	48·7	49·4	58·6	24½	63½	76	"
"	"	"	"	"	"	"	"	"	"	"	" " 700 "				44·2	51·1	27½	67	89	"
"	"	"	"	"	"	"	"	"	"	1⅛	WQ900	68·1	76·2	86·1	99·3	116·0	12½	39	23	"
"	"	"	"	"	"	"	"	"	"	"	" T2hrs.500AC	69·4	72·2	75·1	79·4	85·0	15	47	21	"

En 41

MECHANICAL PROPERTIES AT ROOM TEMPERATURE (Continued)

Process of Manu-facture	Chemical Composition (%) C	Si	Mn	S	P	Ni	Cr	Mo	Al	Section Size dia. (in.)	Heat Treatment °C	Tensile 0·05% P.S. (0·05Rp) tons/sq.in.	0·1% P.S. (0·1Rp)	0·2% P.S. (0·2Rp)	0·5% P.S. (0·5Rp)	Y.P. (Rs)	T.S. (Rm)	El % (A)	R.A. % (Z)	Izod ft.lb.	Ref.
B.E.A.	0·32	0·45	0·57	0·009	0·018	0·24	1·65	0·20	1·04	1⅛	WQ900 T2hrs.550AC	63·5	64·5	66·0	68·5		73·6	17	55½	39	2
"	"	"	"	"	"	"	"	"	"	"	" " 600 "	55·3	55·8	56·5	58·1		66·0	20½	59½	63	"
"	"	"	"	"	"	"	"	"	"	"	" " 650 "	47·3	47·5	47·5	49·0		59·1	23½	63½	79	"
"	"	"	"	"	"	"	"	"	"	"	" " 675 "				43·9		55·8	25	65	88	"
"	"	"	"	"	"	"	"	"	"	"	" " 700 "				39·9		51·4	28	67½	93	"
"	"	"	"	"	"	"	"	"	"	2½	WQ900 T2hrs.400AC	73·4	79·7	86·6	95·3		102·4	11½	39½	14	"
"	"	"	"	"	"	"	"	"	"	"	" " 500 "	68·6	73·6	77·5	82·0		87·6	14½	41	16	"
"	"	"	"	"	"	"	"	"	"	"	" " 550 "	64·5	66·1	67·9	70·1		75·0	17	49½	31	"
"	"	"	"	"	"	"	"	"	"	"	" " 600 "	56·5	57·9	59·4	61·5		68·3	20	54	55	"
"	"	"	"	"	"	"	"	"	"	"	" " 650 "	46·8	47·3	47·8	48·6		58·7	23	61½	83	"
"	"	"	"	"	"	"	"	"	"	"	" " 700 "				37·5		50·0	30½	68½	104	"
"	"	"	"	"	"	"	"	"	"	4½	WQ900 T2hrs.400AC	55·0	61·9	70·9	82·6		91·5	11	38	13	"
"	"	"	"	"	"	"	"	"	"	"	" " 500 "	50·6	57·6	66·8	73·0		80·4	14	39	15	"
"	"	"	"	"	"	"	"	"	"	"	" " 550 "	50·2	57·2	62·7	67·7		73·1	15	45½	19	"
"	"	"	"	"	"	"	"	"	"	"	" " 600 "	48·3	57·4	54·2	57·0		63·0	19	52	53	"
"	"	"	"	"	"	"	"	"	"	"	" " 650 "	43·7	44·9	45·7	48·2		57·1	23½	61½	80	"
"	"	"	"	"	"	"	"	"	"	"	" " 700 "			39·2	42·4		51·4	26	65½	113	"
"	"	"	"	"	"	"	"	"	"	6 (O)	WQ900 T2hrs.400AC	55·0	60·7	67·0	74·6		83·0	13½	43	14	"
"	"	"	"	"	"	"	"	"	"	" (M)	" " " "	41·4	51·5	57·7	66·0		75·5	15	41	9	"
"	"	"	"	"	"	"	"	"	"	" (O)	" " 500 "	52·1	58·0	62·9	67·8		77·5	15½	35½	13	"
"	"	"	"	"	"	"	"	"	"	" (M)	" " " "	47·4	52·4				74·6	15	37½	8	"
"	"	"	"	"	"	"	"	"	"	" (O)	" " 550 "	52·7	54·9	61·5	65·5		73·7	15½	38½	14	"
"	"	"	"	"	"	"	"	"	"	" (M)	" " " "	45·1	50·5	55·6	61·0		71·3	15	35½	7	"
"	"	"	"	"	"	"	"	"	"	" (O)	" " 600 "	45·2	47·9	50·2	52·5		61·0	16½	44½	35	"
"	"	"	"	"	"	"	"	"	"	" (M)	" " " "	44·4	47·0	49·3	51·8		61·2	16	46½	12	"
"	"	"	"	"	"	"	"	"	"	" (O)	" " 650 "	43·4	44·7	45·6	46·8		58·2	21	56	70	"
"	"	"	"	"	"	"	"	"	"	" (M)	" " " "	42·6	43·7	44·6	46·0		57·3	21	55	36	"
"	"	"	"	"	"	"	"	"	"	" (O)	" " 700 "	36·0	36·6	36·7	37·0		50·7	26½	62	91	"
"	"	"	"	"	"	"	"	"	"	" (M)	" " " "	35·3	35·9	36·2	37·0		50·9	26½	60½	46	"
	0·33	0·23	0·60				1·67	0·20	1·19	3¾	OQ900 T640					46·0	59·0	20	47	24	2
	0·35		0·55				1·56	0·20	0·99	2½	OQ900 T700					43·8	55·3	23		77	5
	"		"				"	"	"	"	" T750					37·7	49·1	25		90	"
	0·37	0·37	0·59				1·62	0·18	1·18	1⅝	OQ900 T650					52·0	64·0	21		51	2
	0·37	0·31	0·60				1·62	0·20	1·22	1⅜	OQ900 T650					48·0	60·0	21½		52	"
	0·37	0·28	0·63				1·70	0·19	1·11	3⅜	OQ910 T660					47·0	60·3	21	49½	63	"
	0·37	0·41	0·60				1·62	0·20	1·21	6¼	OQ900 T610					48·2	60·3	17½	47	20	"
	0·37	0·35	0·56				1·44	0·18	0·82	6½	OQ900 T620					42·0	54·0	20½	53	34	"

MECHANICAL PROPERTIES AT ROOM TEMPERATURE (Continued)

Process of Manu-facture	ASTM Grain Size	Chemical Composition (%) C	Si	Mn	S	P	Ni	Cr	Mo	Al	Section Size dia. (in.)	Heat Treatment °C	Tensile (tons/sq.in.) 0·05% P.S. (0·05Rp)	0·1% P.S. (0·1Rp)	0·2% P.S. (0·2Rp)	0·5% P.S. (0·5Rp)	Y.P. (Rs)	T.S. (Rm)	El % (A)	R.A. % (Z)	Izod ft.lb.	Ref.
		0·38	0·35	0·58				1·59	0·21	1·11	$2\frac{1}{4}$	OQ900 T630					50·0	63·0	22	47	46	2
		0·38		0·58				1·51	0·23	0·93	$2\frac{1}{2}$	OQ900 T700					43·9	54·2	24		68	5
		0·38						1·62	0·19	0·92	$2\frac{1}{2}$	OQ900 T650					46·2	56·4	22		59	1
		0·39	0·21	0·51				1·46	0·16	1·39	$2\frac{1}{4}$	OQ900 T630					50·0	63·0	21	$44\frac{1}{2}$	39	2
		0·39						1·66	0·23	1·14	$\frac{3}{4}$	OQ900 T650					52·4	60·4	$24\frac{1}{2}$		63	1
		"						"	"	"	$2\frac{9}{16}$	" "					46·8	57·4	20		56	"
		0·39						1·68	0·25	0·92	$2\frac{1}{2}$	OQ900 T670					43·2	55·0	$20\frac{1}{2}$		47	"
B.E.A.	Actual 7	0·40	0·27	0·46	0·011	0·016	0·20	1·58	0·25	1·01	1	N925 OQ900 T650					55·1	65·4	$21\frac{1}{2}$	61	53	4
		0·40		0·55				1·63	0·23	1·05	$1\frac{1}{8}$	OQ900 T700					43·6	56·4	24		85	5
		0·40						1·54	0·23	1·00	$1\frac{1}{8}$	OQ900 T650					50·4	60·0	$22\frac{1}{2}$		63	1
		0·40						1·63	0·19	1·16	$2\frac{1}{2}$	" "					47·8	59·0	21		59	"
		0·41		0·54				1·54	0·21	0·98	$2\frac{7}{8}$	OQ900 T650					49·6	59·0	16		37	5
		0·41		0·54				1·62	0·24	1·01	$2\frac{1}{2}$	OQ900 T700					47·6	59·2	22		68	"
		0·41						1·73	0·18	0·97	$3\frac{1}{2}$	OQ900 T650					47·2	58·4	21		58	1
		0·42						1·71	0·28	1·19	$\frac{3}{4}$	OQ900 T650					54·6	62·2	23		56	1
		"						"	"	"	$1\frac{1}{4}$	" "					48·6	61·2	$23\frac{1}{2}$		60	"
		0·43		0·57				1·64	0·20	1·05	$2\frac{3}{4}$	OQ900 T650					49·8	61·2	17		56	5
		0·43						1·61	0·25	0·94	1	OQ900 T650					52·0	62·2	22		56	1
		0·43						1·74	0·23	1·03	$2\frac{5}{8}$	" "					47·6	61·4	18		46	"

En 41

EFFECT OF TEMPERING ON CASE HARDNESS

Typical hardness of nitrided case after tempering
at various temperatures for 1 hr.

(Ref. 1)

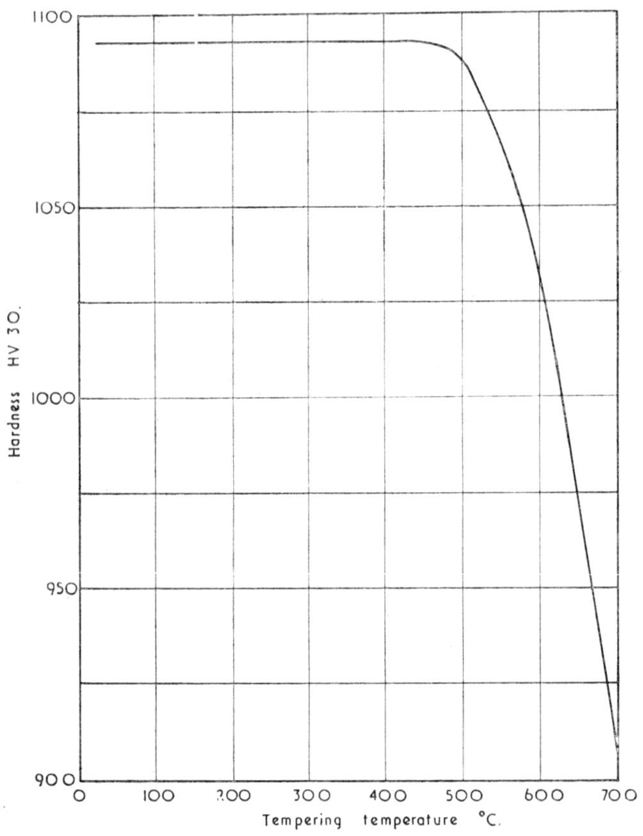

Fig. 41.9

EFFECT OF TEMPERING ON THE MECHANICAL PROPERTIES AT ROOM TEMPERATURE

Chemical Composition (%)

C	Si	Mn	S	P	Ni	Cr	Mo	Al
0·32	0·45	0·57	0·009	0·018	0·24	1·65	0·20	1·04

Basic Electric Arc.

En 41A

$1\frac{1}{8}$ in. dia. bars. Oil quenched 900°C. Tempered 2hrs. Air cool.

(Ref. 2)

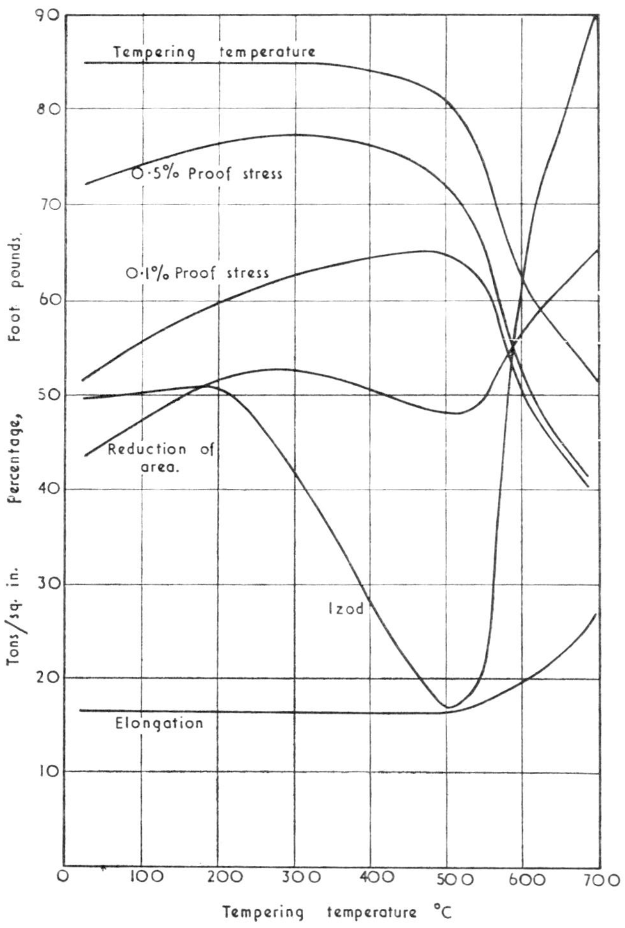

Fig. 41. 10

Typical curves for En 41B

$1\frac{1}{8}$ in. dia. bars. Oil quenched 900°C. Tempered as indicated.

(Ref. 2)

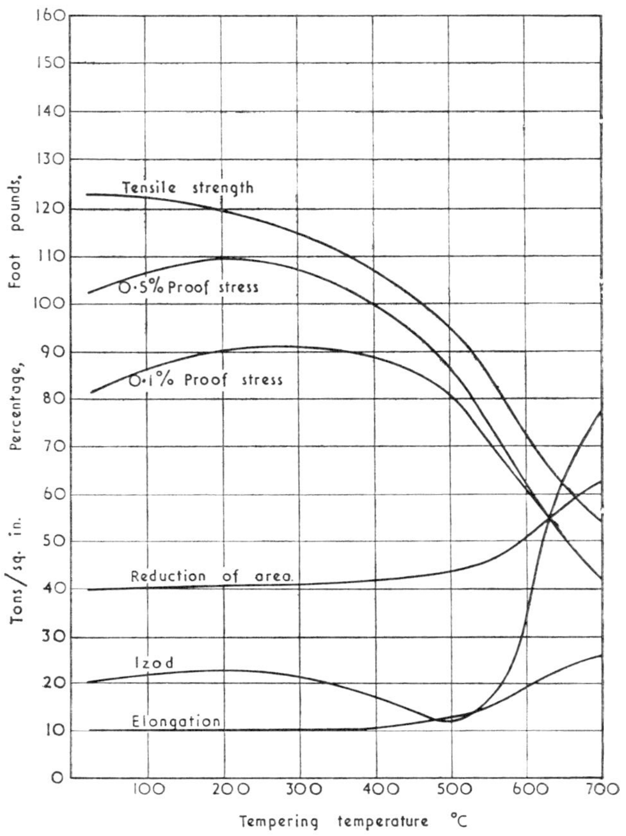

Fig. 41.11

EFFECT OF SECTION SIZE ON THE MECHANICAL PROPERTIES AT ROOM TEMPERATURE

	Chemical Composition (%)								
	C	Si	Mn	S	P	Ni	Cr	Mo	Al
En 41A	0·32	0·45	0·57	0·009	0·018	0·24	1·65	0·20	1·04

Basic Electric Arc.

Oil quenched 900°C. Tempered 600°C. Air cool. (Ref. 2)

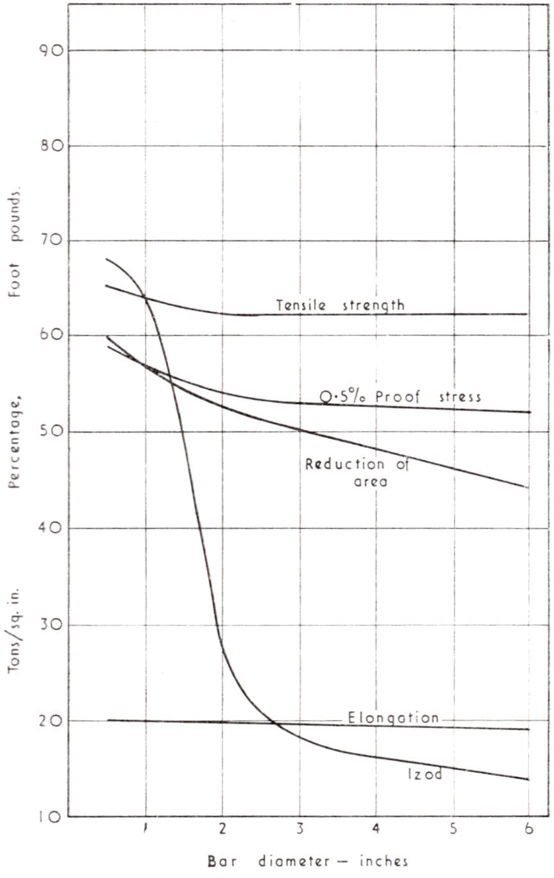

Fig. 41.12

Oil quenched 900°C. Tempered 650°C. Air cool. (Ref. 2)

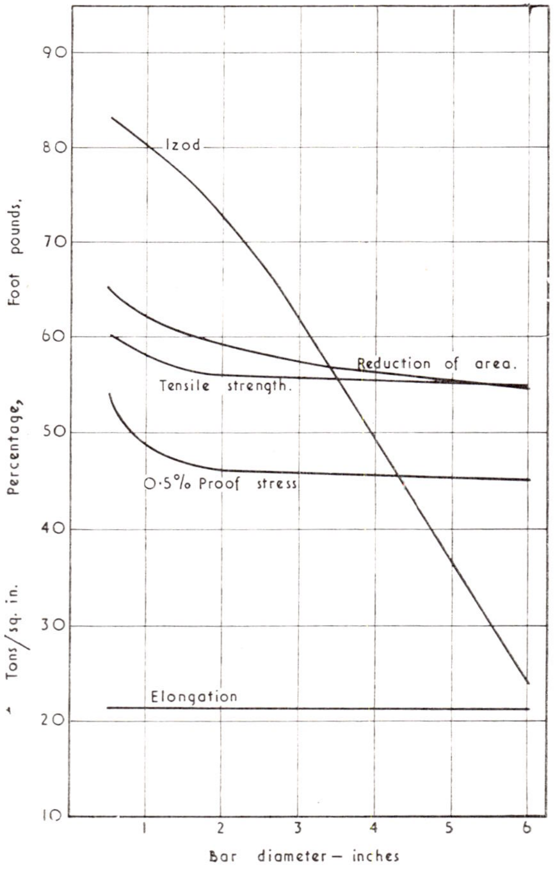

Fig. 41.13

Oil quenched 900°C. Tempered 700°C. Air cool (Ref. 2)

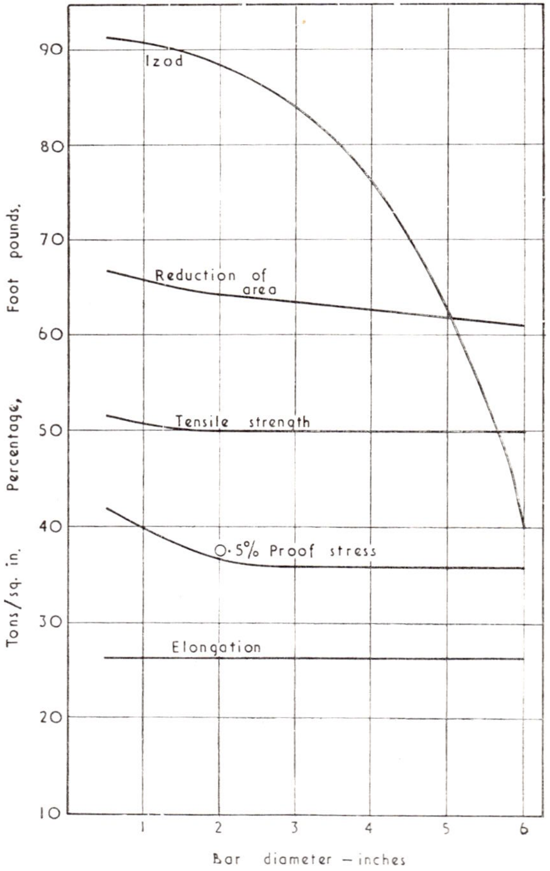

Fig. 41.14

EFFECT OF SECTION SIZE ON THE MECHANICAL PROPERTIES AT ROOM TEMPERATURE (Continued)

Typical curves for En 41B

Oil quenched 900°C. Tempered 600°C. Air cool. (Ref. 2)

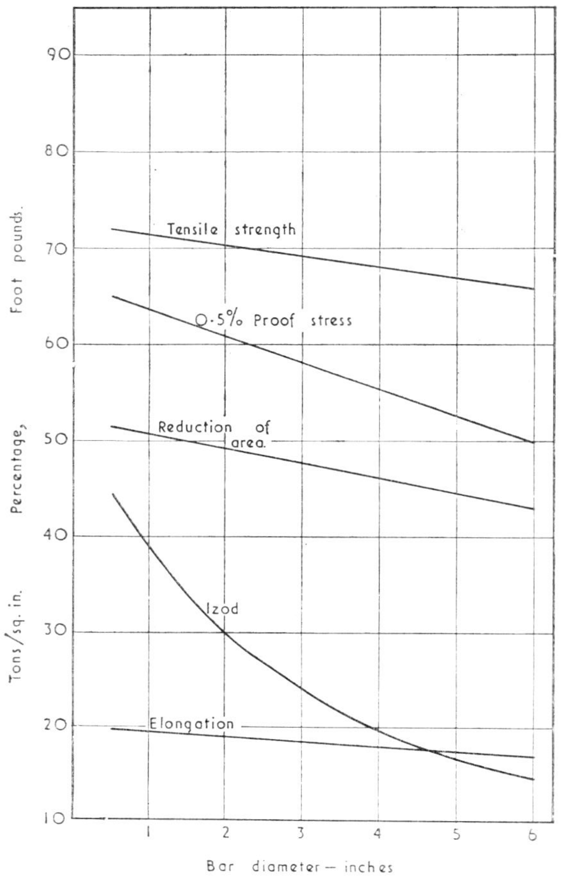

Fig. 41.15

Oil quenched 900°C. Tempered 650°C. Air cool. (Ref. 2)

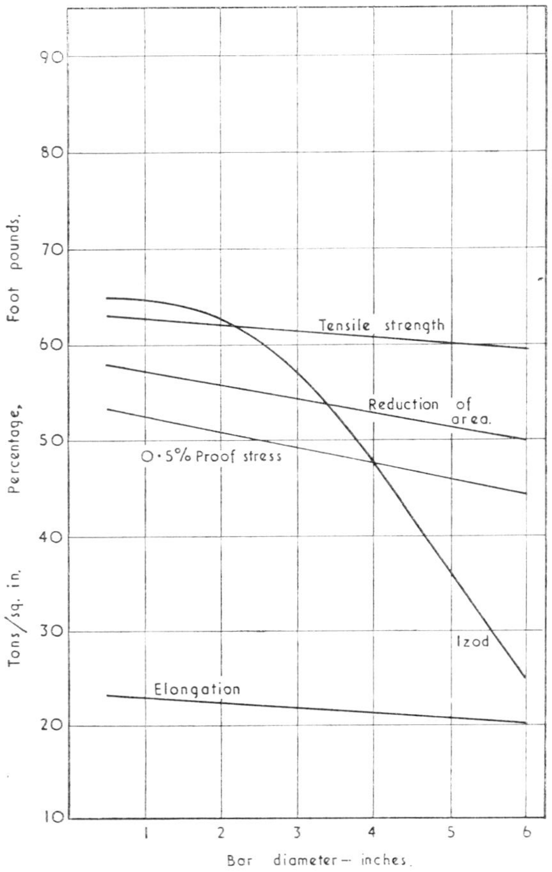

Fig. 41.16

Oil quenched 900°C. Tempered 700°C. Air cool. (Ref. 2)

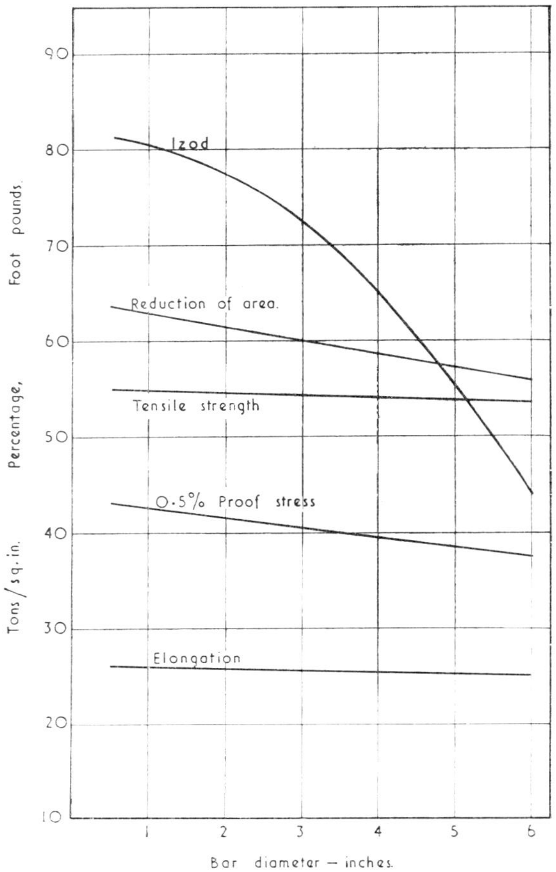

Fig. 41.17

MECHANICAL PROPERTIES AT LOW TEMPERATURE

Chemical Composition (%)									Heat Treatment °C	Test Temperature °C	Tensile							Ref.
C	Si	Mn	S	P	Ni	Cr	Mo	Al			0·01% P.S. (0·01Rp)	0·1% P.S. (0·1Rp)	0·2% P.S. (0·2Rp)	0·5% P.S. (0·5Rp)	T.S. (Rm)	El % (A)	R.A. % (Z)	
											tons/sq.in.							
0·28	0·24	0·47	0·005	0·016	0·22	1·55	0·24	1·07	OQ900 T650+72 hrs. 515	20	40·0	44·4	45·4	46·2	54·0	21½	62	6
"	"	"	"	"	"	"	"	"	" " "	— 50	42·4	46·4	47·6	48·8	58·5	22	57	"
"	"	"	"	"	"	"	"	"	" " "	—120	48·0	52·0	53·2	54·2	66·0	24	54	"
"	"	"	"	"	"	"	"	"	" " "	—196	60·0	67·8	69·2	70·4	80·2	22½	44½	"
0·41	0·22	0·43	0·012	0·025	0·27	1·54	0·30	1·04	OQ900 T650+72 hrs. 515	20	56·4	61·6	62·5	63·6	71·0	18	54½	"
"	"	"	"	"	"	"	"	"	" " "	— 50	57·2	61·6	63·2	64·6	74·0	19½	52	"
"	"	"	"	"	"	"	"	"	" " "	—120	63·6	70·0	71·6	73·4	83·2	19½	47½	"
"	"	"	"	"	"	"	"	"	" " "	—196	72·4	84·0	86·0	87·2	96·8	19½	39	"

MECHANICAL PROPERTIES AT HIGH TEMPERATURE

Chemical Composition (%)						
C	Si	Mn	Ni	Cr	Mo	Al
0·30	0·09	0·42	0·67	1·21	0·18	1·30

$\frac{3}{4}$ in. dia. bar. Oil quenched 840°C. Tempered 650°C.
(Ref. 6)

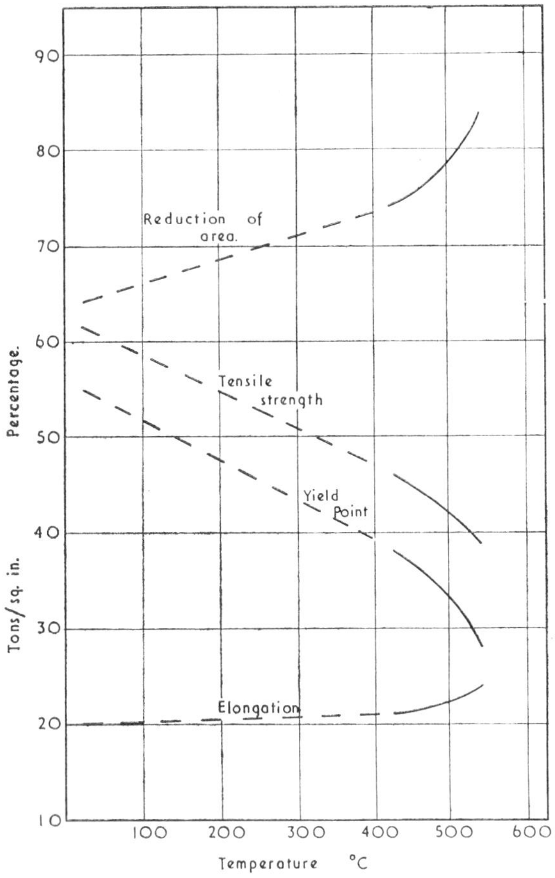

Fig. 41.18

CREEP PROPERTIES

Chemical Composition (%)							Section Size dia. (in.)	Heat Treatment °C	Test Tempera-ture °C	Tensile				Stress tons/sq.in.	Creep rate between 24 to 72hr. in/in.	Ref.
C	Si	Mn	Ni	Cr	Mo	Al				0·5% P.S. (0·5Rp)	T.S. (Rm)	El % (A)	R.A. % (Z)			
										tons/sq.in.						
0·28	0·30	0·52	0·21	1·63	0·21	0·99	$1\frac{1}{8}$	OQ900 T650	500					7·5	$2{\cdot}0\times10^{-6}$	1
"	"	"	"	"	"	"	"	" "	"					5·0	$1{\cdot}1\times10^{-6}$	"
0·46	0·27	0·59	0·07	1·72	0·33	1·01	$1\frac{1}{8}$	OQ900 T650 3hrs. AC	400	58·4	67·6	$16\frac{1}{2}$	53	16·5	$1{\cdot}0\times10^{-6}$	"
"	"	"	"	"	"	"	"	" " "	500					5·5	$1{\cdot}0\times10^{-6}$	"
"	"	"	"	"	"	"	"	" " "	600					2·2	$1{\cdot}0\times10^{-6}$	"

FATIGUE PROPERTIES

Chemical Composition (%)							Section Size dia. (in.)	Heat Treatment °C	Tensile						Fatigue Limit (tons/sq.in.)			Ref.
C	Si	Mn	Ni	Cr	Mo	Al			0·1% P.S. (0·1Rp)	0·5% P.S. (0·5Rp)	T.S. (Rm)	El % (A)		R.A. % (Z)	Rotating Bending		Reversed Axial	
									tons/sq.in.			$4\sqrt{A}$	5D		Plain	Notched	Plain	
0·33		0·70		1·50		1·0	65mm	OH and T			54				±30	±24*		8
"		"		"		"	"	OH and T nitrided 45 hrs. 520							±43	* Semi-circular notch 0·5 mm radius		"
0·30	0·43	0·41		1·43		1·12		OQ920 T650			51·2		22	67	±27			9
"	"	"		"		"		" " +48 hrs. 500 (not nitrided)			50·5		22	66	±26			"
"	"	"		"		"		" " +48 hrs. 500 (nitrided)			53·0				±35			"
0·35	0·22	0·54	0·20	1·67	0·20	0·91	1·06	OQ955 T590 stabilized 3 hrs. 540			72	16		49	±34			10
"	"	"	"	"	"	"	"	" " nitrided 90 hrs. 510			72·5	6		8	±37·5			"
0·39	0·23	0·61		1·63	0·18	1·10	1¼	OQ900 T650	47·6	50·2	57·5		22½	59	±24·9			7
0·44		0·34		1·71		1·33	1	OQ850 T630+90 hrs. 500	42·7	53·2	62·0	19			±31·9		±22·9	11
"		"		"		"	"	" " nitrided 90 hrs. 500	42·5	50·5	64·3				±38·4		±23·4	"
0·25	0·26	0·23		1·40		1·11		OH and T not nitrided			46·0		25½	63	±25	±20(b)		12
"	"	"		"		"		" " "			"		"	"		±13(e)		"
"	"	"		"		"		" " "			"		"	"		±14·5(f)		"
"	"	"		"		"		" " nitrided 36 hrs.			"		"	"	±35	±35(a)		"
"	"	"		"		"		" " "								±31·5(b)		"
"	"	"		"		"		" " "								±35(c)		"
"	"	"		"		"		" " "								±32(d)		"
"	"	"		"		"		" " "								±36·5(e)		"
"	"	"		"		"		" " "								±23(f)		"

* Notches: (a) 0·5 mm radius fillet at end of test portion.
(b) 0·2 " " " " " " " "
(c) 1 mm radius notch 0·55 mm deep; O.D. 8·5 mm; root dia. 7·4 mm.
(d) 1 mm " " 2·3 " " O.D. 12 " " " 7·4 mm.
(e) 0·3 mm " " 0·3 " " O.D. 8 " " " 7·4 mm.
(f) 0·3 mm " " 1·8 " " O.D. 11 " " " 7·4 mm.

CORROSION FATIGUE

Chemical Composition (%)				Heat Treatment °C	Rotating Bending Fatigue Strength (tons/sq.in.)		Ref.
C	Cr	Mo	Al		in air	in River Tees water 10^8 cycles	
0·28	1·58	0·33	0·87	H and T stress relieved 550 +90 hrs. 500 (not nitrided)	±33	<±5	13
"	"	"	"	H and T stress relieved 550 +90 hrs. 500 (nitrided)	±37	±25	"

Chemical Composition (%)						Section Size dia. (in.)	Heat Treatment °C	Tensile				Rotating Bending Fatigue Strength (tons/sq.in.) 10^8 cycles at 2,200 r.p.m.				Ref.
C	Si	Mn	Cr	Mo	Al			Y.P. (Rs) tons/sq.in.	T.S. (Rm) tons/sq.in.	El % (A)	R.A. % (Z)	in air	in steam jet in air	in steam at 60 lb/sq.in., 149°C	in steam at 220 lb/sq.in. 371°C	
0·36	0·27	0·51	1·50	0·18	1·20	1	OQ930 T650 not nitrided	44·3	55·1	21	63	±33		±28·5	±22·5	14
							" " nitrided 24 hrs. 500	48·2	56·2	13½	33½	±40·5	±32·5	±31	±26	"

LIST OF REFERENCES

1. Private correspondence with Firth-Brown Ltd.
2. Private correspondence with the English Steel Corporation Ltd.
3. G. Delbart, A. Constant and A. Clerc. Courbes de transformation des aciers de fabrication française. Institut de Recherches de la Sidérurgie (IRSID).
4. Symposium on the Hardenability of Steel. Organized by the Technical Advisory Committee of the Special and Alloy Steels Committee (Ministry of Supply) and The Iron & Steel Institute. Special Report No. 36 of The Iron & Steel Institute, 1946.
5. B.S. 971: Commentary on the British Standard Wrought Steels En Series. British Standards Institution, London, 1950.
6. War Office, unpublished report.
7. Private correspondence with The United Steel Companies Ltd.
8. H. Wiegand. Nitriding. *Revue de Métallurgie*, 1948, Vol. 45, p. 105.
9. O. Hengstenberg and R. Mailänder. Fatigue Properties of Nitrided Steels. *Z. Ver. Dtsch. Ing.*, 1930, Vol. 74, p. 1126.
10. N. L. Mochel. Fatigue Tests on Nitrided Steels. *Proceedings of the American Society for Testing Materials*, 1930, Vol. 30, pt. 2, pp. 406—410.
11. H. Sutton. Fatigue Properties of Nitrided Steels. *Metal Treatment*, 1936, Vol. 2, No. 5, p. 89.
12. R. Mailänder. Fatigue Strength of Nitrided Specimens. *Z. Ver. Dtsch. Ing.*, 1933, Vol. 77, p. 271.
13. N. P. Inglis and G. F. Lake. Corrosion Fatigue Tests of Nitrided and Nickel Plated Steels in River Tees Water. *Trans. Faraday Soc.*, 1932, Vol. 28, p. 715.
14. T. S. Fuller. Endurance Properties of Steel in Steam. *Proceedings of the American Society for Testing Materials*, 1931, Vol. 19, p. 97.

CARBON SPRING STEEL

SPECIFICATION

CHEMICAL COMPOSITION (%)

	C	Si	Mn	S	P
En 42 [1]	0·70—0·85	0·10—0·40	0·55—0·75	0·050 max.	0·050 max.
En 42B [2]	0·60—0·70	0·35 max.	0·55—0·80	0·050 max.	0·050 max.
En 42C [2]	0·70—0·80	0·35 max.	0·55—0·80	0·050 max.	0·050 max.
En 42D [2]	0·80—0·90	0·35 max.	0·55—0·80	0·050 max.	0·050 max.
En 42E [3]	0·60—0·75	0·35 max.	0·30—0·60	0·050 max.	0·050 max.
En 42F [3]	0·60—0·75	0·35 max.	0·60—0·90	0·050 max.	0·050 max.
En 42G [3]	0·75—0·90	0·35 max.	0·30—0·60	0·050 max.	0·050 max.
En 42J [3]	0·75—0·90	0·35 max.	0·60—0·90	0·050 max.	0·050 max.

[1] Laminated springs to this specification shall conform to the dimensional requirements in Appendix C BS 970.
[2] Steel of this type may be required in the form of wire for oil hardened and tempered springs (see BS 1429).
[3] Steel of this type may be required in the form of cold rolled strip and flat wire for hardening and tempering (see BS 1449).

MECHANICAL PROPERTIES

No mechanical properties are specified for these steels.

RELATED SPECIFICATIONS

United Kingdom				America		France	Germany			Sweden	U.S.S.R.	
En No.	BS Air	DTD	BS	SAE	AISI	AFNOR	Werkstoff No.	Name	DIN	SIS	GOST	Mark
42	S.513	5A 215 239A		 1074 1080	 C1074 C1080	 XC80f					1050—60	80
42B		215	1429	1064	C1064	XC65f	1.0761	C67	17222	14.1770	1050—60	65
42C	S.513	5A 215 239A	1429	1074	C1074						1050—60	75
42D		215	1429	1080 1084	C1080 C1084		1.0783	M85	17222		1050—60	85
42E		215	1449	1064	C1064		1.0763	M75	17222	14.1770		
42F		215	1449	1065 1070	C1065 C1070	XC65f XC70f	1.0761 1.0773	C67 C75	17222 17222		1050—60 1050—60	65 70
42G	S.513	239A	1449	1078	C1078		1.0763 1.0783	M75 M85	17222 17222	 14.1780		
 42J	 S.513	 239A	 1449	 1080	 C1080	 XC80f	1.0773	C75	17222		1050—60 1050—60	80 85

APPLICATIONS

These steels are mainly used for springs for general engineering when the stress conditions are not severe. The choice of steel is based on the stress requirement of the springs and whether the steel is to be oil-hardened and tempered (high manganese) or water hardened and tempered (low manganese).

Typical applications are for laminated springs especially for railway vehicles; support and foundation springs; hardened and tempered small springs, clips and washers; springs for rifle magazines and cartridge chargers, knives, saws, printers' rules, hand tools, and agricultural and motor vehicle components.

WELDING

See Introduction (List of Abbreviations) for key to the symbols used in this table.

Specification	Welding Process					
	MA	G	IG	S	F	B
En 42	f	f	f	m	f	j
En 42B, C & D	m	f	m	f	b	j
En 42E & F	dx	dx	e	g	b	j
En 42G & J	f	f	f	f	g	

Reproduced by kind permission of H.M.S.O.

MACHINABILITY

In the softened condition the machinability of these steels is about 50 to 60% of that of mild steel (En 3).

HOT WORKING AND HEAT TREATMENT TEMPERATURES

Forging, rolling and stamping	1100°C, finish above 800°C.
Annealing*	760°C to 840°C, furnace cool.
Normalizing*	760°C to 840°C, air cool.
Softening	650°C to 700°C, air or furnace cool.
Hardening*	760°C to 840°C, oil or water quench, according to the manganese content.
Tempering	400°C to 500°C.

* Use the higher part of the range for the lower carbon contents and vice-versa.

PHYSICAL PROPERTIES

SPECIFIC GRAVITY (d)

Chemical Composition (%)								Heat Treatment °C	Specific Gravity at 20°C	Ref.
C	Si	Mn	S	P	Ni	Cr	Mo			
0·80	0·13	0·32	0·009	0·008	0·13	0·11	<0·01	A800	7·85	1
0·80								—	7·84	2
0·90								Normalized	7·82	3

SPECIFIC HEAT (c)

Chemical Composition (%)								Heat Treatment °C	Mean Specific Heat cal/g deg. C										Ref.
C	Si	Mn	S	P	Ni	Cr	Mo		20 to 100°C	20 to 200°C	20 to 300°C	20 to 400°C	20 to 500°C	20 to 600°C	20 to 700°C	20 to 800°C	20 to 900°C	20 to 1000°C	
0·80	0·13	0·32	0·009	0·008	0·13	0·11	<0·01	A800	0·116	0·122	0·126	0·130	0·135	0·141	0·146	0·169	0·167	0·166	1
0·84	0·13	0·24	0·014	0·014				—	0·119	0·123	0·127	0·131	0·136	0·142	0·150	0·174	0·171	0·169	"

MEAN COEFFICIENT OF THERMAL EXPANSION (α)

Chemical Composition (%)								Heat Treatment °C	10^6 × Mean Coefficient of Thermal Expansion per deg. C										Ref.
C	Si	Mn	S	P	Ni	Cr	Mo		20 to 100°C	20 to 200°C	20 to 300°C	20 to 400°C	20 to 500°C	20 to 600°C	20 to 700°C	20 to 800°C	20 to 900°C	20 to 1000°C	
0·80	0·13	0·32	0·009	0·008	0·13	0·11	<0·01	A800	11·2	11·8	12·6	13.5	13·8	14·3	14.9	13·9	15·3	15·8	1

THERMAL CONDUCTIVITY (k)

Chemical Composition (%)								Heat Treatment °C	Thermal Conductivity cal/cm s deg. C												Ref.
C	Si	Mn	S	P	Ni	Cr	Mo		0°C	20°C	100°C	200°C	300°C	400°C	500°C	600°C	700°C	800°C	900°C	1000°C	
0·80	0·13	0·32	0·009	0·008	0·13	0·11	<0·01	A800	0·119	0·119	0·115	0·108	0·099	0·091	0·084	0·078	0·072	0·058	0·061	0·064	1
0·84	0·13	0·24	0·014	0·014				—	0·122	0·121	0·117	0·110	0·100	0·092	0·085	0·079	0·073	0·058	0·060	0·064	"

ELECTRICAL RESISTIVITY (ϱ)

Chemical Composition (%)								Heat Treatment °C	Electrical Resistivity microhm-cm												Ref.
C	Si	Mn	S	P	Ni	Cr	Mo		0°C	20°C	100°C	200°C	300°C	400°C	500°C	600°C	700°C	800°C	900°C	1000°C	
0·80	0·13	0·32	0·009	0·008	0·13	0·11	<0·01	A800	17·0	18·0	23·2	30·8	39·8	50·5	62·8	77·2	93·5	112·9	116·4	119·1	1
"	"	"	"	"	"	"	"	WQ950, ¼″ dia.		33·9											4
"	"	"	"	"	"	"	"	" 1″ dia.		23·9											"
0·83	0·25	0·69	0·015	0·018	0·22	0·05		—		17·5(25°C)											5
0·84	0·13	0·24	0·014	0·014				—	16·5	17·6	22·8	30·3	39·2	49·8	62·2	76·7	92·7	112·9	116·6	119·7	1

YOUNG'S MODULUS (E); SHEAR MODULUS (G); POISSON'S RATIO (σ)

Chemical Composition (%)							Section Size dia. (in.)	Heat Treatment °C	Test Temperature °C	Young's Modulus tons/sq.in.		Shear Modulus tons/sq.in.		Poisson's Ratio*		Ref.
C	Si	Mn	S	P	Ni	Cr				static	dynamic	static	dynamic	static	dynamic	
0·64	0·21	0·76	0·025	0·025				C.D+Oil tempered	RT	13,150		5170		0·272		6
								" " +350 ½hr.	"	13,100		5100		0·284		"
0·68	0·26	0·65	0·013	0·020				C.D+Oil tempered	"	13,350		5180		0·289		"
								" " +350 ½hr.	"	13,450		5130		0·311		"
0·69	0·24	0·87	0·029	0·029	0·08	0·07	1	As rolled	18	13,350						7
"	"	"	"	"	"	"	"	" "	— 60	13,700						"
"	"	"	"	"	"	"	"	" "	—120	13,750						"
"	"	"	"	"	"	"	"	" "	—152	13,850						"
"	"	"	"	"	"	"	"	" "	—185	13,850						"
"	"	"	"	"	"	"	"	" "	—197	13,950						"
"	"	"	"	"	"	"	"	A820	18	13,450						"
"	"	"	"	"	"	"	"	"	— 60	13,800						"
"	"	"	"	"	"	"	"	"	—197	14,150						"
"	"	"	"	"	"	"	"	OQ820 T650	15	13,750						"
"	"	"	"	"	"	"	"	" "	— 60	14,050						"
"	"	"	"	"	"	"	"	" "	—120	14,350						"
"	"	"	"	"	"	"	"	" "	—196	14,400						"
0·60	0·21	0·77	0·007	0·001	0·08	0·09		OQ950 T500	RT			5270				6
0·66	0·19	0·56						OQ820 T600	"	13,250						8
0·82	0·25	0·41	0·019	0·026			0·4	OQ900 T500	RT			5180				6
								" T550	"			5180				"
0·88	0·17	0·88						N850	"	13,300						8
0·90								Normalized	"	13,200	13,500	5140	5,240	0·285	0·292	3
0·78	0·18	0·63	0·044	0·035	0·07	0·04		OH and T				5200				9

* Calculated.

TRANSFORMATION CHARACTERISTICS

ISOTHERMAL TRANSFORMATION

Chemical Composition (%)		
C	Si	Mn
0·64	0·22	0·68

McQuaid Ehn grain size 6—9 ASTM. Austenitizing Temperature 870°C.

(Ref. 10)

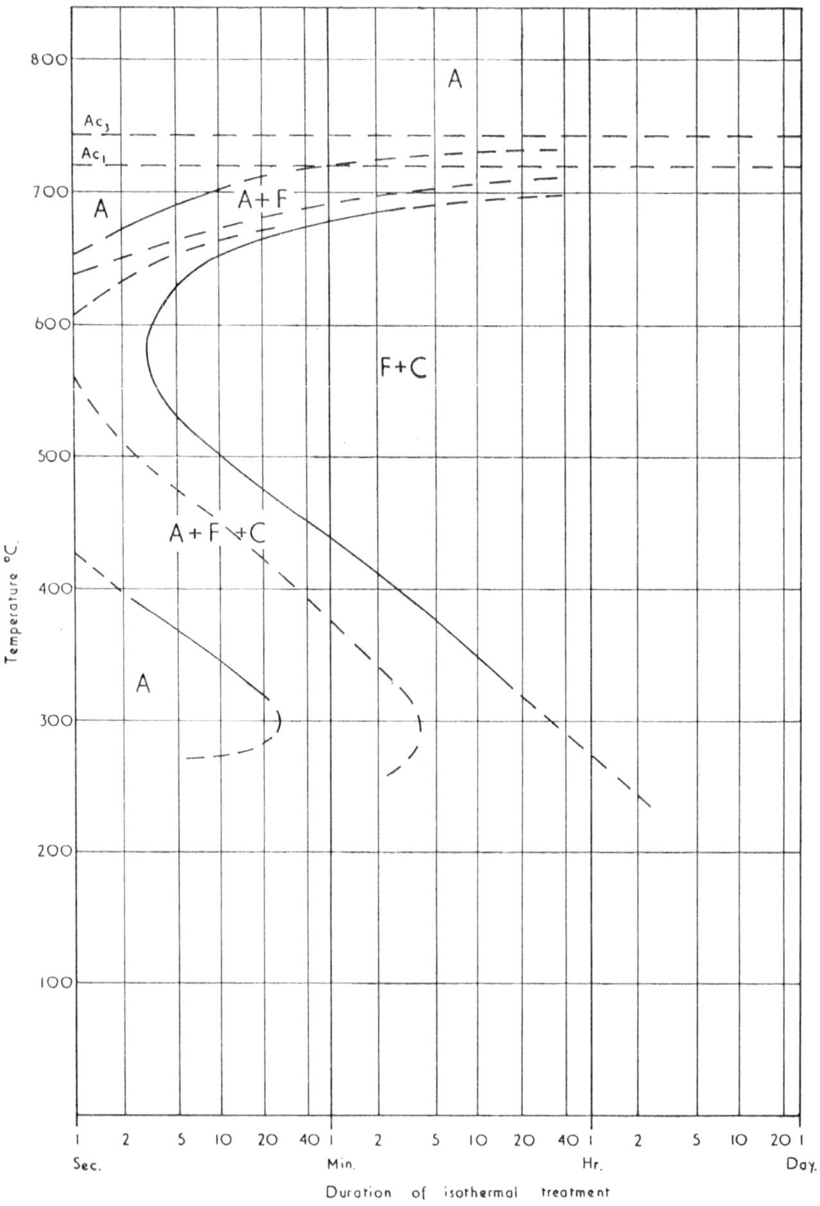

Fig. 42.1

ISOTHERMAL TRANSFORMATION (Continued)

Chemical Composition (%)						
C	Si	Mn	S	P	Ni	Cr
0·75	0·24	0·75	0·010	0·012	0·43	0·06

Pre-treatment annealed.
McQuaid Ehn grain size 6—7 (ASTM).
Actual grain size 12 (ASTM).

Austenitizing Temperature 800°C, 15 min.

(Ref. 11)

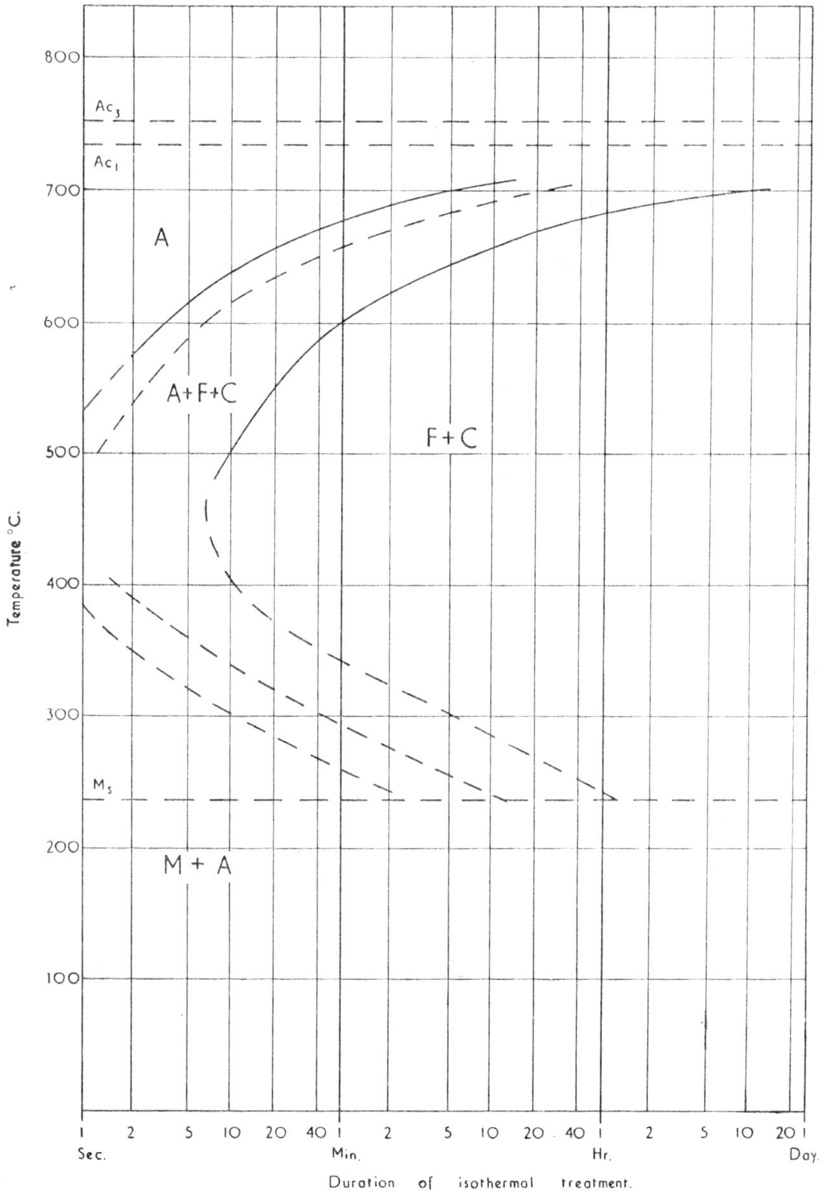

Fig. 42.2

En 42

ISOTHERMAL TRANSFORMATION (Continued)

Chemical Composition (%)							
C	Si	Mn	S	P	Ni	Cr	Mo
0·75	0·33	0·70	0·016	0·017	0·20	0·17	0·02

Basic Electric Arc.
Bars rolled to 4 S.W.G. with intermediate annealing at 780°C. Rolled to tape 0·03 in. × 0·1 in.
McQuaid Ehn grain size 5—6 (ASTM).
As quenched grain size 5—6 (ASTM).

Austenitizing Temperature 800°C, 30 min.

(Ref. 12)

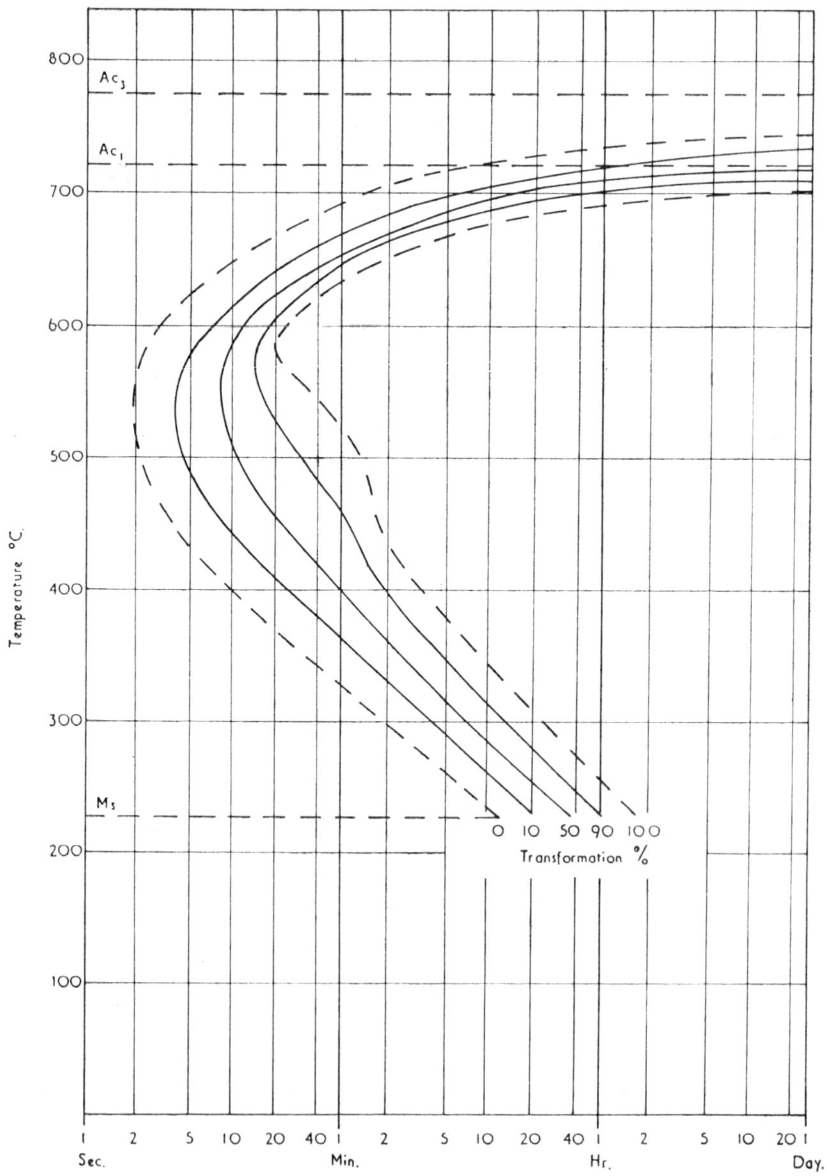

Fig. 42.3

ISOTHERMAL TRANSFORMATION (Continued)

Chemical Composition (%)				
C	Si	Mn	S	P
0·79	0·21	0·76	0·030	0·026

McQuaid Ehn grain size 6 (ASTM).

Austenitizing Temperature 900°C.

(Ref. 10)

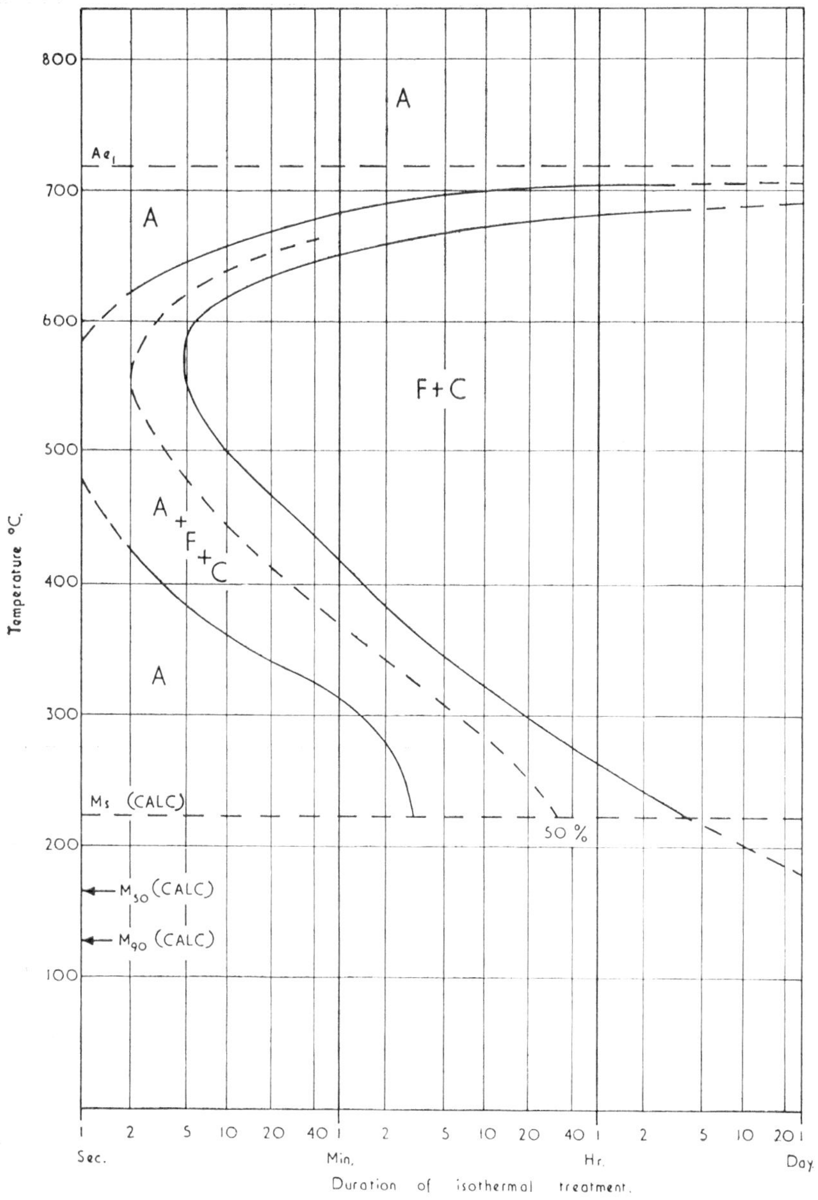

Fig. 42.4

HARDENABILITY

END QUENCH HARDENABILITY DATA

Curve No.	Chemical Composition (%)							Method of Manufacture	Austenite Grain Size (ASTM)
	C	Si	Mn	S	P	Ni	Cr		
A	0·80	0·18	0·63	0·038	0·039	0·14	Trace	A.O.H.	5
B	0·75	0·14	0·75	0·034	0·030	0·21	0·08	B.O.H.	5

Pre-treatment N. 790°C.
Austenitizing Temperature 780°C, 20 min.

(Ref. 13)

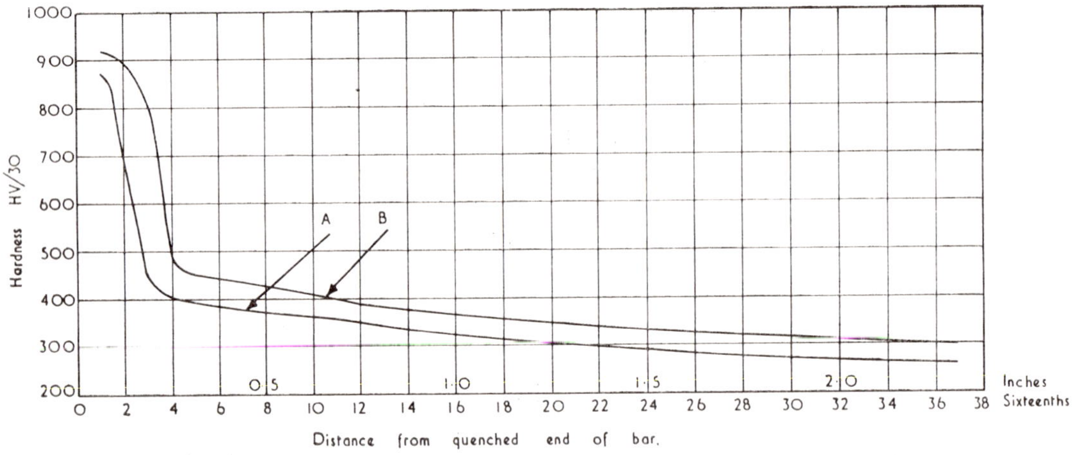

Fig. 42.5

Chemical Composition (%)						
C	Si	Mn	S	P	Ni	Cr
0·78	0·18	0·63	0·044	0·035	0·07	0·04

Acid Open Hearth.
Austenitizing Temperature 820°C.

(Ref. 9)

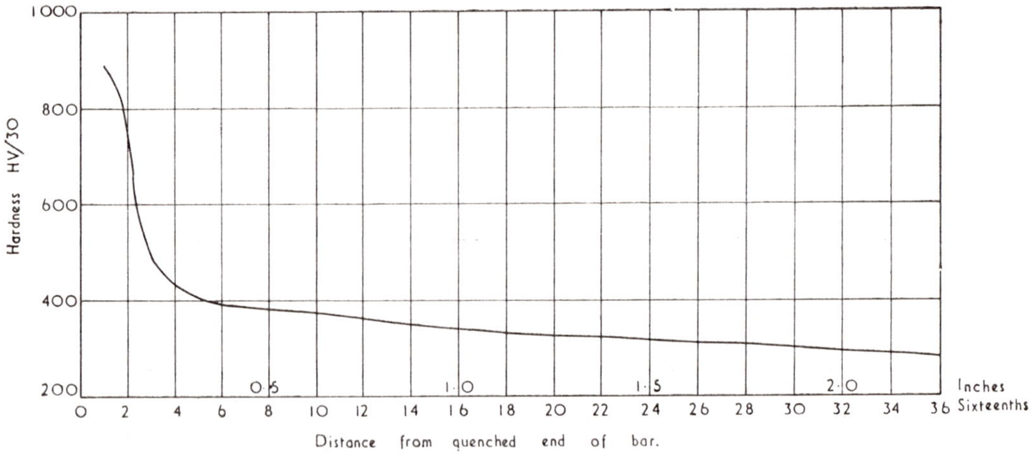

Fig. 42.6

END QUENCH HARDENABILITY DATA (Continued)

Chemical Composition (%)						
C	Si	Mn	S	P	Ni	Cr
0·75	0·24	0·75	0·010	0·012	0·43	0·06

Pre-treatment, annealed.
McQuaid Ehn grain size 6—7 (ASTM).
End quenched from 800°C, 15 min.

(Ref. 11)

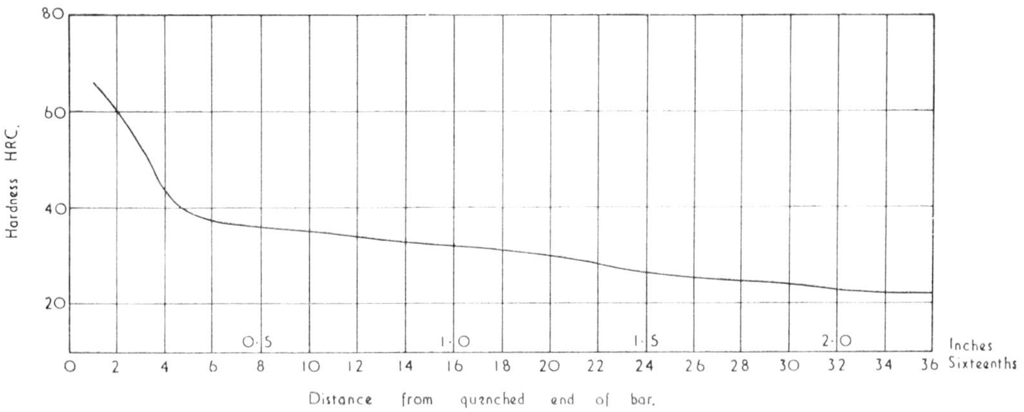

Fig. 42.7

En 42

MECHANICAL PROPERTIES AT ROOM TEMPERATURE

Process of Manufacture	ASTM Grain Size	C	Si	Mn	S	P	Ni	Cr	Section Size dia. (in.)	Heat Treatment °C	0·01% P.S. (0·01Rp) tons/sq. in.	0·1% P.S. (0·1Rp) tons/sq. in.	0·2% P.S. (0·2Rp) tons/sq. in.	Y.P. (Rs) tons/sq. in.	T.S. (Rm) tons/sq. in.	El % (A)	R.A. % (Z)	Izod Impact Value ft.lb.	Ref.
A.O.H.		0·60	0·12	0·75	0·035	0·033			Spring	H and T				63·0	76·0	9½	15	10	14
		0·60	0·21	0·77	0·007	0·011	0·08	0·09		OQ950 T500	57·0				79·0				6
		0·62	0·06	0·62	0·062	0·063	0·15			Normalized				23·1	46·5	19½	23	3	14
		0·64	0·21	0·76	0·025	0·025			0·104	CD & tempered	88·4	98·0			107·2	5½(on6″)	48½		6
		"	"	"	"	"			"	" "+350, ½hr.	89·9	96·2			104·6	4¼ "	54½		"
		"	"	"	"	"			"	"	77·8	87·3			94·2	5 "	54		"
		"	"	"	"	"			"	" +350. ½hr.	78·8	85·5			96·5	5½ "	47		"
		0·66	0·19	0·56					bar	N820			27·0		53·3	19	33		8
		"	"	"					"	A820			27·0		52·4	22	33		"
		"	"	"					"	OQ820 T600			37·4		62·2	19	42		"
		0·69	0·05	0·33		0·025			1⅛	N820				25·0	51·7	13	18½		15
		"	"	"		"			"	OQ820 T650WQ				35·0	55·4	18½	36½		"
		0·75	0·26	0·52						A800				25·4	50·8	18	33		16
		"	"	"						OQ825 T550				57·1	66·5	13	49		"
		"	"	"						OQ825 T650				45·1	50·3	14	52		"
	5—7 (1—4)	0·85	0·13	0·76	0·027	0·012			1	As Rolled				37·5	62·9	12	17	5	17
	"	"	"	"	"	"			"	A787				24·3	40·0	24½	45	4½	"
	"	"	"	"	"	"			½	N898				35·9	67·2	12½	27½	6½	"
	"	"	"	"	"	"			1	"				33·9	65·4	11	20½	5	"
	"	"	"	"	"	"			2	"				31·2	62·9	10½	17	4½	"
	"	"	"	"	"	"			4	"				28·8	59·7	10½	15½	4½	"
	"	"	"	"	"	"			½	OQ820 T482				56·0	82·1	12	34½	7	"
	"	"	"	"	"	"			"	" T538				54·2	75·5	15	38½	7½	"
	"	"	"	"	"	"			"	" T593				47·8	67·9	17	43½	8	"
	"	"	"	"	"	"			1	OQ815 T204				63·4	84·8	12	35	10	"
	"	"	"	"	"	"			"	" T315				63·8	84·4	12	35	10	"
	"	"	"	"	"	"			"	" T426				61·6	83·5	13	36	10	"
	"	"	"	"	"	"			"	" T482				50·2	81·0	13	36	7½	"
	"	"	"	"	"	"			"	" T538				46·2	74·1	15	37½	8	"
	"	"	"	"	"	"			"	" T593				43·3	67·0	16½	40½	8½	"
	"	"	"	"	"	"			"	" T650				38·8	57·1	21	50	12	"
	"	"	"	"	"	"			"	" T704				31·2	52·2	24	51	12	"

MECHANICAL PROPERTIES AT ROOM TEMPERATURE

(Continued)

Process of Manufacture	ASTM Grain Size	Chemical Composition (%)								Section Size dia. (in.)	Heat Treatment °C	Tensile							Izod Impact Value ft.lb.	Ref.
		C	Si	Mn	S	P	Ni	Cr	Mo			0·01% P.S. (0·01Rp)	0·1% P.S. (0·1Rp)	0·2% P.S. (0·2Rp)	Y.P. (Rs)	T.S. (Rm)	El % (A)	R.A. % (Z)		
												tons/sq.in.								
	5—7 (1—4)	0·85	0·13	0·76	0·027	0·012				2	OQ815 T482				49·1	80·4	12½	37½	6½	17
		"	"	"	"	"				"	" T538				45·8	73·0	15	38	7	"
		"	"	"	"	"				"	" T593				39·1	62·6	17½	42	7	"
		"	"	"	"	"				4	OQ815 T482				46·4	76·5	11½	28½	5½	"
		"	"	"	"	"				"	" T538				40·1	70·1	11½	24½	4½	"
		"	"	"	"	"				"	" T593				33·5	60·0	15½	33	6	"
		0·88	0·17	0·88						bar	N850			34·4		67·9	15	27		8
		0·89	0·32	0·28	0·012	0·022	0·09	0·09	0·06		Normalized		28·0			58·0	14	27		7

En 42

EFFECT OF TEMPERING ON THE MECHANICAL PROPERTIES AT ROOM TEMPERATURE

Chemical Composition (%)				
C	Si	Mn	S	P
0·85	0·13	0·76	0·027	0·012

ASTM Grain Size 80% 5—7, 20% 1—4.
1 in. dia. bars. Oil quenched 815°C.
Tempered as indicated.

(Ref. 17)

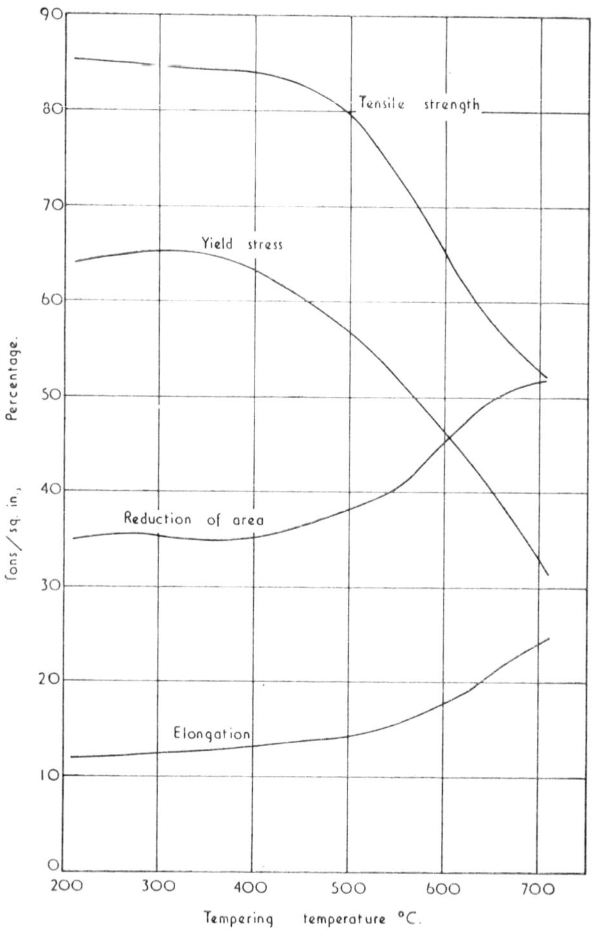

Fig. 42.8

EFFECT OF SECTION SIZE ON THE MECHANICAL PROPERTIES AT ROOM TEMPERATURE

Chemical Composition (%)				
C	Si	Mn	S	P
0·85	0·13	0·76	0·027	0·012

Normalized 898°C.

(Ref. 17)

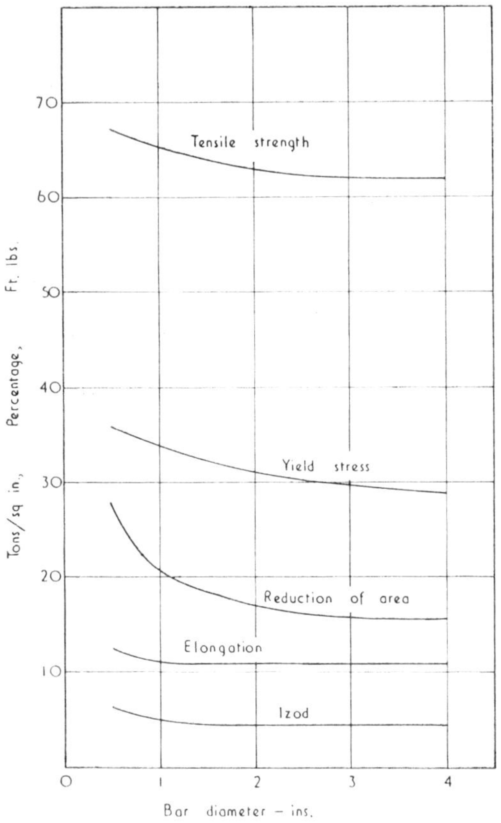

Fig. 42.9

Oil quenched 820°C. Tempered 482°C.

(Ref. 17)

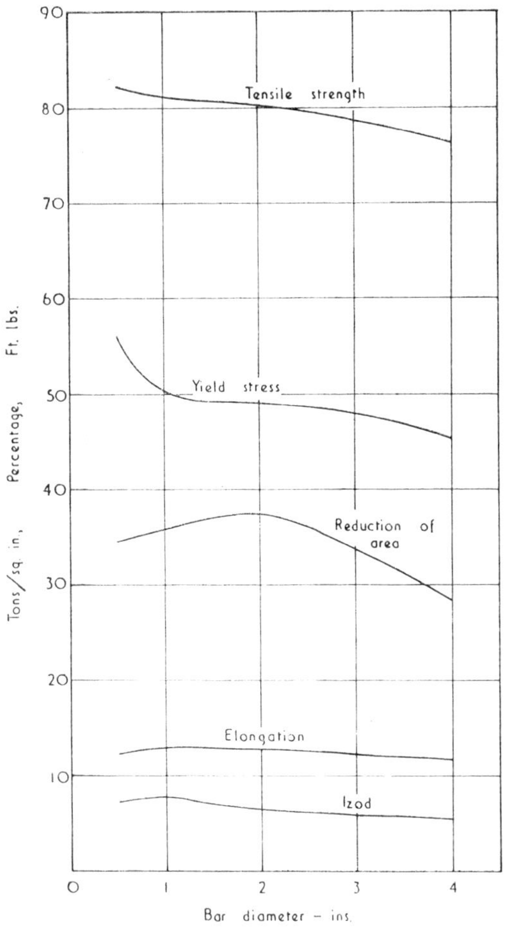

Fig. 42.10

EFFECT OF SECTION SIZE ON THE MECHANICAL PROPERTIES AT ROOM TEMPERATURE (Continued)

Oil quenched 820°C. Tempered 538°C.

(Ref. 17)

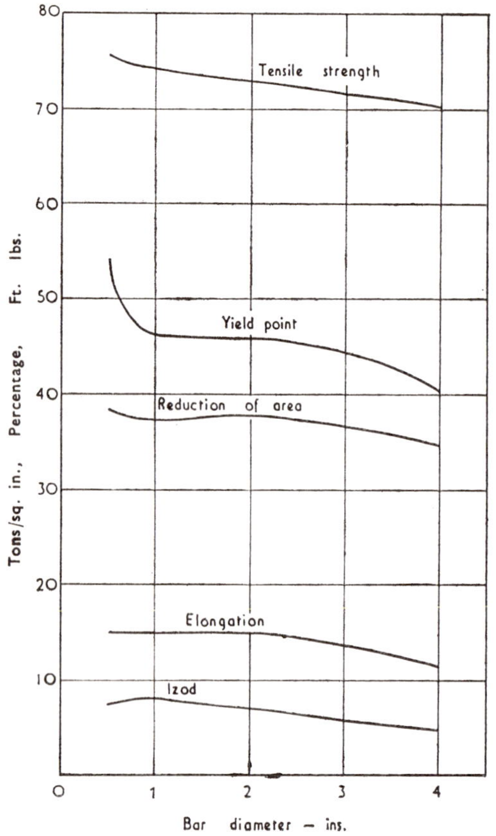

Fig. 42.11

Oil quenched 820°C. Tempered 593°C.

(Ref. 17)

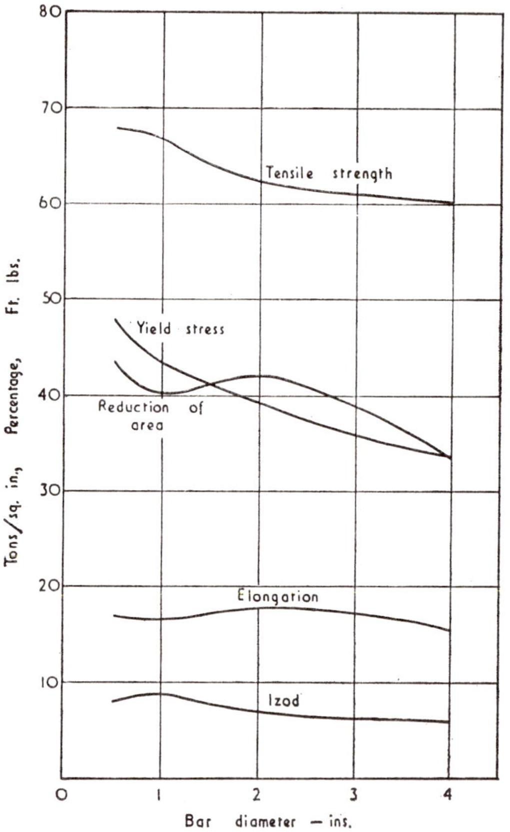

Fig. 42.12

MECHANICAL PROPERTIES AT LOW TEMPERATURE

ASTM Grain Size	Chemical Composition (%) C	Si	Mn	S	P	Ni	Cr	Mo	Section Size dia. (in.)	Heat Treatment °C	Test Temperature °C	Tensile 0·1% P.S. (0·1Rp) tons/sq.in.	0·2% P.S. (0·2Rp) tons/sq.in.	0·5% P.S. (0·5Rp) tons/sq.in.	Y.P. (Rs) tons/sq.in.	T.S. (Rm) tons/sq.in.	El % (A)	R.A. % (Z)	Charpy V Notch ft.lb.	Ref.
	0·69	0·24	0·087	0·029	0·029	0·08	0·07	0·02	1	As rolled	18				32·9	60·5	18½	39		7
	"	"	"	"	"	"	"	"	"	"	— 60				38·5	65·2	21½	33		"
	"	"	"	"	"	"	"	"	"	"	—120				45·8	71·8	18	22		"
	"	"	"	"	"	"	"	"	"	"	—152	54·2		54·4	54·4	76·5	17½	26		"
	"	"	"	"	"	"	"	"	"	"	—185	65·6		66·8		80·5	4½	5½		"
	"	"	"	"	"	"	"	"	"	"	—197	72·0				78·0	1	1		"
	"	"	"	"	"	"	"	"	1	A820	18				27·5	49·7	26	50½		"
	"	"	"	"	"	"	"	"	"	"	— 60				31·6	56·1	26	41½		"
	"	"	"	"	"	"	"	"	"	"	—123				43·6	66·8	22½	31½		"
	"	"	"	"	"	"	"	"	"	"	—197				67·6	71·2	2½	3		"
	"	"	"	"	"	"	"	"	1	OQ820 T650 (1hr.)	15	38·4		39·4	39·4	60·3	18	47½		"
	"	"	"	"	"	"	"	"	"	" " "	— 60				47·2	70·0	21	46½		"
	"	"	"	"	"	"	"	"	"	" " "	—120				62·5	75·0	22½	42½		"
	"	"	"	"	"	"	"	"	"	" " "	—196			66·0	66·5	66·6	3½	4½		"
5—7	0·76	0·20	0·70	0·033	0·019	0·03	0·05			H and T	20					87*	(41RC)		16	18
"	"	"	"	"	"	"	"			"	0								15	"
"	"	"	"	"	"	"	"			"	20					68*	(30RC)		24	"
"	"	"	"	"	"	"	"			"	0								22	"
3	0·80	0·23	0·75	0·030	0·017	0·02	0·06			"	20					86*	(40RC)		14	"
"	"	"	"	"	"	"	"			"	0								13	"
"	"	"	"	"	"	"	"			"	20					70*	(31RC)		23	"
"	"	"	"	"	"	"	"			"	0								20	"

* Estimated.

MECHANICAL PROPERTIES IN TORSION

Chemical Composition (%)								Section Size dia. (in.)	Heat Treatment °C	Test Temperature °C	Tensile						Torsion				Ref.
C	Si	Mn	S	P	Ni	Cr	Mo				L.P. (tons/sq. in.)	0·1% P.S. (0·1Rp) (tons/sq. in.)	Y.P. (Rs) (tons/sq. in.)	T.S. (Rm) (tons/sq. in.)	El % (A)	R.A. % (Z)	E.L. (tons/sq.in.)	Y.P. (tons/sq.in.)	Apparent max. Shear Stress (tons/sq.in.)	Probable max. Shear Stress (tons/sq.in.)	
0·60	0·12	0·75	0·035	0·033				Spring	H and T	RT			63·0	76·0	9½	15		35·0	62·7	47·0	14
0·60	0·21	0·77	0·007	0·011	0·08	0·09			OQ950 T500	"	57·0(0·01%PS)			79·0			37·0		67·0	50·3	6
0·62	0·06	0·62	0·062	0·063	0·15				Normalized	"	16·5		23·1	46·5	19½	23	14·5		44·3	33·2	14
0·64	0·21	0·76	0·025	0·025				0·104	CD & tempered	"	83·4(0·01%PS)	98·0		107·2	5½(on 6")	48½	56·7		91·7	68·8	6
"	"	"	"	"				"	" " +350 ½ hr.	"	89·9(")	96·2		104·6	4½(")	54½	60·8		91·7	68·8	"
"	"	"	"	"				"	" "	"	77·8(")	87·3		94·2	5 (")	54	49·2		84·9	63·7	"
"	"	"	"	"				"	" " +350 ½ hr.	"	78·8(")	85·5		96·5	5½(")	47	51·5		84·9	63·7	"
0·69	0·24	0·87	0·029	0·029	0·08	0·07	0·02	¾	A820 1 hr.	RT			28·6	49·7	26	50½		16·4	42·4	31·8	7
"	"	"	"	"	"	"	"	"	" "	— 10								16·5	45·4	34·0	"
"	"	"	"	"	"	"	"	"	" "	— 20								18·5	44·3	33·2	"
"	"	"	"	"	"	"	"	"	" "	— 40								18·5	45·0	33·7	"
"	"	"	"	"	"	"	"	"	" "	— 60			33·4	56·1	26	41½		20·2	47·1	35·3	"
"	"	"	"	"	"	"	"	"	" "	— 96								23·5	49·6	37·2	"
"	"	"	"	"	"	"	"	"	" "	—120			45·7	66·8	22½	31½		26·9	51·2	38·4	"
"	"	"	"	"	"	"	"	"	" "	—147								32·8	58·0	43·5	"
"	"	"	"	"	"	"	"	"	" "	—196			69·0	71·2	2½	3		42·8	60·1	45·1	"
0·78	0·18	0·63	0·044	0·035	0·07	0·04		1⅛	Normalized	RT	13·5									32·7	9
"	"	"	"	"	"	"			Annealed	"	9·8									31·8	"
"	"	"	"	"	"	"			N, OQ and T400	"	33·2									49·6	"
"	"	"	"	"	"	"			" " T500	"	29·1									47·7	"
"	"	"	"	"	"	"			" " T600	"	24·1									43·2	"
0·82	0·25	0·41	0·019	0·026				0·4	OQ900 T500	"							33·0		69·3	52·0	6
"	"	"	"	"				"	" T550	"							31·0		64·1	48·1	"
0·85	0·14	0·76	0·060	0·032	0·42	0·19		1·15	OQ825 T450		(457 HV)		85·5	98·9	8½	19			81·0	60·8	19
"	"	"	"	"	"	"		"	" T350		(557 HV)			110*					93·1	69·8	"

* Estimated value.

FATIGUE PROPERTIES

Chemical Composition (%)							Section Size dia. (in.)	Heat Treatment °C	Tensile				Fatigue Limit tons/sq.in.		Ref.
C	Si	Mn	S	P	Ni	Cr			Y.P. (Rs) tons/sq.in.	T.S. (Rm) tons/sq.in.	El % (A) 5D $4\sqrt{A}$	R.A. % (Z)	Rotating Bending	Reverse Axial	
0·65	0·18	0·64	0·025	0·037				A800	29·2	50·8	17		±20·7		16
0·75	0·26	0·52						A800	25·4	50·8	18	33	±19·0		"
"	"	"						OQ825 T550	57·1	66·5	13	49	±27·3		"
"	"	"						OQ825 T650	45·1	50·3	14	52	±22·8		"
0·85	0·14	0·76	0·060	0·032	0·42	0·19	1·15	OQ825 T450	85·5	98·9	8½	19	±49·5		19
"	"	"	"	"	"	"	"	" T350		110*			±58·5		"
0·87							2¼ × ¾	N810 A760	(HRB86)	38*			±17·8	±13·4	20
"							"	OQ810 T540	(HB100)	54*			±26·3	±26·2	11

* Estimated.

LIST OF REFERENCES

1. *Physical Constants of Some Commercial Steels at Elevated Temperatures*. Edited by the British Iron and Steel Research Association, Metallurgy Division, Thermal Treatment Sub-committee. Butterworth's Scientific Publications, London, 1953.
2. Private correspondence with the English Steel Corporation Ltd.
3. M. F. Markham. Measurement of Elastic Constants by the Ultrasonic Pulse Method. *British Journal of Applied Physics*, Supplement No. 6, Physics of Non-destructive Testing, 1957, p.S62.
4. R. W. Powell. Effect of Quenching on Conductivity and Resistivity *JISI*, 1956, Vol. 184, Sept., pt. 1, p. 7.
5. G. Bullock. Thermal Changes in Steels as shown by Resistivity. *JISI*, 1956, Vol. 183, Aug. No. 2, p. 362.
6. Private correspondence with the Spring Manufacturers' Research Association.
7. War Office, unpublished report.
8. H. H. Abram. The Value of Young's Modulus for Steel. *JISI*, 1934, Vol. 129, No. 1, p. 325.
9. Private correspondence with the London Polytechnic.
10. Atlas of Isothermal Transformation Diagrams. The United States Steel Corporation. Second Edition, 1957.
11. G. Delbart, A. Constant, and A. Clerc. Courbes de transformation des aciers de fabrication française. Institut de Recherches de la Sidérurgie (IRSID).
12. Atlas of Isothermal Transformation Diagrams of the B.S. En Steels. Special Report No. 56 of The Iron & Steel Institute, 1956.
13. Symposium on the Hardenability of Steel. Organized by the Technical Advisory Committee of the Special and Alloy Steels Committee (Ministry of Supply) and The Iron & Steel Institute. Special Report No. 36 of The Iron & Steel Institute, 1946.
14. Private correspondence with Firth-Brown Ltd.
15. Private correspondence with Edgar Allen and Co. Ltd.
16. R. Cazaud (translated by A. J. Fenner). Fatigue of Metals. Chapman and Hall Ltd., 1953.
17. *Modern Steels and Their Properties—Carbon and Alloy Steel Bars*. The Bethlehem Steel Company. Sixth Edition, 1961.
18. H. W. Gillett. Impact Resistance and Tensile Properties of Metals at Sub-atmospheric Temperatures. Project No. 13 of the joint A.S.M.E—A.S.T.M. Research Committee on the Effect of Temperature on the Properties of Metals.
19. A. S. Kenneford and G. C. Ellis. A Comparison of Six Spring Steels. *JISI*, 1950, Vol. 164, No. 1, p. 265.
20. R. D. France. Endurance Testing of Steels: Comparison of Rotating Beam and Axially Loaded Specimens. *Proceedings of the American Society for Testing Materials*, 1931, Vol. 31, pt. II, pp. 176—193.

CARBON SPRING STEEL—"50" CARBON STEEL

SPECIFICATION

CHEMICAL COMPOSITION (%)

	C	Si	Mn	S	P	
En 43	0·45—0·60	0·10—0·40	0·60—0·80	0·050 max.	0·050 max.	Springs, bars for
En 43G*	0·45—0·60	0·35 max.	0·30—0·60	0·050 max.	0·050 max.	water hardening
En 43J*	0·45—0·60	0·35 max.	0·60—0·90	0·050 max.	0·050 max.	and tempering
En 43A	0·45—0·55	0·05—0·35	0·70—1·00	0·060 max.	0·060 max.	Bars and billets
En 43B	0·45—0·50	0·05—0·35	0·70—1·00	0·060 max.	0·060 max.	for general
En 43C	0·50—0·55	0·05—0·35	0·70—1·00	0·060 max.	0·060 max.	engineering
						components
En 43D	0·60—0·65	0·05—0·35	0·40—0·60	0·060 max.	0·060 max.	
En 43E	0·65—0·70	0·05—0·35	0·70—0·90	0·060 max.	0·060 max.	

* Steel of this type may be required in the form of cold rolled strip or flat wire for hardening and tempering (see BS 1449).

MECHANICAL PROPERTIES

No mechanical properties are specified for En 43, En 43G, En 43J, En 43B, En 43C, En 43D, and En 43E which are supplied to composition only.

	En 43A			Cold Drawn
	Normalized	Hardened and Tempered		
		R	S	
Limiting ruling section, in.	$1\frac{1}{8}$	2	$1\frac{1}{8}$	2
Tensile strength, tons/sq.in., min. (Rm)	40	45	50	45—60
Yield stress[1] tons/sq.in., min. (Rs)	21	30	33	—
0·2% proof stress [2], tons/sq.in., min. (0·2Rp)	—	28	31	—
Elongation, %, min. (A)	18	18	18	12
Brinell hardness No. [3] (HB)	179/229	201/255	223/277	201/277

(1) Not to be used as acceptance tests except by special arrangement between the purchaser and manufacturer.
(2) When specifically requested in the enquiry or order.
(3) For information purposes only. Not contractual part of the specification.

RELATED SPECIFICATIONS

United Kingdom		America		France	Germany			Sweden	U.S.S.R.	
En No.	BS	AISI	SAE	AFNOR	Werkstoff No	Name	DIN	SIS	GOST	Mark
43	1449	C1050 C1055	1050 1055	XC55				14.1606	1050—60 1050—60	50 55
43A		C1049 C1050	1049 1050						1050—60	50G
43B		C1046	1046	XC48	1.0503	C45	17200 17240			
43C		C1050	1050		1.0505	C53	17222		1050—60	50G
43D					1.0601	C60	17200 17222	14.1655	1050—60	60
43E		C1065	1065	XC65	1.0603	C67	17222		1050—60	65
43G	1449							14.1650		
43J	1449	C1050 C1055	1050 1055	XC55					1050—60 1050—60	50 50G

APPLICATIONS

En 43, En 43G and En 43J are used for springs for general engineering when the stress conditions are not severe, particularly where water hardening is preferred to oil hardening. Typical applications are similar to those given for **En 42.**

Steels to the **En 43** specifications are used also in the oil or water hardened and tempered condition for automobile and general engineering components.

Typical uses are:

En 43A, En 43B, En 43C: for crankshafts, sprockets, and small arms parts.

En 43D, En 43E: for keys, clips, and other parts requiring high surface hardness and resistance to wear.

En 43, En 43A and **En 43C** overlap and are very like **En 9** and **En 9K** and may be used for similar applications (see En 9, Vol. I).

WELDING

See Introduction (List of Abbreviations) for key to symbols used in this table.

	Welding Process					
	MA	G	IG	S	F	B
En 43A	e	bx	g	f	b	j
En 43G and En 43J	dx	dx	e	g	b	j

Reproduced by kind permission of H.M.S.O.

MACHINABILITY

In the softened condition the machinability of this steel is about 50—60% of that for mild steel (En 3).

HOT WORKING AND HEAT TREATMENT TEMPERATURE

Forging, rolling and stamping	1150°C, finish above 850°C.
Annealing	810°C—840°C, furnace cool.
Softening	650°C—700°C, air or furnace cool.
Normalizing	810°C—840°C, air cool.
Hardening	810°C—840°C, oil or water quench (depending on the carbon and manganese contents.)
Tempering (for springs)	400°C—500°C, air cool.
(for engineering components)	550°C—650°C, air cool.

PHYSICAL PROPERTIES

(see also En 9, Vol. 1)

SPECIFIC GRAVITY (d)

Chemical Composition (%)			Heat Treatment °C	Specific Gravity at 20°C	Ref.
C	Si	Mn			
0·55	0·23	0·38	Normalized	7·85	1
"	"	"	OQ830 T600	7·83	"
0·55	0·28	0·63	OQ830 T640	7·86	2
0·55			Softened	7·85	3
0·70			"	7·84	"

SPECIFIC HEAT (c)

Chemical Composition (%)			Heat Treatment °C	Specific Heat cal/deg. C						Ref.
C	Si	Mn		20 to 100°C	20 to 200°C	20 to 300°C	20 to 400°C	20 to 500°C	20 to 600°C	
0·55	0·28	0·65	OQ830 T600		0·114	0·123	0·131	0·138	0·147	2
0·55			Softened	0·115						3

MEAN COEFFICIENT OF THERMAL EXPANSION (α)

Chemical Composition (%)			Heat Treatment °C	$10^6 \times$ Mean Coefficient of Expansion per deg.C						Ref.
C	Si	Mn		20 to 100°C	20 to 200°C	20 to 300°C	20 to 400°C	20 to 500°C	20 to 600°C	
0·55			Softened	11·5						3
0·55	0·28	0·65	OQ830 T640	11·6	12·3	13·2	13·8	14·3	14·8	2
0·59	0·25	0·92	Softened	11·1		12·9			14·6	4
0·70			Softened	10·8						3

En 43

ELECTRICAL RESISTIVITY (ϱ)

Chemical Composition (%) C	Si	Mn	Heat Treatment °C	Electrical Resistivity microhm-cm 20°C	25°C	100°C	200°C	300°C	400°C	500°C	600°C	Ref.
0·49	0·15	0·69	—		16·0							5
0·55			Softened	18								3
0·55	0·28	0·65	OQ830 T640	19·9		24·5	31·8	40·8	50·8	60·6	76·5	2

THERMAL CONDUCTIVITY (k)

Chemical Composition (%) C	Si	Mn	Heat Treatment °C	Thermal Conductivity cal/cm s deg. C 20°C	100°C	200°C	300°C	400°C	500°C	600°C	Ref.
0·55			Softened	0·12							3
0·55	0·28	0·65	OQ830 T640		0·118	0·112	0·105	0·096	0·089	0·080	2

YOUNG'S MODULUS (E) AND SHEAR MODULUS (G)

Chemical Composition (%) C	Si	Mn	S	P	Section Size dia. (in.)	Heat Treatment °C	Young's Modulus (Static) tons/sq.in.	Shear Modulus (Static) tons/sq.in.	Ref.
0·47	0·10	0·48	0·050	0·041	—	Patented 970CD 88%	12,500		14
0·46	0·12	0·50	0·048	0·041	—	A700 12 hrs.	13,600		"
0·46	0·09	0·51	0·037	0·026	$3 \times \frac{3}{8}$	N820	13,400		6
"	"	"	"	"	"	WQ810 T500	13,550		"
"	"	"	"	"	"	" T550	13,500		"
0·60	0·21	0·77	0·007	0·012	$3 \times \frac{3}{8}$	N850	13,500		"
"	"	"	"	"	"	OQ950 T400	13,200	5,250	"
"	"	"	"	"	"	" T450	13,050	5,250	"
"	"	"	"	"	"	" T500	13,400	5,290	"
"	"	"	"	"	"	" T550	13,100	5,270	"
0·51	0·11	0·67	0·033	0·054	0·160	As rolled		5,310	7
"	"	"	"	"	"	OQ850 T350		5,090	"
"	"	"	"	"	"	" T380		5,090	"
"	"	"	"	"	"	" T420		5,090	"
"	"	"	"	"	"	" T460		5,270	"

MAGNETIC PROPERTIES

See En 9, Vol. 1.

TRANSFORMATION CHARACTERISTICS

(See also En 9, Vol. 1)

ISOTHERMAL TRANSFORMATION

Chemical Composition	
C	Mn
0·50	0·91

McQuaid Ehn grain size 7—8 (ASTM).
Austenitizing Temperature 910°C.

(Ref. 8)

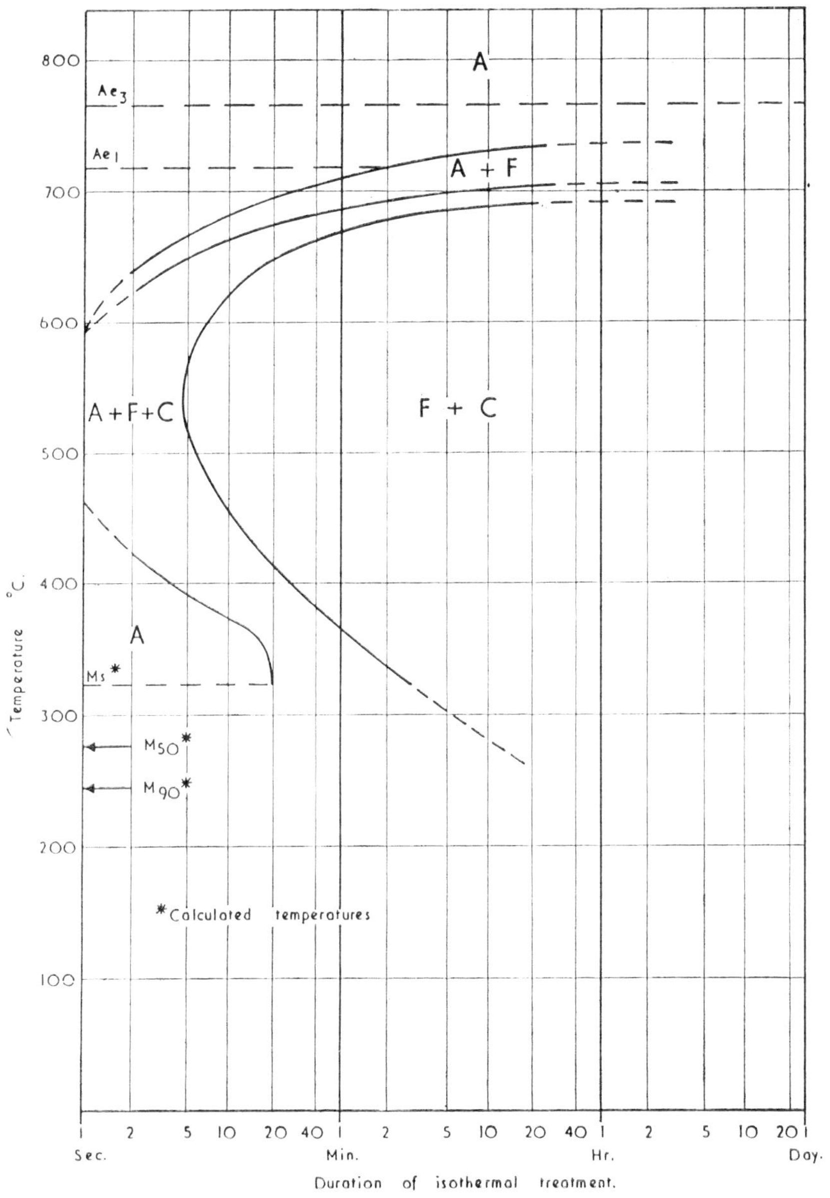

Fig. 43.1

HARDENABILITY

(See also data for En 9, Vol. 1)

END QUENCH HARDENABILITY DATA

Chemical Composition (%)	
C	Mn
0·46	0·99

McQuaid Ehn grain size 7 (ASTM).
Austenitizing Temperature 843°C.

(Ref. 8)

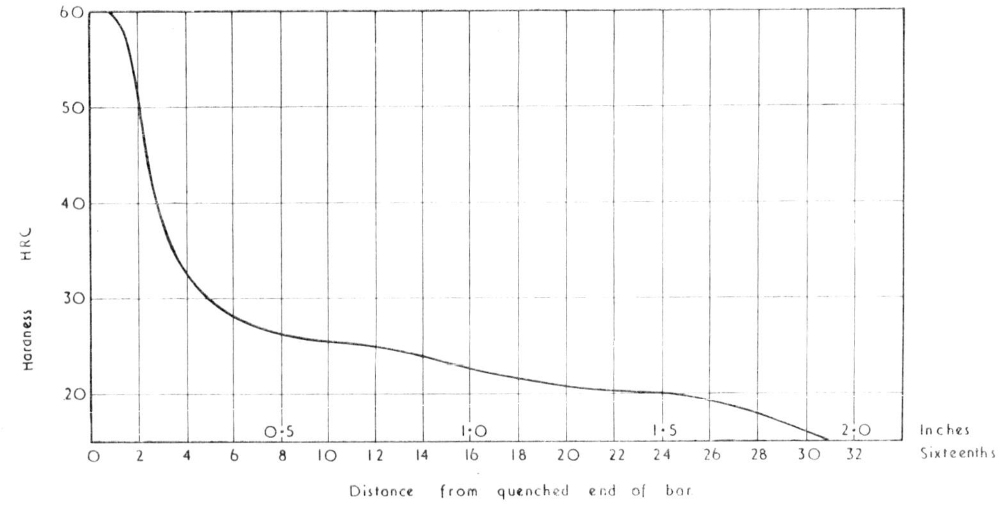

Fig. 43.2

Chemical Composition (%)						
C	Si	Mn	S	P	Ni	Cr
0·51	0·23	0·81	0·040	0·040	0·10	0·03

Acid Open Hearth.
Pre-treatment, bars normalized 820°C.
Austenitizing Temperature 820°C.

(Ref. 9)

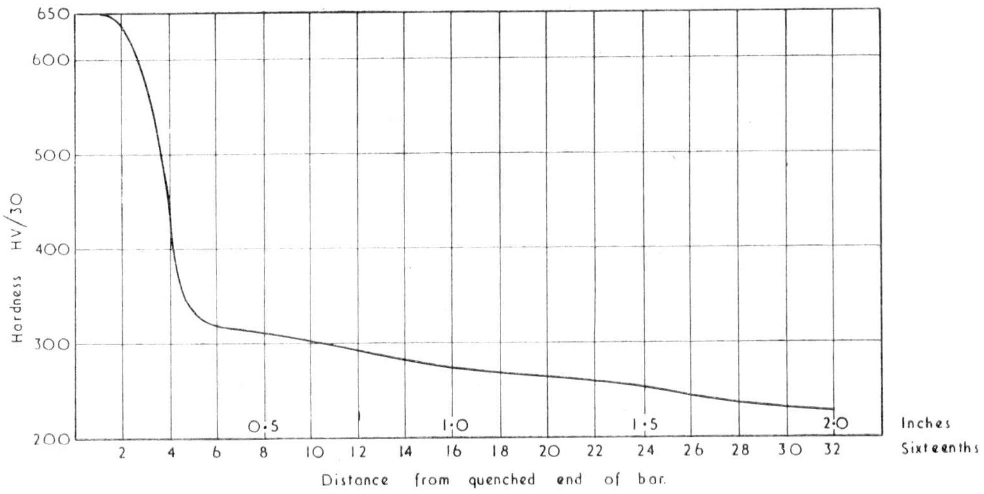

Fig. 43.3

MECHANICAL PROPERTIES AT ROOM TEMPERATURE

(See also En 9, Vol. 1)

Process of Manufacture	Chemical Composition (%)								Section Size dia. (in.)	Heat Treatment °C	Tensile					Charpy Keyhole KgM/sq. cm.	Ref.
	C	Si	Mn	S	P	Ni	Cr	Mo			Y.P. (Rs) tons/sq.in.	T.S. (Rm) tons/sq.in.	El % (A)	R.A. % (Z)	Izod ft.lb.		
B.O.H.	0·45	0·23	0·73	0·028	0·024	0·22			2¼	Softened		41·8	26				10
"	"	"	"	"	"	"			2⅛	"		41·8	26				"
	0·45	0·25	0·75						2¾	N830	28·0	46·0	24	36			3
	0·45	0·17	0·84						2⅛×1¼	OQ830 T630	31·0	50·0	30	56			"
	0·45	0·32	0·78	0·020	0·025				11/16	OQ870	38·0	58·0	18	50	22		11
	"	"	"	"	"				"	" T500	39·0	56·0	21	57	40		"
	"	"	"	"	"				"	" T600	35·0	51·0	24	61	53		"
	"	"	"	"	"				"	WQ870		120·0	3	4	3		"
	"	"	"	"	"				"	" T500	54·0	66·0	18	53	53		"
	"	"	"	"	"				"	" T600	44·0	55·0	23	60	76		"
	"	"	"	"	"				1⅛	N870	27·0	44·0	27	54½	31	3	"
	"	"	"	"	"				"	OQ870	35·8	56·0	20	49	27	3½	"
	"	"	"	"	"				"	" T300	37·5	56·0	21	51	23	3	"
	"	"	"	"	"				"	" T400	37·0	55·0	21	52	27	—	"
	"	"	"	"	"				"	" T500	36·0	55·0	21	54	32	3½	"
	"	"	"	"	"				"	" T600	32·0	50·0	24	59	39	4½	"
	"	"	"	"	"				"	" T700	28·0	45·0	27	65	44	5	"
	"	"	"	"	"				"	WQ870	47·0	67·0	12	28	14	2	"
	"	"	"	"	"				"	" T300	43·0	65·0	15	38	16	2½	"
	"	"	"	"	"				"	" T400	46·0	64·0	17	47	19	3	"
	"	"	"	"	"				"	" T500	42·0	59·0	21	55	34	3½	"
	"	"	"	"	"				"	" T600	36·0	52·0	25	62	48	5	"
	"	"	"	"	"				"	" T700	31·0	45·0	28	67	59	6	"
	"	"	"	"	"				2⅛	OQ870	33·9	52·0	23	54	24		"
	"	"	"	"	"				"	" T500	34·0	52·0	23	54	31		"
	"	"	"	"	"				"	" T600	29·0	47·0	26	59	39		"
	"	"	"	"	"				"	WQ870	37·0	56·0	18	43	16		"
	"	"	"	"	"				"	" T500	36·0	54·0	23	54	31		"
	"	"	"	"	"				"	" T600	31·0	49·0	26	60	44		"
	"	"	"	"	"				3	OQ870	31·9	49·0	23	55	26		"
	"	"	"	"	"				"	" T500	31·0	49·0	23	55	30		"
	"	"	"	"	"				"	" T600	27·0	45·0	26	59	39		"
	"	"	"	"	"				"	WQ870	30·0	52·0	19	47	24		"
	"	"	"	"	"				"	" T500	34·0	51·0	24	56	31		"
	"	"	"	"	"				"	" T600	29·0	48·0	26	59	44		"

MECHANICAL PROPERTIES AT ROOM TEMPERATURE (Continued)

Process of Manufacture	ASTM Grain Size	Chemical Composition (%) C	Si	Mn	S	P	Ni	Cr	Mo	Section Size dia. (in.)	Heat Treatment °C	Tensile Y.P. (Rs) tons/sq.in.	T.S. (Rm) tons/sq.in.	El % (A)	R.A. % (Z)	Izod ft. lb.	Charpy Keyhole KgM/sq. cm.	Ref.
		0·46		0·68						$1\frac{1}{8}$	N850	28·0	44·5	$25\frac{1}{2}$				12
		0·46	0·19	0·91						2	OQ830 T620	31·0	49·0	27	54			3
B.O.H.		0·46	0·24	0·81	0·039	0·027	0·17			6	Normalized	23·6	43·4	$25\frac{1}{2}$				10
B.O.H.		0·46	0·29	0·86	0·035	0·025	0·24			2	Softened		42·2	27				"
														$l=11\cdot3\sqrt{A}$				
		0·46	0·09	0·51	0·037	0·026	0·03			$3\times\frac{3}{8}$	N820	25·0	40·2	21	39			7
		"	"	"	"	"	"			"	WQ810 T400	62·0	83·3	4				"
		"	"	"	"	"	"			"	" T450	59·0	71·5	$5\frac{1}{2}$				"
		"	"	"	"	"	"			"	" T500	53·5	63·6	$7\frac{1}{2}$				"
		"	"	"	"	"	"			"	" T550	50·0	59·3	9				"
														$4\sqrt{A}$				
		0·47		0·80						$1\frac{1}{8}$	OQ840 T640	35·2	51·6	$25\frac{1}{2}$				12
B.O.H.	7 to 8	0·47	0·16	0·89	0·028	0·029	0·17			$1\frac{1}{8}$	Normalized	26·2	45·4	27				10
B.O.H.		0·47	0·23	0·82	0·031	0·022	0·21			5	Normalized	24·8	45·2	23				"
B.O.H.		0·47	0·26	0·92	0·042	0·038	0·21			$1\frac{3}{8}$	Softened		48·2	23				"
B.O.H.		0·47	0·24	0·82	0·032	0·032	0·16			$1\frac{7}{8}$	Softened		46·2	22				"
B.O.H.	5 to 7	0·47	0·14	0·76	0·028	0·021	0·25			$2\frac{1}{4}$	Softened		40·0	27				"
		0·48	0·15	0·76						$3\frac{1}{4}$	N830	30·0	46·0	23	39			3
B.O.H.		0·48	0·22	0·79	0·027	0·024	0·21			6	Normalized	23·0	44·0	24				10
B.O.H.		0·48	0·25	0·81	0·033	0·030	0·20			$1\frac{5}{8}$	Softened		45·2	23				"
B.O.H.		0·48	0·25	0·80	0·032	0·018	0·15			2	Softened		45·0	23				"
B.O.H.		0·48	0·25	0·83	0·020	0·022	0·18			$2\frac{1}{8}$	Softened		43·0	26				"
B.O.H.		0·48	0·25	0·80	0·032	0·018	0·15			$2\frac{1}{4}$	Softened		43·6	25				"
		0·49		0·75						$1\frac{1}{8}$	OQ840 T620	34·1	52·0	21·5				12
		0·49	0·16	0·79						$1\frac{1}{8}$	OQ830 T620	34·0	53·0	22	46			3
B.O.H.	6 to 7	0·49	0·16	0·95	0·033	0·027	0·22			5	Normalized	25·4	45·8	26				10
B.O.H.		0·49	0·23	0·83	0·029	0·019	0·19			6	Normalized	24·6	46·0	23				"
B.O.H.		0·49	0·25	0·82	0·036	0·027	0·21			$1\frac{3}{8}$	Softened	22·4	41·4	28				"
B.O.H.		0·49	0·21	0·84	0·037	0·024	0·26			$1\frac{3}{8}$	Softened		44·0	26				"
		"	"	"	"	"	"			$1\frac{1}{2}$	"	20·8	39·2	29				"
		0·50		0·70						$1\frac{1}{2}$	Cold drawn		52·3	$14\frac{1}{2}$				12
		0·50	0·24	0·79						$2\frac{3}{4}$	N830	27·0	49·0	21	$30\frac{1}{2}$			3
		0·50	0·19	0·75						$3\frac{1}{2}\times2\frac{1}{2}$	OQ330 T630	33·0	53·0	23	47			"
B.O.H.		0·50	0·25	0·78	0·030	0·022	0·15			$3\frac{3}{4}$	Normalized	25·2	47·2	23				10
B.O.H.		0·50	0·22	0·77	0·038	0·032	0·18			4	Normalized	24·0	46·6	23				"
B.O.H.		0·50	0·22	0·70	0·043	0·025				$1\frac{1}{4}$	Softened		46·8	19				"
B.O.H.		0·50	0·19	0·74	0·031	0·027	0·15			$1\frac{1}{4}$	Softened	22·0	43·6	26				"

MECHANICAL PROPERTIES AT ROOM TEMPERATURE

(Continued)

Process of Manufacture	ASTM Grain Size	Chemical Composition (%)								Section Size dia. (in.)	Heat Treatment °C	Tensile					Charpy Keyhole KgM/sq. cm.	Ref.
		C	Si	Mn	S	P	Ni	Cr	Mo			Y.P. (Rs)	T.S. (Rm)	El % (A)	R.A. % (Z)	Izod ft.lb.		
												tons/sq.in.						
B.O.H.		0·51	0·24	0·70	0·026	0·022	0·15			2½	Softened	21·2	43·4	23				10
		0·51	0·11	0·67	0·033	0·054		0·07		0·160	As received		52					13
		"	"	"	"	"		"		"	OQ850 T350		104	12	39			"
		"	"	"	"	"		"		"	" T380		97					"
		"	"	"	"	"		"		"	" T420		88		46			"
		"	"	"	"	"		"		"	" T460		74	18	48			"
		0·52		0·75						1⅛	N840	27·3	45·9	23½				12
B.O.H.	6—7	0·52	0·18	0·90	0·032	0·034	0·22			1⅛	OQ840 T600	33·6	53·8	22				10
B.O.H.		0·52	0·25	0·78	0·026	0·022	0·18			1⅞	Softened		48·5	22				"
B.O.H.	5—8	0·52	0·17	0·88	0·018	0·022	0·24			1⅜	Softened		48·8	20				"
B.O.H.	6—8	0·52	0·23	0·79	0·038	0·029	0·23			2⅛	Softened		49·2	20				"
B.O.H.		0·53	0·26	0·82	0·043	0·031	0·20			2¼	Softened		46·8	23				10
B.O.H.		0·54	0·25	0·70	0·035	0·025	0·18			2	Softened		46·8	23				10
B.O.H.		0·54	0·23	0·70	0·028	0·019	0·20			1⅛	Softened	21·6	44·4	24				"
"		"	"	"	"	"	"			2	"		46·8	22				"

EFFECT OF TEMPERING ON THE MECHANICAL PROPERTIES AT ROOM TEMPERATURE

(See also En 9, Vol. 1)

Chemical Composition (%)				
C	Si	Mn	S	P
0·45	0·32	0·78	0·020	0·025

$1\frac{1}{8}$ in. dia. bars, quenched and tempered as indicated.

Oil quenched 870°C. (Ref. 11)

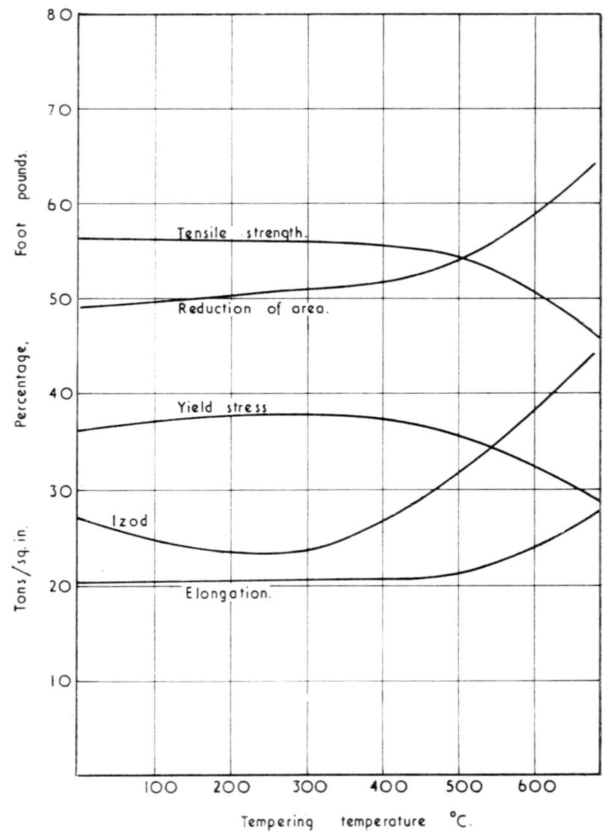

Fig. 43.4

Water quenched 870°C. (Ref. 11)

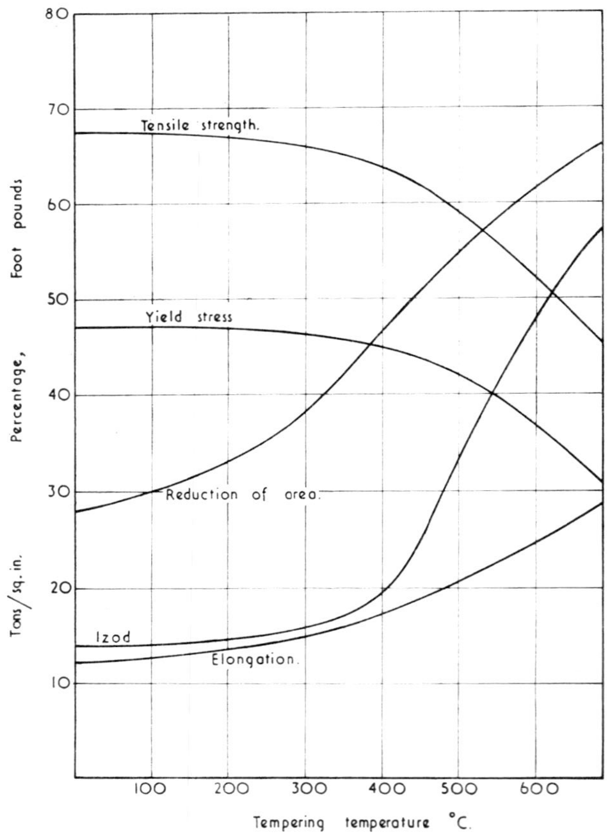

Fig. 43.5

EFFECT OF SECTION SIZE ON THE MECHANICAL PROPERTIES AT ROOM TEMPERATURE

Chemical Composition (%)

C	Si	Mn	S	P
0·45	0·32	0·78	0·020	0·025

Oil quenched 870°C. Tempered 500°C.

(Ref. 11)

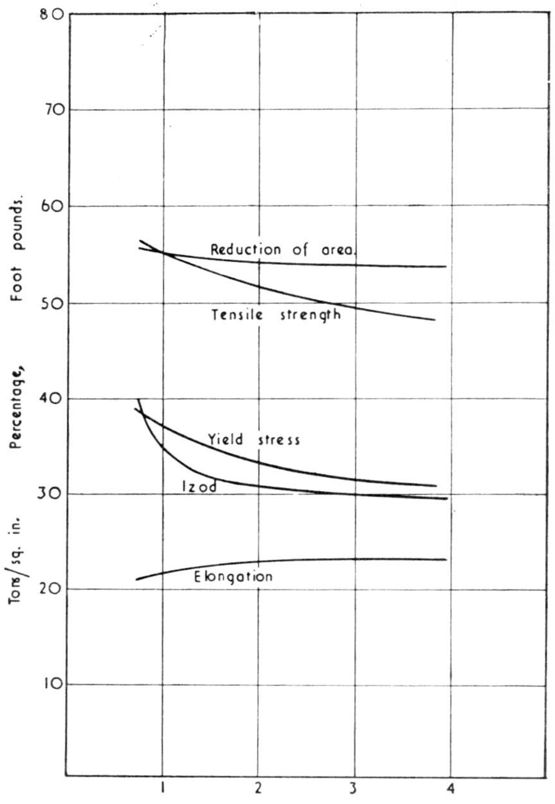

Fig. 43.6

Oil quenched 870°C. Tempered 600°C.

(Ref. 11)

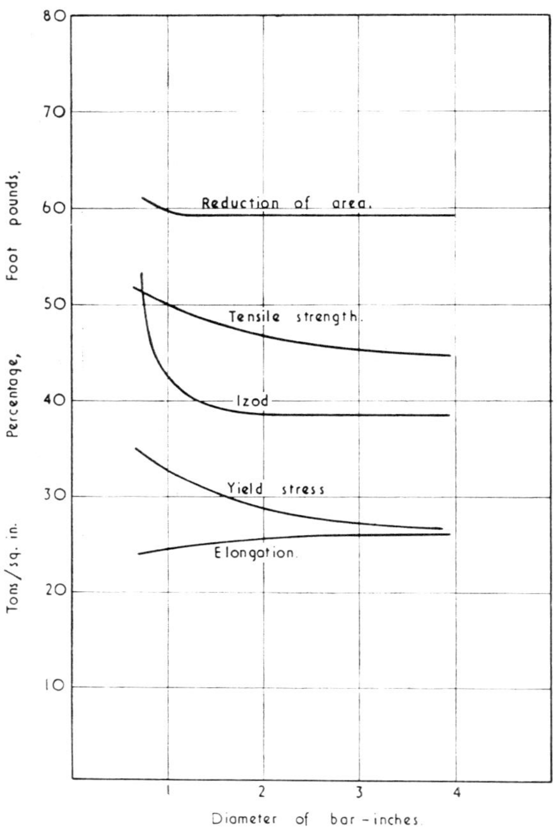

Fig. 43.7

EFFECT OF SECTION SIZE ON THE MECHANICAL PROPERTIES AT ROOM TEMPERATURE

(Continued)

Chemical Composition (%)				
C	Si	Mn	S	P
0·45	0·32	0·78	0·020	0·025

Water quenched 870°C. Tempered 500°C.

(Ref. 11)

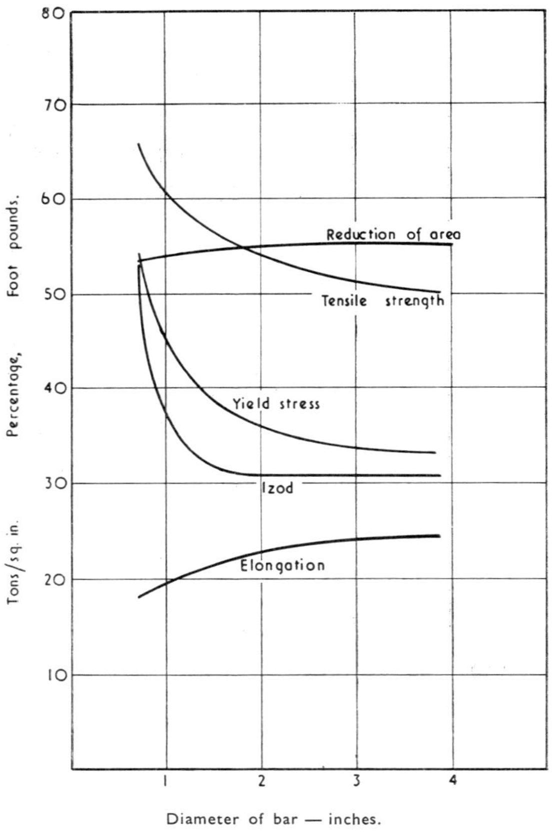

Fig. 43.8

Water quenched 870°C. Tempered 600°C.

(Ref. 11)

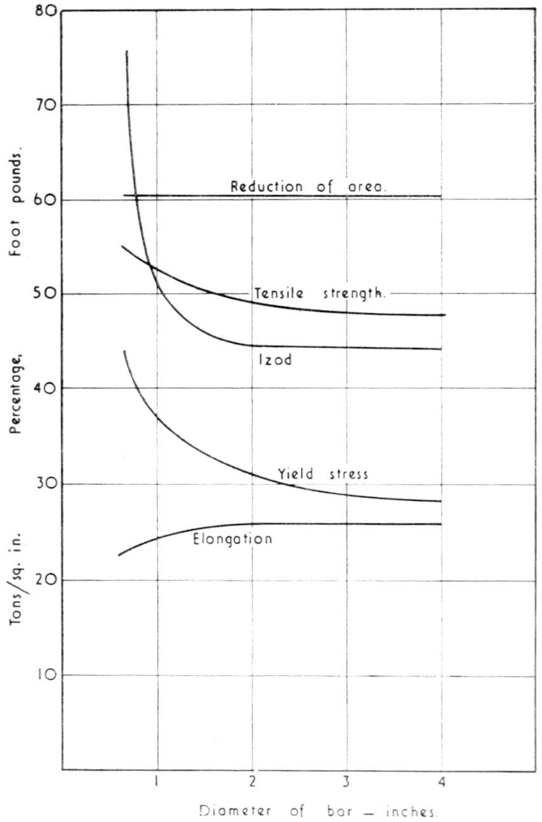

Fig. 43.9

MECHANICAL PROPERTIES AT LOW TEMPERATURE

(See also En 9, Vol. 1)

Chemical Composition (%)							Section Size dia. (in.)	Heat Treatment °C	Test Temperature °C	Tensile							D.V.M. Notched Impact KgM/sq. cm.	Ref.
C	Si	Mn	S	P	Ni	Cr				0·1% P.S. (0·1Rp) tons/sq.in.	0·2% P.S. (0·2Rp) tons/sq.in.	0·5% P.S. (0·5Rp) tons/sq.in.	Y.P. (Rs) tons/sq.in.	T.S. (Rm) tons/sq.in.	El % (A) l = 5d	R.A. % (Z)		
0·46	0·43	0·55	0·033	0·018	0·12	0·43	30 mm	WQ832 T2 hrs. 580AC	RT		40·8			51·3	23½	64	12·6	15
"	"	"	"	"	"	"	"	" " "	—78				47·4	60·3	24½	61	6·0	"
0·59	0·34	0·46	0·040	0·017	0·10	0·05	30 mm	WQ820 T2 hrs. 540AC	RT				40·5	60·7	18½	50		"
"	"	"	"	"	"	"	"	" " "	—78				46·2	67·8	19	48		"
0·46	0·12	0·50	0·048	0·041			0·044 wire	A700 12 hrs.	20	27·7		29·5		38·2		73		14
"	"	"	"	"			"	" "	—30	29·4		31·6		41·3		72½		"
"	"	"	"	"			"	" "	—60	30·6		34·0		43·4		66½		"
0·47	0·10	0·48	0·050	0·041			0·044 wire	Patented 970 CD 88%	20	59·2		81·6		85·0		47		"
"	"	"	"	"			"	" " "	—30	60·5		83·9		88·0		44		"
"	"	"	"	"			"	" " "	—60	61·2		88·3		91·5		44½		"

MECHANICAL PROPERTIES IN TORSION

Chemical Composition (%)							Section Size dia. (in.)	Heat Treatment °C	Test Temperature °C	Tensile				Torsion		Ref.
C	Si	Mn	S	P	Ni	Cr				Y.P. (Rs) tons/sq.in.	T.S. (Rm) tons/sq.in.	El % (A)	R.A. % (Z)	Apparent max. Shear Stress tons/sq.in.	Probable max. Shear Stress tons/sq.in.	
0·51	0·11	0·67	0·033	0·054		0·07	0·160	As received	RT		52			41·0	30·8	13
"	"	"	"	"		"	"	OQ850 T460	"		74	18	48	59·0	34·3	"
0·60	0·21	0·77	0·007	0·012	0·08	0·09	3 × ⅜	OQ950 T400	RT	93·4	101·8		2	85·2	63·9	6
"	"	"	"	"	"	"	"	" T450	"	83·5	91·9	5	10½	73·5	55·1	"
"	"	"	"	"	"	"	"	" T500	"	72·5	78·5	7	17	67·3	51·0	"
"	"	"	"	"	"	"	"	" T550	"	67·9	74·3	6½	21	60·0	45·6	"

FATIGUE PROPERTIES

Chemical Composition (%)							Section Size dia. (in.)	Heat Treatment °C	Test Tempera-ture °C	Tensile				Fatigue Limit tons/sq.in. Rotating bending		Reversed Axial	Ref.
C	Si	Mn	S	P	Ni	Cr				Y.P. (Rs) tons/sq.in.	T.S. (Rm) tons/sq.in.	El % (A)	R.A. % (Z)	Plain	Notched		
0·45							¾	N877 A800	RT		29*			±16·0		±12·8	16
"							"	OQ830 T596	"		36*			±20·6		±20·4	"
0·47							2¼ × ¾	N880 A800	"		33*			±17·2		=13·8	"
"							"	OQ830 T538	"		40*			±26·1		=22·1	"
												$l=11{\cdot}3\sqrt{A}$					
0·46	0·09	0·51	0·037	0·026	0·03		3 × ⅜	WQ810 T400	RT	62·0	83·3	4		±36			7
"	"	"	"	"	"		"	" T450	"	59·0	71·5	5½		±35 to 38			"
"	"	"	"	"	"		"	" T500	"	53·5	63·6	7½		±36			"
"	"	"	"	"	"		"	" T550	"	50·0	59·3	9		±33			"
0·46	0·12	0·50	0·048	0·041			0·044 wire	A700 (12 hrs.)	20	29·5	38·2		73	±16·8			14
"	"	"	"	"			"	" "	—30	31·6	41·3		72½	±17·0			"
"	"	"	"	"			"	" "	—60	34·0	43·4		66½	±17·3			"
0·47	0·10	0·48	0·050	0·041			0·044 wire	Patented 970 CD 88%	20	81·6	85·0		47	±26·5			"
"	"	"	"	"			"	" " "	—30	83·9	88·0		44	±27·0			"
"	"	"	"	"			"	" " "	—60	88·3	91·5		44½	±29·5			"
												l=5D					
0·46	0·34	0·55	0·033	0·018	0·12	0·43	30 mm	WQ830 T580 2 hrs. AC	RT		51·3	23½	64	±24·1	±7·0†		15
"	"	"	"	"	"	"	"	" " "	—78	47·4	60·3	24½	61	±28·0	±7·0†		"
0·59	0·34	0·46	0·040	0·017	0·10	0·05	30 mm	WQ820 T540 2 hrs. AC	RT	40·5	60·7	18½	50	±25·4	±8·9†		"
"	"	"	"	"	"	"	"	" " "	—78	46·2	67·8	19	48	±31·8	±10·2†		"
												$l=11{\cdot}3\sqrt{A}$					
0·60	0·21	0·77	0·007	0·012	0·08	0·09	3 × ⅜	OQ950 T400	RT	93·4	101·8			±47			6
"	"	"	"	"	"	"	"	" T450	"	83·5	91·9	5	10½	±39 to 41			"
"	"	"	"	"	"	"	"	" T500	"	72·5	78·5	7	17	±38			"
"	"	"	"	"	"	"	"	" T550	"	67·9	74·3	6½	21	±37·5			"
0·52		0·63								23·6	41·7			±14·1 (in air)			17
"		"								"	"			±8·8 (in brine, 10^7 cycles)			"

* Estimated values.

† Notch 60°V, 0·1 mm root radius, 0·5 mm deep, O.D. 8 mm.

LIST OF REFERENCES

1. Private correspondence with Firth-Brown Ltd.
2. Private correspondence with The United Steel Companies Ltd.
3. Private correspondence with the English Steel Corporation Ltd.
4. *International Critical Tables of Numerical Data, Physics, Chemistry and Technology*. Prepared by the National Research Council, U.S.A. Editor-in-Chief, E. W. Washburn, McGraw Hill. First Edition, 1926.
5. G. Bullock. Effect of Quenching on Conductivity and Resistivity. *JISI* 1956, Vol. 183, Aug., No. 2, p. 362.
6. G. A. Hankins, D. Hansen and G. W. Ford. The Mechanical Properties of Four Heat-treated Spring Steels. *JISI*, 1926, Vol. 114, No. 2, p. 265.
7. G. A. Hankins and G. W. Ford. The Mechanical and Metallurgical Properties of Spring Steels as revealed by Laboratory Tests. *JISI*, 1929, Vol. 119, No. 1, p. 217.
8. Atlas of Isothermal Transformation Diagrams. The United States Steel Corporation. Second Edition, 1957.
9. Private correspondence with the London Polytechnic.
10. Private correspondence with The Park Gate Iron and Steel Co. Ltd.
11. Report of the Steel Research Committee founded by the Institution of Automobile Engineers and the Society of Motor Manufacturers and Traders as an Investigation to establish the Mechanical Properties of the British Standard Wrought Automobile Steels, Report 75, British Engineering Standards Association, 1920.
12. B.S. 971: Commentary on the British Standard Wrought Steels En Series. British Standards Institution, London, 1950.
13. A. Swan, H. Sutton and W. D. Douglas. Investigation of Aircraft Engine Valve Springs. *Proceedings of the Institution of Mechanical Engineers*, 1931, Vol. 120, p. 261.
14. War Office, unpublished report.
15. A. Pomp and M. Hempel. Fatigue Strength and Notch Coefficient of Unalloyed and Alloy Constructional Steel at +20° and —78°C. *Arch. für das Eisenhüttenwesen*, 1950, Vol. 21, Jan.-Feb., p. 53.
16. R. D. France. Endurance Testing of Steels: Comparison of Rotating Beam and Axially Loaded Specimens. *Proceedings of the American Society for Testing Materials*, 1931, Vol. 31, pt. II, pp. 176—193.
17. S. L. Hoyt. *Metals and Alloys Data Book*. Reinhold, New York, 1952.

CARBON SPRING STEEL (HIGHER CARBON CONTENT)

SPECIFICATION

CHEMICAL COMPOSITION (%)

	C	Si	Mn	S	P
En 44	0·90—1·20	0·30 max.	0·45—0·70	0·050 max.	0·050 max.
En 44B [1]	0·90—1·00	0·35 max.	0·40—0·70	0·050 max.	0·050 max.
En 44C [1]	1·00—1·20	0·35 max.	0·40—0·70	0·050 max.	0·050 max.
En 44D [2]	0·90—1·05	0·35 max.	0·30—0·70	0·050 max.	0·050 max.
En 44E [2]	1·05—1·25	0·35 max.	0·30—0·70	0·050 max.	0·050 max.

(1) Steel of this type may be required in the form of wire for oil hardened and tempered springs (see BS 1429: Annealed steel wire for oil hardened and tempered springs).
(2) Steel of this type may be required in the form of cold rolled strip and flat wire for hardening and tempering (see BS 1449: Steel plate, sheet and strip).

MECHANICAL PROPERTIES

No mechanical properties are specified for these steels.

RELATED SPECIFICATIONS

United Kingdom		America		Germany			Sweden
En No.	BS	SAE	AISI	Werkstoff No.	Name	DIN	SIS
44	1423	1095	C1095				
44B	1423 1429	1095	C1095	1.1274	Mk101	17222	
44C	1429						
44D	1407 1423 1449	1095	C1095	1.1274	Mk101	17222	14.1880
44E	1407 1449						14.1885

APPLICATIONS

This steel is used for coil and spiral springs, especially of large size, such as are employed in railway suspensions; feeler gauges, machine knives, piston rings, saws, cutting tools, chisels, hand tools, etc.

WELDING

See Introduction (List of Abbreviations) for key to symbols used in this table.

	Welding Process					
	MA	G	IG	S	F	B
En 44	f	f	f	f	f	j
En 44B	m	f	m	f	f	j
En 44C	m	f	m	f	f	j
En 44D	f	f	f	f	f	j
En 44E	f	f	f	f	f	j

Reproduced by kind permission of H.M.S.O.

MACHINABILITY

The machinability of this steel in the softened condition is about 50% of that for mild steel (En 3).

HOT WORKING AND HEAT TREATMENT TEMPERATURES

Forging, rolling and stamping	1100°C, finish above 850°C.
Annealing	800°C to 820°C, furnace cool.
Softening	680°C to 720°C, air or furnace cool.
Hardening	780°C to 820°C, oil quench.
Tempering	400°C to 500°C, air cool.

PHYSICAL PROPERTIES

SPECIFIC GRAVITY (d)

Chemical Composition (%)								Heat Treatment °C	Specific Gravity at 20°C	Ref.
C	Si	Mn	S	P	Ni	Cr	Mo			
0·90								Annealed	7·83	1
1·20								Softened	7·83	2
1·22	0·16	0·35	0·015	0·009	0·13	0·11	0·01	A800	7·83	3

SPECIFIC HEAT (c)

Chemical Composition (%)								Heat Treatment °C	Mean Specific Heat cal/g deg. C										Ref.
C	Si	Mn	S	P	Ni	Cr	Mo		20 to 100°C	20 to 200°C	20 to 300°C	20 to 400°C	20 to 500°C	20 to 600°C	20 to 700°C	20 to 800°C	20 to 900°C	20 to 1000°C	
1·22	0·16	0·35	0·015	0·009	0·13	0·11	0·31	A800	0·115	0·122	0·125	0·129	0·133	0·138	0·147	0·169	0·167	0·165	3

MEAN COEFFICIENT OF THERMAL EXPANSION (α)

Chemical Composition (%)								Heat Treatment °C	$10^6 \times$ Mean Coefficient of Expansion per deg. C										Ref.
C	Si	Mn	S	P	Ni	Cr	Mo		20 to 100°C	20 to 200°C	20 to 300°C	20 to 400°C	20to 500°C	20 to 600°C	20 to 700°C	20 to 800°C	20 to 900°C	20 to 1000°C	
1·22	0·16	0·35	0·015	0·009	0·13	0·11	0·01	A800	10·7	11·4	12·2	13·0	13·7	14·3	14·8	14·4	16·5	17·0	3
1·28	0·19	0·37						—	11·0		12·0			14·1					4

ELECTRICAL RESISTIVITY (ϱ)

Chemical Composition (%)								Section Size dia. (in.)	Heat Treatment °C	Electrical Resistivity microhm-cm												Ref.
C	Si	Mn	S	P	Ni	Cr	Mo			0°C	20°C	100°C	200°C	300°C	400°C	500°C	600°C	700°C	800°C	900°C	1000°C	
1·22	0·16	0·35	0·015	0·009	0·13	0·11	0·01		A800	18·4	19·6	25·2	33·3	43·0	54·0	66·5	80·2	96·4	115·2	119·6	122·6	3
,,	,,	,,	,,	,,	,,	,,	,,	¼	WQ950		32·1											,,
,,	,,	,,	,,	,,	,,	,,	,,	1	WQ950		35·8											,,

THERMAL CONDUCTIVITY (k)

Chemical Composition (%)								Heat Treatment °C	Thermal Conductivity cal/cm s deg. C												Ref.
C	Si	Mn	S	P	Ni	Cr	Mo		0°C	20°C	100°C	200°C	300°C	400°C	500°C	600°C	700°C	800°C	900°C	1000°C	
1·22	0·16	0·35	0·015	0·009	0·13	0·11	0·01	A800	0·108	0·108	0·107	0·102	0·096	0·089	0·083	0·076	0·068	0·057	0·059	0·062	3

THERMAL CONDUCTIVITY (Continued)

Chemical Composition (%)					Heat Treatment °C	Thermal Conductivity cal/cm s deg. C											Ref.
C	Si	Mn	S	P		90°C	100°C	120°C	140°C	160°C	180°C	190°C	200°C	230°C	260°C	270°C	
0·93	0·11	0·16	0·007	0·014	WQ790	0·078		0·080(115°C)									5
"	"	"	"	"	" T150	0·082(88°C)		0·083(116°C)									"
"	"	"	"	"	" T200	0·085			0·085(141°C)								"
"	"	"	"	"	" T300			0·098(115°C)			0·096(178°C)			0·096(233°C)			"
"	"	"	"	"	" T400		0·104(96°C)						0·101(196°C)			0·102(273°C)	"
"	"	"	"	"	" T600		0·106(96°C)					0·103(185°C)				0·103	"
"	"	"	"	"	Annealed		0·107(104°C)						0·106(202°C)			0·104(271°C)	"
1·14	0·12	0·21	0·026	0·020	WQ780	0·077(85°C)		0·078(116°C)									5
"	"	"	"	"	" T150	0·080(85°C)		0·081(114°C)									"
"	"	"	"	"	" T200	0·084(92°C)							0·084(197°C)				"
"	"	"	"	"	" T300	0·095(94°C)				0·094(157°C)						0·094(267°C)	"
"	"	"	"	"	" T400		0·099(95°C)					0·099(188°C)				0·097(272°C)	"
"	"	"	"	"	" T600	0·102(88°C)					0·099(184°C)				0·099(263°C)		"
"	"	"	"	"	Annealed	0·103					0·102				0·100(263°C)		"

YOUNG'S MODULUS (E); SHEAR MODULUS (G); POISSON'S RATIO (σ)

Chemical Composition (%)					Section Size dia. (in.)	Heat Treatment °C	Test Tempera-ture °C	Young's Modulus tons/sq.in.		Shear Modulus tons/sq.in.		Poisson's Ratio tons/sq.in.		Ref.
C	Si	Mn	S	P				Static	Dynamic*	Static	Dynamic*	Static	Dynamic*	
0·89	0·08	0·43	0·044	0·038		As received				5,190				6
"	"	"	"	"		OQ800 T460				5,220				"
"	"	"	"	"		OQ740 T460				5,190				"
0·90						Annealed		13,200	13,500	5,100	5,240	0·285	0·292	1
0·92	0·27	0·42	0·012	0·010	0·104	Cold drawn		12,100		5,220				7
"	"	"	"	"	"	CD ½ hr. 350		12,900		5,280				"
1·0					½	OQ780			13,000	4,980	5,010		0·297	8
"					"	" T100			13,000	4,980	5,020		0·296	"
"					"	" T200			13,100	5,040	5,070		0·295	"
"					"	" T400			13,450	5,180	5,200		0·293	"
"					"	" T500			13,550	5,200	5,250		0·292	"
"					"	" T600			13,600	5,260	5,200		0·288	"
"					"	" T650			13,600	5,310	5,270		0·288	"
"					"	OQ870			13,100	5,040	5,070		0·295	"
"					"	" T100			13,150	5,050	5,080		0·295	"
"					"	" T200			13,300	5,100	5,140		0·295	"
"					"	" T300			13,500	5,170	5,200		0·294	"
"					"	" T400			13,600	5,220	5,250		0·292	"
"					"	" T500			13,700	5,260	5,300		0·290	"
"					"	" T600			13,800	5,360	5,350		0·288	"
"					"	" T650			13,800	5,400	5,370		0·288	"
1·04	0·16	0·38				A820		13,270						9
"	"	"				N820		13,270						"
"	"	"				OQ820 T600		13,270						"
1·06	0·22	0·30	0·011	0·005		As rolled	25	13,450†		5,320†		0·265†		10
"	"	"	"	"		"	150	12,900†		5,140†		0·255†		"
"	"	"	"	"		"	260	12,600†		4,980†		0·265†		"
"	"	"	"	"		"	480	12,300†		4,880†		0·260†		"
"	"	"	"	"		"	590	10,750†		4,260†		0·260†		"

* Ultrasonic pulse method.
† Combined bending and twisting method.

MAGNETIC PROPERTIES

Chemical Composition (%)				Heat Treatment °C	Brinell Hardness HB	Magnetic induction B (Gauss) for magnetizing field H (Oersted)=			Ref.
C	Si	Mn	Ni			10	50	200	
1·03	0·026	0·32	0·19	H & T	383	2,140	13,300	—	11
"	"	"	"	N850	318	2,250	12,900	17,250	"
"	"	"	"	A850	257	2,150	12,800	17,100	"
1·09	0·18	0·49	0·11	H & T	395	2,700	12,750	17,000	"
"	"	"	"	N850	335	1,300	11,600	16,600	"
"	"	"	"	A850	257	2,000	13,200	16,600	"

TRANSFORMATION CHARACTERISTICS

ISOTHERMAL TRANSFORMATION

Chemical Composition (%)						
C	Si	Mn	S	P	Ni	Cr
0·93	0·11	0·31	0·010	0·012	0·20	0·12

Open Hearth.
Actual grain size 12 (ASTM).
Pre-treatment, annealed.

Austenitizing Temperature 800°C, 15 min.

(Ref. 12)

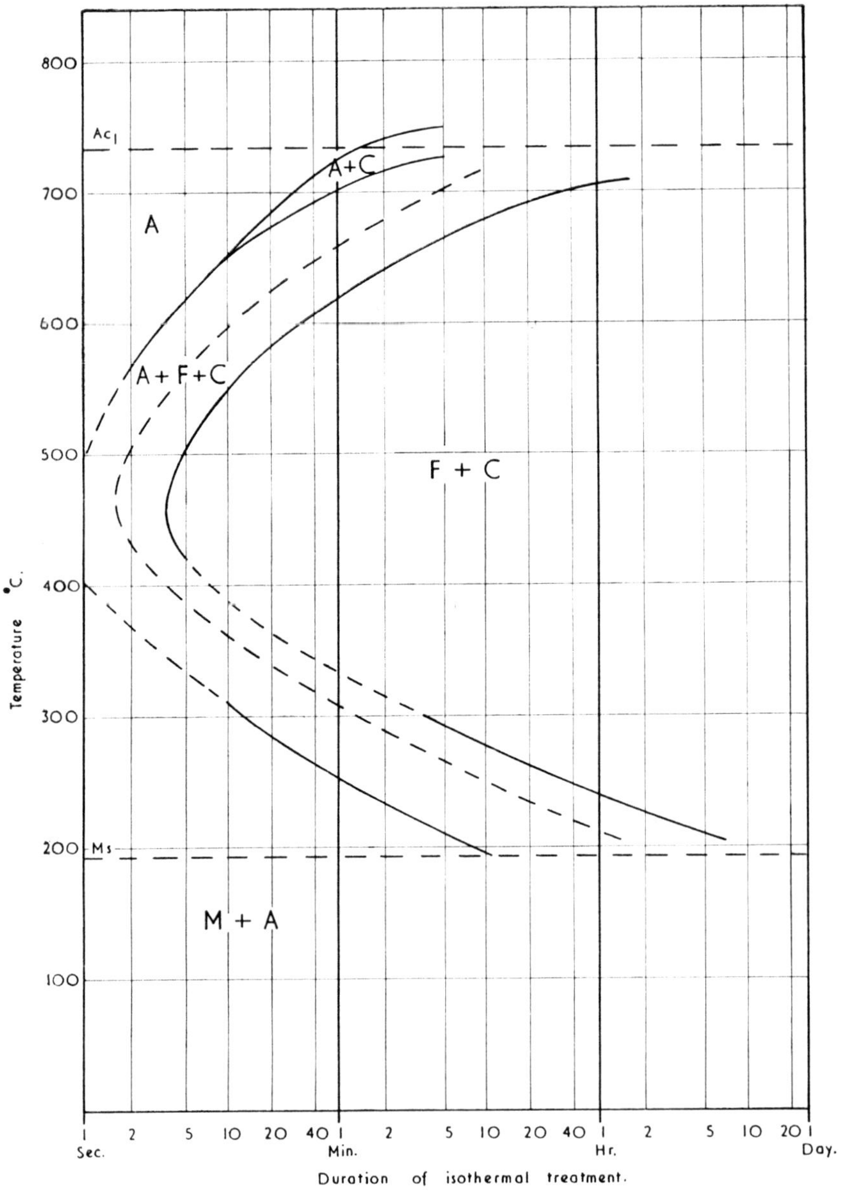

Fig. 44.1

ISOTHERMAL TRANSFORMATION (Continued)

Chemical Composition (%)							
C	Si	Mn	S	P	Ni	Cr	Mo
0·96	0·32	0·55	0·012	0·013	0·08	0·11	0·01

Basic Electric Arc.
McQuaid Ehn grain size 4—5 (ASTM).
As quenched grain size 5 (ASTM).
Bars hot rolled to 4 S.W.G., cold drawn to 11 S.W.G., with intermediate annealing at 780°C. Rolled to tape 0·03 in. × 0·1 in.

Austenitizing Temperature 780°C, 30 min.

(Ref. 13)

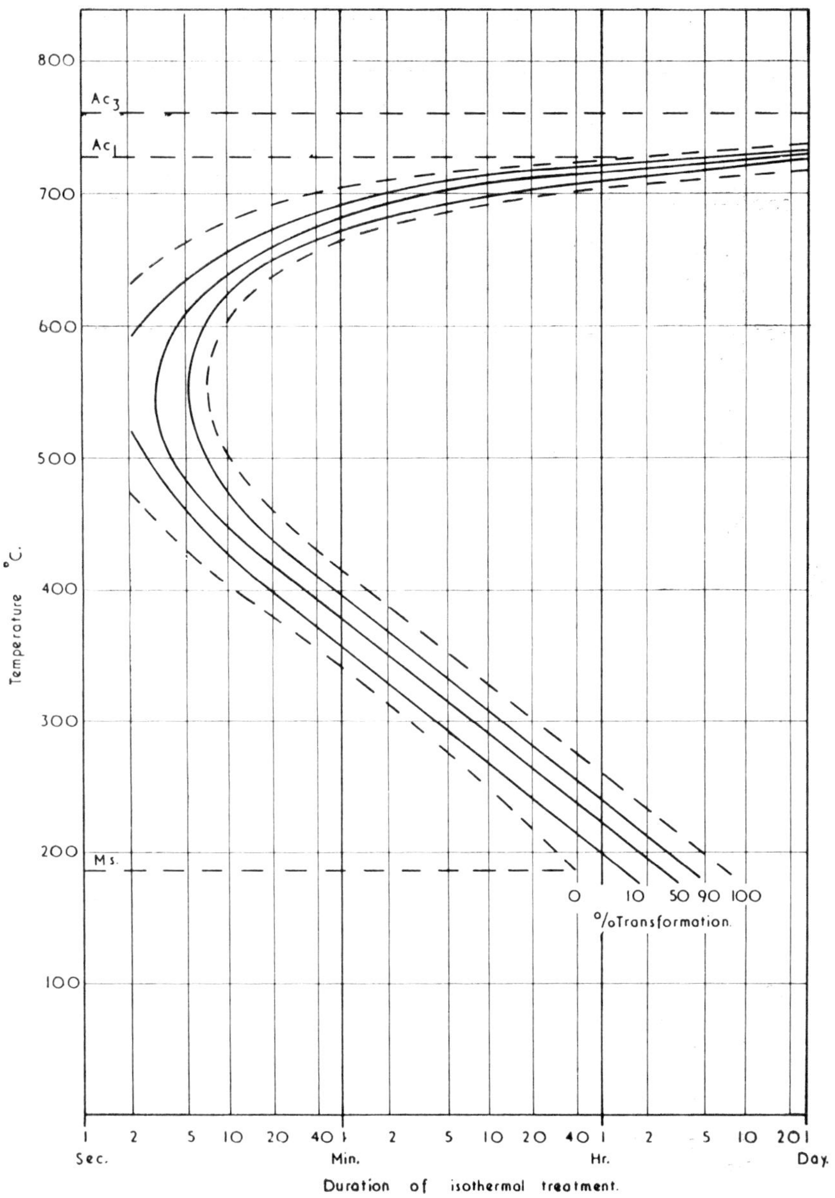

Fig. 44.2

En 44

ISOTHERMAL TRANSFORMATION (Continued)

Chemical Composition (%)	
C	Mn
1·13	0·30

Austenite grain size 7—8 (ASTM).

Austenitizing Temperature 910°C.

(Ref. 14)

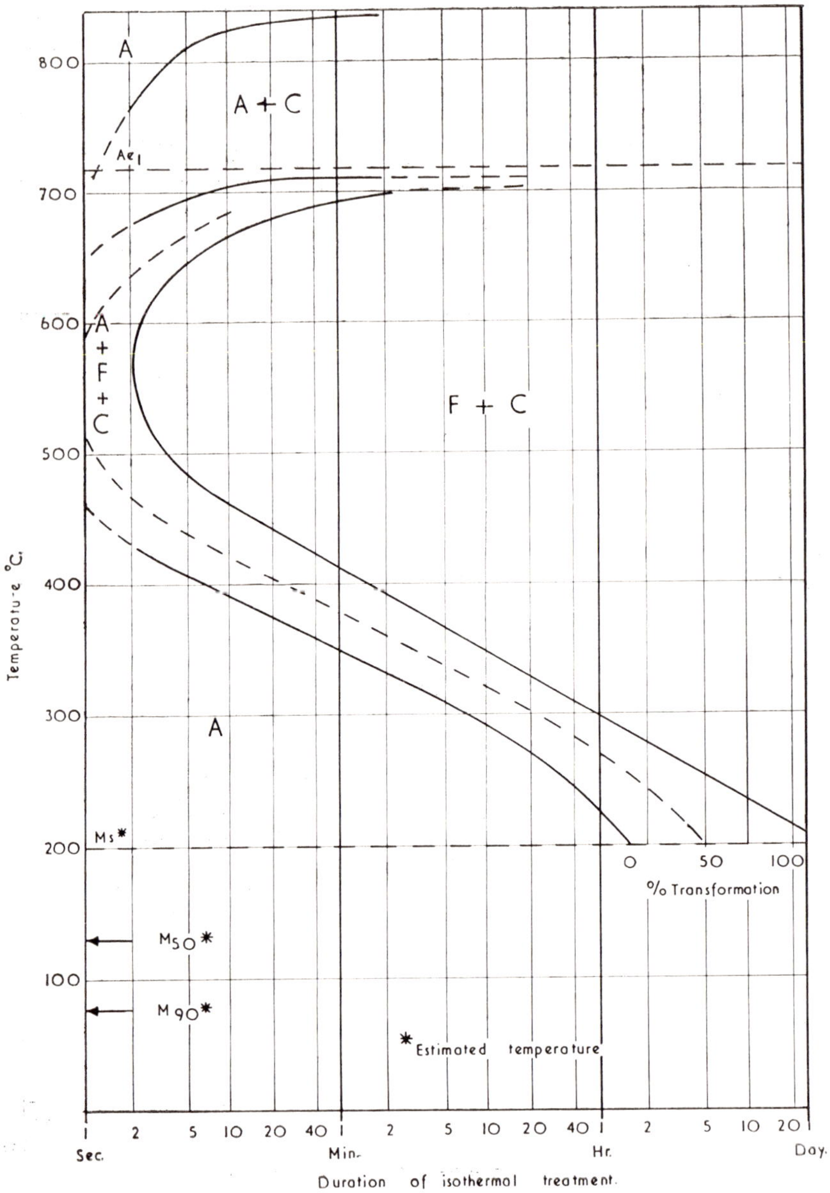

Fig. 44.3

HARDENABILITY

END QUENCH HARDENABILITY DATA

Chemical Composition (%)						
C	Si	Mn	S	P	Ni	Cr
0·93	0·11	0·31	0·010	0·012	0·20	0·12

Open Hearth.
Actual grain size 12 (ASTM).
Pre-treatment, annealed.
End quenched from 800°C.

(Ref. 12)

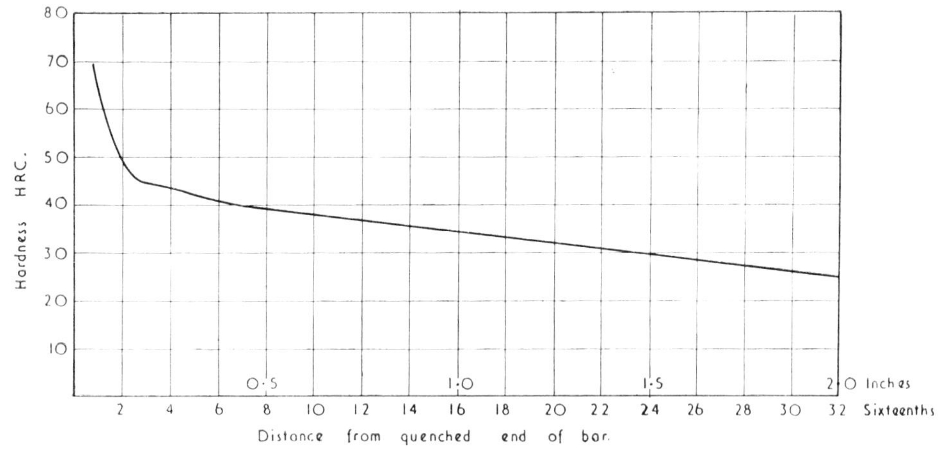

Fig. 44.4

Chemical Composition (%)							
C	Si	Mn	S	P	Ni	Cr	Mo
0·96	0·32	0·55	0·012	0·013	0·08	0·11	0·01

Basic Electric Arc.
McQuaid Ehn grain size 4—5 (ASTM).
As quenched grain size 5 (ASTM).

Austenitizing Temperature 780°C, 30 min.

(Ref. 13)

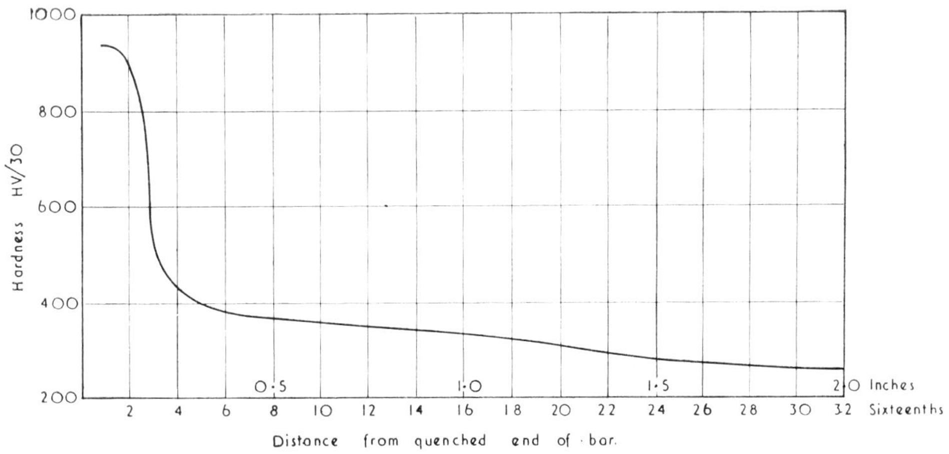

Fig. 44.5

En 44

END QUENCH HARDENABILITY (Continued)

Chemical Composition (%)	
C	Mn
1·17	0·28

Austenite grain size 7 (ASTM).
Austenitizing Temperature 927°C.

(Ref. 14)

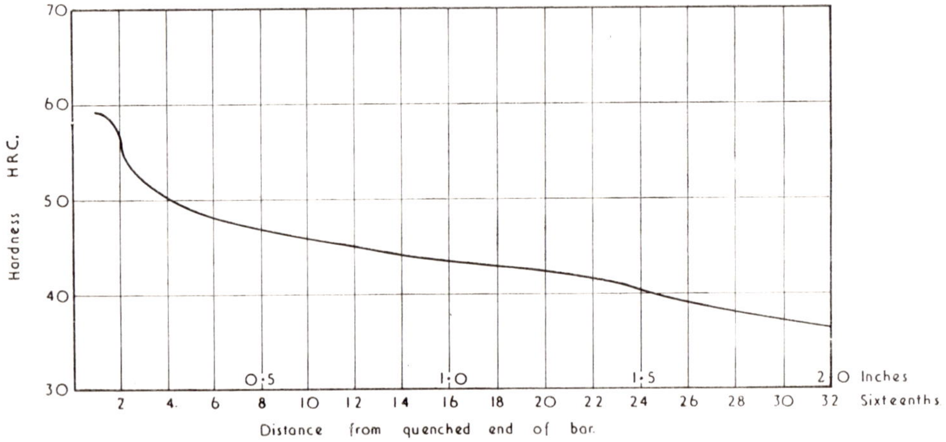

Fig. 44.6

Chemical Composition (%)						
C	Si	Mn	S	P	Ni	Cr
0·99	0·23	0·64	0·050	0·032	0·12	0·07

Acid Open Hearth.
Pre-treatment, normalized 820°C.
End quenched from 780°C.

(Ref. 15)

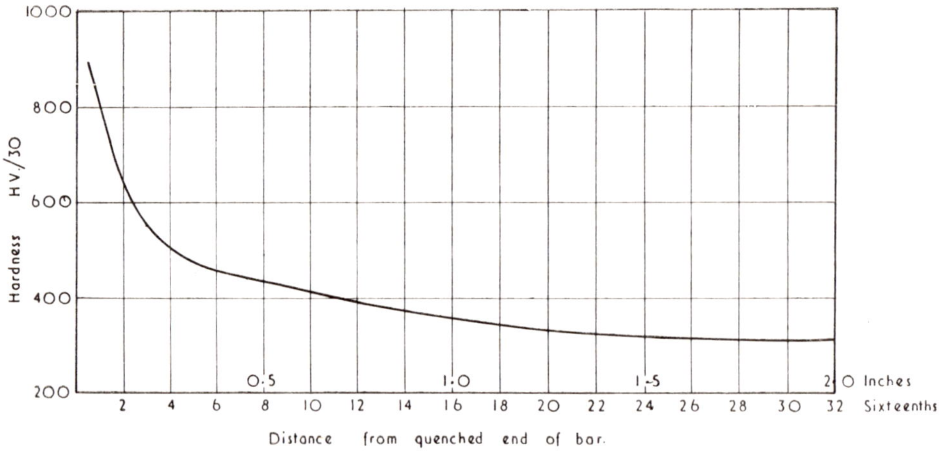

Fig. 44.7

HARDNESS TRAVERSE CURVES

Chemical Composition (%)				
C	Si	Mn	Ni	Cr
0·98	0·14	0·59	0·07	0·05

Bars oil quenched from 820°C.

(Ref. 7)

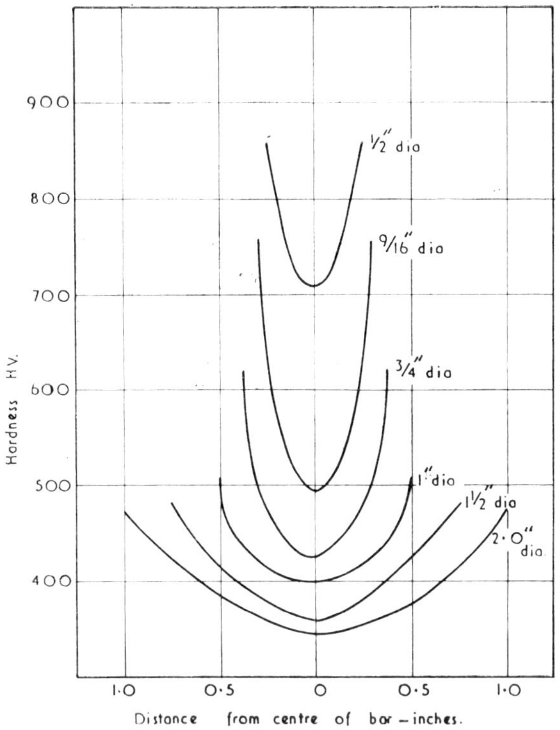

Fig. 44.8

MECHANICAL PROPERTIES AT ROOM TEMPERATURE

Process of Manufacture	ASTM Grain Size	Chemical Composition (%)								Section Size dia. (in.)	Heat Treatment °C	Tensile								Izod ft.lb.	Ref.
		C	Si	Mn	S	P	Ni	Cr	Mo			0·01% P.S. (0·01Rp)	0·1% P.S. (0·1Rp)	0·2% P.S. (0·2Rp)	0·5% P.S. (0·5Rp)	Y.P. (Rs)	T.S. (Rm)	El % (A)	R.A. % (Z)		
												tons/sq.in.									
		0·89	0·08	0·43	0·044	0·038		0·39		0·160	As received						64				6
		"	"	"	"	"		"		"	OQ800 T460						92				"
		"	"	"	"	"		"		"	OQ740 T460						99				"
		0·90	0·11	0·55	0·036	0·031				1¼	N850				32·4		59·4	11	15½	8	11
A.O.H.		0·92	0·27	0·42	0·012	0·019				0·104	Cold drawn	50·5	87·0				127·0	2	52		7
"		"	"	"	"	"				"	CD+½ hr. 350	53·6	89·8				125·8	5	46		"
		0·93	0·03	0·38	0·045	0·017				2×⅞	AC871 FC788					14·9	37·5	25	37		16
		"	"	"	"	"				"	OQ788 T649 AC					30·2	51·3	23	39½		"
		"	"	"	"	"				"	" T454 AC						84·0	10	29½		"
		0·95		0·40	0·05	0·04					As rolled					37·5	63·4	9	18	3	11
		"		"	"	"					Annealed					23·7	44·6	21	42	5	"
		"		"	"	"					OQ830 T310					52·7	80·4	11	30	5	"
	5—7(1—4)	0·96	0·20	0·40	0·029	0·012				1	As rolled					37·5	63·5	9	17		17
	"	"	"	"	"	"				"	A 787°C					24·5	42·5	13	20½	2	"
	"	"	"	"	"	"				½	N898					35·9	67·4	12½	27½	4½	"
	"	"	"	"	"	"				1	"					32·4	65·6	9½	13½	4	"
	"	"	"	"	"	"				2	"					25·7	59·1	9	13½	3	"
	"	"	"	"	"	"				4	"					25·6	57·3	10	14	3	"
	"	"	"	"	"	"				½	OQ800 T482					51·8	82·1	13	35½	4½	"
	"	"	"	"	"	"				"	" T538					45·3	74·3	15½	40	5½	"
	"	"	"	"	"	"				"	" T593					38·8	63·4	17½	43	6	"
	"	"	"	"	"	"				1	OQ800 T204					53·6	83·5	10	30	5	"
	"	"	"	"	"	"				"	" T315					52·7	81·7	10	30	5	"
	"	"	"	"	"	"				"	" T426					50·5	78·6	12	32	6	"
	"	"	"	"	"	"				"	" T482					45·7	78·4	10	23½	5½	"
	"	"	"	"	"	"				"	" T538					42·5	71·0	13	32½	6	"
	"	"	"	"	"	"				"	" T593					35·3	62·4	17	39	6	"
	"	"	"	"	"	"				"	" T648					35·7	57·6	21	47	6	"
	"	"	"	"	"	"				"	" T704					33·0	48·7	26	54	6	"
	"	"	"	"	"	"				2	OQ800 T482					43·9	74·9	12	30	4½	"
	"	"	"	"	"	"				"	" T538					41·3	67·4	13½	31½	4	"
	"	"	"	"	"	"				"	" T593					34·5	60·0	18½	43½	4½	"
	"	"	"	"	"	"				4	OQ800 T482					41·5	73·7	12	17½	4	"
	"	"	"	"	"	"				"	" T538					35·7	66·1	12	22	4½	"
	"	"	"	"	"	"				"	" T593					29·3	58·0	17	34½	5	"

MECHANICAL PROPERTIES AT ROOM TEMPERATURE

(Continued)

Process of Manufacture	ASTM Grain Size	Chemical Composition (%) C	Si	Mn	S	P	Ni	Cr	Mo	Section Size dia. (in.)	Heat Treatment °C	Tensile 0·01% P.S. (0·01Rp) tons/sq.in.	0·1% P.S. (0·1Rp)	0·2% P.S (0·2Rp)	0·5% P.S. (0·5Rp)	Y.P. (Rs)	T.S. (Rm)	El % (A)	R.A. % (Z)	Izod ft.lb.	Ref.
	5–7–(14)	0·96	0·20	0·40	0·029	0·012				½	WQ787 T482					60·5	85·5	12½	31½	4½	17
	,,	,,	,,	,,	,,	,,				,,	,, T538					49·5	76·8	12½	44	6½	,,
	,,	,,	,,	,,	,,	,,				,,	,, T593					44·2	64·3	17	45	6	,,
	,,	,,	,,	,,	,,	,,				1	WQ787 T204					67·9	96·9	10	31	5	,,
	,,	,,	,,	,,	,,	,,				,,	,, T315					67·0	95·1	11	33	5	,,
	,,	,,	,,	,,	,,	,,				,,	,, T426					61·6	89·3	12	36	5	,,
	,,	,,	,,	,,	,,	,,				,,	,, T482					54·0	81·2	13	37½	4½	,,
	,,	,,	,,	,,	,,	,,				,,	,, T538					45·7	73·7	16	41½	5	,,
	,,	,,	,,	,,	,,	,,				,,	,, T593					43·1	63·8	16½	43½	5½	,,
	,,	,,	,,	,,	,,	,,				,,	,, T648					37·5	54·9	20	47	5	,,
	,,	,,	,,	,,	,,	,,				,,	,, T704					33·0	46·4	23	50	5	,,
	,,	,,	,,	,,	,,	,,				2	WQ787 T482					50·5	80·3	12½	34	4	,,
	,,	,,	,,	,,	,,	,,				,,	,, T538					44·0	69·1	15½	39	4½	,,
	,,	,,	,,	,,	,,	,,				,,	,, T593					40·2	62·5	17½	43½	5	,,
	,,	,,	,,	,,	,,	,,				4	WQ787 T482					42·2	74·7	12½	31½	3	,,
	,,	,,	,,	,,	,,	,,				,,	,, T538					36·2	67·0	15½	35½	4	,,
	,,	,,	,,	,,	,,	,,				,,	,, T593					34·8	58·6	18½	41	4½	,,
	7	0·98	0·20	0·47	0·035	0·012		0·02	0·12	1⅝	OQ845 T454			58·0			84·8	11	30		18
	,,	,,	,,	,,	,,	,,		,,	,,	,,	,, T510			53·6			82·6	13	32		,,
	,,	,,	,,	,,	,,	,,		,,	,,	,,	,, T566			48·2			77·2	16	36		,,
	,,	,,	,,	,,	,,	,,		,,	,,	,,	,, T621			40·2			67·0	18	38		,,
		0·98	0·10	0·66	0·040	0·036				⅝	OQ800 T475					83·2	92·0	10	27½	4	19
		1·04	0·16	0·38						Bar	A820			26·0			55·0	10	12		9
		,,	,,	,,						,,	N820			28·6			55·2	15	22		,,
		,,	,,	,,						,,	OQ820 T600			44·0			60·8	14	22		,,
		1·20	0·19	(0·25)	0·021	0·021				2×1	N860					27·1	52·2	8	11½		16
		,,	,,	,,	,,	,,				,,	OQ799 T460					58·0	80·4	9	15		,,
		,,	,,	,,	,,	,,				,,	OQ885						98·2	1	2½		,,
		,,	,,	,,	,,	,,				,,	,, T204 AC						100·0	2	3		,,
		,,	,,	,,	,,	,,				,,	,, T343 AC						100·2	3½	6		,,
		,,	,,	,,	,,	,,				,,	,, T496 AC						96·6	3½	3½		,,
		,,	,,	,,	,,	,,				,,	,, T615 AC					43·2	61·8	9½	12		,,
		,,	,,	,,	,,	,,				,,	,, T760 AC					31·9	52·7	14½	17½		,,

EFFECT OF TEMPERING ON THE MECHANICAL PROPERTIES AT ROOM TEMPERATURE

Oil quenched 800°C.

(Ref. 17)

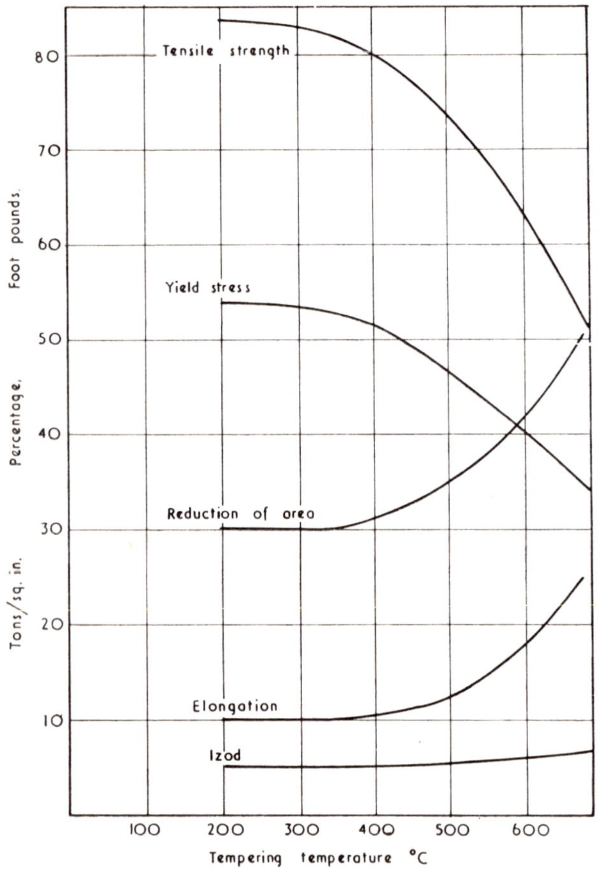

Fig. 44.9

Chemical Composition (%)				
C	Si	Mn	S	P
0·96	0·20	0·40	0·029	0·012

Austenite grain size 50% 5—7 (ASTM).
50% 1—4 (ASTM).
1 in. dia. bars hardened and tempered as indicated.

Water quenched 788°C.

(Ref. 17)

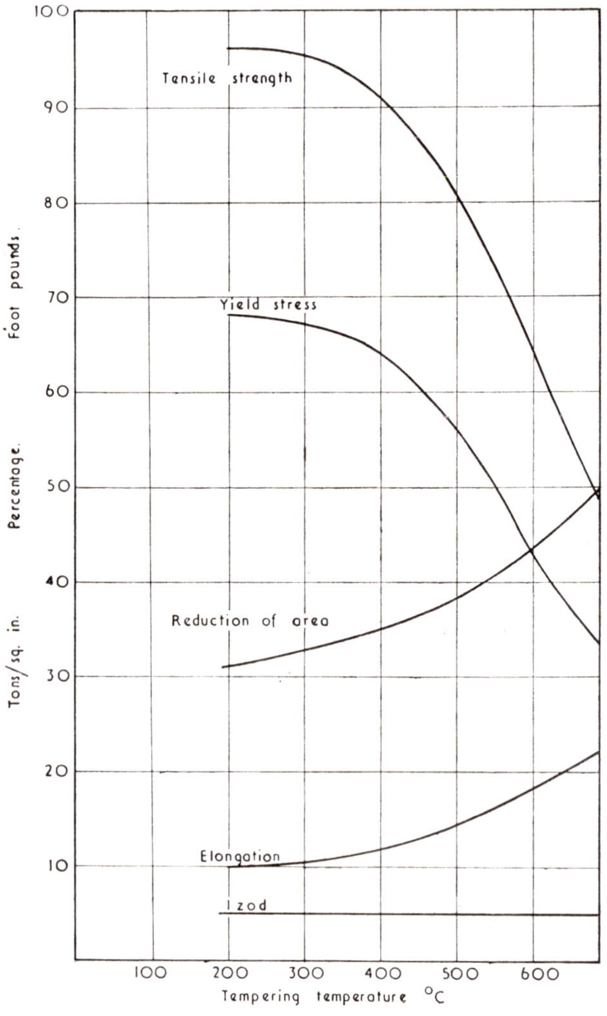

Fig. 44.10

EFFECT OF SECTION SIZE ON THE MECHANICAL PROPERTIES AT ROOM TEMPERATURE

Chemical Composition (%)

C	Si	Mn	S	P
0·96	0·20	0·40	0·029	0·012

Normalized 898°C. (Ref. 17)

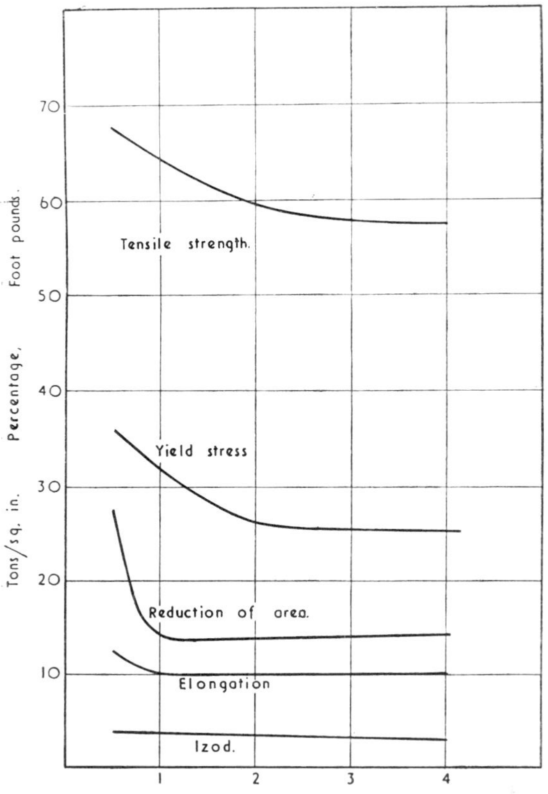

Fig. 44.11

Oil quenched 800°C. Tempered 482°C. (Ref. 17)

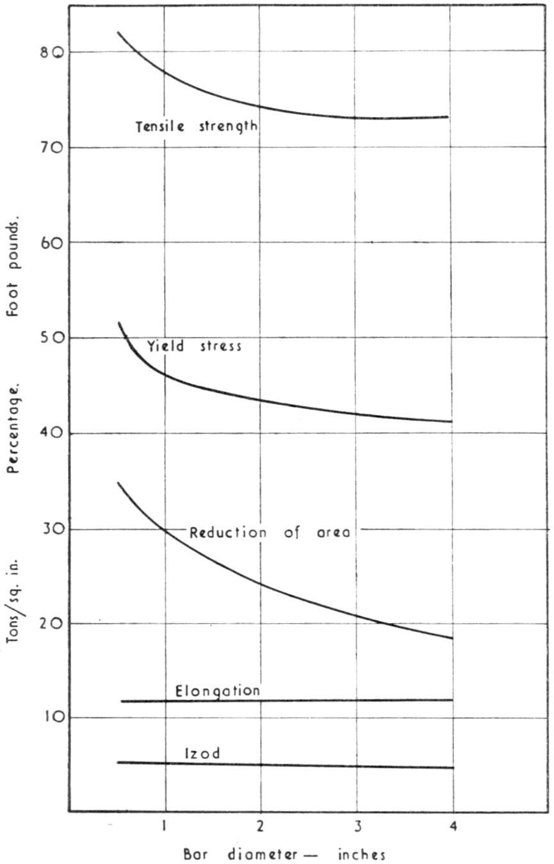

Fig. 44.12

Oil quenched 800°C. Tempered 538°C. (Ref. 17)

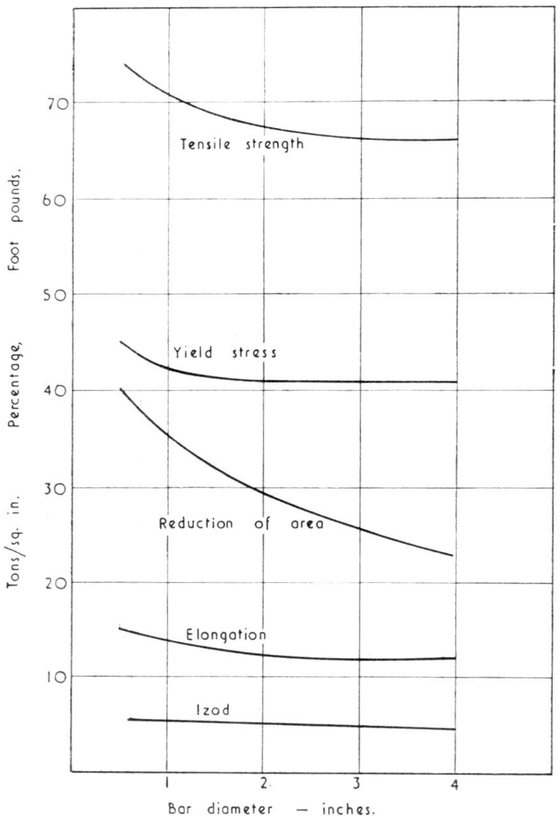

Fig. 44.13

EFFECT OF SECTION SIZE ON THE MECHANICAL PROPERTIES AT ROOM TEMPERATURE (Continued)

Chemical Composition (%)				
C	Si	Mn	S	P
0·96	0·20	0·40	0·029	0·012

Oil quenched 800°C. Tempered 593°C. (Ref. 17)

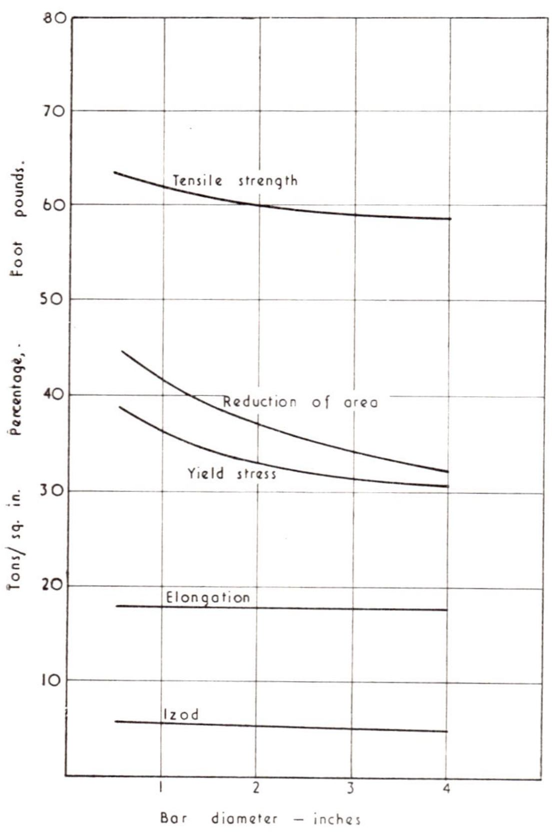

Fig. 44.14

Water quenched 787°C. Tempered 482°C. (Ref. 17)

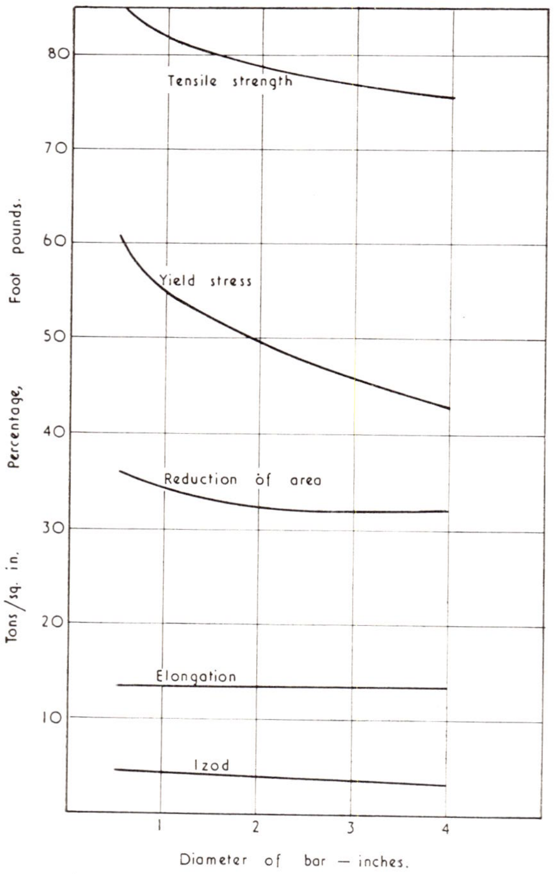

Fig. 44.15

Water quenched 787°C. Tempered 538°C. (Ref. 17)

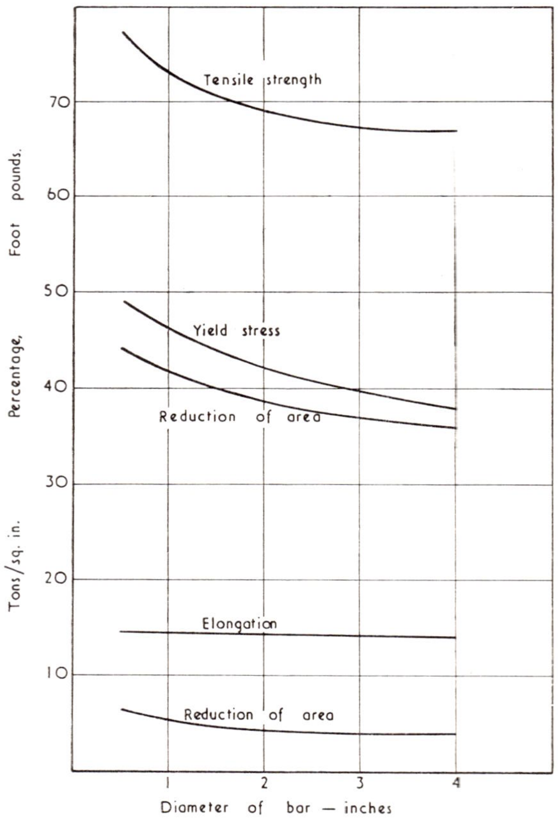

Fig. 44.16

EFFECT OF SECTION SIZE ON THE MECHANICAL PROPERTIES AT ROOM TEMPERATURE
(Continued)

Water quenched 787°C. Tempered 593°C.

(Ref. 17)

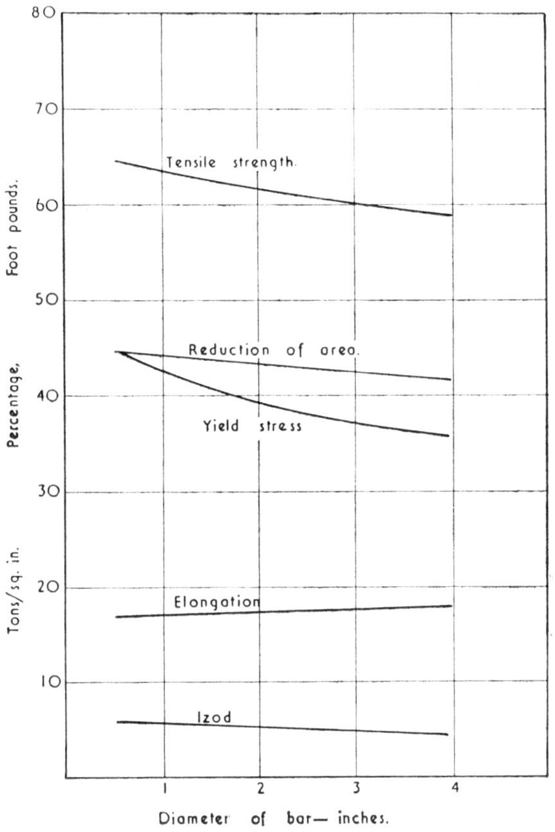

Fig. 44.17

MECHANICAL PROPERTIES AT HIGH TEMPERATURE

Chemical Composition (%)		Heat Treatment	Test Tempera-ture °C	Tensile				Ref.
				0·5% P.S. (0·5Rp)	T.S. (Rm)	El % (A)	R.A. % (Z)	
				tons/sq.in.				
C	Mn							
0·90	0·70	Normalized	RT	32	64·5	14	20	21
"	"	"	250		63·5	10	18	"
"	"	"	450		39	38	77	"
"	"	"	650		18·5	60	82	"
"	"	"	850		4	84	99	"

MECHANICAL PROPERTIES IN TORSION

Chemical Composition (%)						Section Size dia. (in.)	Heat Treatment °C	Tensile					Torsion			Ref.
C	Si	Mn	S	P	Cr			0.1% P.S. (0.1Rp)	Y.P. (Rs)	T.S. (Rm)	El % (A)	R.A. % (Z)	Y.P.	App. Max. Shear Stress	Prob. Max. Shear Stress	
								tons/sq.in.					tons/sq.in.			
0·89	0·08	0·43	0·044	0·038	0·09	0·160	As received			64				50	37·5	6
”	”	”	”	”	”	”	OQ800 T460			92				70	52·5	”
”	”	”	”	”	”	”	OQ740 T460			99				66	49·5	”
0·90	0·11	0·55	0·036	0·031		1¼	N 850		32·4	59·4	11	15½	18·8	47·6	35·7	11
											on 6″					
0·92	0·27	0·42	0·012	0·019		0·104	Cold drawn	87·0		127·0	2	52		112·5	84·4	7
”	”	”	”	”		”	CD+½ hr. 350	89·8		125·8	5	46		111·7	83·8	”
0·95						¾	N815		25·5	53·2	17	56	18·7	50·7	38·0	20
1.2	0·19	(0·25)	0·021	0·021		2×1	N860		27·1	52·2	8	11½	17·7			16

FATIGUE PROPERTIES

Chemical Composition (%)					Section Size dia. (in.)	Heat Treatment °C	Tensile				Fatigue Limit (tons/sq.in.)			Ref.
C	Si	Mn	S	P			Y.P. (Rs) tons/sq.in.	T.S. (Rm) tons/sq.in.	El % (A)	R.A. % (Z)	Rotating Bending	Reverse Axial	Torsional	
0·89	0·08	0·43	0·044	0·038	0·160	As received		64					15·7±9·4	6
"	"	"	"	"	"	OQ800 T460		92						"
"	"	"	"	"	"	OQ740 T460		99					15±9	"
0·92	0·27	0·42	0·012	0·019	0·104	CD+½ hr. 350		125·8	5	46	±37·0		19·25±19·25	7
0·93	0·03	0·38	0·045	0·17	2×⅞	N871 A788	14·9	37·5	25	37	±13·6		±17·3	16
"	"	"	"	"	"	OQ788 T649 AC	30·2	51·3	23	39½			±12·9	"
"	"	"	"	"	"	OQ788 T454		84·0	10	29½	±43·7		±23·2	"
0·98	0·10	0·66	0·040	0·036	⅝	OQ800 T475	83·2	92·0	10	27½		0±39·0		19
"	"	"	"	"	"	" "	"	"	"	"		20±35·0		"
"	"	"	"	"	"	" "	"	"	"	"		40±27·5		"
"	"	"	"	"	"	" "	"	"	"	"		60±18·5		"
1·2	0·19	0·25	0·021	0·021	2×1	N860	27·1	52·2	8	11½	±22·3		±10·9	16
"	"	"	"	"	"	OQ799 T460 AC	58·0	80·4	9	15	±41·1		±21·4	"
"	"	"	"	"	"	OQ885		98·2	1	2½	±46·9			"
"	"	"	"	"	"	" T204 AC		100·0	2	3	±47·8			"
"	"	"	"	"	"	" T343 "		100·2	3½	6	±48·2			"
"	"	"	"	"	"	" T496 "		96·6	3½	3½	±46·0			"
"	"	"	"	"	"	" T615 "	43·2	61·8	9½	12	±30·4			"
"	"	"	"	"	"	" T760 "	31·9	52·7	14½	17½	±23·9			"

LIST OF REFERENCES

1. M. F. Markham. Measurement of Elastic Constants by the Ultrasonic Pulse Method. *British Journal of Applied Physics*, Supplement No. 6, Physics of Non-destructive Testing, 1957, p. S62.
2. Private correspondence with the English Steel Corporation Ltd.
3. *Physical Constants of Some Commercial Steels at Elevated Temperatures*. Edited by the British Iron and Steel Research Association, Metallurgy Division, Thermal Treatment Sub-committee. Butterworth's Scientific Publications, London, 1953.
4. *International Critical Tables of Numerical Data, Physics, Chemistry and Technology*. Prepared by the National Research Council, U.S.A. Editor-in-Chief, E. W. Washburn. McGraw-Hill. First Edition, 1926.
5. British Iron and Steel Research Association, unpublished data.
6. A. Swan, H. Sutton and W. D. Douglas. Investigation of Aircraft Engine Valve Springs. *Proceedings of the Institution of Mechanical Engineers*, 1931, Vol. 120, p. 261.
7. Private correspondence with the Spring Manufacturers' Research Association.
8. A. F. C. Brown, A. G. Cole and M. F. Markham. Variation of Elastic Constants of Hardened 1 per cent Carbon Steels on Tempering. *Nature*, 1957, Vol. 180, Dec. 7, p. 1254.
9. H. H. Abram. The Value of Young's Modulus for Steel. *JISI*, 1934, Vol. 129, No. 1, p. 325.
10. F. Garofalo, P. R. Malenock and G. V. Smith. The Influence of Temperature on the Elastic Constants of Some Commercial Steels. American Society for Testing and Materials, Symposium on the Determination of Elastic Constants, June 25, 1952. A.S.T.M. Special Publication No. 129.
11. Private correspondence with Firth-Brown Ltd.
12. G. Delbart, A. Constant and A. Clerc. Courbes de transformation des aciers de fabrication française. Institut de Recherches de la Sidérurgie (IRSID).
13. Atlas of Isothermal Transformation Diagrams of the B.S. En Steels. Prepared by the Metallurgy Division, British Iron and Steel Research Association. Special Report No. 56 of The Iron & Steel Institute, 1956.
14. Atlas of Isothermal Transformation Diagrams. The United States Steel Corporation. Second Edition, 1957.
15. Private correspondence with the London Polytechnic.
16. H. F. Moor and T. M. Jasper. Fatigue of Metals. *Bulletin* of the University of Illinois, Engineering Experimental Station, 1923, No. 136.
17. *Modern Steels and Their Properties. Carbon and Alloy Steel Bars*. The Bethlehem Steel Company. Sixth Edition, 1961.
18. E. T. Bittner. Alloy Spring Steels. *Transactions Amer. Soc. Metals*, 1948, No. 40, pp. 263—280.
19. Private correspondence with The United Steel Companies Ltd.
20. M. C. Fetzer. Torsional and Tensile Properties of Steels. *Steel*, 1956, Vol. 118, June 24, pp. 92—94, 119—123.

SILICON–MANGANESE SPRING STEEL

SPECIFICATION

CHEMICAL COMPOSITION (%)

	C	Si	Mn	S	P
En 45	0·50—0·60	1·50—2·00	0·70—1·00	0·050 max.	0·050 max.
En 45A	0·55—0·65	1·70—2·00	0·70—1·00	0·050 max.	0·050 max.

Steel of this type may also be required in the form of wire for oil hardened and tempered springs (See BS 1429).

MECHANICAL PROPERTIES

No mechanical properties are specified for these steels.

RELATED SPECIFICATIONS

United Kingdom		America		France	Germany			Sweden	U.S.S.R.	
En No.	BS	SAE	AISI	AFNOR	Werkstoff No.	Name	DIN	SIS	GOST	Mark
45	24 pt 3A 1429	9255	9255	55S7	1.0970	55Si7	17221 17222	14.2090	2052—53 2052—53	55S2 55SG
45A	24 pt 3B 1429	9260	9260		1.0971	65Si7	17221 17222		2052—53 2052—53 2052—53	60S2 60SG 60S2A

APPLICATIONS

An oil hardening steel of higher hardenability suitable for more highly stressed applications than the plain carbon spring steels. En 45 is suitable for laminated springs of all types, volute springs, and coil springs with bar diameters not exceeding 1 in. En 45A is suitable for heavier coil or volute springs, particularly coil springs made from bars over 1 in. diameter.

The steel is also suitable for the manufacture of torsion bars up to about 1—1¼ in. diameter.

WELDING

See Introduction (List of Abbreviations) for key to symbols used in this table.

	Welding Process					
	MA	G	IG	S	F	B
En 45	f	f	f	f	b	j
En 45A	f	f	f	f	b	j

Reproduced by kind permission of H.M.S.O.

MACHINABILITY

In the softened condition the machinability of this steel is approximately 45% of that for mild steel (En 3).

HOT WORKING AND HEAT TREATMENT TEMPERATURES

Forging, rolling and stamping	1150°C, finish above 850°C.
Normalizing	870°C—930°C, air cool.
Annealing	840°C—900°C, furnace cool.
Softening (sub-critical annealing)	680°C—720°C, air or furnace cool.
Hardening	870°C—930°C, oil quench.
Tempering	400°C—550°C.

PHYSICAL PROPERTIES

SPECIFIC GRAVITY (d)

Chemical Composition (%)							Heat Treatment °C	Specific Gravity at 20°C	Ref.
C	Si	Mn	S	P	Ni	Cr			
0·49	1·98	0·90	0·047	0·044	0·16	0·04	Annealed 900	7·72	1
0·52	1·95	1·05	—	—	—	0·05	OQ860 T480	7·73	2

SPECIFIC HEAT (c)

Chemical Composition (%)							Heat Treatment °C	Mean Specific Heat cal/g deg. C										Ref.
C	Si	Mn	S	P	Ni	Cr		20 to 100°C	20 to 200°C	20 to 300°C	20 to 400°C	20 to 500°C	20 to 600°C	20 to 700°C	20 to 800°C	20 to 900°C	20 to 1000°C	
0·49	1·98	0·90	0·047	0·044	0·16	0·04	A900	0·119	0·121	0·125	0·129	0·134	0·141	0·149	0·164	0·162	0·161	1

MEAN COEFFICIENT OF THERMAL EXPANSION (α)

Chemical Composition (%)							Heat Treatment °C	$10^6 \times$ Mean Coefficient of Thermal Expansion per deg. C										Ref.
C	Si	Mn	S	P	Ni	Cr		20 to 100°C	20 to 200°C	20 to 300°C	20 to 400°C	20 to 500°C	20 to 600°C	20 to 700°C	20 to 800°C	20 to 900°C	20 to 1000°C	
0·49	1·98	0·90	0·047	0·044	0·16	0·04	A900	11·3	12·3	13·1	13·5	13·9	14·2	14·5	13·7	13·8	14·6	1

ELECTRICAL RESISTIVITY (ϱ)

Chemical Composition (%)							Heat Treatment °C	Electrical Resistivity microhm-cm												Ref.
C	Si	Mn	S	P	Ni	Cr		0°C	20°C	100°C	200°C	300°C	400°C	500°C	600°C	700°C	800°C	900°C	1000°C	
0·49	1·98	0·90	0·047	0·044	0·16	0·04	A900	41·9	42·9	47·0	52·9	60·1	68·5	78·8	91·1	105·7	117·3	119·7	122·3	1
0·52	1·95	1·05	—	—	—	0·05	OQ860 T480		45·0											2

THERMAL CONDUCTIVITY (k)

Chemical Composition (%)							Heat Treatment °C	Thermal Conductivity cal/cm s deg. C												Ref.
C	Si	Mn	S	P	Ni	Cr		0°C	20°C	100°C	200°C	300°C	400°C	500°C	600°C	700°C	800°C	900°C	1000°C	
0·49	1·98	0·90	0·047	0·044	0·16	0·04	A900	0·060	0·062	0·068	0·072	0·074	0·074	0·074	0·072	0·067	0·060	0·061	0·063	1
0·52	1·95	1·05	—	—	—	0·05	OQ860 T480		0·045											2

YOUNG'S MODULUS (E); SHEAR MODULUS (G)

Chemical Composition (%)					Section Size dia. (in.)	Heat Treatment °C	Young's Modulus Static	Shear Modulus Static	Ref.
C	Si	Mn	S	P			tons/sq.in.	tons/sq.in.	
0·54	1·90	0·99	0·026	0·032	0·160	OQ950 T420		5,160	3
0·51	1·95	0·94	0·021	0·021	3 × $\frac{3}{8}$	N950	13,500		4
,,	,,	,,	,,	,,	,,	OQ950 T450	13,350	5,180	,,
,,	,,	,,	,,	,,	,,	,, T500	13,450	5,160	,,
,,	,,	,,	,,	,,	,,	,, T550	13,500	5,130	,,
,,	,,	,,	,,	,,	,,	,, T600	13,400	5,180	,,
,,	,,	,,	,,	,,	,,	WQ950 T450	13,300	5,130	,,
,,	,,	,,	,,	,,	,,	,, T500	13,350	5,130	,,
,,	,,	,,	,,	,,	,,	,, T550	13,550	5,160	,,
,,	,,	,,	,,	,,	,,	,, T600	13,400	5,130	,,

TRANSFORMATION CHARACTERISTICS

ISOTHERMAL TRANSFORMATION

Chemical Composition (%)							
C	Si	Mn	S	P	Ni	Cr	Mo
0·55	1·74	0·87	0·037	0·038	0·16	0·10	0·02

Acid Open Hearth.
McQuaid Ehn grain size 4 to 5 (ASTM).
As quenched grain size 7 to 8 (ASTM).
Bars hot rolled to 4 S.W.G., cold drawn to 11 S.W.G., with intermediate annealing at 780°C.
Austenitizing Temperature 915°C, 30 min.

(Ref. 5)

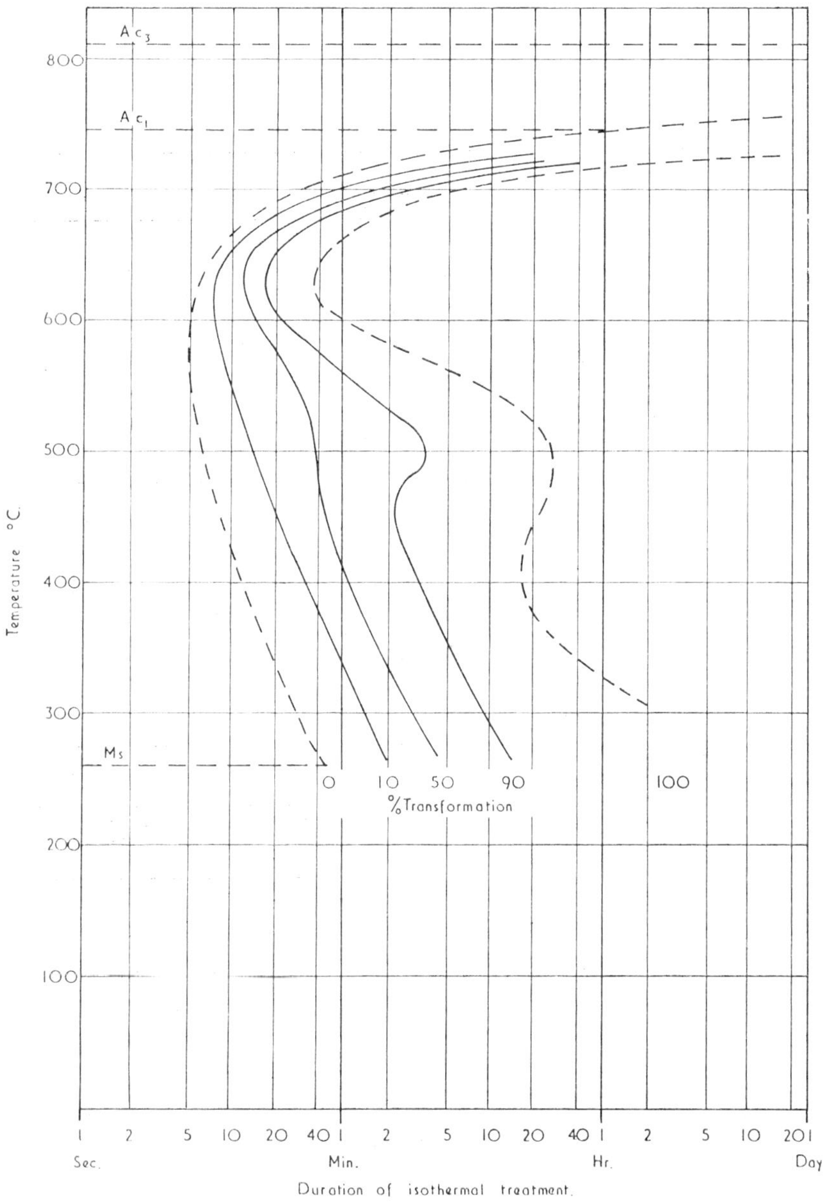

Fig. 45.1

En 45

ISOTHERMAL TRANSFORMATION DIAGRAMS (Continued)

Chemical Composition (%)			
C	Si	Mn	Cr
0·62	2·01	0·82	0·07

McQuaid Ehn grain size 6 to 7 (ASTM).

Austenitizing Temperature 872°C.

(Ref. 6)

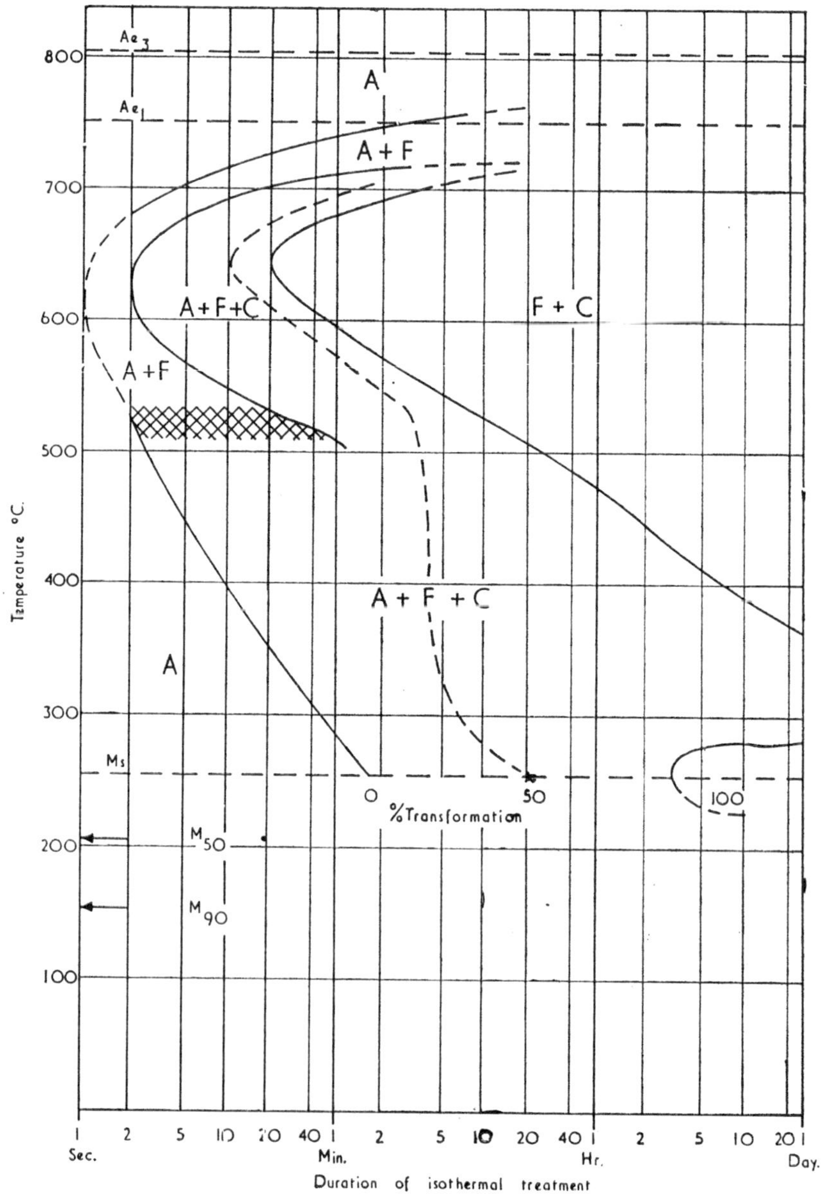

Fig. 45.2

ISOTHERMAL TRANSFORMATION DIAGRAMS (Continued)

Chemical Composition (%)			
C	Si	Mn	Cr
0·62	2·01	0·95	0·15

McQuaid Ehn grain size 6 to 7.

Austenitizing Temperature 872°C.

(Ref. 6)

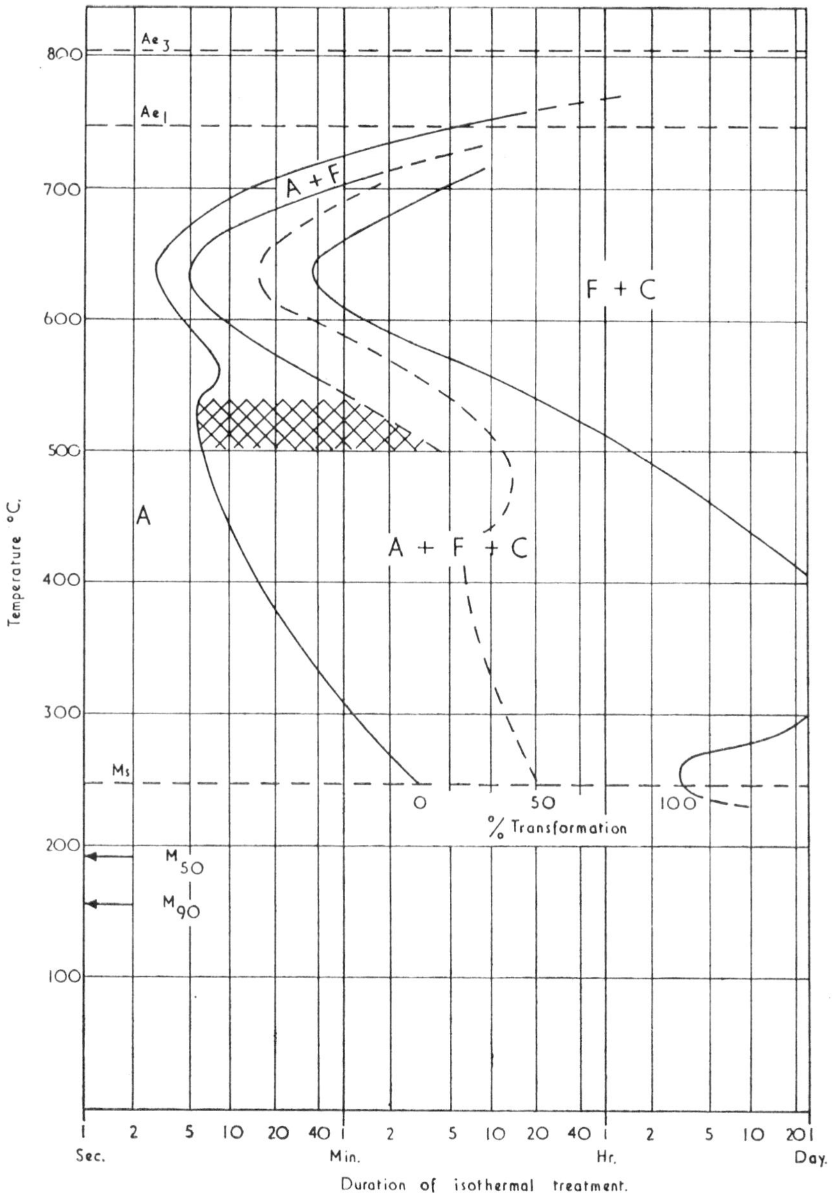

Fig. 45.3

HARDENABILITY

END QUENCH HARDENABILITY DATA

Curve No.	Chemical Composition (%)							Process	Austenitic Grain Size (ASTM)
	C	Si	Mn	S	P	Ni	Cr		
A	0·56	0·82	0·78	0·036	0·045	0·16	0·06	A.O.H.	6
B	0·57	1·75	0·88	0·039	0·037	0·20	0·10	"	6
C	0·58	1·80	0·84	0·037	0·040	0·22	0·10	"	5
D	0·58	1·84	0·98	0·041	0·046	0·14	0·06	"	6
E	0·59	1·80	0·89	0·031	0·040	0·28	0·09	"	6
F	0·59	1·89	1·00	0·031	0·041	0·14	0·07	"	5

Bars normalized 910°C.
Austenitizing Temperature 910°C, 20 min

(Ref. 7)

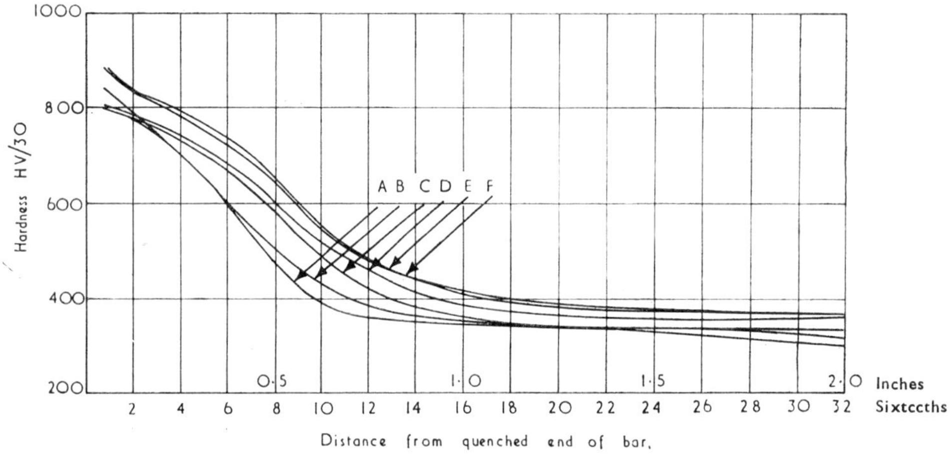

Fig. 45.4

Curve No.	Chemical Composition (%)								McQuaid Ehn Grain Size (ASTM)
	C	Si	Mn	S	P	Ni	Cr	Mo	
G	0·56	1·95	1·03	0·040	0·030	0·26	0·07	—	5
H	0·60	1·78	0·84	0·040	0·035	0·18	0·09	0·06	—
J	0·56	1·61	0·90	0·041	0·032	0·25	0·09	0·05	—
K	0·56	1·81	0·82	0·044	0·034	0·19	0·08	0·06	—
L	0·54	1·59	0·69	0·045	0·033	0·26	0·09	0·05	—

1¼ in. dia. bars, oil quenched 950°C, tempered 450°C.
Austenitizing Temperature 950°C.

(Ref. 7)

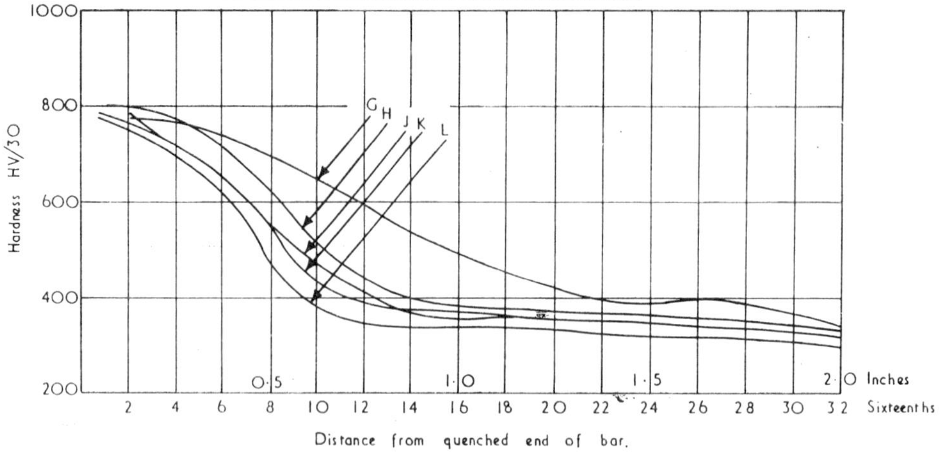

Fig. 45.5

END QUENCH HARDENABILITY DATA (Continued)

Chemical Composition (%)							
C	Si	Mn	S	P	Ni	Cr	Mo
0·56	2·02	0·94	0·045	0·036	0·19	0·10	0·12

Acid Open Hearth. McQuaid Ehn grain size 3—4 (some 2 and 5). Bars normalized 900°C. End quenched from 950°C. Tempered as indicated.

(Ref. 7)

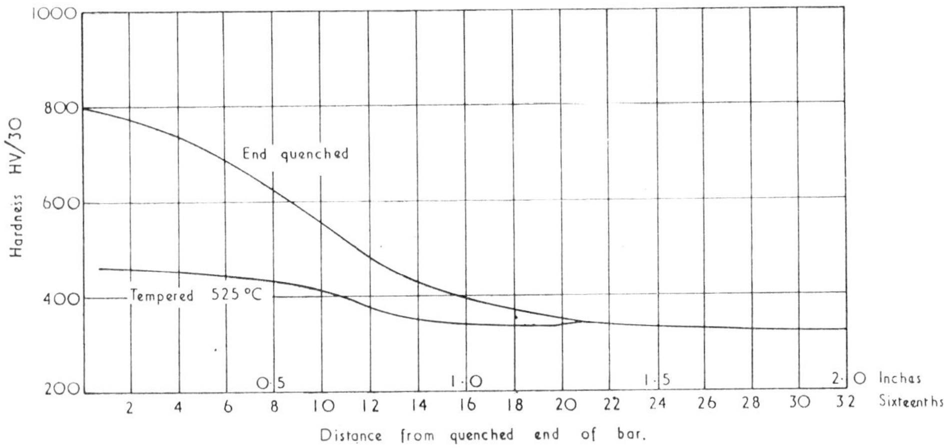

Fig. 45.6

Curve No.	Process of Manufacture	Chemical Composition (%)								Original Bar dia. (in.)	ASTM Grain Size Actual
		C	Si	Mn	S	P	Ni	Cr	Mo		
M	B.E.A.	0·57	1·97	1·22	0·008	0·016	0·24	0·11	0·02	$1\frac{3}{4}$	7 (some 8)
N	A.O.H.	0·53	2·02	0·95	0·039	0·029	0·25	0·18	0·03	$1\frac{1}{16}$	7 (some 6 and 8)

Bars normalized 925°C.
End quenched 900°C. Tempered as indicated.

(Ref. 7)

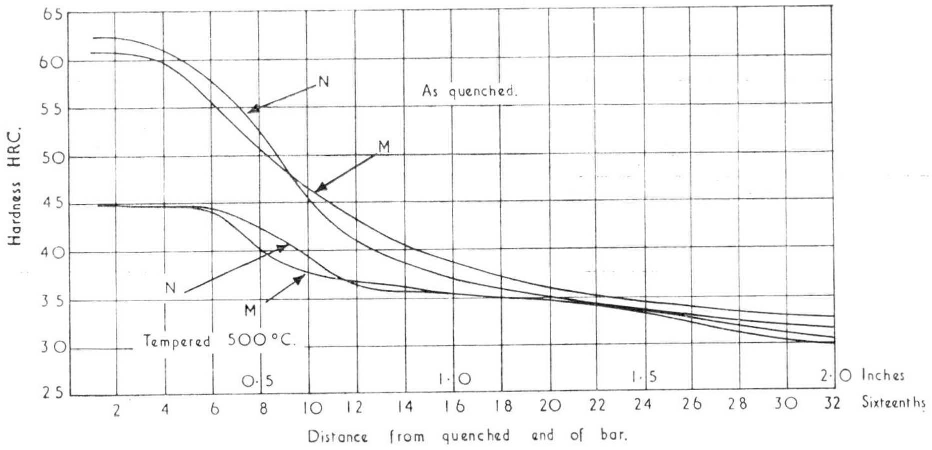

Fig. 45.7

En 45

END QUENCH HARDENABILITY BAND

These curves have been constructed from the results obtained on a large number of casts of En 45.

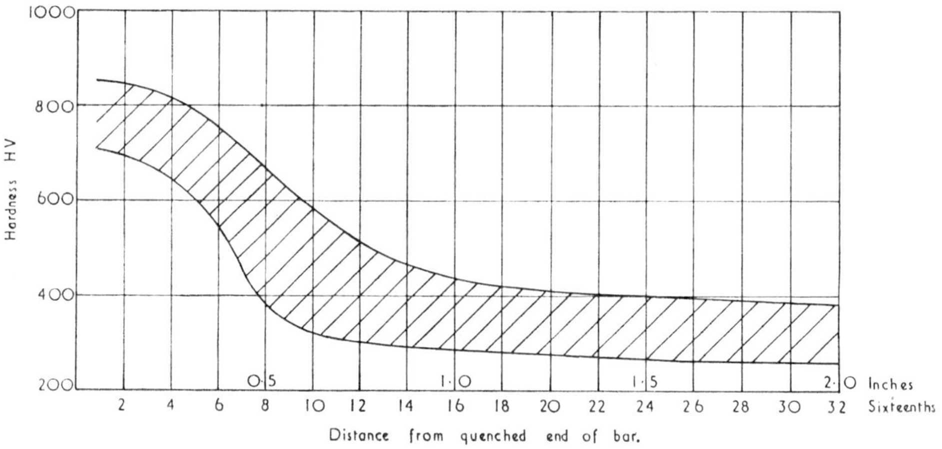

Fig. 45.8

HARDNESS TRAVERSE CURVES

Chemical Composition (%)							
C	Si	Mn	S	P	Ni	Cr	Mo
0·56	2·02	0·94	0·045	0·036	0·19	0·10	0·02

Acid Open Hearth.
McQuaid Ehn grain size 3—4 (some 2 and 6) (ASTM).
As quenched grain size, 900°C 3—4 (ASTM).
950°C 3 (ASTM).

Oil quenched 900°C. (Ref. 7)

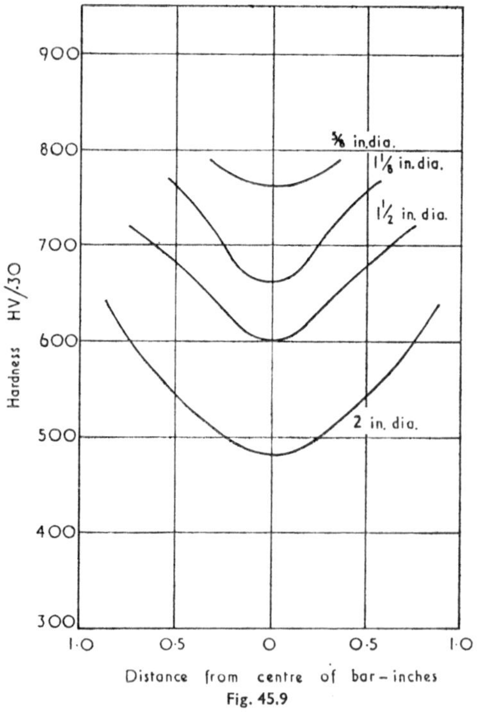

Fig. 45.9

Oil quenched 950°C. (Ref. 7)

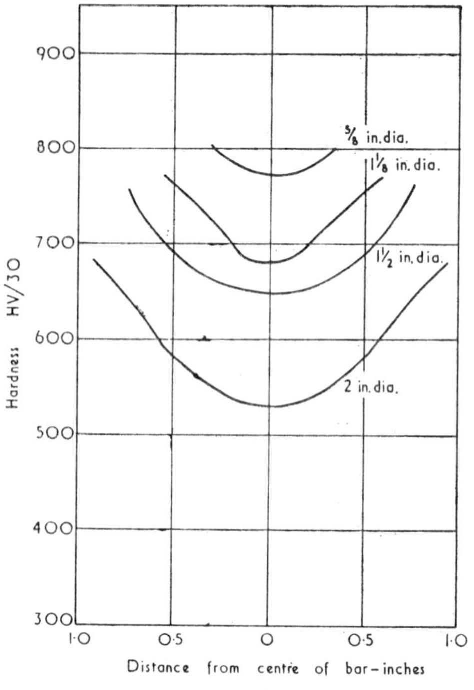

Fig. 45.10

HARDNESS TRAVERSE CURVES (Continued)

(For details of steel used see page 118.)

Oil quenched 900°C. Tempered 525°C.
(Ref. 7)

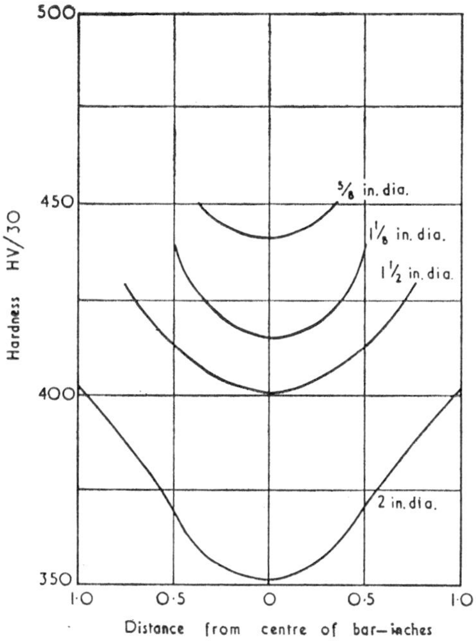

Fig. 45.11

Oil quenched 950°C. Tempered 525°C.
(Ref. 7)

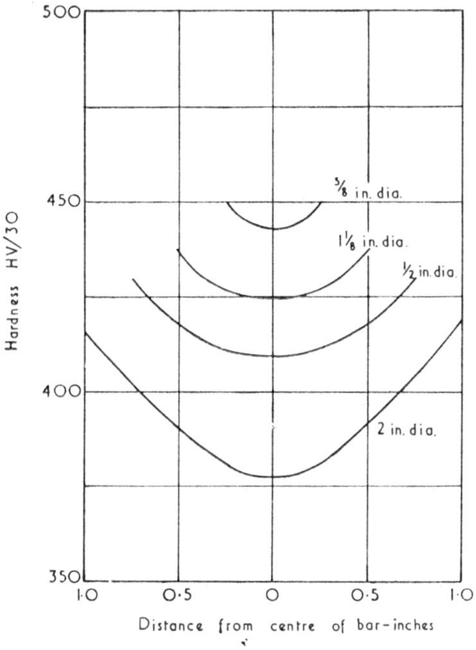

Fig. 45.12

HARDNESS TRAVERSE CURVES (Continued)

Chemical Composition (%)		
C	Si	Mn
0.57	1.98	0.82

Oil quenched 900°C.
Tempered as indicated (1 in. dia).
(Ref. 8)

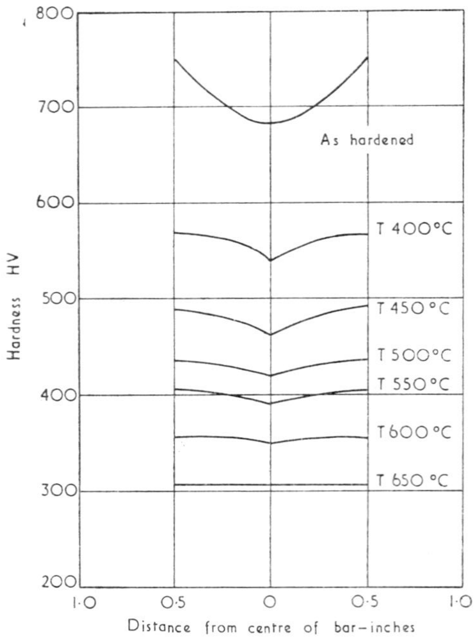

Fig. 45.13

Oil quenched 900°C.
Tempered as indicated (2 in. dia.).
(Ref. 8)

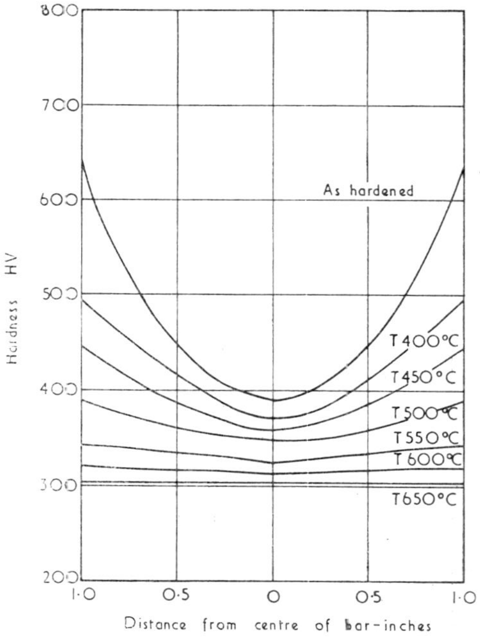

Fig. 45.14

Oil quenched 900°C.
(Ref. 9)

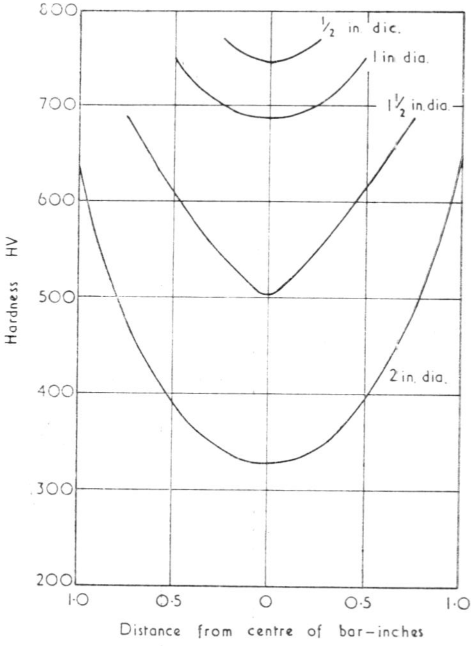

Fig. 45.15

MECHANICAL PROPERTIES AT ROOM TEMPERATURE

Process of Manufacture	Chemical Composition (%)							Section Size dia. (in.)	Heat Treatment °C	Tensile						Ref.
	C	Si	Mn	S	P	Ni	Cr			0·1% P.S. (0·1Rp)	0·2% P.S. (0·2Rp)	0·5% P.S. (0·5Rp)	T.S. (Rs)	El % (A)		
										tons/sq.in.				l=4√A	l=2″	
A.O.H.	0·50	2·00	0·99	0·025	0·026	0·19	0·07	$\frac{3}{16}$ thick	OQ925 T450	87·0	96·2	100·0	107·0	11	7½	9
"	"	"	"	"	"	"	"	"	" T500	81·5	85·9	88·1	94·6	13½	10½	"
"	"	"	"	"	"	"	"	"	" T525	77·9	78·5	80·1	87·1	16	10½	"
"	"	"	"	"	"	"	"	"	" T550	73·8	74·3	75·5	82·5	16½	12	"
"	"	"	"	"	"	"	"	¼ thick	OQ925 T450	93·3	96·1	99·1	105·6	10	9½	"
"	"	"	"	"	"	"	"	"	" T500	84·9	85·7	88·8	94·5	12	10	"
"	"	"	"	"	"	"	"	"	" T525	78·7	79·8	81·5	88·9	14½	12½	"
"	"	"	"	"	"	"	"	"	" T550	72·9	73·3	74·6	82·7	13½	13½	"
"	"	"	"	"	"	"	"	$\frac{5}{16}$ thick	OQ925 T450	84·5	92·8	94·4	104·9	12	10	"
"	"	"	"	"	"	"	"	"	" T500	83·3	84·9	87·2	94·1	11	10	"
"	"	"	"	"	"	"	"	"	" T525	72·5	76·5	78·9	85·9	12½	12	"
"	"	"	"	"	"	"	"	"	" T550	70·0	70·4	72·3	80·1	17	15	"
"	"	"	"	"	"	"	"	⅜ thick	OQ925 T450	93·6	96·6	102·0	107·0	12½	11	"
"	"	"	"	"	"	"	"	"	" T500	81·6	83·8	85·1	93·5	12½	10½	"
"	"	"	"	"	"	"	"	"	" T525	76·8	78·8	79·4	88·9	15	12½	"
"	"	"	"	"	"	"	"	"	" T550	61·7	62·7	63·5	82·2	15	13½	"
"	"	"	"	"	"	"	"	$\frac{7}{16}$ thick	OQ925 T450	89·6	94·0	97·3	104·2	10	10	"
"	"	"	"	"	"	"	"	"	" T500	85·6	87·0	88·8	94·6	11½	10	"
"	"	"	"	"	"	"	"	"	" T525	78·8	80·4	81·1	89·6	11½	12	"
"	"	"	"	"	"	"	"	"	" T550	72·0	76·5	78·6	85·9	13½	12	"
"	"	"	"	"	"	"	"	½ thick	OQ925 T450	95·6	98·8	100·6	109·2	—	10	"
"	"	"	"	"	"	"	"	"	" T500	84·4	86·0	88·1	95·0	—	10	"
"	"	"	"	"	"	"	"	"	" T525	82·3	84·3	86·3	92·3	—	9½	"
"	"	"	"	"	"	"	"	"	" T550	71·4	73·7	75·3	83·5	—	14	"
"	"	"	"	"	"	"	"	$\frac{9}{16}$ thick	OQ925 T450	95·7	96·4	101·8	109·1	9½	9½	"
"	"	"	"	"	"	"	"	"	" T500	81·8	84·0	86·4	94·4	11½	12	"
"	"	"	"	"	"	"	"	"	" T525	79·8	81·6	82·3	90·8	12½	13	"
"	"	"	"	"	"	"	"	"	" T550	70·4	71·2	71·9	86·7	12½	12	"
"	"	"	"	"	"	"	"	⅝ thick	OQ925 T450	87·7	88·2	88·7	107·2	10	12	"
"	"	"	"	"	"	"	"	"	" T500	82·4	85·0	86·3	93·0	11	12½	"
"	"	"	"	"	"	"	"	"	" T525	75·9	77·8	79·9	91·6	12½	14	"
"	"	"	"	"	"	"	"	"	" T550	75·2	75·8	77·2	82·9	13	14½	"

MECHANICAL PROPERTIES AT ROOM TEMPERATURE

(Continued)

Process of Manu-facture	ASTM Grain Size: Mc Quaid Ehn	ASTM Grain Size: Actual	Chemical Composition (%): C	Si	Mn	S	P	Ni	Cr	Mo	Section Size dia. (in.)	Heat Treatment °C	Tensile: 0·05% P.S. (0·05Rp) tons/sq.in.	0·2% P.S. (0·2Rp) tons/sq.in.	Y.P. (Rs) tons/sq.in.	T.S. (Rm) tons/sq.in.	El % (A)	R.A % (Z)	Izod ft.lb	Ref.
A.O.H.	—	7(6,8)	0·53	2·02	0·95	0·039	0·029	0 25	0·18	0·03	1	N925 OQ900 T500			77·9	89·6	11½	26½	7	7
"			0·53	1·85	0·80						1¾	N900		32·4		62·5	8			10
"			"	"	"						"	N900 T625 AC		35·5		62·4	9			"
"			0·53	1·92	0·95						⅜ thick	OQ900 T480			87·7	96·5	6	29	10	"
"			"	"	"						"	" T560			73·1	80·3	6½		10	"
"			"	"	"						"	" T630			61·0	68·0	13	42	14	"
A.O.H.			0·54	1·98	0·85	0·037	0·038				⅝	OQ920 T300 AC	95·2		105·8	110·0	9	24		11
"			"	"	"	"	"				"	" T350 "	86·5		93·9	99·3	11	26		"
"			"	"	"	"	"				"	" T450 "	83·1		90·4	97·0	13	27½		"
"			"	"	"	"	"				"	" T500 "	64·1		69·6	78·0	17	33		"
A.O.H.	3—4	3—4	0·56	2·02	0·94	0·045	0·036	0·19	0·10	0·02	⅝	OQ900 T525 AC			86·2	94·6	13	22½	5	7
"	"	"	"	"	"	"	"	"	"	"	1⅛ (O)	" " "			79·7	88·4	14½	19½	4	"
"	"	"	"	"	"	"	"	"	"	"	" (C)	" " "			79·2	86·2	11	12½	2	"
"	"	"	"	"	"	"	"	"	"	"	1½ (O)	" " "			75·4	86·4	12	14	4	"
"	"	"	"	"	"	"	"	"	"	"	" (C)	" " "			74·2	84·4	12	14	3	"
"	"	"	"	"	"	"	"	"	"	"	2 (O)	" " "			76·3	83·4	15½	20	6	"
"	"	"	"	"	"	"	"	"	"	"	" (M)	" " "			71·8	79·7	14½	15	5	"
"	"	"	"	"	"	"	"	"	"	"	" (C)	" " "			62·9	77·1	14½	15	5	"
"	3—4	3	"	"	"	"	"	"	"	"	⅝	OQ950 T525 AC			89·4	94·5	14½	22½	4	"
"	"	"	"	"	"	"	"	"	"	"	1⅛(O)	" " "			82·8	89·9	13½	20	3	"
"	"	"	"	"	"	"	"	"	"	"	" (C)	" " "			81·8	87·0	14½	17½	2	"
"	"	"	"	"	"	"	"	"	"	"	1½ (O)	" " "			78·7	87·0	13½	19	4	"
"	"	"	"	"	"	"	"	"	"	"	" (C)	" " "			76·9	84·5	11	15	3	"
"	"	"	"	"	"	"	"	"	"	"	2 (O)	" " "			76·3	83·0	13½	21	4	"
"	"	"	"	"	"	"	"	"	"	"	" (M)	" " "			76·5	80·4	13½	17	4	"
"	"	"	"	"	"	"	"	"	"	"	" (C)	" " "			69·8	76·8	11	13½	3	"
			0·56	1·82	0·83	0·040	0·039				0·12×0·14	OQ900 T530			83·2	91·6				12
B.E.A.			0·57	1·97	1·22	0·008	0·016	0·24	0·11	0·02	1	N925 OQ900 T500			77·1	88·9	12	28½	6	7

MECHANICAL PROPERTIES AT ROOM TEMPERATURE

(Continued)

Process of Manufacture	ASTM Grain Size	Chemical Composition (%)								Section Size dia. (in.)	Heat Treatment °C	Tensile						Izod ft.lb.	Ref.
		C	Si	Mn	S	P	Ni	Cr	Mo			0·05% P.S. (0·05Rp)	0·2% P.S. (0·2Rp)	Y.P. (Rs)	T.S. (Rm)	El % (A)	R.A. % (Z)		
												tons/sq.in.							
	6 to 8	0·58	2·00	0·78	0·024	0·020	0·08	0·08		1	A843			34·4	51·3	22	45		13
	,,	,,	,,	,,	,,	,,	,,	,,		,,	N899			40·2	59·4	17	38		,,
	,,	,,	,,	,,	,,	,,	,,	,,		,,	OQ885 T204			128·1	136·0	2	4	3	,,
	,,	,,	,,	,,	,,	,,	,,	,,		,,	,, T260			123·2	131·6	3	6	3	,,
	,,	,,	,,	,,	,,	,,	,,	,,		,,	,, T316			116·1	125·0	4	10	3	,,
	,,	,,	,,	,,	,,	,,	,,	,,		,,	,, T371			107·1	116·1	5	16	3	,,
	,,	,,	,,	,,	,,	,,	,,	,,		,,	,, T427			96·0	104·0	8	22	6	,,
	,,	,,	,,	,,	,,	,,	,,	,,		,,	,, T482			82·6	91·5	11	27	7	,,
	,,	,,	,,	,,	,,	,,	,,	,,		,,	,, T538			71·4	81·2	14	32	8	,,
	,,	,,	,,	,,	,,	,,	,,	,,		,,	,, T593			61·6	71·4	17	37	15	,,
	,,	,,	,,	,,	,,	,,	,,	,,		,,	,, T649			52·7	64·3	20	42	22	,,
	,,	,,	,,	,,	,,	,,	,,	,,		,,	,, T704			45·5	58·0	22	50	25	,,
		0·63	1·94	0·82			0·14	0·14	0·03	0·525	OQ870 T515		82·5		91·6	12½	31	Charpy V 9·2	14
	Actual 6	0·63	2·16	0·75	0·031	0·015	0·15	0·30		1⅝	OQ870 T454		93·7		107·1	10	26		15
	,,	,,	,,	,,	,,	,,	,,	,,		,,	,, T510		78·1		96·0	11	28		,,
	,,	,,	,,	,,	,,	,,	,,	,,		,,	,, T566		71·4		82·6	14	31		,,
	,,	,,	,,	,,	,,	,,	,,	,,		,,	,, T621		60·3		75·9	18	37		,,

EFFECT OF TEMPERING ON THE MECHANICAL PROPERTIES AT ROOM TEMPERATURE

Chemical Composition (%)				
C	Si	Mn	Ni	Cr
0·58	2·0	0·78	0·08	0·08

1 in. dia. bars. Oil quenched 885°C. Tempered as shown.

(Ref. 13)

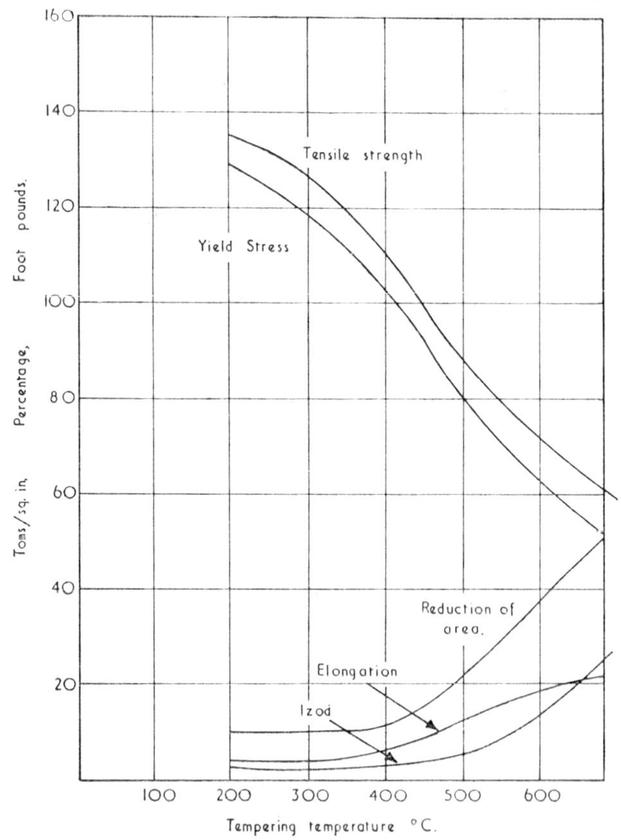

Fig. 45.16

MECHANICAL PROPERTIES IN TORSION

Chemical Composition (%)							Section Size dia. (in.)	Heat Treatment °C	Tensile					Torsion			Ref.
C	Si	Mn	S	P	Ni	Cr			0·2% P.S. (0·2Rp)	Y.P. (Rs)	T.S. (Rm)	El % (A)	R.A. % (Z)	Y.P.	Apparent Shear Strength	Probable Shear Strength	
									tons/sq.in.					tons/sq.in.			
0·54	1·90	0·99	0·026	0·032		0·05	0·160	As received			55·0				44	33	3
,,	,,	,,	,,	,,		,,	,,	OQ950 T420			103·5				81	60·8	,,
												l=11·3√A					
0·54	1·95	0·94	0·021	0·021			3 × ⅜	OQ950 T450		102·7	108·1	4	20½		86·5	64·8	4
,,	,,	,,	,,	,,			,,	,, T500		83·2	91·3	8	21½		77·5	58·1	,,
,,	,,	,,	,,	,,			,,	,, T550		72·0	79·6	10	25½		72·6	54·5	,,
,,	,,	,,	,,	,,			,,	,, T600		63·5	72·5	12½	30½		55·8	41·8	,,
,,	,,	,,	,,	,,			,,	WQ870 T450		109·0	112·4	4	28		88·8	66·6	,,
,,	,,	,,	,,	,,			,,	,, T500		85·0	91·1	8	22		80·4	60·3	,,
,,	,,	,,	,,	,,			,,	,, T550		77·0	84·6	8½	25½		69·3	52·0	,,
,,	,,	,,	,,	,,			,,	,, T600		69·5	77·7	10½	26½		66·9	50·2	,,
0·59	1·64	1·10	0·049	0·040	0·28	0·18	1·15	OQ850 T425	109·5		120·2	10	19		97·3	73·0	16
,,	,,	,,	,,	,,	,,	,,	,,	,, T505	86·2		95·6	11½	26½		80·3	60·2	,,

FATIGUE PROPERTIES

ASTM Grain Size	Chemical Composition (%)							Section Size dia. (in.)	Heat Treatment °C	Tensile					Fatigue Limit tons/sq.in.					Ref.
															Rotating Bending			Reversed Axial	Torsional	
															Plain		Notched			
	C	Si	Mn	S	P	Ni	Cr			0·2% P.S. (0·2Rp)	Y.P. (Rs)	T.S. (Rm)	El % (A)	R.A. % (Z)	In Air	In Brine	In Air			
										tons/sq.in.										
													$l=11{\cdot}3\sqrt{A}$							
	0·54	1·95	0·94	0·021	0·021			3 × ⅜	OQ950 T450		102·7	108·1	4	20½	±47·8					4
	"	"	"	"	"			"	" T500		33·2	91·3	8	21½	±46·0					"
	"	"	"	"	"			"	" T550		72·0	79·6	10	25½	±44·0					"
	"	"	"	"	"			"	" T600		53·5	72·5	12½	30½	±35·4					"
	"	"	"	"	"			"	WQ870 T450		109·0	112·4	4	28½	±50·0					"
	"	"	"	"	"			"	" T500		35·0	91·1	8	22	±47·0					"
	"	"	"	"	"			"	" T550		77·0	84·6	8½	25½	±41·4					"
	"	"	"	"	"			"	" T600		69·5	77·7	10½	26½	±39·0					"
	"	"	"	"	"			0·2	N950						±19·0					17
	"	"	"	"	"			"	OQ950 T500						±46·0					"
	0·54	1·90	0·99	0·026	0·032	0·05		0·160	As received			55							22·5±13·5	3
	"	"	"	"	"	"		"	OQ950 T420			103·5							23·75±14·25	"
													$4\sqrt{A}$							
	0·55	1·82	0·83					⅝	OQ900 T530		83·2	91·6	13	31				0±41·5		10
	"	"	"					"	" "		"	"	"	"				20±30·0		"
	"	"	"					"	" "		"	"	"	"				40±25·5		"
	"	"	"					"	" "		"	"	"	"				60±19·0		"
	0·56	1·90	0·79						N871		39·7	64·3	18	45	±32·2	±22·3				18
7—8	0·58	1·98	1·02	0·025	0·029			½	OQ880 T550		72·8	82·5	15½	30	±43·0		±18·2*			19
	0·59	1·64	1·10	0·049	0·040	0·28	0·18	1·15	OQ850 T425	109·5		120·2	10	19	±58·5					16
	"	"	"	"	"	"	"	"	" T505	86·2		95·6	11½	26½	±49·2					"
6—8	0·60	1·63	0·95	0·033	0·033			½	OQ875 T575		74·7	81·7	18	43½	±46·5		±21·0*			19

* Notch 55°V, 0·008 in. root radius, 0·020 in. deep, test dia. 0·335 in., root dia. 0·295 in., Neuber stress concentration factor 2·85.

LIST OF REFERENCES

1. *Physical Constants of Some Commercial Steels at Elevated Temperatures.* Edited by the British Iron and Steel Research Association, Metallurgy Division, Thermal Treatment Sub-Committee. Butterworth's Scientific Publications, London, 1953.
2. Private correspondence with Firth-Brown Ltd.
3. A. Swan, H. Sutton and W. D. Douglas. Investigation of Aircraft Engine Valve Springs. *Proceedings of the Institution of Mechanical Engineers*, 1931, Vol. 120, p. 261.
4. G. A. Hankins, D. Hanson and G. W. Ford. The Mechanical Properties of Four Heat-treated Spring Steels. *JISI*, 1926, Vol. 114, No. 2, p. 265.
5. Atlas of Isothermal Transformation Diagrams of the B.S. En Steels. Prepared by the Metallurgy Division, British Iron and Steel Research Association. Special Report No. 56 of The Iron & Steel Institute, 1956.
6. Atlas of Isothermal Transformation Diagrams. The United States Steel Corporation. Second Edition, 1957.
7. Symposium on the Hardenability of Steel. Organized by the Technical Advisory Committee of the Special and Alloy Steels Committee (Ministry of Supply) and The Iron and Steel Institute. Special Report No. 36 of The Iron & Steel Institute, 1946.
8. Private correspondence with the English Steel Corporation Ltd.
9. Private correspondence with Brown Bayley Steels Ltd.
10. Private correspondence with The United Steel Companies Ltd.
11. Private correspondence with Edgar Allen and Co. Ltd.
12. Private correspondence with the Spring Manufacturers' Research Association.
13. *Modern Steels and Their Properties, Carbon, and Alloy Steel Bars.* The Bethlehem Steel Company. Sixth Edition, 1961.
14. R. S. Archer, J. Z. Briggs, C. M. Loeb. *Molybdenum Steels, Irons, Alloys.* Climax Molybdenum Co. Ltd., 1947.
15. E. T. Bittner. Alloy Spring Steels. *Transactions of the Amer. Soc. Metals*, 1948, No. 40, pp. 263—280.
16. A. S. Kenneford and G. C. Ellis. A Comparison of Six Spring Steels. *JISI*, 1950, Vol. 164, No. 1, p. 265.
17. G. A. Hankins and M. L. Becker. Effect of Surface Conditions Produced by Heat Treatment on the Fatigue Resistance of Spring Steels. *JISI*, 1931, Vol. 124, No. 2, p. 387.
18. S. L. Hoyt. *Metals and Alloys Data Book.* Reinhold, New York, 1952.
19. W. E. Bardgett and F. Garside. Effect of Variation in Silicon Content on the Properties of Silicon–Manganese Spring Steel. *Metallurgia*, 1949, Vol. 39, Feb., pp. 171—177.

SILICON-MANGANESE SPRING STEEL

SPECIFICATION

CHEMICAL COMPOSITION (%)

C	Si	Mn	S	P
0·35—0·45	1·50—2·00	0·70—1·00	0·050 max.	0·050 max.

MECHANICAL PROPERTIES

No mechanical properties are specified for this steel.

APPLICATIONS

En 46 is used for applications similar to those of En 45 but where water hardening is preferred to oil hardening.

WELDING

(See Introduction (List of Abbreviations) for key to symbols used in this table.)

	Welding Process					
	MA	G	IG	S	F	B
En 46	f	f	f	f	b	j

Reproduced by kind permission of H.M.S.O.

MACHINABILITY

In the softened condition the machinability is approximately 45% of that for mild steel (En 3).

HOT WORKING AND HEAT TREATMENT TEMPERATURES

Forging, rolling and stamping	1150°C, finish above 850°C.
Normalizing	860°C to 920°C, air cool.
Annealing	860°C to 920°C, furnace cool.
Softening (sub-critical annealing)	680°C to 720°C, air or furnace cool.
Hardening	860°C to 920°C, water quench.
Tempering	400°C to 550°C, air cool.

PHYSICAL PROPERTIES

No data are available for the physical properties of En 46 Steel. Data for specific gravity, specific heat, coefficient of expansion, electrical resistivity and thermal conductivity will be similar to those given for En 45.

HARDENABILITY

END QUENCH HARDENABILITY DATA

Chemical Composition (%)				
C	Si	Mn	S	P
0·49	1·98	0·90	0·047	0·044

Acid Open Hearth.
Bars normalized 900°C. End quenched 900°C.

(Ref. 1)

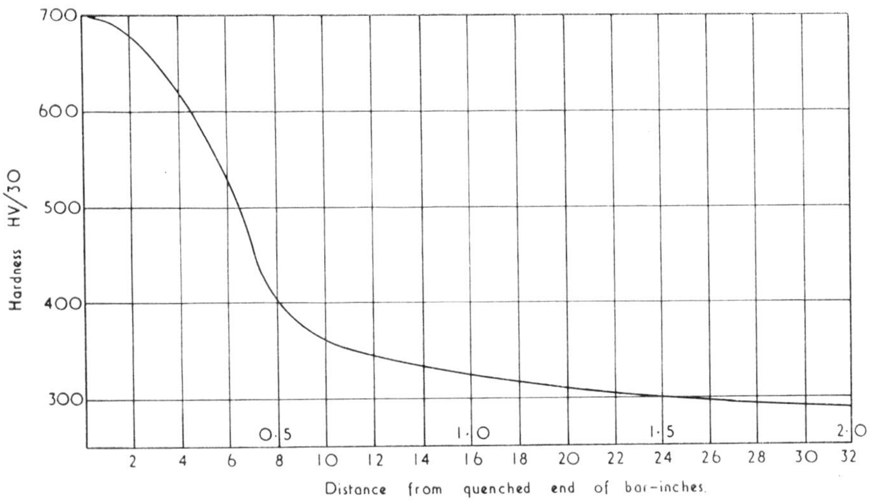

Fig. 46.1

En 46

MECHANICAL PROPERTIES AT ROOM TEMPERATURE

Chemical Composition (%)			Section Size dia. (in.)	Heat Treatment °C	Tensile					Izod ft.lb.	Ref.
C	Si	Mn			0·3% P.S. (0·3Rp) tons/sq.in.	Y.P. (Rs) tons/sq.in.	T.S. (Rm) tons/sq.in.	El % (A)	R.A. % (Z)		
0·34	1·89	0·85	½	WQ890 T500	69·3		76·8	20	48		2
"	"	"	"	" T550	62·7		70·0	21	50½		"
"	"	"	"	" T600	56·0		63·5	25	53½		"
"	"	"	"	" T640	51·0		58·5	29	57		"
"	"	"	1	WQ890 T500	54·0		68·8	17	38	25	"
"	"	"	"	" T550	50·0		63·0	22	48	45	"
"	"	"	"	" T600	45·8		57·3	25	54	64	"
"	"	"	"	" T640	42·6		53·0	27	57½	—	"
"	"	"	2	WQ890 T400	44·6		63·3	18	34	11	"
"	"	"	"	" T450	44·2		62·0	19	38	14	"
"	"	"	"	" T500	41·6		58·2	22	43½	17	"
"	"	"	"	" T550	39·0		54·5	25	50	17	"
"	"	"	½	OQ890 T500	59·5		72·4	20	42		"
"	"	"	"	" T550	54·0		66·4	20½	49		"
"	"	"	"	" T600	49·0		60·7	24½	54		"
"	"	"	"	" T640	46·0		56·9	29	57		"
"	"	"	1	OQ890 T500	48·2		65·0	19	39	24	"
"	"	"	"	" T550	43·5		59·1	22½	47	35	"
"	"	"	"	" T600	40·6		54·2	26	53½	48	"
"	"	"	"	" T640	39·0		51·3	27½	58½	60	"
"	"	"	2	OQ890 T400	40·0		54·6	23	50½	16	"
"	"	"	"	" T450	40·0		54·6	25½	51	15½	"
"	"	"	"	" T500	39·8		54·6	25½	51½	13	"
"	"	"	"	" T550	37·8		54·2	25½	52	9	"
0·36	1·65	0·94	1¾ × ¾	N860		30·2	48·3	28	53	20	3
"	"	"	"	OQ1000 T2hrs. 550		43·0	58·6	19	46	25	"
"	"	"	"	OQ900 " 500		43·1	60·4	21	47	17	"
"	"	"	"	" " 550		40·1	57·0	22	52	29	"
"	"	"	"	" " 600		41·1	57·0	20	47	24	"
"	"	"	"	OQ820 " 550		35·6	53·3	25	56	17	"

EFFECT OF TEMPERING TEMPERATURE ON HARDNESS

Chemical Composition (%)					Section Size dia. (in.)	Heat Treatment °C	Brinell Hardness (HB) Tempering Temperature °C								Ref.
C	Si	Mn	S	P			20	300	350	400	450	500	550	600	
0·36	1·65	0·94			¾×5×5	OQ900						300	270	270	3
0·38	1·90	0·85				WQ900	578			461	395	360			4
0·39	1·98	0·88	0·027	0·041	¼×1×3	WQ820	375	385	380	360	335	310			5
"	"	"	"	"	⅜×3×12	WQ1000	530			495	415	375	330		"

EFFECT OF TEMPERING ON THE MECHANICAL PROPERTIES AT ROOM TEMPERATURE

Chemical Composition (%)		
C	Si	Mn
0·34	0·89	0·85

½ in. dia. bars. Quenched 890°C. Tempered as indicated.

Water quench.

(Ref. 2)

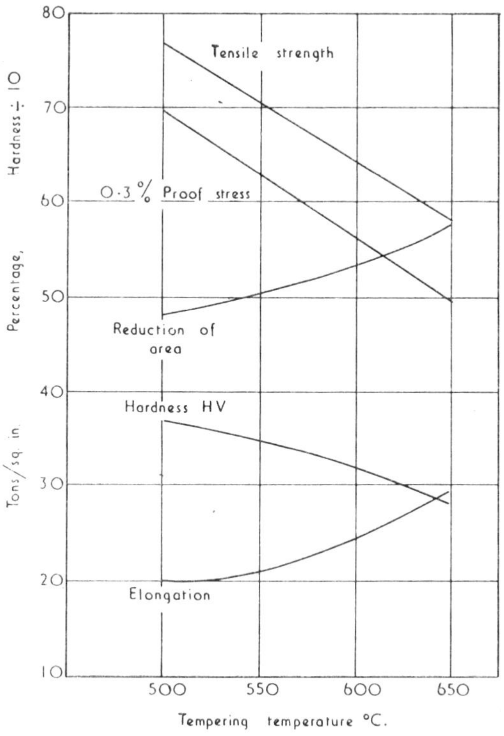

Fig. 46.2

Oil quench.

(Ref. 2)

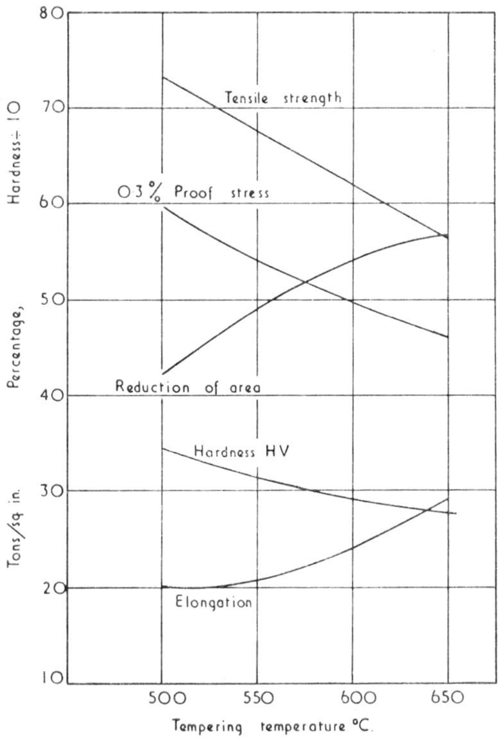

Fig. 46.3

EFFECT OF TEMPERING ON THE MECHANICAL PROPERTIES AT ROOM TEMPERATURE (Continued)

(Chemical Composition as on page 131)

Water quench. (Ref. 2) Oil quench. (Ref. 2)

1 in. dia. bars quenched 890°C. Tempered as indicated.

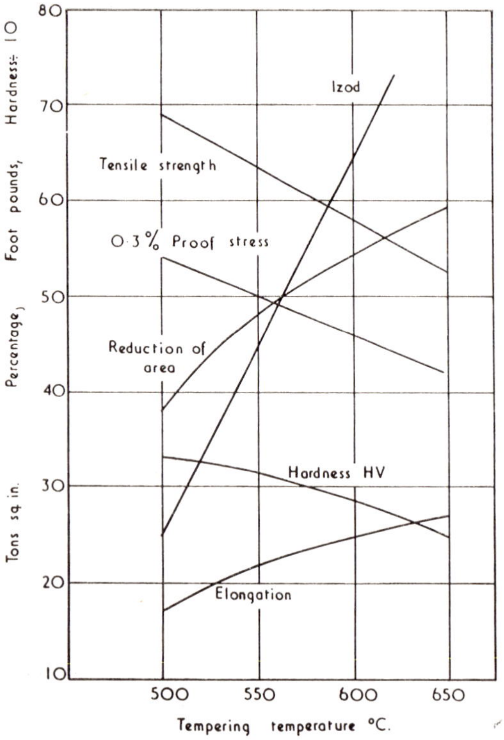

Fig. 46.4

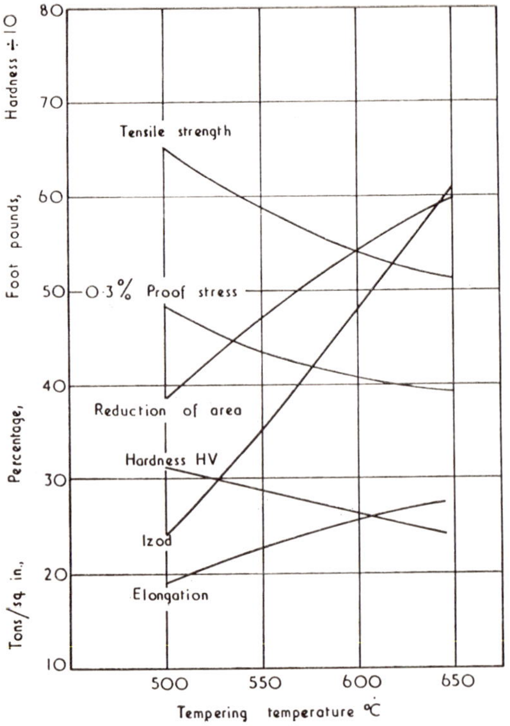

Fig. 46.5

2 in. dia. bars quenched 890°C. Tempered as indicated.

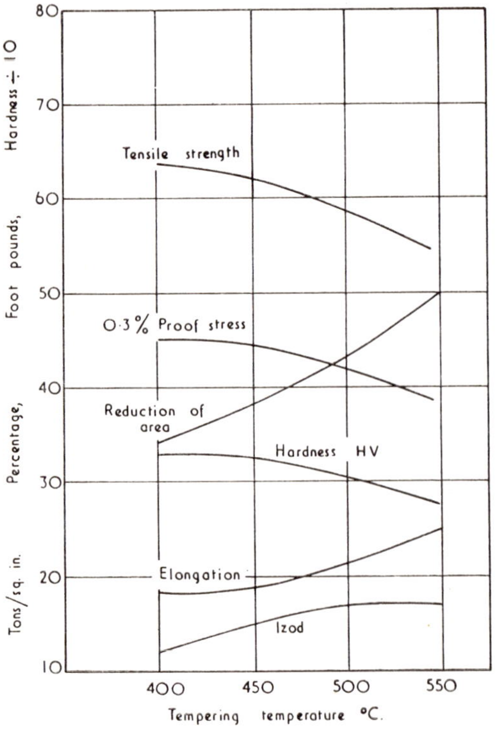

Fig. 46.6

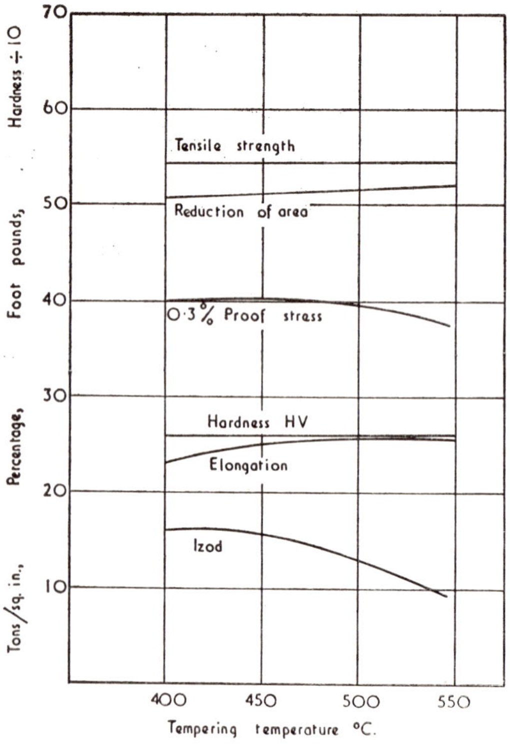

Fig. 46.7

LIST OF REFERENCES

1. Private correspondence with the London Polytechnic.
2. War Office, unpublished report.
3. G. Burns. The Properties of Some Silico-Manganese Steels. *JISI*, 1932, Vol. 125, No. 1, p. 363.
4. B.S. 971: Commentary on the British Standard Wrought Steels En Series. British Standards Institution, London, 1950.
5. S. L. Hoyt. *Metals and Alloys Data Book*. Reinhold, New York, 1952.

1 PER CENT CHROMIUM-VANADIUM SPRING STEEL

SPECIFICATION

CHEMICAL COMPOSITION (%)

C	Si	Mn	S	P	Cr	V
0·45—0·55	0·50 max.	0·50—0·80	0·050 max.	0·050 max.	0·80—1·20	0·15 min.

MECHANICAL PROPERTIES

No mechanical properties are specified for this steel.

RELATED SPECIFICATIONS

United Kingdom		America		France	Germany			Sweden	U.S.S.R.	
En No.	BS	SAE	AISI	AFNOR	Werkstoff No.	Name	DIN	SIS	GOST	Mark
47	1429	6150	6150	50CV4	1.8159	50CrV4	17221 17222 17225	14.2230	4543—57	50khFA

APPLICATIONS

An oil hardening steel which has deeper hardenability than the silicon-manganese En 45 type but, because of its lower carbon content, does not develop such a high surface hardness in small sections. It is not as susceptible to decarburization as the silicon–manganese type. The steel is suitable for high duty volute and leaf springs and for the heavier engine valve springs.

WELDING

(See Introduction (List of Abbreviations) for key to symbols used in this table.)

	Welding Process					
	MA	G	IG	S	F	B
En 47	m	b	m	f	b	j

MACHINABILITY

In the softened condition the machinability is approximately 45% of that for mild steel (En 3).

HOT WORKING AND HEAT TREATMENT TEMPERATURES

Forging, rolling and stamping	1150°C, finish above 850°C.
Annealing	850°C to 900°C, furnace cool.
Normalizing	850°C to 900°C, air cool.
Softening (sub-critical annealing)	680°C to 730°C, air or furnace cool.
Hardening	850°C to 900°C, oil quench.
Tempering	400°C to 550°C, air cool.

PHYSICAL PROPERTIES

(See also data for En 50)

SPECIFIC GRAVITY (d)

Chemical Composition	Heat Treatment	Specific Gravity at 20°C	Ref.
(within specification)	Softened	7·83	1

SPECIFIC HEAT (c)

Chemical Composition	Heat Treatment	Mean Specific Heat cal/g deg. C	Ref.
		20 to 100°C	
(within specification)	Softened	0·115	1

Values quoted for En 18 and En 19 (Vol. 1) will be very similar to those for En 47 up to approximately 600°C.

MEAN COEFFICIENT OF THERMAL EXPANSION (α)

Chemical Composition (%)							Heat Treatment °C	$10^6 \times$ Mean Coefficient of Expansion per deg. C.							Ref.
C	Si	Mn	S	P	Cr	V		20 to 100°C	20 to 200°C	20 to 300°C	20 to 400°C	20 to 500°C	20 to 600°C	20 to 700°C	
0·53	0·15	0·80	0·020	0·015	1·02	0·17	A860 3 hrs.	12·4	12·8	13·4	13·9	14·2	14·5	14·7	2
"	"	"	"	"	"	"	OQ860 T410 3 hrs.	11·8	12·4	13·1	13·6	13·9	14·1	14·2	"
"	"	"	"	"	"	"	" T650 10 hrs.	12·3	12·7	13·4	13·9	14·3	14·7	14·8	"
(within specification)							Softened	12·0							1

ELECTRICAL RESISTIVITY (ϱ)

Chemical Composition	Heat Treatment	Electrical Resistivity microhm-cm	Ref.
(within specification)	Softened	23	1

THERMAL CONDUCTIVITY (k)

Chemical Composition	Heat Treatment	Thermal Conductivity cal/cm s deg. C.	Ref.
(within specification)	Softened	0·12	1

YOUNG'S MODULUS (E); SHEAR MODULUS (G)

Chemical Composition (%)								Heat Treatment °C	Young's Modulus Static	Shear Modulus Static	Ref.
C	Si	Mn	S	P	Ni	Cr	V		tons/sq.in.	tons/sq.in.	
0·55	0·29	0·68	Tr	0·006	0·10	1·16	0·27	N850	13,500		3
"	"	"	"	"	"	"	"	OQ850 T400	13,550	5,250	"
"	"	"	"	"	"	"	"	" T475	13,400	5,180	"
"	"	"	"	"	"	"	"	" T550	13,650	5,200	"
"	"	"	"	"	"	"	"	" T600	13,700	5,270	"

TRANSFORMATION CHARACTERISTICS

ISOTHERMAL TRANSFORMATION

Chemical Composition (%)								
C	Si	Mn	S	P	Ni	Cr	Mo	V
0·51	0·27	0·72	0·020	0·021	0·15	0·94	0·05	0·20

High Frequency.
Mc Quaid Ehn grain size 7 (ASTM).
As quenched grain size 7 (ASTM).
Bars hot rolled to 4 S.W.G, cold drawn to 11 S.W.G., with intermediate annealing at 800°C.

Austenitizing Temperature 875°C, 30 min.

(Ref. 4)

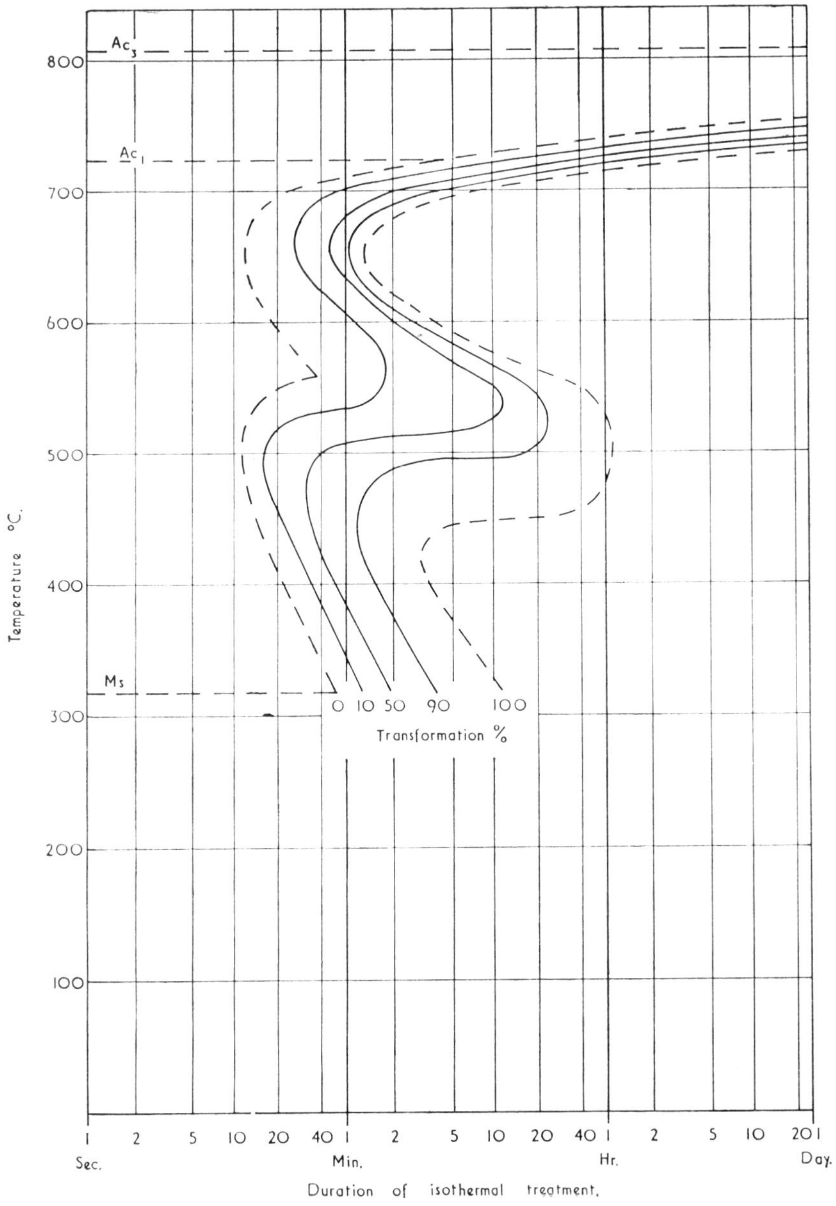

Fig. 47.1

ISOTHERMAL TRANSFORMATION (Continued)

Chemical Composition (%)			
C	Mn	Cr	V
0·53	0·67	0·93	0·18

McQuaid Ehn grain size 9 (ASTM).
Austenitizing Temperature 843°C.

(Ref.5)

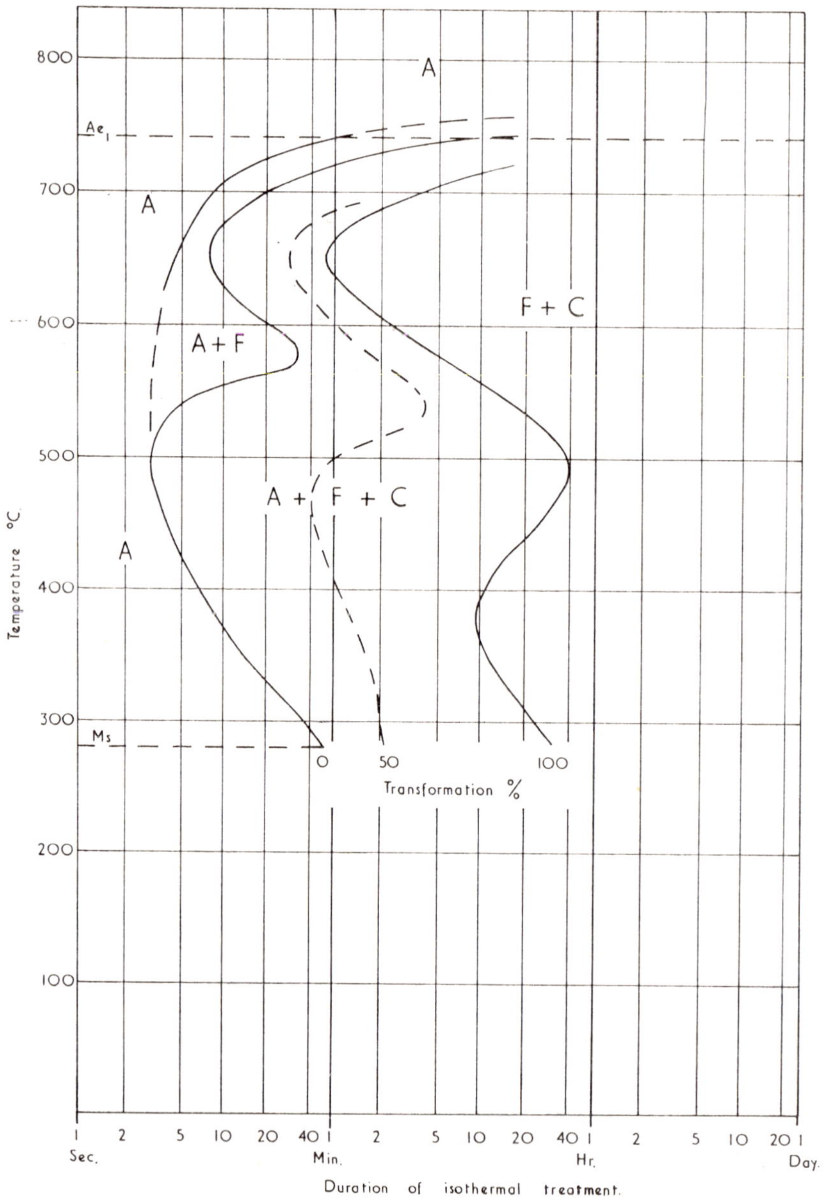

Fig. 47.2

HARDENABILITY

(See also En 50)

END QUENCH HARDENABILITY DATA

Chemical Composition (%)								
C	Si	Mn	S	P	Ni	Cr	Mo	V
0·51	0·27	0·72	0·020	0·021	0·15	0·94	0·05	0·20

High Frequency.
McQuaid Ehn grain size 7 (ASTM).
As quenched grain size 7 (ASTM).
End quenched from 875°C.

(Ref. 4)

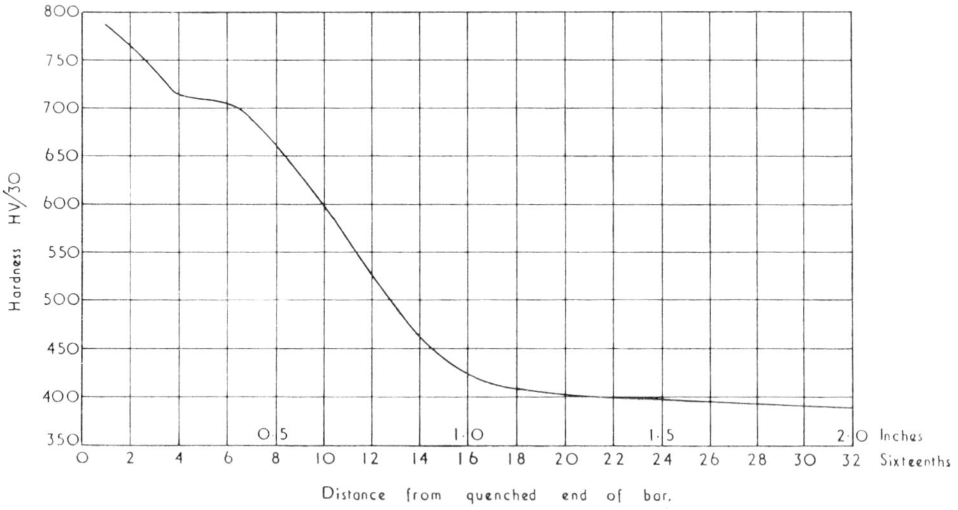

Fig. 47.3

Chemical Composition (%)			
C	Mn	Cr	V
0·53	0·81	0·92	0·16

McQuaid Ehn grain size 9 to 10 (ASTM).
End quenched from 843°C.

(Ref. 5)

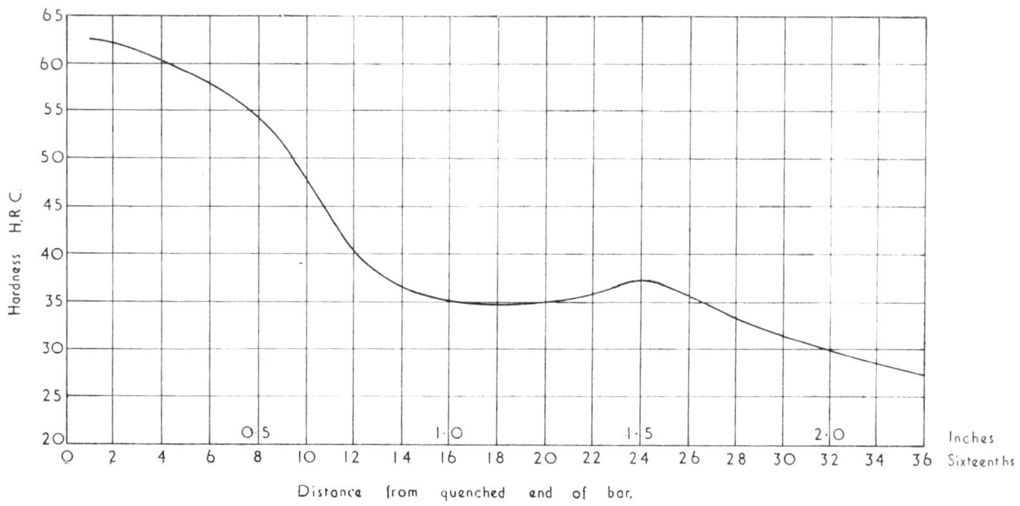

Fig. 47.4

En 47

END QUENCH HARDENABILITY BAND

These curves were compiled from data published in the S.A.E. *Handbook*, 1960.

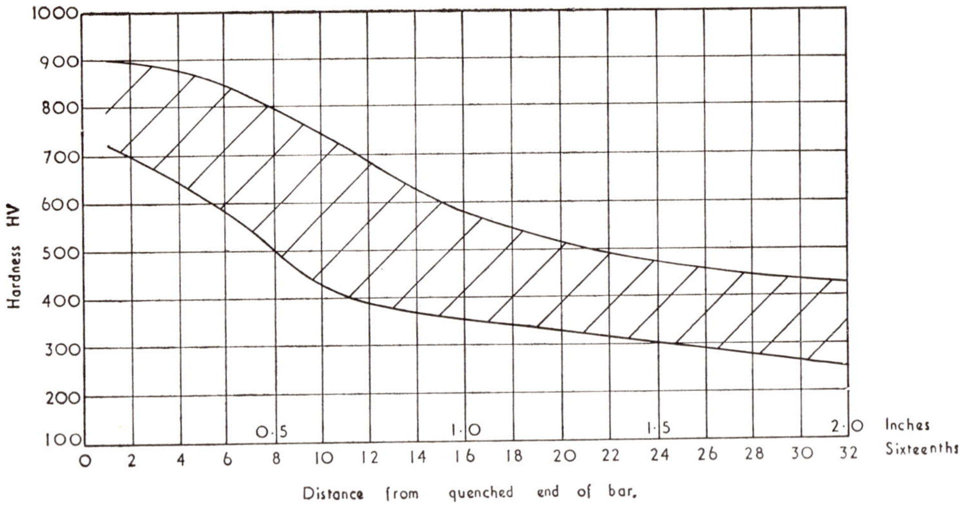

Fig. 47.5

HARDNESS TRAVERSE DATA

Chemical Composition (%)						
C	Si	Mn	Ni	Cr	Mo	V
0·50	0·41	0·65	0·18	1·01	0·05	0·18

Bars oil quenched from 875°C.

(Ref. 1)

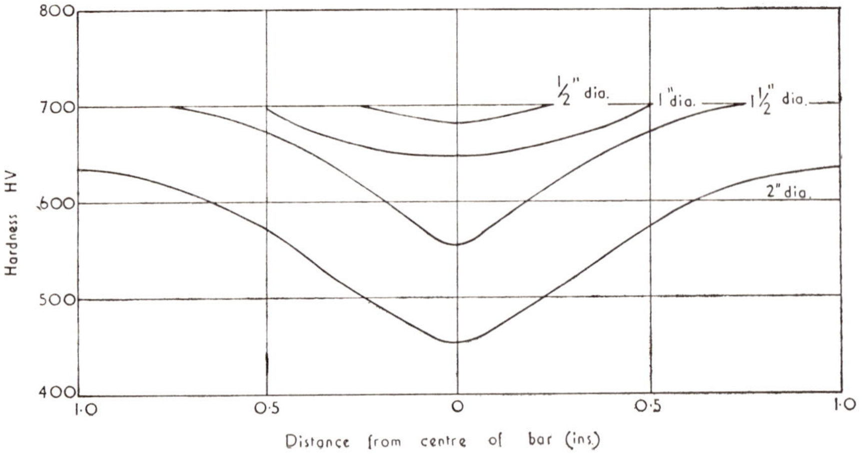

Fig. 47.6

HARDNESS TRAVERSE DATA (Continued)

(For Chemical composition, see the opposite page.)

Oil quenched 900°C. Tempered as indicated.
1 in. dia. bars. (Ref. 1)

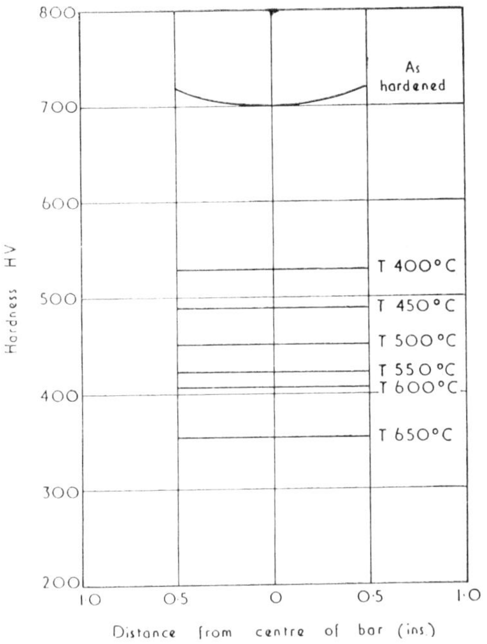

Fig. 47.7

Oil quenched 900°C. Tempered as indicated.
2 in. dia. bars. (Ref. 1)

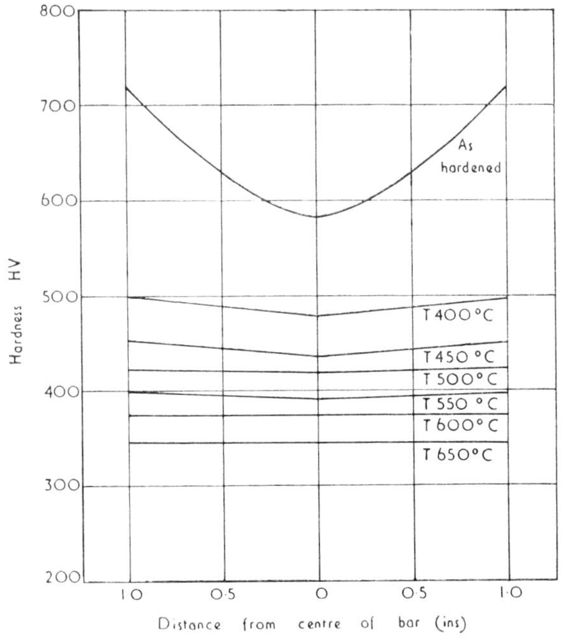

Fig. 47.8

MECHANICAL PROPERTIES AT ROOM TEMPERATURE

(See also En 50)

Process of Manu-facture	ASTM Grain Size		Chemical Composition (%)									Section Size dia. (in.)	Heat Treatment °C	Tensile					Izod ft.lb.	Ref.
	McQuaid Ehn	As Quench-ed	C	Si	Mn	S	P	Ni	Cr	Mo	V			0·2% P.S. (0·2Rp) tons/sq.in.	Y.P. (Rs) tons/sq.in.	T.S. (Rm) tons/sq.in.	El % (A)	R.A. % (Z)		
B.E.A.	7	7 to 8	0·45	0·19	0·65	0·017	0·016	0·23	1·31	0·07	0·20	½ (C)	N870 OQ870 T540 T620 OQ		76·0	79·5	18	49	41	6
"	"	" " "	"	"	"	"	"	"	"	"	"	" (M)	" " " " "		76·0	78·5	18½	54	38	"
"	"	" " "	"	"	"	"	"	"	"	"	"	1¼ (C)	" " " " "		69·5	73·5	20	58	45	"
"	"	" " "	"	"	"	"	"	"	"	"	"	" (M)	" " " " "		71·5	75·0	18	55	40	"
"	"	" " "	"	"	"	"	"	"	"	"	"	1¾ (C)	" " " " "		63·5	69·4	19	53½	44	"
"	"	" " "	"	"	"	"	"	"	"	"	"	" (M)	" " " " "		67·2	71·9	20	56	42	"
"	"	" " "	"	"	"	"	"	"	"	"	"	" (O)	" " " " "		66·4	70·4	20	58	41	"
"	"	" " "	"	"	"	"	"	"	"	"	"	3 (C)	" " " " "		58·9	66·2	17	54½	38	"
"	"	" " "	"	"	"	"	"	"	"	"	"	" (M)	" " " " "		59·0	67·2	20	57	35	"
"	"	" " "	"	"	"	"	"	"	"	"	"	" (O)	" " " " "		58·4	66·1	20	58½	37	"
			0·48	0·35	0·81	0·029	0·011	0·33	1·00		0·18	1⅝	OQ845 T454	91·5		100·4	13	47		7
			"	"	"	"	"	"	"		"	"	" T510	82·6		89·3	14	48		"
			"	"	"	"	"	"	"		"	"	" T566	75·9		80·4	15	49½		"
			"	"	"	"	"	"	"		"	"	" T621	67·0		73·7	18	51		"
	6 to 8		0·49	0·29	0·78	0·016	0·012	0·18	1·00	0·05	0·17	0·565	N870 OQ845 T204		120·6	140·6	7	16	14	8
	" " "		"	"	"	"	"	"	"	"	"	"	" " T260		114·0	131·9	8	29	11	"
	" " "		"	"	"	"	"	"	"	"	"	"	" " T316		107·2	120·6	9½	34	11	"
	" " "		"	"	"	"	"	"	"	"	"	"	" " T371		99·6	110·0	10	37	9	"
	" " "		"	"	"	"	"	"	"	"	"	"	" " T427		92·9	99·6	10	40	14	"
	" " "		"	"	"	"	"	"	"	"	"	"	" " T482		85·3	90·7	11	43	23	"
	" " "		"	"	"	"	"	"	"	"	"	"	" " T538		77·7	81·7	12	46	30	"
	" " "		"	"	"	"	"	"	"	"	"	"	" " T593		68·7	73·2	15	50	39	"
	" " "		"	"	"	"	"	"	"	"	"	"	" " T649		58·5	63·0	18	55	65	"
	" " "		"	"	"	"	"	"	"	"	"	"	" " T704		47·8	52·2	22	62	87	"
	5 to 6(2 to 4)		0·51	0·35	0·80	0·015	0·014	0·11	0·95	0·01	0·18	1	A816		26·7	43·2	23	48½	20	"
	" " "		"	"	"	"	"	"	"	"	"	½	N870		41·5	63·1	20½	63	27	"
	" " "		"	"	"	"	"	"	"	"	"	1	"		39·8	60·7	22	61	26	"
	" " "		"	"	"	"	"	"	"	"	"	2	"		33·6	57·9	20½	56½	20	"
	" " "		"	"	"	"	"	"	"	"	"	4	"		29·9	57·1	18	49½	17	"

MECHANICAL PROPERTIES AT ROOM TEMPERATURE

(See also En 50)

(Continued)

Process of Manufacture	ASTM Grain Size	Chemical Composition (%) C	Si	Mn	S	P	Ni	Cr	Mo	V	Section Size dia. (in.)	Heat Treatment °C	Tensile 0·2% P.S. (0.2Rp) tons/sq.in.	Y.P. (Rs) tons/sq.in.	T.S. (Rm) tons/sq.in.	El % (A)	R.A %. (Z)	Izod ft.lb.	Ref.
	5 to 6 (2 to 4)	0·51	0·35	0·80	0·015	0·014	0·11	0·95	0·01	0·18	½	OQ843 T538		79·4	80·1	14½	49½	26	8
	" "	"	"	"	"	"	"	"	"	"	1	" "		74·9	77·5	14½	48	25	"
	" "	"	"	"	"	"	"	"	"	"	2	" "		64·8	74·1	14½	46½	19	"
	" "	"	"	"	"	"	"	"	"	"	4	" "		56·7	67·6	16	48½	17	"
	" "	"	"	"	"	"	"	"	"	"	½	OQ845 T593		70·8	71·4	16½	52½	39	"
	" "	"	"	"	"	"	"	"	"	"	1	" "		67·2	70·7	16	53	34	"
	" "	"	"	"	"	"	"	"	"	"	2	" "		58·8	66·2	17½	55	34	"
	" "	"	"	"	"	"	"	"	"	"	4	" "		48·4	58·0	19	55½	38	"
	" "	"	"	"	"	"	"	"	"	"	½	OQ845 T649		63·2	65·6	18	54	44	"
	"	"	"	"	"	"	"	"	"	"	1	" "		57·8	63·1	18½	56½	57	"
	" "	"	"	"	"	"	"	"	"	"	2	" "		52·0	59·7	19½	57½	52	"
	" "	"	"	"	"	"	"	"	"	"	4	" "		42·2	54·2	21	59½	45	"
		0·53		0·84				1·12		0·20	1⅛	OQ875 T500		92·7	95·0	12	44	15	9
		0·55	0·29	0·68	Tr	0·006	0·10	1·16		0·27	3 × ⅜	N850			69·6	$l = 11{\cdot}3\sqrt{A}$ 10½	47		3
		"	"	"	"	"	"	"		"	"	OQ850 T400		105·7	111·6	5½	12		"
		"	"	"	"	"	"	"		"	"	" T475		103·5	105·3	5	14		"
		"	"	"	"	"	"	"		"	"	" T500		81·7	84·3	7½	32		"
		"	"	"	"	"	"	"		"	"	" T600		75·6	79·5	7½	31		"

EFFECT OF TEMPERING ON HARDNESS

Chemical Composition (%) C	Si	Mn	S	P	Ni	Cr	V	Section Size dia. (in.)	Heat Treatment °C	Brinell Hardness (HB) Tempering Temperature °C 20	250	300	350	400	450	500	550	600	650	Ref.
0·45		0·64				1·13	0·175		OQ900	595				482	454	462	406	386	377	10
0·55	0·29	0·68	Tr	0·006	0·10	1·16	0·27	3 × ⅜	OQ850	640	620	578	540	512	477	466	444	400		3

EFFECT OF TEMPERING ON THE MECHANICAL PROPERTIES AT ROOM TEMPERATURE

Chemical Composition (%)								
C	Si	Mn	S	P	Ni	Cr	Mo	V
0·49	0·29	0·78	0·016	0·012	0·18	1·00	0·05	0·17

McQuaid Ehn grain size 6—8 (ASTM).
0.565 in. dia. bars normalized 870°C.
Oil quenched 845°C. Tempered as indicated.

(Ref. 8)

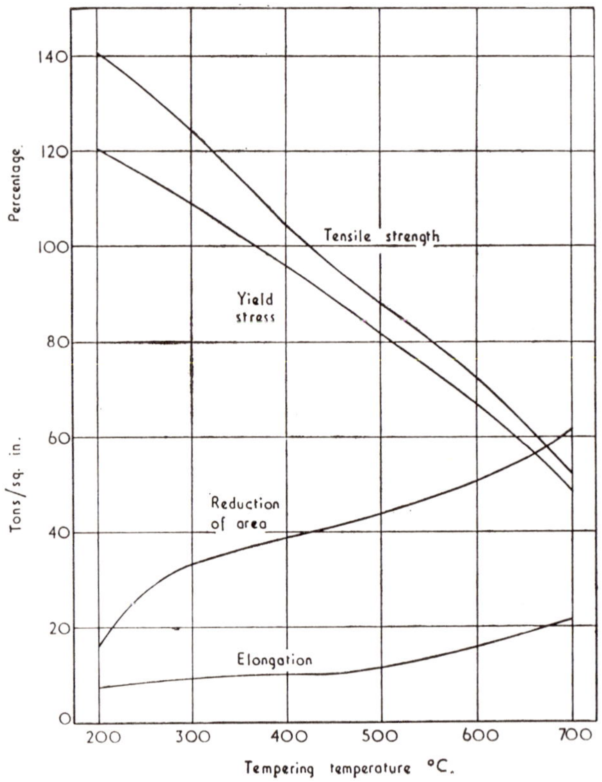

Fig. 47.9

EFFECT OF SECTION SIZE ON THE MECHANICAL PROPERTIES AT ROOM TEMPERATURE

Chemical Composition (%)								
C	Si	Mn	S	P	Ni	Cr	Mo	V
0·51	0·35	0·80	0·015	0·014	0·11	0·95	0·01	0·18

McQuaid Ehn grain size 5—6 (some 2—4).

Normalized 870°C.

(Ref. 8)

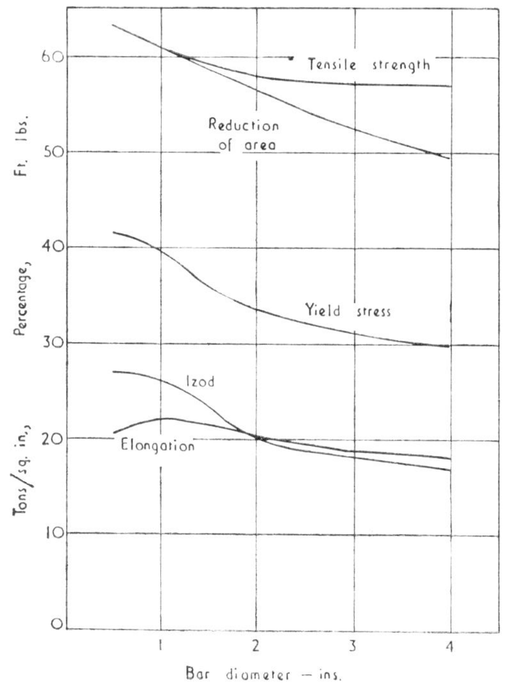

Fig. 47.10

Oil quenched 845°C. Tempered 538°C.

(Ref. 8)

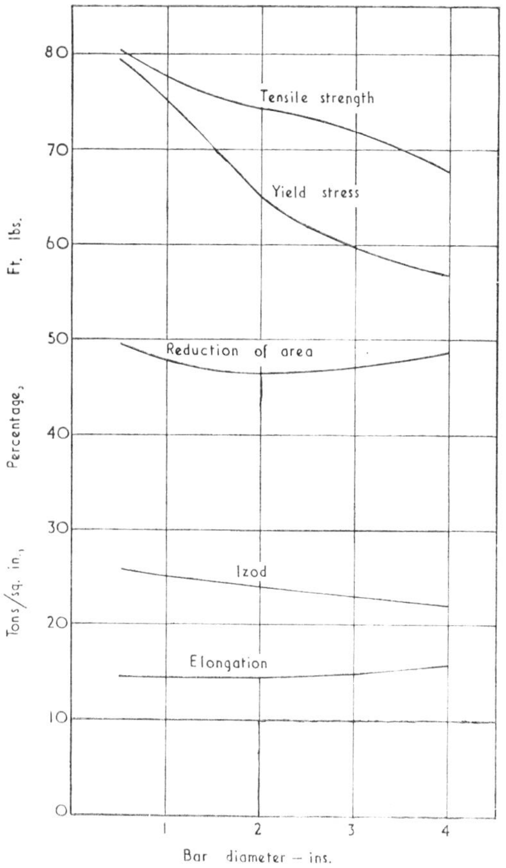

Fig. 47.11

En 47

EFFECT OF SECTION SIZE ON MECHANICAL PROPERTIES (Continued)

Chemical Composition %								
C	Si	Mn	S	P	Ni	Cr	Mo	V
0·51	0·35	0·80	0·015	0·014	0·11	0·95	0·01	0·18

Oil quenched 845°C. Tempered 593°C. (Ref. 8)

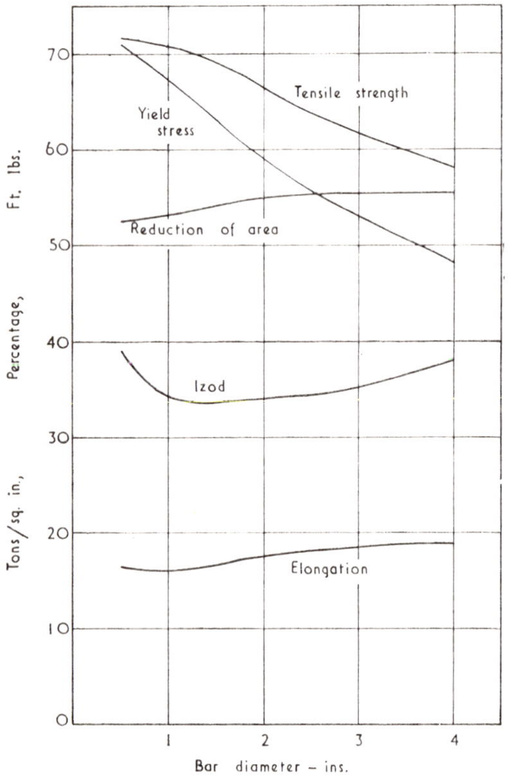

Fig. 47.12

Oil quenched 845°C. Tempered 649°C. (Ref. 8)

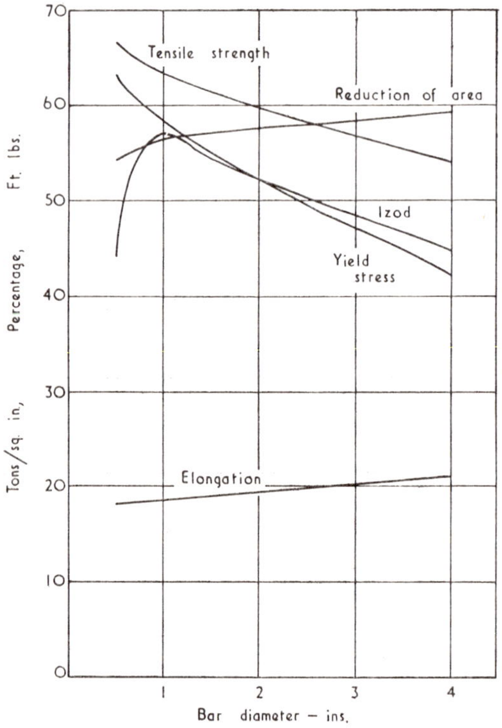

Fig. 47.13

MECHANICAL PROPERTIES AT LOW TEMPERATURE

Chemical Composition (%)									Section Size dia.	Heat Treatment °C	Test Temperature °C	Tensile						D.V.M. Impact Value Kg M/sq.cm	Ref.
C	Si	Mn	S	P	Ni	Cr	Mo	V				0·2% P.S. (0·2Rp) tons/sq.in.	0·5% P.S. (0·5Rp) tons/sq.in.	Y.P. (Rs) tons/sq.in.	T.S. (Rm) tons/sq.in.	El % (A) l=5D	R.A. % (Z)		
0·50	0·24	0·63	0·016	0·029	0·14	1·05		0·18	30 mm	WQ850 T550 2hrs. AC	20		50·5	52·1	58·5	18½	60	11·8	11
”	”	”	”	”	”	”		”	”	” ” ” ”	—78		56·8	59·0	67·6	19½	54	4·6	”
0·54	0·27	0·61	0·017	0·005	0·23	1·10	0·08	0·17	30 mm	OQ840 T580 2hrs. AC	20		52·4	54·0	58·7	20½	59	11·4	”
”	”	”	”	”	”	”	”	”	”	” ” ” ”	—78		58·9	61·1	67·0	21	55	5·5	”

IMPACT TRANSITION TEMPERATURE DATA

Chemical Composition (%)					Heat Treatment °C	Charpy Keyhole Notch, ft.lb. — Test Temperature °C							Ref.
C	Si	Mn	Cr	V		R.T.	—4	—18	—32	—46	—60	—68	
0·45	0·17	0·69	0·99	0·20	N871 OQ857 T427	14		14		13	14	13	12
”	”	”	”	”	” ” T538	25		19		15	14	13	”
”	”	”	”	”	N885 OQ857 T316	11	11	12		9	7		”
”	”	”	”	”	” ” T427	14	15	12		12	11		”
0·51	0·18	0·68	1·09	0·18	” ” T538	19	17	17	14	13	13	12	”
”	”	”	”	”	” ” T649	34	29	30	27	25	22	22	”

MECHANICAL PROPERTIES IN TORSION

Chemical Composition (%)								Section Size dia. (in.)	Heat Treatment °C	Tensile		Torsion		Ref.
C	Si	Mn	S	P	Ni	Cr	V			Y.P. (Rs) tons/sq.in.	T.S. (Rm) tons/sq.in.	Apparent max. Shear Stress tons/sq.in.	Probable max. Shear Stress tons/sq.in.	
0·55	0·29	0·68	Tr	0·006	0·10	1·16	0·27	3′ × ⅜	OQ850 T400	105·7	111·6	91·4	68·6	3
”	”	”	”	”	”	”	”	”	” T475	103·5	105·3	78·0	58·5	”
”	”	”	”	”	”	”	”	”	” T550	81·7	84·3	68·3	51·2	”
”	”	”	”	”	”	”	”	”	” T600	75·6	79·5	66·1	49·6	”

FATIGUE PROPERTIES

Chemical Composition (%)								Section Size dia. (in.)	Heat Treatment °C	Test Tempera-ture °C	Tensile					Fatigue Limit tons/sq. in.				Ref.
C	Si	Mn	S	P	Ni	Cr	V				0·2% P.S. (0·2Rp)	Y.P. (Rs)	T.S. (Rm)	El (A) %	R.A. (Z) %	Rotating Bending		Reversed Axial	Tor-sional	
											tons/sq.in.					Plain	Notched			
0·49	0·22	0·66	0·041	0·031		1·11	0·15	⅝	OQ870 T500			85·0	87·0	16	47			0±43·5		9
"	"	"	"	"		"	"	"	" "			"	"	"	"			20±33·5		"
"	"	"	"	"		"	"	"	" "			"	"	"	"			40±30·0		"
"	"	"	"	"		"	"	"	" "			"	"	"	"			60±22·5		"
														l=5D						
0·54	0·40	0·70	0·016	0·015		1·08	0·27		OQ875 T500			91·1	93·7	7	36	±41·0				13
"	"	"	"	"		"	"		" T600			78·5	85·1	9	38	±35·6				"
0·54	0·24	0·63	0·016	0·029	0·14	1·05	0·18	30 mm	WQ850 T550	20	50·5	52·1	58·5	18½	60	±25·4	±7·0*			11
"	"	"	"	"	"	"	"	"	" "	—78	56·8	59·0	67·6	19½	54	±27·3	±7·0*			"
0·54	0·27	0·61	0·017	0·005	0·23	1·10	0·17	30 mm	OQ840 T580	20	52·4	54·0	58·7	20½	59	±30·5	±7·6*			"
"	"	"	"	"	"	"	"	"	" "	—78	58·9	61·1	67·0	21	55	±32·4	±8·9*			"
0·55	0·29	0·68	Tr	0·006	0·10	1·16	0·27	0·20	OQ850 T600							±42·0				14
"	"	"	"	"	"	"	"	3×⅜	OQ850 T400			105·7	111·6	5½	12	±47·0				3
"	"	"	"	"	"	"	"	"	" T475			103·5	105·3	5	14	±43·5				"
"	"	"	"	"	"	"	"	"	" T550			81·7	84·3	7½	32	±43·5				"
"	"	"	"	"	"	"	"	"	" T600			75·6	79·5	7½	31	±42·5				"
"	"	"	"	"	"	"	"	"	OQ850 T600		(apparent shear strength 66 tons/sq.in.)								0±25	15
"	"	"	"	"	"	"	"	"	" "		" " " " "								28±22	"
"	"	"	"	"	"	"	"	"	" "		" " " " "								43±16	"

*Notch 60°V, 0·1 mm root radius, 0·5 mm deep, bar dia. 8 mm, root dia., 7 mm.

LIST OF REFERENCES

1. Private correspondence with the English Steel Corporation Ltd.
2. P. Hidnert. Thermal Expansion of Some Chromium-Vanadium Steels. *Journal of Research of the National Bureau of Standards*, 1940, Vol. 24, Jan., pp. 25—30.
3. G. A. Hankins, D. Hanson and G. W. Ford. The Mechanical Properties of Four Heat Treated Spring Steels. *JISI*, 1926, Vol. 114, No. 2, p. 265.
4. Atlas of Isothermal Transformation Diagrams of the B. S. En Steels. Special Report No. 56 of The Iron & Steel Institute, 1956.
5. Atlas of Isothermal Transformation Diagrams. The United States Steel Corporation. Second Edition, 1957.
6. Symposium on the Hardenability of Steel. Organized by the Technical Advisory Committee of the Special and Alloy Steel Committee (Ministry of Supply) and The Iron and Steel Institute. Special Report No. 36 of The Iron & Steel Institute, 1946.
7. E. T. Bittner. Alloy Spring Steels. *Transactions of the Amer. Soc. Metals*, 1948, No. 40, pp. 263—280.
8. *Modern Steels and Their Properties — Carbon and Alloy Steel Bars*, The Bethlehem Steel Company. Sixth Edition, 1961.
9. Private correspondence with The United Steel Companies Ltd.
10. B. S. 971 : Commentary on the British Standard Wrought Steels En Series. British Standards Institution, London, 1950.
11. A. Pomp and M. Hempel. Fatigue Strength and Notch Coefficient of Unalloyed and Alloy Constructional Steel at +20° and —78°C. *Arch. für das Eisenhüttenwesen*, 1950, Vol. 21, Jan/Feb., p. 53.
12. H. W. Gillett. Impact Resistance and Tensile Properties of Metals at Sub-atmospheric Temperatures. Project No. 13 of the joint ASME-ASTM Research Committee on the Effect of Temperature on the Properties of Metals.
13. R. Cazaud (translated by A. J. Fenner). *Fatigue of Metals*. Chapman and Hall Ltd., 1953.
14. G. A. Hankins and M. L. Becker. Effect of Surface Conditions Produced by Heat-treatment on the Fatigue Resistance of Spring Steels. *JISI*, 1931, Vol. 124, No. 2, p. 387.
15. G. A. Hankins. D.S.I.R. Engineering Research Special Report No. 9. H.M.S.O., 1928.

1 PER CENT CHROMIUM SPRING STEEL (En 48)
SILICON-CHROMIUM SPRING STEEL (En 48A)

SPECIFICATION

CHEMICAL COMPOSITION (%)

	C	Si	Mn	S	P	Cr
En 48	0·45—0·55	0·10—0·50	0·50—0·80	0·050max.	0·050max.	1·00—1·40
En 48A	0·50—0·60	1·35—1·65	0·60—0·90	"	"	0·55—0·85

MECHANICAL PROPERTIES

No mechanical properties are specified for these steels.

RELATED SPECIFICATIONS

United Kingdom	America		France	U.S.S.R.	
En No.	SAE	AISI	AFNOR	GOST	Mark
48	5147	5147	50C4	4543—57	50ch
48A		9254	55SC7		

APPLICATIONS

En 48 is used for applications similar to those of En 47. En 48A is used for applications similar to those of En 45.

WELDING

(See Introduction (List of Abbreviations) for key to symbols used in this table.)

	Welding Process					
	MA	G	IG	S	F	B
En 48*	d	cx	g	e	b	
En 48A	f		f	f	b	

* The symbols given are those quoted for En 18, which differs only slightly from En 48.

MACHINABILITY

In the softened condition the machinability of En 48 and En 48A is approximately 45% of that for mild steel (En 3).

HOT WORKING AND HEAT TREATMENT TEMPERATURES

	En 48	*En 48A*
Forging, rolling and stamping	1150°C, finish above 850°C.	
Annealing	830°C—880°C, furnace cool.	860°C—920°C, furnace cool.
Normalizing	830°C—880°C, air cool.	860°C—920°C, air cool.
Softening	670°C—700°C, air or furnace cool.	680°C—720°C, air or furnace cool.
Hardening	830°C—880°C, oil quench.	860°C—920°C, oil quench.
Tempering	400°C—550°C, air cool.	400°C—550°C, air cool.

PHYSICAL PROPERTIES

Few data are available for the physical properties of En 48 and En 48A. Data for specific gravity, specific heat, coefficient of expansion, will be very similar to En 18 and En 45 for En 48 and En 48A respectively.

SPECIFIC GRAVITY (d)

	Chemical Composition (%)						Heat Treatment °C	Specific Gravity at 20°C	Ref.
	C	Si	Mn	S	P	Cr			
En 48	0·46	0·21	0·93	0·034	0·028	0·95	A 720 AC	7·83	1
"	"	"	"	"	"	"	OQ850 T550	7·82	"
"	"	"	"	"	"	"	" T600	7·82	"
"	"	"	"	"	"	"	" T650	7·83	"
"	"	"	"	"	"	"	" T680	7·83	"

YOUNG'S MODULUS (E); SHEAR MODULUS (G)

	Chemical Composition (%)						Heat Treatment °C	Young's Modulus	Shear Modulus	Ref.
	C	Si	Mn	S	P	Cr		tons/sq.in.		
En 48	0·45	0·12	0·69	0·010	0·014	1·14	N830	13500		2
"	"	"	"	"	"	"	OQ830 T350	13300	5070	"
"	"	"	"	"	"	"	" T425	13300	5110	"
"	"	"	"	"	"	"	" T500	13400	5090	"
"	"	"	"	"	"	"	" T575	13500	5160	"
"	0·46	0·21	0·93	0·034	0·028	0·95	A720 AC	13200		1
"	"	"	"	"	"	"	OQ850 T550	13200		"

MECHANICAL PROPERTIES AT ROOM TEMPERATURE

	Chemical Composition (%)								Section Size dia. (in.)	Heat Treatment °C	Tensile							Izod ft.lb.	Ref.
	C	Si	Mn	S	P	Ni	Cr	Mo			0·1% P.S (0·1Rp)	0·2% P.S. (0·2Rp)	0·5% P.S. (0·5Rp)	Y.P. (Rs)	T.S. (Rm)	El % (A)	R.A. % (Z)		
											tons/sq.in.								
En 48	0·45	0·12	0·69	0·010	0·014		1·14		3 × ⅜	N830				40·0	53·9	l=11·3 √A	50		2
”	”	”	”	”	”		”		”	OQ830 T350					102·7	12½	1		”
”	”	”	”	”	”		”		”	” T425				95·0	100·3	4½	17½		”
”	”	”	”	”	”		”		”	” T500				79·5	86·0	6½	30½		”
”	”	”	”	”	”		”		”	” T575				70·0	74·4	8½	35½		
																l=4√A			”
”	0·45	0·29	0·91			0·04	0·95		1⅛	OQ850 T625	50·5		51·6		60·7	22	60		1
En 48	0·46	0·27	0·91			0·20	0·95		1¾	OQ850 T650				42·8	52·6	25	58	68	1
”	0·46	0·21	0·93	0·034	0·028	0·17	0·95		1⅛	A720 AC	21·9		25·3		47·6	26	54		”
”	”	”	”	”	”	”	”		”	OQ850 T550	63·0		70·0		75·0	17	44		”
”	”	”	”	”	”	”	”		”	” T600	58·6		64·1		70·5	20	51		”
”	”	”	”	”	”	”	”		”	” T650	49·8		54·0		61·5	22	60		”
”	”	”	”	”	”	”	”		”	” T680	41·2		46·0		55·5	25	63		”
”	0·46	0·28	1·01			0·20	0·99		1⅛	OQ870 T600				63·8	70·0	19	50	36	”
”	”	”	”			”	”		”	” T650				57·2	64·0	20	55	56	”
”	0·46	0·24	0·92			0·15	0·97		1¼	OQ850 T550				71·2	75·6	15	41	14	”
En 48 (SAE5150)	(within specification)								1	As rolled				44·6	60·7	14	50		3
”	”								”	A788				30·4	46·4	22	57		”
”	”								”	N885				44·2	60·3	15	50		”
”	”								”	OQ815/845 T316				105·0	116·0	3	10		”
”	”								”	” T371				99·1	107·0	9	27		”
”	”								”	” T427				90·2	98·7	10	30		”
”	”								”	” T482				76·8	84·8	13	38		”
”	”								”	” T538				67·0	75·0	16	45		”
”	”								”	” T593				58·0	66·5	19	51		”
”	”								”	” T649				51·3	59·4	21	57		”
”	”								”	” T704				45·1	53·6	22	61		”
																l=10″			
En 48A	0·57	1·60	0·76	0·017	0·014		0·77		0·097	H & T				87·5	110·3	7½	47		4
”	”	”	”	”	”		”		”	”				84·8	116·1	8	50		”
”	”	”	”	”	”		”		”	”				109·8	125·4	9½	49		”
																l=4 √A			
En 48A	0·60	1·50	0·69	0·030	0·023	0·42	0·65		1·15	OQ850 T425		103·4			114·4	11½	40		5
”	”	”	”	”	”	”	”		”	” T510		85·6			94·6	12	37		”

EFFECT OF TEMPERING ON THE MECHANICAL PROPERTIES AT ROOM TEMPERATURE

Typical curve for steel to En 48 composition (SAE 5150).
1 in. dia. bars. Oil quenched 815—845°C.

(Ref. 3)

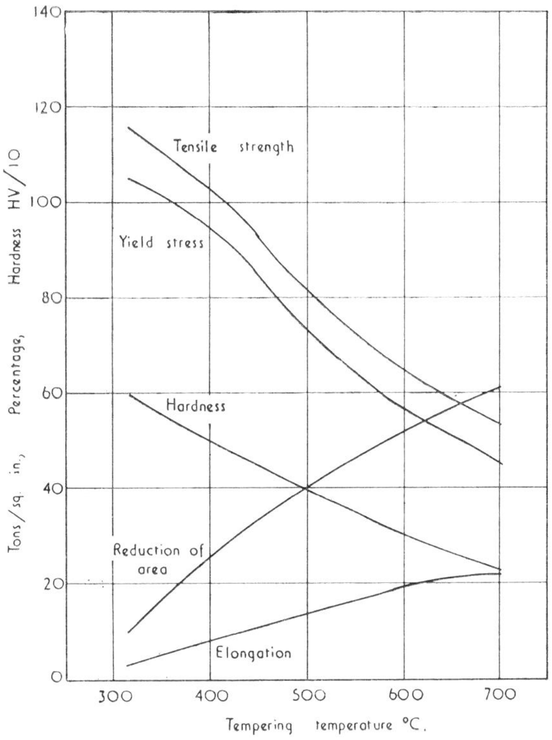

Fig. 48.1

MECHANICAL PROPERTIES AT HIGH TEMPERATURE

	Chemical Composition (%)					Section Size dia. (in.)	Heat Treatment °C	Test Temperature °C	Tensile					Ref.
	C	Si	Mn	Ni	Cr				0·1% P.S. (0·1Rp) tons/sq.in.	0·5% P.S. (0·5Rp) tons/sq.in.	T.S. (Rm) tons/sq.in.	El % (A)	R.A. % (Z)	
En 48	0·45	0·29	0·91	0·04	0·95	1⅛	OQ850 T625	RT	50·5	51·6	60·7	22	60	1
	”	”	”	”	”	”	” ”	250	42·9	56·4	70·2	13½	27	”
	”	”	”	”	”	”	” ”	300	45·0	51·8	60·7	22	60	”
	”	”	”	”	”	”	” ”	400	39·2	44·6	52·3	28	74	”
	”	”	”	”	”	”	” ”	500	19·6	27·8	38·5	33½	84	”
	”	”	”	”	”	”	” ”	600	6·5	11·0	22·1	50	94	”

TORSION AND FATIGUE PROPERTIES

	Chemical Composition (%)						Section Size dia. (in.)	Heat Treatment °C	Tensile				Torsion		Rotating Bending Fatigue Limit	Ref.
	C	Si	Mn	S	P	Cr			Y.P. (tons/sq.in.)	T.S. (Rm) (tons/sq.in.)	El % (A)	R.A. % (Z)	Apparent max. Shear Stress (tons/sq.in.)	Probable max. Shear Stress (tons/sq.in.)	tons/sq.in.	
											$l=11\cdot3\sqrt{A}$					
En 48	0·45	0·12	0·69	0·010	0·041	1·14	3 × 3/8	OQ830 T350		102·7		1	89·6	67·2	±50	2
"	"	"	"	"	"	"	"	" T425	95·0	100·3	4 1/2	17 1/2	83·0	62·3	±45	"
"	"	"	"	"	"	"	"	" T500	79·5	86·0	6 1/2	30 1/2	70·2	52·7	±39	"
"	"	"	"	"	"	"	"	" T575	70·0	74·4	8 1/2	35 1/2	64·3	48·2	±38	"
											$l=4\sqrt{A}$					
En 48A	0·60	1·50	0·69	0·030	0·042	0·65	1·15	OQ850 T425	103·4(0·2%PS)	114·4	11 1/2	40	94·4	70·8	±46·5	5
"	"	"	"	"	"	"	"	" T510	85·6(")	94·6	12	37	81·8	61·4	±41·5	"

LIST OF REFERENCES

1. Private correspondence with Firth-Brown Ltd.
2. G. A. Hankins and G. W. Ford. The Mechanical and Metallurgical Properties of Spring Steels as revealed by Laboratory Tests. *JISI*, 1929, Vol. 119, No. 1, p. 217.
3. S. L. Hoyt. *Metals and Alloys Data Book*. Reinhold, New York, 1952.
4. H. J. Elmendorf. An Investigation of Tempered Chromium-silicon Spring Steels. *Transactions of the Amer. Soc. Metals*. 1948, Vol. 40, pp. 281—301.
5. A. S. Kenneford and G. C. Ellis. A Comparison of Six Spring Steels. *JISI*, 1950, Vol. 164, No. 1, p. 265.

CARBON STEELS FOR HARD DRAWN WIRE

SPECIFICATION

CHEMICAL COMPOSITION (%)

	C	Si	Mn	S	P
En 49A	0·40—0·85	0·30 max.	1·00 max.	0·050 max.	0·050 max.
En 49B	0·45—0·85	0·30 max.	1·00 max.	0·050 max.	0·050 max.
En 49C	0·55—0·85	0·30 max.	0·75 max.	0·040 max.	0·040 max.
En 49D	0·65—0·85	0·30 max.	0·75 max.	0·040 max.*	0·040 max.*

* By agreement between purchaser and manufacturer steel to En 49D may be supplied with sulphur and phosphorus contents each 0·030 % max.

MECHANICAL PROPERTIES

No mechanical properties are specified for these steels.

RELATED SPECIFICATIONS

United Kingdom			America		France	Germany			Sweden	U.S.S.R.	
En No.	BS	DTD	SAE	AISI	AFNOR	Werkstoff No.	Name	DIN	SIS	GOST	Mark
49A	1408	5A 215	1045 1050 1060 1070 1080	C1045 C1050 C1060 C1070 C1080	XC42 XC45 XC48 XC55 XC65 XC70 XC80	1.0503	C45	17200 17240			
49B	1408	5A 215	1045 1050 1060 1070 1080	C1045 C1050 C1060 C1070 C1080	XC45 XC48 XC55 XC65 XC70 XC80						

(Continued)

RELATED SPECIFICATIONS (Continued)

United Kingdom			America		France	Germany			Sweden	U.S.S.R.	
En No.	BS	DTD	SAE	AISI	AFNOR	Werkstoff No.	Name	DIN	SIS	GOST	Mark
49C	1408	5A	1060 1070 1080	C1060 C1070 C1080	XC55 XC65 XC70 XC80	1.0601	C60	17200 17222			
49D	1408	5A	1070 1080	C1070 C1080	XC65 XC70 XC80	1.0605	C75	17222 17223	14.1770	1435-54 1435-54	U7A U7

APPLICATIONS

The En 49 specifications cover a wide range of steels which are usually supplied in the cold drawn condition for the manufacture of springs. En 49A is used for general purposes where the stresses are relatively low, e.g., clips, fasteners, etc.

En 49B is used for general purposes for static applications such as upholstery springs, etc.

En 49C is used for medium duty applications such as machinery and some valve springs.

En 49D is used for applications requiring higher stress than En e.g., 49C, engine valve springs.

WELDING

The welding of cold drawn wires is not recommended.

MACHINABILITY

The machinability of these steels in the softened condition is about 50% of that for mild steel (En 3).

HOT WORKING AND HEAT TREATMENT TEMPERATURES

Where applicable the temperatures recommended for En 49 will be the same as those given for En 8, En 9 (Vol. 1), and En 42, En 43, depending upon the carbon content.

0·35—0·45% C	refer to En 8.
0·45—0·60% C	refer to En 43.
0·50—0·60% C	refer to En 9.
0·70—0·85% C	refer to En 42.

PHYSICAL PROPERTIES

Few data are available for the physical properties of En 49 in the cold drawn condition. Data for steels in the heat treated condition can be obtained by reference to En 8, En 9 (Vol. 1), and En 42, En 43 depending upon the carbon content.

YOUNG'S MODULUS (E); SHEAR MODULUS (G)

Chemical Composition (%)					Section Size dia. (in.)	Heat Treatment °C	Young's Modulus tons/sq.in.	Shear Modulus tons/sq.in.	Ref.
C	Si	Mn	S	P					
0·64	0·23	0·75	0·013	0·023	0·104	OH T350	13,350		1
0·74	0·16	0·48	0·015	0·024	0·104	Patented CD T350	13,000		"
0·82	0·24	0·64	0·027	0·010	0·104	" " T350	13,000		"
0·72	0·19	0·62	0·038	0·029	0·044	Patented CD	12,600		2
0·73	0·19	0·62	0·041	0·030	0·044	A700 (12hrs.)	13,100		"
0 60	0·22	0·71	0·045	0·045	wire	Cold drawn	13,250	5,150	3
0·71	0·12	0·59	0·010	0·040	0·160	Cold drawn		4,910	4

TRANSFORMATION CHARACTERISTICS

See also En 8, En 9 (Vol. 1) and En 42, En 43 (Vol. 3).

ISOTHERMAL TRANSFORMATION

Chemical Composition (%)	
C	Mn
0·63	0·87

McQuaid Ehn grain size 5—6 (ASTM).
Austenitizing Temperature 816°C.

(Ref. 5)

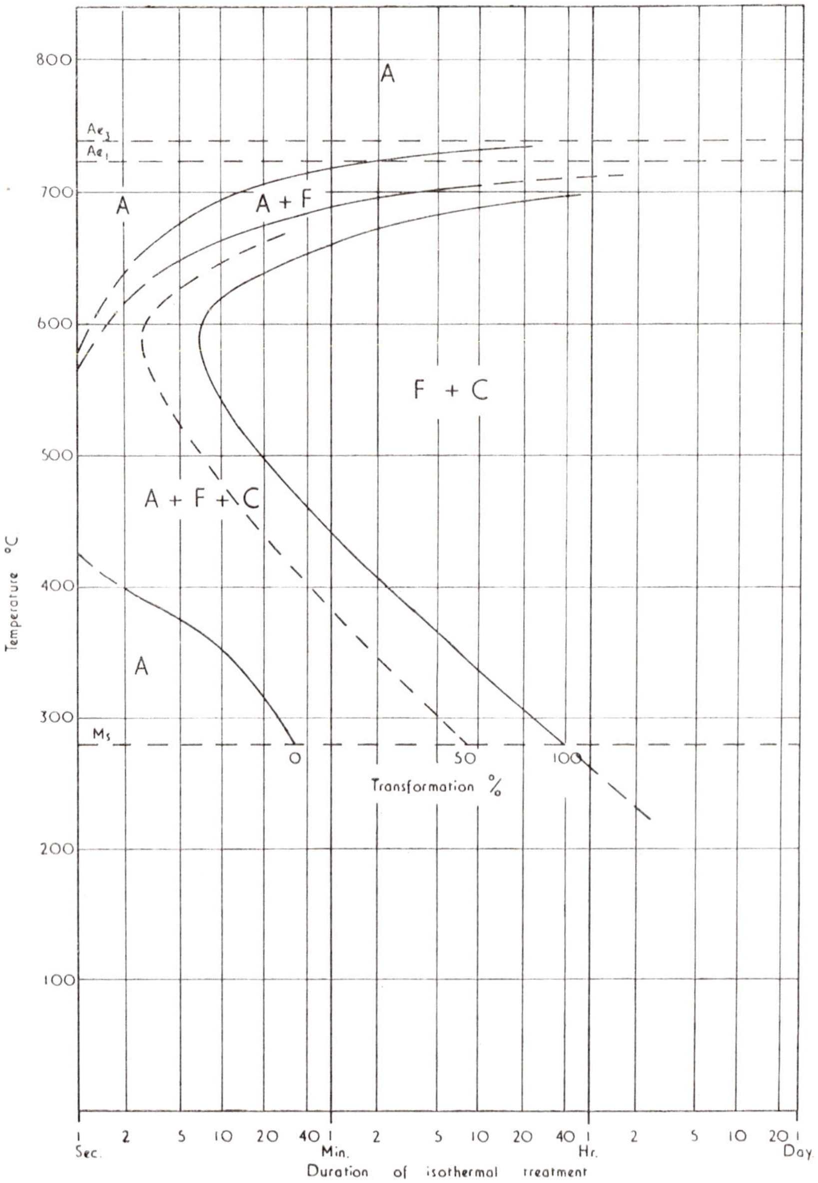

Fig. 49.1

HARDENABILITY

See also En 8, En 9 (Vol. 1) and En 42, En 43 (Vol. 3).

END QUENCH HARDENABILITY DATA

Chemical Composition (%)	
C	Mn
0·63	0·87

McQuaid Ehn grain size 5—6 (ASTM).
End quenched from 816°C.

(Ref. 5)

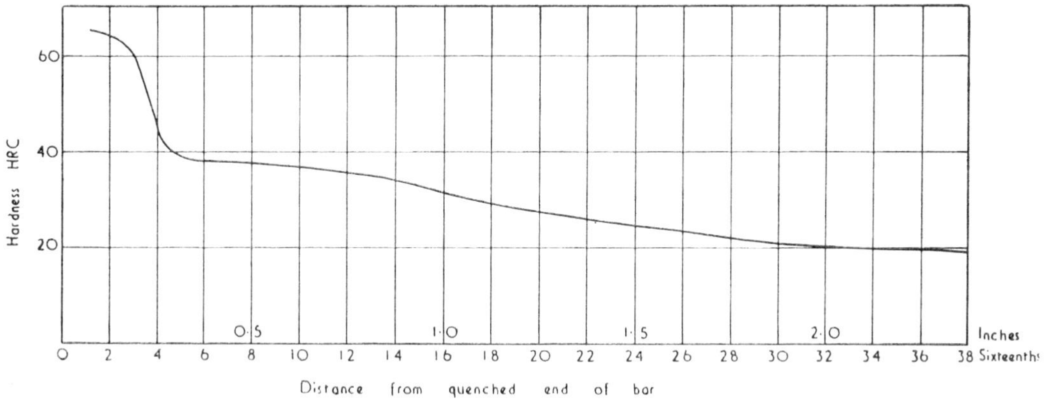

Fig. 49.2

En 49

MECHANICAL PROPERTIES AT ROOM TEMPERATURE

Process of Manufacture	Chemical Composition (%)					Section Size dia. (in.)	Heat Treatment °C	Tensile					Ref.
	C	Si	Mn	S	P			0·1% P.S. (0·1Rp)	0·2% P.S. (0·2Rp)	0·5% P.S. (0·5Rp)	T.S. (Rm)	El % (A)	
								tons/sq.in.					
A.O.H.	0·64	0·23	0·75	0·013	0·023	0·104	OH and T	95·7	96·8	97·5	103·0		1
"	"	"	"	"	"	"	" " T350 $\frac{1}{2}$hr.	97·3	99·3	101·6	103·0		"
												on 10″	
	0·65	0·23	0·72	0·024	"	0·192	Air patented				60·7	12	6
	"	"	"	"	0·032	0·166	" CD 1 hole				72·5	3	"
	"	"	"	"	"	0·144	" " 2 "				77·2	3	"
	"	"	"	"	"	0·125	" " 3 "				83·0	$2\frac{1}{2}$	"
	"	"	"	"	"	0·108	" " 4 "				90·8	$3\frac{1}{2}$	"
	"	"	"	"	"	0·093	" " 5 "				94·2	3	"
	"	"	"	"	"	0·080	" " 6 "				103·8	3	"
	"	"	"	"	"	0·069	" " 7 "				109·0	2	"
	"	"	"	"	"	0·060	" " 8 "				116·1	$2\frac{1}{2}$	"
	"	"	"	"	"	0·052	" " 9 "				124·1	2	"
	"	"	"	"	"	0·045	" " 10 "				134·8	2	"
	"	"	"	"	"	0·039	" " 11 "				148·2	$1\frac{1}{2}$	"
	"	"	"	"	"	0·034	" " 12 "				165·6	$1\frac{1}{2}$	"
	"	"	"	"	"	0·192	Lead patented				69·2	10	"
	"	"	"	"	"	0·166	" +CD, 1 hole				79·7	4	"
	"	"	"	"	"	0·144	" " 2 "				84·4	$3\frac{1}{2}$	"
	"	"	"	"	"	0·125	" " 3 "				90·4	$2\frac{1}{2}$	"
	"	"	"	"	"	0·108	" " 4 "				99·1	3	"
	"	"	"	"	"	0·093	" " 5 "				102·8	$3\frac{1}{2}$	"
	"	"	"	"	"	0·080	" " 6 "				116·0	3	"
	"	"	"	"	"	0·069	" " 7 "				122·8	$2\frac{1}{2}$	"
	"	"	"	"	"	0·060	" " 8 "				131·7	2	"
	"	"	"	"	"	0·052	" " 9 "				147·4	2	"
	"	"	"	"	"	0·045	" " 10 "				163·0	2	"
	"	"	"	"	"	0·039	" " 11 "				175·2	2	"
	"	"	"	"	"	0·034	" " 12 "				192·8	$1\frac{1}{2}$	"
A.O.H.	0·74	0·16	0·48	0·015	0·024	0·104	Patented & CD	76·0	85·4	96·9	105·0		1
"	"	"	"	"	"	"	" " +T350 $\frac{1}{2}$hr.	87·3	89·1	91·1	104·0		"
A.O.H.	0·82	0·24	0·64	0·027	0·010	0·144	Patented & CD	72·4	80·6	92·1	99·0		"
"	"	"	"	"	"	"	" " +T350 $\frac{1}{2}$hr.	81·2	83·9	86·3	97·0		"

Chemical Composition (%)					Section Size dia. (in.)	Heat Treatment °C	Tensile				Ref.
C	Si	Mn	S	P			Y.P. (Rs)	T.S. (Rm)	El % (A)	R.A. % (Z)	
							tons/sq.in.				
0·82	0·12	0·84	0·050	0·012	0·252	OQ843 T316	116·1	—	3	—	7
"	"	"	"	"	"	" T371	102·7	112·9	7	35	"
"	"	"	"	"	"	" T427	85·7	96·0	9	42	"
"	"	"	"	"	"	" T482	71·4	83·9	11	43	"
"	"	"	"	"	"	" T538	61·2	71·9	12	44	"
"	"	"	"	"	"	" T593	51·3	60·3	16	49	"
"	"	"	"	"	"	" T649	44·6	52·7	19	50	"
"	"	"	"	"	"	" T704	38·0	45·1	22	55	"

EFFECT OF TEMPERING ON THE MECHANICAL PROPERTIES AT ROOM TEMPERATURE

Chemical Composition (%)				
C	Si	Mn	S	P
0·82	0·12	0·84	0·050	0·012

0·252 in. dia. bar.
Oil quenched 843°C. Tempered as indicated.

(Ref. 7)

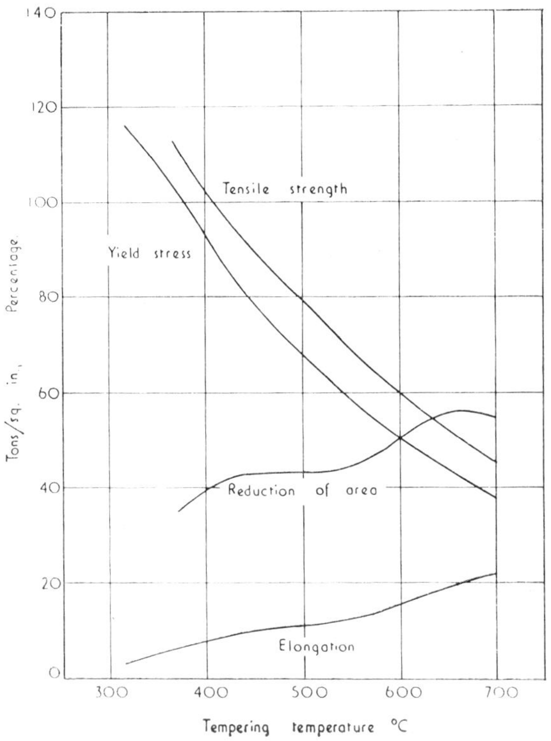

Fig. 49.3

MECHANICAL PROPERTIES AT LOW TEMPERATURE

Chemical Composition (%)					Section Size dia. (in.)	Heat Treatment °C	Test Temper-ature °C	Tensile					Ref.
C	Si	Mn	S	P				0·1% P.S. (0·1Rp) tons/sq.in.	0·5% P.S. (0·5Rp) tons/sq.in.	T.S. (Rm) tons/sq.in.	El % (A)	R.A. % (Z)	
0·4					3·6 mm		20			69	3½	55	8
"					"		—20			71	4	53	"
"					"		—60			74	4	52	"
0·44	0·16	0·72			2·03 mm		20			90·5	2½	62	"
"	"	"			"		—20			93·6	2	62	"
"	"	"			"		—50			95·1	2	61	"
0·68	0·13	0·70			4·03 mm		20			112	3½	53	"
"	"	"			"		—20			117	3	51	"
"	"	"			"		—50			119	2½	50	"
0·80					3·6 mm		20			110	2	43	"
"					"		—20			114	2	41½	"
"					"		—60			117	2	40	"
0·46	0·12	0·50	0·048	0·041	0·044	A700 12hrs.	20	27·7	29·5	38·2		73	9
"	"	"	"	"	"	"	—30	29·4	31·6	41·3		72½	"
"	"	"	"	"	"	"	—60	30·6	34·0	43·4		66½	"
0·47	0·10	0·48	0·050	0·041	0·044	Patented 950/980+CD88%	20	59·2	81·6	85·0		47	"
"	"	"	"	"	"	" "	—30	60·5	83·9	88·0		44	"
"	"	"	"	"	"	" "	—60	61·2	88·3	91·5		44½	"
0·54	0·16	0·68	0·034	0·028	0·043	Patented electrogalvanized +CD	20	71·0	96·4	98·9		54½	"
"	"	"	"	"	"		—30	75·1	99·6	103·3		19½	"
"	"	"	"	"	"		—60	80·1	103·5	106·2		50½	"
0·73	0·19	0·62	0·041	0·030	0·044	A700 12hrs.	20	40·0	41·1	46·5		60½	"
"	"	"	"	"	"	"	—30	40·8	43·5	48·7		56	"
"	"	"	"	"	"	"	—60	41·7	46·2	50·4		57½	"
0·72	0·19	0·62	0·038	0·029	0·044	Patented 950/980 CD88%	20	92·5	120·0	121·6		58½	"
"	"	"	"	"	"	" "	—30	93·0	124·5	126·0		48½	"
"	"	"	"	"	"	" "	—60	93·5	130·5	132·0		46	"
0·77	0·13	0·51	0·041	0·042	0·043	Patented electro-galvanized & CD.	20	89·8	119·0	127·5		50½	"
"	"	"	"	"	"	" "	—30	97·2	128·8	130·4		45½	"
"	"	"	"	"	"	" "	—60	105·7	134·2	137·1		47½	"

MECHANICAL PROPERTIES IN TORSION

Chemical Composition (%)					Section Size dia. (in.)	Heat Treatment °C	Tensile Strength (Rm) tons/sq.in.	Apparent max. Shear Strength tons/sq.in.	Probable max. Shear Strength tons/sq.in.	Ref.
C	Si	Mn	S	P						
0·64	0·23	0·75	0·013	0·023	0·104	OH and T	103	89	67	1
"	"	"	"	"	"	" +T350 ½hr.	103	87	65	"
0·74	0·16	0·48	0·015	0·024	0·104	Patented CD	105	89	67	"
"	"	"	"	"	"	" "+T350 ½hr.	104	86	64·5	"
0·82	0·24	0·64	0·027	0·010	0·144	Patented CD	99	70	52·5	"
"	"	"	"	"	"	" "+T350 ½hr.	97	73	55	"
0·71	0·12	0·59	0·010	0·040	0·160	Cold drawn	89	62	46·5	4

FATIGUE PROPERTIES

Chemical Composition (%)					Section Size dia. (in.)	Heat Treatment °C	Test Temperature °C	Tensile Strength (Rm) tons/sq.in	Rotating Bending Fatigue Limit tons/sq.in.			Ref.
C	Si	Mn	S	P					5×10⁴ cycles	5×10⁵ cycles	10⁷ cycles	
0·46	0·12	0·50	0·048	0·041	0·044	A700 12hrs.	20	38·2	±26·2	±20·8	±16·8	1
"	"	"	"	"	"	"	—30	41·3	±26·8	±21·1	±17·0	"
"	"	"	"	"	"	"	—60	43·4	±27·3	±21·7	±17·3	"
0·47	0·10	0·48	0·050	0·041	0·044	Patented CD 88%	20	85·0	±40·6	±28·5	±26·5	"
"	"	"	"	"	"	" " "	—30	88·0	±41·4	±28·5	±27·0	"
"	"	"	"	"	"	" " "	—60	91·5	±41·6	±30·5	±29·5	"
0·54	0·16	0·68	0·034	0·028	0·043	Patented electro-galvanized	20	98·9	±42·0	±29·0	±28·5	"
"	"	"	"	"	"	" "	—30	103·3	±39·5	±27·0	±27·0	"
"	"	"	"	"	"	" "	—60	106·2	±40·5	±25·5	±23·0	"
0·72	0·19	0·62	0·038	0·029	0·044	Patented CD 88%	20	121·6	±50·0	±35·5	±33·0	"
"	"	"	"	"	"	" "	—30	126·0	±48·6	±34·8	±34·0	"
"	"	"	"	"	"	" "	—60	122·0	±47·2	±35·2	±35·5	"
0·73	0·19	0·62	0·041	0·030	0·044	A700 12hrs.	20	46·5	±29·8	±23·8	±19·5	"
"	"	"	"	"	"	" "	—30	48·7	±28·8	±23·3	±19·5	"
"	"	"	"	"	"	" "	—60	50·4	±28·8	±22·3	±20·5	"
0·77	0·13	0·51	0·041	0·042	0·043	Patented electro-galvanized	20	127·5	±47·5	±33·0	±31·5	"
"	"	"	"	"	"	" "	—30	130·4	±43·4	±30·5	±30·0	"
"	"	"	"	"	"	" "	—60	137·1	±43·6	±28·5	±28·0	"
0·79	0·25	0·49	0·011	0·024	0·144	Patented CD41·5%		100·3			±24·6	10
"	"	"	"	"	0·128	" CD69·5%		103·0			±26·3	"
"	"	"	"	"	0·116	" CD75%		108·6			±28·3	"
"	"	"	"	"	0·104	" CD80%		115·6			±30·9	"
"	"	"	"	"	0·092	" CD84·2%		124·5			±35·4	"
"	"	"	"	"	0·080	" CD88%		143·0			±38·1	"
"	"	"	"	"	0·144	" CD41½% T200		103·0			±26·4	"
"	"	"	"	"	0·128	" CD69½% "		106·5			±23·7	"
"	"	"	"	"	0·116	" CD75% "		110·1			±27·4	"
"	"	"	"	"	0·104	" CD80% "		120·5			±33·0	"
"	"	"	"	"	0·092	" CD84.2%		128·1			±40·1	"
"	"	"	"	"	0·080	" CD88%		140·0			±42·8	"

CORROSION FATIGUE

Chemical Compositio (%)					Section Size dia. (in.)	Heat Treatment °C	Tensile Strength (Rm)	Rotating Bending Fatigue Limit tons/sq.in. (3,000 rev/min.)			Ref.
C	Si	Mn	S	P				In Air	In Tap Water	In 3% Salt Water	
0·48	0·25	0·50	0·024	0·036	Bar	Normalized	45·5	±18·2	±11·3	± 7·5	11
"	"	"	"	"	"	Hot galvanized (30—60 μ)		±15·0	±16·3	±15·0	"
"	"	"	"	"	"	Elect. zinc (5—10 μ)		±17·5	±17·5	±15·0	"
"	"	"	"	"	"	Elect. cadmium (5—10 μ)		±18·7	±16·3	± 8·1	"
"	"	"	"	"	"	Tin coating (5—20 μ)		±17·5	±12·5	± 7·5	"
"	"	"	"	"	"	Nickel coating (20 μ)		±17·5	± 9·4	± 7·5	"
"	"	"	"	"	"	Chrome plating (20 μ Ni+ 0·25/1·25 μ Cr)		±17·5	±14·4	± 9·4	"

Chemical Composition (%)					Section Size dia. (in.)	Heat Treatment °C	Tensile Strength (Rm)	Rotating Bending Fatigue Limit tons/sq.in.			Ref.
									In sea water		
C	Si	Mn	S	P				In Air	10^7 cycles	10^8 cycles	
0·80	0·17	0·57	0·039	0·024	$\frac{3}{8}$	Lead Patented CD from 0·5″ dia.	90·5	±32	±4	±2·5	12

TORSIONAL FATIGUE

Chemical Composition (%)					Section Size dia. (in.)	Heat Treatment °C	Tensile Strength	Torsional Fatigue Limit	Ref.
C	Si	Mn	S	P			tons/sq.in.		
0·71	0·12	0·59	0·010	0·040	0·160	Cold drawn	89	16·9±10·1	4

LIST OF REFERENCES

1. Private correspondence with the Spring Manufacturers' Research Association.
2. R. W. Nichols. The Mechanical Properties of Carbon Steel Wire at Low Temperature. *JISI*, 1956, Vol. 182, No. 1, p. 337.
3. British Iron and Steel Research Association, unpublished data.
4. A. Swan, H. Sutton and W. D. Douglas. Investigation of Aircraft Engine Valve Springs. *Proceedings of the Institution of Mechanical Engineers*, 1931, Vol. 120, p. 261.
5. Atlas of Isothermal Transformation Diagrams. The United States Steel Corporation. Second Edition, 1957.
6. V. J. Vitelli. The Effect of Heat Treatment and Cold Work on the Mechanical Properties of High Carbon Steel Wire. *Wire and Wire Products*, 1962, Vol. 33, Dec., pp. 1763—1766.
7. H. Muir, B. L. Averbach and M. Cohen. The Elastic Limit and Yield Behaviour of Hardened Steels. *Transactions of the Amer. Soc. Metals*, 1955, Vol. 47.
8. A. Pomp and A. Krish. Mechanical Properties of Rope Wire in the Temperature Range +20° to —50°C. *Mitteilungen aus dem Kaiser-Wilhelm Inst. für Eisenforschung*, 1937, Vol. 19, No. 7, pp. 97—103.
9. War Office, unpublished report.
10. E. T. Gill, R. Goodacre. Some Aspects of the Fatigue Properties of Patented Steel Wires. *JISI*, 1935, Vol. 132, p. 143.
11. O. Forsman and Lundin. The Influence of Different Surface Coatings on the Fatigue Strength of Steel. *Proceedings of the First World Metallurgical Congress, Oct. 13—19, 1951. Amer. Soc. Metals*, 1952.
12. A. J. Gould and V. R. Evans. The Effect of Shot-peening upon the Corrosion Fatigue of a High Carbon Steel. *JISI*, 1948, Vol. 160, No. 3, p. 164.

CHROMIUM-VANADIUM STEEL FOR WIRE FOR VALVE SPRINGS

SPECIFICATION

CHEMICAL COMPOSITION (%)

C	Si	Mn	S	P	Cr	V
0·40—0·50	0·10—0·35	0·50—0·70	0·040max.	0·040max.	1·00—1·50	0·15min.

MECHANICAL PROPERTIES

No mechanical properties are specified for this steel.

RELATED SPECIFICATIONS

United Kingdom			America		France	Germany			Sweden	U.S.S.R.	
En No.	BS	DTD	SAE	AISI	AFNOR	Werkstoff No.	Name	DIN	SIS	GOST	Mark
50	1429	4A	6150	6150	50CV4	1.7561	42CV6	1654 17200	14.2230	2052—53 4543—57	50KhFA 50KhFA
						1.8159	50CV4	17221 17222 17225			

APPLICATIONS

This steel has similar applications to those of En 47. It is used for engine valve springs for high duty service, e.g., aircraft engines.

WELDING

The welding characteristics are similar to those given for En 47.

MACHINABILITY

In the softened condition the machinability is approximately 45% of that for mild steel (En 3).

HOT WORKING AND HEAT TREATMENT TEMPERATURES

Forging, rolling and stamping	1150°C, finish above 850°C.
Annealing	850°C—900°C, furnace cool.
Normalizing	850°C—900°C, air cool.
Softening (sub-critical annealing)	680°C—730°C, air or furnace cool.
Hardening	850°C—900°C, oil quench.
Tempering	400°C—550°C, air cool.

PHYSICAL PROPERTIES

SPECIFIC GRAVITY (d)

Chemical Composition (%)								Heat Treatment °C	Specific Gravity at 20°C	Ref.
C	Si	Mn	S	P	Ni	Cr	V			
0·41	0·23	0·71	0·006	0·017	—	1·27	0·28	OQ and T	7·82	1
0·46	0·17	0·57	—	—	0·15	1·40	0·18	OQ850 T490	7·83	2

SPECIFIC HEAT (c)

Chemical Composition (%)						Heat Treatment °C	Specific Heat cal/g deg. C 20—100°C	Ref.
C	Si	Mn	Ni	Cr	V			
0·46	0·17	0·57	0·15	1·40	0·18	OQ850 T490	0·115	2

MEAN COEFFICIENT OF THERMAL EXPANSION (α)

(See also En 47)

Chemical Composition (%)						Heat Treatment °C	$10^6 \times$ Mean Coefficient of Expansion per deg. C						Ref.
C	Si	Mn	Ni	Cr	V		20to 100°C	20to 200°C	20to 300°C	20to 400°C	20to 500°C	20to 600°C	
0·46	0·17	0·57	0·15	1·40	0·18	OQ850 T490	12·6						2
(within specification)						—	—	12·8	13·4	13·9	14·2	14·5	3
,, ,,						—	12·0						4

ELECTRICAL RESISTIVITY (ϱ)

Chemical Composition (%)						Heat Treatment °C	Electrical Resistivity microhm-cm 20°C	Ref.
C	Si	Mn	Ni	Cr	V			
0·46	0·17	0·57	0·15	1·40	0·18	OQ850 T490	23·0	2

THERMAL CONDUCTIVITY (k)

Chemical Composition (%)						Heat Treatment °C	Thermal Conductivity cal/cm s deg. C				Ref.
C	Si	Mn	Ni	Cr	V		20°C	200°C	400°C	600°C	
0·46	0·17	0·57	0·15	1·40	0·18	OQ850 T490	0·110				2
(within specification)								0·102	0·098	0·095	3

YOUNG'S MODULUS (E); SHEAR MODULUS (G); POISSON'S RATIO (σ)

(See also En 47)

Chemical Composition (%)							Heat Treatment °C	Young's Modulus tons/sq.in.		Shear Modulus tons/sq.in.		Poisson's Ratio	Ref.
C	Si	Mn	S	P	Cr	V		Static	Dynamic	Static	Dynamic		
0·41	0·23	0·71	0·006	0·017	1·27	0·28	OQ850 T700	13,550	13,800	5,400	5,360	0,287	1
(within specification)							OQ and T			5,270			3

TRANSFORMATION CHARACTERISTICS

(See also En 47)

ISOTHERMAL TRANSFORMATION

Chemical Composition (%)			
C	Mn	Cr	V
0·45	0·70	1·00	0·20

Austenitizing Temperature 900°C.

(Ref. 5)

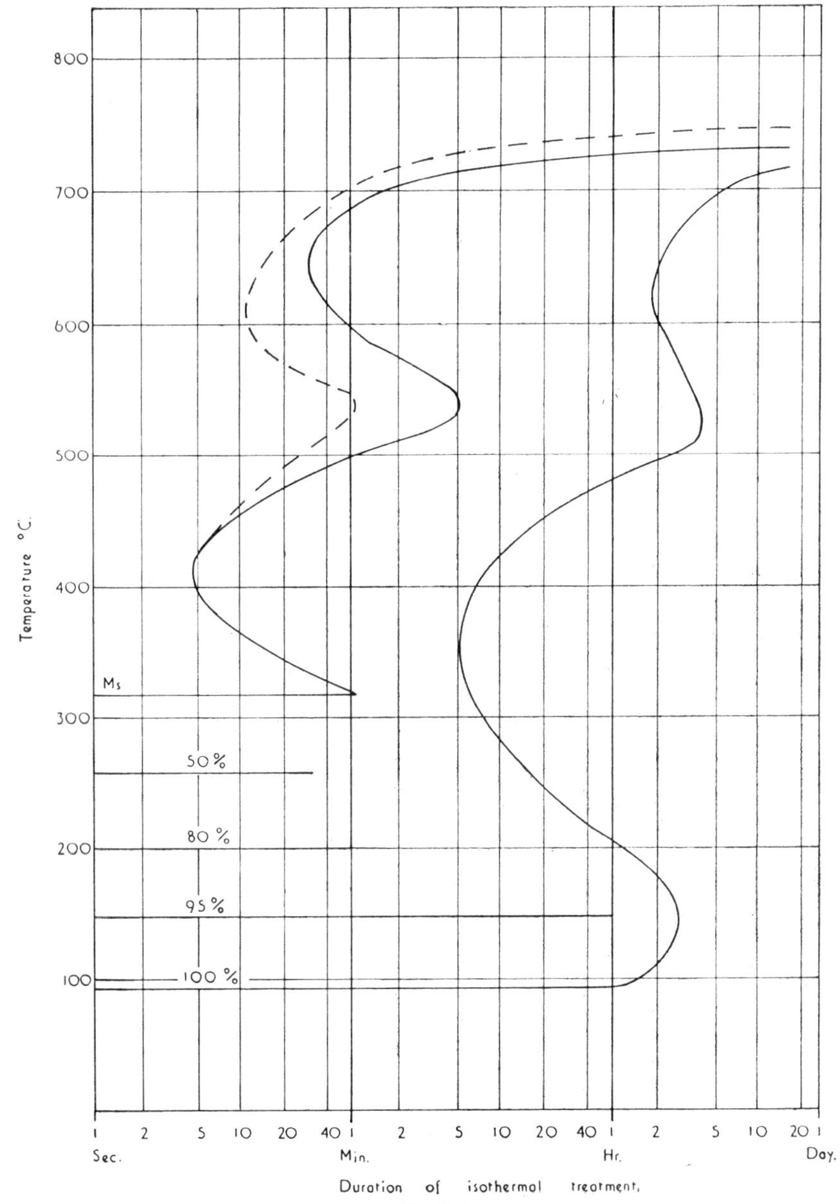

Fig. 50.1

HARDENABILITY

(See also En 47)

END QUENCH HARDENABILITY DATA

Chemical Composition (%)								
C	Si	Mn	S	P	Ni	Cr	Mo	V
0·45	0·19	0·65	0·017	0·016	0·23	0·31	0·07	0·20

Basic Electric Arc.
McQuaid Ehn grain size 7 (ASTM).
As quenched grain size 7—8 (ASTM).
Bars normalized 870°C. End quenched from 870°C.

(Ref. 6)

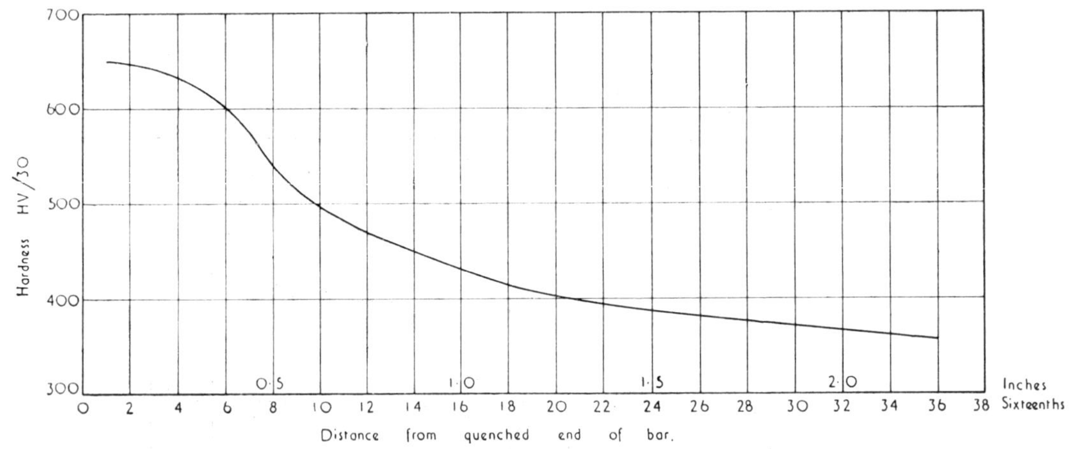

Fig. 50.2

HARDNESS TRAVERSE CURVES

Chemical Composition (%)								
C	Si	Mn	S	P	Ni	Cr	Mo	V
0·45	0·19	0·65	0·017	0·016	0·23	1·31	0·07	0·20

Basic Electric Arc.
McQuaid Ehn grain size 7 (ASTM).
As quenched grain size 7—8 (ASTM).
Bars normalized 870°C. Oil quenched 870°C.
Tempered 540°C and 620°C.

Oil quenched 870°C. (Ref. 6)

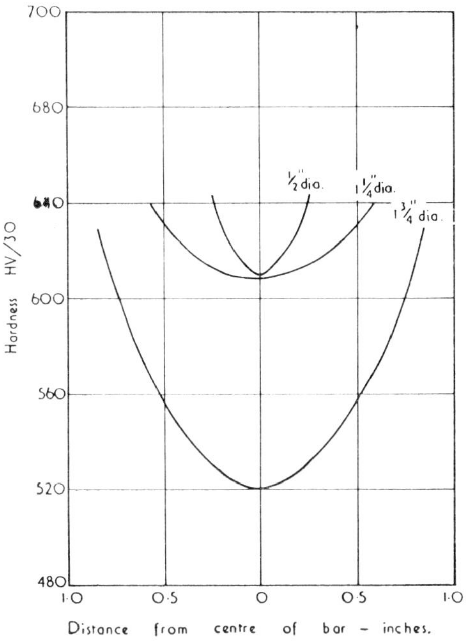

Fig. 50.3

Oil quenched 870°C. Tempered 540°C. (Ref. 6)

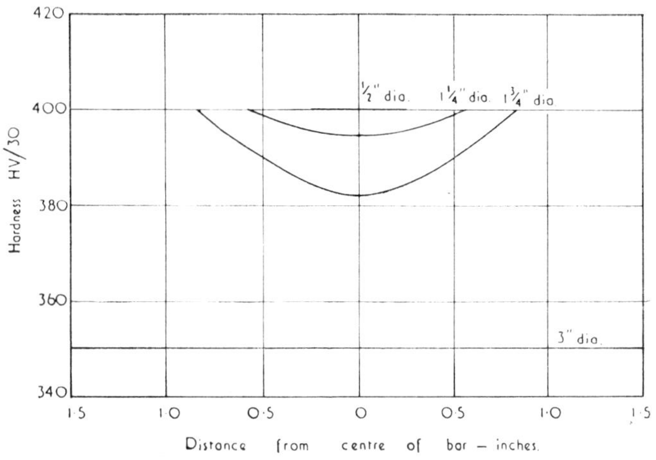

Fig. 50.4

Oil quenched 870°C. Tempered 620°C. (Ref. 6)

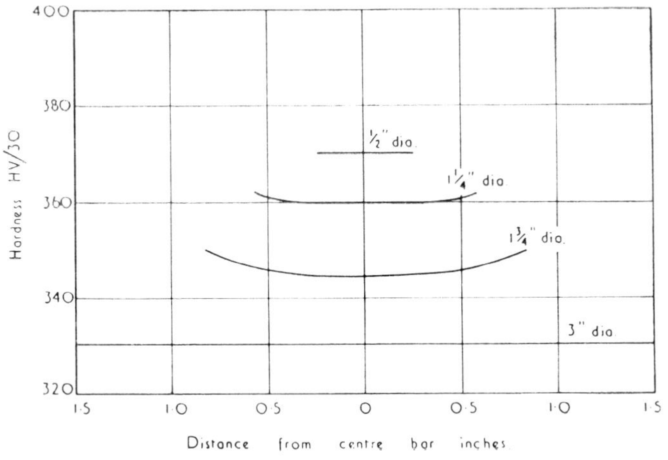

Fig. 50.5

En 50

MECHANICAL PROPERTIES AT ROOM TEMPERATURE

(See also En 47)

Process of Manu-facture	Chemical Composition (%)								Section Size dia. (in.)	Heat Treatment °C	Tensile				Izod ft.lb.	Ref.
	C	Si	Mn	S	P	Ni	Cr	V			Y.P. (Rs) tons/sq.in.	T.S. (Rm) tons/sq.in.	El % (A)	R.A. % (Z)		
	0·40	0·14	0·65				1·33	0·18	1 1/8	OQ850 T500	65·9	87·7	15	49 1/2	14	7
	"	"	"				"	"	"	" T600	60·9	65·1	21	57	33	"
	"	"	"				"	"	"	" T700	47·0	53·7	26 1/2	66	99	"
	0·42		0·75				1·04	0·15	1 1/8	OQ870 T400	69	83·0	13 1/2	55	16	8
	"		"				"	"	"	" T500	75	80·9	11	54	22	"
	"		"				"	"	"	" T600	63	70·5	15	56	19	"
	"		"				"	"	"	" T650	59	65·0	18 1/2	57 1/2	77	"
	"		"				"	"	"	" T700	45	52·2	25 1/2	61	90	"
	"		"				"	"	"	" T750	40	46·2	30 1/2	64	117	"
	0·44		0·65			0·25	1·20	0·17	5/8	OQ850 T400	93·6	107·6	11 1/2		10	9
	"		"			"	"	"	"	" T430	88·0	97·6	15		19	"
	"		"			"	"	"	"	" T460	86·0	93·6	15		19	"
B.E.A.	0·45	0·22	0·62	0·009	0·018	0·24	1·17	0·19	1 5/8	OQ850 T440	64·0	72·0	15	49	—	2
"	"	"	"	"	"	"	"	"	"	OQ880 T440	66·8	73·6	15	48 1/2	8	"
B.E.A.	0·45	0·35	0·64	0·006	0·013	0·25	1·18	0·18	0·505	OQ870 T200	128·0	135·0	12	32	12	"
"	"	"	"	"	"	"	"	"	"	OQ850	107·5	144·0	—	—	5	"
"	"	"	"	"	"	"	"	"	"	OQ850 T200	115·0	137·5	—	—	9	"
"	"	"	"	"	"	"	"	"	"	OQ830	124·0	139·5	—	—	6	"
"	"	"	"	"	"	"	"	"	"	OQ830 T200	122·0	134·0	—	—	10	"

MECHANICAL PROPERTIES AT ROOM TEMPERATURE

(Continued)

Process of Manu-facture	ASTM Grain Size		Chemical Composition (%)									Section Size dia. (in.)	Heat Treatment °C	Tensile				Izod ft.lb.	Ref.
	Mc Quaid Ehn	Actual	C	Si	Mn	S	P	Ni	Cr	Mo	V			Y.P. (Rs) tons/sq.in.	T.S. (Rm) tons/sq.in.	E l % (A)	R.A. % (Z)		
B.E.A.	7	7—8	0·45	0·19	0·65	0·017	0·016	0·23	1·31	0·70	0·20	½ (C)	N870 OQ870 T540 T620 OQ	76·0	79·5	18	49	41	6
"	"	"	"	"	"	"	"	"	"	"	"	(M)	" " " " "	76·0	78·5	18½	54	38	"
"	"	"	"	"	"	"	"	"	"	"	"	1¼ (C)	" " " " "	69·5	73·5	20	58	45	"
"	"	"	"	"	"	"	"	"	"	"	"	(M)	" " " " "	71·5	75·0	18	55	40	"
"	"	"	"	"	"	"	"	"	"	"	"	1¾ (C)	" " " " "	63·5	69·4	19	53½	44	"
"	"	"	"	"	"	"	"	"	"	"	"	(M)	" " " " "	67·2	71·9	20	56	42	"
"	"	"	"	"	"	"	"	"	"	"	"	(O)	" " " " "	66·4	70·4	20	58	41	"
"	"	"	"	"	"	"	"	"	"	"	"	3 (C)	" " " " "	58·9	66·2	17	54½	38	"
"	"	"	"	"	"	"	"	"	"	"	"	(M)	" " " " "	59·0	67·2	20	57	35	"
"	"	"	"	"	"	"	"	"	"	"	"	(O)	" " " " "	58·4	66·1	20	58½	37	"
			0·46	0·18	0·45			0·19	1·45		0·19	—	OQ850 T450	89·5	107·3	8½	24	14	2

EFFECT OF TEMPERING ON THE MECHANICAL PROPERTIES AT ROOM TEMPERATURE

Chemical Composition (%)			
C	Mn	Cr	V
0·42	0·75	1·04	0·15

1⅛ in. dia. bars. Oil quenched 870°C.

(Ref. 8)

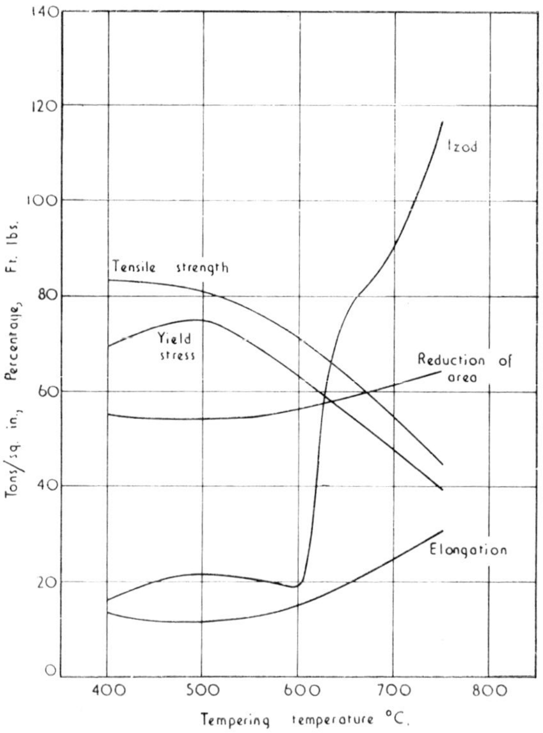

Fig. 50.6

MECHANICAL PROPERTIES AT HIGH TEMPERATURE

Process of Manu-facture	Chemical Composition (%) C	Si	Mn	S	P	Ni	Cr	V	Section Size dia. (in.)	Heat Treatment °C	Test Temper-ature °C	Tensile 0·1% P.S. (0·1Rp) tons/sq.in.	0·5% P.S. (0·5Rp) tons/sq.in.	Y.P. (Rs) tons/sq.in.	T.S. (Rm) tons/sq.in.	El % (A)	R.A. % (Z)	Ref.
A.O.H.	0·43	0·22	0·59	0·039	0·034	0·32	1·14	0·18	¾	OH850 T650	RT		55·5		61·0	21½	61½	2
"	"	"	"	"	"	"	"	"	"	" "	100		57·5		62·6	20	63	"
"	"	"	"	"	"	"	"	"	"	" "	200		53·1		59·6	15½	61½	"
"	"	"	"	"	"	"	"	"	"	" "	300		49·2		55·6	20	59	"
"	"	"	"	"	"	"	"	"	"	" "	400		46·6		56·3	15½	61½	"
"	"	"	"	"	"	"	"	"	"	" "	500		27·0		32·6	26½	74½	"
"	"	"	"	"	"	"	"	"	"	" "	600				24·8	30½	87	"
"	"	"	"	"	"	"	"	"	"	" "	700				11·6	52	92	"
"	"	"	"	"	"	"	"	"	"	" "	800				6·7	68	95	"
"	"	"	"	"	"	"	"	"	"	" "	900				4·7	81½	96	"
"	"	"	"	"	"	"	"	"	"	" "	1000				3·1	73½	97	"
	0·43	0·35	0·59				1·06	0·18	¾	OQ840 T590	RT			65	70	20	58	10
	"	"	"				"	"	"	" "	100			(64)	(68)	(20½)	(59)	"
	"	"	"				"	"	"	" "	200			(62)	(65)	(20½)	(60)	"
	"	"	"				"	"	"	" "	300			(58)	(61)	(21)	(64)	"
	"	"	"				"	"	"	" "	400			50	54	21	73	"
	"	"	"				"	"	"	" "	500			28	45	23	84	"
	"	"	"				"	"	"	" "	600			11	29	32	91	"
	"	"	"				"	"	"	" "	650			5	18	40	93	"
	(within specification)								0·160	OQ and T	20	82·5			90	15		3
	" "								"	" "	200	80·5			88·0	17		"
	" "								"	" "	300	73·0			83·5	19		"
	" "								"	" "	400	58·0			70·5	19		"
	" "								"	" "	500	27·0			51·5	24		"

CREEP PROPERTIES

Chemical Composition (%)	Section Size dia. (in.)	Heat Treatment °C	Time hrs.	Strain %	Creep Stress tons/sq. in. Test Temperature °C 300	350	400	Ref.
(within specification)	0·357	OQ and T	100	0·1	54	32	—	3
" "	"	" "	"	0·5	—	—	31	"
" "	"	" "	"	1·0	—	—	36	"
" "	"	" "	300	0·1	52	30	—	"
" "	"	" "	"	0·5	—	—	29	"

En 50

MECHANICAL PROPERTIES IN TORSION

Process of Manufacture	Chemical Composition (%) C	Si	Mn	S	P	Ni	Cr	V	Section Size dia. (in.)	Heat Treatment °C	Tensile Y.P. (Rs) tons/sq.in.	T.S. (Rm) tons/sq.in.	El % (A)	R.A. % (Z)	Torsion Y.P.	Apparent max. Shear Strength tons/sq.in.	Probable max. Shear Strength tons/sq.in.	Ref.
B.E.A.	0·41	0·23	0·71	0·006	0·017		1·27	0·28		OQ850 T700	44·1	48·7	26½	66	30·8	46·2	34·7	11
	0·42	0·24	0·62	0·047	0·035	0·38	0·95	0·18		OQ850 T325	98·8	111·8	12½	45½		90·0	67·5	12
	"	"	"	"	"	"	"	"		" T450	85·5	90·3	14	52		74·3	55·7	"
A.O.H.	0·46	0·19	0·64	0·028	0·031	0·19	1·48	0·33		OQ850 T650	48·6	56·5	22	60½	36·8	52·0	39·0	2
	0·53	0·28	0·77	0·020	0·020		1·22	0·18	0·128	OQ840 T420		97·5				73·0	54·8	13

FATIGUE PROPERTIES

Process of Manufacture	Chemical Composition (%) C	Si	Mn	S	P	Ni	Cr	V	Section Size dia. (in.)	Heat Treatment °C	Tensile Y.P. (Rs) tons/sq.in.	T.S. (Rm) tons/sq.in.	Torsion Y.P.	Apparent max. Shear Stress tons/sq.in.	Probable max. Shear Stress tons/sq.in.	Fatigue limit tons/sq.in. Rotating Bending plain	Rotating Bending notched	Torsional plain	Torsional notched	Ref.
B.E.A.	0·41	0·23	0·71	0·006	0·017		1·27	0·28		OQ850 T700	44·1	48·7	30·8	46·2	34·7	±27·8	±14·0(1)	±16·7	±10·4(2)	11
	0·42	0·24	0·62	0·047	0·035		0·95	0·18		OQ850 T325	98·8	111·8		90·0	67·5	±55·0				12
	"	"	"	"	"			"		" T400	85·5	90·3		74·3	55·7	±46·2				"
A.O.H.	0·46	0·19	0·57			0·15	1·40	0·18		OQ850 T490	82·5	87·0				±42·0				2
	0·53	0·28	0·77	0·020	0·020		1·22	0·18	0·128	OQ840 T420		97·5		73·0	54·8			23·75± 14·25		13

(1) Notch 55°V, 0·02 in. deep, O.D.O. 34 in. dia., root radius 0·00035 in.
(2) " " " " " " " 0·005 in.

LIST OF REFERENCES

1. M. F. Markham. Measurement of Elastic Constants by the Ultrasonic Pulse Method. *British Journal of Applied Physics*, Supplement No. 6, Physics of Non-destructive Testing, 1957, p. S62.
2. Private correspondence with Firth-Brown Ltd.
3. Private correspondence with Rolls-Royce Ltd.
4. Private correspondence with the English Steel Corporation Ltd.
5. Data Sheet, Crucible Steel Co. of America, Dec., 1947.
6. Symposium on the Hardenability of Steel. Organized by the Technical Advisory Committee of the Special and Alloy Steels Committee (Ministry of Supply) and The Iron and Steel Institute. Special Report No. 36 of The Iron and Steel Institute, 1946.
7. Private correspondence with Hadfields Ltd.
8. L. Aitchison and W. I. Pumphrey. *Engineering Steels*. MacDonald and Evans, London, 1953.
9. B.S. 971: Commentary on the British Standard Wrought Steels, En Series. British Standards Institution, London, 1950.
10. War Office, unpublished report.
11. H. J. Gough, Engineering Steels under Combined Cyclic and Static Stresses, *Proceedings of the Institution of Mechanical Engineers*, 1949, Vol. 160, pp. 417—440.
12. A. S. Kenneford and G. C. Ellis. A Comparison of Six Spring Steels. *JISI*, 1950, Vol. 164. No. 1, p. 265.
13. A. Swan, H. Sutton and W. D. Douglas. Investigation of Aircraft Engine Valve Springs. *Proceedings of the Institution of Mechanical Engineers*, 1931, Vol. 120, p. 261.

3 PER CENT NICKEL VALVE STEEL

SPECIFICATION

CHEMICAL COMPOSITION (%)

C	Si	Mn	S	P	Ni	Cr
0·25—0·35	0·10—0·35	0·35—0·75	0·050 max.	0·050 max.	2·75—3·25	0·30 max.

MECHANICAL PROPERTIES

Izod impact value, ft. lb. min.	40
Brinell hardness number, HB, max.	229

APPLICATIONS

This steel is of similar composition to that of En 21 (Volume 2). En 51 is specifically designated as a valve steel and is suitable for inlet valves and low duty exhaust valves in internal combustion engines where the temperature conditions are not liable to cause excessive scaling.

WELDING

(See Introduction for key to symbols used in this table.)

	Welding Process					
	MA	G	IG	S	F	B
En 51	d	cx	g	e	b	h

Reproduced by kind permission of H.M.S.O.

MACHINABILITY

The machinability of this steel when heat treated to 45 to 55 tons/sq.in. is approximately 50 to 55% of that for mild steel (En 3). When heat treated to 55 to 65 tons/sq.in. the machinability is 45 to 50% of that for mild steel. In the softened condition the machinability is approximately 60% of that for mild steel.

HOT WORKING AND HEAT TREATMENT TEMPERATURES

Forging, rolling and stamping	1180°C, finish above 850°C.
Annealing	830°C—850°C, furnace cool.
Softening (subcritical annealing)	600°C—650°C, air cool.
Normalizing	830°C—860°C, air cool.
Hardening	830°C—860°C, oil (or water quench for thin sections or C in the lower range).
Tempering	550°C—650°C, air cool, oil or water quench.

MECHANICAL AND PHYSICAL PROPERTIES

The mechanical and physical properties of this steel are the same as those given for En 21. For information concerning all the properties of En 51 steel reference should be made to the data given for En 21 in Volume 2.

SILICON–CHROMIUM VALVE STEEL

SPECIFICATION

CHEMICAL COMPOSITION (%)

C	Si	Mn	S	P	Ni	Cr
0·40—0·50	3·00—3·75	0·30—0·60	0·040max.	0·040max.	0·50max.	7·50—9·50

MECHANICAL PROPERTIES

The Brinell hardness of hardened and tempered test bars shall be 255—293.

RELATED SPECIFICATIONS

United Kingdom	Germany		USSR	
En No.	Werkstoff No.	Name	GOST	Mark
52	1.4718	X45CrSi93	5632—61	4Ch9S2

APPLICATIONS

A silicon-chromium valve steel suitable for applications requiring good hot strength and resistance to creep up to 700°C. Typical applications are inlet and exhaust valves subjected to heavier duties and higher temperatures than those suitable for En 51. Valve seats, nuts and bolts for use at moderately elevated temperatures.

WELDING

(See Introduction for key to symbols used in this table.)

	Welding Process					
	MA	G	IG	S	F	B
En 52				m	b	m

MACHINABILITY

The machinability of this steel when hardened and fully tempered and in the softened condition is approximately 40% of that for mild steel (En 3).

HOT WORKING AND HEAT TREATMENT TEMPERATURES

Forging, rolling and stamping	1150°C, finish above 900°C.
Annealing (softening)	800°C— 850°C, furnace or air cool.
Hardening	1000°C—1050°C, air cool or oil quench.
Tempering	650°C— 850°C, air cool or oil quench.

PHYSICAL PROPERTIES

SPECIFIC GRAVITY (d)

Chemical Composition (%)						Heat Treatment °C	Specific Gravity at 20°C	Ref.
C	Si	Mn	Ni	Cr	Mo			
0·42	3·5	0·50	0·19	7·5		OQ950 T700	7·60	1
0·44	3·9	0·52	0·15	8·1		OQ950 T650	7·60	"
0·50	(3·0)			7·5	added		7·63	2
(Within specification)							7·61	3

SPECIFIC HEAT (c)

Chemical Composition (%)					Heat Treatment °C	Mean Specific Heat cal/g. deg. C					Ref.
C	Si	Mn	Cr	Mo		20 to 100°C	20 to 200°C	20 to 400°C	20 to 600°C	20 to 800°C	
0·45	3·40	0·50	8·0		Softened	0·115					1
0·50	(3·0)		7·5	added			0·126	0·136	0·149	0·164	2

MEAN COEFFICIENT OF THERMAL EXPANSION (α)

Chemical Composition (%)						Heat Treatment °C	$10^6 \times$ Mean Coefficient of Thermal Expansion per deg. C									Ref.
C	Si	Mn	Ni	Cr	Mo		20 to 100°C	20 to 200°C	20 to 300°C	20 to 400°C	20 to 500°C	20 to 600°C	20 to 700°C	20 to 800°C	20 to 900°C	
0·45	3·30	—		8·50				12·6		12·9		13·4		13·7		4
0·50	3·04	0·37	0·16	8·28		Softened	12·6	12·8	12·9	12·9	13·2	13·5	13·6	13·7	13·8	1
0·50	(3·0)			7·5	added			12·1	12·6	12·9	13·3	13·6	13·9	14·3		2
(Within specification)							12·6			12·9		13·4	13·6			3

En 52

THERMAL CONDUCTIVITY (k)

Chemical Composition (%)						Heat Treatment °C	Thermal Conductivity cal/cm sec. deg. C										Ref.
C	Si	Mn	Ni	Cr	Mo		20°C	100°C	200°C	300°C	400°C	500°C	600°C	700°C	800°C	900°C	
0·42	3·5	0·50	0·19	7·5		OQ950 T650	0·030									0·075	1
0·45	3·5			9·0				0·030						0·060			4
0·45	3·4	0·50		8·0		Softened	0·053									0·075	1
0·50	(3·0)			7·5	added			0·042	0·047	0·051	0·055	0·058	0·059	0·060	0·060		2

ELECTRICAL RESISTIVITY (ϱ)

Chemical Composition (%)						Heat Treatment °C	Electrical Resistivity microhm-cm				Ref.
C	Si	Mn	Ni	Cr	Mo		20°C	100°C	200°C	900°C	
0·42	3·5	0·50	0·19	7·5		OQ950 T650	75				1
0·45	3·4	0·50		8·0			80			110	"
0·50	(3·0)			7·5	added	Softened	77	80	84		2

YOUNG'S MODULUS (E)

Chemical Composition (%)					Heat Treatment °C	Young's Modulus (dynamic*) tons/sq.in.						Ref.
C	Si	Mn	Ni	Cr		20°C	200°C	400°C	600°C	800°C	950°C	
0·43	3·40	0·48	0·29	7·7	OQ990 T750	13,700	13,100	12,000	10,600	7,800	6,600	1

* Vibration method.

MECHANICAL PROPERTIES AT ROOM TEMPERATURE

Chemical Composition (%)						Section Size dia. (in.)	Heat Treatment °C	Tensile						Izod ft.lb.	Ref.
C	Si	Mn	Ni	Cr	Mo			0·1% P.S. (0·1Rp) Tons/sq.in.	0·5% P.S. (0·5Rp) Tons/sq.in.	Y.P. (Rs) Tons/sq.in.	T.S. (Rm) Tons/sq.in.	El % (A)	R.A. % (Z)		
0·42	3·50	0·50	0·19	7·50		¾	OQ950 T650			44·4	60·7	25	33½		1
0·44	3·90	0·52	0·15	8·10		1⅛	OQ950 T700	45·0	53·2		66·2	16	32		1
0·44	3·30	0·36		9·18		1½	OQ1050 T850				56·0	23		8	5
0·45	3·32	0·41	0·50	7·70		1	OQ1050 T850			46·3	59·2	28½	63	14	2
0·46	3·30	0·43		9·20		1⅛	OQ1050 T850				57·6	25		10	5
0·48	3·54	0·59	0·03	9·16			A843				61	23	47		6
0·50	(3·0)			7·5	added		H & T				58·9	27½	59	18	2
0·51	3·16	0·38	0·16	8·26			OQ950 T750		46·9		60·2	26½	55		1

EFFECT OF TEMPERING ON HARDNESS

Chemical Composition (%)		
C	Si	Cr
0·45	3·50	9·0

$1\frac{1}{8}$ in. dia. bars. Oil quenched from 1050°C. Tempered as indicated for one hour.

(Ref. 4)

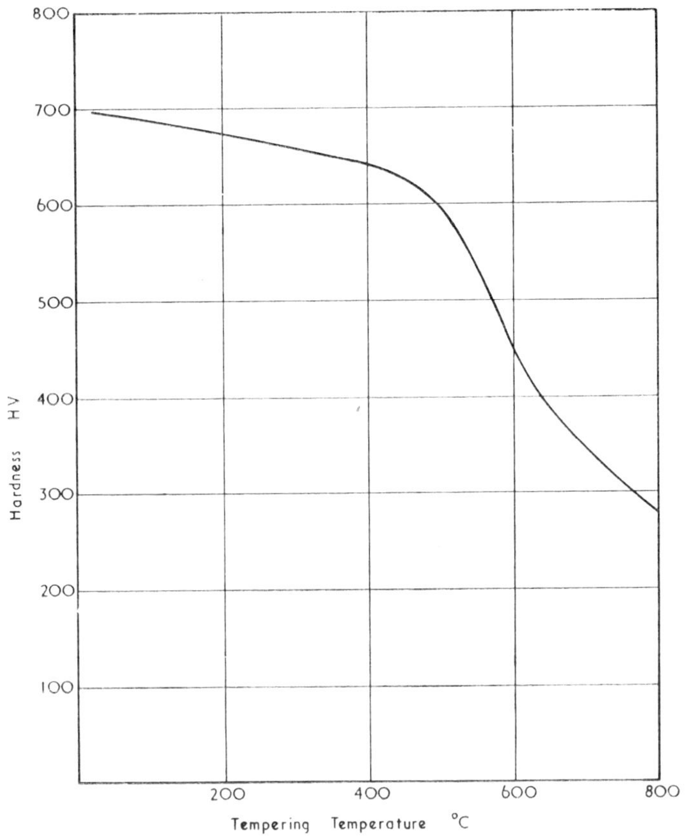

Fig. 52.1

MECHANICAL PROPERTIES AT HIGH TEMPERATURE

Chemical Composition (%)							Section Size dia. (in.)	Heat Treatment °C	Test Temperature °C	Tensile							Izod ft.lb.	Ref.
C	Si	Mn	S	P	Ni	Cr				0·1% P.S. (0·1Rp)	0·2% P.S. (0·2Rp)	0·5% P.S. (0·5Rp)	Y.P. (Rs)	T.S. (Rm)	El % (A)	R.A. % (Z)		
										Tons/sq.in.								
0·42	3·50	0·50	0·013	0·030	0·19	7·47	¾	OQ950 T650	RT			44·4		60·7	25	53½		1
,,	,,	,,	,,	,,	,,	,,	,,	,, ,,	400	27·4		30·6		46·9	27	60		,,
,,	,,	,,	,,	,,	,,	,,	,,	,, ,,	500	21·0		22·8		33·6	40	75		,,
,,	,,	,,	,,	,,	,,	,,	,,	,, ,,	600	9·8		11·4		17·6	60½	93		,,
,,	,,	,,	,,	,,	,,	,,	,,	,, ,,	700					9·6	—	—		,,
,,	,,	,,	,,	,,	,,	,,	,,	,, ,,	800					6·0	66	99		,,
,,	,,	,,	,,	,,	,,	,,	,,	,, ,,	900					3·1	—	—		,,
								Strain rate	0·085 in./min to Y.P. then 0·125 in./min on 2 in. gauge length									
0·43	3·23	0·33	0·020	0·021	0·08	8·43		H & T	RT		47·8			62·3	15½	47		7
,,	,,	,,	,,	,,	,,	,,		,,	600		12·1			22·5	45½	90		,,
0·45	3·30	0·50				8·50		OQ1050 T850	20					60	26		9	4
,,	,,	,,				,,		,, ,,	600					20	57			,,
,,	,,	,,				,,		,, ,,	700					9	67			,,
,,	,,	,,				,,		,, ,,	800					2	80			,,
								Strain rate	0·028 in./min.									
0·48	3·54	0·59			0·03	9·16		A843	21					61	23	47		6
,,	,,	,,			,,	,,		,,	399					49	18	45		,,
,,	,,	,,			,,	,,		,,	482					39	25	60		,,
,,	,,	,,			,,	,,		,,	538					31	28	70		,,
,,	,,	,,			,,	,,		,,	590					20	34	88		,,
,,	,,	,,			,,	,,		,,	677					11	46	98		,,
									815					4	66	99		,,
0·51	3·04	0·37			0·16	8·28	¾	OQ1075 T900	RT				46·0	54·0	5	5		1
,,	,,	,,			,,	,,	,,	,, ,,	800				,,	4·0	100	99		,,

CREEP PROPERTIES

Chemical Composition (%)							Heat Treatment °C	Test Temperature °C	Stress tons/sq.in. to give creep rate of 10^{-6} per hr. from 24 to 72 hours	Ref.
C	Si	Mn	S	P	Ni	Cr				
0·40	3·88	0·35			0·16	8·19	OQ950 T750	400	19·0	1
"	"	"			"	"	" "	500	4·0	"
"	"	"			"	"	" "	600	1·0	"
"	"	"			"	"	" "	700	0·13	"
"	"	"			"	"	" "	800	0·18	"
0·51	3·16	0·38			0·16	8·26	OQ950 T750			
"	"	"			"	"	" "	400	19·0	1
"	"	"			"	"	" "	500	2·5	"
0·43	3·38	0·48	0·008	0·014	0·29	7·66	OQ975 T700	650	Stress tons/sq.in. to give fracture in 4 mins. 12·0	1

FATIGUE PROPERTIES

Chemical Composition (%)					Section Size dia. (in.)	Heat Treatment °C	Test Temperature °C	Tensile				Rotating Bending Fatigue Limit tons/sq.in.	Ref.
C	Si	Mn	Ni	Cr				0·1% P.S. (0·1Rp) Tons/sq. in.	0·5% P.S. (0·5Rp) Tons/sq. in.	T.S. (Rm) Tons/sq. in.	El % (A)		
0·44	3·9	0·52	0·15	8·1	$1\frac{1}{8}$	OQ950 T700	RT	45·0	53·2	66·2	16	±24·0	1

LIST OF REFERENCES

1. Private correspondence with Firth-Vickers Stainless Steels Ltd.
2. Private correspondence with The United Steel Companies Ltd.
3. Private correspondence with Low Moor Alloy Steelworks Ltd.
4. Private correspondence with Jessop-Saville Ltd.
5. British Standard 971: Commentary on the British Standard Wrought Steels, En Series. British Standards Institution, London, 1950.
6. C. L. Clarke and A. E. White. The Stability of Metals at Elevated Temperatures. *University of Michigan Engineering Research Bulletin*, No. 11, Nov., 1928.
7. F. Korber and A. Pomp. Comparative Investigations on Heat-resisting at High Temperatures. *Mitt. K.-Wilh.-Inst. Eisenforsch.*, vol. 18, No. 18, 1936, 247—251.

SILICON–CHROMIUM VALVE STEEL

SPECIFICATION

CHEMICAL COMPOSITION (%)

C	Si	Mn	S	P	Ni	Cr
0·55—0·65	1·40—1·70	0·30—0·60	0·050 max.	0·050 max.	0·50 max.	5·75—6·75

MECHANICAL PROPERTIES

Izod impact value, ft. lb. min.	12
Brinell hardness number, HB.	235—285

APPLICATIONS

This steel, which is of similar type to En 52, is not used to any great extent. Typical applications are inlet and exhaust valves for automobile and low powered aircraft engines. The steel has a lower scaling resistance above 700°C than En 52 steel.

WELDING AND MACHINABILITY

Similar to that of En 52.

HOT WORKING AND HEAT TREATMENT TEMPERATURES

Forging, rolling and stamping	1100°C, finish above 900°C.
Annealing (softening)	800°C— 850°C, furnace cool.
Hardening	950°C—1000°C, air cool or oil quench.
Tempering	720°C— 800°C, air cool or oil quench.

MECHANICAL AND PHYSICAL PROPERTIES

Because of the small usage of this steel few data are available on the physical and mechanical properties. The following values have, however, been obtained on steel supplied to the En 53 specification.

MEAN COEFFICIENT OF THERMAL EXPANSION (α)

Chemical Composition (%)	$10^6 \times$ Mean Coefficient of Thermal Expansion per deg. C				Ref.
	20 to 200°C	20 to 400°C	20 to 600°C	20 to 700°C	
Within specification	12·2	13·0	13·6	13·9	1

MECHANICAL PROPERTIES AT ROOM AND HIGH TEMPERATURE

Chemical Composition (%)				Section Size dia. (in.)	Heat Treatment °C	Test Temperature °C	Tensile		Izod ft.lb.	Ref.
C	Si	Mn	Cr				T.S. tons/sq.in.	El %		
0·59	1·58	0·60	5·76	$1\frac{1}{8}$	AH980 T780	RT	54·0	25	25	2
0·59	1·54	0·52	6·06	$1\frac{5}{8}$	AH980 T780	RT	55·2	24	23	"
0·60	1·55		5·80		AH980 T780	RT	55	24	15	1
"	"		"		" "	600	20	50		"
"	"		"		" "	700	8	100		"
						800	4	140		"

LIST OF REFERENCES

1. Private correspondence with Jessop-Saville Ltd.
2. British Standard 971: Commentary on the British Standard Wrought Steels, En Series. British Standards Institution, London, 1950.

HIGH NICKEL–CHROMIUM–TUNGSTEN VALVE STEEL

SPECIFICATION

CHEMICAL COMPOSITION (%)

En No.	C	Si	Mn	S	P	Ni	Cr	Mo*	W	Nb*
54	0·35—0·50	1·00—2·50	1·50 max.	0·045 max.	0·045 max.	10·0 min.	12·00—16·0		2·00—4·00	
54A	0·37—0·47	1·00—2·00	0·50—0·80	0·045 max.	0·045 max.	13·0—15·0	13·0—15·0	0·40—0·60	2·2—3·0	0·16—0·22

* Where specified in order.

MECHANICAL PROPERTIES

	En 54	En 54A
Izod impact value ft.lb. min.	15	15
Brinell hardness number, HB, max.	302	269

RELATED SPECIFICATIONS

United Kingdom	
En No.	BS Air
54	S111
54A	S111

APPLICATIONS

An austenitic chromium-nickel-tungsten valve steel suitable for heat-resisting duties requiring strength and resistance to scaling at temperatures up to 800°C. Typical uses are high duty exhaust valves for aircraft, motor car and motor cycle engines, furnace impellers, bars and carriers, bolts and stays, tube bending mandrels, tube piercing points and screw conveyors where hard facing is employed to give wear resistance.

WELDING

(See Introduction for key to symbols used in this table.)

	Welding Process					
	MA	G	IG	S	F	B
En 54				m	a	m
En 54A				m	a	m

Reproduced by kind permission of H.M.S.O.

MACHINABILITY

In the softened condition the machinability is approximately 30% of that of mild steel (En 3).

HOT WORKING AND HEAT TREATMENT TEMPERATURES

Forging, rolling and stamping	1150°C, finish above 900°C.
Softening	950°C—1020°C, cool quickly in air, oil or water.

This steel is not hardenable by heat treatment.

PHYSICAL PROPERTIES

SPECIFIC GRAVITY (d)

Chemical Composition (%)						Heat Treatment °C	Specific Gravity at 20°C	Ref.
C	Si	Mn	Ni	Cr	W			
0·40	0·97	0·41	10·3	13·6	3·53	AC950	8·00	1
0·42	1·50	0·70	14·0	14·0	2·60	AC950	8·00	"
0·44	1·10	0·71	14·1	14·0	2·05	AC950	7·92	2
0·51	1·37	0·59	13·5	13·6	2·17	AC1050	7·97	3

SPECIFIC HEAT (c)

Chemical Composition (%)						Heat Treatment °C	Mean specific heat cal/g.deg. C				Ref.
C	Si	Mn	Ni	Cr	W		20 to 200°C	20 to 400°C	20 to 600°C	20 to 800°C	
0·51	1·37	0·59	13·5	13·6	2·17	AC1050	0·122	0·126	0·130	0·133	3

MEAN COEFFICIENT OF THERMAL EXPANSION (α)

Chemical Composition (%)						Heat Treatment °C	10^6 × Mean Coefficient of Thermal Expansion per deg. C										Ref.
C	Si	Mn	Ni	Cr	W		20 to 100°C	20 to 200°C	20 to 300°C	20 to 400°C	20 to 500°C	20 to 600°C	20 to 700°C	20 to 800°C	20 to 900°C	20 to 1000°C	
0·40	0·97	0·41	10·3	13·6	3·53	AC950	17		17		18		18		19		1
0·40	1·50	1·00	13·5	14·0	2·50			17·1		18·3		18·9		19·3		20·0	4
0·42	1·50	0·70	14·0	14·0	2·60	AC950	17		18		18		19		20		1
0·42	1·30		13·0	13·0	2·50			17·3		18·3		18·9	19·1	19·2			5
0·43	1·44	0·68	14·3	13·4	2·58	WQ1050	17·1	17·6	17·9	18·2	18·5	18·8	19·1	19·3	19·6	20·0	3
0·50	2·25	0·80	14·5	13·5	2·00		16·5	16·7	16·8	16·9	17·1	17·3	17·4	17·8	18·2		6
0·51	1·37	0·59	13·5	13·6	2·17	AC1050				17·4		18·9		18·3			3
(Within En 54A specification)						AC950		17·3		18·2	18·6	19·0					7

THERMAL CONDUCTIVITY (k)

Chemical Composition (%)						Heat Treatment °C	Thermal Conductivity cal/cm sec. deg. C											Ref.
C	Si	Mn	Ni	Cr	W		20°C	100°C	200°C	300°C	400°C	500°C	600°C	700°C	800°C	900°C	1000°C	
0·40	0·97	0·41	10·3	13·6	3·53	AC950	0·033										0·075	1
0·48	1·22	1·44	12·23	12·85	2·88	—	0·048	0·048	0·048	0·050	0·053	0·056	0·059	0·061	0·064	0·068	0·073	8
(within En 54 specification)							0·032											2
(within En 54A specification)						AC950	0·038		0·050		0·063							7

ELECTRICAL RESISTIVITY (ρ)

Chemical Composition (%)						Heat Treatment °C	Electrical Resistivity microhm-cm					Ref.
C	Si	Mn	Ni	Cr	W		20°C	100°C	200°C	900°C	1000°C	
0·40	0·97	0·41	10·3	13·6	3·53	AC950	78				109	1
0·43	1·44	0·68	14·3	13·4	2·58	WQ1050	83·8	88·3	95·0			3
0·51	1·37	0·59	13·5	13·6	2·17	AC1050	83·5					"
(Within En 54 specification)							79					2
(Within En 54 specification)						AC950	78					7
(Within En 54 specification)							72			127		5

YOUNG'S MODULUS (E), SHEAR MODULUS (G), POISSON'S RATIO (σ)

Chemical Composition (%)	Test Temperature °C	Young's Modulus (dynamic*) tons/sq.in.	Shear Modulus tons/sq.in.	Poisson's Ratio	Ref.
Within specification	20		5·350	0·29	7
Within specification	20	13,100			1
	200	12,100			"
	400	11,100			"
	600	10,000			"
	800	8,900			"

* Vibration Method

MAGNETIC PROPERTIES

This steel is virtually non-magnetic in the fully softened condition.

MECHANICAL PROPERTIES AT ROOM TEMPERATURE

Process of Manu-facture	Chemical Composition (%)									Section Size dia. (in.)	Heat Treatment °C	Tensile							Izod ft.lb.	Ref.
	C	Si	Mn	S	P	Ni	Cr	W	Nb			0·1% P.S. (0·1Rp) tons/sq. in.	0·2% P.S. (0·2Rp) tons/sq. in.	0·5% P.S. (0·5Rp) tons/sq. in.	Y.P. (Rs) tons/sq. in.	T.S. (Rm) tons/sq. in.	El % (A)	R.A. % (Z)		
	0·40	0·97	0·41			10·3	13·6	3·53			AC950				36·0	63·0	23½	24		1
	0·42	1·30				13·0	13·0	2·50			AC950					53·6	31	35		4
	0·42	1·04	0·54			14·1	12·7	2·84			As rolled				36·6	54·3	42	50½		"
	"	"	"			"	"	"			AC1050				35·7	58·5	28½	40½		"
	0·42	1·50	0·70			14·0	14·0	2·50			AC850				35·0	50·0	28		15	1
	0·42	1·38	0·80	0·012	0·025	13·0	13·0	2·36	1·56		AC950		25·3			53·1	30½	26		6
B. E. A	0·43	1·48	0·58	0·028	0·020	11·0	13·8	2·2		1⅛	AC850		30·8			69·0	25½	27	44	9
	0·44	1·10	0·71			14·1	14·0	2·05		1⅛	AC950				39·0	54·6	39½	47½	38	1
	0·44	1·58	0·64	0·005	0·022	13·8	14·1	2·54		—	AC950	20·4	22·6	25·2		49·0	40	53		"
	0·45	1·45	0·96			13·7	13·4	2·42		1⅛	AC950					52·0	30		35	10
	0·51	1·37	0·59			13·5	13·6	2·17		¾	AC1050				29·3	54·1	32½	44½	43	3

MECHANICAL PROPERTIES AT LOW TEMPERATURE

Chemical Composition (%)									Heat Treatment °C	Test Temperature °C	Tensile				Ref.
C	Si	Mn	S	P	Ni	Cr	W	Nb			0·2% P.S. (0·2Rp) tons/sq.in.	T.S. (Rm) tons/sq.in.	El % (A)	R.A. % (Z)	
0·42	1·38	0·80	0·012	0·025	13·0	13·0	2·36	1·56	AC950	+25	25·3	53·1	30½	26	6
"	"	"	"	"	"	"	"	"	"	—50	30·7	60·6	19	21	"
"	"	"	"	"	"	"	"	"	"	—196	44·2	73·0	15	17	"

Chemical Composition (%)									Heat Treatment °C	Izod Impact Values ft. lbs. at							
C	Si	Mn	S	P	Ni	Cr	W	Nb		20°C	0	—20	—40	—60	—80	—100	—120
0·42	1·38	0·80	0·012	0·025	13·0	13·0	2·36	1·56	AC950	41	38	35	32	29	26	23	20

MECHANICAL PROPERTIES AT HIGH TEMPERATURE

Chemical Composition (%)								Section Size dia. (in.)	Heat Treatment °C	Test Temperature °C	Tensile							Izod ft.lb.	Ref.
C	Si	Mn	S	P	Ni	Cr	W				0·1% P.S. (0·1Rp) tons/sq.in.	0·2% P.S. (0·2Rp) tons/sq.in.	0·5% P.S. (0·5Rp) tons/sq.in.	Y.P. (Rs) tons/sq.in.	T.S. (Rm) tons/sq.in.	El % (A)	R.A. % (Z)		
0·40	0·97	0·41	0·016	0·016	10·3	13·6	3·53		AC950	15				36·0	63·0	23½	24		1
"	"	"	"	"	"	"	"		"	100				37·5	53·5	25	33		"
"	"	"	"	"	"	"	"		"	200				35·3	49·4	21	28½		"
"	"	"	"	"	"	"	"		"	300				32·6	43·4	22½	33		"
"	"	"	"	"	"	"	"		"	400				29·4	42·7	21	33		"
"	"	"	"	"	"	"	"		"	500				26·0	39·5	23½	46		"
"	"	"	"	"	"	"	"		"	600				19·6	34·1	25	54½		"
"	"	"	"	"	"	"	"		"	700				14·2	28·2	28	61		"
"	"	"	"	"	"	"	"		"	800				9·1	17·2	40	74		"
"	"	"	"	"	"	"	"		"	900				6·4	12·0	51	79		"
"	"	"	"	"	"	"	"		"	1000				2·6	6·0	70	81		"
										(Rate of loading 1 ton per sq. in. per min.)									
0·42	1·26	0·61			13·3	13·9	2·27	⅝											
"	"	"			"	"	"	"	AC1050	700					23·2	32½	62		3
"	"	"			"	"	"	"	"	800					17·6	19½	36		"
"	"	"			"	"	"	"	"	900					6·5	70	65		"
								"	(Strain rate 0·04 in. per min.)										
0·43	1·48	0·58	0·028	0·020	11·0	13·8	2·22	1⅛	AC850	120		30·8			69·0	25½	27	44	9
"	"	"	"	"	"	"	"	"	"	550		26·0			42·5	21½	45		"
"	"	"	"	"	"	"	"	"	"	600		25·0			38·2	21½	47		"
"	"	"	"	"	"	"	"	"	"	650		22·5			31·7	24	52½		"
"	"	"	"	"	"	"	"	"	"	700		17·5			25·0	29	57½		"
"	"	"	"	"	"	"	"	"	"	750		13·0			20·9	34	59½		"
"	"	"	"	"	"	"	"	"	"	800		10·8			17·7	33½	58		"
"	"	"	"	"	"	"	"	"	"	850		9·0			14·0	32	56		"
										(Rate of loading 1 ton per sq. in. per min.)									

MECHANICAL PROPERTIES AT HIGH TEMPERATURE

(Continued)

Chemical Composition (%)								Section Size dia. (in.)	Heat Treatment °C	Test Temperature °C	Tensile							Charpy V ft.lb.	Ref.
C	Si	Mn	S	P	Ni	Cr	W				0·1% P.S. (0·1Rp)	0·2% P.S. (0·2Rp)	0·5% P.S. (0·5Rp)	Y.P. (Rs)	T.S. (Rm)	El % (A)	R.A. % (Z)		
											tons/sq. in.								
0·44	1·58	0·64			13·8	14·1	2·54	¾	AC950	RT	20·4	22·6	25·2		49·0	40	53		1
,,	,,	,,			,,	,,	,,	,,	,,	100	19·5	21·9	24·3		44·9	38½	54		,,
,,	,,	,,			,,	,,	,,	,,	,,	200	18·7	20·3	22·1		42·6	38½	53		,,
,,	,,	,,			,,	,,	,,	,,	,,	300	18·1	19·4	21·4		42·5	36	52		,,
,,	,,	,,			,,	,,	,,	,,	,,	400	16·7	18·0	19·6		41·1	36	51		,,
,,	,,	,,			,,	,,	,,	,,	,,	500	16·8	17·8	19·2		39·0	34½	52		,,
,,	,,	,,			,,	,,	,,	,,	,,	600	15·3	16·3	17·9		34·3	34½	52		,,
,,	,,	,,			,,	,,	,,	,,	,,	700	14·0	15·2	16·3		29·1	43	56½		,,
,,	,,	,,			,,	,,	,,	,,	,,	800	10·7	11·8	12·8		20·2	61½	65½		,,
,,	,,	,,			,,	,,	,,	,,	,,	900	—	—	—		12·3	84	74½		,,
,,	,,	,,			,,	,,	,,	,,	,,	1000	—	—	—		8·0	102½	77½		,,
0·50	2·25	0·80	0·010	0·016	14·5	13·5	2·0		As received	20				38·6	57·5	45	51	52	6
,,	,,	,,	,,	,,	,,	,,	,,		,,	250				33·2	50·7	36	50	53	,,
,,	,,	,,	,,	,,	,,	,,	,,		,,	350				31·2	51·5	41	51	54	,,
,,	,,	,,	,,	,,	,,	,,	,,		,,	450				29·5	48·5	38	54	55	,,
,,	,,	,,	,,	,,	,,	,,	,,		,,	550				26·5	45·7	36	52	59	,,
,,	,,	,,	,,	,,	,,	,,	,,		,,	650				24·5	44·6	27	35	60	,,
,,	,,	,,	,,	,,	,,	,,	,,		,,	750				21·8	33·3	16	25	62	,,
,,	,,	,,	,,	,,	,,	,,	,,		,,	850				16·2	20·9	15	37	64	

MECHANICAL PROPERTIES AT HIGH TEMPERATURE

Chemical Composition (%)					
C	Si	Mn	Ni	Cr	W
0·45	0·95	0·62	14·32	13·67	2·18

1 in. dia. bars as rolled condition.

(Ref. 6)

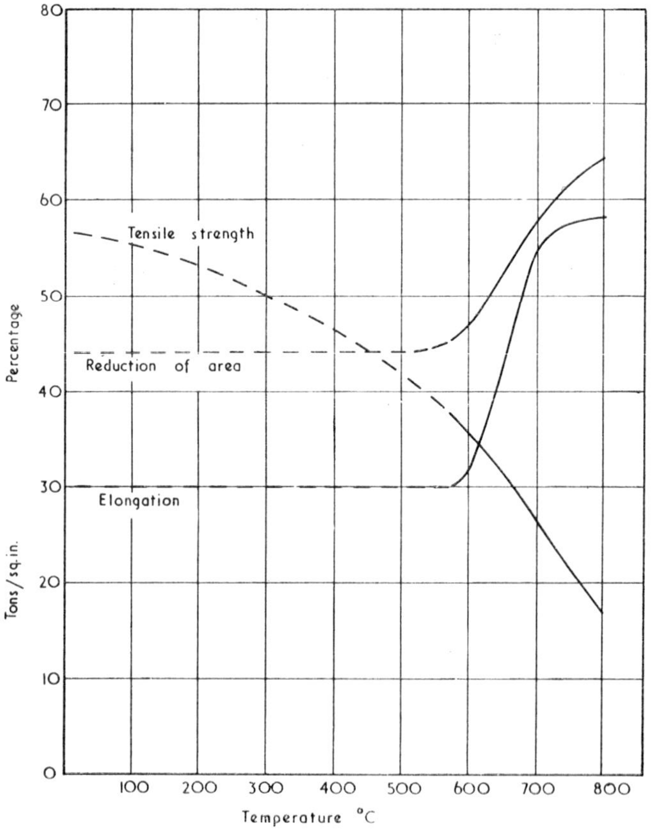

Fig. 54.1

CREEP PROPERTIES

Chemical Composition (%)								Heat Treatment °C	Test Temperature °C	Tensile				Stress tons/sq.in. to give strain rate of 10^{-6}in./in./hr from 24 to 72 hrs.	Ref.
C	Si	Mn	S	P	Ni	Cr	W			Y.P. (Rs) tons/sq.in.	T.S. (Rm) tons/sq.in.	El % (A)	R.A. % (Z)		
0·40	0·97	0·41	0·016	0·016	10·3	13·6	3·53	AC950	400	29·4	42·7	21	33	18	1
"	"	"	"	"	"	"	"	"	500	26·0	39·5	23½	46	8	"
"	"	"	"	"	"	"	"	"	600	19·6	34·1	25	54½	3·5	"
"	"	"	"	"	"	"	"	"	800	14·2	28·2	28	61	1·5	"
"	"	"	"	"	"	"	"	"	800	9·1	17·2	40	74	0·35	"
"	"	"	"	"	"	"	"	"	900	6·4	12·0	51	79	0·08	"
"	"	"	"	"	"	"	"	"	1000	2·6	6·0	70	81	0·03	"

Chemical Composition (%)	Section Size dia. (in.)	Heat Treatment °C	Test Temperature °C	Stress to give stated strain rate in 100 hrs.		Stress to give stated strain rate in 300 hrs.		Ref.
				0·1%	0·2%	0·1%	0·2%	
(Within En 54A specification)	1⅛	AC950	450	26·0	30·0	24·5	29·0	7
"	"	"	500	18·5	21·0	17·2	20·5	"
"	"	"	550	12·5	14·2	11·2	13·5	"
"	"	"	600	8·0	9·5	7·0	8·5	"
"	"	"	650	5·1	6·5	4·5	5·5	"

FATIGUE PROPERTIES

Chemical Composition (%)						Section Size dia. (in.)	Heat Treatment °C	Test Temperature °C	Tensile (tons/sq.in.)							Izod ft.lb.	Rotary Bending Endurance Limit tons/sq.in.		Ref.
C	Si	Mn	Ni	Cr	W				0·1% P.S. (0·1Rp)	0·2% P.S. (0·2Rp)	0·5% P.S. (0·5Rp)	Y.P. (Rs)	T.S. (Rm)	El % (A)	R.A. % (Z)		10^7 cycles	10^8 cycles	
0·64	1·10	0·71	1·41	14·0	2·05	1⅛	AC950	RT				39·0	54·6	39½	47½	38	±20	—	1
(Within En 54A specification)						1⅛	AC950	20	28·3	30·7	34·0		58·2	27			±30·0	—	7
”						”	”	200	23·0	24·9	27·5		48·0	23			±28·5	—	”
”						”	”	400	22·5	24·5	27·0		47·0	24			±27·0	±26·5	”
”						”	”	500	21·6	23·1	25·4		44·2	27			±24·5	±23·5	”
”						”	”	600	18·8	20·6	22·9		38·5	23			±21·0	±18·0	”
”						”	”	700					30·0	29			±16·5	±13·5	”

LIST OF REFERENCES

1. Private correspondence with Firth-Vickers Stainless Steels Ltd.
2. Private correspondence with Low Moor Alloy Steelworks Ltd.
3. Private correspondence with The United Steel Companies Ltd.
4. Private correspondence with Darwins Ltd.
5. Private correspondence with Jessop-Saville Ltd.
6. War Office unpublished report.
7. Private correspondence with Rolls-Royce Ltd.
8. E. Maurer. The Heat Conductivity of Chromium Steels at High Temperatures. *Archiv für das Eisenhüttenwesen*, 1936, vol. 10, Oct., pp. 145—154.
9. Private correspondence with Edgar Allen and Co. Ltd.
10. British Standard 971: Commentary on the British Standard Wrought Steels, En Series. British Standards Institution, London, 1950.

HIGH CHROMIUM–NICKEL–TUNGSTEN VALVE STEEL

SPECIFICATION

CHEMICAL COMPOSITION (%)

C	S	Mn	S	P	Ni	Cr	W
0·18–0·45	1·00–2·50	1·00 max.	0·045 max.	0·045 max.	6·00–12·0	17·0 min.	2·00–4·00

MECHANICAL PROPERTIES

Izod impact valve, ft.lb., min.	20
Brinell hardness number, HB max.	302

APPLICATIONS

This highly-alloyed chromium nickel steel possesses good strength at elevated temperatures together with a high degree of scaling resistance at temperatures up to 950°C. It has a better resistance to attack from leaded fuels than En 54 and better scale resistance above 900°C, but at valve-operating temperatures there is no significant difference.

Typical uses are similar to those given for En 54, but where higher temperatures than 800°C are involved, e.g. valves, superheater supports, furnace, jet engine and industrial gas turbine parts subjected to high temperatures in service.

WELDING

The welding characteristics will be similar to those given for En 54.

MACHINABILITY

The machinability in the softened condition is approximately 30% of that for mild steel (En 3).

HOT WORKING AND HEAT TREATMENT TEMPERATURES

Forging, rolling and stamping	1150°C, finish above 900°C.
Softening	950°C—1100°C, rapid cool.

This steel cannot be hardened by heat treatment.

PHYSICAL PROPERTIES

SPECIFIC GRAVITY (d)

Chemical Composition (%)						Heat Treatment °C	Specific Gravity at 20°C	Ref.
C	Si	Mn	Ni	Cr	W			
0·20	1·50	0·40	11·5	23·0	3·00	Softened	7·90	1
0·22	1·00	0·70	11·5	22·5	2·70	"	7·90	2
0·23	1·65	0·41	11·8	23·2	2·98	"	7·90	1
0·34	1·39	0·46	10·9	21·4	3·16	"	7·90	"
0·34	1·24	0·69	7·6	19·2	3·98	"	7·80	"
0·37	1·27	0·73	9·5	21·0	2·20	"	7·89	3

SPECIFIC HEAT (c)

Chemical Composition (%)						Heat Treatment °C	Mean Specific Heat cal/g. deg. C				Ref.
C	Si	Mn	Ni	Cr	W		20 to 200°C	20 to 400°C	20 to 600°C	20 to 800°C	
0·37	1·27	0·73	9·5	21·0	2·20	Softened	0·120	0·124	0·127	0·131	3

MEAN COEFFICIENT OF THERMAL EXPANSION (α)

Chemical Composition (%)						Heat Treatment °C	$10^6 \times$ Mean Coefficient of Thermal Expansion per deg. C										Ref.
C	Si	Mn	Ni	Cr	W		20 to 100°C	20 to 200°C	20 to 300°C	20 to 400°C	20 to 500°C	20 to 600°C	20 to 700°C	20 to 800°C	20 to 900°C	20 to 1000°C	
0·23	1·65	0·41	11·8	23·2	2·98	Softened	15		16		16		17		18		1
0·34	1·24	0·69	7·61	19·2	3·98	"	16		17		17		18		18		"
0·37	1·27	0·73	9·5	21·0	2·20	"		16·4		17·4		18·2		18·9		19·8	3

THERMAL CONDUCTIVITY (k)

Chemical Composition (%)						Heat Treatment °C	Thermal Conductivity cal/cm sec. deg. C											Ref.
C	Si	Mn	Ni	Cr	W		20°C	100°C	200°C	300°C	400°C	500°C	600°C	700°C	800°C	900°C	1050°C	
0·19	0·92	1·00	11·6	22·5	3·01	AC1050	(0·030)	0·039	0·043	0·047	0·051	0·054	0·057	0·059				4
0·20	1·5	0·40	11·5	23·0	3·0	Softened	0·030										0·070	1
0·22	1·0	0·70	11·5	22·5	2·70	"		0·035	0·040	0·043	0·047	0·052	0·057	(0·060)	(0·062)	(0·065)		2
0·40	1·52	0·65	8·1	20·0	2·12	"		0·034	0·038	0·042	0·046	0·050	0·053	0·056	0·059			3

Values in parentheses are estimated.

ELECTRICAL RESISTIVITY (ϱ)

Chemical Composition (%)						Heat Treatment °C	Electrical Resistivity microhm-cm										Ref.
C	Si	Mn	Ni	Cr	W		20°C	100°C	200°C	300°C	400°C	500°C	600°C	700°C	800°C	1050°C	
0·19	0·92	1·00	11·6	22·5	3·01	AC1050	83·7	89·0	95·2	100·5	105·4	109·7	113·0	115·5	118·0		4
0·23	1·65	0·41	11·8	23·2	2·98	Softened	87									125	1
0·34	1·24	0·69	7·6	19·2	3·98	"	85									125	"
0·37	1·27	0·73	9·5	21·0	2·20	"	79·4	84·6	91·3								3

YOUNG'S MODULUS (E)

Chemical Composition (%)						Young's Modulus, tons/sq. in.										Method	Ref.
C	Si	Mn	Ni	Cr	W	20°C	100°C	200°C	300°C	400°C	500°C	600°C	700°C	800°C	1000°C		
0·21	0.65	0·56	11·8	23·0	3·10	12,800		11,900		10,800		9,800		8,900	7,600	Vibration	1
0·22	1·00	0·70	11·5	22·5	2·70	12,720	12,380	11,860	11,340	10,840	10,340	9,870	9,400	8,860			2
0·32	1·35	0·52	7·97	20·3	3·85	13,200		12,200		11,000		9,900		8,800	7,400	Vibration	1

MAGNETIC PROPERTIES

This steel is virtually non-magnetic in the softened condition.

MECHANICAL PROPERTIES AT ROOM TEMPERATURE

Chemical Composition (%)						Section Size dia. (in.)	Heat Treatment °C	Tensile							Izod ft.lb.	Ref.
C	Si	Mn	Ni	Cr	W			0·1% P.S. (0·1Rp)	0·2% P.S. (0·2Rp)	0.5% P.S. (0·5Rp)	Y.P. (Rs)	T.S. (Rm)	El % (A)	R.A. % (Z)		
								tons/sq.in.								
0·20	1·60	0·40	11·5	23·0	2·75	1⅛	WQ1100	22·0	24·2	27·1		53·5	49	56		1
0·23	1·65	0·41	11·8	23·2	2·98	1⅛	AC1050	18·3	21·1	25·7		46·4	32	34		"
0·30	1·50	0·60	8·0	20·0	4·0		AC1050	22·4	25·6	29·9		54·0	38½	36		"
0·34	1·24	0·69	7·6	19·2	3·98		AC1050	32·2	34·5	37·6		64·7	33½	35		"
0·34	1·39	0·46	10·9	21·4	3·16	1⅛	AC950	20·0		38·0		57·0	25	35	40	"
0·39	1·36	0·66	8·5	20·2	2·01	1	AC1050					54·4	31		46	5
0·45	1·31	0·57	9·1	19·6	2·03	¾	AC1050				31·6	56·8	32	42	36	3
0·33	1·88	0·42	11·29	22·2	2·99		As rolled				42·2	54·6	29	31	47	1
"	"	"	"	"	"		AC900				31·4	51·0	30	40	35	"
"	"	"	"	"	"		AC1050				27·6	49·4	38½	43½	39	"
"	"	"	"	"	"		" +2hrs. 800AC				26·8	48·6	37	46	27	"
0·40	1·31	0·41	12·36	22·8	3·52		As rolled				41·8	56·0	18	30½	12	1
							AC900				38·6	53·2	22½	30½	13	"
							AC1050				35·2	53·6	28	36½	29	"
							" +2hrs. 800AC				34·0	53·6	22	38	33	"
0·43	1·48	0·49	7·0	21·1	3·69		WQ1000									
"	"	"	"	"	"		" +1,000 hrs. 750/800				42·5	56·5	40	49	56	6
"	"	"	"	"	"		" " 800/850				44·5	53·5	21	14	4	"
"	"	"	"	"	"		" " 850/900				43·5	54·0	30½	20	5½	"
											40·2	52·7	29	31	7	"

MECHANICAL PROPERTIES AT HIGH TEMPERATURE

Chemical Composition (%)							Heat Treatment °C	Test Temperature °C	Tensile							Ref.
C	Si	Mn	Ni	Cr	W	Mo			0·1% P.S (0·1Rp)	0·2% P.S. (0·2Rp)	0·5% P.S. (0·5Rp)	Y.P. (Rs)	T.S. (Rm)	El % (A)	R.A. % (Z)	
									tons/sq.in.							
0·20	1·5	0·40	11·5	23·0		3·0	AC1050	20	18·3	21·1	25·7		46·4	32	34	1
"	"	"	"	"		"	"	100	15·6	18·2	22·2		41·4	28	34	"
"	"	"	"	"		"	"	200	14·2	16·5	20·3		41·2	33½	49	"
"	"	"	"	"		"	"	300	14·0	16·0	19·4		39·8	37½	47	"
"	"	"	"	"		"	"	400	14·0	16·0	19·5		39·5	34½	42	"
"	"	"	"	"		"	"	500	13·9	15·8	19·2		36·6	17½	22	"
"	"	"	"	"		"	"	600	13·2	14·4	15·9		29·7	33½	36½	"
"	"	"	"	"		"	"	700					23·2	36	34	"
"	"	"	"	"		"	"	800					14·9	47	41½	"
"	"	"	"	"		"	"	900					8·4	44½	36	"
"	"	"	"	"		"	"	1000					4·5	76	48	"

Rate of straining: 0·0085 in./min. to Y.P. then at 0·125 in./min.

MECHANICAL PROPERTIES AT HIGH TEMPERATURE (Continued)

Chemical Composition (%)							Section Size dia. (in.)	Heat Treatment °C	Test Temperature °C	Tensile							Ref.
C	Si	Mn	Ni	Cr	W	Mo				0·1% P.S. (0·1Rp)	0·2% P.S. (0·2Rp)	0·5% P.S. (0·5Rp)	Y.P. (Rs)	T.S. (Rm)	El % (A)	R.A. % (Z)	
										tons/sq.in.							
0·22	1·00	0·70	11·5	22·5	2·70			AC950	20	19·6		24·6		47·5	41½	50	2
"	"	"	"	"	"			"	100	17·0		21·6		43·0	40	53	"
"	"	"	"	"	"			"	200	15·0		19·0		39·0	38½	56½	"
"	"	"	"	"	"			"	300	14·8		18·3		38·0	34	54	"
"	"	"	"	"	"			"	400	14·6		17·9		37·5	32	50	"
"	"	"	"	"	"			"	500	13·0		15·8		34·5	33	49	"
"	"	"	"	"	"			"	600	12·3		14·6		30·5	33½	48	"
"	"	"	"	"	"			"	700	10·0		12·0		24·5	43½	60	"
"	"	"	"	"	"			"	800	6·3		9·8		17·0			"
"	"	"	"	"	"			"	900					7·0			"
									Rate of loading 1 ton/sq. in./min.								
0·29	1·57	0·46	11·4	22·3	3·30		¾	AC1050	20			28·5		49·5	44	51	1
"	"	"	"	"	"		"	"	400	11·1		18·0		39·6	36½	50	"
"	"	"	"	"	"		"	"	500	13·0		16·4		36·9	37½	41	"
"	"	"	"	"	"		"	"	550	12·6		16·0		35·5	32	43	"
"	"	"	"	"	"		"	"	600	12·4		15·3		31·6	35	47	"
"	"	"	"	"	"		"	"	700	10·9		13·3		23·2	48	56½	"
									Rate of loading 1 ton/sq.in./min.								
0·31	1·49	0·87	7·50	20·4	2·04		¾	WQ1100	RT	22·0	24·2	27·1		53·5	49½	56½	1
"	"	"	"	"	"		"	"	200	16·3	18·1	19·7		41·6	44	60	"
"	"	"	"	"	"		"	"	400	15·1	16·9	18·8		41·1	45½	56½	"
"	"	"	"	"	"		"	"	600	13·4	14·7	16·2		34·9	40	53½	"
"	"	"	"	"	"		"	"	700	12·0	13·4	14·7		27·2	49	59	"
"	"	"	"	"	"		"	"	800	8·2	9·9	10·8		16·2	61½	47	"
"	"		"	"	"		"	"	1000					6·7	65½	45	"
									Rate of loading 1 ton/sq.in./min								
0·32	1·89	0·47	11·54	22·1	3·27		¾	As forged	15					50·4	29½	44½	1
"	"	"	"	"	"		"	"	400	31·0		34·3		45·1	25	47½	"
"	"	"	"	"	"		"	"	500	29·0		32·4		43·7	24	40½	"
"	"	"	"	"	"		"	"	600	25·0		28·9		38·5	26½	38½	"
"	"	"	"	"	"		"	"	700	17·0		20·3		27·4	34½	56½	"
"	"	"	"	"	"		"	"	800					18·7	41½	65½	"

MECHANICAL PROPERTIES AT HIGH TEMPERATURE (Continued)

Chemical Composition (%)							Section Size dia. (in.)	Heat Treatment °C	Test Temperature °C	Tensile							Ref.
C	Si	Mn	Ni	Cr	W	Mo				0·1% P.S. (0·1Rp)	0·2% P.S. (0·2Rp)	0·5% P.S. (0·5Rp)	Y.P. (Rs)	T.S. (Rm)	El % (A)	R.A. % (Z)	
										tons/sq.in.							
0·32	1·89	0·47	11·54	22·1	3·27		3/4	AC1050	15			27·0		48·0	34½	48	1
"	"	"	"	"	"		"	"	400	14·5		18·1		39·5	34½	51	"
"	"	"	"	"	"		"	"	600	13·6		16·3		33·0	36	45	"
"	"	"	"	"	"		"	"	800	6·2		8·3		15·0	60	68½	"
"	"	"	"	"	"		"	"	900					8·7	68	62½	"
								Rate of loading 1 ton/sq.in/. min.									
0·31	1·43	0·58	7·4	19·9	4·0		1	AC850	20		21·5			51·7	42	48½	7
"	"	"	"	"	"		"	"	550		16·3			34·5	28	47	"
"	"	"	"	"	"		"	"	600		14·5			30·8	29½	50½	"
"	"	"	"	"	"		"	"	650		12·5			26·9	36½	58	"
"	"	"	"	"	"		"	"	700		11·4			22·4	41	63	"
"	"	"	"	"	"		"	"	750		10·0			17·9	46	65½	"
"	"	"	"	"	"		"	"	800		7·7			13·2	50	64	"
"	"	"	"	"	"		"	"	850		6·5			9·8	45	58½	"
								Rate of loading 1 ton/sq.in·/min.									
0·34	1·24	0·69	7·57	19·1	3·98		3/4	AC1050	20	32·2	34·5	37·6		64·7	33½	35	1
"	"	"	"	"	"		"	"	100	28·3	30·3	32·8		54·1	29½	43	"
"	"	"	"	"	"		"	"	200	26·8	28·6	30·8		51·2	30½	47	"
"	"	"	"	"	"		"	"	300	24·3	26·1	28·4		44·9	27	46½	"
"	"	"	"	"	"		"	"	400	22·4	24·3	26·8		41·8	25	45	"
"	"	"	"	"	"		"	"	500	23·1	24·8	26·9		41·4	25½	42½	"
"	"	"	"	"	"		"	"	600	19·1	20·8	22·9		36·5	23	38½	"
"	"	"	"	"	"		"	"	700					28·5	36	58½	"
"	"	"	"	"	"		"	"	800					19·5	37½	62	"
"	"	"	"	"	"		"	"	900					13·2	45½	55	"
"	"	"	"	"	"		"	"	1000					8·9	44	58½	"
								Rate of loading 1 ton/sq.in./min.									
0·44	1·54	0·60	8·1	19·6	2·29		3/8	AC950	RT				33·8	58·4	36	47	3
"	"	"	"	"	"		"	AC1050	700					19·5	94½	65	"
"	"	"	"	"	"		"	"	800					12·7	54	57	"
"	"	"	"	"	"		"	"	900					6·9	76	60	"
								Rate of straining 0·08 in./min.									

CREEP PROPERTIES

Chemical Composition (%) C	Si	Mn	Ni	Cr	W	Section Size dia. (in.)	Heat Treatment °C	Test Temperature °C	Tensile 0·1% P.S. (0·1Rp) tons/sq.in.	0·2% P.S. (0·2Rp) tons/sq.in.	0·5% P.S. (0·5Rp) tons/sq.in.	T.S. (Rm) tons/sq.in.	El % (A)	R.A. % (Z)	Stress tons/sq.in. to give creep rate of 10^{-6} in./in./hr from 24 to 72 hrs.	Ref.
0·34	1·24	0·69	7·6	19·1	3·98	¾	AC1050	400	22·4	24·3	26·8	41·8	25	45	25	1
"	"	"	"	"	"	"	"	500	23·1	24·8	26·9	41·4	25½	42½	10	"
"	"	"	"	"	"	"	"	600	19·1	20·8	22·9	36·5	23	38½	3·5	"
"	"	"	"	"	"	"	"	700				28·5	36	58½	1·75	"
"	"	"	"	"	"	"	"	800				19·5	37½	62	0·54	"
"	"	"	"	"	"	"	"	900				13·2	45½	55	0·13	"
"	"	"	"	"	"	"	"	1000				8·9	44	58½	0·04	"

Chemical Composition (%) C	Si	Mn	Ni	Cr	W	Test Temperature °C	Stress tons/sq.in. for creep strain of 10^{-4}/hr	10^{-5}/hr	10^{-6}/hr	Ref.
0·22	1·0	0·7	11·5	22·5	2·7	600	13·0			2
(Nominal composition)						650	9·0	(5·6)	(3·4)	"
						700	6·3	3·5	(2·1)	"
						750	3·7	2·1	(1·2)	"
						800	2·1	1·0		"

STRESS RUPTURE DATA

Chemical Composition (%) (nominal)					
C	Si	Mn	Ni	Cr	W
0·22	1·0	0·7	11·5	22·5	2·7

Broken curves have been extrapolated.

(Ref. 2)

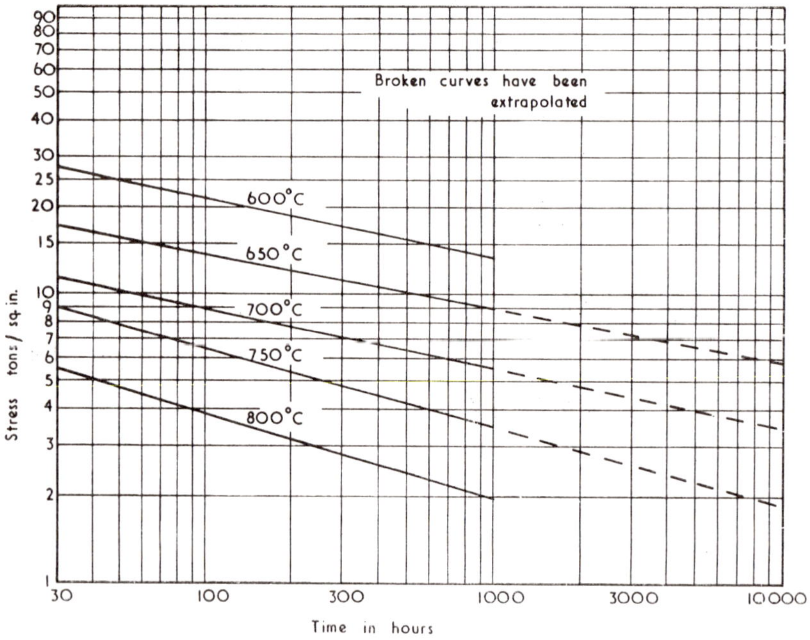

Fig. 55.1

FATIGUE PROPERTIES

Chemical Composition (%)						Section Size dia. (in.)	Heat Treatment °C	Test Temperature °C	Tensile					Rotating Bending Fatigue (Endurance) Limit tons/sq.in.		Ref.
C	Si	Mn	Ni	Cr	W				0·1% P.S. (0·1Rp) tons/sq.in.	0·5% P.S. (0·5Rp) tons/sq.in.	T.S. (Rm) tons/sq.in.	El % (A)	R.A. % (Z)	4×10^7 cycles	10^8 cycles	
0·22	1·0	0·7	11·5	22·5	2·7		AC950	700	10·0	12·0	24·5	43½	60	±12·8	±12·0	2
"	"	"	"	"	"		"	800	6·3	9·8	17·0			±8·2	±7·0	"
"	"	"	"	"	"		"	870						±5·6	±4·9	"
"	"	"	"	"	"		"	900			7·0			±5·6	±4·9	"
0·34	1·39	0·46	10·9	21·4	3·16	1⅛	AC950	RT	20·0	38·0	57·0	25	35	±21·0 (10^7 cycles)		1

LIST OF REFERENCES

1. Private correspondence with Firth-Vickers Stainless Steels Ltd.
2. Private correspondence with Jessop-Saville Ltd.
3. Private correspondence with The United Steel Companies Ltd.
4. Private correspondence with De Havilland Aircraft Co. Ltd.
5. British Standard 971: Commentary on the British Standard Wrought Steels, En Series. British Standards Institution, London, 1950.
6. J. R. Handforth. Metallurgical Problems arising from Internal Combustion Engine Valves. *JISI*, vol. 2, 1932, 93—131.
7. Private correspondence with Edgar Allen and Co. Ltd.

CHROMIUM RUST-RESISTING STEEL

SPECIFICATION

CHEMICAL COMPOSITION (%)

	C	Si	Mn	S	P	Ni	Cr
En 56A	0·12 max.	1·00 max.	1·00 max.	0·045 max.	0·045 max.	1·00 max.	12·0—14·0
En 56B	0·12—0·18	1·00 max.	1·00 max.	0·045 max.	0·045 max.	1·00 max.	12·0—14·0
En 56C	0·18—0·25	1·00 max.	1·00 max.	0·045 max.	0·045 max.	1·00 max.	12·0—14·0
En 56D	0·25—0·35	1·00 max.	1·00 max.	0·045 max.	0·045 max.	1·00 max.	12·0—14·0

MECHANICAL PROPERTIES

Property	Hardened and Tempered Condition[1]							
	P		R		S		T	X
Limiting ruling section, in.	6	2	6	2	4	2	2	$1\frac{1}{8}$
Tensile strength tons/sq. in. min. (Rm)	35	35	45	45	50	50	55	75
Yield stress [2] tons/sq. in. min. (Rs)	25	25	34	34	38	38	44	63
0·2% Proof stress [3] tons/sq. in. min. (0·2Rp)	23	23	32	32	36	36	41	59
Elongation per cent., min. (A)	25	25	20	20	16	16	12	8
Izod impact value, ft. lb. min.	25	45	20	25	10	20	10	
Brinell hardness number[4], HB	152/207	152/207	201/255	201/255	223/277	223/277	248/302	341 min

[1] The selection of En 56A, En 56B, En 56C, or En 56D is governed by the ruling section and the tensile range required and unless otherwise agreed will be at the discretion of the steelmaker.

[2] Not to be used as acceptance values except by special arrangement between the purchaser and the manufacturer.

[3] When specifically requested in the enquiry or order.

[4] For information only; not contractual part of the specification.

RELATED SPECIFICATIONS

United Kingdom				America		France	Germany			Sweden	USSR	
En No.	BS	BS Air	DTD	SAE	AISI	AFNOR	Werkstoff No.	Name	DIN	SIS	GOST	Mark
56A	1449 1554 2056	2S.61	97B 161A	51410	410	Z12C13	1.4006	X10Cr13		14·2302		
56B	1449 1554 2056		97B 161A 203B	51410	410		1.4021	X20Cr13			5632—61	1Kh13
56C	1449 1554 2056	2S.62		51420	420	Z20C13	1.4021	X20Cr13	17224	14·2303	5632—61	2Kh13
56D	1449 1554 2056		271	51420	420	Z30C13	1.4034	X40Cr13			5632—61	3Kh13

APPLICATIONS

These steels are suitable for use under mildly corrosive conditions. A wide range of mechanical properties is obtainable depending upon the type of steel, heat treatment and section size. The steels have good strength and resistance to oxidation up to 750°C although poor creep strength above 600°C. All are magnetic at normal temperatures.

En 56A and En 56B steels are very ductile in the annealed condition and can be readily formed by cold rolling, stamping, pressing and drawing, etc. Typical uses are: Aircraft rivets, split pins, compressor and steam turbine blades and shroudings, airframe parts, printing machine components, lightly stressed pins, nuts and bolts and general engineering fittings, motor-car fittings, golf club heads, corrosion and heat-resisting components for the oil industry where low coefficient of expansion is an advantage. En 56C: Typical uses are: Hydraulic steam and oil refinery pump and valve parts, turbine blading, automobile and aircraft fittings, garden tools, gauges, propeller shafts, piston rods, ball and roller bearing races, high-tensile nuts and bolts, scissors, surgical and dental instruments. En 56D: Used mainly for components requiring high hardness combined with corrosion resistance. Typical uses: Table cutlery, penknife blades, hand and machine knives, surgical and dental instruments, springs and components for hydraulic, steam and oil refinery pumps and valves.

It should be noted that the corrosion resistance of these steels is at its best in the fully-hardened condition but use in this state is limited to the manufacture of cutting blades, wear-resistant surfaces, etc., since the impact resistance and ductility are limited. Toughness is highest if the steels are tempered at 650°C or above and the corrosion resistance is only marginally inferior to that in the fully-hardened state, although varying with the carbon content. Tempering in the range 350°C to 650°C should be avoided, as serious loss of corrosion and stress-corrosion resistance can be caused and the toughness is adversely affected.

WELDING

(See Introduction for key to symbols used in this table.)

	Welding Process					
	MA	G	IG	S	F	B
En 56A	d*	f	e	e	b	k
En 56B	f	f	e	f	b	k
En 56C	f	f	e	f	b	k
En 56D	f	f	e	f	b	k

Reproduced by kind permission of H.M.S.O.

* The use of austenitic electrodes should be considered in the light of the specific application.

MACHINABILITY

In the softened condition the machinability of these steels is approximately 35 to 55% of that for mild steel (En 3), decreasing with increase in carbon content. Free machining grades, En 56AM, En 56BM, En 56CM, En 56DM are available with enhanced machinability.

HOT WORKING AND HEAT TREATMENT TEMPERATURES

Forging, rolling and stamping	1100°C, finish above 900°C; retarded cooling is advisable.
Annealing	800°C— 950°C, furnace cool.
Softening	700°C— 750°C, air cool.
Hardening	950°C—1020°C, air, oil (or water quench for low carbon steels or thin sections).
Tempering (for general engineering purposes)	650°C— 750°C, cool in air, oil or water.
(for cutlery)	150°C— 180°C, cool in air, oil or water.
(for springs)	400°C— 450°C, cool in air, oil or water.

Note: Tempering in the range 350°C—650°C causes a reduction in resistance to corrosion and stress-corrosion.

PHYSICAL PROPERTIES

SPECIFIC GRAVITY (d)

Type	Chemical Composition (%)							Heat Treatment °C	Specific Gravity at 20°C	Ref.
	C	Si	Mn	S	P	Ni	Cr			
En 56A	0·03	0·24	0·38	0·033	0·020	0·20	13·0	A850	7·72	1
"	0·07	0·14	0·29	0·025	0·012	0·11	11·8	A750	7·73	2
"	0·08	0·10	0·12			0·15	13·5	AC950 T750	7·725	3
"	0·10	0·20	0·20				13·5	A750	7·73	"
"	0·10	0·20	0·32			0·18	13·6	OQ940 T700	7·725	"
"	0·11	0·11	0·38			0·21	13·7	OQ950 T550	7·725	"
"	0·12	0·92	0·22			0·24	13·8	A975	7·728	"
"	"	"	"			"	"	OQ1000	7·726	"
"	"	"	"			"	"	" T250	7·726	"
"	"	"	"			"	"	" T820	7·727	"
En 56B	0·13	0·17	0·25	0·024	0·018	0·14	13·0	AC960 T750	7·745	4
"	0·16	0·29	0·45	0·019	0·019	—	13·1	Softened	7·723	5
"	0·18	0·29	0·31			0·22	13·8	AC1000 T150	7·725	3
En 56C	0·22						13·0	H & T750	7·70	1
"	0·23	0·25	0·30				13·0		7·73	6
"	0·25	0·25	0·25				13·5	Softened	7·75	3
En 56D	0·27	0·35	0·26			0·29	13·1	OQ940 T700	7·74	3
"	0·27	0·18	0·28	0·022	0·022	0·20	13·7	AH960 T750	7·741	4
"	0·32		0·30				13·0		7·74	7
"	0·33	0·25	0·35	0·030	0·030		13·0		7·73	8

SPECIFIC HEAT (c)

Type	Chemical Composition (%)							Heat Treatment °C	Mean Specific Heat cal/g.deg. C										Ref
	C	Si	Mn	S	P	Ni	Cr		20 to 100°C	20 to 200°C	20 to 300°C	20 to 400°C	20 to 500°C	20 to 600°C	20 to 700°C	20 to 800°C	20 to 900°C	20 to 1000°C	
En 56A	0·03	0·24	0·38	0·033	0·020	0·20	13·0			0·119	0·123	0·130	0·137	0·146	0·156				1
"	0·10						13·0			0·117		0·129		0·142					"
"	0·10	0·20	0·20				13·5	Softened	0·115										3
En 56B	0·13	0·17	0·25	0·024	0·018	0·14	13·0	AC960 T750	0·111	0·117	0·121	0·127	0·133	0·141	0·150	0·155	0·158	0·157	4
En 56D	0·27	0·18	0·28	0·022	0·022	0·20	13·7	AC960 T750	0·111	0·116	0·121	0·126	0·132	0·140	0·151	0·159	0·162	0·162	4

MEAN COEFFICIENT OF THERMAL EXPANSION (α)

Type	Chemical Composition (%)							Heat Treatment °C	10^6 × Mean Coefficient of Thermal Expansion per deg. C										Ref.
	C	Si	Mn	S	P	Ni	Cr		20 to 100°C	20 to 200°C	20 to 300°C	20 to 400°C	20 to 500°C	20 to 600°C	20 to 700°C	20 to 800°C	20 to 900°C	20 to 1000°C	
En 56A	0·03	0·24	0·38	0·033	0·020	0·20	13·0	A850	10·3	10·7	11·1	11·4	11·8	12·1	12·3				1
"	0·05	0·71	0·84			0·24	12·8	A750	10·4	10·9	11·3	11·7	12·1	12·3	12·5				3
"	0·07	0·14	0·29	0·025	0·012	0·11	11·8	H & T	10·7	10·6	10·9	11·2	11·5	11·9	12·7	12·7			2
"	0·10						13·0	Softened		10·9		11·7		12·3		12·8			"
En 56B	0·12	0·25	0·30				13·0		10·6	10·7	11·0	11·4	12·0	12·4	12·6	12·8			6
"	0·13	0·17	0·25	0·024	0·018	0·14	13·0	AC960 T750	10·2	10·8	11·2	11·6	11·9	12·2	12·4	12·6	10·8	11·8	4
En 56C	0·22						13·0	H & T750		10·9		11·7		12·3		12·8			1
En 56D	0·27	0·18	0·28	0·022	0·022	0·20	13·7	AC960 T750	10·1	10·8	11·2	11·6	11·9	12·3	12·5	12·2	10·7	12·2	4
"	0·30	0·11	0·18				13·1	A760	10·3	10·7		11·5		12·1					9
"	"	"	"				"	H & T	9·8	9·8		9·9		11·2					10
"	0·32						13·0		10·0		11·0			12·2					"
"	0·35	0·30	0·25			0·20	13·1	Softened	10·7	11·0	11·4	11·6	11·8	12·1					"

THERMAL CONDUCTIVITY (k)

Type	Chemical Composition (%)							Heat Treatment °C	Thermal Conductivity cal/cm sec. deg. C											Ref.
	C	Si	Mn	S	P	Ni	Cr		20°C	100°C	200°C	300°C	400°C	500°C	600°C	700°C	800°C	900°C	1000°C	
En 56A	0·03	0·38	0·24	0·033	0·020	0·20	13·0	A850		0·060	0·060	0·063	0·064	0·064	0·064	0·064				1
"	0·07	0·02	0·04				13·1	—	0·061	0·061	0·061	0·061	0·061	0·061	0·061	0·061	0·061	0·061	0·061	11
"	0·07	0·09	0·09			0·23	12·0	A845		0·060	0·062	0·064	0·066							12
"	0·07	0·14	0·29	0·025	0·012	0·11	11·8	H & T	0·059						0·066					2
En 56B	0·13	0·17	0·25	0·024	0·018	0·14	13·0	AC960 T750	0·064	0·066	0·066	0·067	0·066	0·065	0·063	0·061	0·060	0·064	0·066	4
"	0·14	0·12	0·19			0·70	14·6	A845		0·058	0·059	0·060		0·062						12
"	0·16	0·26	0·46			0·50	13·3		0·053	0·053	0·054	0·055	0·056	0·057	0·058	0·060	0·063	0·067	0·070	11
En 56C	0·25	0·25	0·25				13·5	Softened	0·058						0·062					3
En 56D	0·27	0·18	0·28	0·022	0·022	0·20	13·7	AC960 T750	0·061	0·063	0·065	0·066	0·066	0·065	0·064	0·061	0·060	0·064	0·066	4
"	0·27	0·27	0·29			0·37	13·6	OQ940 T725	0·060	0·060	0·061	0·061	0·061	0·061	0·061	0·059	0·056			"
"	En 56D Composition								0·059		0·064		0·070	0·073	0·076					13

ELECTRICAL RESISTIVITY (ϱ)

Type	Chemical Composition (%)							Heat Treatment °C	Electrical Resistivity microhm-cm												Ref.
	C	Si	Mn	S	P	Ni	Cr		0°C	20°C	100°C	200°C	300°C	400°C	500°C	600°C	700°C	800°C	900°C	1000°C	
En 56A	0·03	0·24	0·38	0·033	0·020	0·20	13·0	A750		54·6	60·2	71·2									1
,,	,,	,,	,,	,,	,,	,,	,,	A850		53·5	61·9	69·8									,,
,,	,,	,,	,,	,,	,,	,,	,,	OQ980 T750		53·5	61·9	69·8									,,
,,	0·08	0·10	0·12			0·15	13·5	AC950 T750		55·0											3
,,	0·10	0·30	0·50				13·3	OQ970 T750		56·0						107·0					,,
,,	0·10						13·0			52·9	60·5	69·0									1
En 56A (B)	0·12	0·92	0·22			0·24	13·8	A975		51·4											3
,,	,,	,,	,,			,,	,,	OQ1000		61·4											,,
,,	,,	,,	,,			,,	,,	,, T250		58·7											,,
,,	,,	,,	,,			,,	,,	,, T820		56·6											,,
En 56B	0·13	0·17	0·25	0·024	0·018	0·14	13·0	AC960 T750	48·6	50·6	58·4	67·9	76·9	85·4	93·8	102·1	110·3	116·0	115·0	117·0	4
,,	,,	,,	,,	,,	,,	,,	,,	WQ950		56·1											14
En 56C	0·22						13·0	H & T750		53·0	61·0	70·2									1
,,	,,						,,	H & T200		60·0	67·2	76·2									,,
,,	0·25	0·25	0·25				13·5	Softened		57·0						106·0					3
En 56D	0·27	0·18	0·28	0·022	0·022	0·20	13·7	AC960 T750	50·3	52·2	59·5	68·4	76·9	85·3	93·5	101·5	109·9	115·7	114·6	117·9	4
,,	,,	,,	,,	,,	,,	,,	,,	WQ950 ($\frac{1}{4}$in. dia.)		62·3											14
,,	,,	,,	,,	,,	,,	,,	,,	,, (1in. dia.)		60·2											,,
,,	0·27	0·27	0·29			0·37	13·6	OQ940 T725			63·4	71·8	80·2	88·5	97·0	105·6	114·1	122·7			4

YOUNG'S MODULUS (E), SHEAR MODULUS (G)

Type	Chemical Composition (%)							Heat Treatment °C	Test Temperature °C	Young's Modulus tons/sq.in.		Shear Modulus tons/sq.in.	Ref.
	C	Si	Mn	S	P	Ni	Cr			Static	Dynamic	Static	
56A	0·07	0·14	0·29	0·025	0·012	0·11	11·8	OQ980 T680	20		14,050		2
,,	,,	,,	,,	,,	,,	,,	,,	,,	100		13,800		,,
,,	,,	,,	,,	,,	,,	,,	,,	,,	200		13,380		,,
,,	,,	,,	,,	,,	,,	,,	,,	,,	300		12,850		,,
,,	,,	,,	,,	,,	,,	,,	,,	,,	400		12,250		,,
,,	,,	,,	,,	,,	,,	,,	,,	,,	500		11,460		,,
,,	,,	,,	,,	,,	,,	,,	,,	,,	600		10,600		,,
,,	,,	,,	,,	,,	,,	,,	,,	,,	700		9,450		,,
,,	,,	,,	,,	,,	,,	,,	,,	,,	800		7,900		,,
56A	0·08	0·22	0·47	0·016	0·011	0·14	12·0	A816	24	13,300		5,420	15
,,	,,	,,	,,	,,	,,	,,	,,	,,	149	12,400		5,110	,,
,,	,,	,,	,,	,,	,,	,,	,,	,,	260	11,850		4,930	,,
,,	,,	,,	,,	,,	,,	,,	,,	,,	371	11,300		4,690	,,
,,	,,	,,	,,	,,	,,	,,	,,	,,	482	10,650		4,460	,,
,,	,,	,,	,,	,,	,,	,,	,,	,,	593	8,450		3,640	,,
56A	0·10	0·30	0·50				13·3	Softened	20	13,900			3
,,	,,	,,	,,						200	13,100			,,
,,	,,	,,	,,						400	12,000			,,
,,	,,	,,	,,						600	10,600			,,
56A	0·10	0·20	0·32			0·18	13·6	OQ970 T700	RT	13,400			,,
56B	0·15	0·11	0·14	0·014	0·018	0·27	13·5	OQ950 T650	RT	14,100			16
,,	,,	,,	,,	,,	,,	,,	,,	OQ1000 T650	RT	13,800			,,
56B	0·16	0·29	0·45	0·019	0·019		13·1		93		14,200		17
,,	,,	,,	,,	,,	,,		,,		204		13,800		,,
,,	,,	,,	,,	,,	,,		,,		316		13,300		,,
,,	,,	,,	,,	,,	,,		,,		427		12,650		,,
,,	,,	,,	,,	,,	,,		,,		538		11,800		,,
,,	,,	,,	,,	,,	,,				649		11,400		,,
56B	0·16	0·13	0·13			0·26	13·6	OQ1000 T625	20	13,820			3
	0·18	0·29	0·31			0·22	13·8	AH1000 T150	RT	13,400			,,
56C	0·21	0·28	0·46			0·28	13·9	Softened	RT	13,850			7
	,,	,,	,,			,,	,,	OQ950 T650		13,550			,,
56D	0·27	0·35	0·26			0·29	13·1	OQ940 T700	RT	13,700			3
,,	0·31	0·31	0·27			0·18	14·2	OQ950 T650	RT	14,400			16

YOUNG'S MODULUS (E), SHEAR MODULUS (G) (Continued)

Type	Chemical Composition (%)							Heat Treatment °C	Test Temperature °C	Young's Modulus tons/sq. in.		Shear Modulus tons/sq.in.	Ref.
	C	Si	Mn	S	P	Ni	Cr			Static	Dynamic	Static	
56D	0·31	0·31	0·27			0·18	14·2	OQ1000 T650	RT	14,300			16
”	0·32	0·21	0·21			0·81	12·7	OQ990 T750	20		14,200		3
”	”	”	”			”	”	”	200		13,400		”
”	”	”	”			”	”	”	400		12,300		”
”	”	”	”			”	”	”	600		10,800		3
”	”	”	”			”	”	”	750		9,000		”
	Within En 56D specification								20	13,300			13
”	”	”	”						200	12,850			”
”	”	”	”						400	12,150			”
”	”	”	”						500	11,100			”
”	”	”	”						600	10,300			”

MAGNETIC PROPERTIES

Type	Chemical Composition (%)							Heat Treatment °C	Magnetic Induction B (Gauss) for magnetizing field H (Oersted) =											For H max. = 250		Ref.
	C	Si	Mn	S	P	Ni	Cr		5	10	20	40	50	60	100	150	200	250	400	Brem	H_c	
56A	0·03	0·38	0·24	0·033	0·020	0·20	13·0	AC750	5,700	8,000	11,000	13,000	13,800	14,300	15,200	15,900	16,450	17,100		11,300	8	1
”	”	”	”	”	”	”	”	A850	6,600	11,200	12,650	13,800	14,150	14,400	15,200	16,050	16,900	17,700		9,800	4	”
”	”	”	”	”	”	”	”	OQ980 T750	2,100	4,200	8,150	12,700	13,500	14,000	15,200	16,100	16,700	17,100		8,000	9	”
56B	0·15	0·97	0·28			0·22	14·1	—	2,500	5,600	8,100	10,100	10,750	11,250	12,500	13,300	13,700		14,500			”
56D	0·31	0·05	0·22			0·11	14·5	AC900 (HB445)		250			2.050				10,600					3
”	”	”	”			”	”	A900 (HB 191)		2,300			12,500				16,050					”

MAGNETIC PROPERTIES (Continued)

Type	Chemical Composition (%)					Heat Treatment °C	Magnetic Properties for H max. = 400 Oersted				Ref.
	C	Si	Mn	Ni	Cr		B max. (Gauss)	Max. Permeability (μ max)	Brem (Gauss)	H_c (Oersted)	
56A or B	0·12	0·92	0·22	0·24	13·8	OQ1000	14,900	110	6,100	37·5	3
"	"	"	"	"	"	" T250	13,940	110	8,280	37·5	"
"	"	"	"	"	"	" T820	15,200	170	7,070	19	"
"	"	"	"	"	"	A975	17,170	500	6,100	7·5	"
En 56C	0·22	0·40	0·30		13·0	OQ960 T700	17,000	310	8,100	16	3
En 56D	0·30	0·40	0·30		13·0	OQ980 T180	12,500	75	6,500	58	"

Type En 56D (0·30%C, 13·0%Cr)			Ref.
Heat Treatment °C	Magnetic Properties for H max. = 400		
	Brem (Gauss)	H_c (Oersted)	
WQ 860	7,050	54	3
" 820	7,500	46	"
" 840	6,980	52	"
" 860	7,290	54·5	"
" 880	6,840	56	"
" 900	6,770	59	"
" 950	6,580	63·5	"
" 1000	6,220	69	"
" 1050	5,260	67	"
AC950 WQ840	6,870	50	"
" WQ880	6,700	54	"
" WQ950	6,440	58½	"
WQ800 A950 OQ900	6,900	61½	"
WQ820 A950 OQ950	6,640	63	"
WQ840 A950 OQ1000	6,040	69	"
WQ860 T250 ½ hr.	7,200	55	"
WQ880 "	6,850	59½	"
WQ900 "	6,620	63½	"

En 56

TRANSFORMATION CHARACTERISTICS

ISOTHERMAL TRANSFORMATION

En 56A

Chemical Composition (%)							
C	Si	Mn	S	P	Ni	Cr	Mo
0·11	0·45	0·49	0·005	0·012	0.13	12·00	0·020

Basic Electric Arc.
Actual Grain Size 10 (ASTM).
Austenitizing Temperature 1000°C, 30 min.

(Ref. 18)

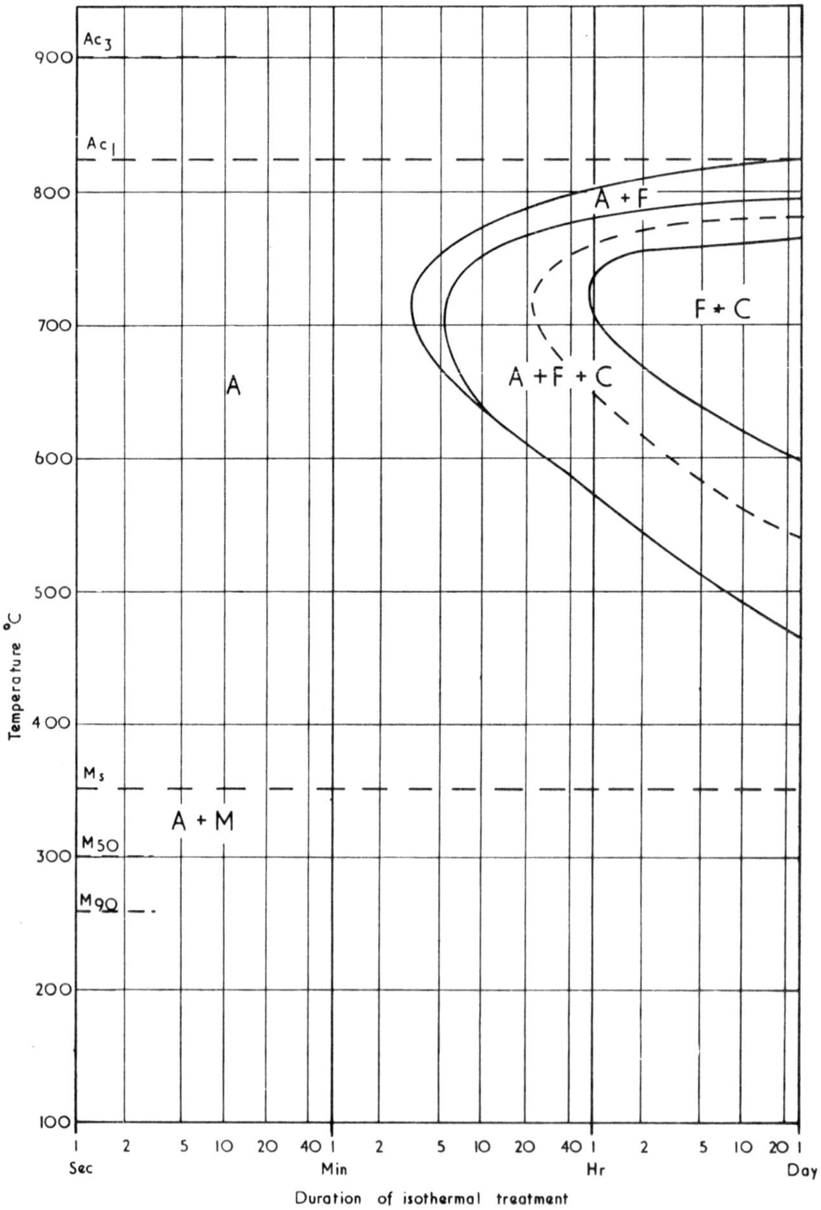

Fig. 56.1

ISOTHERMAL TRANSFORMATION (Continued)

En 56A

Chemical Composition (%)				
C	Si	Mn	Ni	Cr
0·11	0·37	0·44	0·16	12·2

Actual Grain Size 6—7 (ASTM).
Austenitizing Temperature 1800°F (982°C).

(Ref. 19)

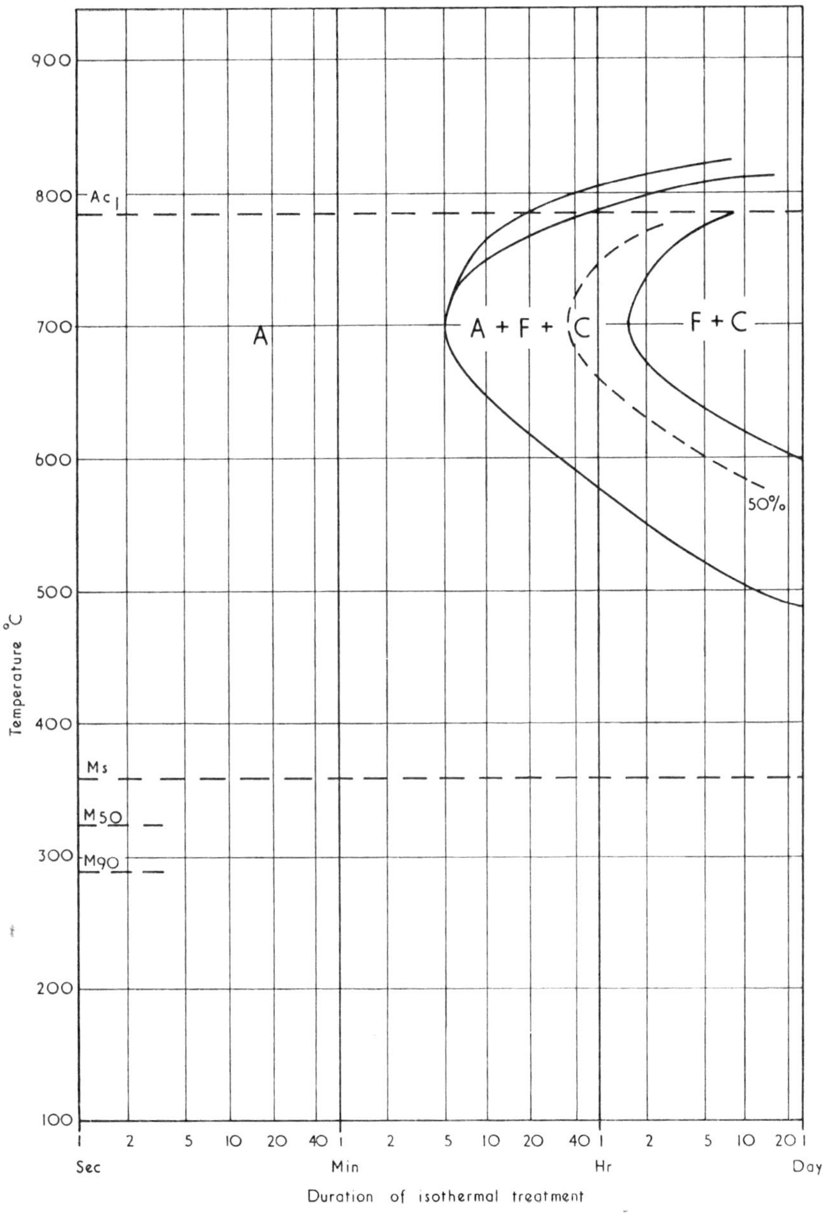

Fig. 56.2

En 56

ISOTHERMAL TRANSFORMATION (Continued)

En 56C

Chemical Composition (%)							
C	Si	Mn	S	P	Ni	Cr	Mo
0·24	0·37	0·27	0·010	0·021	0·32	13·3	0·06

Basic Electric Arc.
As quenched Grain Size 7 (ASTM).
Bars hot rolled to ⅜ in. dia., cold drawn to 11 SWG with intermediate annealing at 600°C.
Austenitizing Temperature 960°C, 30 min.

(Ref. 20)

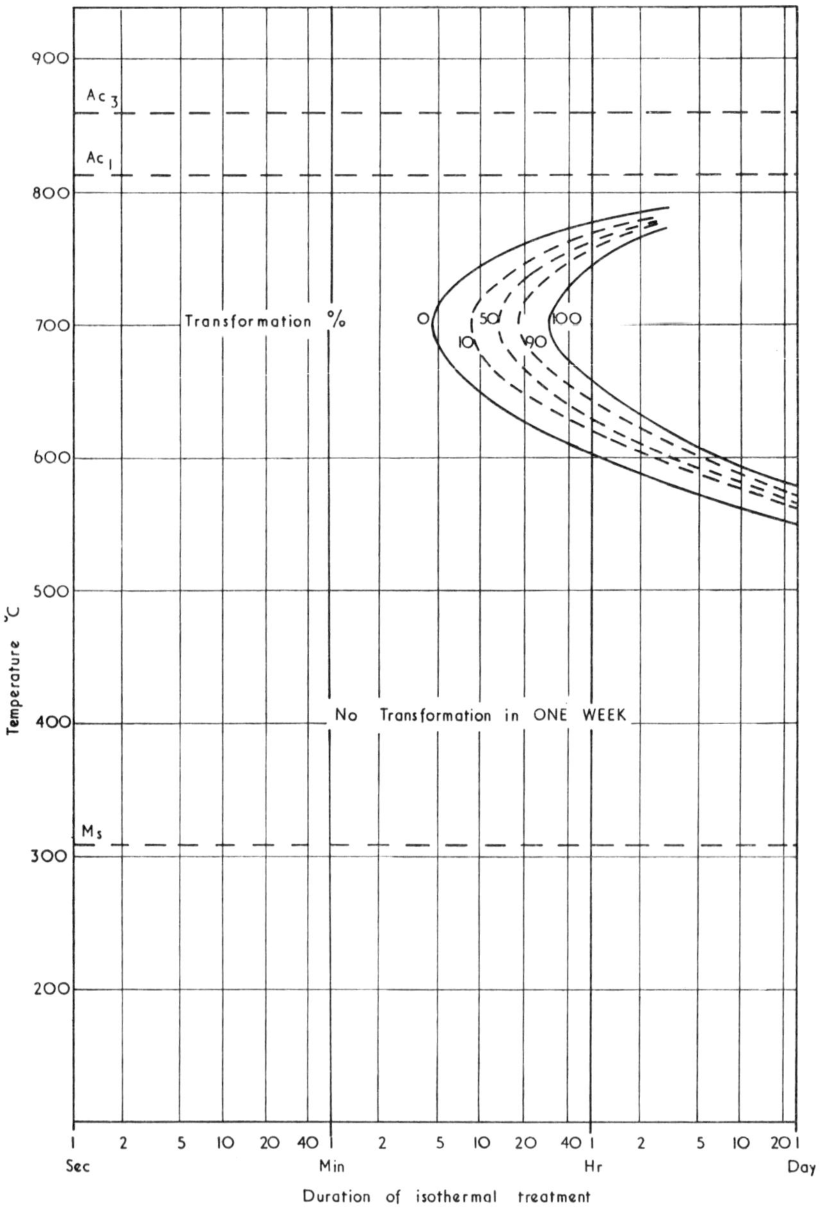

Fig. 56.3

ISOTHERMAL TRANSFORMATION (Continued)

En 56C

Chemical Composition (%)				
C	S	Mn	Ni	Cr
0·21	0·37	0·25	0·36	13·2

Austenitizing Temperature 1000°C, 20 min.

(Ref. 21)

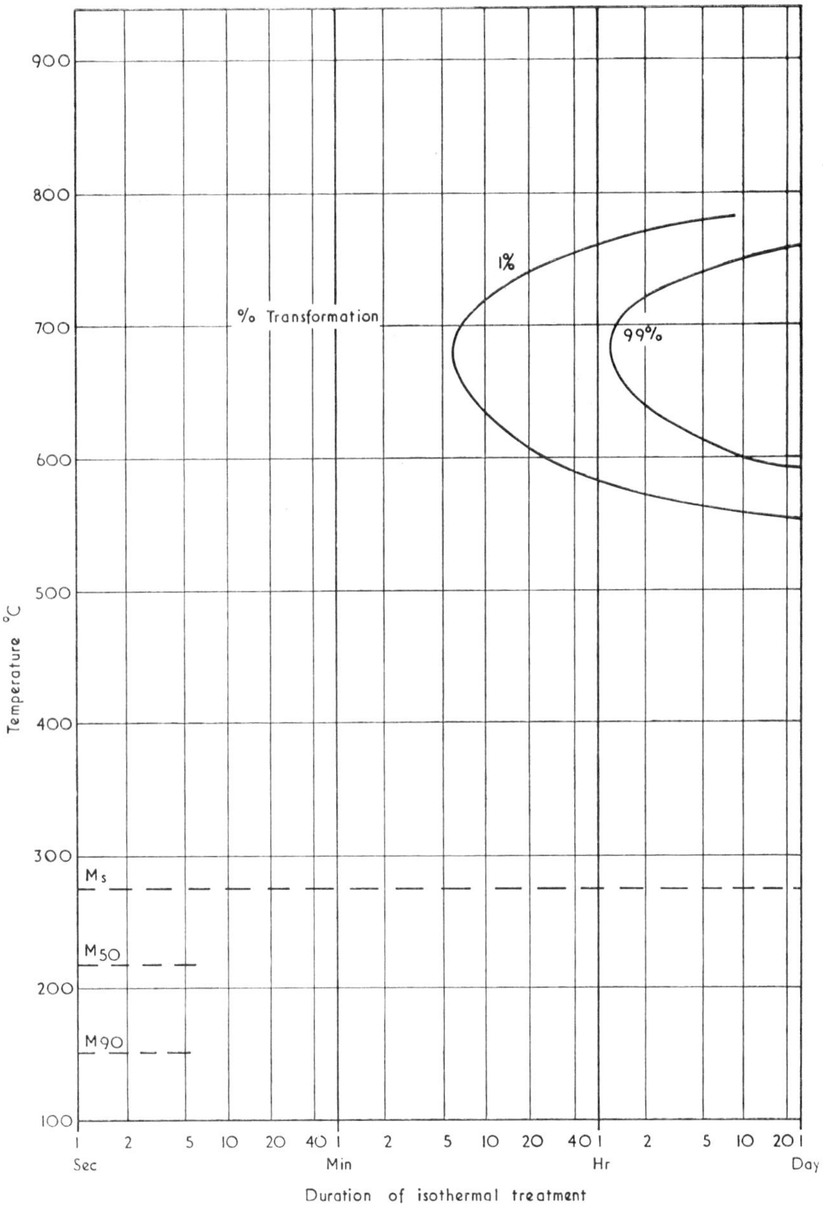

Fig. 56.4

En 56

ISOTHERMAL TRANSFORMATION (Continued)

En 56D

Chemical Composition (%)						
C	Si	Mn	S	P	Ni	Cr
0·29	0·85	0·40	0·005	0·023	0·18	12·3

Basic Electric Arc.
Actual Grain Size 10 (ASTM).
Austenitizing Temperature 1000°C, 30 min.

(Ref. 18)

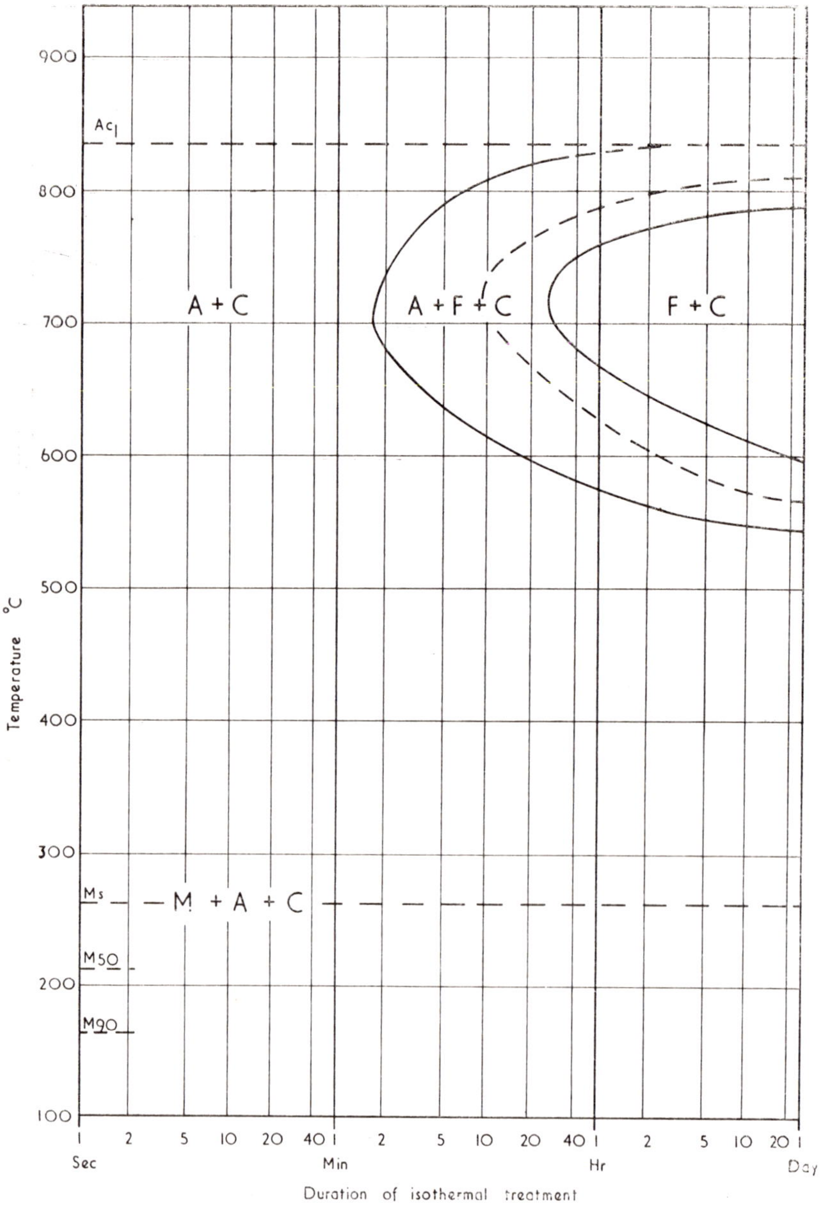

Fig. 56.5

HARDENABILITY

END QUENCH HARDENABILITY DATA

En 56A

Chemical Composition (%)							
C	Si	Mn	S	P	Ni	Cr	Mo
0·11	0·45	0·49	0·005	0·012	0·13	12·0	0·020

Basic Electric Arc.
Actual Grain Size 10 (ASTM).
End quenched from 1000°C.

(Ref. 18)

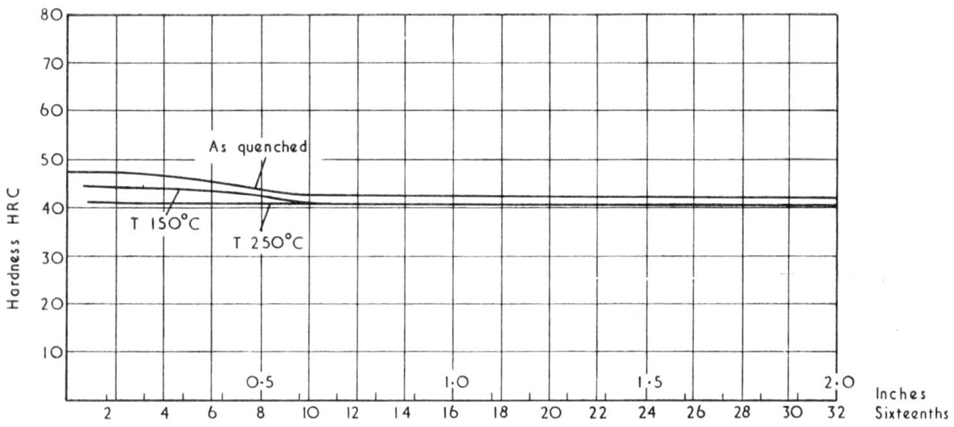

Fig. 56.6

En 56A

Chemical Composition (%)				
C	Si	Mn	Ni	Cr
0·11	0·37	0·44	0·16	12·2

Actual Grain Size 6—7 (ASTM).
End quenched from 1800°F (982°C).

(Ref. 19)

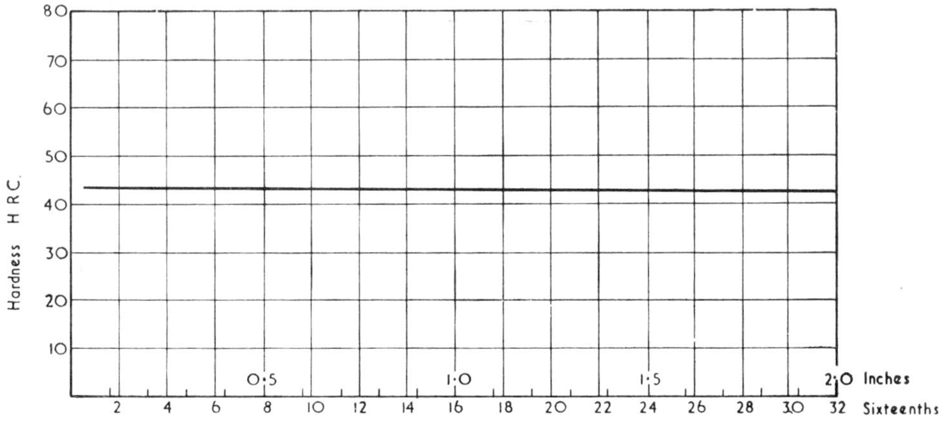

Fig. 56.7

END QUENCH HARDENABILITY DATA (Continued)

En 56C

Chemical Composition (%)							
C	Si	Mn	S	P	Ni	Cr	Mo
0·24	0·37	0·27	0·010	0·021	0·32	13·3	0·06

Basic Electric Arc.
As quenched Grain Size 7 (ASTM).
End quenched from 960°C.

(Ref. 20)

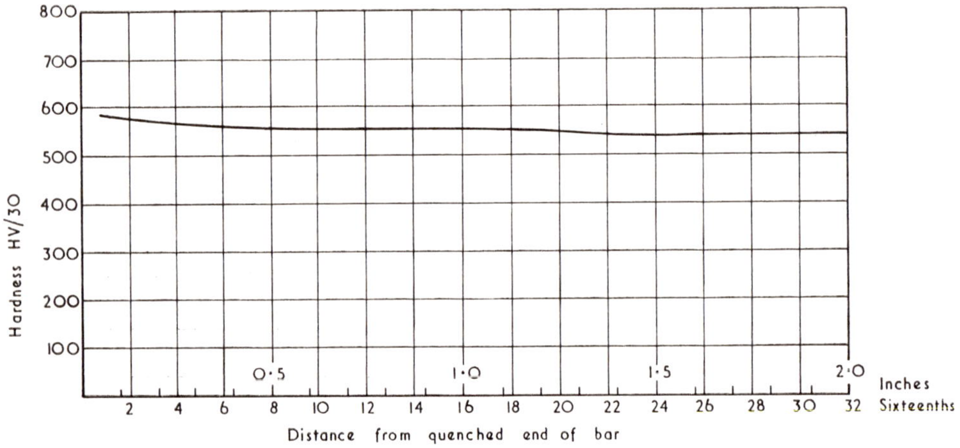

Fig. 56.8

En 56D

Chemical Composition (%)						
C	Si	Mn	S	P	Ni	Cr
0·29	0·85	0·40	0·005	0·023	0·18	12·32

Basic Electric Arc.
Actual Grain Size 10 (ASTM).
End quenched from 1000°C.

(Ref. 18)

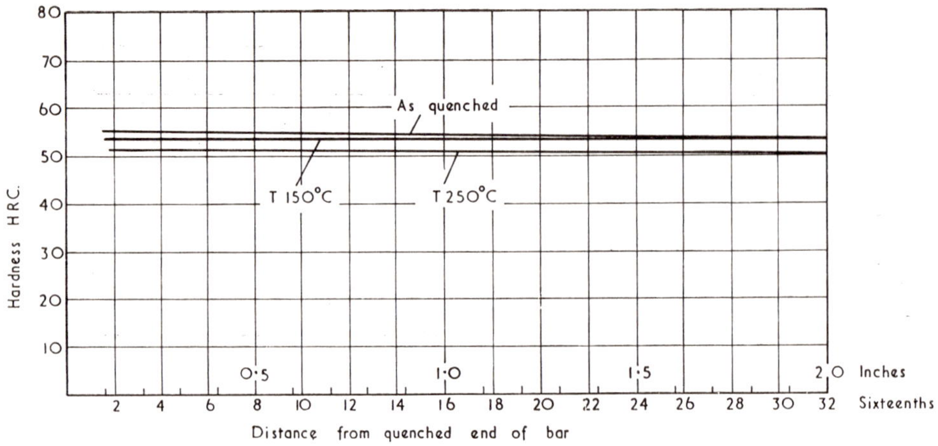

Fig. 56.9

MECHANICAL PROPERTIES AT ROOM TEMPERATURE

Type	Chemical Composition (%)							Section Size dia. (in.)	Heat Treatment °C	Tensile							Izod ft.lb.	Ref.
	C	Si	Mn	S	P	Ni	Cr			0·1% P.S. (0·1Rp) tons/sq.in.	0·2% P.S. (0·2Rp) tons/sq.in.	0·5% P.S. (0·5Rp) tons/sq.in.	Y.P. (Rs) tons/sq.in.	T.S. (Rm) tons/sq.in.	El % (A)	R.A. % (Z)		
En 56A	0·06	0·11	0·18			0·28	13·4	1⅛	AC1000				60·2	76·8	10	26	21	3
"	"	"	"			"	"	"	OQ1000 T2hrs. 200				75·0	82·4	7	13½	15	"
"	"	"	"			"	"	"	" " 400				61·8	75·2	18	58	15	"
"	"	"	"			"	"	"	" " 500				56·8	64·5	15	59	14	"
"	"	"	"			"	"	"	" " 600				45·0	51·1	19	54½	82	"
"	"	"	"			"	"	"	" " 700				32·3	41·3	26½	66	110	"
"	"	"	"			"	"	"	" " 750				25·6	35·9	31	66½	108	"
"	"	"	"			"	"	"	AC950 SC				15·8	31·2	39	71½	68	"
"	"	"	"			"	"	"	A850 SC				15·9	30·1	40½	75	102	"
"	"	"	"			"	"	"	OQ1000 T2hrs. 800 T750WQ				25·1	36·7	36	75	114	"
En 56A	0·06	0·07	0·18			0·20	12·60	1⅛	As rolled				67·8	76·0	4	9½		3
"	"	"	"			"	"	"	AC1000				59·5	73·8	10½	21	26	"
"	"	"	"			"	"	"	OQ1000				71·2	78·5	4	10	10	"
"	"	"	"			"	"	"	" T1hr. 500				73·0	78·5	21	61½	8	"
"	"	"	"			"	"	"	" " 750				21·6	32·8	35	77½	110	"
"	"	"	"			"	"	"	" " 800				23·5	34·4	35	77	110	"
"	"	"	"			"	"	"	A750 3hrs.SC				25·0	35·7	33½	77	100	"
"	"	"	"			"	"	"	A850 " "				12·6	27·9	39½	78	87	"
"	"	"	"			"	"	"	OQ927 T593				43·6	49·6	23½	50	75	"
En 56A	0·07	0·14	0·29	0·025	0·012	0·11	11·8	1⅛	OQ980 T680	34·8		38·3		46·0	29	67		2
En 56A	0·08	0·08	0·16			0·93	11·5	1⅛	As rolled				68·2	84·2	5½	11½		3
"	"	"	"			"	"	"	AC1000				63·4	79·7	2½	8½	24	"
"	"	"	"			"	"	"	OQ1000				75·2	85·2	1½	5	15	"
"	"	"	"			"	"	"	OQ1000 T1hr. 500AC				77·2	86·1	19	57	7	"
"	"	"	"			"	"	"	" T1hr. 750 "				25·4	37·6	28	73	100	"
"	"	"	"			"	"	"	" T800 T750				43·2	51·5	20	61	41	"
"	"	"	"			"	"	"	A750 3hrs.SC				31·5	39·5	29	79½	104	"
"	"	"	"			"	"	"	AC850				75·4	83·0	4½	8½	13	"
"	"	"	"			"	"	"	OQ927 T593				48·1	53·4	23	66½	40	"

MECHANICAL PROPERTIES AT ROOM TEMPERATURE (Continued)

Type	Chemical Composition (%) C	Si	Mn	S	P	Ni	Cr	Section Size dia. (in.)	Heat Treatment °C	Tensile 0·1% P.S. (0·1Rp) tons/sq.in.	0·2% P.S. (0·2Rp) tons/sq.in.	0·5% P.S. (0·5Rp) tons/sq.in.	Y.P. (Rs) tons/sq.in.	T.S. (Rm) tons/sq.in.	El % (A) 4√A	El % (A) 5d	R.A. % (Z)	Izod ft.lb.	Ref.
En 56A	0·08	0·20	0·40			0·11	12·3	1⅛					30·6	38·5	29½			100	2
En 56A	0·08	0·18	0·25	0·015	0·017	0·30	12·1	1	AC1000 T2hrs. 525 T2hrs. 540	54·2	59·8	63·7		65·6	21½	15½	64·6		3
En 56A	0·08	0·08	0·06				13·3	1⅛	AC975				28·2	57·6	10½		18	7	"
"	"	"	"				"	"	OQ975 T2hrs. 200				56·6	63·7	7½		15	8	"
"	"	"	"				"	"	" " 400				51·4	65·0	18½		54½	8	"
"	"	"	"				"	"	" " 500				51·2	62·7	22		64½	10	"
"	"	"	"				"	"	" " 600				33·2	40·0	27½		74½	97	"
"	"	"	"				"	"	" " 700				22·0	33·0	35		78	114	"
"	"	"	"				"	"	" " 750				20·0	31·9	36		80	99	"
"	"	"	"				"	"	AC950 SC				11·6	26·9	42		72½		"
"	"	"	"				"	"	A850 "				13·2	25·3	43		75½		"
"	"	"	"				"	"	OQ1000 T800 T2hrs. 750 WQ				17·1	29·8	38½		80½	106	"
En 56A	0·08					0·46	13·3	1⅛	AH950 T700 WQ				34·4	40·7	32		69	99	23
En 56A	0·08					0·80	13·6	1⅛	AH950 T700 WQ				39·0	45·0	29		63½	87	"
En 56A	0·09	0·23	0·47			0·11	12·5	¾	OQ1000 T1hr. 750				33·8	42·7	32		73	104	3
"	"	"	"			"	"	"	" " 1 month 500				33·5	42·9	32		72	91	"
"	"	"	"			"	"	"	" " 6 months "				34·2	42·4	30		72	102	"
"	"	"	"			"	"	"	" " 12 " "				32·3	42·2	32		75½	98	"
En 56A	0·09	0·12	0·20			0·23	13·1	1⅛	AC850 T750			18·4		30·9	36		72	100	"
En 56A	0·09	0·27	0·26			0·14	13·5	1⅛	OQ960 T740 WQ			25·5		37·4	32		74	110	"
En 56A	0·09	0·28	0·47			0·35	14·0	1⅛	OQ960 T750 WQ			25·9		37·8	31		75	107	"
En 56A	0·09	0·14	0·28			0·12	13·6	1⅛	AC750			23·9		36·7	34		73½	103	"
En 56A	0·09	0·30	0·43	0·008	0·020	0·27	13·7	9/16	AC1050 T1hr. 650	33·6	34·8	36·3		45·7	31	24	71		"
En 56A	0·09					0·10	13·7	1⅛	AC950 T700 WQ				29·6	37·1	32½	69	63½	101	23
En 56A	0·10	0·20	0·20				13·5	1⅛	A850				16·9	31·5	37½		76½	100	1
"	"	"	"				"	"	AC1000				58·0	75·0	8		15	12	"
"	"	"	"				"	"	OQ1000				77·0	80·6	2½		6½	5	"
"	"	"	"				"	"	" T500				69·0	82·3	16		50½	4	"
"	"	"	"				"	"	" T750				24·5	37·2	33		75	98	"
En 56A	0·10	0·18	0·20				13·1	1⅛	AC850 AC750			14·8		27·5	40		77	90	"

MECHANICAL PROPERTIES AT ROOM TEMPERATURE (Continued)

Type	Chemical Composition (%) C	Si	Mn	S	P	Ni	Cr	Section Size dia. (in.)	Heat Treatment °C	Tensile 0·1% P.S. (0·1Rp) tons/sq.in.	0·2% P.S. (0·2Rp) tons/sq.in.	0·5% P.S. (0·5Rp) tons/sq.in.	Y.P. (Rs) tons/sq.in.	T.S. (Rm) tons/sq.in.	El % (A)	R.A. % (Z)	Izod ft.lb.	Ref.
En 56A	0·11	0·89					12·5	1	OQ960 T700				30·4	44·0	25	52	70	23
En 56A	0·11	0·13				0·55	12·4	1⅜	OQ950 T750				30·6	41·2	29		92	22
"	"	"				"	"	3¼	OQ950 T730				32·8	41·2	29		70	"
En 56A 56B	0·12	0·92	0·22			0·24	13·8	1⅛	A975 SC			18·7		30·3	37½	72½	2	3
"	"	"	"			"	"	"	OQ1000 T750 WQ			25·6		38·4	29	58	26	"
"	"	"	"			"	"	"	" " SC			25·4		39·2	32	67	24	"
"	"	"	"			"	"	"	" T820 WQ			21·5		36·2	31	54½	32	"
"	"	"	"			"	"	"	" " SC			22·8		38·1	34	67	13	"
En 56A 56B	0·12		0·3				13·0		A750				17·0	30·0	38	77	100	7
"	"		"				"		OQ950 T700				25·0	38·0	35	75	98	"
En 56A 56B	0·12	0·20	0·25				12·9	1⅛	OQ1000 T750			24·0		37·9	32	70	80	3
En 56B	0·13	0·31	0·29			0·33	13·5	1⅛	OQ950 T730 WQ			32·4		43·0	30	73	119	3
En 56B	0·14	0·28	0·34	0·016	0·015	0·19	13·3	1·7	OQ970 T500 WQ				70·0	83·0	17	58	18	7
"	"	"	"	"	"	"	"	"	" T600 "				50·0	60·0	20	63	40	"
"	"	"	"	"	"	"	"	"	" T700 "				34·0	46·0	26	67	94	"
En 56B	0·15	0·11	0·14	0·014	0·018	0·25	13·5		OQ950 T600 1hr. AC		50·5			59·8	21	63	27	16
"	"	"	"	"	"	"	"		" T650 " "		39·5			49·4	24	66	39	"
"	"	"	"	"	"	"	"		" T700 " "		32·4			43·6	28	69	107	"
"	"	"	"	"	"	"	"		OQ1000 T600 " "		47·0			57·0	22	63		"
"	"	"	"	"	"	"	"		" T650 " "		40·0			50·0	25	65		"
"	"	"	"	"	"	"	"		" T700 " "		32·5			43·8	29	67		
En 56B	0·16	0·52	0·27			0·31	13·0	1⅛	OQ950				89·6	94·4	7	12	36	23
"	"	"	"			"	"	"	" T500 WQ				84·0	94·0	20	59½	22	"
"	"	"	"			"	"	"	" T600 "				44·0	54·8	24½	65	50	"
"	"	"	"			"	"	"	" T650 "				38·8	48·4	24	67	89	"
"	"	"	"			"	"	"	" T700 "				37·2	44·8	28½	71	100	"
"	"	"	"			"	"	"	" T750 "				36·8	43·4	31	67	120	"

En 56

MECHANICAL PROPERTIES AT ROOM TEMPERATURE (Continued)

Type	Chemical Composition (%) C	Si	Mn	S	P	Ni	Cr	Section Size dia. (in.)	Heat Treatment °C	Tensile (tons/sq.in.) 0·1% P.S. (0·1Rp)	0·2% P.S. (0·2Rp)	0·5% P.S. (0·5Rp)	Y.P. (Rs)	T.S. (Rm)	El % (A)	R.A. % (Z)	Izod ft.lb.	Ref.
En 56B	0·16	0·28	0·20			0·51	13·0	1⅛	AH950 T300 WQ					88·5	10½	33	35	23
"	"	"	"			"	"	"	" T600 "				44·8	54·8	23	60½	78	"
"	"	"	"			"	"	"	" T650 "				40·0	48·0	24	62	98	"
"	"	"	"			"	"	"	" T700 "				39·3	46·2	26	62	96	"
"	"	"	"			"	"	"	" T750 "				34·4	44·2	31	66	93	"
En 56B	0·16	1·00	0·23			0·27	13·8	1⅛	OQ1000				51·6	64·0	2	6	3	3
"	"	"	"			"	"	"	" T200 AC				42·4	56·4	7½	10	7	"
"	"	"	"			"	"	"	" T300 "				48·9	63·7	9½	18	6½	"
"	"	"	"			"	"	"	" T400 "				48·2	63·5	13	30½	5	"
"	"	"	"			"	"	"	" T500 "				49·3	64·4	16	36½	3½	"
"	"	"	"			"	"	"	" T600 "				37·1	44·5	25	59	23	"
"	"	"	"			"	"	"	" T700 "				32·3	41·6	27	62½	60	"
"	"	"	"			"	"	"	" T800 "				22·3	35·8	32½	59	35	"
"	"	"	"			"	"	"	" T875 "				21·1	33·2	33	57	25	"
"	"	"	"			"	"	"	A975 SC				18·6	31·8	42	73½		"
En 56B	0·16	0·13	0·13			0·26	13·6		OQ1000 T625		53·0			63·0	22	62		16
En 56B	0·17	0·29	0·50			0·12	12·6	1⅛	OQ980 T750				36·4	47·0	27½		90	22
En 56B	0·17	0·94	0·34				14·0	"	OQ1000 T820 AC				25·5	40·1	32		26	3
"	"	"	"				"	"	" T820 WQ				25·7	39·9	31		40	"
"	"	"	"				"	"	" T750 AC				27·8	43·4	31		39	"
En 56B 56C	0·18	0·27	0·41			0·11	13·1	1⅛	OQ980 T200 OQ					94·3	4	6½	16	1
"	"	"	"			"	"	"	" T400 "				90·7	95·5	16	49½	5	"
"	"	"	"			"	"	"	" T500 "				85·8	98·7	18½	56½	7	"
"	"	"	"			"	"	"	" T550 "				80·5	82·6	20	49½	33	"
"	"	"	"			"	"	"	" T600 "				57·8	62·3	20½	68	17	"
"	"	"	"			"	"	"	" T650 "				51·0	56·0	26	62½	28	"
"	"	"	"			"	"	"	" T675 "				48·2	53·3	26	65	76	"
"	"	"	"			"	"	"	" T700 "				44·0	50·6	26	68	84	"
"	"	"	"			"	"	"	" T725 "				40·4	48·2	28½	68	89	"
"	"	"	"			"	"	"	" T750 "				38·3	46·6	30½	68	94	"

MECHANICAL PROPERTIES AT ROOM TEMPERATURE (Continued)

Type	Chemical Composition (%) C	Si	Mn	S	P	Ni	Cr	Section Size dia. (in.)	Heat Treatment °C	Tensile 0·1% P.S. (0·1Rp) tons/sq.in.	0·2% P.S. (0·2Rp) tons/sq.in.	0·5% P.S. (0·5Rp) tons/sq.in.	Y.P. (Rs) tons/sq.in.	T.S. (Rm) tons/sq.in.	El % (A) 4√A	El % (A) 5d	R.A. % (Z)	Izod ft.lb.	Ref.
En 56C	0·19	0·52	0·44	0·010	0·010	0·38	13·2	1⅛	OQ900 T690				43·8	54·4	25		66	78	2
En 56C	0·20	0·14	0·20			0·42	13·3	1⅛	AC750			31·5		44·2	37½		63	84	3
En 56C	0·20	0·28	0·49			0·15	12·6	1⅛	OQ980 T750				38·7	48·7	28			68	22
"	"	"	"			"	"	2¼	" "				33·7	44·7	29			89	"
En 56C	0·20	0·26	0·25			0·42	14·2	1⅛	OQ975 T3hrs. 200			99·8		107·7	14		43	11	3
"	"	"	"			"	"	"	" " 500					103·7	13		39	6	"
"	"	"	"			"	"	"	" " 600			47·6		60·3	18½		53½	18	"
"	"	"	"			"	"	"	" " 700			33·1		45·4	29½		67	73	"
"	"	"	"			"	"	"	" " 750			29·6		43·1	30		68½	107	"
En 56C	0·21	0·41	0·42	0·012	0·017	0·24	12·7	1·7	OQ970 T550 WQ				66·0	84·0	16		55	16	7
"	"	"	"	"	"	"	"	"	" T600 "				58·0	70·0	18		58	22	"
"	"	"	"	"	"	"	"	"	" T700 "				37·0	52·0	24		63	63	"
En 56C	0·21	0·38	0·61	0·013	0·007	0·10	13·0	⅝	OQ900 T690				44·0	53·6	22		64	37	2
En 56C	0·21	0·52	0·49	0·010	0·014	0·28	13·4	1	OQ900 T670				41·6	51·6	25		66	76	"
En 56C	0·21					0·30	13·3	1⅛	AH950 T700 WQ				30·0	44·0	26		61½	38	23
"	"																		
En 56C	0·22	0·23	0·45			0·14	13·6	2¼	OQ1000 T730				43·1	49·4	25			81	22
En 56C	0·22	0·28	0·28			0·14	12·4	2	OQ875 T600 WQ				44·6	52·6	23		66½	70	3
"	"	"	"			"	"	1⅜ sq.	" " "				41·2	48·6	25½		70	70	"
"	"	"	"			"	"	2 sq.	OQ900 "				44·8	52·8	21		63½	70	"
En 56C	0·22					0·74	12·8	1⅛	AH950 T700 WQ				38·0	48·4	25½		60½	58	23
En 56C	0·22	0·40	0·30				13·0	1⅛	OQ960 T500				90·7	98·2	15		44		3
"	"	"	"				"	"	" T550				72·8	79·7	17		50½		"
"	"	"	"				"	"	" T600				51·4	60·5	20		56		"
"	"	"	"				"	"	" T650				43·3	52·3	22		62½		"
"	"	"	"				"	"	" T700				41·3	49·1	26		64½		"
"	"	"	"				"	"	" T750				36·6	45·9	30		70½		"
En 56C	0·23	0·34	0·37	0·023	0·012	0·24	12·7	1¾	OQ900 T720				38·0	49·0	26		64	76	2
En 56C	0·23	0·24	0·61	0·010	0·021	0·29	13·0	¾	OQ1000 2hrs. 550 AC 2hrs. 600 AC	50·0	52·1	54·3		63·6	23	17½	57		3
En 56C	0·24	0·20	0·30			0·20	13·0	1⅛	OQ975 T750 AC			30·4		42·4	28½		65	70	3
En 56C	0·24	0·31	0·24			0·46	13·3	1⅛	OQ970 T600 WQ			44·1		55·6	20½		57	18	"
En 56C	0·24	0·61	0·35			0·19	14·2	1⅛	OQ970 T760 WQ			34·8		47·4	28		66½	100	"

MECHANICAL PROPERTIES AT ROOM TEMPERATURE (Continued)

Type	Chemical Composition (%) C	Si	Mn	S	P	Ni	Cr	Section Size dia. (in.)	Heat Treatment °C	Tensile 0·1% P.S. (0·1Rp) tons/sq.in.	0·2% P.S. (0·2Rp)	0·5% P.S. (0·5Rp)	Y.P. (Rs)	T.S. (Rm)	El % (A)	R.A. % (Z)	Izod ft.lb.	Ref.
En 56C 56D	0·25	0·15	0·28			0·44	14·0	4¼	OQ1000 T640				48·3	56·0	24		32	22
En 56C 56D	0·25	0·21	0·22			0·20	12·2	1⅛	AH925 T700 WQ				43·6	53·5	22½	60½	77	23
"	"	"	"			"	"	"	" " AC								47	"
En 56D	0·26	0·35	0·44	0·019	0·018	0·19	14·0	¾	AH980 T1hr. 660	43·1	44·7			58·2	24½	64	36	1
"	"	"	"	"	"	"	"	1	" " "	43·1	44·8			58·0	25	62½	35	"
"	"	"	"	"	"	"	"	1¼	" " "	43·0	45·2			57·9	26	62½	40	"
"	"	"	"	"	"	"	"	1½	" " "	43·4	44·9	47·2		58·6	25	61	37	"
"	"	"	"	"	"	"	"	1	OQ980 " "	42·7	44·4			57·5	27	65½	38	"
"	"	"	"	"	"	"	"	1¼	" " "	43·0	44·5			57·4	24½	62½	35	"
"	"	"	"	"	"	"	"	1½	" " "	43·2	45·0			58·1	26½	64	37	"
En 56D	0·26	0·36	0·27			0·27	13·0	⅞	AH920 T725 AC				38·4	48·6	28	67	70	3
"	"	"	"			"	"	½	OQ875 T600 WQ				50·0	56·0	22½	62½	21	"
"	"	"	"			"	"	13/16	" T600 WQ					51·6	25	63½	54	"
En 56D	0·26	0·36	0·27			0·31	13·7	1 1/16	OQ970 T630	41·0	43·4	46·6		56·0	19	58		3
En 56D	0·26	0·32	0·32			0·24	13·6	1 3/16	OQ970 T630	40·9	43·7	46·6		56·6	19	57		"
En 56D	0·26	0·25	0·24			0·30	13·6	2¼	OQ875 T600 WQ				37·6	49·8	25	63½	57	"
En 56D	0·27	0·23	0·29			0·28	13·2	1½	OQ970 T700				39·2	48·0	27		91	22
"	"	"	"			"	"	1¾	" "				40·0	48·8	25		80	"
"	"	"	"			"	"	1⅞	" T650				45·2	53·6	22		51	"
"	"	"	"			"	"	4½	OQ1000 T700				36·8	46·0	30		48	"
"	"	"	"			"	"	5⅞	" T650				45·6	54·0	22		23	"
En 56D	0·27	0·71	0·31			0·23	13·9	1⅛	OQ975 T780 WQ			33·2		44·3	27	68	104	3
En 56D	0·27	0·30	0·31			0·37	13·2	1⅛	OQ960 T710 WQ			36·6		45·8	28½	67	99	"
En 56D	0·27	0·34	0·36			0·32	13·5	1 1/16	OQ970 T630	42·6	45·1	48·2		58·4	19½	57		"
En 56D	0·27	0·35	0·33			0·34	13·4	0·74	OQ920 T600 WQ				41·8	48·6	24	66	70	"
En 56D	0·27	0·26	0·23			0·36	13·6	1¼	OQ970 T630	40·3	43·2	49·0		55·8	19	55		"
En 56D	0·28	0·10	0·22	0·03	0·026	0·20	12·6	1⅛	OQ950 T650 WQ				39·8	49·0	22	55	48	3
"	"	"	"	"	"	"	"	"	" T250				95·0	106·7	7	29½	8	

MECHANICAL PROPERTIES AT ROOM TEMPERATURE (Continued)

Type	Chemical Composition (%) C	Si	Mn	S	P	Ni	Cr	Section Size dia. (in.)	Heat Treatment °C	Tensile 0·1% P.S. (0·1Rp) tons/sq.in.	0·2% P.S. (0·2Rp) tons/sq.in.	0·5% P.S. (0·5Rp) tons/sq.in.	Y.P. (Rs) tons/sq.in.	T.S. (Rm) tons/sq.in.	El % (A)	R.A. % (Z)	Izod ft.lb.	Ref.
En 56D	0·28	0·27	0·27			0·38	13·1	2	OQ875 T600 WQ				47·0	55·2	20	61½	22	3
”	”	”	”			”	”	”	” T650				37·8	49·0	28½	61½	47	”
”	”	”	”			”	”	”	OQ950 T650 ”				37·0	50·3	24	60½	44	”
”	”	”	”			”	”	”	OQ975 T600				52·6	61·3	20½	53½	10	”
”	”	”	”			”	”	”	” T650				44·2	53·8	21	54½	25	”
”	”	”	”			”	”	”	WQ975 T650				44·4	55·8	20	52	22	”
”	”	”	”			”	”	”	” T675				34·0	47·7	29	66	85	”
”	”	”	”			”	”	”	AC975 T650				34·8	52·6	22	58	26	”
”	”	”	”			”	”	”	OQ1000 T650				39·2	52·7	25	59	20	”
”	”	”	”			”	”	”	A975 SC				19·6	41·7	30½	63½	50	”
En 56D	0·28	0·30	0·35	0·017	0·018	0·20	12·9	1⅛	OQ950 T680				36·0	56·8	35	59½	30	2
”	”	”	”	”	”	”	”	”	” T700				38·4	49·2	25½	61½	37	”
En 56D	0·29	0·79	0·27			0·23	13·7		H & T			35·0		48·7	28	65½		16
En 56D	0·29	0·21	0·21			0·28	13·5	1⅛	AC750			34·4		46·4	28	64	76	3
En 56D	0·29	0·30	0·28			0·35	13·0	¾	OQ975 T750			36·0		48·0	29	59		”
En 56D	0·30		0·30				13·0	1⅛	OQ950 T100 SC				95	120	4	12	5	3
”	”		”				”	”	OQ950 T300 ”				90	100	8	20	8	”
”	”		”				”	”	” T500 ”				83	89	10	39		”
”	”		”				”	”	” T550 ”				64	68	12	47	13	”
”	”		”				”	”	” T600 ”				42	49	23	60		”
”	”		”				”	”	” T650 ”				42	49	22	57	44	”
”	”		”				”	”	” T700 ”								72	”
”	”		”				”	”	” T100 WQ				95	120	4	12	5	”
”	”		”				”	”	” T300 ”				90	100	8	20	8	”
”	”		”				”	”	” T500 ”				81	84	7	39	7	”
”	”		”				”	”	” T550 ”				61	66	16	52	15	”
”	”		”				”	”	” T600 ”				45	50	21	59	19	”
”	”		”				”	”	” T650 ”				43	49	22	59	54	”
”	”		”				”	”	” T700 ”				36	44	26	66	86	”
En 56D	0·30	0·20	0·20				13·0	1⅛	OQ960 T180					108				3
En 56D	0·30	0·67	0·31			0·32	13·6	1⅛	OQ975 T780 WQ			35·2		48·8	26	63½	37	”

MECHANICAL PROPERTIES AT ROOM TEMPERATURE (Continued)

Type	Chemical Composition (%)							Section Size dia. (in.)	Heat Treatment °C	Tensile							Izod ft.lb.	Ref.
	C	Si	Mn	S	P	Ni	Cr			0·1% P.S. (0·1Rp)	0·2% P.S. (0·2Rp)	0·5% P.S. (0·5Rp)	Y.P. (Rs)	T.S. (Rm)	El % (A)	R.A. % (Z)		
										tons/sq.in.								
En 56D	0·31	0·56	0·20			0·60	13·0		OQ1000 T700		44·4			56·3	22		58	16
En 56D	0·31	0·31	0·27	0·018	0·016	0·18	14·2		OQ950 T600 AC			54·0		63·4	18	55	29	"
"	"	"	"	"	"	"	"		" T650 "			43·0		55·6	23	60	37	"
"	"	"	"	"	"	"	"		" T700 "			35·0		48·7	28	67	63	"
"	"	"	"	"	"	"	"		OQ1000 T600 "			54·0		67·7	19	50		"
"	"	"	"	"	"	"	"		" T650 "			45·0		59·4	23	56		"
"	"	"	"	"	"	"	"		" T700 "			37·0		52·2	26	62		"
En 56D	0·34	0·28	0·39			0·31	13·5	1	OQ960 T660				45·2	53·6	22		58	22
"	"	"	"			"	"	1⅛	" T725				40·0	48·0	29		102	"
"	"	"	"			"	"	1½	" T660				46·0	54·4	22		35	"
"	"	"	"			"	"	2⅛	" T650				46·0	54·4	24		35	"
"	"	"	"			"	"	4¾	OQ970 T650				42·8	51·6	26		30	"
En 56D	0·35	0·30	0·40				13·5		H & lightly T				94·7	100·8	10	37		24

EFFECT OF NITRIDING ON THE SURFACE HARDNESS

Type	Chemical Composition (%)					Heat Treatment °C		Effect of Nitriding for 90 hrs. at °C				Ref.
	C	Si	Mn	Ni	Cr			500—510	550—560	600—610	650—660	
En 56D	0·28	0·18	0·25	0·35	13·4	OQ950 T600	Hardness HV at surface	725—850	760—750	720—735	510—550	25
							Max. hardness HV below surface	1,050—1,100	1,020—1,030	770—780	600—610	
							Depth of case (in.)	0·012	0·015	0·018	0·025	

Note: Nitriding reduces the corrosion resistance.

EFFECT OF TEMPERING ON THE MECHANICAL PROPERTIES AT ROOM TEMPERATURE

(Ref. 3)

En 56A

Chemical Composition (%)				
C	Si	Mn	Ni	Cr
0·06	0·11	0·18	0·28	13·4

1⅛ in. dia. bars. Oil quenched 1000°C. Tempered as indicated.

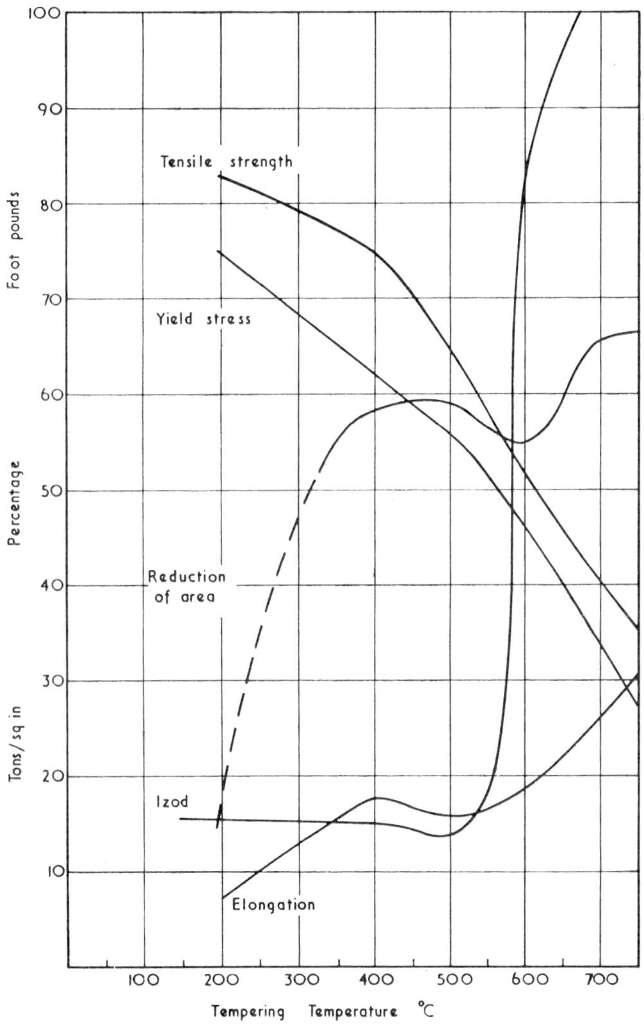

Fig. 56.10

EFFECT OF TEMPERING ON THE MECHANICAL PROPERTIES AT ROOM TEMPERATURE

(Continued)

En 56B

Chemical Composition (%)						
C	Si	Mn	S	P	Ni	Cr
0·14	0·23	0·56	0·020	0·016	0·27	12·6

35 mm. dia. bars. Oil quenched 950°C. Tempered as indicated.

(Ref. 2)

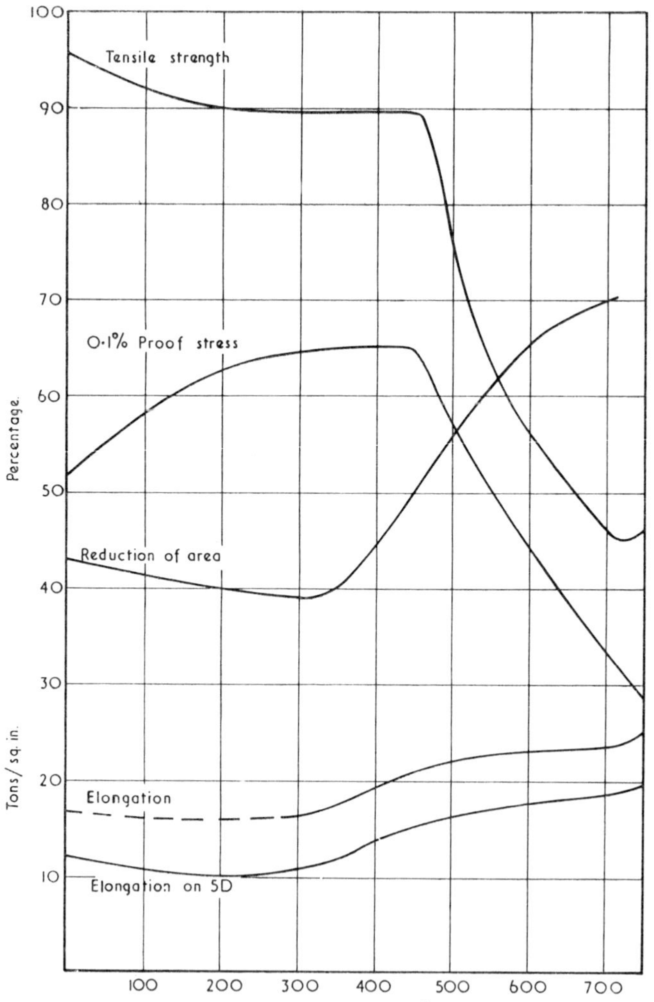

Fig. 56.11

EFFECT OF TEMPERING ON THE MECHANICAL PROPERTIES AT ROOM TEMPERATURE

(Continued)

En 56B· En 56C

Chemical Composition (%)				
C	Si	Mn	Ni	Cr
0·18	0·27	0·41	0·11	13·1

$1\frac{1}{8}$ in. dia. bars. Oil quenched 980°C. Tempered as indicated.

(Ref. 1)

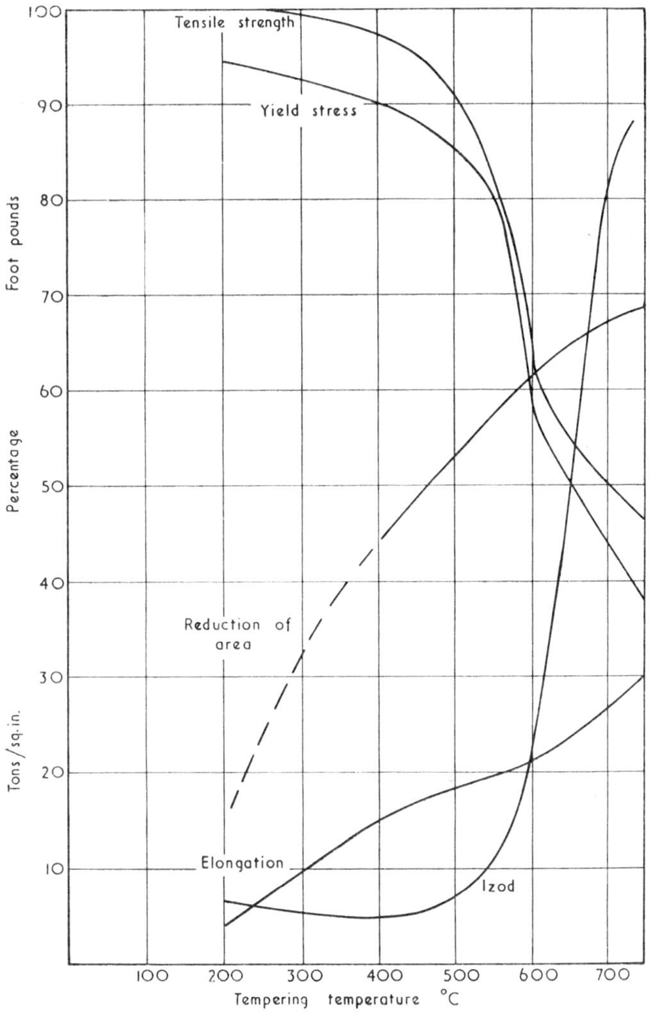

Fig. 56.12

EFFECT OF TEMPERING ON THE MECHANICAL PROPERTIES AT ROOM TEMPERATURE

(Continued)

En 56D

Chemical Composition (%)							
C	Si	Mn	S	P	Ni	Cr	Mo
0·27	0·21	0·28	0·014	0·011	0·25	12·9	0·05

Basic Electric Arc.
1⅛ in. dia. bars. Oil quenched 950°C. Tempered as indicated.

(Ref. 2)

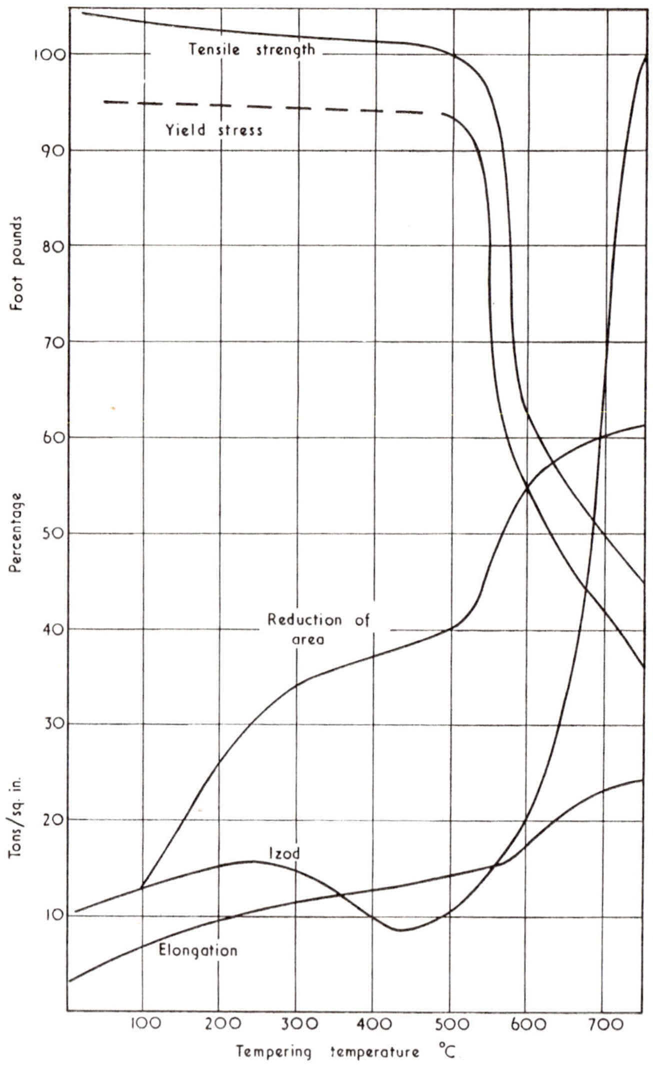

Fig. 56.13

MECHANICAL PROPERTIES AT LOW TEMPERATURE

Type	Chemical Composition (%) C	Si	Mn	S	P	Ni	Cr	Heat Treatment °C	Test Temperature °C	Tensile 0·2% P.S. (0·2Rp) tons/sq.in.	Y.P. (Rs) tons/sq.in.	T.S. (Rm) tons/sq.in.	El % (A)	R.A. % (Z)	Notched Impact ft.lb. Charpy Keyhole	Ref.
En 56A	0·07	0·15	0·20	0·025	0·011	0·08	13·8	Annealed	RT			29·5	40	76	39	26
"	"	"	"	"	"	"	"	"	—40			33·7	45	73	25	"
"	"	"	"	"	"	"	"	OQ1010 T2hr.550 T1½ hrs. 557	RT			54·9	22½	66	28	"
"	"	"	"	"	"	"	"	" " "	— 40			58·0	25	67	26	"
En 56A	0·11	0·25	0·50	0·020	0·020		12·5	AH & Temp.	RT						27	27
"	"	"	"	"	"		"	"	— 18						25	"
"	"	"	"	"	"		"	"	— 32						24	"
"	"	"	"	"	"		"	"	— 46						21	"
"	"	"	"	"	"		"	"	— 73						4	"
															Izod	
En 56A or B	(Within specification)							Annealed	RT		38·8	49·1	21	68	85	28
"	" "							"	0		39·7	51·3	24	69	40	"
"	" "							"	— 40		40·2	54·5	23	64	25	"
"	" "							"	— 62		42·0	57·1	22	60	25	"
"	" "							"	—196		66·1	70·5	10	11	5	"
En 56C	0·21	0·31	0·61	0·010	0·018	0·55	13·1	OQ950 T250	+ 25	70·7		96·0	18½	50½	See fig 56·14	29
"	"	"	"	"	"	"	"	" "	— 50	80·9		102·3	18½	57½		"
"	"	"	"	"	"	"	"	" "	—120	86·8		109·1	18½	51		"
"	"	"	"	"	"	"	"	" "	—135	87·0		106·4	3	3		"
"	"	"	"	"	"	"	"	" "	—150	90·2		106·5	0	1		"
"	"	"	"	"	"	"	"	" "	—195	101·2		111·7				"
																"
En 56D	0·30	0·25	0·26			0·30	13·7	H & T	RT			46·6	33½	67		3
"	"	"	"			"	"	"	+0·5			46·6	35	66		"
"	"	"	"			"	"	"	— 30			49·0	37½	66		"
"	"	"	"			"	"	"	— 60			52·0	36	65½		"
"	"	"	"			"	"	"	— 90			53·7	35	65		"
En 56D	0·35	0·30	0·40				13·5	H & lightly T	RT		94·7	100·8	10	37		24
"	"	"	"				"	" "	—183		94·1	116·0	3	3½		"

(See also data on page 239)

IMPACT TRANSITION TEMPERATURE DATA

En 56C

Chemical Composition (%)						
C	Si	Mn	S	P	Ni	Cr
0·21	0·31	0·61	0·010	0·018	0·55	13·1

Oil quenched 950°C. Tempered 250°C.
(See also Mechanical Properties at Low Temperature (page 237) for tensile test data.)

(Ref. 29)

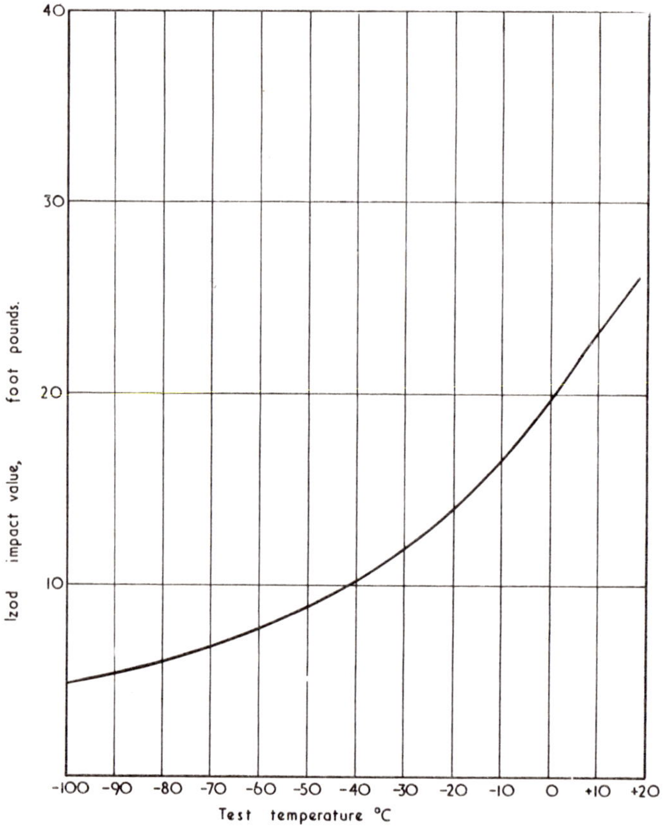

Fig. 56.14

IMPACT TRANSITION TEMPERATURE DATA (Continued)

En 56D

C	Si	Mn	Ni	Cr	Heat Treatment °C	Test Temperature °C	Tensile Y.P. (Rs) tons/sq.in.	Tensile T.S. (Rp) tons/sq.in.	Tensile El % (A)	Tensile R.A. % (Z)	Ref.
0·32	0·21	0·39	0·20	12·77	OQ950 T550	+ 20	59·2	67·4	22	54·5	39
"	"	"	"	"	" "	0	62·1	70·1	20	53	"
"	"	"	"	"	" "	— 20	62·7	71·5	20	51	"
"	"	"	"	"	" "	— 40	61·8	71·4	17	52	"
"	"	"	"	"	" "	— 60	63·5	74·4	17	49	"
"	"	"	"	"	" "	— 80	64·0	76·5	17	46	"
"	"	"	"	"	" "	—100	71·4	78·4	17	45	"
"	"	"	"	"	OQ950 T650	+ 20	46·2	54·4	26½	66	"
"	"	"	"	"	" "	0	47·7	54·7	26½	67	"
"	"	"	"	"	" "	— 20	46·1	56·6	27	65½	"
"	"	"	"	"	" "	— 40	48·0	57·9	23	63	"
"	"	"	"	"	" "	— 60	50·9	60·6	24	62	"
"	"	"	"	"	" "	— 80	52·5	62·1	24	61	"
"	"	"	"	"	" "	—100	56·0	63·8	23	60	"
"	"	"	"	"	OQ950 T700	+ 20	39·4	48·0	37	66	"
"	"	"	"	"	" "	0	41·6	49·6	29½	66	"
"	"	"	"	"	" "	— 20	42·9	51·4	30½	63	"
"	"	"	"	"	" "	— 40	42·6	52·8	27½	63	"
"	"	"	"	"	" "	— 60	44·8	54·4	27	60	"
"	"	"	"	"	" "	— 80	46·1	56·5	27	58½	"
"	"	"	"	"	" "	—100	48·3	58·3	26	58	"

1¾ in., dia. bars, oil quenched. Tempered as indicated.

(Ref. 39)

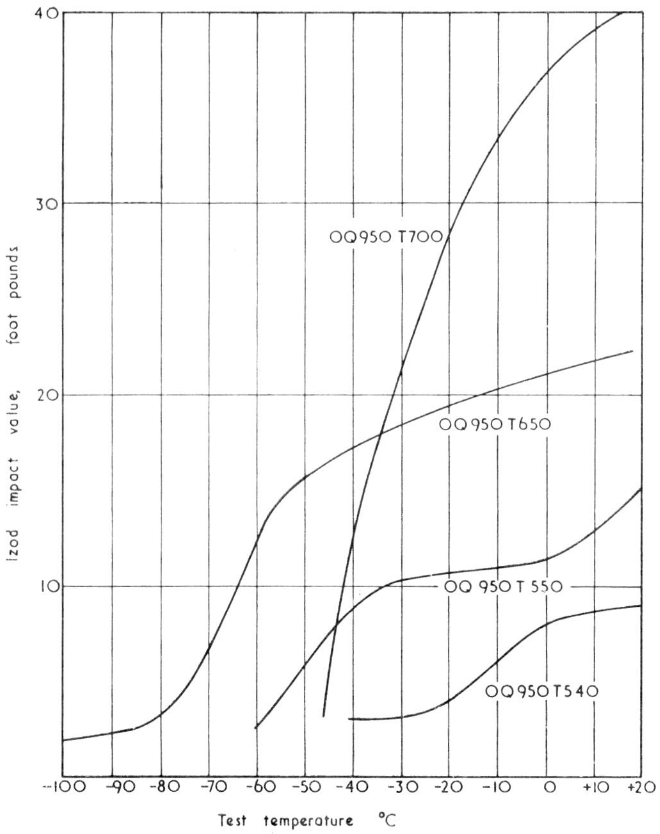

Fig. 56.15

MECHANICAL PROPERTIES AT HIGH TEMPERATURE

Type	Chemical Composition (%)							Section Size dia. (in.)	Heat Treatment °C	Test Temperature °C	Tensile						Izod ft.lb.	Ref.
	C	Si	Mn	S	P	Ni	Cr				0·1% P.S. (0·1Rp)	0·2% P.S. (0·2Rp)	0·5% P.S. (0·5Rp)	T.S. (Rm)	El % (A)	R.A. % (Z)		
											tons/sq.in.							
En 56A	0·07	0·14	0·29	0·025	0·012	0·11	11·8	1⅛	OQ980 T680	20	34·8		38·3	46·0	29	67		2
"	"	"	"	"	"	"	"	"	" "	100	32·6		36·0	42·5	26	66½		"
"	"	"	"	"	"	"	"	"	" "	200	30·4		33·3	39·0	23½	64½		"
"	"	"	"	"	"	"	"	"	" "	300	29·7		32·9	37·1	21	62½		"
"	"	"	"	"	"	"	"	"	" "	400	28·7		32·4	35·7	19	61½		"
"	"	"	"	"	"	"	"	"	" "	500	22·3		26·5	29·3	24	67		"
"	"	"	"	"	"	"	"	"	" "	600	12·0		15·2	18·5	39	84		"
								(Rate of loading: 1 ton/sq. in. min.)										
En 56A	0·10	0·30	0·50				13·3	1⅛	OQ970 T740	20	23·2	25·1	26·9	39·3	33½	72½		3
"	"	"	"				"	"	" "	100	20·6	22·1	23·4	34·5	34½	79½		"
"	"	"	"				"	"	" "	200	19·8	21·4	23·5	32·3	32½	76		"
"	"	"	"				"	"	" "	300	20·4	21·4	22·7	32·7	29½	74½		"
"	"	"	"				"	"	" "	400	18·8	20·1	21·3	29·9	28	74½		"
"	"	"	"				"	"	" "	500	16·2	17·6	19·3	24·9	33	77½		"
"	"	"	"				"	"	" "	600	7·4	9·25	11·3	19·8	50½	90½		"
								(Rate of straining 0·0085 in./min. to Y. P. then ⅛ in./min.)										
En 56A	0·11	0·09	0·43	0·016	0·017	0·27	12·4	1⅛	OQ970 T750 WQ	20		28·4		40·5	31½	74		7
"	"	"	"	"	"	"	"	"	" " "	300		25·0		33·4	22½	70		"
"	"	"	"	"	"	"	"	"	" " "	400		24·9		32·5	22½	69		"
"	"	"	"	"	"	"	"	"	" " "	450		24·1		30·8	23	68½		"
"	"	"	"	"	"	"	"	"	" " "	500		22·0		28·1	24	72½		"
"	"	"	"	"	"	"	"	"	" " "	550		17·1		24·0	26½	79½		"
"	"	"	"	"	"	"	"	"	" " "	600		11·9		18·4	30	86		"
"	"	"	"	"	"	"	"	"	" " "	650		6·9		12·0	39	91½		"
								(Rate of loading 1 ton/sq. in. min.)										
En 56C	0·21	0·39	0·30	0·026	0·016	0·28	13·5	1⅛	OQ950 T750 WQ	20		35·0		48·0	29	68		7
"	"	"	"	"	"	"	"	"	" " "	350		31·7		41·9	21	64½		"
"	"	"	"	"	"	"	"	"	" " "	400		30·5		39·6	20	64		"
"	"	"	"	"	"	"	"	"	" " "	450		29·5		37·6	21	64		"
"	"	"	"	"	"	"	"	"	" " "	500		26·0		32·1	22½	66½		"
"	"	"	"	"	"	"	"	"	" " "	550		20·0		25·7	25	76		"
"	"	"	"	"	"	"	"	"	" " "	600		13·2		22·5	28½	81½		"
"	"	"	"	"	"	"	"	"	" " "	650		6·5		13·0	40½	91		"
								(Rate of loading 1 ton/sq. in. min.)										

MECHANICAL PROPERTIES AT HIGH TEMPERATURE (Continued)

Type	Chemical Composition (%) C	Si	Mn	S	P	Ni	Cr	Section Size dia. (in.)	Heat Treatment °C	Test Temperature °C	Tensile 0·1% P.S. (0·1Rp) tons/sq.in.	0·2% P.S. (0·2Rp) tons/sq.in.	0·5% P.S. (0·5Rp) tons/sq.in.	T.S. (Rm) tons/sq.in.	El % (A)	R.A. % (Z)	Izod ft.lb.	Ref.
En 56C	0·22	0·40	0·30				13·0	1⅛	OQ960 T700	20	29·5	31·4	34·0	45·6	27	60½		3
”	”	”	”				”	”	” ”	100	27·2	29·0	31·4	41·5	29½	67½		”
”	”	”	”				”	”	” ”	200	25·9	27·6	29·6	38·4	27	62½		”
”	”	”	”				”	”	” ”	300	25·2	26·9	28·6	36·9	25	60		”
”	”	”	”				”	”	” ”	400	23·6	25·6	28·0	34·9	21½	58½		”
”	”	”	”				”	”	” ”	500	19·2	21·2	23·4	27·3	29	65½		”
”	”	”	”				”	”	” ”	600	9·7	12·0	15·0	18·8	44	80		”
								(Rate of straining 0·0085 in./min. to Y.P. then ⅛ in./min.)										
En 56D	0·30						13·0		H & lightly T	300				108·2	9	13		30
”	”						”		”	400				101·5	4	6½		”
”	”						”		”	500				87·0	9	27½		”
”	”						”		”	600				32·0	19	70		”
”	”						”		”	700				14·0	33	73		”
En 56D	(Within specification)							1¼	OH & T	20	52·5	54·0	56·0	62·0	23			13
”		”	”					”	”	300	44·0	46·0	47·5	62·5	18			”
”		”	”					”	”	400	37·0	40·5	42·0	47·0	17			”
”		”	”					”	”	500	21·0	23·5	27·0	38·5	21			”
”		”	”					”	”	550	14·0	16·5	19·0	32·5	25			”
”		”	”					”	”	600	10·0	12·0	14·0	22·5	30			”

STRESS RUPTURE AND CREEP PROPERTIES

Type	Chemical Composition (%)							Section Size dia. (in.)	Heat Treatment °C	Test Temperature °C	Stress (tons/sq.in.) to produce minimum creep rate of				Ref.
	C	Si	Mn	S	P	Ni	Cr				10^{-4} per hr.	10^{-5} per hr.	10^{-6} per hr.	10^{-7} per hr.	
En 56A	0·07	0·14	0·29	0·025	0·012	0·11	11·8	1⅛	OQ980 T680	400	34*	23·5	16·0	11*	2
"	"	"	"	"	"	"	"	"	" "	500	16*	10·6	7·0	4·7	"
"	"	"	"	"	"	"	"	"	" "	550	9*	6·1	4·1	2·8*	"
"	"	"	"	"	"	"	"	"	" "	600	4·7	2·9	1·8	1·1*	"
En 56A	0·09	0·18	0·47			0·18	12·40		H & T (45 tons/sq.in.)	593			2·23	1·65	31
"	"	"	"			"	"		" "	648			1·16	0·67	"
"	"	"	"			"	"		" "	734			0·62	0·40	"

* Approximate values

Type	Chemical Composition (%)					Heat Treatment °C	Test Temperature °C	Stress (tons/sq.in) for rupture in		Ref.
	C	Si	Mn	Ni	Cr			10,000 hrs	100,000 hrs.	
En 56A	0·11	0·30	0·45	0·14	13·2	Softened	540	5·8	4·5	31
"	"	"	"	"	"	"	590	2·3	1·8	"
"	"	"	"	"	"	"	650	0·9	0·7	"

STRESS RUPTURE AND CREEP PROPERTIES (Continued)

Type	Chemical Composition (%)				Heat Treatment °C	Test Temperature °C	Stress (tons/sq.in.) for 1% creep in 8 years	Ref.
	C	Si	Mn	Cr				
En 56A	0·12	0·50	0·50	14·0	Annealed (31 tons/sq. in.)	480	3·8	32
"	"	"	"	"	"	540	2·0	"
"	"	"	"	"	"	595	0·9	"
"	"	"	"	"	"	650	0·45	"
"	"	"	"	"	"	705	0·24	"

Type	Chemical Composition (%)		Test Temperature °C	Stress (tons/sq.in.) for 1% creep in		Stress (tons/sq.in.) for rupture in		Ref.
	C	Cr		10,000 hrs.	100,000 hrs.	10,000 hrs.	100,000 hrs.	
En 56A or B	0·15 max.	11·5—13·0	480	9·8		15·2	11·6	33
"	"	"	540	5·9	4·5	8·5	5·8	"
			595	2·45	1·65	4·5	2·1	"
			650	0·94	0·58			"
			705	0·44	0·36			"

Type	Chemical Composition (%)					Section Size dia. (in.)	Heat Treatment °C	Test Temperature °C	Stress (tons/sq.in.) for creep rate of 10^{-6} between 24 & 72 hrs.	Ref.
	C	Si	Mn	Ni	Cr					
En 56C	0·22	0·40	0·30		13·0	$\frac{3}{4}$	OQ960 T700	400	21	3
"	"	"	"		"	"	" "	450	12	"
"	"	"	"		"	"	" "	500	5·5	"
"	"	"	"		"	"	" "	550	2·8	"
"	"	"	"		"	"	" "	600	1·0	"
En 56D	0·29	0·30	0·28	0·35	13·0	$\frac{3}{4}$	OQ975 T750	300	25	3
"	"	"	"	"	"	"	" "	400	14·5	"
"	"	"	"	"	"	"	" "	500	5·7	"

Type	Chemical Composition (%)					Section Size dia. (in.)	Heat Treatment °C	Test Temperature °C	Stress (tons/sq.in.) to produce creep rate of				Ref.
	C	Si	Mn	Ni	Cr				0·01 per day	0·001 per day	0·0001 per day	0·00001 per day	
En 56D	0·29	0·79	0·27	0·23	13·7	$\frac{5}{8}$	H & T	300		42·0	39·1	31·5	34
"	"	"	"	"	"	"	"	400	29·8	26·1	22·5	19·5	"
"	"	"	"	"	"	"	"	500	14·1	11·6	8·5	4·5	"
"	"	"	"	"	"	"	"	600	4·4	2·35	1·2	0·8 (estd.)	"

STRESS RUPTURE AND CREEP PROPERTIES (Continued)

Type	Chemical Composition (%)								Section Size dia. (in.)	Heat Treatment °C	Test Temperature °C	Stress (tons/sq.in.) for creep of				Ref.
	C	Si	Mn	S	P	Ni	Cr	Mo				0·1% in 100 hr	0·2% in 100 hr	0·1% in 300 hr	0·2% in 300 hr	
En 56D	(Within specification)								1¼	H & T (HB = 300)	300	37·6	41·5	36·5	41·0	13
"	"								"	"	350	29·5	33·5	28·2	32·0	"
"	"								"	"	400	18·0	25·7	16·0	21·5	"
"	"								"	"	450	8·8	13·0	7·5	11·2	"
"	"								"	"	500	6·5	9·4	4·5*	7·0*	"
"	"								"	"	550	2·0*	3·5	1·5*	3·0*	"

* Approximate values.

Type	Chemical Composition (%)								Section Size dia. (in.)	Heat Treatment °C	Test Temperature °C	Stress (tons/sq.in.) forrupture in				Ref.
	C	Si	Mn	S	P	Ni	Cr	Mo				1 hr	10 hr	100 hr	300 hr	
En 56D	(Within specification)								1¼	H & T (HB = 300)	400		44·0*	40·0*	38·5	13
"	"								"	"	450		29·5*	26·0	24·5	"
"	"								"	"	500	30·0*	25·0*	20·0*	18·5	"

* Approximate values.

MECHANICAL PROPERTIES IN TORSION

Type	Chemical Composition (%)							Section Size dia. (in.)	Heat Treatment °C	Test Temperature °C	Tensile				Torsion			Ref.
	C	Si	Mn	S	P	Ni	Cr				Y.P. (Rs)	T.S. (Rm)	El % (A)	R.A. % (Z)	Y.P.	Shear Apparent max.	Strength Probable max.	
											tons/sq.in.				tons/sq.in.			
En 56A	0·03						13·5	3		RT		36·8	16	26½		36·0	27·0	35
En 56D	0·28	0·10	0·22	0·030	0·026	0·20	12·6	1⅛	OQ950 T650WQ	RT	39·8	49·0	22	55	26·1	43·9	32·9	3
"	"	"	"	"	"	"	"	"	" T250		95·0	106·7	7	29½	50·3	76·5	57·4	"
En 56D	(Within specification)							1¼	H & T	RT	56·0	62·0	23			54·7	41·0	13

FATIGUE PROPERTIES

Type	Chemical Composition (%)							Section Size dia. (in.)	Heat Treatment °C	Test Temperature °C	Tensile				Rotating Bending Fatigue Limit tons/sq.in. 10^6 cycles			Ref.
	C	Si	Mn	S	P	Ni	Cr				Y.P. (Rs) tons/sq.in.	T.S. (Rm) tons/sq.in.	El % (A)	R.A. % (Z)	In Air	In Fresh Water	In Sea Water	
En 56A	0·03						13·5	3		RT		36·8			±17·8			35
En 56A	0·08						12·2		A985 SC	RT		27·5	41	67½	±17·4	±13·4		37
"	"						"		WQ985 T650	RT		40·4	26½	75	±21·8	±13·8		"
En 56A	0·08	0·10	0·12			0·15	13·5	1⅛	AC950 T750 AC	RT	24·0	33·9	32	72	±14·0			3
En 56A	0·08	0·23	0·17	0·025	0·018	0·25	12·6		OH & T		25·4	36·4	33	73½	±19·0			36
En 56A	0·09	0·23	0·47			0·11	12·5		OQ1000 T750	RT	30·4	41·0	31	72½	±21·0			3
"	"	"	"			"	"		" "	427	26·0	31·0	26½	71½	±14·0			"
En 56A	0·09	0·20	0·25	0·030	0·030		13·0			RT	30·0	35·0	30	60	±15·5			8
En 56A	0·10	0·30	0·50				13·3		OQ1000 T750	RT	24·5	37·2	33	75	±15·5			3
En 56A	0·10	0·20	0·32			0·18	13·6		OQ940 T700	RT	31·0	40·8	29	71	±17·5			"
En 56A	0·10						13·3		WQ985 T540	RT		78·5	19	61	±29·0	± 9·8		37
"	"						"		" T595	"		55·6	22	70	±29·5	±11·8		"
"	"						"		" T650	"		46·7	23	70	±25·2	±14·7		"
"	"						"		A875 SC	"		36·0	34½	73½	±20·3	±11·8		"
En 56A	0·11	0·11	0·38			0·21	13·7		OQ950 T550	"	43·5	55·2	19	57	±24·5			3
En 56A	0·11	0·15	0·34			0·11	12·9		Annealed	RT		40·0	35½	44	±25·0	±17·0	±12	37
En 56A, 56B	0·12	0·18	0·40			0·45	12·7		H & T	"		50·2	23½	67	±24·0	±17·5		"
En 56B, 56C	0·18	0·29	0·31			0·22	13·8	Strip	AH1000 T150	RT	76·5	83·4	7		±43*			3
En 56C	0·21	0·28	0·46			0·28	13·9	1⅛	OQ950 T650 WQ	RT	41·2	55·9	20	57	±30			7
En 56C	0·23	0·35	0·25	0·030	0·030		13·0			RT	40·0	48·0	25	55	±22			8
En 56C, 56D	0·25	0·67	0·45	0·021	0·017		13·8		OQ950 T500 RT	RT	82·6	101·6		33	±41			38
"	"	"	"	"	"				" T600 "	"	63·5	72·5		49	±37·5			"
En 56D	0·27	0·35	0·26			0·29	13·1	1¼	OQ940 T700 WQ	RT	38·5	50·2	25	61½	±21·0			"
															10^8 Cycles	10^7 Cycles		
En 56D			(Within specification)					1¼	H & T	20	56·0	62·0	23		±34·5	±34·5		13
"			"					"	"	300	47·5	52·5	18		±31·7	±32·0		"
"			"					"	"	400	42·0	47·0	17		±28·0	±28·7		"
"			"					"	"	500	27·0	38·5	21		±18·5	±22·5		"
"			"					"	"	550	19·0	32·5	25		±14·5	±18·0		"
"			"					"	"	600	14·0	22·5	30		±12·5	±15·0		"

* Bending Fatigue

FATIGUE PROPERTIES (Continued)

Type	Chemical Composition (%)							Section Size dia. (in.)	Heat Treatment °C	Test Temperature °C	Tensile				Axial Fatigue Limit tons/sq.in. 10^7 rev.	Ref.
	C	Si	Mn	S	P	Ni	Cr				Y.P. (Rs) tons/sq.in.	T.S. (Rm) tons/sq.in.	El % (A)	R.A. % (Z)		
En 56C	0·21	0·47	0·53	0·013	0·026	0·29	13·1	$\frac{7}{8}$	OQ1000 T650 WQ	RT	42·7	54·2	27	66	±25·5	3
"	"	"	"	"	"	"	"	"	" " "	"	"	"	"	"	21·65 ±21·65	"
"	"	"	"	"	"	"	"	"	" " "	"	"	"	"	"	13·5 ±13·5	"
"	"	"	"	"	"	"	"	$\frac{3}{4}$	" " "	"	46·6	56·4	24	62	21·5 ±21·5	"
"	"	"	"	"	"	"	"	"	" " "	"	"	"	"	"	±27·0	"
En 56C	0·22	0·40	0·30				13·0		OQ960 T700	RT	41·3	49·1	26	69	±22·0	"

LIST OF REFERENCES

1. Private correspondence with The United Steel Companies Ltd.
2. Private correspondence with Jessop-Saville Ltd.
3. Private correspondence with Firth-Vickers Stainless Steels Ltd.
4. Physical Constants of some Commercial Steels at Elevated Temperatures. Butterworths Scientific Publications, London, 1953.
5. Private correspondence with the National Physical Laboratory.
6. Private correspondence with Richard Thomas and Baldwins Ltd.
7. Private correspondence with Edgar Allen and Co. Ltd.
8. Private correspondence with Samuel Osborn and Co. Ltd.
9. W. H. Souder and P. Hidnert. Thermal Expansion of Nickel, Monel metal, Stellite, Stainless steel, and Aluminium. United States Bureau of Standards, Scientific Paper No. 426. 1921.
10. International Critical Tables of Numerical Data, Physics, Chemistry and Technology. Prepared by the National Research Council, U.S.A. Editor-in-Chief: E. W. Washburn, 1st Edition, McGraw-Hill, 1926.
11. E. Maurer. The Heat Conductivity of Chromium Steels at High Temperature. *Archiv für das Eisenhüttenwesen*, 1936, vol. 10, Oct., 145—154.
12. S. M. Shelton. Thermal Conductivity of some Irons and Steels over the Temperature Range 100 to 500°C. Bureau of Standards *Journal of Research*, vol. 12, April, 1934, 441—450.
13. Private correspondence with Rolls-Royce Ltd.
14. R. W. Powell. Effect of Quenching on Conductivity and Resistivity. *JISI*, vol. 184, Pt. 1, Sept., 1956, 7.
15. F. Garofalo, R. P. Malenock and G. V. Smith. Elastic Constants of some Commercial Steels. American Soc. for Testing Materials. Symposium on the Determination of Elastic Constants. ASTM Special Publication No. 129, June 25, 1952.
16. H. H. Abram. The Value of Young's Modulus for Steel. *JISI*, vol. 129, No. 1, 1934, 325.
17. W. H. Hill, K. D. Shimmin, B. A. Wilcox. Elevated Temperature Dynamic Moduli of Metallic Materials. *Proceedings* ASTM, vol. 61, 1961, 890—906.
18. G. Delbart, A. Constant and A. Clerc. Courbes de Transformation des Aciers de Fabrication Française. Institut de Recherches de la Sidérurgie (IRSID).
19. Atlas of Isothermal Transformation Diagrams. United States Steel Corporation. Second printing, 1957.
20. Atlas of Isothermal Transformation Diagrams of the B. S. En Steels. Iron and Steel Institute Special Report No. 56, 1956.
21. Molinder. Atlas of Isothermal Transformation Diagrams of several Swedish Stainless Steels. Communication from Jernkontorets Tech. Council, vol. 185, No. 203, 1954, 263—290.
22. British Standard 971, Commentary on the British Standard Wrought Steels, En Series. British Standards Institution, 1950.
23. J. H. G. Monypenny. *Stainless Iron and Steel*. 2nd edition. Chapman and Hall, London, 1931.
24. J. Strauss. Metals and Alloys for Industrial Applications requiring Extreme Stability. *Transactions*, American Society for Steel Treating, vol. 16, Aug., 1929, 191—226.
25. B. Jones. Further Experiments of the Nitrogen-hardening of High-chromium and Austenitic Steels. *JISI*, vol. 136, No. 2, 1937, 169—185.
26. Russell and Welckes. ASME/ASTM Symposium 1931, 658—680.
27. H. W. Gillett. Impact Resistance and Tensile Properties of Metals at Sub-atmospheric temperatures. Project 13 of the Joint ASME-ASTM Research Committee on the Effect of Temperature on the Properties of Metals.
28. *Stainless Steel Handbook*. United States Steel Corporation, 1957.
29. War Office unpublished report.
30. D. G. Clark. Recent Developments in Stainless Steel. *Yearbook* of the American Iron and Steel Institute, 1925, 262—299.
31. F. H. Norton. *Creep of Steel at High Temperatures*. 1st edition. McGraw-Hill, London, 1929, pp. 36, 41 and 65.
32. Second Report on Comparative High-Temperature Tension Tests of Metals at different Laboratories. *Proceedings*, ASTM, 1930, vol. 30, Pt. 1, 96—109.

33. Technical Data-Stainless Steels. Atlas Gibson Steels S. A. and Atlas Steels, Canada.

34. H. J. Tapsell, Johnson and W. J. Clenshaw. DSIR Engineering Research Special Report No. 18, 1932, H.M.S.O.

35. R. R. Moore. Resistance of Metals to repeated Static and Impact Stresses. *Proceedings*, ASTM, 1924, vol. 24, Pt. 2, 547—573.

36. J. W. Cuthbertson. *The Load-deflection Fatigue Test. A critical investigation of its application to steels.* ISI, Carnegie Scholarship Memoirs, 1935, vol. 24, 1—50.

37. D. J. McAdam. Stress-strain Cycle Relationship and Corrosion Fatigue of Metals. *Proceedings*, ASTM, 1926, vol. 26, Pt. 2, 224—254.

38. R. Cazaud. Fatigue of Metals. Translated by A. J. Fenner. Chapman and Hall Ltd., London, 1953.

39. Private correspondence with Imperial Chemical Industries Ltd.

CHROMIUM RUST-RESISTING STEEL (FREE MACHINING)

SPECIFICATION

CHEMICAL COMPOSITION (%)

	En 56AM	En 56BM	En 56CM	En 56DM
C	0·12 max.	0·12—0·18	0·18—0·25	0·25—0·35
Si	1·00 max.	1·00 max.	1·00 max.	1·00 max.
Mn	1·50 max.	1·50 max.	1·50 max.	1·50 max.
S	0·75 max.	0·75 max.	0·75 max.	0·75 max.
P	0·045 max.	0·045 max.	0·045 max.	0·045 max.
Ni	1·00 max.	1·00 max.	1·00 max.	1·00 max.
Cr	12·0—14·0	12·0—14·0	12·0—14·0	12·0—14·0
Mo*	0·60 max.	0·60 max.	0·60 max.	0·60 max.
Se*	0·60 max.	0·60 max.	0·60 max.	0·60 max.
Zr*	0·60 max.	0·60 max.	0·60 max.	0·60 max.
Pb*	0·35 max.	0·35 max.	0·35 max.	0·35 max.

* Total 1·00% max.

MECHANICAL PROPERTIES

	Hardened and Tempered Condition		
	P	R	S
Limiting ruling section, in.	6	6	4
Tensile strength, tons/sq.in. min. (Rm).	35	45	50
Elongation, %, min, (A)	20	15	12
Izod impact value, ft, lb., min.	25	20	
Brinell hardness number, [1], HB.	152/207	201/255	223/277

[1] For information purposes only. Not contractual part of the specification.

RELATED SPECIFICATIONS

United Kingdom		America	
En No.	BS Air	SAE	AISI
En 56AM		51416 51416Se	416 416Se
En 56BM		51416 51416Se	416 416Se
En 56CM	S124		
En 56DM		51420F	

APPLICATIONS

These steels are used for applications requiring resistance to corrosion similar to that for En 56 where easy machining is needed, but where transverse properties are not so important. Typical applications are: Nuts and bolts, printing machine components, air-frame components, chemical and food processing plant, shafts, steam valves and where intricate machining is part of the component manufacture.

Additions of S and Se are made to improve the machining properties but cause a reduction in corrosion resistance, partially offset by the Mo or Zr additions.

WELDING

See Introduction for key to symbols used in this table.

	Welding Process					
	MA	G	IG	S	F	B
En 56AM	f	f	f	f	f	k
En 56BM	f	f	f	f	f	k
En 56CM	f	f	f	f	f	k
En 56DM	f	f	f	f	f	k

Reproduced by kind permission of H.M.S.O.

MACHINABILITY

The machinability of these steels is approximately 85% of that for mild steel (En 3).

HOT WORKING AND HEAT TREATMENT TEMPERATURES

Forging, rolling and stamping	1150°C, finish above 950°C. Retarded cooling is advisable.
Annealing	800°C— 900°C, furnace cool.
Softening (subcritical annealing)	650°C— 750°C, air cool.
Hardening	950°C—1020°C, air cool or oil quench (or water quench for low C steels or thin sections).
Tempering	Up to 750°C, depending on properties required.

Note: Corrosion and stress-corrosion resistance are reduced by tempering in the range 450°C—650°C.

PHYSICAL PROPERTIES

The physical properties of the free-machining grades are generally similar to those of the non-free machining grades of En 56 of similar composition apart from the free-machining additions.

The following data have been obtained on steel to En 56CM composition.

MEAN COEFFICIENT OF THERMAL EXPANSION (α)

Type	Chemical Composition (%)							$10^6 \times$ Mean Coefficient of Thermal Expansion per deg. C									Ref.
	C	Si	Mn	S	P	Cr	Zr	20 to 100°C	20 to 200°C	20 to 300°C	20 to 400°C	20 to 500°C	20 to 600°C	20 to 700°C	20 to 800°C	20 to 900°C	
En 56CM	0·20	0·50	0·85	0·27	0·037	13·8	0·11	10·0	10·0	10·4	10·5	10·7	10·9	11·3	11·5	10·6	1

TRANSFORMATION CHARACTERISTICS

ISOTHERMAL TRANSFORMATION

En 56AM

Chemical Composition (%)								
C	Si	Mn	S	P	Ni	Cr	Zr	N
0·12	0·74	0·79	0·190	0·017	0·25	12·82	0·08	0·037

Actual Grain Size 7—9 (ASTM).
Austenitizing Temperature 1800°F (982°C).

(Ref. 2)

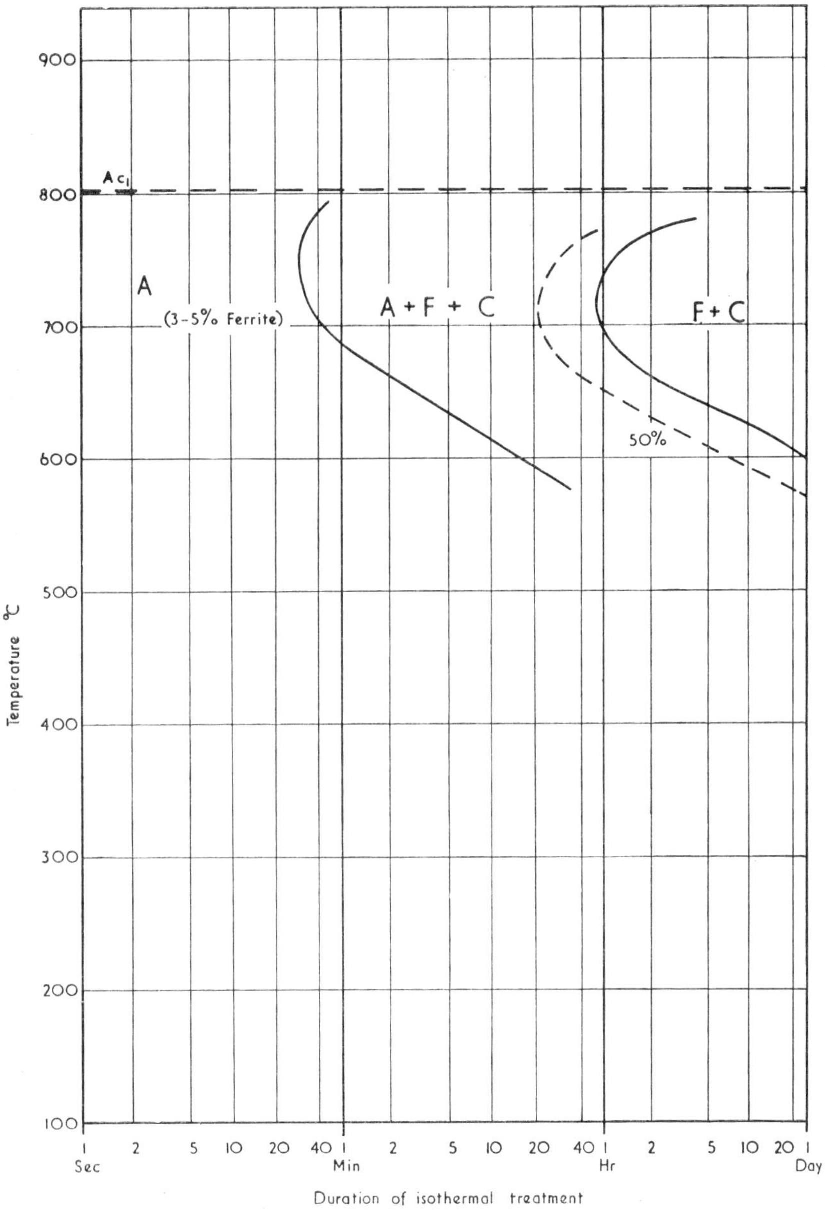

Fig. 56M.1

MECHANICAL PROPERTIES AT ROOM TEMPERATURE

Type	Chemical Composition (%)								Section Size dia. (in.)	Heat Treatment °C	Tensile					Izod ft.lb.	Ref.
	C	Si	Mn	S	P	Ni	Cr	Se			0·5% P.S. (0·5Rp)	Y.P. (Rs)	T.S. (Rm)	El % (A)	R.A. % (Z)		
											tons/sq.in.						
En 56AM	0·08						11·3	0·23	1⅛	OQ950 T740 WQ		24·0	34·8	36	72	81	3
En 56AM	0·08						13·3	0·21	1⅛	OQ950 T740 WQ		26·4	35·6	33	73½	80	"
En 56AM	0·09						13·8	0·24	1⅛	OQ950 T740 WQ		25·6	36·0	34	70	80	"
En 56AM	0·10						13·6	0·18	1⅛	OQ950 T740 WQ		24·4	35·2	32	70	78	"
En 56AM	0·11	0·40	0·81	0·50		0·17	13·7			OQ950 T500 WQ		71·4	77·6	16	39½	17	4
"	"	"	"	"		"	"			" T600 "		42·4	52·2	17½	48½	22	"
"	"	"	"	"		"	"			" T700 "		32·2	44·4	22	55	29	"
"	"	"	"	"		"	"			" T750 "		28·8	41·7	23½	55½	35	"
En 56AM	0·11	0·28	0·75	0·50			13·0			OQ970 T500 WQ		64·9	74·5	16	39	35	"
"	"	"	"	"			"			" T600 "		41·6	49·3	18½	50	46	"
"	"	"	"	"			"			" T700 "		31·4	41·6	24	57	51	"
"	"	"	"	"			"			" T750 "		27·4	38·7	27	53½	62	"
								Mo									
En 56AM, BM	0·12	0·60	1·20	0·23			13·5	0·28	1⅛	OQ1000 T750		28·0	41·0	28	65	52	3
En 56BM	0·13	0·25	0·58	0·47	0·014	0·19	13·8	0·42	1⅛	Softened 750 AC	26·5		39·8	32	60½	85	"
En 56BM	0·14	0·16	0·39	0·52	0·016	0·32	14·1		1⅛	OH1000 T600 1hr. AC		49·0	55·4	18	50	79	"
"	"	"	"	"	"	"	"		"	" T650 " "		43·0	49·4	20	52	101	"
"	"	"	"	"	"	"	"		"	" T700 " "		29·8	47·0	21	53	100	"
"	"	"	"	"	"	"	"		"	" T750 " "		32·4	42·4	24½	55	95	"
								Mo									
En 56BM, CM	0·18	0·60	1·23	0·23			13·5	0·28	1⅛	OQ1000 T700		28·9	44·5	25½	54	36	3
En 56CM	0·21	0·39	0·58	0·54	0·016	0·24	14·5	0·46	1⅛	Softened 750 AC	33·4		44·5	26½	51½	73	"

MECHANICAL PROPERTIES AT HIGH TEMPERATURE

Type	Chemical Composition (%) C	Si	Mn	S	P	Cr	Zr	Heat Treatment °C	Test Temperature °C	Tensile Y.P. (Rs) tons/sq.in.	T.S. (Rm) tons/sq.in.	El % (A)	R.A. % (Z)	Charpy V Notch ft.lb.	Ref.
En 56CM	0·20	0·50	0·85	0·27	0·037	13·8	0·11	OQ1000 T740 Long.	20	37·1	48·0	23	57	32	11
"	"	"	"	"	"	"	"	" " Trans.	"	35·7	46·7	11	14	4	"
"	"	"	"	"	"	"	"	" " Long.	250	31·6	47·7	19	55	32	"
"	"	"	"	"	"	"	"	" " Trans.	"	32·5	41·3	10	15	7	"
"	"	"	"	"	"	"	"	" " Long.	350	32·1	41·7	17	52	29	"
"	"	"	"	"	"	"	"	" " Trans.	"	31·6	41·0	9	12	7	"
"	"	"	"	"	"	"	"	" " Long.	450	30·2	39·0	18	53	27	"
"	"	"	"	"	"	"	"	" " Trans.	"	30·2	38·3	8	10	6	"
"	"	"	"	"	"	"	"	" " Long.	550	25·3	31·3	23	62	27	"
"	"	"	"	"	"	"	"	" " Trans.	"	25·7	30·8	11	16	8	"
"	"	"	"	"	"	"	"	" " Long.	650	19·0	22·4	32	76	31	"
"	"	"	"	"	"	"	"	" " Trans.	"	18·7	21·3	19	46	8	"
"	"	"	"	"	"	"	"	" " Long.	750	7·0	8·6	41	89	31	"
"	"	"	"	"	"	"	"	" " Trans.	"	5·6	7·5	37	81	7	"
"	"	"	"	"	"	"	"	" " Long.	850	2·1	4·2	67	92	47	"
"	"	"	"	"	"	"	"	" " Trans.	"	2·0	4·1	67	91	13	"

FATIGUE PROPERTIES

Type	Chemical Composition (%) C	Si	Mn	S	P	Ni	Cr	Mo	Heat Treatment °C	Test Temper ature °C	Tensile 0·2% P.S. (0·2Rp) tons/sq.in.	Y.P. (Rs) tons/sq.in.	T.S. (Rm) tons/sq.in.	El % (A)	R.A. % (Z)	Rotating Bending Fatigue Strength (10^7) cycles tons/sq.in.	Ref.
En 56AM	0·10	0·40	0·40	0·30	0·020	0·25	12·7	0·35	Annealed	RT	20·3		33·5	30	61½	±19·2	
"	"	"	"	"	"	"	"	"	H & T	"	36·8		46·9	20	57	±24·8	5
En 56CM	0·18	0·60	1·30	0·23			13·0	0·28	OQ1000 T700	RT		28·9	44·5	25½	54	±24·7	3

LIST OF REFERENCES

1. War Office unpublished data.
2. Atlas of Isothermal Transformation Diagrams (Supplement). United States Steel Corporation, 1953.
3. Private correspondence with Firth-Vickers Stainless Steels Ltd.
4. Private correspondence with Edgar Allen and Co. Ltd.
5. S. P. Watkins. The Free Machining Stainless Steels. *Metal Progress*, vol. 39, April, 1941, 452—459.

MARTENSITIC CHROMIUM–NICKEL RUST-RESISTING STEEL

SPECIFICATION

CHEMICAL COMPOSITION (%)

C	Si	Mn	S	P	Ni	Cr
0·25 max.	0·10—1·00	1·00 max.	0·045 max.	0·045 max.	1·00—3·00	15·5—20·0

MECHANICAL PROPERTIES

Limiting ruling section., in.	6	2½
Tensile strength tons/sq.in. min. (Rm).	55	55
Yield stress [1] tons/sq.in. min. (Rs).	44	44
0·2% proof stress [2] tons/sq.in. min. (0·2Rp).	41	41
Elongation, %, min.	15	15
Izod impact valve, ft.lb., min.	15	25
Brinell hardness [3] number, min., HB.	248	248

[1] Not to be used as acceptance values except by special arrangement between the purchaser and the manufacturer.
[2] When specifically requested in the enquiry or order.
[3] For information purposes only. Not contractual part of the specification.

RELATED SPECIFICATIONS

United Kingdom				America		France	Germany			Sweden	USSR	
En No.	BS	BS-Air	DTD	SAE	AISI	AFNOR	Werkstoff No.	Name	DIN	SIS	GOST	Mark
57	1449 1554 2056	2S80		51431	431	Z15CN16-2	1.4057 1.2787	X.22Cr	Ni 17	14.2321	5632-61 5632-61	1Kh17N2 2KhI7N2

APPLICATIONS

A high tensile steel with a corrosion resistance between the plain chromium steels of En 56 type and the austenitic chromium-nickel steels of En 58. It has also a high resistance to electrolytic corrosion which renders it suitable where it is in contact with other metals. It is not suitable for applications involving sea water or conditions where selective corrosion of its duplex structural constituents may occur.

The steel is usually used in the hardened and tempered condition to 55 to 65 tons/sq.in. It is frequently used at strengths of 75 tons/sq.in. and can be hardened to 90 tons/sq.in. Typical application are pump shafts, spindles and valves working in high-pressure steam or for hydraulic work; highly-stressed aircraft and general engineering fittings, leather rollers, paper machine parts and high tensile nuts and bolts.

WELDING

(See Introduction for key to symbols used in this table.)

	Welding Process					
	MA	G	IG	S	F	B
En 57	f	f	f	f	b	k

Reproduced by kind permission of H.M.S.O.

MACHINABILITY

In the softened condition the machinability is approximately 50% of that for mild steel (En 3).

HOT WORKING AND HEAT TREATMENT TEMPERATURES

Forging, rolling and stamping	1120°C, finish above 900°C. Retarded cooling is advisable.
Softening	630°C—660°C, air or furnace cool.
Hardening	950°C—1020°C, air or oil quench.
Tempering (for normal conditions)	600°C—700°C, cool in air or oil.
(for wear resistance)	350°C—450°C, cool in air or oil.

For maximum corrosion resistance temper at 650°C minimum.

PHYSICAL PROPERTIES

SPECIFIC GRAVITY (d)

Chemical Composition (%) C	Si	Mn	Ni	Cr	Heat Treatment °C	Specific Gravity at 20°C	Ref.
0·11	0·13	0·15	1·90	17·9	Softened 670	7·73	1
0·11		0·42	1·98	17·7	Softened	7·71	2
0·12	0·29	0·15	2·75	17·7	OQ960 T520	7·73	1
0·14	0·30	0·38	2·40	19·0	OQ960 T625	7·70	"
0·15			1·80	17·0	OQ970 T650	7·74	2
0·16	0·40	0·30	2·50	16·5	Softened	7·70	1
0·16	0·35	0·75	1·88	16·3		7·70	3
0·20		1·00	1·25	18·0		7·70	4
(Within specification)						7·72	5

SPECIFIC HEAT (c)

Chemical Composition (%) C	Si	Mn	Ni	Cr	Heat Treatment °C	Mean Specific Heat cal/g. deg. C 20 to 100°C	20 to 200°C	20 to 300°C	20 to 400°C	20 to 500°C	20 to 600°C	20 to 700°C	Ref.
0·11		0·42	1·98	17·7	Softened		0·122	0·128	0·135	0·142	0·150	0·159	2
0·16	0·25	0·30	2·00	17·0		0·120							6
0·16	0·40	0·30	2·50	16·5		0·115							1

MEAN COEFFICIENT OF THERMAL EXPANSION (α)

Chemical Composition (%) C	Si	Mn	Ni	Cr	Heat Treatment °C	$10^6 \times$ Mean Coefficient of Thermal Expansion per deg. C 20 to 100°C	20 to 200°C	20 to 300°C	20 to 400°C	20 to 500°C	20 to 600°C	20 to 700°C	Ref.
0·11		0·42	1·98	17·7	Softened		10·9	11·3	11·6	11·7	11·7	11·85	2
0·14	0·30	0·38	2·40	19·0	OQ960 T625	10·4	11·1	11·3	11·4	11·6			1
"	"	"	"	"	A800	10·4	10·9	11·1	11·2	11·3			"
0·16	0·35	0·75	1·88	16·3		10·5	10·9	11·3	11·6	12·0	12·4		3
(Within specification)						10·3		11·0		11·7			5

ELECTRICAL RESISTIVITY (ρ)

Chemical Composition (%)					Heat Treatment °C	Electrical Resistivity microhms-cm						Ref.
C	Si	Mn	Ni	Cr		20°C	100°C	200°C	300°C	400°C	500°C	
0·11		0·42	1·98	17·8	AC650	68·5	74·3	81·4				2
"		"	"	"	OQ970 T550	68·8	76·2	83·0				"
"		"	"	"	OQ970 T650	69·3	83·0	82·5				"
0·12	0·29	0·15	2·75	17·7	OQ960 T520	75·0						"
0·14	0·30	0·38	2·40	19·0	OQ960 T625	72·0	78·1	85·1	91·3	97·3	102·8	1
	(Within specification)				Softened	69·4						3
	"		"		Hardened	73·1						"

THERMAL CONDUCTIVITY (k)

Chemical Composition (%)					Heat Treatment °C	Thermal Conductivity cal/cm sec. deg. C					Ref.
C	Si	Mn	Ni	Cr		20°C	100°C	400°C	500°C	600°C	
0·15			1·8	17·0	H & T650		0·049	0·060		0·063	2
0·16	0·40	0·30	2·5	16·5			0·045		0·058		1
	(Within specification)						0·048				7
	"		"			0·045	0·049	0·060			5

YOUNG'S MODULUS (E)

Chemical Composition (%)					Heat Treatment °C	Test Temperature °C	Young's Modulus (tons/sq.in) (Static Method)	Ref.
C	Si	Mn	Ni	Cr				
0·16	0·40	0·30	2·5	16·5		RT	13,700	1
0·12	0·30	0·23	2·46	18·3	OQ960 T520	RT	13,500	"
0·16	0·15	0·35	1·80	17·1	OH & T	20	12,700	11
"	"	"	"	"	"	250	12,500	"
"	"	"	"	"	"	400	11,500	"
"	"	"	"	"	"	550	9,200	"
	(Within specification)					RT	12,900	7
	"	"				RT	12,500	8
	"	"			H & T550	RT	13,600	5
	"	"			H & T650	RT	13,600	"

MAGNETIC PROPERTIES

Chemical Composition (%)					Heat Treatment °C	Magnetic Induction B (Gauss) for Magnetizing Field H (Oersted)											B_{rem} (Gauss)	H_c (Oersted)	Ref.
C	Si	Mn	Ni	Cr		H = 5	10	20	40	50	60	100	150	200	250	400			
																	H max. = 400		
0·14	0·30	0·38	2·40	19·0	A800	1,100	2,100	3,800	6,550	7,650	8,550	10,650	11,900	12,600		14,300	6,150	17	1
"	"	"	"	"	H & T625	650	1,800	4,250	8,500	9,500	10,250	11,900	12,850	13,450		14,300	6,400	19	"
																	H max. = 250		
0·11		0·42	1·98	17·7	AC650	800	1,800	5,850	12,400	13,100	13,500	14,400	14,800	15,100	15,400		10,380	13·8	2
"		"	"	"	OQ970 T550	700	2,000	6,900	9,950	12,050	12,600	14,000	15,000	15,800	16,450		8,270	13·8	"
"		"	"	"	OQ970 T650	800	2,600	7,700	11,200	12,000	12,600	13,900	14,750	15,100	15,250		9,270	15·6	"

Chemical Composition (%)					Heat Treatment °C	B for H = 250	Permeability μ max.	B_{rem} (Gauss)	H_c (Oersted)	Ref.
C	Si	Mn	Ni	Cr						
0·16	0·40	0·50	2·50	16·5	OQ975 T650	14,600	190	8,350	30	1

MECHANICAL PROPERTIES AT ROOM TEMPERATURE

Chemical Composition (%)							Section Size dia. (in.)	Heat Treatment °C	Tensile							Izod ft.lb.	Ref.
C	Si	Mn	S	P	Ni	Cr			0·1% P.S. (0·1Rp)	0·2% P.S. (0·2Rp)	0·5% P.S (0·5Rp)	Y.P. (Rs)	T.S. (Rm)	El % (A)	R.A. % (Z)		
									tons/sq.in.								
0·09					2·08	17·8	1⅛	AH950 T300 WQ				49·6	58·8	20	56	74	9
"					"	"	"	" T400 "				49·6	58·8	20	57	67	"
"					"	"	"	" T500 "				46·0	57·8	22	59½	91	"
"					"	"	"	" T600 "				34·4	43·2	28½	62	85	"
"					"	"	"	" T650 "				33·4	42·5	30	62	86	"
"					"	"	"	" T700 "					44·6	28	62	82	"
"					"	"	"	" T750 "					46·5	27½	59½	79	"
0·09	0·17	0·18			2·42	18·4	1¼	AH950 T475AC				46·8	52·8	20	49½	7	1
"	"	"			"	"	"	" T500 "				44·8	50·0	19	51	17	"
"	"	"			"	"	"	OQ950 T475 WQ				52·0	58·4	19	51	3	"
"	"	"			"	"	"	" T500 "				44·0	50·0	24	59½	49	"
"	"	"			"	"	"	" T490 "				48·8	54·4	21	52	35	"
0·09	0·19	0·23			2·53	18·4	½	OQ950 T500 WQ				50·4	56·7	22½	65	67	"
"	"	"			"	"	"	" T550 "				40·9	49·7	25	66½	67	"
"	"	"			"	"	"	" T600 "				37·8	47·6	27	69	70	"
"	"	"			"	"	1⅛	OQ950 T550 WQ				41·2	50·6	27	65	66	"
"	"	"			"	"	"	" T575 "				40·4	49·4	27	66	58	"
"	"	"			"	"	"	" T600 "				38·2	48·0	28	61½	52	"
"	"	"			"	"	"	" T625 "				35·2	48·0	29	65	54	"
"	"	"			"	"	"	" T650 "				33·6	47·6	27	60½	60	"
"	"	"			"	"	"	" T675 "				28·4	48·4	25	62½	66	"
"	"	"			"	"	"	" T700 "				32·8	47·2	25	62½	71	"
"	"	"			"	"	"	" T725 "				34·0	48·8	26	65	66	"
"	"	"			"	"	1¼	OQ950 T510 WQ				44·8	51·2	26	65	90	"
"	"	"			"	"	"	WQ950 T520 WQ				43·2	49·6	26	65	92	"
"	"	"			"	"	2	OQ950 T550 WQ				44·0	53·2	24	58	61	"
0·11	0·15	0·13			1·84	17·73	1¼ sq.	As rolled				60·0	63·6	3	8½	13	1
"	"	"			"	"	1¼	AC950 T500				44·0	50·4	22	56	63	"
"	"	"			"	"	"	" T550				38·8	44·8	23	54½	69	"
"	"	"			"	"	"	OQ950 T475 WQ				55·6	60·0	21½	54½	19	"
"	"	"			"	"	"	" T500 "				47·6	52·8	21	52	57	"
"	"	"			"	"	3sq.	AC950				53·2	58·0	7	15	8	"
"	"	"			"	"	"	OQ950				53·2	59·6	5½	10½	4	"

MECHANICAL PROPERTIES AT ROOM TEMPERATURE (Continued)

Chemical Composition (%)							Section Size dia. (in.)	Heat Treatment C°	Tensile							Izod ft.lb.	Ref.
C	Si	Mn	S	P	Ni	Cr			0·1% P.S. (0·1Rp)	0·2% P.S. (0·2Rp)	0·5% P.S. (0·5Rp)	Y.P. (Rs)	T.S. (Rm)	El % (A)	R.A. % (Z)		
									tons/sq.in.								
0·11		0·42			1·98	17·8	1⅛	FC 600				45·1	50·7	23½	59	63	2
"		"			"	"	"	FC 650				39·0	46·6	24½	59	79	"
"		"			"	"	"	OQ980 T200 OQ				72·5	80·6	16	36	29	"
"		"			"	"	"	" T400 "				74·1	78·0	18½	57	18	"
"		"			"	"	"	" T500 "				70·9	76·0	20	53	18	"
"		"			"	"	"	" T550 "				53·1	60·3	24	60	76	"
"		"			"	"	"	" T575 "				53·2	53·9	22	65	76	"
"		"			"	"	"	" T600 "				47·5	52·8	24	60	66	"
"		"			"	"	"	" T625 "				45·0	51·8	24½	60	68	"
"		"			"	"	"	" T650 "				43·8	50·7	28½	65	64	"
0·11		0·43			1·35	17·2	⅝	OQ980 T550				56·4	66·5	19½	57	30	"
"		"			"	"	1⅛	" "				54·6	65·7	19	55	38	"
"		"			"	"	2½	" "				55·3	64·0	17½	52½	21	"
"		"			"	"	⅝	" T600				49·2	55·6	20	57	53	"
"		"			"	"	1⅛	" "				48·5	55·5	20	55	54	"
"		"			"	"	2½	" "				44·8	55·7	19½	55	57	"
"		"			"	"	⅝	" T650				44·5	52·4	21	57	58	"
"		"			"	"	1⅛	" "				43·1	52·0	22	57	56	"
"		"			"	"	2½	" "				42·4	50·7	21½	52½	48	"
0·11					2·18	18·0		AC950 T650 WQ				36·0	45·3	27½	62	86	9
0·12	0·40	0·38			2·33	18·5	0·525 hex.	OQ950 T550 AC				59·0	69·7	17½	56	53	1
"	"	"			"	"	⅝	OQ960 T510 AC					60·4	21	61½	50	"
"	"	"			"	"	"	" T525 "				48·7	56·7	23	63	60	"
"	"	"			"	"	1⅛	OQ950 T500 1hr. AC	49·5		58·0		66·4	20	53	AC WQ	"
																56 59	
"	"	"			"	"	"	" T525 " "				51·9	59·2	19	56	73 75	"
"	"	"			"	"	"	" T550 " "				46·3	54·5	22	56	79 71	"
"	"	"			"	"	"	" T750 " "				42·4	51·6	25	62	30 50	"
0·12	0·19	0·31			2·67	17·8	¼	OQ1050 T2hrs. 520				77·0	82·0	30	55	23	"
"	"	"			"	"	"	" " 530				62·2	66·7	30	60	22	"
"	"	"			"	"	"	" " 540				62·2	66·6	30	57½	24	"
"	"	"			"	"	"	" " 550				60·7	64·4	31	60	22	"
"	"	"			"	"	"	" T½hr. 520				74·6	84·5	13½	42½	7	"

MECHANICAL PROPERTIES AT ROOM TEMPERATURE (Continued)

Chemical Composition (%)							Section Size dia. (in.)	Heat Treatment C°	Tensile							Izod ft.lb.	Ref.
C	Si	Mn	S	P	Ni	Cr			0·1% P.S. (0·1Rp)	0·2% P,S. (0·2Rp)	0·5% P.S. (0·5Rp)	Y.P. (Rs)	T.S. (Rm)	El % (A)	R.A. % (Z)		
									tons/sq.in.								
0·12	0·19	0·31			2·67	17·8	¼	OQ1050 T½hr. 570				55·3	62·5	20	60½	21	1
"	"	"			"	"	"	" " 580				57·9	64·1	21½	60½	22	"
"	"	"			"	"	"	OQ960 T½hr. 500				68·1	75·5	19½	55	10	"
"	"	"			"	"	"	" " 520				61·7	66·6	19½	58	11	"
"	"	"			"	"	"	AC960 T½hr. 540				62·7	70·0	19½	59½	17	"
"	"	"			"	"	"	" " 560				58·1	64·5	17	49½	14	"
"	"	"			"	"	"	OQ900 T2hrs. 520				57·0	58·8	35	62½	22	"
"	"	"			"	"	"	" " 530				52·4	55·3	31	62½	40	"
"	"	"			"	"	"	AC750 OQ960 T2hr.525				62·7	64·7	31	62½	18	"
"	"	"			"	"	"	AC850 " " "				59·8	63·2	31	60	20	"
"	"	"			"	"	"	AC1050 " " "				56·3	61·3	30	57½	22	"
"	"	"			"	"	½	A820 AC960 T520				67·5	74·1	20	52½	11	"
"	"	"			"	"	"	" " T530				61·3	65·5	19½	58	13	"
"	"	"			"	"	"	" OQ960 T520				66·3	74·4	25	58	12	"
"	"	"			"	"	"	" " T530				63·3	67·3	19½	60½	11	"
"	"	"			"	"	"	A850 AC960 T520				67·6	72·6	20½	55	14	"
"	"	"			"	"	"	" " T530				61·0	64·8	21½	56½	15	"
"	"	"			"	"	"	" OQ960 T520				68·0	72·2	19½	60½	10	"
"	"	"			"	"	"	" " T530				59·9	65·3	20	62	16	"
0·13	0·23	0·33			2·36	18·0	1⅛	OQ950 T1hr. 500AC				62·4	68·6	21	53	52	1
"	"	"			"	"	"	" " " WQ								81	"
"	"	"			"	"	"	" " 525 "				51·2	58·0	23	56½	80	"
"	"	"			"	"	"	" " " WQ								92	"
"	"	"			"	"	"	" " 550 "				47·0	54·0	25	58		"
"	"	"			"	"	"	" " " WQ								77	"
"	"	"			"	"	"	" " 750 "				40·2	51·8	25½	62½	67	"
"	"	"			"	"	"	" " " WQ								69	"
"	"	"			"	"	1⅜	OQ950 T525 AC	48·8		53·0		59·9	17	53	64	"
"	"	"			"	"	2⅞	" " "	49·0		55·6		62·0	15	49	23	"
0·13	0·23	0·29			2·61	17·1	1	OQ975 T400	63·4		73·4		84·7	19½	56		"
"	"	"			"	"	"	" T500	60·9		67·5		77·8	19½	59		"
"	"	"			"	"	"	" T550	49·9		57·0		67·7	22½	62		"
"	"	"			"	"	"	" T620			54·0		60·8	20	59	54	"
"	"	"			"	"	"	" T650	39·4		47·0		58·8	23	62		"

MECHANICAL PROPERTIES AT ROOM TEMPERATURE (Continued)

Chemical Composition (%)							Section Size dia. (in.)	Heat Treatment °C	Tensile							Izod ft.lb.	Ref.
C	Si	Mn	S	P	Ni	Cr			0·1% P.S. (0·1Rp) tons/sq.in.	0·2% P.S. (0·2Rp) tons/sq.in.	0·5% P.S. (0·5Rp) tons/sq.in.	Y.P. (Rs) tons/sq.in.	T.S. (Rm) tons/sq.in.	El % (A)	R.A. % (Z)		
0·14	0·21	0·24			2·48	17·3	$1\frac{3}{4}$	OQ1000 T620	40·2	43·7	46·8		54·8	$23\frac{1}{2}$	$62\frac{1}{2}$		1
0·14	0·40	0·70	0·019	0·013	2·07	16·1	$2\frac{1}{2}$(M)	OQ1000 T610	43·6				59·2	20	55	36	10
0·14		0·68			2·35	17·1	1	OQ1000 T610 WQ				48·5	57·6	23		69	5
0·15	0·13	0·27			2·47	16·0	$1\frac{3}{16}$	OQ1000 WQ750 T650AC	42·2	49·7	52·8		62·2	18		55	1
0·15		0·23			2·50	16·9	$1\frac{1}{8}$	OQ1000 T620 WQ				49·2	56·8	22		56	5
"		"			"	"	$1\frac{1}{2}$	" " "				52·4	59·2	21		53	"
"		"			"	"	$3\frac{1}{4}$	" " "				50·8	58·4	21		50	"
"		"			"	"	4	" " "				50·4	58·0	20		45	"
0·16	0·20	0·23			2·67	17·8	15/32	AC960 T520				72·5	75·5	16	$53\frac{1}{2}$	22	1
"	"	"			"	"	"	" T530				69·1	74·3	18	52	18	"
"	"	"			"	"	"	" T540				63·5	66·6	20	$55\frac{1}{2}$	26	"
"	"	"			"	"	"	OQ960 T520				75·5	78·0	18	$54\frac{1}{2}$	12	"
"	"	"			"	"	"	" T530				75·4	78·0	18	$54\frac{1}{2}$	12	"
"	"	"			"	"	"	" T540				66·6	68·7	$19\frac{1}{2}$	$55\frac{1}{2}$	24	"
"	"	"			"	"	0·44 hex.	AC 960 T530				69·0	73·5	18	$51\frac{1}{2}$	15	"
"	"	"			"	"	"	" T540				59·0	63·8	18	$54\frac{1}{2}$	22	"
"	"	"			"	"	"	OQ960 T530				75·5	78·0	$19\frac{1}{2}$	56	15	"
"	"	"			"	"	"	" T540				65·8	69·0	$18\frac{1}{2}$	58	20	"
"	"	"			"	"	$\frac{1}{2}$	960 Cool in sand T530				63·5	67·4	$19\frac{1}{2}$	55	22	"
"	"	"			"	"	"	" " " " T540				61·5	66·4	$19\frac{1}{2}$	$54\frac{1}{2}$	17	"
"	"	"			"	"	"	960 Q in molten lead T530				69·7	80·2	$13\frac{1}{2}$	49	18	"
"	"	"			"	"	"	" " " T540				61·4	67·5	20	58	22	"
"	"	"			"	"	"	960 FC to 450 AC T530				56·7	61·0	$20\frac{1}{2}$	56	19	"
"	"	"			"	"	"	" " " T540				56·0	60·3	20	$60\frac{1}{2}$	24	"
"	"	"			"	"	"	" FC to 300 OQ T530				57·9	60·3	$19\frac{1}{2}$	$60\frac{1}{2}$	21	"
"	"	"			"	"	"	" " " T540				55·5	60·2	$21\frac{1}{2}$	$60\frac{1}{2}$	26	"
0·16					2·30	17·2		AC950 T650 WQ				38·8	49·0	24	$61\frac{1}{2}$	71	9
0·16	0·27	0·26			2·47	16·4	$1\frac{3}{16}$	OQ1000 T620 WQ	40·0	44·0	49·0		58·0	20	$60\frac{1}{2}$		1
0·16		0·65			2·40	17·3	$1\frac{1}{8}$	OQ1000 T610 WQ				50·8	60·4	21		61	5
0·16		0·59			2·11	18·3	2sq.	OQ1000 T545 WQ				45·8	56·0	20		59	"

MECHANICAL PROPERTIES AT ROOM TEMPERATURE (Continued)

Chemical Composition (%)							Section Size dia. (in.)	Heat Treatment °C	Tensile							Izod ft.lb.	Ref.
C	Si	Mn	S	P	Ni	Cr			0·1% P.S. (0·1Rp) tons/sq.in.	0·2% P.S. (0·2Rp) tons/sq.in.	0·5% P.S. (0·5Rp) tons/sq.in.	Y.P. (Rs) tons/sq.in.	T.S. (Rm) tons/sq.in.	El % (A)	R.A. % (Z)		
0·16	0·33	0·38			2·42	17·0	1	OQ975 T400	68·0		78·6		89·6	15½	48½	10	1
"	"	"			"	"	"	" T500	69·3		75·8		84·7	17	49½	10	"
"	"	"			"	"	"	" T550	59·2		65·6		74·4	17½	53	11	"
"	"	"			"	"	"	" T620			52·2		57·6	18	44½	18	"
"	"	"			"	"	"	" T650	40·8		47·3		59·8	21½	55½	32	"
0·16	0·40	0·50			2·50	16·5	1	OQ975 T400				75	85	15	45	40	1
"	"	"			"	"	"	" T500				60	70	18	50	49	"
"	"	"			"	"	"	" T550				45	57	22	55	56	"
0·17	0·31	0·55	0·024	0·017	2·55	16·4	¾	OQ1000 T600				54·0	64·5	21½	58½	16	1
"	"	"	"	"	"	"	"	" "+3 months 450				65·2	74·3	17	42	3	"
"	"	"	"	"	"	"	"	" "+6 " "				60·8	70·8	18½	43	7	"
"	"	"	"	"	"	"	"	" "+12 " "				64·0	73·8	16	39	5	"
0·17	0·47	0·60	0·026	0·012	2·14	16·8	2¾(M)	OQ950 T630	47·2				62·0	21	59½	50	10
0·17	0·14	0·27			2·42	15·9	1¼	OQ1000 T620	40·6	43·9	48·0		57·8	22	61		1
0·18	0·74	0·73	0·031	0·035	1·90	17·0	1 13/16 (C)	OQ950 T620	45·0				56·0	22	59½	53	10
0·18		0·60			2·22	17·0	1⅛	OQ1000 T610 WQ				50·0	60·0	22		69	5
0·18	0·18	0·23			2·45	16·7	1⅜	OQ1000 T620 WQ	37·2	41·1	45·6		56·6	21½	59		1
0·18	0·48	0·73			2·02	16·6	1 1/16	OQ950 T620 WQ	45·4	48·6	52·6		63·0	20	54		"
0·21					2·46	17·6		AC950 T650 WQ				39·4	54·8	21	53½	55	9
0·21	0·52	0·72	0·012	0·010	1·86	16·2	1¼ (C)	OQ950 T620	48·0				59·2	21	59½	52	10
0·21		0·75			1·60	17·4	1⅛	OQ975 T580	44·8				60·8	18½			5
"		"			"	"	2½	" "	49·6				59·8	21		36	"
"		"			"	"	4	" "	41·6				55·6	19		26	"
"		"			"	"	5¼	" "	40·4				57·8	17½		21	"
0·22	0·22	0·60	0·022	0·023	2·01	15·8		OQ950 T250		71·5			96·0	16	36½	20	11
0·23	0·42	0·72	0·025	0·019	2·30	16·2	5¼(M)	OQ950 T630	50·0				61·2	18	55	28	10

EFFECT OF TEMPERING ON THE MECHANICAL PROPERTIES AT ROOM TEMPERATURE

Chemical Composition (%)			
C	Mn	Ni	Cr
0·15	0·50	1·80	16·70

1⅛ in. dia. bars. Oil quenched 1000°C. Tempered as indicated.

(Ref. 5)

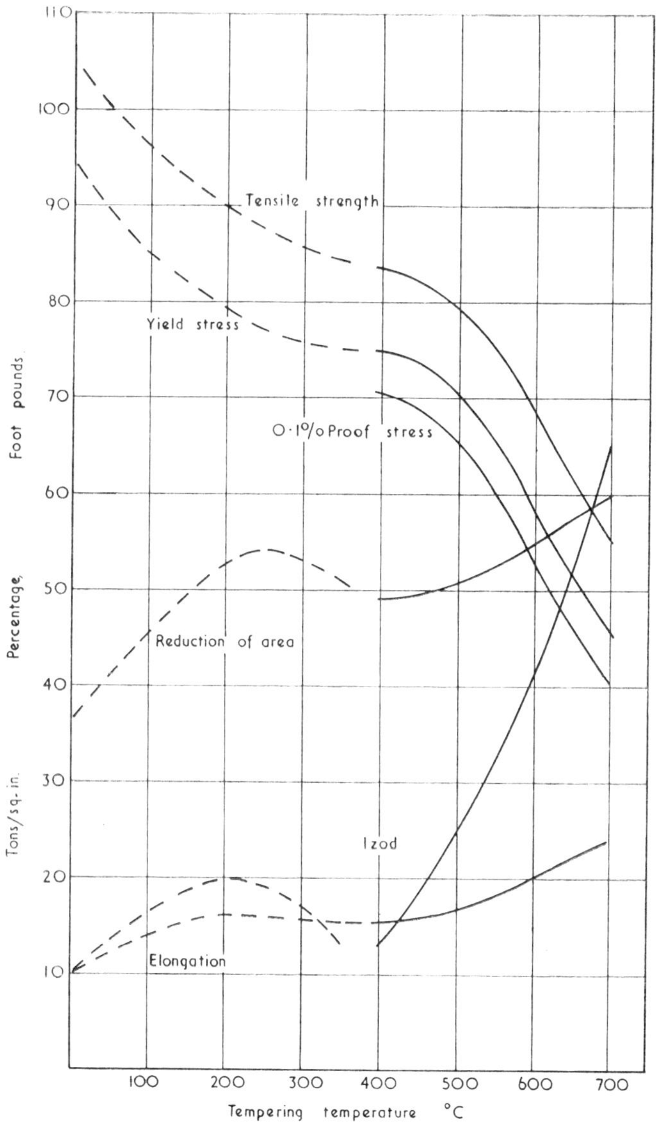

Fig. 57.1

MECHANICAL PROPERTIES AT LOW TEMPERATURE

Chemical Composition (%)								Section Size dia. (in.)	Heat Treatment °C	Test Temperature °C	Tensile							Notched Impact	Ref.
C	Si	Mn	S	P	Ni	Cr	Mo				0·1% P.S. (0·1Rp)	0·2% P.S. (0·2Rp)	0·5% P.S. (0·5Rp)	Y.P. (Rs)	T.S. (Rm)	El % (A)	R.A. % (Z)		
											tons/sq.in.								
0·11					1·72	16·3			OQ980 T455	+20				70·0	88·7	12	60	Charpy (ft.lb.) 23	12
"					"	"			" "	—40								19	"
"					"	"			" "	—75				79·8	97·0	14	60	12	"
0·14	0·20	0·25			2·61	17·0			OQ970 T630	RT					61·2	24	54		1
"	"	"			"	"			" "	+0·5					62·7	25	54		"
"	"	"			"	"			" "	—26					63·8	22	56		"
"	"	"			"	"			" "	—55					67·2	23	54		"
"	"	"			"	"			" "	—88					68·2	21	44		"
0·17	0·43	0·29	0·010	0·018	2·32	16·7		¾	OQ1000 T550 AC	+ 20	57·4	61·3	65·4		74·0	19	58	See Fig. 57.3	5
"	"	"	"	"	"	"		"	" "	— 78	60·4	64·8	69·1		81·0	22	51		"
"	"	"	"	"	"	"		"	" "	—196	77·0	81·6	87·4		100·3	6	26		"
"	"	"	"	"	"	"		"	" T650AC	+ 20	45·3	48·5	52·2		64·4	22	58		"
"	"	"	"	"	"	"		"	" " "	— 78	50·1	53·4	56·8		72·0	27	54		"
"	"	"	"	"	"	"		"	" " "	—196	68·0	72·0	76·2		92·0	20	34		"
0·18	0·33	0·37	0·023	0·016	2·56	16·8		1 1/16	OQ & T	RT	46·4		53·6		60·0	22½	58½	Charpy V (ft.lb.) 44	1
"	"	"	"	"	"	"		"	"	—78	56·3		62·2		68·0	24	53	17	"
0·22	0·22	0·60	0·022	0·023	2·01	15·8			OQ950 T250	+ 25		71·5			96·0	16	36½		11
"	"	"	"	"	"	"			" "	— 50		72·9			100·6	12	29½		"
"	"	"	"	"	"	"			" "	—120		71·2			109·0	11½	14½		"
"	"	"	"	"	"	"			" "	—150		88·5			110·8	8	10½		"
"	"	"	"	"	"	"			" "	—195		99·0			111·0	1	1		"

IMPACT TRANSITION TEMPERATURE DATA

Chemical Composition (%)						
C	Si	Mn	S	P	Ni	Cr
0·22	0·22	0·60	0·022	0·023	2·01	15·8

Bars oil quenched 950°C. Tempered 250°C.

(Ref. 11)

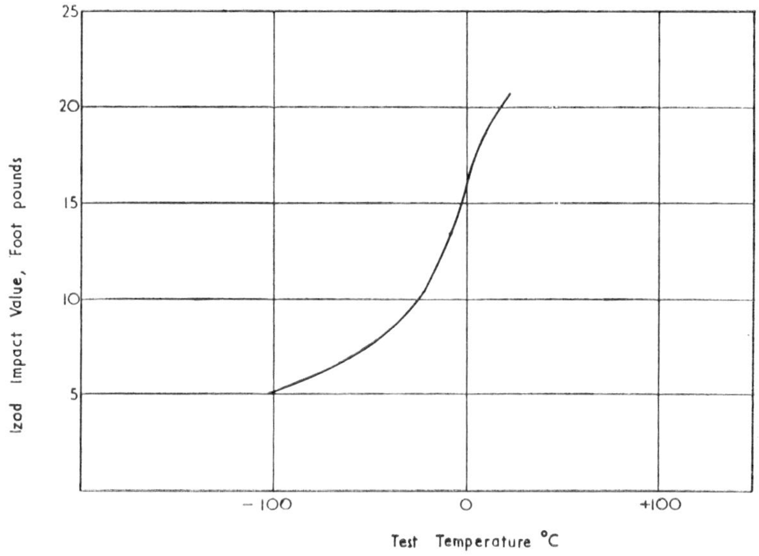

Fig. 57.2

Chemical Composition (%)						
C	Si	Mn	S	P	Ni	Cr
0·17	0·43	0·29	0·010	0·018	2·32	16·7

⅝ in. dia. bars. Oil quenched 1000°C. Tempered as indicated.

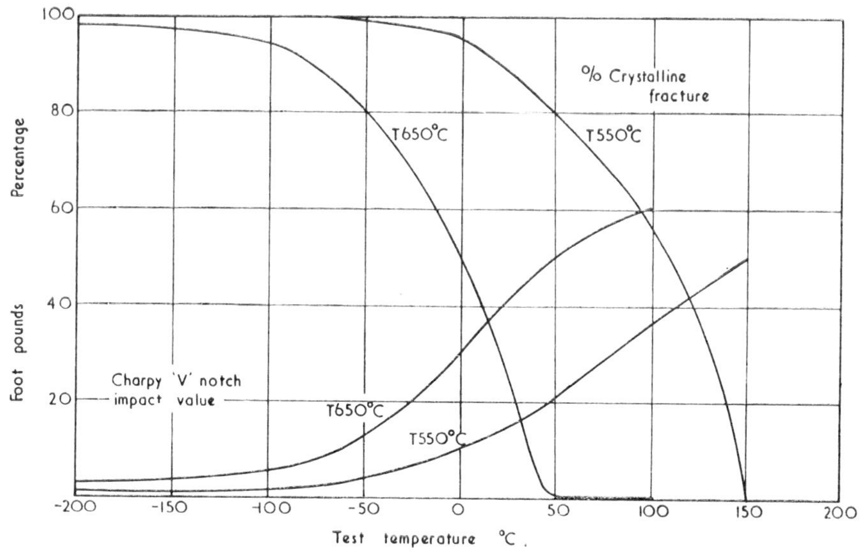

Fig. 57.3

(See Mechanical Properties at Low Temperature, page 268, for tensile test results.)

MECHANICAL PROPERTIES AT HIGH TEMPERATURE

Chemical Composition (%)							Section Size dia. (in.)	Heat Treatment °C	Test Temperature °C	Tensile						Ref.
C	Si	Mn	S	P	Ni	Cr				0·1% P.S. (0·1Rp) tons/sq.in.	0·2% P.S. (0·2Rp) tons/sq.in.	0·5% P.S. (0·5Rp) tons/sq.in.	T.S. (Rm) tons/sq.in.	El % (A)	R.A. % (Z)	
0·11	0·22	0·32			2·64	18·1	¾	OQ970 T600	RT	37·2		43·0	54·9	24	66	1
"	"	"			"	"	"	" "	100	35·6		42·7	47·5	23	58½	"
"	"	"			"	"	"	" "	200	36·5		39·6	47·1	23	64	"
"	"	"			"	"	"	" "	300	42·8		47·7	51·9	16	58½	"
"	"	"			"	"	"	" "	400	37·6		40·8	44·5	19½	63	"
"	"	"			"	"	"	" "	500	23·0		27·5	33·8	25½	64	"
"	"	"			"	"	"	" "	600				21·6	40	83½	"
"	"	"			"	"	"	" "	700				13·7	56	81½	"
"	"	"			"	"	"	" "	800				7·9	74½	85	"
"	"	"			"	"	"	" "	900				4·1	93½	83½	"
"	"	"			"	"	"	" "	1000				2·7	136	84	"
"	"	"			"	"	"	OQ950 T530	500	23·9		32·0	37·5	25½	46½	"
"	"	"			"	"	"	" "	600	8·8		13·7	22·0	37½	84½	"
"	"	"			"	"	"	" "	700				10·7	69	92	"
"	"	"			"	"	"	" "	800				7·1	99	88½	"
0·14	0·30	0·50	0·024	0·029	2·60	17·5	9/16	OQ1000 T620 WQ	RT	47·0	49·5	51·5	58·6	23	63	1
"	"	"	"	"	"	"	"	" " "	100	43·3	46·8	50·2	55·2	23	63	"
"	"	"	"	"	"	"	"	" " "	200	40·5	43·5	46·0	52·2	21	63	"
"	"	"	"	"	"	"	"	" " "	300	44·8	47·2	49·0	54·2	19	58½	"
"	"	"	"	"	"	"	"	" " "	400	44·8	46·8	47·4	55·2	16	44	"
"	"	"	"	"	"	"	"	" " "	500	35·2	37·6	40·3	44·8	23	59½	"
"	"	"	"	"	"	"	"	" " "	600	18·2	20·8	22·7	25·3	38	83	"
0·16	0·15	0·34	0·019	0·033	1·80	17·1	1	H & T	20	41·0	44·2	47·6	56·8	22		13
"	"	"	"	"	"	"	"	"	250	36·3	39·0	41·7	48·3	18		"
"	"	"	"	"	"	"	"	"	400	33·4	36·9	40·7	48·5	16		"
"	"	"	"	"	"	"	"	"	550	17·0	20·0	23·2	29·6	29		"
(Within specification)							1¼	H & T	20	47·5	48·3	49·2	56·0	22	56	14
"							"	"	200	45·0			52·5			"
"							"	"	300	42·0			50·0			"
"							"	"	400	35·3	37·2	39·3	44·0			"
"							"	"	500	16·5	18·6	22·3	29·5			"

CREEP PROPERTIES

Chemical Composition (%)					Section Size dia. (in.)	Heat Treatment °C	Test Temperature °C	Stress (tons/sq.in.) to give creep strain of 10^{-6} from 24 to 72 hrs.	Ref.
C	Si	Mn	Ni	Cr					
0·11	0·22	0·32	2·64	18·1	¾	OQ950 T600	400	16·0	1
"	"	"	"	"	"	" "	450	6·5	"
"	"	"	"	"	"	" "	500	3·5	"

(See also Mechanical Properties at High Temperature, page 270.)

Chemical Composition (%)					Section Size dia. (in.)	Heat Treatment °C	Test Temperature °C	Time hrs.	Stress (tons/sq.in.) to produce creep strains of				Ref.
C	Si	Mn	Ni	Cr					0·1%	0·2%	0·5%	1·0%	
0·16	0·15	0·34	1·80	17·1	1	H & T	400	10	26·8				13
"	"	"	"	"	"	"	"	100	18·9	23·9	(31·0)		"
"	"	"	"	"	"	"	"	300	15·2	20·5	27·5		"
"	"	"	"	"	"	"	"	1000	11·0	16·6	23·6	26·8	"
"	"	"	"	"	"	"	500	10	4·3	5·8	8·3	9·8	"
"	"	"	"	"	"	"	"	100	2·1	3·0	4·4	5·7	"
"	"	"	"	"	"	"	"	300	1·3	2·0	2·9	3·8	"
"	"	"	"	"	"	"	"	1000	(0·8)	1·0	1·8	(2·4)	"
									Figures in brackets are extrapolated.				

(See also Mechanical Properties at High Temperature, page 270.)

MECHANICAL PROPERTIES IN TORSION

Chemical Composition (%)					Heat Treatment °C	Tensile					Torsion			Ref.
C	Si	Mn	Ni	Cr		0·1% P.S. (0·1Rp) tons/sq.in.	0·5% P.S. (0·5Rp) tons/sq.in.	T.S. (Rm) tons/sq.in.	El (A) %	R.A. (Z) %	Y.P. tons/sq.in.	Apparent max. Shear Stress tons/sq.in.	Probable max. Shear Stress tons/sq.in.	
0·16	0·40	0·50	2·5	16·5	OQ975 T650		45·0	57·0	22	55	28·4	47·6	35·7	1
(Within specification)					H & T	40·0	43·5	55·0				48·7	36·5	5

FATIGUE PROPERTIES

Chemical Composition (%)					Section Size dia. (in.)	Heat Treatment °C	Test Temperature °C	Tensile				El (A) %	R.A. (Z) %	Rotating Bending Fatigue Limit tons/sq.in. (10⁷ cycle)		Ref.
C	Si	Mn	Ni	Cr				0·1% P.S. (0·1Rp)	0·5% P.S. (0·5Rp)	Y.P. (Rs)	T.S. (Rm)			Plain	Notched[1]	
								tons/sq.in.								
0·12	0·29	0·15	2·75	17·7		H and T	RT	39·0	42·7		56·8	19½	56	±23·5		1
0·15	0·33	0·50	2·09	16·0		OQ950 T600 WQ	"			46·6	58·1	21	60½	±32·2		4
0·16	0·40	0·50	2·50	16·5		OQ975 T650 WQ	"			45·0	57·0	22	55	±24·0		1
0·18	0·33	0·37	2·56	16·8	11/16	H and T	RT	46·4	53·6		60·0	22½	58½	±35·0	±16·5	"
"	"	"	"	"	"	"	—78	56·3	62·2		68·0	24	53	±39·0	±22·0	"

[1] Neuber Stress Concentration Factor 2.2.

Chemical Composition (%)					Heat Treatment °C	Tensile				Fatigue Limit tons/sq.in.				Ref.
C	Si	Mn	Ni	Cr		Y.P. (Rs)	T.S. (Rm)	El % (A)	R.A. % (A)	Plain Bending		Reversed	Axial	
						tons/sq.in.				In Air	Lanoline Coated	In Air	Lanoline Coated	
0·25	0·52	0·68	1·16	17·1	OQ950 T630 WQ	41·1	54·4	22½	53½	±32·7	±31·2	±28·3	±28·5	15

In Sea Water

Rotating Bending		Reversed Axial	
Applied Stress tons/sq.in.	Cycles to Failure $\times 10^6$	Applied Stress tons/sq.in.	Cycles to Failure $\times 10^6$
±20	2·18	±32	0·13
±18	3·30	±28	0·30
±16	3·52	±24	0·88
±14	17·69	±20	1·20
±13·5	24·29 (UB)	±19	4·12
±13	31·26	±18	13·14
		±16·5	6·39
		±15·5	50·33

LIST OF REFERENCES

1. Private correspondence with Firth-Vickers Stainless Steels Ltd.
2. Private correspondence with The United Steel Companies Ltd.
3. Private correspondence with Samuel Osborn and Co. Ltd.
4. Private correspondence with Edgar Allen and Co. Ltd.
5. Private correspondence with International Nickel Ltd.
6. Private correspondence with Richard Thomas and Baldwins Ltd.
7. Technical Data—Stainless Steels. Atlas Gibsons Steels S. A., and Atlas Steels, Canada.
8. *Stainless Steel Handbook*. United States Steel Corporation, 1957.
9. J. H. G. Monypenny. *Stainless Iron and Steel*. 2nd edition. Chapman and Hall, London, 1931.
10. Private correspondence with Jessop-Saville Ltd.
11. War Office, unpublished report.
12. H. W. Gillett. Impact Resistance and Tensile Properties of Metals at Sub-atmospheric Temperatures. Project 13 of the Joint ASME-ASTM Research Committee on the Effect of Temperature on the Properties of Metals. *Proceedings*, ASTM, vol. 41, 1941.
13. Private correspondence with the Royal Aircraft Establishment, Farnborough.
14. Private correspondence with Rolls-Royce Ltd.
15. H. J. Gough and D. G. Sopwith. Some Comparative Corrosion-fatigue Tests employing two types of Stressing Action. *JISI*, vol. 127, No. 1, 1934, 301—332.

AUSTENITIC CHROMIUM–NICKEL RUST, ACID AND HEAT-RESISTANT STEEL

SPECIFICATION

CHEMICAL COMPOSITION (%)

	En 58A	En 58B	En 58C	En 58D‡	En 58E	En 58F	En 58G	En 58H	En 58J
C	0·16 max.	0·15 max.	0·15 max.	0·16 max.	0·08 max.	0·15 max	0·15 max.	0·12 max.	0·12 max.
Si	0·20 min.	0·20 min.	0·20 min.	0·20 min.	0·20 min.	0·20 min.	0·20 min.	0·20 min.	0·20 min.
Mn	2·00 max.	2·00 max.	2·00 max.	2·00 max.	2·00 max.	2·00 max.	2·00 max.	2·00 max.	2·00 max.
S	0·045 max.	0·045 max.	0·045 max.	0·045 max.	0·045 max.	0·045 max.	0·045 max.	0·045 max.	0·045 max.
P	0·045 max.	0·045 max.	0·045 max.	0·045 max.	0·045 max.	0·045 max.	0·045 max.	0·045 max.	0·045 max.
Ni	7·0—10·0*	7·0—10·0*	9·0—12·0	11·0—14·0	8·0—11·0	7·0—10·0*	9·0—12·0	8·0—12·0	8·0—12·0
Cr	17·0—20·0*	17·0—20·0*	17·0—20·0	11·0—14·0	17·5—20·0	17·0—20·0*	17·0—20·0	17·0—20·0	17·0—20·0
Mo								1·5—2·5	2·5—3·5
Ti		4×C min.	4×C min.					opt.*	opt.†
Nb						8×C min.	8×C min.	opt.†	opt.†

* The sum of the nickel and chromium contents shall be not less than 25.0%.
† These elements are present at the option of the manufacturer.
‡ Where steel is required for deep drawing or spinning. The sum of nickel and chromium contents shall be not less than 23.0%.

En 58 M

Free-machining modifications of the above steels are supplied to a composition agreed between the purchaser and the supplier.

MECHANICAL PROPERTIES

Limiting ruling section, in.	6
Tensile strength, tons/sq. in., min. (Rm).	35
Yield stress (1) tons/sq. in., min. (Rs).	12
Elongation, %, min.	30
Izod impact value, ft. lb., min.	50

(1) Not to be used as acceptance values except by special arrangement between the purchaser and the manufacturer.

Material to specifications En 58B, 58C, 58F and 58G, and to specifications En 58H and 58J when stabilized with titanium or niobium, shall satisfy a special ringing and bend test, which shall be carried out as follows:

Each bend test piece shall be not larger than ½ in. diameter or thickness.

The test piece shall be heated for 30 min at a temperature of 650°C. and cooled in air and shall then be immer sedfor 72 hrs. in a boiling solution having the following composition:

111 grammes copper sulphate ($CuSO_4 \cdot 5H_2O$)
98 grammes sulphuric acid (sp. gr. 1.84)

made up to 1 litre with distilled water.

Precautions should be taken during boiling to prevent concentration due to evaporation.

Each test piece shall then be dropped on a metal or stone surface and must emit a clear metallic ring. The test pieces shall then be bent through 90° over a radius of three times the diameter or thickness of the test piece and shall withstand this treatment without cracking.

RELATED SPECIFICATIONS

United Kingdom				America		France	Germany		Sweden	USSR	
En No.	BS	BS Air	DTD	SAE	AISI	AFNOR	Werk-stoff No.	Name	SIS	GOST	Mark
58A	1449			30302	302	Z 12CN18—10	1.4300	X12 Cr Ni 18 8		5632—61	Kh18 N9E
	1554								14·2330		
	2056								14·2331		
58B	1449	S110	189A	30321	321	Z10CNT18—10	1.4541	X10 Cr Ni Ti 18 9	14·2337	5632—61	Kh18 N10T
	1554	S520	712A							5632—61	06Kh18 N9T
	2056	S521	743								
		T55									
		T58									
58C	1449	S110		30321	321		1.4541	X10 Cr Ni Ti 18 9		5632—61	OKh18 N10T
	1554	S520	712A						14·2337	5632—61	Kh18 N10T
	2056	S521	743							5632—61	OKh18 N12T
		T55				Z10CNT18—10				5632—61	Kh18 N12T
		T58									
58D	1449					Z10CN12—12	1.4307	X8 Cr Ni 12 12			
	1554										
	2056										
58E	1449		734	30304	304	Z6CN18—10	1.4301	X5 Cr Ni 18 9	14·2333	5632—61	OKh18 N11
	1554										
	2056										
58F	1449	S110	189A	30347	347	Z10CNNb18—10	1.4550	X10 Cr Ni Nb 18 9	14·2338		
	1554	S520	712A	30348	348						
	2056	S521	743								
		T55									
		T58									
58G	1449	S110		30347	347	Z10CNNb18—10	1.4550	X10 Cr Ni Nb 18 9	14·2338	5632—61	OKh18 N12B
	1554	S520	743	30348	348						
	2056	S521									
		T55									
		T58									
58H	1449	S110		30316	316		1.4571	X10 Cr Ni Mo Ti 18 10	14·2340		
	1554	S520									
	2056	S521	712A								
58J	1449	S110		30316	316	Z8CNDNb18—12	1.4580	X10 Cr Ni Mo Nb 18 10	14·2342		
	1554	S520		30317	317	Z6CND18—12	1.4436	X5 Cr Ni Mo 18 12			
	2056	S521									
		T55									
		T58									
58M	1449			30303	303					5632—61	Kh18 N10E
				30303Se	303Se						

APPLICATIONS

Because of the high chromium and nickel contents the steels of the En 58 series have a very high resistance to corrosion and are used where this property is of paramount importance. All the steels are austenitic and are virtually non-magnetic although some which are either low in carbon or contain stabilizing additions of titanium or niobium may contain some delta ferrite, and in consequence may exhibit a slight degree of magnetic susceptibility, sufficient to enable them to be attracted by a strong magnet.

The steels cannot be hardened by heat treatment, and if they are required in a hardened condition this can only be achieved by cold work. All have a high capacity for cold work-hardening.

Steels without stabilizing additions of titanium and niobium may be susceptible to intercrystalline corrosion if they are heated in the range 350 to 900°C or if they are allowed to cool slowly within this range, e.g. after welding. This does not apply to unstabilized steels of very low carbon content below 0·03%; even so, they are not recommended for service involving exposure to temperatures above 300°C.

Steels containing molybdenum have an increased resistance to attack from phosphoric, acetic and dilute sulphuric acid solutions, to sulphurous waters and waters containing halogen salts including, in the last-named case, increased resistance to pitting attack.

The steels have a high resistance to scaling. Those with stabilizing additions have high strength and resistance to creep at temperatures up to 800°C and are frequently used in steam plants and for furnace parts. Steels with high nickel contents, e.g. En 58C and En 58G, are preferred for such applications.

The stabilized steels do not take a very high polish (the niobium-stabilized steels being better in this respect than the titanium-stabilized). If a very high polish is desired, an unstabilized quality should be used, a low carbon variety being adopted if the parts are fabricated by welding and no subsequent heat treatment is possible.

En 58D is a quality having exceptionally high ductility and somewhat less capacity for work-hardening than the other steels. It is therefore suited for applications where fabrication is by deep drawing, pressing or spinning. With its lower chromium content than the remainder of the En 58 series it has a somewhat lower corrosion resistance and has now to a considerable extent been replaced by a low-carbon type with a high nickel content such as En 58E.

Typical applications are:

En 58A: Parts for chemical plant, cooking utensils, dairy equipment, shop and office fittings, surgical instruments, hospital equipment, photographic equipment, plates for plastic laminates, domestic sinks, drawing boards, bath fittings, table tops, plates, decorative trim both inside and outside buildings and motor cars.

En 58B and C: Chemical and oil refinery plant, dairy and brewing equipment, cooking utensils, aircraft construction, steam, hydraulic and mechanical engineering components where the parts are fabricated by welding or are subjected to hot-working operations during fabrication, e.g. hot pressing or flanging, or are operated after fabrication either constantly or intermittently at elevated temperatures up to 800°C in which case En 58C is preferred.

En 58D: Shop and office fittings, spoons, forks, fishknives, ashtrays and domestic and decorative applications involving deep drawing and spinning.

En 58E: As En 58A, but including components fabricated by welding from sheet not greater than $\frac{3}{16}$ in. thick. It is also replacing En 58D for deep-drawn and cold-spun components, pressed and spun saucepans, mixing bowls, cream separators, hospital bowls, oil separators, refrigerator expansion units, washing machines.

En 58F and G: Applications similar to those for En 58B and C, especially where cross welding is essential. En 58G is preferred for service at high temperature, filler wires for gas welding and wires for metallic-arc electrodes. For chemical plant, especially for storage and transport of concentrated nitric acid, gas turbine parts up to 700°C and general heat-resisting components for temperatures up to 800°C En 58G is preferred. For large welded sections, however, En 58J is now preferred, since it is less susceptible to cracking in the weld.

En 58H and J: Parts of chemical and industrial plant exposed to phosphoric and acetic acids and solutions containing halogen salts and sulphites, papermaking, textile printing machinery, dye vats, surgical and medical equipment, plant for the manufacture of plastics and synthetic fibres, chemical and petroleum products, marine parts in contact with sea water.

En 58M: The free-machining grades are used for similar duties to the corresponding grades without the free-machining additions where difficult machining operations have to be carried out, e.g. nuts, bolts, screws, skids, aircraft fittings, pump parts in contact with brine, valve bodies where the sulphur addition lessens the tendency to seizure. The addition of sulphur for free-machining purposes reduces the corrosion resistance in some environments but a little molybdenum may be added to offset this.

WELDING

(See Introduction for key to symbols used in this table.)

	MA	G	IG	S	F	B	Remarks
En 58A	a	a	a	a	a	h	See Notes 2 and 3.
En 58B	a	a	a	a	a	h	See Note 3.
En 58C	a	a	a	a	a	h	See Note 3.
En 58D	a	a	a	a	a	h	See Note 1 and 3.
En 58E	a	a	a	a	a	h	See Note 1 and 3.
En 58F	a	a	a	a	a	h	See Note 3.
En 58G	a	a	a	a	a	h	See Note 3.
En 58J	a	a	a	a	a	h	See Note 3.

Reproduced by kind permission of H.M.S.O.

Notes

1. Post-welding heat treatment *may be* necessary to restore the corrosion resistance of welded joints.
2. Post-welding heat treatment *is* necessary to restore the corrosion resistance of welded joints.
3. Hard-drawn wire should not be welded or brazed.

MACHINABILITY

The machinability is approximately 30% of that for mild steel (En 3).
The free-machining grades have a machinability approximately 75% of that for mild steel (En 3).

HOT WORKING AND HEAT TREATMENT TEMPERATURES

Forging, rolling and stamping	1200°C, finish above 900°C.
Softening	950°C—1150°C, quickly cool in air, oil or water.

Cooling after softening should be as quick as possible, depending on the section, thus thin sections may be air cooled or oil quenched, thick sections water quenched. The stabilized grades do not need cooling so quickly as the non-stabilized.

PHYSICAL PROPERTIES

SPECIFIC GRAVITY (d)

Type	Chemical Composition (%)								Heat Treatment °C	Specific Gravity at 20°C	Ref.
	C	Si	Mn	Ni	Cr	Mo	Ti	Nb			
En 58E	0·03			8·0	18·0				Softened	7·87	1
En 58E	0·05	0·50	0·40	10·0	19·0				"	7·93	2
En 58E	0·06	0·50	1·40	8·25	18·0				"	7·92	3
En 58E	0·07	"	0·45	8·93	17·3				OQ1090	7·92	4
En 58E	0·08	0·68	0·37	8·14	19·1				WQ1100	7·916	5
En 58A	0·10	0·50	1·40	8·25	18·0				Softened	7·90	3
En 58A	0·12	0·24	0·36	7·96	17·7				"	7·916	2
En 58A	0·13	0·23	0·45	7·99	17·8				"	7·919	"
En 58A	0·13	0·41	0·47	7·81	17·0				"	7·911	"
En 58A	0·14	0·23	0·27	8·06	17·8				"	7·921	"
En 58B	0·06			8·0	18·0		added		Softened	7·89	1
En 58B	0·07	0·50	1·40	8·25	18·0		0·35		"	7·93	3
En 58B	0·13	0·50	0·33	8·22	18·6		0·61		AC1100	7·96	2
En58B	0·07	0·50	0·90	9·0	17·0		0·40		Softened	7·93	6
En 58D	0·07		0·90	12·0	12·0				Softened	8·00	6
En 58D	0·08			12·5	13·0				"	8·00	2
En 58D	0·10	0·30	0·30	12·5	12·5				"	8·01	1
En 58F(G)	0·08	0·60	0·80	9·5	18·0			0·85	Softened	7·93	2
En 58F	0·10			8·5	18·5			1·0	"	7·93	7
En 58G(F)	0·12	0·60	0·40	10·0	18·0			1·2	"	7·93	2
En 58G(F)	0·09	0·40	0·52	9·98	18·2			0·85	AC1050	7·933	"
En 58G	0·09	0·60	1·50	11·0	17·5			0·90	Softened	7·94	"
En 58G	0·11	0·50	1·30	12·0	18·7			1·25	AC1050	7·92	8
En 58G	0·13	0·60	1·20	12·0	18·0			1·30	"	7·94	"
En 58H	0·06	0·50	1·40	10·2	18·0	1·75	0·75		Softened	7·93	
En 58J	0·07	0·30	0·50	8·0	18·0	2·75			Softened	7·96	2
En 58J	0·07	0·40	1·50	9·0	17·5	2·75	0·30		"	7·96	"
En 58J	0·08			12·0	19·0	3·5			"	7·95	9

SPECIFIC HEAT (c)

Type	Chemical Composition (%)								Heat Treatment °C	Mean Specific Heat cal/g.deg. C										Ref.
	C	Si	Mn	Ni	Cr	Mo	Ti	Nb		20 to 100°C	20 to 200°C	20 to 300°C	20 to 400°C	20 to 500°C	20 to 600°C	20 to 700°C	20 to 800°C	20 to 900°C	20 to 1000°C	
En 58A(E)	0·03			8·0	18·0				Softened		0·125		0·129		0·131		0·135			1
En 58E(A)	0·06			10·0	19·0				”		0·123		0·126		0·131		0·135			”
En 58A(E)	0·08	0·68	0·37	8·14	18·1				WQ1100	1·121	0·125	0·126	0·128	0·131	0·135	0·137	0·139	0·141	0·142	5
En 58A	0·10			8·0	18·0						0·122		0·122		0·132		0·141			3
En 58B	0·06			8·0	18·0		added		Softened		0·126		0·128		0·132		0·138			1
En 58B	0·07	0·50	1·40	8·25	18·0		0·35		”	0·120										3
En 58C(B)	0·07	0·50	0·90	9·0	17·0		0·40		”	0·120										6
En 58B	0·12	0·60	0·40	8·0	18·0		0·60		”	0·120										2
En 58D	0·07	0·50	0·90	12·0	12·0				Softened	0·119										6
En 58D	0·08			12·5	13·0				”		0·123		0·126		0·131		0·135			1
En 58D	0·10	0·20	0·20	12·5	12·5				”	0·117										10
En 58F(G)	0·06			9·0	19·0			added	Softened		0·126		0·128		0·132		0·138			1
En 58G	0·09	0·60	1·50	11·0	17·5			0·90	”	0·120										2
En 58G	0·08	0·52	1·30	11·1	18·3			0·86	”	0·115	0·118	0·121	0·123	0·126	0·128	0·130	0·132	0·134		11
En 58H	0·08			9·0	19·0	1·75	added		Softened		0·120		0·128		0·133		0·134			1
En 58J	0·08			12·0	19·0	3·5			Softened		0·122		0·124		0·128		0·134			9
En 58J	0·07	0·30	0·50	8·0	18·0	2·75			”	0·120										2

MEAN COEFFICIENT OF THERMAL EXPANSION (α)

Type	Chemical Composition (%)								Heat Treatment °C	$10^6 \times$ Mean Coefficient of Thermal Expansion per deg. C.										Ref.
	C	Si	Mn	Ni	Cr	Mo	Ti	Nb		20 to 100°C	20 to 200°C	20 to 300°C	20 to 400°C	20 to 500°C	20 to 600°C	20 to 700°C	20 to 800°C	20 to 900°C	20 to 1000°C	
En 58E(A)	0·03			8·0	18·0				Softened		17·5		18·4		18·7		19·4			1
En 58E	0·06			10·0	18·0				”		18·0		18·7		19·3		19·8			”
En 58E	0·06	0·45	0·41	9·9	18·4				AC1100	16·1	16·8	17·5	17·9	18·2	18·6	19·0	19·3	19·6		2
En 58E		Within specification								17·0		18·0		18·6			19·6			12
En 58E(A)	0·06	0·50	1·40	8·25	18·0				Softened	17·0	17·5		18·4		18·7		19·4			3
En 58E(A)	0·07	0·53	0·65	8·49	19·19				Softened	15·9	16·6	17·1	17·6	18·0						13
En 58A(E)	0·07		0·45	8·93	17·35				OQ1090	17·2	17·4	17·8	18·1	18·4	18·7	18·9				4
En 58A(E)	0·08			9·0	18·0				Softened		18·0		18·5		19·2		19·9		20·4	1
En 58A(E)	0·08	0·68	0·37	8·14	19·1				WQ1100	15·1	16·7	17·3	17·8	18·3	18·6	18·9	19·2	19·3	19·5	5
En 58A	0·10			8·0	18·0				Softened +336 days at 500	17·7	18·0	18·2	18·6	19·0	19·2					”
En 58A		Within specification							Softened	17·0		17·8		18·4			19·7			12
En 58B	0·06			8·0	18·0		added		Softened		17·2		18·3		19·0		19·6			1
En 58B	0·08			8·0	19·0		added		”		16·9		17·8		18·5		19·2			”
En 58B	0·07	0·50	1·40	8·25	18·0		0·35		”	17·0	17·4		18·0		19·0		19·6			3
En 58B(C)		Within specification							”	16·9		17·7		18·3			19·6			12
En 58B(C)	0·10	0·40	0·30	9·0	18·0		0·60		”	17·0										2
En 58B	0·11	0·82	0·42	8·83	19·2		0·55		”	16·5	17·1	17·7	18·0	18·3	18·7	19·1	19·3	19·6		2
En 58D	0·08			12·5	13·0				Softened		18·0		18·7		19·3		19·8			1
En 58D	0·08			12·5	12·5				”	17·8										2
En 58D		Within specification							”	18·0		18·3		19·0			19·8			12
En 58F(G)	0·06			9·0	19·0			added	Softened		17·2		18·3		19·0		19·6			1
En 58G	0·06			10·0	18·0			”	”		17·5		18·1		18·9		19·7		20·3	8
En 58G	0·07	0·56	1·74	11·3	18·6			0·77	”	16·3	16·8	17·3	17·7	18·1						13
En 58G	0·11	0·50	1·30	12·0	18·75			1·25	”	17·0	17·2	17·4	17·6	17·9	18·6	18·8				8
En 58G	0·12	0·65	1·20	12·0	18·0			1·25	”		16·5		18·0		18·5		19·0			14
En 58G	0·12	0·60	1·50	11·0	17·5			1·2	”	17		18		18		19				2
En 58H	0·06	1·50	1·40	10·25	18·0	1·75		0·25	Softened	16·9	17·0		17·6		18·3		19·0			3
En 58H	0·06	0·56	1·77	12·70	17·8	2·38			”	16·2	16·8	17·2	17·6	17·9						13
En 58H	0·08			9·0	19·0	1·75	added		”		16·8		17·6		18·3		19·0			1
En 58H		Within specification							”	16·5		17·2		18·0			19·0			12
En 58H		”	”						”	17		17		18		18				8

MEAN COEFFICIENT OF THERMAL EXPANSION (α) (Continued)

Type	Chemical Composition (%)								Heat Treatment °C	10^6 × Mean Coefficient of Thermal Expansion per deg. C										Ref.
	C	Si	Mn	Ni	Cr	Mo	Ti	Nb		20 to 100°C	20 to 200°C	20 to 300°C	20 to 400°C	20 to 500°C	20 to 600°C	20 to 700°C	20 to 800°C	20 to 900°C	20 to 1000°C	
En 58J	0·06	0·18	0·37	8·05	18·0	2·81			WQ1150	15·7	16·2	16·7	17·0	17·3	17·6	18·0	18·4	18·7		2
En 58J	0·06	0·50	1·40	10·25	18·0	2·8			Softened	16·9	17·0		17·6		18·3		19·0			3
En 58J	0·08			9·0	19·0	2·75			"		16·8		17·6		18·3		19·0			9
En 58J	0·08			12·0	19·0	3·5			"		16·8		17·3		18·1		18·7			"
En 58J	0·12	0·50		9·5	18·5	2·75			"	16		18		19		20				8
En 58J	Within specification								"	16·5		17·3		18·0			19·4			12
En 58J	(AISI 316 Composition)								"	15·5	16·1	16·8	17·5	18·0	18·3	18·6	18·8	19·0	19·2	15

The coefficients of expansion of the free-cutting qualities of En 58M are similar to those of the corresponding grades with similar composition apart from the free-cutting additions.

COEFFICIENT OF THERMAL EXPANSION AT LOW TEMPERATURES

Type	Chemical Composition (%)								10^{-6} × Coefficient of Thermal Expansion per deg. C						Ref.
	C	Si	Mn	Ni	Cr	Mo	Ti	Nb	0 to 20°C	—50 to 20°C	—100 to 20°C	—150 to 20°C	—200 to 20°C	—250 to 20°C	
En 58E (A)*	0·054	0·42	0·49	8·82	18·5				14·7	14·1	13·3	12·5	11·6	10·6	16
En 58E (A)	0·07	0·53	0·65	8·49	19·2				16·0	15·2	14·4	13·7	13·2		13
En 58E	0·05	0·85	0·67	10·1	18·6				15·6	14·8	14·4	13·9	13·7		2
En 58B	0·10			8·0	18·0		0·6			(—30 to 20°C) 14·2					"
En 58G	0·7	0·56	1·74	11·3	18·6			0·77	15·9	15·5	15·0	14·3	13·0		13
En 58G (F)	Not stated (AISI 347 Composition)									16·0	15·4	14·7			15
En 58H	0·06	0·58	1·77	12·7	17·8	2·38			14·7	14·4	13·8	13·2	12·6		13
En 58J	(AISI 316 Composition)									15·2	15·0	14·8			15

* Cold rolled. All others were in the softened condition.

THERMAL CONDUCTIVITY (k)

Type	Chemical Composition (%)								Heat Treatment °C	Thermal Conductivity cal/cm sec. deg. C												Ref.
	C	Si	Mn	Ni	Cr	Mo	Ti	Nb		0°C	20°C	100°C	200°C	300°C	400°C	500°C	600°C	700°C	800°C	900°C	1000°C	
En 58E (A)	0·03			8·0	18·0				Softened			0·038	0·042	0·045	0·049	0·051	0·054	0·058				1
En 58E (A)	0·06			10·0	19·0				"			0·034	0·038	0·041	0·045	0·048	0·052	0·055	0·058			"
En 58A (E)	0·08	0·68	0·37	·14	19·1				WQ 1100	0·038	0·038	0·039	0·041	0·044	0·048	0·052	0·057	0·061	0·064	0·064	0·067	5
En 58A (E)	0·07		0·27	·10	18·6				8hrs. 735 SC			0·039	0·042	0·045	0·048	0·052						17
En 58A	Within specification								Softened		0·036	0·039			0·049							12
En 58A	0·11		0·19	9·21	18·5				WQ 1120			0·039	0·042	0·045	0·048	0·051						17
En 58E	Within specification								Softened		0·038											12
En 58A	0·12	0·13		9·60	18·4				AC 1100			0·032	0·036	0·040	0·044	0·049	0·053	0·057	0·061	0·065		18
En 58A	0·15	0·19	0·26	8·04	17·9							0·039	0·042	0·045	0·049	0·052	0·056	0·059	0·062			5
En 58B	0·06			8·0	18·0		added		Softened			0·038			0·049		0·054	0·058				1
En 58B	0·08			8·0	19·0		added		"			0·037			0·050		0·059		0·065			"
En 58B (C)	Within specification										0·037	0·039			0·049				0·059			12
En 58B (C)	0·07	0·47	0·59	9·12	18·1		0·34		"			0·038	0·042	0·046	0·049	0·053						17
En 58D	0·08			12·5	13·0				Softened			0·034	0·038	0·041	0·045	0·048	0·052	0·055	0·058			1
En 58D	0·10	0·30	0·80	12·5	12·5				"		0·037	0·040										2
En 58F	0·08	0·62	1·23	8·85	18·7		0·14	0·99	AC 1050		0·035	0·038	0·042	0·046	0·049	0·052	0·055	0·058	0·061	0·064		19
En 58F (G)	Within specification										0·037	0·039			0·049							12
En 58G	0·06	0·47	1·51	11·3	17·6			0·83	"		0·036	0·040	0·044	0·048	0·051	0·054	0·057	0·060	0·062			"
En 58G	0·07	0·70	1·80	11·2	18·0			0·77	Softened	0·046		0·049	0·053	0·056	0·061	0·066	0·070	0·075	0·081	0·087		20
En 58G (F)	AISI No. 347 Composition								"	0·034	0·035	0·037	0·040	0·044	0·047	0·050	0·056	0·060	0·063	0·065		15
En 58G	0·11	0·50	1·30	12·0	18·7			1·25	"			0·036	0·040	0·044	0·048	0·052	0·058	0·064				8
En 58H	0·08			9·0	19·0	1·75	added		Softened			0·038	0·043	0·047	0·051	0·055	0·059	0·062	0·065			1
En 58H	0·08			12·0	19·0	3·50			"			0·038			0·049		0·057		0·063			9
En 58H (J)	AISI No. 316 Composition								"	0·031	0·032	0·034	0·038	0·041	0·044	0·044	0·051	0·054	0·057	0·060	0·062	15
En 58H	Within specification										0·038	0·040			0·049							12
En 58J	" "										0·039	0·040			0·049							12

THERMAL CONDUCTIVITY AT LOW TEMPERATURE

Type	Chemical Composition (%)							Heat Treatment °C	Thermal Conductivity cal/cm. sec. deg. C				Ref.
	C	Si	Mn	Ni	Cr	Ti	Nb		—150°C	—100°C	—50°C	0°C	
En 58G (F)	AISI No. 347 Composition							Softened	0·029	0·031	0·033	0·034	15
En 58H (J)	AISI No. 316 "							"	0·026	0·027	0·029	0·031	"

ELECTRICAL RESISTIVITY (ϱ)

Type	Chemical Composition (%)								Heat Treatment °C	Electrical Resistivity microhm-cm												Ref.
	C	Si	Mn	Ni	Cr	Mo	Ti	Nb		0°C	20°C	100°C	200°C	300°C	400°C	500°C	600°C	700°C	800°C	900°C	1000°C	
En 58A (E)	0·08	0·68	0·37	8·14	19·1				WQ1100	69·4	71·0	77·6	85·0	91·6	97·6	102·6	107·2	111·1	114·1	117·1	119·6	5
”	”	”	”	”	”				WQ1100		70·7											21
”	”	”	”	”	”				WQ950		70·9											”
En 58A	0·08			8·0	18·0				Softened		70·0	76·7	83·9		96·6		106·5		113·7		118·0	1
En 58E	0·06			10·0	19·0				”		73·0	79·4	87·5									”
En 58A	0·10	0·66	0·30	8·04	18·2				”	70·0	71·6	77·6	84·5	91·1	97·0	102·3	106·9	110·7	114·0	116·9	119·6	2
En 58A	0·10			10·5	17·0				”		73·0								119·5			8
En 58A	0·15	0·19	0·26	8·0	17·9				”			74·3			95·0			109·4				2
En 58B	0·12	0·60	0·40	8·0	18·0		0·6		Softened	71·7	73·4	80·1	88·0	95·2	101·5	107·0	111·9	116·0	119·5	122·3	124·2	2
En 58B	0·07			8·5	18·0		added		”		72·0	78·0				105·0		115·0				22
En 58C	0·06	0·38	1·87	12·0	17·2		0·60		WQ1050	72·5	74·3	79·2	86·3	93·4	99·7	104·8	109·4	113·4	117·0	119·8	122·2	23
En 58D	0·08			12·5	13·0				Softened		73·0	79·4	87·5									1
En 58D	0·10	0·30	0·30	12·5	12·5				”		70·0	77·0										”
En 58D	0·10	0·30	0·80	12·5	12·5				”	68·8	70·0	77·2	85·0	91·7	97·8	102·4						2
En 58D	not stated								”		72·0	78·0										12
En 58F	0·08	0·62	1·23	8·85	18·7		0·14	0·99	AC1050		73·4	79·9	86·6	93·0	98·9	104·8	108·7	112·4	115·8	118·2		19
En 58F	Composition not stated								Softened		72·0	78·0				105·0		115·0				22
En 58G	0·06	0·47	1·51	11·3	17·6			0·83	”		72·7	78·8	86·0	92·5	98·4	103·5	108·3	111·6	114·8			19
En 58G	0·07	0·70	1·80	11·2	18·0			0·77	”	79·7		84·8	90·1	93·5	97·2	100·0	101·7	103·5	105·3	107·2		20
En 58G	Composition not stated								”		73·0								119·5			8
En 58H	0·08			9·0	19·0	1·75	added		Softened		77·0	83·2	89·5									1
En 58H	0·08	0·95	0·29	10·0	18·2	1·11			Softened	74·6	77·2	81·1	87·2	92·9	98·2	103·1	107·2	110·7	113·9	116·4	118·8	2
En 58J	0·06	0·72	0·47	8·43	18·5	3·57	0·42		AC1050		77·4											2
En 58J	0·07	0·40	1·50	9·0	17·5	2·75	0·3		Softened		78·0	84·0										”
En 58J	0·13	0·85	0·35	8·16	18·2	3·45			Softened	76·8	78·2	84·1	90·9	97·9	102·2	106·9	112·0	114·5	117·4	120·2	122·6	2

ELECTRICAL RESISTIVITY AT LOW TEMPERATURES (ϱ)

Type	Chemical Composition (%)						Heat Treatment °C	Electrical Resistivity microhm-cm					Ref.
	C	Si	Mn	Ni	Cr	Ti		—183°C	—80°C	—40°C	—20°C	0°C	
En 58C	0·06	0·38	1·87	12·0	17·2	0·64	WQ1050	56·6	66·3	69·5	74·5	72·5	23

YOUNG'S MODULUS (E), SHEAR MODULUS (G), POISSON'S RATIO (σ)

Type	Chemical Composition (%)								Heat Treatment °C	Test Temperature °C	Young's Modulus tons/sq.in.		Shear Modulus tons/sq.in.		Poisson's Ratio		Ref.
	C	Si	Mn	Ni	Cr	Mo	Ti	Nb			Static	Dynamic	Static	Dynamic	Static	Dynamic	
En 58E	0·03	0·56	0·73	9·9	18·5				Softened	20		12,800*		4,950*		0·297*	24
En 58E(A)	0·06	0·39	0·38	9·56	18·71				WQ1,040	20	12,650		5,090		0·245		25
"	"	"	"	"	"				"	100	12,350		4,850		0·285		"
"	"	"	"	"	"				"	200	12,050		4,650		0·295		"
"	"	"	"	"	"				"	300	11,750		4,470		0·315		"
"	"	"	"	"	"				"	400	11,100		4,240		0·310		"
"	"	"	"	"	"				"	500	10,200		3,980		0·280		"
"	"	"	"	"	"				"	600	9,900		3,820		0·295		"
"	"	"	"	"	"				"	700	9,200		3,570		0·285		"
"	"	"	"	"	"				"	800	8,100		3,210		0·275		"
En 58E	0·05	0·25	0·43	10·56	18·1				Softened	RT			4,910				26
En 58E	0·06	0·41	1·73	10·18	18·4				CD ¼ hard	"			4,730				"
En 58E	0·07	0·37	1·68	10·36	18·9				CD ½ hard	"			4,390				"
En 58E	(Within specification)								Softened		13,000						12
En 58A	0·07	0·53	0·76	7·95	18·0				Cold rolled	"	11,950						1
"	"	"	"	"	"				" +½hr. 100	"	12,500						"
"	"	"	"	"	"				" " 200	"	12,800						"
"	"	"	"	"	"				" " 300	"	12,000						"
"	"	"	"	"	"				" " 400	"	12,650						"
"	"	"	"	"	"				" " 500	"	12,600						"
En 58A	0·08	0·46	0·65	8·02	18·0				Cold rolled	"	11,750						"
"	"	"	"	"	"				" +½hr. 100	"	11,750						"
"	"	"	"	"	"				" " 200	"	13,150						"
"	"	"	"	"	"				" " 300	"	12,250						"
"	"	"	"	"	"				" " 400	"	12,800						"
"	"	"	"	"	"				" " 500	"	13,100						"
En 58A	(Within specification)								Softened		12,900						12
En 58B	0·06	0·70	0·57	8·40	18·5		0·48		AC1050	20		12,900‡					10
"	"	"	"	"	"		"		"	100		12,500					"
"	"	"	"	"	"		"		"	200		11,900					"
"	"	"	"	"	"		"		"	300		12,000					"
"	"	"	"	"	"		"		"	400		10,550					"
"	"	"	"	"	"		"		"	500		9,950					"
"	"	"	"	"	"		"		"	600		9,450					"
"	"	"	"	"	"		"		"	700		9,000					"
"	"	"	"	"	"		"		"	800		8,550					"
"	"	"	"	"	"		"		"	900		8,050					"
"	"	"	"	"	"		"		"	1000		7,300					"

* Ultrasonic Pulse Method
‡ Vibration Method

YOUNG'S MODULUS (E), SHEAR MODULUS (G), POISSON'S RATIO (σ) (Continued)

Type	Chemical Composition (%)								Heat Treatment °C	Test Temperature °C	Young's Modulus tons/sq.in.		Shear Modulus tons/sq.in.		Poisson's Ratio		Ref.
	C	Si	Mn	Ni	Cr	Mo	Ti	Nb			Static	Dynamic	Static	Dynamic	Static	Dynamic	
En 58C	0·05	0·50	0·56	11·21	18·2		0·60		WQ1040	20	12,500		4,910		0·270(calc.)		25
"	"	"	"	"	"		"		"	100	12,250		4,800		0·255 "		"
"	"	"	"	"	"		"		"	200	11,500		4,620		0·245 "		"
"	"	"	"	"	"		"		"	300	11,100		4,350		0·265 "		"
"	"	"	"	"	"		"		"	400	10,700		4,160		0·285 "		"
"	"	"	"	"	"		"		"	500	10,350		3,980		0·300 "		"
"	"	"	"	"	"		"		"	600	9,700		3,750		0·295 "		"
"	"	"	"	"	"		"		"	700	9,050		3,680		0·230 "		"
"	"	"	"	"	"		"		"	800	8,050		3,390		0·190 "		"
En 58B	0·12	0·60	0·40	8·0	18·0		0·60		Softened	20		13,000‡					2
"	"	"	"	"	"		"		"	400		10,500‡					"
"	"	"	"	"	"		"		"	600		9,400‡					"
"	"	"	"	"	"		"		"	800		8,200‡					"
En 58B	0·15	0·87	0·43	8·30	18·6		1·10		AC1050	20	12,500	12,700‡					"
"	"	"	"	"	"		"		"	200	11,600	11,700‡					"
"	"	"	"	"	"		"		"	400	10,500	10,500‡					"
"	"	"	"	"	"		"		"	600	9,300	9,400‡					"
"	"	"	"	"	"		"		"	800		8,200‡					"
En 58C (B)	0·07	0·90	0·70	9·04	17·5		0·40		WQ1050	RT		12,800*		4,950*		0·294*	24
"	(Within specification)								Softened	RT	12,900						12
En 58D				"	"				"	"	12,900						12
En 58D	0·05	0·41	0·71	12·70	12·60				Softened	RT		13,000*		5,050*		0·287*	24
En 58G	0·08	0·61	1·27	10·83	18·0			0·69	AC 1040	20	12,800		4,910			0·30(calc.)	25
"	"	"	"	"	"			"	"	100	12,300		4,780			0·295 "	"
"	"	"	"	"	"			"	"	200	11,900		4,550			0·305 "	"
"	"	"	"	"	"			"	"	300	11,450		4,370			0·31 "	"
"	"	"	"	"	"			"	"	400	10,900		4,160			0·31 "	"
"	"	"	"	"	"			"	"	500	10,300		3,890			0·33 "	"
"	"	"	"	"	"			"	"	600	9,650		3,670			0·315 "	"
"	"	"	"	"	"			"	"	700	9,200		3,420			0·345 "	"
"	"	"	"	"	"			"	"	800	7,800		3,070			0·275 "	"
En 58G	0·15	0·50	0·70	10·0	18·0			1·0	Softened	RT		13,100‡					2
"	"	"	"	"	"			"	"	200		12,100‡					"
"	"	"	"	"	"			"	"	500		10,900‡					"
"	"	"	"	"	"			"	"	600		10,300‡					"
"	"	"	"	"	"			"	"	700		9,700‡					"
"	"	"	"	"	"			"	"	800		9,130‡					"
"	"	"	"	"	"			"	"	900		8,600‡					"

* Ultrasonic Pulse Method
‡ Vibration Method

YOUNG'S MODULUS (E), SHEAR MODULUS (G), POISSON'S RATIO (σ) (Continued)

Type	Chemical Composition (%)								Heat Treatment °C	Test Temperature °C	Young's Modulus tons/sq.in.		Shear Modulus tons/sq.in.		Poisson's Ratio		Ref.
	C	Si	Mn	Ni	Cr	Mo	Ti	Nb			Static	Dynamic	Static	Dynamic	Static	Dynamic	
En 58G	0·06	0·45	1·28	10·64	18·0			0·78	Softened	RT	12,400		4,820		0·280		26
"	0·04	0·42	1·25	10·95	17·6			0·53	CR ¼ hard	RT	11,700		4,620		0·295		"
"	0·06	0·59	1·37	10·62	18·5			0·63	" ½ hard	RT	11,560		4,500		0·309		"
" (F)	(Within specification)								Softened	RT	12,900						12
En 58J	0·07	0·48	1·88	9·84	17·4	2·55			WQ1040	20	12,650		5,050		0·255		25
"	"	"	"	"	"	"			"	100	12,350		4,900		0·26		"
"	"	"	"	"	"	"			"	200	12,000		4,680		0·28		"
"	"	"	"	"	"	"			"	300	11,650		4,450		0·31		"
"	"	"	"	"	"	"			"	400	11,250		4,260		0·32		"
"	"	"	"	"	"	"			"	500	10,550		4,040		0·305		"
"	"	"	"	"	"	"			"	600	10,050		3,810		0·32		"
"	"	"	"	"	"	"			"	700	9,350		3,570		0·31		"
"	"	"	"	"	"	"			"	800	8,500		3,380		0·26		"
En 58H	(Within specification)								Softened	RT	13,200						12
En 58J				"	"				"	"	13,200						"

* Ultrasonic pulse method.
† Vibration method.

MAGNETIC PROPERTIES

En 58 steels, which are completely austenitic in structure, are virtually non-magnetic, the magnetic permeability (μ) being quoted generally between 1·005 and 1·030. Some, however, may contain free ferrite. These are of very low carbon or high silicon content, or contain titanium or niobium if not compensated by increased nickel or manganese. In such cases the steels may possess a slight degree of magnetizability sufficient for them to be feebly attracted by a strong hand magnet. The magnetic permeability of these ferrite-containing steels may be as high as 1·5 or even 2·0.

Cold working tends to increase the magnetic permeability; thus En 58A, which in the fully softened condition has a permeability of 1·005 to 1·03 has, when cold worked 10%, a permeability of 1·08 to 1·13 (Ref. 2).

The following are typical values of the permeability of the individual steels in the softened condition:

Steel Type	Magnetic Permeability					
	Ref. 2	Ref. 10	Ref. 27	Ref. 28	Ref. 1	Ref. 12
En 58A	1·005/1·03	1·0/1·1	1·02		1·00	1·005/1·03
En 58B	1·5/2·0	1·0/1·2	1·02		1·0/1·01	1·03/2·0
En 58C		1·0/1·2	1·02			1·03/2·0
En 58D	1·05/1·03				1·0/1·016	1·005/1·03
En 58E	1·005/1·03	1·0/1·1	1·02		1·00	1·005/1·03
En 58F	1·005/1·03	1·0/1·2	1·02			1·005/1·03
En 58G		1·0/1·2	1·02		1·02/1·04	1·005/1·03
En 58H	1·005/1·03	1·0/1·2	1·02		2·2/4·4 (Ti added)	1·005/1·03
En 58J	1·05	1·1/1·3	1·02		1·0	1·05

The following actual values have been reported (Ref. 2) on a steel containing 0·07% C, 8·08% Ni and 18·18% Cr in the fully softened condition.

Magnetic Permeability (μ) for H (Oersted) = 25	100	200
1·048	1·045	1·043

MECHANICAL PROPERTIES AT ROOM TEMPERATURE

Type	Chemical Composition (%) C	Si	Mn	Ni	Cr	Mo	Ti	Nb	Section Size dia. (in.)	Heat Treatment °C	Tensile 0·1% P.S (0·1Rp) tons/sq.in.	0·2% P.S. (0·2Rp) tons/sq.in.	0·5% P.S (0·5Rp) tons/sq.in.	Y.P. (Rs) tons/sq.in.	T.S. (Rm) tons/sq.in.	El % (A)	R.A. % (Z)	Izod ft.lb.	Ref.
En 58E(A)	0·03			8·0	18·0					Softened			18·4		45·8	55		108	1 "
En 58E	0·05	0·51	0·63	7·80	17·9				$\frac{3}{4}$	WQ1050				17·4	44·6	53		106	29
En 58E	0·05	0·50	0·40	10·0	19·0				$1\frac{1}{8}$	WQ1050				15·0	40·0	60	65	80	2
En 58E	0·05	0·51	0·70	10·08	18·1				$\frac{5}{16}$	WQ1050	14·5	15·1	16·0		39·1	$77\frac{1}{2}$	$77\frac{1}{2}$		"
En 58E	0·06			10·0	19·0					Softened			15·5		42·5	62		102	1
En 58A	0·07	0·50	0·90	8·5	17·0					AC1060			16·0		42·0	66			6
En 58A	0·07	0·76	0·53	7·95	18·0					Cold drawn	56·0				95·8	$6\frac{1}{2}$	$53\frac{1}{2}$		1
"	"	"	"	"	"					" $+\frac{1}{2}$hr. 100	57·0				97·4	$3\frac{1}{2}$	$52\frac{1}{2}$		"
"	"	"	"	"	"					" " 200	59·3				103·0	$3\frac{1}{2}$	51		"
"	"	"	"	"	"					" " 300	71·5				101·3	3	51		"
"	"	"	"	"	"					" " 400	69·5				105·0	3	42		"
"	"	"	"	"	"					" " 500	69·0				101·6	$3\frac{1}{2}$	$51\frac{1}{2}$		"
En 58A	0·08	0·65	0·46	8·02	18·0					Cold drawn	43·1				78·6	22	55		"
"	"	"	"	"	"					" $+\frac{1}{2}$hr. 100	46·8				76·9	$20\frac{1}{2}$	60		"
"	"	"	"	"	"					" " 200	47·0				78·0	$15\frac{1}{2}$	56		"
"	"	"	"	"	"					" " 300	51·2				77·0	$10\frac{1}{2}$	65		"
"	"	"	"	"	"					" " 400	52·5				78·3	$11\frac{1}{2}$	$64\frac{1}{2}$		"
"	"	"	"	"	"					" " 500	51·4				74·3	17	$67\frac{1}{2}$		"
En 58E	0·08	0·26	0·12	11·41	17·9				$1\frac{1}{8}$	As forged				30·0	42·8	46	63		2
"	"	"	"	"	"				"	2hrs. 800AC				25·2	42·8	43	60		"
"	"	"	"	"	"				"	$\frac{1}{2}$hr. 900AC				18·4	39·3	55	63		"
"	"	"	"	"	"				"	" 1130 "				16·5	36·6	68	$89\frac{1}{2}$		"
"	"	"	"	"	"				"	" " " +2hrs. 800				17·2	36·4	62	68		"
En 58E(E)	0·08			10·5	18·0					WQ1025				18·0	38·4	73	$75\frac{1}{2}$	103	30
En 58A(E)	0·08			9·0	19·0								16·0		43·5	63		105	1
En 58A	0·10	0·40	0·30	8·0	18·0				2	Softened				20·0	42·0	60	63	95	9
En 58A	0·10			8·0	18·0					Softened				22·7	44·7	$59\frac{1}{2}$	69	120	31
En 58A	0·10	0·35	0·47	9·38	18·8				$\frac{5}{8}$	WQ1040				16·1	42·4	72	$77\frac{1}{2}$		32
"	"	"	"	"	"					" CD 49%				71·9	78·1	14	69		
"	"	"	"	"	"					" CD 71%					107·1	4	59		"
En 58A	0·12	0·65	0·86	8·62	18·0				$1\frac{1}{8}$	AC1050				24·1	46·2	45		118	29
En 58A	0·12	0·50	0·50	10·0	18·0				$1\frac{1}{8}$	WQ1100				21·0	42·8	61	$69\frac{1}{2}$	120	33

MECHANICAL PROPERTIES AT ROOM TEMPERATURE (Continued)

Type	Chemical Composition (%)								Section Size dia. (in.)	Heat Treatment °C	Tensile							Izod ft.lb.	Ref.
	C	Si	Mn	Ni	Cr	Mo	Ti	Nb			0·1% P.S. (0·1Rp) tons/sq.in.	0·2% P.S. (0·2Rp) tons/sq.in.	0·5% P.S. (0·5Rp) tons/sq.in.	Y.P. (Rs) tons/sq.in.	T.S. (Rm) tons/sq.in.	El % (A)	R.A. % (Z)		
En 58A	0·12	0·30		8·20	18·0					WQ1000				13·5	58·2	63½	58½		34
"	"	"		"	"					WQ1000/1050				16·8	54·8	56	58	119	"
"	"	"		"	"					WQ1100				12·0	53·8	58½	52		"
En 58A	0·14	0·39	0·60	8·75	17·6					Softened	19·0				47·0	63	78		35
En 58A	0·14	0·34	0·17	8·06	17·6					1hr. 600 WQ				19·3	52·5	32	25	117	34
"	"	"	"	"	"					2hrs. 700 "				20·1	62·0	50	52	92	"
"	"	"	"	"	"					24hrs. 700 "				18·3	61·5	36	53	53	"
"	"	"	"	"	"					½hr. 800 "				20·5	59·4	40	35	99	"
"	"	"	"	"	"					½hr. 1000 "				19·0	57·9	31	23	96	"
"	"	"	"	"	"					¼hr. 1200 "				14·0	50·5	54	36	103	"
En 58C (B)	0·05	0·62	0·39	9·15	19·0		0·29		⅝	WQ1038				16·1	38·4	45	79	(Charpy V) 182	36
En 58B	0·06			8·0	18·0		added			Softened			15·6		45·8	56		Izod 103	1
En 58C(B)	0·07	0·42	0·81	9·84	17·5		0·46		⅜	WQ1100				16·4	39·6	52	70		8
En 58C	0·07	0·54	0·90	10·3	19·4		0·50		1	WQ1100				18·4	41·2	54	64	103	"
En 58B	0·08	0·46	0·21	8·08	18·5		1·03		1	As rolled				32·8	43·4	42½	63½		2
"	"	"	"	"	"		"		"	AC800				29·2	42·6	42½	66		"
"	"	"	"	"	"		"		"	AC900				28·6	42·0	42½	66		"
"	"	"	"	"	"		"		"	AC1150				21·2	41·2	45	67	118	"
"	"	"	"	"	"		"		"	" +2hrs. 800				19·4	39·7	43	69	116	"
En 58C	0·08	0·86	0·50	10·2	19·1		0·48		1¼	WQ1050				19·8	42·6	49		108	29
En 58C(B)	0·09	0·67	0·84	9·48	17·5		0·47		1⅜	WQ1100				16·4	39·6	52	70	98	8
En 58B	0·09	1·06	0·71	8·07	18·48		0·93		1	As rolled				32·1	54·7	43	58		2
"	"	"	"	"	"		"		"	2hrs. 800 AC				28·0	51·9	40	65½		"
"	"	"	"	"	"		"		"	AC900				24·0	50·3	45½	65½	116	"
"	"	"	"	"	"		"		"	AC1150				25·0	48·3	42½	62	93	"
"	"	"	"	"	"		"		"	" +2hrs. 850				31·5	51·2	39	64		"
En 58B	0·11	0·68	0·33	8·12	18·6		0·81		¾	AC1050	13·4	15·7	19·2		40·4	53½	67		2

MECHANICAL PROPERTIES AT ROOM TEMPERATURE (Continued)

Type	Chemical Composition (%) C	Si	Mn	Ni	Cr	Mo	Ti	Nb	Section Size dia. (in.)	Heat Treatment °C	Tensile 0·1% P.S. (0·1Rp) tons/sq.in.	0·2% P.S. (0·2Rp) tons/sq.in.	0·5% P.S. (0·5Rp) tons/sq.in.	Y.P. (Rs) tons/sq.in.	T.S. (Rm) tons/sq.in.	El % (A)	R.A. % (Z)	Izod ft.lb.	Ref.
En 58B	0·11	0·66	0·46	8·90	18·9		0·81		1⅛	FC 1050				18·0	39·2	54	63		2
"	"	"	"	"	"		"		"	AC1050				20·0	38·4	53	63		"
"	"	"	"	"	"		"		"	" " +2hrs. 600				20·0	39·4	53½	64		"
"	"	"	"	"	"		"		"	" " " 650				20·0	38·5	53	62½		"
"	"	"	"	"	"		"		"	" " " 700				20·0	38·9	52½	63		"
"	"	"	"	"	"		"		"	" " " 750				20·0	38·9	52	63½		"
"	"	"	"	"	"		"		"	" " " 800				18·0	38·2	55	64½		"
"	"	"	"	"	"		"		"	" " " 850				18·5	38·3	55	64		"
"	"	"	"	"	"		"		"	" " " 900				20·0	38·4	55	64½		"
"	"	"	"	"	"		"		"	" " " 950				19·0	38·1	55	64½		"
En 58C	0·12	0·96	1·24	12·00	18·7		0·49		¾	AC1050	17·5		19·8		41·6	54	63½	109	"
"	"	"	"	"	"		"		"	" T650 14 days	18·2		20·5		42·0	48½	61½	92	"
"	"	"	"	"	"		"		"	" " 3 months	18·0		19·9		42·4	44½	54	52	"
"	"	"	"	"	"		"		"	" " 6 "	16·5		19·4		42·7	41½	47	35	"
"	"	"	"	"	"		"		"	" " 2½ yrs.	15·8		18·9		41·0	35½	37	19	"
En 58B	0·12	0·70	0·66	7·95	17·2		0·64		1⅜	AC1050				21·2	46·7	44		99	29
En 58B	0·13	0·50	0·33	8·22	18·6		0·61		1⅛	AC1150	11·7		16·7		42·6	53½	68½		2
En 58D	0·07	0 50	0·90	12·0	12·0					1050 Air Blast Cool			18·0		38·0	50			6
En 58D	0·08			12·5	13·0					Softened 1000			15·2		37·4	64½		111	1
En 58D	0·08	0·08	0·26	11·96	12·3				1⅛	As rolled				23·7	41·3	53½	63		2
"	"	"	"	"	"				"	800 2hrs. AC				23·0	41·2	40	58½		"
"	"	"	"	"	"				"	900 20 min. AC				16·4	38·4	52	53½		"
"	"	"	"	"	"				"	1125 20 min. AC				15·3	38·0	61	53½		"
"	"	"	"	"	"				"	" " +2hrs. 800AC				18·2	41·3	46	60		"
En 58D	0·10	0·10	0·23	12·64	12·4				1⅛	As rolled				22·4	38·1	58	72		"
"	"	"	"	"	"				"	AC950				16·0	36·2	62	73		"
"	"	"	"	"	"				"	AC1050				14·8	36·2	65	75		"
"	"	"	"	"	"				"	AC1100				13·6	35·8	63	73		"
"	"	"	"	"	"				"	WQ1050				16·4	35·1	66	75		"
"	"	"	"	"	"				"	" +½hr. 650				15·2	35·7	61	73		"
En 58D	0·10	0·30	0·32	12·2	12·4				2	AC1050				16·7	36·0	65		108	29

MECHANICAL PROPERTIES AT ROOM TEMPERATURE (Continued)

Type	Chemical Composition (%)								Section Size dia. (in.)	Heat Treatment °C	Tensile							Izod ft.lb.	Ref.
	C	Si	Mn	Ni	Cr	Mo	Ti	Nb			0·1% P.S. (0·1Rp) tons/sq.in.	0·2% P.S. (0·2Rp) tons/sq.in.	0·5% P.S. (0·5Rp) tons/sq.in.	Y.P. (Rs) tons/sq.in.	T.S. (Rm) tons/sq.in.	El % (A)	R.A. % (Z)		
En 58D	0·13			11·9	11·2					WQ1000/1050				15·2	42·8	80	76½	110	30
En 58F	0·06			8·95	18·3			0·65		AC1050				16·1	38·8	59	69	104	37
"	"			"	"			"		WQ1050				18·7	39·7	57	68	117	"
En 58G	0·06			11·1	18·0			0·57		AC1050				15·2	36·2	72	74	110	"
"	"			"	"			"		WQ1050				15·8	37·0	68	72	119	"
En 58G (F)	0·06			9·0	19·0			added		Softened			17·0		43·0	56		100	2
En 58F	0·07	0·50	1·40	8·25	18·0			0·80		Softened				18·0	45·0	50	60	120	3
En 58F	0·07	0·60	0·63	8·3	17·9			1·22	1⅛	WQ1050				16·8	42·9	46		106	29
En 58G (F)	0·07			9·18	17·8			0·75		WQ1150				15·6	41·0	59	67	120	37
En 58G (F)	0·07			9·42	18·3			0·51		WQ1150				16·1	40·0	60	72	120	"
En 58G	0·09	0·60	1·50	11·0	17·5			0·90	1⅛	WQ1050	14·4	15·9	17·6		40·6	59	65	80	2
En 58G (F)	0·11	0·50	0·41	9·50	17·8			1·22	¾	AC1050	13·7		17·0		39·9	57	65	117	"
"	"	"	"	"	"			"	"	" +1 month 650	16·2		19·8		41·0	56½	63½	116	"
"	"	"	"	"	"			"	"	" +3 " "	15·8		18·9		39·9	54½	63½	96	"
"	"	"	"	"	"			"	"	" +1 year "			18·7		40·7	50	57½	60	"
"	"	"	"	"	"			"	"	" +2½ " "	14·1		18·2		40·2	48	53	38	"
"	"	"	"	"	"			"	"	" +5 " "			21·0		41·9	38	50	33	"
En 58G	0·12	0·60	1·5	11·0	17·5			1·20	¾	AC1050	17·0		20·7		45·0	55	62	101	"
"	"	"	"	"	"			"	"	" +2 days 600	18·0		21·4		44·3	48	54	91	"
"	"	"	"	"	"			"	"	" +2 weeks "	18·1		22·3		46·0	52	57½	89	"
"	"	"	"	"	"			"	"	" +3 months "	17·5		21·6		43·5	47½	57½	106	"
"	"	"	"	"	"			"	"	" +2 days 650	17·7		20·8		44·5	52	59	88	"
"	"	"	"	"	"			"	"	" +2 weeks "	17·6		21·4		44·3	50	54	76	"
"	"	"	"	"	"			"	"	" +3 months "	14·1		18·3		42·0	54	60½	101	"
"	"	"	"	"	"			"	"	" +12 months "	15·9		19·5		42·5	51	54	57	"
En 58G	0·13	0·42	1·37	11·2	17·8			1·30	¾	AC1050+10% CW	32·4		35·0		45·5	42	58½	99	"
"	"	"	"	"	"			"	"	" " +2wks. 650	29·5		32·0		47·4	39½	56	81	"
"	"	"	"	"	"			"	"	" " +3 mths.650	27·9		31·2		45·7	40	57½	81	"
"	"	"	"	"	"			"	"	" " +2 yr. 650	25·4		29·4		44·2	41	52½	41	"
"	"	"	"	"	"			"	"	" " +5 yr. "	24·9		28·7		43·7	34½	45	25	"
"	"	"	"	"	"			"	"	AC1050+20% CW	41·8		44·0		49·5	33½	53	66	"
"	"	"	"	"	"			"	"	" " +2wks 630	37·0		38·9		50·6	35	53	63	"
"	"	"	"	"	"			"	"	" " +3mths 630	34·5		37·8		49·7	32	52½	53	"
"	"	"	"	"	"			"	"	" " +1yr. 630	30·4		34·5		48·0	34	45	23	"
"	"	"	"	"	"			"	"	" " +5yr. 630	16·0		21·8		41·8	32	30	18	"

MECHANICAL PROPERTIES AT ROOM TEMPERATURE (Continued)

Type	Chemical Composition (%)								Section Size dia. (in.)	Heat Treatment °C	Tensile							Izod ft.lb.	Ref.
	C	Si	Mn	Ni	Cr	Mo	Ti	Nb			0·1% P.S. (0·1Rp)	0·2% P.S. (0·2Rp)	0·5% P.S. (0·5Rp)	Y.P. (Rs)	T.S. (Rm)	El % (A)	R.A. % (Z)		
											tons/sq.in.								
En 58H	0·06	0·60	0·38	8·22	18·4	1·44			1⅛	WQ1050				22·4	41·4	48		118	29
En 58H	0·06	0·70	0·47	8·10	17·0	1·56			1⅛	As rolled	20·0				52·0	47	58	107	7
”	”	”	”	”	”	”			”	WQ950	16·2				50·0	38	48½	100	”
”	”	”	”	”	”	”			”	WQ1050	17·5				51·8	46½	54½	91	”
”	”	”	”	”	”	”			”	WQ1150	16·0				50·0	49½	58	107	”
”	”	”	”	”	”	”			”	WQ1250	15·0				49·2	49½	58	91	”
En 58H	0·06	0·50	1·40	10·25	18·0	1·75		0·25		Softened				18·5	45·0	45	60	120	3
En 58H	0·09	0·60	0·38	8·22	18·3	1·44			1¼	WQ1050		15·0		16·8	41·6	64	70½	115	7
En 58J	0·06	0·72	0·47	8·43	18·50	3·57	0·42		1¼	As rolled				35·4	46·6	35½	65	104	2
”	”	”	”	”	”	”	”		”	AC1050				27·6	42·6	36	55	115	”
”	”	”	”	”	”	”	”		”	” +2hrs. 600				29·4	45·8	38	56½	20	”
”	”	”	”	”	”	”	”		”	” +2hrs. 800				30·4	45·2	15½	16	10	”
”	”	”	”	”	”	”	”		”	AC1150				27·2	41·2	46½	69½	104	”
En 58J	0·07	0·49	0·62	8·57	18·3	2·83			2½	AC1050				20·8	43·6	45		106	29
En 58J	0·07	0·42	0·93	8·93	18·6	3·36			⅝	Softened				23·4	46·0	42	72		32
En 58J	0·07	0·48	0·32	8·08	18·3	2·84			1⅛	AC1000				26·6	44·5	47½	57		7
”	”	”	”	”	”	”			”	WQ1000				23·0	43·2	54½	64½		”
”	”	”	”	”	”	”			”	AC1100				19·2	40·7	50½	67½		”
”	”	”	”	”	”	”			”	WQ1100				20·6	41·2	60	75½		”
En 58J	0·08	0·56	1·74	9·90	17·7	2·84	0·42		1⅛	WQ1050	19·0				40·4	54	74	117	38
En 58J	0·08			9·0	19·0	2·75				Softened		17·8			47·7	51		108	1
En 58J	0·08			12·0	19·0	3·50				”		17·0			46·6	50		105	”
En 58J	0·09	0·66	0·39	8·25	18·0	2·74			1⅛	As rolled	37·4		42·7		58·2	38	65	118	7
”	”	”	”	”	”	”			”	WQ1050	16·0		19·3		42·2	60	74	117	”
”	”	”	”	”	”	”			”	” +1hr. 600AC	16·6		20·2		43·6	57	69½	118	”
”	”	”	”	”	”	”			”	” +1hr. 700 ”	14·2		18·2		42·8	50½	63½	82	”
”	”	”	”	”	”	”			”	” +100hrs. 700 AC			21·8		42·6	45½	61½	71	”
”	”	”	”	”	”	”			”	” +1hr. 800 ”	13·2		17·0		44·3	49	61	82	”
”	”	”	”	”	”	”			”	” +1hr. 900 ”	13·2		17·0		44·6	48½	60	78	”
”	”	”	”	”	”	”			”	” +1hr. 1000 ”	13·2		17·5		42·2	60	71	116	”

MECHANICAL PROPERTIES AT ROOM TEMPERATURE (Continued)

Type	Chemical Composition (%)								Section Size dia. (in.)	Heat Treatment °C	Tensile							Izod ft.lb.	Ref.
	C	Si	Mn	Ni	Cr	Mo	Ti	Nb			0·1% P.S. (0·1Rp)	0·2% P.S. (0·2Rp)	0·5% P.S. (0·5Rp)	Y.P. (Rs)	T.S. (Rm)	El % (A)	R.A. % (Z)		
											tons/sq.in.								
En 58J	0·12	0·28	0·48	8·78	18·6	3·25			1⅛	As forged				42·0	50·9	31	61		2
”	”	”	”	”	”	”			”	AC800				46·3	53·8	12	13		”
”	”	”	”	”	”	”			”	AC900				34·5	47·0	23	23		”
”	”	”	”	”	”	”			”	AC1000				28·8	45·7	33	41	27	”
”	”	”	”	”	”	”			”	AC1150				28·1	45·2	46	68	103	”
”	”	”	”	”	”	”			”	AC1000 +2hrs. 800				29·7	46·3	17	15	15	”
”	”	”	”	”	”	”			”	” 2hrs. 600				34·2	46·5	17	15	15	”
”	”	”	”	”	”	”			”	AC1100 +2hrs. 800				26·7	42·7	20	20	17	”
”	”	”	”	”	”	”			”	” 2hrs. 600				31·3	45·7	17	17	17	”
En 58J	0·12	0·42	0·41	9·32	18·0	2·60			1⅛	As rolled				37·2	51·9	42	66		”
”	”	”	”	”	”	”			”	AC800				33·0	50·2	41	57		”
”	”	”	”	”	”	”			”	AC900				33·8	50·6	42	58		”
”	”	”	”	”	”	”			”	AC1150				17·6	43·2	63	71½	119	”
”	”	”	”	”	”	”			”	” +2hrs. 800				16·8	42·8	61½	63	112	”

Type	Chemical Composition (%)											Section Size dia (in.)	Heat Treatment °C	Tensile							Izod ft.lb.	Ref.
	C	Si	Mn	S	P	Ni	Cr	Mo	Ti	Se	Zr			0·1% P.S. (0·1Rp)	0·2% P.S. (0·2Rp)	0·5% P.S. (0·5Rp)	Y.P. (Rs)	T.S. (Rm)	El % (A)	R.A. % (Z)		
														tons/sq.in.								
En 58AM (EM)	0·08	0·40	0·60	0·300	0·020	8·5	18·5						Softened	12·9	14·5			37·9	55	60	73(L)15(T)	39
En 58AM (EM)	0·08	0·50	0·60	0·120	0·020	8·5	18·5			0·30			”	12·9	15·4			38·9	60	65	100(L)26(T)	”
En 58BM	0·12	0·66	0·54	0·290	0·022	7·76	18·5	0·65	0·51			1⅛	AC1050			18·3		47·4	53½	42	51	2
En 58CM	0·12	0·60	1·20	0·300		10·0	18·0	0·30	0·60		0·25	1⅛	AC1050				15·8	38·8	53	54	50	”

MECHANICAL PROPERTIES AT LOW TEMPERATURE

Type	Chemical Composition (%) C	Si	Mn	Ni	Cr	Mo	Ti	Nb	Section Size dia. (in.)	Heat Treatment °C	Test Temperature °C	Tensile 0·1% P.S. (0·1Rp)	0·2% P.S. (0·2Rp)	0·5% P.S. (0·5Rp)	Y.P. (Rs)	T.S. (Rm)	El % (A)	R.A. % (Z)	Notched Impact ft.lb.	Ref.
												tons/sq.in.								
																			Charpy V	
En 58E	0·02	0·60	1·17	9·96	20·8	0·18				Softened	—196								68	43
"	"	"	"	"	"	"				"	—252								67	"
																			Charpy K	
En 58E (A)	0·05			8·80	18·5					Cold Rolled	20				66·1	93·8	12	54	25	40
"	"			"	"					" "	— 78				87·0	108·5	34	63	35	"
"	"			"	"					" "	—127								38	"
"	"			"	"					" "	—192					134·0	29	44	30	"
"	"			"	"					" "	—253								26	"
																			DVM Kg. M/sq.c.	
En 58E	0·05			9·0	18·7					WQ1030	20				13·7	41·6	62	70	28·6	41
"	"			"	"					"	— 75				14·6	66·0	37	62	—	"
"	"			"	"					"	—180				20·6	100·7	33	54	24·5	"
																			Charpy V	
En 58E	0·05			10·0	18·0					Softened	RT								97	2
"	"			"	"					"	—83								97	"
																			Charpy V	
En 58E	0·06	0·60	1·55	10·0	18·4	0·38				Softened	—196								90	43
"	"	"	"	"	"	"				"	—252								87	"
En 58A (E)	0·07	0·50	0·64	8·63	18·2				5/8	WQ1030	RT				16·1	44·3	56	78		42
"	"	"	"	"	"				"	"	—78				22·4	80·4	36	69½		"
"	"	"	"	"	"				"	Sub-zero —78	RT				15·2	44·7	55½	77½		"
"	"	"	"	"	"				"	Hot rolled	RT				36·6	52·7	37½	73½		"
"	"	"	"	"	"				"	" "	—78				34·9	82·2	36½	67		"
"	"	"	"	"	"				"	Sub-zero —78	RT				36·6	53·1	38½	72½		"
"	"	"	"	"	"				"	Cold drawn	RT				41·1	67·8	22	62		"
"	"	"	"	"	"				"	" "	—78				46·0	87·5	29	67		"
"	"	"	"	"	"				"	Sub-zero —78	RT				40·6	68·0	23	65		"
																			Izod	
En 58E (A)	AISI 302 Composition									Softened	RT				17·0	41·1	68	78	110	28
"	" " "									"	—0				17·8	54·5	65	76	110	"
"	" " "									"	—40				21·4	64·7	60	73	110	"
"	" " "									"	—62				22·3	71·9	57	70	110	"
"	" " "									"	—196				25·0	104·9	40	59	110	"
En 58E (A)	AISI 304 Composition									Softened	RT				15·6	39·3	62	71	110	28
"	" " "									"	0				15·2	58·0	55	68	110	"
"	" " "									"	—40				15·2	69·2	47	64	110	"

MECHANICAL PROPERTIES AT LOW TEMPERATURE (Continued)

Type	Chemical Composition (%)								Section Size dia. (in.)	Heat Treatment °C	Test Temperature °C	Tensile							Izod ft.lb.	Ref.
	C	Si	Mn	Ni	Cr	Mo	Ti	Nb				0·1% P.S. (0·1Rp)	0·2% P.S. (0·2Rp)	0·5% P.S. (0·5Rp)	Y.P. (Rs)	T.S. (Rm)	El % (A)	R.A. % (Z)		
												tons/sq.in.								
En 58E (A)	AISI 304 Composition									Softened	—62				15·2	75·9	39	63	110	28
,,	,, ,, ,,									,,	—196				17·4	109·8	32	55	110	,,
En 58A	0·10			7·98	18·2					Softened	RT					45·5	82½	74		2
,,	,,			,,	,,					,,	—28					70·6	65	69		,,
,,	,,			,,	,,					,,	—60					80·1	54	67½		,,
,,	,,			,,	,,					,,	—89					84·2	52	67½		,,
En 58A	0·11		0·5	10·1	18·3					WQ1095	RT				17	41	54	74	114	12
,,	,,		,,	,,	,,					,,	—20				19	51	63	73		,,
,,	,,		,,	,,	,,					,,	—60				22	61	61	74	118 (—65°)	,,
,,	,,		,,	,,	,,					,,	—96				24	69	46	70		,,
,,	,,		,,	,,	,,					,,	—116				32	80	49	63	108 (—120°)	,,
,,	,,		,,	,,	,,					,,	—180				33	100	46	55	100	,,
En 58A	0·12	0·43	0·24	8·1	18·8					WQ1150	RT				25·9	52·4	56	53½		44
,,	,,	,,	,,	,,	,,					,,	—253				55·8	119·8	25	30½		,,
En 58A	0·12			8·0	18·0					Softened	RT				17·2	39·7	83	82		2
,,	,,			,,	,,					,,	—70				28·2	64·7	63	76		,,
En 58A	0·13	0·33	0·43	10·7	17·3					AC1050	+25		12·6			40·0	72½	82		35
,,	,,	,,	,,	,,	,,						—120		22·2			81·3	70½	68½		,,
,,	,,	,,	,,	,,	,,						—150		26·7			86·5	54	66½		,,
,,	,,	,,	,,	,,	,,						—196		33·3			96·5	32	25		,,
																			DVM Kg.M/sq.cm.	
En 58A	0·14	0·66	0·56	9·0	17·9					Softened	RT				16·1	44·6	70	74	26·5	41
,,	,,	,,	,,	,,	,,					,,	—180				20·5	100·5	45	42	24·5	,,
En 58A	0·09	0·72	0·23	8·2	19·0	0·09			⅝	WQ1050	+20	13·8	14·5	15·9		42·2	70	69		12
,,	,,	,,	,,	,,	,,	,,			,,	,,	—78	14·5	16·6	20·0		82·5	47	66		,,
,,	,,	,,	,,	,,	,,	,,			,,	,,	—196	14·7	17·1	22·1		113·8	39	56		,,

MECHANICAL PROPERTIES AT LOW TEMPERATURE (Continued)

Type	Chemical Composition (%) C	Si	Mn	Ni	Cr	Mo	Ti	Nb	Section Size dia. (in.)	Heat Treatment °C	Test Temperature °C	Tensile 0·1% P.S. (0·1Rp) tons/sq.in.	0·2% P.S. (0·2Rp) tons/sq.in.	0·5% P.S. (0·5Rp) tons/sq.in.	Y.P. (Rs) tons/sq.in.	T.S. (Rm) tons/sq.in.	El % (A)	R.A. % (Z)	Izod ft.lb.	Ref.
En 58A	0·10	0·35	0·47	9·38	18·82				$\frac{5}{8}$	WQ1038	RT				16·1	42·4	72	$77\frac{1}{2}$		32
"	"	"	"	"	"				"	"	—78				24·5	73·7	80	71		"
"	"	"	"	"	"				"	"	—188				35·7	103·6	$55\frac{1}{2}$	62		"
"	"	"	"	"	"				"	" CD 49%	RT				71·9	78·1	14	69		"
"	"	"	"	"	"				"	" " "	—78					88·4	46	68		"
"	"	"	"	"	"				"	" " "	—188					124·1	41	70		"
"	"	"	"	"	"				"	" CD 71%	RT					107·1	4	59		"
"	"	"	"	"	"				"	" " "	—78					118·7	4	$66\frac{1}{2}$		"
"	"	"	"	"	"				"	" " "	—188				129·5	147·3	$27\frac{1}{2}$	$53\frac{1}{2}$		"
En 58C (B)	0·05	0·62	0·39	9·15	19·0		0·29		$\frac{5}{8}$	WQ1038	RT				16·5	38·8	44	79		42
"	"	"	"	"	"		"		"	"	—78				17·0	63·8	46	74		"
"	"	"	"	"	"		"		"	" Cooled —78	RT				14·7	38·4	46	79		"
"	"	"	"	"	"		"		"	" Cold drawn	RT				38·0	58·7	22	69		"
"	"	"	"	"	"		"		"	" " "	—78				39·3	78·1	32	67		"
"	"	"	"	"	"		"		"	CD & cooled —78	RT				38·4	58·9	21	69		"
En 58C (B)	0·05	0·62	0·39	9·15	19·0		0·29		$\frac{5}{8}$	WQ1038	RT				15·6	38·0				32
"	"	"	"	"	"		"		"	"	—78				22·3	64·3				"
"	"	"	"	"	"		"		"	"	—188				38·0	91·5				"
"	"	"	"	"	"		"		"	Cold drawn	RT				44·6	53·6	38	64		"
"	"	"	"	"	"		"		"	"	—78				45·5	81·7	49	61		"
"	"	"	"	"	"		"		"	"	—188				58·9	113·0	35	60		"
En 58B (C)	0·09	0·66	0·75	9·04	17·9		0·63		$\frac{1}{2}$	AC1050	—75	15·4	17·4	21·0		71·0	51	$64\frac{1}{2}$	74	2
En 58C (B)	0·09	0·99	0·84	9·83	18·5	0·13	0·66			AC1050	RT		17·7			42·5	44	$68\frac{1}{2}$	120	35
"	"	"	"	"	"	"	"			"	—50		19·6			62·2	$50\frac{1}{2}$	67		"
"	"	"	"	"	"	"	"			"	—120		16·7			80·5	44	$61\frac{1}{2}$	120	"
"	"	"	"	"	"	"	"			"	—196		15·9			98·0	45	$58\frac{1}{2}$	94	"
En 58B	0·11	0·71	0·47	8·19	18·2		0·63		$\frac{5}{8}$	WQ1050	+20	13·3	14·9	17·4		40·5	56	69		12
"	"	"	"	"	"		"		"	"	—78	15·0	18·2	22·7		76·5	37	60		"
"	"	"	"	"	"		"		"	"	—196	19·0	22·2	27·0		109·8	35	51		"

MECHANICAL PROPERTIES AT LOW TEMPERATURE (Continued)

Type	Chemical Composition (%)								Section Size dia. (in.)	Heat Treatment °C	Test Temperature °C	Tensile							Izod ft.lb.	Ref.
	C	Si	Mn	Ni	Cr	Mo	Ti	Nb				0·1% P.S. (0·1Rp) tons/sq.in.	0·2% P.S. (0·2Rp) tons/sq.in.	0·5% P.S. (0·5Rp) tons/sq.in.	Y.P. (Rs) tons/sq.in.	T.S. (Rm) tons/sq.in.	El % (A)	R.A. % (Z)		
En 58D	0·11			12·0	12·7				$\frac{1}{4}$	WQ1100	RT				15·2	37·0	58	77	118*	45
"	"			"	"				"	"	—21				15·0	47·3	64	76		"
"	"			"	"				"	"	—60				24·0	59·3	50	72	118*	"
"	"			"	"				"	"	—98				15·0	74·5	43	67		"
"	"			"	"				"	"	—120				19·4	81·2	38	64	98*	"
"	"			"	"				"	"	—180				28·2	105·0	36	50	102*	"
"	"			"	"				$\frac{1}{2}$	"	RT				14·9	34·0	$62\frac{1}{2}$	78		"
"	"			"	"				"	"	—180				35·8	101·9	37	53		"
En 58D	0·11			13·8	12·2				$\frac{1}{4}$	WQ1100	RT				12·1	38·0	$51\frac{1}{2}$	76		45
"	"			"	"				"	"	—20				15·0	42·0	67	80		"
"	"			"	"				"	"	—62				20·8	47·5	60	78		"
"	"			"	"				"	"	—100				22·0	59·5	56	75		"
"	"			"	"				"	"	—121				24·0	69·0	50	71		"
"	"			"	"				"	"	—177				32·0	94·5	$44\frac{1}{2}$	58		"
"	"			"	"				$\frac{1}{2}$	"	RT				14·6	34·7	57	75		"
"	"			"	"				"	"	—180				36·6	92·3	$47\frac{1}{2}$	59		"
En 58F (G)			AISI No 347 Composition							Softened	RT				17·9	37·5	60	72	110	28
"			" " "								0				18·7	46·9	62	72	110	"
"			" " "								—40				19·6	52·2	63	71	110	"
"			" " "								—62				20·1	58·0	57	70	110	"
"			" " "								—196				22·8	87·0	40	64	110	"
En 58G	0·07	0·61	0·64	10·16	18·6			0·93	$\frac{5}{8}$	WQ1038	RT				16·5	42·4				32
"	"	"	"	"	"			"	"	"	—78				22·3	63·4				"
"	"	"	"	"	"			"	"	"	—188				31·2	93·7				"
"	"	"	"	"	"			"	"	" Cold drawn	RT				49·1	70·1	10	54		"
"	"	"	"	"	"			"	"	" " "	—78				72·3	78·6	30	61		"
"	"	"	"	"	"			"	"	" " "	—188				87·0	110·3	38	50		"
En 58G	0·07	0·32	0·42	10·29	19·0			0·95		WQ1038	RT				22·8	41·5	41	75		42
"	"	"	"	"	"			"		"	—78				23·7	65·2	40	69		"
"	"	"	"	"	"			"		" Cooled —78	RT				22·8	41·4	42	75		"
"	"	"	"	"	"			"		WQ1038, Cold drawn	RT				56·2	65·2	14	61		"
"	"	"	"	"	"			"		" " "	—78				59·8	74·1	33	63		"
"	"	"	"	"	"			"		" CD. Cooled —78	RT				57·1	64·7	14	61		"

* Unbroken.

MECHANICAL PROPERTIES AT LOW TEMPERATURE (Continued)

Type	Chemical Composition (%) C	Si	Mn	Ni	Cr	Mo	Ti	Nb	Section Size dia. (in.)	Heat Treatment °C	Test Temperature °C	Tensile (tons/sq.in.) 0·1% P.S. (0·1Rp)	0·2% P.S. (0·2Rp)	0·5% P.S. (0·5Rp)	Y.P. (Rs)	T.S. (Rm)	El % (A)	R.A. % (Z)	Izod ft.lb.	Ref.
En 58G	0·07	0·32	0·42	10·29	19·0			0·95		Hot rolled	RT				43·3	52·2	25	70		42
”	”	”	”	”	”			”		” ”	—78				44·6	69·2	37	70		”
”	”	”	”	”	”			”		” ” Cooled —78	RT				42·9	43·3	26	66		”
En 58G	0·10	0·81	0·42	10·4	17·5	0·12		1·26	5/8	WQ1050	+20	13·7	14·8	16·7		39·9	56	65		12
”	”	”	”	”	”	”		”	”	”	—78	17·8	20·0	24·2		65·0	53	62		”
”	”	”	”	”	”	”		”	”	”	—196	18·4	23·2	30·8		92·0	46	54		”
En 58J	0·05			8·20	18·1	2·90				Softened	RT					45·8	68	76		2
”	”			”	”	”				”	0					49·1	80½	78		”
”	”			”	”	”				”	—29					58·4	60	75		”
”	”			”	”	”				”	—54					68·0	48	70		”
”	”			”	”	”				”	—90					76·0	40			”
En 58J	0·06	0·32	0·31	8·4	18·6	3·50	0·42		5/8	WQ1050	+20	15·1	19·9	25·5		40·8	50	70		12
”	”	”	”	”	”	”	”			”	—78	19·4	26·0	32·5		65·8	48	67		”
”	”	”	”	”	”	”	”			”	—196	37·5	41·3	45·8		96·8	44	60		”
En 58J	0·07	0·42	0·93	8·93	18·6	3·36			5/8	WQ1038	RT				23·4	46·0	42	72		42
”	”	”	”	”	”	”			”	”	—78				32·6	68·7	50	68		”
”	”	”	”	”	”	”			”	” Cooled —78	RT				23·2	45·2	42	72		”
En 58J	0·07	0·42	0·93	8·93	18·6	3·36			5/8	Softened	RT				23·4	46·0	42	72		32
”	”	”	”	”	”	”			”	”	—78				32·6	68·7	50	68		”
”	”	”	”	”	”	”			”	”	—188				51·3	112·5				”
En 58J	0·07	0·48	0·73	7·88	18·4	2·80			”	Cold drawn	RT				52·2	79·2				32
”	”	”	”	”	”	”			”	” ”	—78				60·3	82·6				”
”	”	”	”	”	”	”			”	” ”	—188				66·9	116·0				”
En 58J (H)	AISI No. 316 Composition									Softened	RT				17·0	34·8	60	76	110	28
”	” ” ”									”	0				17·4	40·2	60	75	110	”
”	” ” ”									”	—40				18·3	46·4	59	75	110	”
”	” ” ”									”	—62				19·6	52·7	57	73	110	”
”	” ” ”									”	—196				31·2	80·4	46	64	110	”

IMPACT TRANSITION TEMPERATURE DATA

Type	Chemical Composition (%)								Section Size dia. (in.)	Heat Treatment °C	Charpy V Notch Impact Value ft.lb. at										Ref.
	C	Si	Mn	Ni	Cr	Mo	Ti	Nb			—194°C	—100°C	—78°C	—50°C	—40°C	—20°C	0	+20°C	+30°C	+100°C	
En 58A(E)	0·07	0·50	0·44	8·63	18·2				5/8	WQ1030			123		144	157	158	158		189	42
"	"	"	"	"	"				"	Hot rolled			192		171	170	149	147		147	"
"	"	"	"	"	"				"	Cold drawn			55		62	70	68	61		61	"
En 58C(B)	0·05	0·62	0·39	9·15	19·0		0·29		5/8	WQ1038			159		178	181	182	183			"
"	"	"	"	"	"					Cold rolled			83		87	89	88	91			"
En 58G	0·07	0·32	0·42	10·29	19·0			0·95		WQ1038			122		121	123	120		123		"
"	"	"	"	"	"			"		Hot rolled			72		80	89	86		91		"
"	"	"	"	"	"			"		Cold drawn			37		36	35	34		36		"
En 58J	0·06			8·40	18·1	2·92				AC1050	91	134		158			151	151			2

See also Mechanical Properties at Low Temperature, page 294.

Type	Chemical Composition (%)							Heat Treatment °C	Izod Impact Value ft.lb. at													Ref.
	C	Si	Mn	Ni	Cr	Mo	Ti		—196°C	—180°C	—160°C	—150°C	—140°C	—120°C	—120°C	—80°C	—75°C	—50°C	—40°C	0	RT	
En 58C(B)	0·09	0·99	0·84	9·83	18·5	0·13	0·66	AC1050	94		113		118	120		120			120	120	120	35
En 58J	0·07			8·45	18·0	3·36		Softened		27		29			37		58	90		106	110	46
"	"			"	"	"		20% Cold stretched		10		13			32		44	53		58	58	"

En 58

IMPACT TRANSITION TEMPERATURE DATA (Continued)

Curve	Type	Chemical Composition (%)										Section Size dia. (in.)	Heat Treatment °C
		C	Si	Mn	S	P	Ni	Cr	Mo	Ti	Nb		
A	En 58A	0·09	0·72	0·23	0·010	0·021	8·2	19·0	0·09			$\frac{5}{8}$	WQ1050
B	En 58B	0·11	0·71	0·47	0·019	0·020	8·19	18·2		0·63		$\frac{5}{8}$	WQ1050
C	En 58F(G)	0·10	0·81	0·42	0·007	0·010	10·4	17·5	0·12		1·26	$\frac{5}{8}$	WQ1050

(See also Mechanical Properties at Low Temperature, page 294 for tensile test data.)

(Ref. 12)

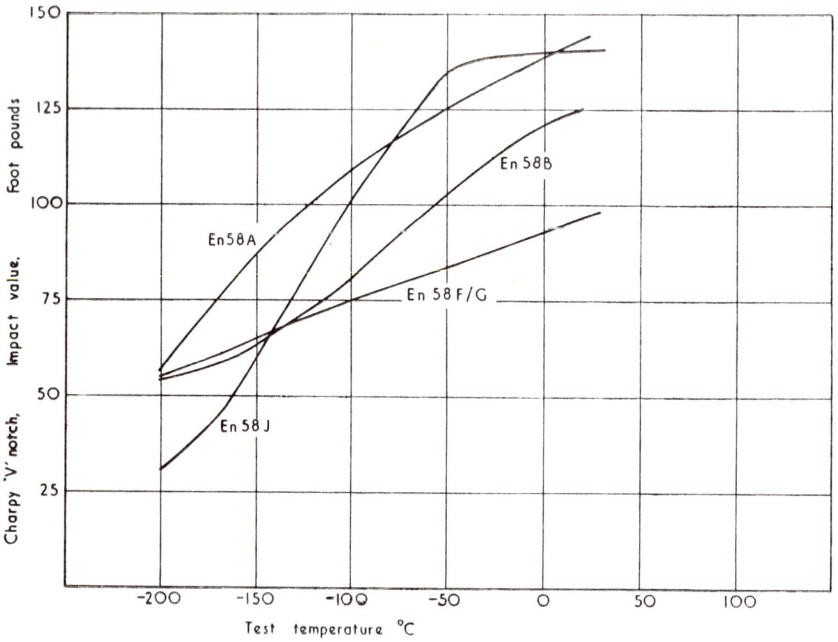

Fig. 58.1

MECHANICAL PROPERTIES AT HIGH TEMPERATURE

Type	Chemical Composition (%)									Section Size dia. (in.)	Heat Treatment °C	Test Temperature °C	Tensile							Izod ft.lb.	Ref.
	C	Si	Mn	Ni	Cr	Mo	Ti	Nb	Se				0·1% P.S. (0·1Rp) tons/sq.in.	0·2% P.S. (0·2Rp) tons/sq.in.	0·5% P.S. (0·5Rp) tons/sq.in.	Y.P. (Rs) tons/sq.in.	T.S. (Rm) tons/sq.in.	El % (A)	R.A. % (Z)		
En 58E (A)	0·04	0·70	0·50	9·74	19·0						Softened	RT				15·5	39·0	61	77		47
"	"	"	"	"	"						"	430				9·6	28·7	46	71		"
"	"	"	"	"	"						"	540				8·7	26·5	43	71		"
"	"	"	"	"	"						"	650				7·6	23·5	46	64		"
"	"	"	"	"	"						"	760				5·4	11·1	38	44		"
"	"	"	"	"	"						"	800				4·2	8·7	34	42		"
																				Charpy K	
En 58E (A)	0·06	0·50	0·61	9·25	17·75						WQ1093	30		13·6			36·8	61½	70	74	51
"	"	"	"	"	"						"	399		7·1			26·8	46	69	59	"
"	"	"	"	"	"						"	482		6·5			25·0	45	69	51	"
"	"	"	"	"	"						"	538		6·2			23·7	44	70	49	"
"	"	"	"	"	"						"	593		5·4			21·7	39	58	55	"
"	"	"	"	"	"						"	649		4·9			19·2	37	44	57	"
"	"	"	"	"	"						"	704		4·9			15·6	35	36		"
"	"	"	"	"	"						"	760		4·7			12·1	31	28		"
"	"	"	"	"	"						"	816		4·5			9·2	30	28		"
"	"	"	"	"	"						"	871					7·9	31	36		"
"	"	"	"	"	"						"	982					4·3	47	42½		"
"	"	"	"	"	"						"	1093					2·2	80	72½		"
En 58E	0·05	0·51	0·70	10·08	18·1					15/16	WQ1050	RT	14·5	15·1	16·0		39·1	77½	77½		2
"	"	"	"	"	"					"	"	100	12·0	12·5	13·4		32·3	73	80		"
"	"	"	"	"	"					"	"	200	10·6	11·0	11·7		29·2	54½	74½		"
"	"	"	"	"	"					"	"	300	9·5	10·0	10·6		29·2	52	70		"
En 58E (A)	AISI No. 304L Composition										Softened	21		15·2			37·3	65	77		48
"	" " "										"	149		10·3			29·5	50	75		"
"	" " "										"	260		8·6			25·8	42	72		"
"	" " "										"	371		7·8			25·1	41	70		"
"	" " "										"	482		7·1			23·1	40	69		"
"	" " "										"	593		6·5			19·6	38	62		"
"	" " "										"	704		5·8			15·0	37	46		"
"	" " "										"	816		5·4			9·4	32	32		"

MECHANICAL PROPERTIES AT HIGH TEMPERATURE (Continued)

Type	Chemical Composition (%) C	Si	Mn	Ni	Cr	Mo	Ti	Nb	Se	Section Size dia. (in.)	Heat Treatment °C	Test Temperature °C	Tensile 0·1% P.S. (0·1Rp) tons/sq.in.	0·2% P.S. (0·2Rp) tons/sq.in.	0·5% P.S. (0·5Rp) tons/sq.in.	Y.P. (Rs) tons/sq.in.	T.S. (Rm) tons/sq.in.	El % (A)	R.A. % (Z)	Izod ft.lb.	Ref.
En 58E (A)	AISI No. 304 Composition										Softened	21		17·0			37·9	63	76		48
,,	,, ,, ,,										,,	149		12·5			32·6	60	75		,,
,,	,, ,, ,,										,,	260		9·8			30·8	55	74		,,
,,	,, ,, ,,										,,	371		9·3			29·5	50	72		,,
,,	,, ,, ,,										,,	482		8·0			26·8	45	69		,,
,,	,, ,, ,,										,,	593		7·1			21·9	40	62		,,
,,	,, ,, ,,										,,	704		6·2			15·9	37	51		,,
,,	,, ,, ,,										,,	816		4·9			9·4	39	45		,,
,,	,, ,, ,,										,,	927		3·1			4·5	46	50		,,
,,	,, ,, ,,										,,	1038		1·3			2·0	63	62		,,
,,	,, ,, ,,										,,	1149		0·7			1·1	77	84		,,
,,	,, ,, ,,										,,	1260		0·5			0·8	103	99		,,
En 58E (A)	0·05	0·60	0·80	10·0	18·5					Transverse from pipe	AC1050	RT	(1) 12·8	13·2	13·8		36·4	92	72		49
,,	,,	,,	,,	,,	,,					,,	,,	,,	(2) 13·9	14·6	15·4		34·6	76	69		,,
,,	,,	,,	,,	,,	,,					,,	,,	300	(3) 6·4	6·7	7·2						,,
,,	,,	,,	,,	,,	,,					,,	,,	,,	(4) 7·1	7·5	8·2						,,
																		1 = 10d			
En 58A	0·11	0·45	0·54	8·21	17·82						AC1120	550					28·4	40	59		50
,,	,,	,,	,,	,,	,,						,,	650					22·4	37	55		,,
,,	,,	,,	,,	,,	,,						,,	760					14·2	44	60½		,,
,,	,,	,,	,,	,,	,,							870					7·3	48	56½		,,
																		1 = 4 A			
En 58A	0·12			8·0	18·0						Softened	15				17·2	39·7	83	82		
,,	,,			,,	,,						,,	500				13·4	30·8	49	73		
,,	,,			,,	,,						,,	600				11·6	22·3	25	28		
,,	,,			,,	,,						,,	700				8·9	18·3	16	20		
,,	,,			,,	,,						,,	800				5·8	10·9	10	13		
,,	,,			,,	,,						,,	900				3·6	6·7	13	14		
,,	,,			,,	,,						,,	1000				2·5	3·8	22	20		

Strain rate (1) To 0.2% P.S. 0·00109/min. Beyond 0·2% P.S. 0·021/min.
,, ,, (2) ,, ,, ,, 0·052 ,, ,, ,, 0·74 ,,
,, ,, (3) ,, ,, ,, 0·00107 ,,
,, ,, (4) ,, ,, ,, 0·042 ,,

MECHANICAL PROPERTIES AT HIGH TEMPERATURE (Continued)

Type	Chemical Composition (%)									Section Size dia. (in.)	Heat Treatment °C	Test Temperature °C	Tensile							Izod ft.lb.	Ref.
	C	Si	Mn	Ni	Cr	Mo	Ti	Nb	Se				0·1% P.S. (0·1Rp) tons/sq.in.	0·2% P.S. (0·2Rp) tons/sq.in.	0·5% P.S. (0·5Rp) tons/sq.in.	Y.P. (Rs) tons/sq.in.	T.S. (Rm) tons/sq.in.	El % (A)	R.A. % (Z)		
En 58A	0·13	0·56	0·13	7·74	17·82					1⅛	AC1200	15				17·5	48·6	50	39		2
"	"	"	"	"	"					"	"	40					47·8	63	70		"
"	"	"	"	"	"					"	"	60					40·8	70½	78		"
"	"	"	"	"	"					"	"	100					35·8	75	83½		"
"	"	"	"	"	"					"	"	200					30·6	65½	82		"
"	"	"	"	"	"					"	"	300					29·4	51	78		"
"	"	"	"	"	"					"	"	400					30·3	52	72		"
"	"	"	"	"	"					"	"	500					29·7	51	69½		"
"	"	"	"	"	"					"	"	600					25·7	37	47½		"
"	"	"	"	"	"					"	"	700					16·5	15½	24		"
"	"	"	"	"	"					"	"	800					7·5	10½	14½		"
"	"	"	"	"	"					"	"	900					5·1	22½	30½		"
"	"	"	"	"	"					"	"	1,000					3·7	56	77		"
En 58C (B)	Not given (AISI No. 321)										Softened	21		14·7			38·0	58	75		48
"		"	"		"						"	149		13·0			30·6	49	76		"
"		"	"		"						"	260		11·6			27·7	43	74		"
"		"	"		"						"	371		10·3			26·6	38	71		"
"		"	"		"						"	482		9·2			25·0	37	70		"
"		"	"		"						"	593		8·5			21·9	43	73		"
"		"	"		"						"	704		7·4			16·5	56	78		"
"		"	"		"						"	806		5·8			9·8	73	85		"
"		"	"		"						"	927		3·8			4·7	60	62		"
"		"	"		"						"	1,038		1·8			2·2	57	60		"
"		"	"		"						"	1,149		0·7			1·1	98	81		"
"		"	"		"						"	1,260		0·4			0·7	98	88		"
En 58B	0·10	0·73	0·44	8·05	18·12		0·56			¾	AC1050	RT	13·1	15·2	18·2		43·4	48	64		2
"	"	"	"	"	"		"			"	"	100	11·9	13·9	16·3		33·2	53	70		"
"	"	"	"	"	"		"			"	"	200	10·6	12·3	14·2		29·4	48	68½		"
"	"	"	"	"	"		"			"	"	300	10·3	11·7	13·5		27·8	49½	65½		"
"	"	"	"	"	"		"			"	"	400	10·4	11·6	13·2		27·8	47	65½		"

MECHANICAL PROPERTIES AT HIGH TEMPERATURE (Continued)

Type	Chemical Composition (%) C	Si	Mn	Ni	Cr	Mo	Ti	Nb	Se	Section Size dia. (in.)	Heat Treatment °C	Test Temperature °C	Tensile 0·1% P.S. (0·1Rp) tons/sq.in.	0·2% P.S. (0·2Rp) tons/sq.in.	0·5% P.S. (0·5Rp) tons/sq.in.	Y.P. (Rs) tons/sq.in.	T.S. (Rm) tons/sq.in.	El % (A)	R.A. % (Z)	Izod ft.lb.	Ref.
En 58B	0·11	0·68	0·33	8·12	18·6		0·81			¾	AC1050	RT	13·4	15·7	19·2		40·4	53½	67		2
"	"	"	"	"	"		"			"	"	100	11·0	13·1	16·2		33·4	52	71½		"
"	"	"	"	"	"		"			"	"	200	10·9	12·6	15·0		30·3	45½	68½		"
"	"	"	"	"	"		"			"	"	300	10·2	12·1	13·7		28·9	41½	66		"
"	"	"	"	"	"		"			"	"	400	10·1	11·5	13·0		28·5	39	64		"
En 58B	0·12	0·95	0·80	8·60	17·1		0·51			¾	AC1050	RT	15·4	16·4	17·9		46·8	54½	64		2
"	"	"	"	"	"		"			"	"	100	12·8	14·0	15·4		36·3	64	74		"
"	"	"	"	"	"		"			"	"	200	11·0	11·8	12·9		31·0	49½	69		"
"	"	"	"	"	"		"			"	"	300	10·4	11·1	12·7		30·3	44	68		"
"	"	"	"	"	"		"			"	"	400	9·4	10·1	11·1		30·7	46½	65½		"
En 58C(B)	0·10			9·6	18·6		0·37				Softened	20	13·7	14·7	16·2		41·3	63	68		8
"	"			"	"		"				"	300	9·5	10·1	11·4		30·2	45½	66½		"
"	"			"	"		"				"	400	9·0	9·7	10·9		29·6	44	66		"
"	"			"	"		"				"	500	8·5	9·2	10·5		28·5	44½	65½		"
En 58B	Not stated									1¼	AC1150	20	16·0	17·5	19·5		45·0	53	65		14
"	"									"	"	400	14·9	16·2	18·0		28·0	36	40		"
"	"									"	"	500	12·2	13·1	14·2		26·5	35	36		"
"	"									"	"	600	9·5	10·5	11·4		21·0	36	65		"
"	"									"	"	700	7·4	8·1	9·0		15·2	50	60		"
"	"									"	"	800	5·0	5·4	6·1		9·5	60	80		"
En 58B	Not stated									Strip 0·64×½	Cold rolled	RT	33·1	38·4	42·9		59·3	32			35
"	"									"	" "	100	30·0	35·6	41·1		48·5	19			"
"	"									"	" "	200	27·0	32·1	37·5		42·9	13			"
"	"									"	" "	300	27·7	32·0	36·8		41·7	10			"
"	"									"	" "	400	26·7	30·7	35·6		41·3	10½			"
"	"									"	CR+4 hrs. 525	RT	38·9	42·2	46·2		57·1	34½			"
"	"									"	" "	100	37·3	40·2	43·0		49·6	19			"
"	"									"	" "	200	34·4	37·1	39·6		44·9	13			"
"	"									"	" "	300	32·7	35·3	37·8		42·2	10			"
"	"									"	" "	400	30·8	33·1	35·6		41·0	11			

MECHANICAL PROPERTIES AT HIGH TEMPERATURE (Continued)

Type	Chemical Composition (%)									Section Size dia. (in.)	Heat Treatment °C	Test Temperature °C	Tensile							Izod ft.lb.	Ref.
	C	Si	Mn	Ni	Cr	Mo	Ti	Nb	Se				0·1% P.S. (0·1Rp) tons/sq.in.	0·2% P.S. (0·2Rp) tons/sq.in.	0·5% P.S. (0·5Rp) tons/sq.in.	Y.P. (Rs) tons/sq.in.	T.S. (Rm) tons/sq.in.	El % (A)	R.A. % (Z)		
En 58B	(Within specification)										Softened 1050	20	13·3	14·2			46·4	6½	71	111	1
"	" "										" "	600	9·7	10·1			23·2	42½	70	78	"
"	" "										" "	700	8·4	8·9			16·3	56½	66	69	"
"	" "										" "	800	5·9	6·2			8·4	92	80	63	"
"	" "										" "	900					5·1	105	97		"
										(Rate of loading 0·5 tons/sq.in./min. to 0·2 P.S. then 0·04 in./min. strain)											
En 58F(G)	0·11	0·50	0·41	9·50	17·84			1·22		¾	AC1050	20	14·4	15·9	17·6		40·6	59	65½		2
"	"	"	"	"	"			"		"	"	100	12·8	14·9	15·3		33·6	53	71½		"
"	"	"	"	"	"			"		"	"	200	11·5	12·4	13·5		30·0	49½	68½		"
"	"	"	"	"	"			"		"	"	300	11·0	11·9	14·1		28·4	44	65½		"
"	"	"	"	"	"			"		"	"	400	10·1	10·8	11·8		28·0	40	65½		"
"	"	"	"	"	"			"		"	"	500	9·5	10·1	11·1		26·1	33½	66		"
"	"	"	"	"	"			"		"	"	550	9·5	10·1	11·0		25·1	38½	65		"
"	"	"	"	"	"			"		"	"	600	9·6	10·1	10·9		24·7	38½	60		"
"	"	"	"	"	"			"		"	"	650	9·5	10·1	10·8		23·1	39	63		"
"	"	"	"	"	"			"		"	"	700	9·2	9·7	10·4		21·5	44½	65½		"
"	"	"	"	"	"			"		"	"	750	9·0	9·4	10·4		18·9	50½	68½		"
En 58G	Not stated									1	Softened	20	13·3	15·2	17·2		40·8	58	65		14
"	"									"	"	400	9·5	10·6	11·7		27·9	38	64		"
"	"									"	"	500	9·6	10·3	11·2		27·0	38	62		"
"	"									"	"	600	9·0	9·6	10·5		24·5	40	62		"
"	"									"	"	700					19·2	46	70		"
En 58G(F)	Not stated (AISI 347 Composition)										Softened	21		17·6			40·6	50	71		48
"	" " " " "										"	149		15·2			33·3	47	75		"
"	" " " " "										"	260		14·3			30·8	41	74		"
"	" " " " "										"	371		14·3			29·9	35	72		"
"	" " " " "										"	482		14·1			28·6	35	69		"
"	" " " " "										"	593		12·7			25·0	39	69		"
"	" " " " "										"	704		10·7			17·9	51	74		"
"	" " " " "										"	816		8·7			10·3	76	92		"
"	" " " " "										"	927		4·7			6·2	68	51		"
"	" " " " "										"	1038		3·6			4·2	68	50		"
"	" " " " "										"	1149		1·1			1·6	61	68		"
"	" " " " "										"	1260		0·6			0·9	66	76		"

MECHANICAL PROPERTIES AT HIGH TEMPERATURE (Continued)

Type	Chemical Composition (%) C	Si	Mn	Ni	Cr	Mo	Ti	Nb	Se	Section Size dia. (in.)	Heat Treatment °C	Test Temperature °C	Tensile 0·1% P.S. (0·1Rp) tons/sq.in.	0·2% P.S. (0·2Rp) tons/sq.in.	0·5% P.S. (0·5Rp) tons/sq.in.	Y.P. (Rs) tons/sq.in.	T.S. (Rm) tons/sq.in.	El % (A)	R.A. % (Z)	Izod ft.lb.	Ref.
En 58G	0·11	0·77	1·28	11·95	18·2			1·35		¾	AC1050	RT	15·4	15·4	17·9		40·8	56	58½		2
"	"	"	"	"	"			"		"	"	100	13·9	14·9	16·1		34·7	49½	60		"
"	"	"	"	"	"			"		"	"	200	12·4	13·2	14·5		31·9	48½	64		"
"	"	"	"	"	"			"		"	"	300	11·7	12·4	13·4		30·2	38½	62		"
"	"	"	"	"	"			"		"	"	400	10·2	11·0	12·3		30·3	40	58½		"
"	"	"	"	"	"			"		"	"	500	10·1	10·6	11·7		29·2	38½	56½		"
"	"	"	"	"	"			"		"	"	600	10·4	11·1	12·1		26·9	38½	58½		"
"	"	"	"	"	"			"		"	"	700	10·0	10·4	11·0		19·9	53	68½		"
En 58G	0·12	0·60	1·5	11·0	17·5			1·20			AC1050	RT	13·3	15·2	17·2		40·8	58½	65½		2
"	"	"	"	"	"			"			"	400	9·5	10·6	11·7		27·9	38½	64		"
"	"	"	"	"	"			"			"	500	9·6	10·3	11·2		27·0	38½	62		"
"	"	"	"	"	"			"			"	550	9·1	10·1	11·0		26·4	41	60		"
"	"	"	"	"	"			"			"	600	9·0	9·6	10·5		24·5	40	62		"
"	"	"	"	"	"			"			"	650	9·9	10·5	11·2		23·9	40	62		"
"	"	"	"	"	"			"			"	700					19·2	46½	70		"
"	"	"	"	"	"			"			"	750					15·5	48	73		"
En 58G	0·11	0·50	1·30	12·0	18·7			1·25			Softened	20	13·5		16·8		38·5	52	60½		8
"	"	"	"	"	"			"			"	100	12·7		15·3		35·0	47	60		"
"	"	"	"	"	"			"			"	200	11·5		13·8		31·5	42	59		"
"	"	"	"	"	"			"			"	300	10·5		12·5		29·5	38	57½		"
"	"	"	"	"	"			"			"	400	10·0		11·8		28·5	36	51		"
"	"	"	"	"	"			"			"	500	9·5		11·0		27·5	35	47		"
"	"	"	"	"	"			"			"	600	9·2		10·7		25·2	37	52½		"
"	"	"	"	"	"			"			"	700	8·8		10·5		21·5	40	60½		"
"	"	"	"	"	"			"			"	800	8·3		9·5		15·0	54	60		"
												Rate of loading 1 ton/sq. in./min.									
													(0·05% P.S.)								
En 58H	Not stated									0·42	WQ1050	RT	14·1				39·7	67½	75		1
"	"	"								"	"	150	11·8				32·4	53½	73		"
"	"	"								"	"	300	11·0			13·5	30·3	32	68		"

MECHANICAL PROPERTIES AT HIGH TEMPERATURE (Continued)

Type	Chemical Composition (%) C	Si	Mn	Ni	Cr	Mo	Ti	Nb	Se	Section Size dia. (in.)	Heat Treatment °C	Test Temperature °C	Tensile 0·1% P.S. (0·1Rp) tons/sq.in.	0·2% P.S. (0·2Rp) tons/sq.in.	0·5% P.S. (0·5Rp) tons/sq.in.	Y.P. (Rs) tons/sq.in.	T.S. (Rm) tons/sq.in.	El % (A)	R.A. % (Z)	Izod ft. lb.	Ref.
En 58H	0·09	0·60	0·38	8·22	18·4	1·44				1⅛	WQ1050	20		15·0			41·6	64	70½		7
"	"	"	"	"	"	"				"	"	400		9·7			30·7	48	64½		"
"	"	"	"	"	"	"				"	"	450		9·5			30·5	49½	64		"
"	"	"	"	"	"	"				"	"	500		9·0			30·0	48½	64½		"
"	"	"	"	"	"	"				"	"	550		8·5			28·8	45½	64½		"
"	"	"	"	"	"	"				"	"	600		8·4			26·3	42	64½		"
"	"	"	"	"	"	"				"	"	650		8·0			22·0	40	62½		"
"	"	"	"	"	"	"				"	"	700		6·9			17·5	39½	60½		"
											Rate of loading 1 ton/sq.in./min.										
En 58H(J)	Not stated AISI 316 Composition										Softened	21		17·2			38·2	60	77		48
"	" " " " "										"	149		13·8			33·9	53	75		"
"	" " " " "										"	260		12·1			32·8	49	73		"
"	" " " " "										"	371		10·7			32·4	47	70		"
"	" " " " "										"	482		9·8			31·7	45	66		"
"	" " " " "										"	593		9·2			29·7	43	62		"
"	" " " " "										"	704		8·7			20·5	42	58		"
"	" " " " "										"	816		8·3			12·3	42	55		"
"	" " " " "										"	927		4·7			6·7	67	60		"
"	" " " " "										"	1038		2·0			3·6	60	47		"
"	" " " " "										"	1149		1·0			1·3	75	55		"
"	" " " " "										"	1260		0·6			0·8	92	83		"
En 58H(J)	0·03 max C. AISI 316 L Composition										Softened	21		14·1			35·6	73	78		48
"	" " " " "										"	149		12·1			31·7	47	73		"
"	" " " " "										"	260		10·6			29·4	43	71		"
"	" " " " "										"	371		9·5			28·6	42	69		"
"	" " " " "										"	482		9·1			27·7	40	66		"
"	" " " " "										"	593		8·5			24·6	43	65		"
"	" " " " "										"	704		7·1			18·3	46	52		"
"	" " " " "										"	816		5·4			11·2	49	46		"

MECHANICAL PROPERTIES AT HIGH TEMPERATURE (Continued)

Type	Chemical Composition (%)									Section Size dia. (in.)	Heat Treatment °C	Test Temperature °C	Tensile							Izod ft.lb.	Ref.
	C	Si	Mn	Ni	Cr	Mo	Ti	Nb	Se				0.1% P.S. (0·1Rp)	0·2% P.S. (0·5Rp)	0·5% P.S. (0·2Rp)	Y.P. (Rs)	T.S. (Rm)	El % (A)	R.A. % (Z)		
													tons/sq.in.								
En 58J	0.06	0·65	0·33	8·27	18·6	2·70				¾	AC1050	400	12·6	14·2	15·9		35·2	44	68		2
”	”	”	”	”	”	”				”	”	500	11·4	13·0	15·0		33·1	45	68		”
”	”	”	”	”	”	”				”	”	600	10·6	11·8	13·1		27·3	41½	70		”
”	”	”	”	”	”	”				”	”	700					19·2	57	68		”
”	”	”	”	”	”	”				”	”	800					12·3	64	60		”
”	”	”	”	”	”	”				”	”	900					7·5	54	41		”
											Rate of loading 1 ton/sq. in./min.										
En 58J	0·06	0·72	0·47	8·43	18·5	3·57				¾	AC1050	RT				27·6	42·6	36	55		2
”	”	”	”	”	”	”				”	”	400	15·1	17·6	19·4		35·7	35½	67		”
”	”	”	”	”	”	”				”	”	500	15·4	19·0	22·3		35·2	33½	65½		”
”	”	”	”	”	”	”				”	”	600					27·4	44	75		”
”	”	”	”	”	”	”				”	”	700					18·8	67	72		”
”	”	”	”	”	”	”				”	”	800					14·0	104	92		”
											Rate of loading 1 ton/sq. in./min.										
En 58J	0·07	0·26	0·36	8·55	18·2	2·82				¾	AC1050	RT	16·5	18·5	21·0		43·9	59	68½		2
”	”	”	”	”	”	”				”	”	66	14·6	16·2	18·3		39·1	57½	72		”
”	”	”	”	”	”	”				”	”	100	12·6	14·2	16·5		37·1	54½	70		”
”	”	”	”	”	”	”				”	”	200	11·1	12·5	14·7		34·5	45½	68		”
”	”	”	”	”	”	”				”	”	300	10·3	11·4	13·1		34·4	43	66		”
”	”	”	”	”	”	”				”	”	400	10·0	11·0	12·5		33·8	43	63		”
”	”	”	”	”	”	”				”	”	500	9·5	10·3	11·8		32·6	43	64		”
											Rate of loading 1 ton/sq. in./min.										
En 58J	0·07	0·41	0·40	8·35	18·5	3·00				¾	AC1050	RT	16·8	19·0	22·2		44·4	56	71½		2
”	”	”	”	”	”	”				”	”	66	14·4	16·3	19·2		38·9	50½	71		”
”	”	”	”	”	”	”				”	”	100	14·3	16·1	18·5		37·9	49½	71½		”
”	”	”	”	”	”	”				”	”	200	12·1	14·1	16·5		34·8	45½	70		”
”	”	”	”	”	”	”				”	”	300	11·0	12·7	14·8		34·3	43	67		”
”	”	”	”	”	”	”				”	”	400	10·8	12·3	14·5		34·1	44	67		”
”	”	”	”	”	”	”				”	”	500	10·7	12·1	13·9		33·0	42½	67½		”
											Rate of loading 1 ton/sq. in./min.										

MECHANICAL PROPERTIES AT HIGH TEMPERATURE (Continued)

Type	Chemical Composition (%)									Section Size dia. (in.)	Heat Treatment °C	Test Temperature °C	Tensile							Izod ft.lb.	Ref.
	C	Si	Mn	Ni	Cr	Mo	Ti	Nb	Se				0·1% P.S. (0·1Rp)	0·2% P.S. (0·2Rp)	0·5% P.S. (0·5Rp)	Y.P. (Rs)	T.S. (Rm)	El % (A)	R.A. % (A)		
													tons/sq.in.								
En58J	0·13	0·73	0·49	9·17	17·2	2·75					AC1100	20	14·5		16·3		38·2	60	77		8
,,	,,	,,	,,	,,	,,	,,					,,	300	8·4		10·0		32·0	50	69½		,,
,,	,,	,,	,,	,,	,,	,,					,,	400	9·0		10·2		32·5	51	69		,,
,,	,,	,,	,,	,,	,,	,,					,,	500	8·0		8·5		30·0	53	66		,,
,,	,,	,,	,,	,,	,,	,,					,,	550	6·4		8·5		28·5	51	67		,,
,,	,,	,,	,,	,,	,,	,,					,,	600	7·5		8·5		28·0	43	62		,,
,,	,,	,,	,,	,,	,,	,,					,,	650	7·0		8·1		25·5	48	61		,,
,,	,,	,,	,,	,,	,,	,,					,,	700	6·8		7·9		20·5	42	58		,,
										Rate of loading 1 ton/sq.in./min.											
En 58AM	0·08	0·48	0·39	8·45	17·6				0·23		Softened	20				23·4	50·0	41½	65		52
,,	,,	,,	,,	,,	,,				,,		,,	425				17·5	37·4	27½	62		,,
,,	,,	,,	,,	,,	,,				,,		,,	480				16·2	35·9	24½	56½		,,
,,	,,	,,	,,	,,	,,				,,		,,	540				14·0	33·1	21	57½		,,

CREEP AND RUPTURE PROPERTIES

Type	Chemical Composition (%)									Heat Treatment °C	Test Temperature °C	Stress (tons/sq.in.) for				Ref.
												Creep Rate		Rupture in		
	C	Si	Mn	Ni	Cr	Mo	Ti	Nb	Se			10^{-6} per hr. (1% in 10,000 hrs.)	10 per hr. (1% in 100,000 hrs.)	1,000 hrs.	10,000 hrs.	
En 58E(A)	0·03	AISI	No.	304 L	Comp.					Softened	482			21·0	18·3	48
,,	max.	,,	,,	,,	,,					,,	538			14·7	12·5	,,
,,	,,	,,	,,	,,	,,					,,	593	4·3	2·9	10·3	7·3	,,
,,	,,	,,	,,	,,	,,					,,	649	3·0		7·1	5·2	,,
,,	,,	,,	,,	,,	,,					,,	704	1·5	1·0	4·5	3·1	,,
,,	,,	,,	,,	,,	,,					,,	760			2·8	1·9	,,
,,	,,	,,	,,	,,	,,					,,	816	0·7	0·1	1·8	1·1	,,
,,	,,	,,	,,	,,	,,					,,	871			1·1	0·7	,,
En 58A(E)	AISI No. 304 Composition									Softened	427	14·9	11·2			48
,,	,,	,,	,,							,,	482	10·7	7·9	21·9	16·9	,,
,,	,,	,,	,,							,,	538	7·9	5·1	15·6	12·1	,,
,,	,,	,,	,,							,,	593	5·1	3·2	10·1	7·6	,,
,,	,,	,,	,,							,,	649	3·1	1·8	6·8	4·5	,,
,,	,,	,,	,,							,,	704	1·7	1·1	4·1	2·6	,,
,,	,,	,,	,,							,,	760	1·1	0·8	2·7	1·7	,,
,,	,,	,,	,,							,,	816	0·7	0·5	1·7	1·1	,,
,,	,,	,,	,,							,,	871	0·3	0·4	1·2	0·9	,,
,,	,,	,,	,,							,,	982			0·7		,,
,,	,,	,,	,,							,,	1093			0·4		,,
												Stress (tons/sq.in.)	Total creep in 10 days			
En 58E(A)	0·04	0·70	0·50	9·74	19·02					As rolled	650	4·5	0·27			47
,,	,,	,,	,,	,,	,,					,,	760	1·3	0·13			,,
,,	,,	,,	,,	,,	,,					,,	870	0·4	0·23			,,
,,	,,	,,	,,	,,	,,					WQ1150	650	4·5	0·16			,,
,,	,,	,,	,,	,,	,,					,,	760	1·3	0·08			,,
,,	,,	,,	,,	,,	,,					,,	870	0·4	0·06			,,

CREEP AND RUPTURE PROPERTIES (Continued)

Type	Chemical Composition (%) C	Si	Mn	Ni	Cr	Mo	Ti	Nb	Se	Heat Treatment °C	Test Temperature °C	Stress (tons/sq.in.) for creep strain of 0·01% in 1000 hrs.	Stress (tons/sq.in.) for creep strain of 0·10% in 1000 hrs.	Stress (tons/sq.in.) for rupture in 100,000 hrs.	Ref.
En 58E(A)	0·06	0·61	0·50	9·25	17·75					WQ1093	538	4·8	8·0		51
”	”	”	”	”	”					”	593	3·5	5·9		”
”	”	”	”	”	”					”	649	1·9	3·7	3·2	”
”	”	”	”	”	”					”	704	1·1	2·5	1·7	”
”	”	”	”	”	”					”	760	0·8	1·8	1·2	”
”	”	”	”	”	”					”	816	0·6	1·3	0·8	”
”	”	”	”	”	”					”	871			0·5	”
”	”	”	”	”	”					”	982			0·2	”
”	”	”	”	”	”					CW, AC927	538	4·2	8·3		”
”	”	”	”	”	”					”	649	1·6	2·8	5·8	”
”	”	”	”	”	”					”	704	1·0	2·1	2·8	”
”	”	”	”	”	”					”	816	0·2	0·6	0·8	”
												(Stress (tons/sq in.) for creep strain of 1% in 100,000 hrs.)			
En 58A	0·12			8·5	18·0					Softened	500	8·2			53
”	”			”	”					”	550	5·6			”
”	”			”	”					”	600	3·7			”
”	”			”	”					”	650	2·4			”
”	”			”	”					”	700	1·6			”
”	”			”	”					”	750	0·9			”
”	”			”	”					”	800	0·5			”
												Stress (tons/sq. in) for 1 % creep strain in 10,000 hrs.		100,000 hrs.	
En 58E(A)	AISI	No. 302	or 304	Composition						Softened	482	10·7		2·9	27
”	”	”	”	”						”	538	7·9		5·1	”
”	”	”	”	”						”	593	5·1		3·2	”
”	”	”	”	”						”	649	3·1		1·8	”
”	”	”	”	”						”	704	1·7		1·1	”
”	”	”	”	”						”	760	1·1		0·8	”
”	”	”	”	”						”	816	0·6		0·5	”
”	”	”	”	”						”	871	0·36			”

CREEP AND RUPTURE PROPERTIES (Continued)

Type	C	Si	Mn	Ni	Cr	Mo	Ti	Nb	Se	Section Size dia. (in.)	Heat Treatment °C	Test Temperature °C	Stress (tons/sq.in.) for Creep rate of 10^{-6} per hr. (1% in 10,000 hrs.)	Stress (tons/sq.in.) for Creep rate of 10^{-7} per hr. (1% in 100,000 hrs.)	Stress (tons/sq.in.) for Rupture in 1,000 hrs.	Stress (tons/sq.in.) for Rupture in 10,000 hrs.	Ref.
En 58C(B)	AISI No. 321 Composition										Softened	427	19·2	16·7	23·7	21·2	48
"	" " "										"	482	12·7	13·4	21·4	19·2	"
"	" " "										"	538	11·4	8·3	18·7	15·2	"
"	" " "										"	593	7·6	5·6	13·2	9·6	"
"	" " "										"	649	4·3	2·9	8·0	5·4	"
"	" " "										"	704	2·3	1·8	4·9	2·9	"
"	" " "										"	760	1·1	0·8	3·1	1·6	"
"	" " "										"	816	0·9	0·6	1·6	1·0	"
"	" " "										"	871	0·7		1·3	0·8	"
En 58C(B)	0·10	1·15	0·69	9·60	18·6		0·37			1⅛	AC1050	650	5·3		9·0		8
"	"	"	"	"	"		"			"	"	700	3·8		6·2		"
"	"	"	"	"	"		"			"	"	750	2·7		4·1		"

Type	C	Si	Mn	Ni	Cr	Mo	Ti	Nb	Se	Section Size dia. (in.)	Heat Treatment °C	Test Temperature °C	Stress (tons/sq.in.) for 1% Creep strain in 1,000 hrs.	5,000 hrs.	10,000 hrs.	30,000 hrs.	Stress (tons/sq.in.) for Rupture in 1,000 hrs	5,000 hrs.	10,000 hrs.	30,000 hrs.	100,000 hrs.	Ref.
En 58C	0·12	0·96	1·24	12·0	18·7		0·49				AC1100	600	12·0	8·0	7·0	5·2	15·5	11·0	9·2	6·6	4·3*	2
"	"	"	"	"	"		"				"	650	7·4	5·0	4·0		9·6	6·6	5·4	3·4*	2·0*	"
"	"	"	"	"	"		"				"	700	5·0	2·8	2·0		6·2	3·8	2·9	1·7*	0·9*	"
"	"	"	"	"	"		"				"	750					4·0	2·1	1·5	0·8*	0·6*	"
"	"	"	"	"	"		"				"	800					2·3	1·1	0·7*			

* Extrapolated.

Type	Chemical Composition (%)	Section Size dia. (in.)	Heat Treatment °C	Test Temperature °C	Stress (tons/sq.in.) for creep of 0·1% in 100 hrs.	300 hrs.	Stress (tons/sq.in.) for rupture in 100 hrs.	Ref.
En 58B	Within specification			500	13·0	12·0		14
"	" "			600	9·2	6·8	15·0	"
"	" "			700	1·2	"	6·7	"

CREEP AND RUPTURE PROPERTIES (Continued)

Type	Chemical Composition (%) C	Si	Mn	Ni	Cr	Mo	Ti	Nb	Section Size dia. (in.)	Heat Treatment °C	Test Temperature °C	Stress (tons/sq.in.) for total creep strain in 100 hrs. of 0·1%	0·2%	0·3%	0·5%	1·0%	in 300 hrs. of 0·1%	0·2%	0·3%	0·5%	1·0%	Stress (tons/sq. in.) for rupture in 100 hrs.	300 hrs.	Ref.
En 58B	Within specification										600	7·6	7·8	8·3	8·6	9·2*	7·2	7·5	8·0	8·3	8·8	15·2	13·5	1
,,	,, ,,										700	4·8	5·9				4·0	5·0	5·5	6·2*	6·7	9·6	8·1	,,
,,	,, ,,										800	2·2	2·7	2·9	3·1*	3·4*	1·7	2·1	2·4	2·5	2·8	4·5	3·8	,,

Type	Chemical Composition (%)	Section Size dia. (in.)	Heat Treatment °C	Test Temperature °C	Stress (tons/sq.in.) for total creep strain of 1% in 10,000 hrs.	100,000 hrs.	Stress (tons/sq.in.) for rupture in 1,000 hrs.	10,000 hrs.	Ref.
En 58C(B)	AISI No. 321 Composition		Softened	482	11·2				27
,,	,, ,, ,,		,,	538	8·2				,,
,,	,, ,, ,,		,,	593	5·8	5·0	12·1	7·1	,,
,,	,, ,, ,,		,,	649	3·6	2·7	7·8	4·4	,,
,,	,, ,, ,,		,,	704	2·1	1·3	4·5	2·7	,,
,,	,, ,, ,,		,,	760	1·1	0·7	2·5	1·6	,,
,,	,, ,, ,,		,,	816	0·4		1·7	1·0	,,
,,	,, ,, ,,		,,	871	0·2				
En 58G(F)	AISI No. 347 Composition		Softened	482	18·7	17·0	25·0	22·8	48
,,	,, ,, ,,		,,	538	14·3	12·3	21·4	17·0	,,
,,	,, ,, ,,		,,	593	10·3	7·6	15·2	11·6	,,
,,	,, ,, ,,		,,	649	7·4	4·2	9·8	7·6	,,
,,	,, ,, ,,		,,	704	4·4	2·2	5·8	4·2	,,
,,	,, ,, ,,		,,	760	2·2	1·0	3·3	2·3	,,
,,	,, ,, ,,		,,	816	0·9	0·7	2·0	1·2	,,
,,	,, ,, ,,		,,	871	0·4	0·2	1·0	0·4	,,

Type	C	Si	Mn	Ni	Cr	Mo	Ti	Nb	Section Size dia. (in.)	Heat Treatment °C	Test Temperature °C	Stress (tons/sq.in.) for minimum creep rate of 10^{-4} per hr.	10^{-5} per hr.	10^{-6} per hr.	10^{-7} per hr.	Ref.
En 58G	0·11	0·50	1·30	12·0	18·7	1·25				Softened	550		20·0	14·0	9·8	8
,,	,,	,,	,,	,,	,,	,,				,,	600	23·0	14·8	9·7	6·4	,,
,,	,,	,,	,,	,,	,,	,,				,,	650	14·9	10·5	7·0	4·8	,,
,,	,,	,,	,,	,,	,,	,,				,,	700	9·0	6·0	3·9	2·6*	,,

* Approximate.

CREEP AND RUPTURE PROPERTIES (Continued)

Type	Chemical Composition (%) C	Si	Mn	Ni	Cr	Mo	Ti	Nb	Section Size dia, (in.)	Heat Treatment °C	Test Temperature °C	Stress (tons/sq.in.) for a creep strain in 100 hrs. of 0·1 %	0·2 %	in 300 hrs. of 0·1%	0·2%	0·5%	1·0%	in 1,000 hrs of 0·1%	0·2%	0·5%	1·0%	Stress (tons/sq. in.) for rupture in 300 hrs.	1,000 hrs.	Ref.
En 58G	Within specification									Softened	600	14·5	16·5	11·9	13·2	15·5*	17·0*	9·6	10·5	12·1	13·4	18·5	14·7	14
"	" "									"	650	9·0	10·2	7·3	8·2	9·3	10·0	4·8	6·6	7·6	8·2	10·7	8·7	"
"	" "									"	700							2·7*	3·2*	3·7*	4·5*	6*	4·8*	"

* Approximate.

Type	C	Si	Mn	Ni	Cr	Mo	Ti	Nb	Heat Treatment °C	Test Temperature °C	Stress for 0·1% creep in 300 hrs.	1,000 hrs.	5,000 hrs.	10,000 hrs.	20,000 hrs.	Ref.
En 58G	0·12	0·60	1·50	11·0	17·5			1·2	Softened	600	12·0	9·5	8·0	6·7		2
"	"	"	"	"	"			"	"	650	7·0	6·0	3·5	2·5	2·0	"
"	"	"	"	"	"			"	"	700		3·0	1·5			"

Type	C	Si	Mn	Ni	Cr	Mo	Ti	Nb	Heat Treatment °C	Test Temperature °C	Stress for 0·2% creep in 300 hrs.	1,000 hrs.	5,000 hrs.	10,000 hrs.	20,000 hrs.	Ref.
"	"	"	"	"	"			"	"	600	13·0	10·5	8·0	7·0	6·0	"
"	"	"	"	"	"			"	"	650	8·0	7·5	4·5	3·5	2·5	"
"	"	"	"	"	"			"	"	700		4·0	2·0	1·5		"

Type	C	Si	Mn	Ni	Cr	Mo	Ti	Nb	Heat Treatment °C	Test Temperature °C	Stress for 0·5% creep in 300 hrs.	1,000 hrs.	5,000 hrs.	10,000 hrs.	20,000 hrs.	Ref.
"	"	"	"	"	"			"	"	600	15·0+	12·0	9·0	8·0	7·5	"
"	"	"	"	"	"			"	"	650	9·0	8·5	6·0	4·5	2·7	"
"	"	"	"	"	"			"	"	700		5·0	2·5	2·0		"

Type	C	Si	Mn	Ni	Cr	Mo	Ti	Nb	Heat Treatment °C	Test Temperature °C	For rupture in 300 hrs.	1,000 hrs.	5,000 hrs.	10,000 hrs.	20,000 hrs.	30,000 hrs.	Ref.
"	"	"	"	"	"			"	"	600	19·0	16·0	12·0	9·5	8·5	7·5	"
"	"	"	"	"	"			"	"	650	12·5	10·0	7·5	6·5	5·0	4·0	"
"	"	"	"	"	"			"	"	700	9·0	7·5	4·5	3·0+	2·5	2·0	"
"	"	"	"	"	"			"	"	750	4·5	3·5	2·2	1·5			"
"	"	"	"	"	"			"	"	800	3·0	2·0	1·2				"

Type	Chemical Composition (%)	Heat Treatment °C	Test Temperature °C	Stress (tons/sq.in.) for total creep strain 1% in 10,000 hrs.	100,000 hrs.	Stress (tons/sq.in.) for rupture in 1,000 hrs.	10,000 hrs.	Ref.
En 58G(F)	AISI No. 347 Composition	Softened	538	8·5				27
"	" "	"	593	6·2	5·0	13·6	9·8	"
"	" " "	"	649	3·7	2·7	7·8	4·9	"
"	" " "	"	704	2·1	1·1	5·0	2·1	"
"	" " "	"	760	1·1		3·3		"
"	" " "	"	816	0·7		2·0		"
"	" " "	"	871	0·4				"

CREEP AND RUPTURE PROPERTIES (Continued)

Type	Chemical Composition (%) C	Si	Mn	Ni	Cr	Mo	Ti	Nb	Section Size dia. (in.)	Heat Treatment °C	Test Temperature °C	Stress (tons/sq. in.) at 1,000 hrs. for creep rate of 10⁻⁵ per hr.	10⁻⁶ per hr.	10⁻⁷ per hr.	Stress for total plastic strain of 0·1 %	1·0 %	Stress (tons/sq.in.) for rupture in 100 hrs.	300 hrs.	1,000 hrs.	3,000 hrs.	10,000 hrs.	Ref.
En 58G	Composition not stated									AC1050	550			19·7			26·0	24·2	22·1	19·3	17·0	1
"	" " "									"	600	13·4	11·4	9·5*	9·7	14·1	19·2	18·1	15·1	12·7	10·5	"
"	" " "									"	650	7·9	5·8	3·9	5·8	8·2*	15·0	12·7	10·6	9·0	7·6	"
"	" " "									"	700						10·7	8·6	6·8	5·6*		"
												* Extrapolated.										
												Stress (tons/sq.in.) for a creep rate of 10⁻⁶ per hr.	10⁻⁷ per hr.				Stress (tons/sq.in.) for rupture in 1,000 hrs.	10,000 hrs.				
En 58H	AISI No. 316 Composition									Softened	482	14·1	8·5				29·5	28·6				43
"	" " "									"	538	10·8	6·6				18·3	12·5				"
"	" " "									"	593	8·1	4·7				14·7	11·1				"
"	" " "									"	649	5·7	2·9				11·0	8·1				"
"	" " "									"	704	3·5	1·7				7·8	5·4				"
"	" " "									"	760	2·0	0·9				5·0	3·1				"
"	" " "									"	816	1·2	0·6				2·9	1·5				"
"	" " "									"	871	0·4	0·3				1·4	0·7				"
"	" " "									"	982						0·5	0·2				"
"	" " "									"	1,093						0·2	0·05				"
												Stress (tons/sq.in.) for 1% total creep strain in 10,000 hrs.	100,000 hrs.				Stress (tons/sq.in.) for rupture in 1,000 hrs.	10,000 hrs.				
En 58H(J)	AISI No. 316 Composition									Softened	538	10·8	6·6									27
"	" " "									"	593	8·1	4·7				14·7					"
"	" " "									"	649	5·7	2·9				11·0	8·1				"
"	" " "									"	704	3·5	1·7				7·8	5·4				"
"	" " "									"	760	2·0	0·9				5·0	3·1				"
"	" " "									"	816	1·2	0·6				2·9	1·5				"
"	" " "									"	871						1·8	0·8				"
												Stress (tons/sq.in.) for minimum creep rate of 10⁻⁶ per hr.										
En 58J	0·12	0·50		9·50	18·5	2·75					600	11·0										8
"	"	"		"	"	"					650	6·0										"
"	"	"		"	"	"					700	2·3										"
"	"	"		"	"	"					750	1·9										"

CREEP AND RUPTURE PROPERTIES (Continued)

Type	C	Si	Mn	Ni	Cr	Mo	Ti	Nb	Section Size dia. (in.)	Heat Treatment °C	Test Temperature °C	Stress (tons/sq.in.) for 1% creep strain in 10,000 hrs.	Ref.
	Chemical Composition (%)												
En 58J	0·07	0·05	0·90	10·00	17·0	3·0	0·4				400	15·5	6
"	"	"	"	"	"	"	"				500	12·8	"
"	"	"	"	"	"	"	"				600	7·8	"
"	"	"	"	"	"	"	"				700	3·0	"
"	"	"	"	"	"	"	"				800	0·8	"

MECHANICAL PROPERTIES IN TORSION

Type	Chemical Composition (%)							Section Size dia. (in.)	Heat Treatment °C	Tensile				Torsion			Ref.
	C	Si	Mn	Ni	Cr	Ti	Nb			Y.P. (Rs) tons/sq.in.	T.S. (Rm) tons/sq.in.	El % (A)	R.A. % (Z)	Y.P. tons/sq.in.	Apparent max. Shear Stress tons/sq.in.	Probable max. Shear Stress tons/sq.in.	
En 58A	Within specification								Softened	18·0	40·0	50	50	9·5	52·0	39·0	2
En 58A	0·07			8·08	18·2					14·4	51·4	54	55	7·5	59·3	44·5	2
En 58A	Within specification							¾	Softened		42·6	66	72½	10·8	43·0	32·3	54
En 58E	0·05	0·25	0·43	10·56	18·1			Tube 0·842 OD 0·600 ID	Softened	17·0	34·0	73		9·1	27·3	24·2	26
En 58E	0·06	0·41	1·73	10·18	18·4			" 1·002 OD 0·875 ID	CD ¼ hard	39·0	49·1	41		25·6	34·7	31·2	"
En 58E	0·07	0·37	1·68	10·36	18·9			1·054 OD 0·813 ID	CD ½ hard	56·7	75·8	13		31·2	44·5	40·0	"
En 58B	Within specification								Softened	18·0	42·0	45	50	8·5	47·9	35·9	2
En 58C	0·07	0·55	1·50	11·24	18·2	0·55		Tube 0·503 OD 0·406 ID	CD ¼ hard		51·3	24		20·9	34·7	31·6	26
"		"	"	"	"	"		" 0·502 OD 0·402 ID	" ½ hard		58·5	14		25·0	36·5	33·2	"
En 58D	Within specification								WQ1050	17·0	35·5	55	73		36·7	27·5	31
En 58G	0·04	0·58	1·42	11·04	17·6		0·58	Tube 0·839 OD 0·636 ID	Softened	22·4	41·9	43		10·4	23·8	21·2	26
"	0·07	0·43	1·52	12·69	18·7		0·87	" 1·004 OD 0·870 ID	CD ¼ hard	35·7	50·4	23½		22·1	35·6	33·3	"
"	0·07	0·41	1·74	12·20	17·7		0·91	" 1·050 OD 0·817 ID	CD ½ hard	60·7	70·5	13½		35·1	41·9	37·6	"

FATIGUE PROPERTIES

Type	Chemical Composition (%) C	Si	Mn	Ni	Cr	Ti	Heat Treatment °C	Test Temperature °C	Tensile (tons/sq.in.) 0·1% P.S. (0·1Rp)	0·2% P.S. (0·2Rp)	Y.P. (Rs)	T.S. (Rm)	El % (A)	R.A. % (Z)	Fatigue Limit (10⁷ cycles) tons/sq.in. Rotating Bending	Reverse Bending	Reversed Axial	Ref.
En 58A	0·08	0·60	0·80	8·75	18·0		Softened 1050	RT			18	40	50	50	±17			2
En 58E	0·05	0·60	0·80	10·0	18·2		" "	RT			15	40	60	65	±17			"
En 58A	0·06	0·50	1·40	8·25	18·0		Softened	RT			15	43	60	60	±17			3
En 58A	0·10	0·50	1·40	8·25	18·0		" "	RT			18	45	50	60	±17·5			"
En 58E(A)	0·05	0·42	0·49	8·82	18·5		Cold rolled	+25		66·4		98·9	12	54½		±49·1		40
"	"	"	"	"	"		"	—78		80·4		107·8	14½	62		±55·8		"
"	"	"	"	"	"		"	—192		84·4		132·4	23½	45½		±69·2		"
En 58A	0·07	0·76	0·53	7·95	18·0		Cold drawn	RT	56·0			95·8	6½	53½	±29·6			1
"	"	"	"	"	"		CD + ½hr. 100	"	57·0			97·4	3½	52½	±29·8			"
"	"	"	"	"	"		" " 200	"	59·3			103·0	3½	51	±29·0			"
"	"	"	"	"	"		" " 300	"	71·5			101·3	3	51	±30·0			"
"	"	"	"	"	"		" " 500	"	69·0			101·6	3½	51½	±29·8			"
En 58A	0·08	0·65	0·46	8·02	18·0		Cold drawn	"	43·1			78·6	22	55	±30·8			"
"	"	"	"	"	"		CD + ½hr. 100	"	46·8			76·9	20½	60	±31·0			"
"	"	"	"	"	"		" " 200	"	47·0			78·0	15½	56	±32·8			"
"	"	"	"	"	"		" " 300	"	51·2			77·0	10½	65	±35·2			"
"	"	"	"	"	"		" " 400	"	52·5			78·3	11½	64½	±35·0			"
"	"	"	"	"	"		" " 500	"	51·4			74·3	17	67½	±32·6			"
En 58A	0·11	0·28	0·36	8·23	18·32		Cold rolled	RT			41·2	66·3	52	36	±23·7		±24·2 (in air)	55
"	"	"	"	"	"		" "	"			"	"	"	"	±17·5		±17·2 (in salt spray)	"
"	"	"	"	"	"		" "	"			"	"	"	"	±16·7 (2 × 10⁷ cycles)		±16·1 " (2 × 10⁷ cycles)	"
"	"	"	"	"	"		" "	"			"	"	"	"	±15·8 (5 × 10⁷ cycles)		±14·8 (5 × 10⁷ cycles)	"
En 58C (B)	0·08	0·80	0·80	9·5	18·0	0·50	Softened 1050	RT			18	42	45	50	±17·5			2
En 58B	Within specification						AC1150	20	16·0		19·5	45·0	53	65	±24·0			14
"	" "						"	400	14·9		18·0	28·0	36	40	±15·5			"
"	" "						"	500	12·2		14·2	26·5	35	36	±14·5			"
"	" "						"	600	9·5		11·4	21·0	36	65	±13·0			"
"	" "						"	700	7·4		9·0	15·2	50	60	±11·5			"
"	" "						"	800	5·0		6·1	9·5	60	80	± 7·0 (approx.)			

FATIGUE PROPERTIES (Continued)

Type	Chemical Composition (%)									Heat Treatment °C	Test Temperature °C	Tensile						Fatigue Limit tons/sq.in.			Ref.
	C	Si	Mn	Ni	Cr	Mo	S	Se	Nb			0·1% P.S. (0·1Rp)	0·2% P.S. (0·2Rp)	Y.P. (Rs)	T.S. (Rm)	El % (A)	R.A. % (Z)	Rotating Bending			
												tons/sq.in.						(10^7 cycles)	(4×0^7 cycles)	(10^8 cycles)	
																					2
En 58F (G)	0·08	0·60	0·80	9·5	18·0				0·85	Softened 1050				17·2	40·8	58	65	±19·5			3
En 58F	0·07	0·50	1·40	8·25	18·0				0·80	Softened				18·0	45·0	50	60	±17·5			14
En 58G	Within specification									Softened	20	13·3		17·2	40·8	58	65	±19			"
"	"			"						"	600	9·0		10·5	24·5	40	62	±13·5			
En 58G	0·11	0·50	1·30	12·0	18·7				1·25	Softened 1050	550								±15·7	±15·5	8
"	"	"	"	"	"				"	" "	600	9·2		10·7	25·2	37	52½		±15·0	±15·5	"
"	"	"	"	"	"				"	" "	650								±13·5	±14·7	"
En 58H	0·07	0·30	1·5	10·0	17·7	2·75				Softened 1050				18·0	40·0	50	50	±17·5			2
En 58AM	0·08	0·50	0·60	8·5	18·5			0·120	0·30	Softened		12·9	15·4		38·9	60	65	±15·4			39
"	0·08	0·40	0·60	8·5	18·5			0·300		"		12·9	14·5		37·9	55	60	±15·6			"

LIST OF REFERENCES

1. Private correspondence with The United Steel Companies Ltd.
2. Private correspondence with Firth-Vickers Stainless Steels Ltd.
3. Private correspondence with Samuel Osborn and Co. Ltd.
4. J. B. Austin and R. H. H. Pierce. Thermal Expansion of Heat-resisting Iron Alloys. *Ind. Eng. Chem.*, 1933, 25, July, 776—779.
5. Physical Constants of Some Commercial Steels at Elevated Temperatures. Butterworths Scientific Publications, London, 1953.
6. Private correspondence with Richard Thomas and Baldwins Ltd.
7. Private correspondence with Edgar Allen and Co. Ltd.
8. Private correspondence with Jessop-Saville Ltd.
9. Private correspondence with Hadfields Ltd.
10. Private correspondence with Brown Bayley Steels Ltd.
11. T. B. Douglas and J. L. Dever. Enthalpy and Specific Heat of Four Corrosion-resistant Alloys at High Temperature. *J. Res. Nat. Bur. Stand.*, 54, Jan. 15—19, 1955.
12. Private correspondence with International Nickel Ltd.
13. D. E. Furman. Thermal Expansion Characteristics of Stainless Steels between —300° and 1000°F, *Transactions*, AIME, 1950, 188, 688—691: *J. Met.*, Apr., 1950.
14. Private correspondence with Rolls-Royce Ltd.
15. Thermal Properties of 13 Metals. ASTM Special Technical Publication No. 227.
16. M. G. Fantana. Investigation of Mechanical and Physical Properties of Aircraft Alloys at Low Temperatures, Pt 2, A. F. Tech. Rep. 5662: PB 130662, Oct., 1948, 125.
17. S. M. Shelton. Thermal Conductivity of Some Irons and Steels over the Temperature Range 100°C to 500°C. *J. Res. Nat. Bur. Stand.*, 12. Apr., 1934, 441—450.
18. L. Silverman. Thermal Conductivity Data presented for various Metals and Alloys up to 900°C. *J. Met.*, 5, May, Section 1, 1953, 631—632.
19. Private correspondence with De Havilland Aircraft Co. Ltd.
20. C. L. Hogan and R. B. Sawyer. The Thermal Conductivity of Metals at High Temperature. *J. Appl. Phys.*, 23, Feb., 1952, 177—180.
21. R. W. Powell and R. P. Tye. The Effect of Oil Quenching and Tempering on the Thermal Conductivities and Electrical Resistivities of Three Steels. *JISI*, vol. 184, p. 10, Sept., 1956.
22. Private correspondence with Low Moor Alloy Steelworks Ltd.
23. R. Kohlhaas and F. Richter. The Temperature Dependency of the Electrical Resistance of Pure Iron and Steel with particular reference to Phase Changes. *Arch. Eisenh.*, 33, May, 1962, 291—299.
24. Private correspondence with the National Physical Laboratory.
25. F. Garofalo, P. R. Malenock and G. V. Smith. The Influence of Temperature on the Elastic Constants of some Commercial Steels, ASTM Symposium on the Determination of Elastic Constants, June 25, 1952. ASTM Special Publication No. 129.
26. C. W. Muhlenburch *et al.* Mechanical Properties in Torsion and Poisson's Ratio for certain Stainless Steel Alloys. *Proceedings*, ASTM, 51, 1951, 832—856.
27. Technical data — Stainless Steels. Atlas Gilson Steels S. A. and Atlas Steels, Canada.
28. *Stainless Steel Handbook*. Allegheny Stainless Steels, New York, U.S.A.
29. British Standard 971, Commentary on the British Standard Wrought Steels, En Series. British Standards Institution, London, 1950.
30. J. H. G. Monypenny. *Stainless Iron and Steel*. Chapman and Hall, 1951.
31. Private correspondence with Darwins Ltd.
32. D. J. McAdam, jun., *et al.* Influence of Low Temperatures on the Mechanical Properties of 18 : 8 chromium–nickel steel. *J. Res. Nat. Bur. Stand.*, 40, May, 1948, 375—392.
33. Private correspondence with Hall and Pickles Ltd.

34. N. B. Pilling. Some Effects of Nickel Content in Austenitic Iron-chromium Nickel Alloys. *Proceedings*, ASTM, Vol. 30, Pt. 2, 1930, p. 284.

35. War Office unpublished data.

36. S. A. Henes and A. M. Tunkals. Welding of Hardenable Steels with High Alloy (austenitic) Electrodes. *Weld, J.*, 23, Oct., 1946, 6905.

37. S. L. Hoyt. *Metal Data*. Reinhold, New York, 1952.

38. B. J. Connolly and G. Boyd. The High Temperature Properties of Four Wrought Steels used in the Chemical Industry. *Proc. I.M.E.*, 172 (31), 1958.

39. S. P. Watkins. The Free-machining Stainless Steels. *Met. Prog.*, 194, 39, 452—458, 500.

40. J. L. Zambrow and M. G. Fontana. Mechanical Properties, including Fatigue of Aircraft Alloys at very low temperatures. *Trans. A.S.M.*, 41. 1949, 480—518.

41. H. J. Weister. Steel for Use at Low Temperature and its Testing. *Stahl und Eisen*, vol. 63, No. 4, Jan., 1943, 64.

42. S. J. Rosenberg. Effect of Low Temperatures on the Properties of Aircraft Metals. *J. Res. Nat. Bur. Stand.*, 25, Dec, 1940, 673—699.

43. H. W. Mischeer and H. J. Nichols. An Investigation of the Low Temperature Impact Properties of Stainless Steel Weldments. *Weld. J.*, 40, Dec., 1961, 5645—5685.

44. W. J. De Hass and R. Hadfield. Effect of Temperature of Liquid Hydrogen on the Tensile Properties of forty-one Specimens of Metal. *Philosophical Transactions*, the Royal Society, Vol. 232, 1934.

45. E. W. Colbeck, W. E. McGillivray and W. R. D. Manning. The Mechanical Properties of some Austenitic Steels at Low Temperature. *The Engineer*, vol. 155, June, 1933, 556—557.

46. N. P. Allen. Symposium on the Mechanical Properties at Low Temperature. Nat. Bureau of Standards Circular 520, May, 1952.

47. R. H. Aborn and E. C. Bain. *Nature of Nickel—Chromium Rustless Steels*. Transactions of the American Society for Steel Treating, vol. 18, 1930, 837—893.

48. Private correspondence with the United States Steel Corporation.

49. Private correspondence with the Royal Aircraft Establishment, Farnborough.

50. M. Schmidt and O. Jungwirth. Hot Brittleness of Austenitic Steels. *Arch. Eisenh.*, 6, June, 1933, 559—562.

51. Digest of Steels for High Temperature Service, Timken Roller Bearing Co. Ltd.

52. Carpenter Steel Co., U.S.A. *Iron Age*, Sept 15, 1932, 404.

53. Data published by Babcock and Wilcox Co., U.S.A.

54. M. C. Fetzer. Torsional and Tensile Properties of Steels. *Steel*, vol. 118, 92—94, June, 1956.

55. H. J. Gough and D. G. Sopwith. The Influence of the Mean Stress of the Cycle on the Resistance of Metals to Corrosion-Fatigue. *JISI*, 135 (1), 1937, 293—313.

CHROMIUM–NICKEL–SILICON VALVE STEEL

SPECIFICATION

CHEMICAL COMPOSITION

C	Si	Mn	S	P	Ni	Cr
0·74—0·84	1·75—2·25	0·20—0·60	0·030 max.	0·030 max.	1·15—1·65	19·0—20·5

MECHANICAL PROPERTIES

The steel shall be capable of local hardening by oil quenching to a Rockwell C hardness of not less than 47 or the equivalent Vickers hardness number 480.

RELATED SPECIFICATIONS

United Kingdom	Germany	
En No.	Werkstoff No.	Name
59	1.4747	X80 CrNiSi20

APPLICATIONS

This steel has a high resistance to scaling and good mechanical properties at temperatures up to 800 to 900°C. It has a high resistance to corrosion where the use of leaded fuels of high octane rating is involved. Its high resistance to sulphur attack also renders it suitable for furnace parts subject to sulphurous atmospheres. Typical uses are roasting furnace parts, engine valves.

WELDING

Welding is not recommended.

MACHINABILITY

The machinability of this steel in the softened condition is approximately 30% of that for mild steel (En 3).

HOT WORKING AND HEAT TREATMENT TEMPERATURES

Forging, rolling and stamping	1100°C, finish above 900°C. Slow cooling advisable.
Softening	750°C, air cool.
Hardening	1020°C—1080°C, air cool or oil quench.
Tempering	700°C—750°C, air cool or oil quench.

PHYSICAL PROPERTIES

SPECIFIC GRAVITY (d)

Chemical Composition (%)					Specific Gravity at 20°C	Ref.
C	Si	Mn	Ni	Cr		
0·78	2·0	0·45	1·35	19·5	7·54	1

MEAN COEFFICIENT OF THERMAL EXPANSION (α)

Chemical Composition (%)					$10^6 \times$ Mean Coefficient of Thermal Expansion per deg. C								Ref.
C	Si	Mn	Ni	Cr	20 to 100°C	20 to 200°C	20 to 300°C	20 to 400°C	20 to 500°C	20 to 600°C	20 to 700°C	20 to 800°C	
0·78	2·0	0·45	1·35	19·5	8·2		10·2		10·8		11·7		1
0·80	2·0		1·50	19·5		11·0		11·7		12·3		12·5	2

THERMAL CONDUCTIVITY (k)

Chemical Composition (%)					Thermal Conductivity at 20°C cal/cm sec. deg. C	Ref.
C	Si	Mn	Ni	Cr		
0·78	2·0	0·45	1·35	19·5	0·027	1

ELECTRICAL RESISTIVITY (ρ)

Chemical Composition (%)					Electrical Resistivity at 20°C microhm-cm	Ref.
C	Si	Mn	Ni	Cr		
0·78	2·0	0·45	1·35	19·5	97	1

YOUNG'S MODULUS (E)

Chemical Composition (%)					Young's Modulus (tons/sq. in.)				Ref.
C	Si	Mn	Ni	Cr	20°C	200°C	300°C	500°C	
0·78	2·0	0·45	1·35	19·5	13,900	13,500	13,200	10,700	1

En 59

TRANSFORMATION CHARACTERISTICS

ISOTHERMAL TRANSFORMATION

Chemical Composition (%)				
C	Si	Mn	Ni	Cr
0·81	2·15	0·55	1·63	19·4

Austenitizing Temperature 1063°C, 30 min.

(Ref. 3)

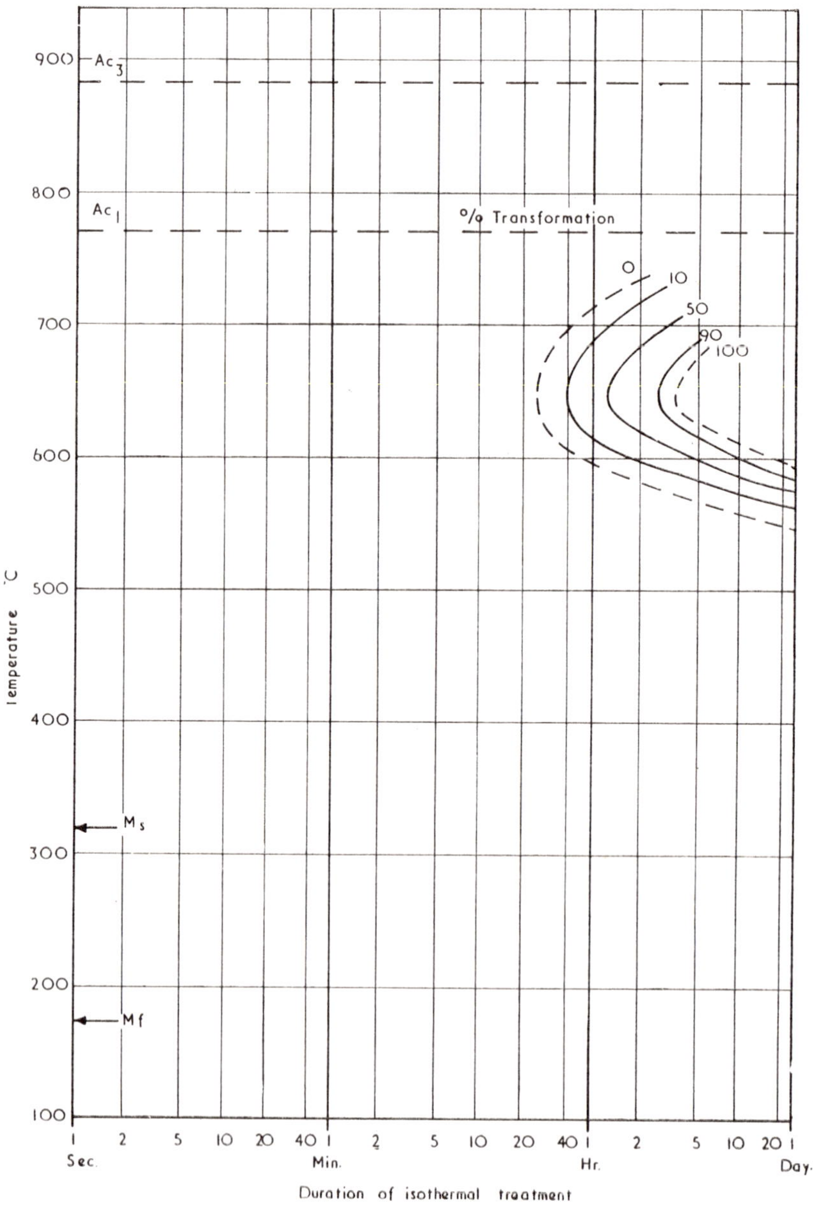

Fig. 59.1

MECHANICAL PROPERTIES AT ROOM AND ELEVATED TEMPERATURES

Chemical Composition (%)					Heat Treatment °C	Test Temperature °C	Tensile						Ref.
C	Si	Mn	Ni	Cr			0·1% P.S. (0·1Rp) tons/sq.in.	0·2% P.S. (0·2Rp) tons/sq.in.	0·5% P.S. (0·5Rp) tons/sq.in.	T.S. (Rm) tons/sq.in.	El % (A)	R.A. % (Z)	
0·78	2·0	0·45	1·35	19·5	AC1050 T750	20	56·8	56·8	56·8	62·7	16	21	1
"	"	"	"	"	" "	100	48·2	48·2	48·2	57·4	16	25	"
"	"	"	"	"	" "	200	42·6	42·8	43·2	54·8	15	22	"
"	"	"	"	"	" "	300	40·3	40·5	40·5	53·8	15	27	"
"	"	"	"	"	" "	400	33·4	34·8	36·8	49·0	15	28	"
"	"	"	"	"	" "	500	25·3	27·9	30·8	37·9	25	39	"
"	"	"	"	"	" "	600	11·8	14·1	16·5	21·1	44	66	"
"	"	"	"	"	" "	700	—	5·1	6·4	10·1	76	85	"
0·80	2·0	0·50	1·50	20·0	OQ1050 T740	20	49·4		58·0	70·2	16	27	2
"	"	"	"	"	" "	200	44·4		51·8	62·6	14	35	"
"	"	"	"	"	" "	400	39·4		48·3	57·0	14½	31½	"
"	"	"	"	"	" "	500	25·3		34·3	44·0	17	35	"
"	"	"	"	"	" "	600	8·2		13·6	24·4	36½	66	"
"	"	"	"	"	" "	700	3·5		5·5	12·7	60	84½	"

HIGH TEMPERATURE IMPACT PROPERTIES

Chemical Composition (%)					Charpy Mesnager Impact Values (Kg. M) at											Ref.
C	Si	Mn	Ni	Cr	RT	110	190	260	330	390	440	520	600	660	750	
0·80	2·0	0·50	1·50	20·0	1·31	1·70	2·10	2·43	2·62	2·73	2·96	3·18	3·41	4·83	5·67	2

LIST OF REFERENCES

1. Private correspondence with Firth-Vickers Stainless Steels Ltd.
2. Private correspondence with Jessop–Saville Ltd.
3. Private correspondence with Firth Brown Ltd.

FERRITIC CHROMIUM RUST-RESISTING STEEL 17 PER CENT CHROMIUM

SPECIFICATION

CHEMICAL COMPOSITION (%)

C	Si	Mn	S	P	Ni	Cr
0·12 max.	1·00 max.	1·00max.	0·045 max.	0·045 max.	0·50 max.	16·0—18·0

MECHANICAL PROPERTIES

No mechanical properties are specified for this steel.

RELATED SPECIFICATIONS

United Kingdom		America		Germany		USSR	
En No.	BS	SAE	AISI	Werkstoff No.	Name	GOST	Mark
60	1449	51430	430	1.4016 1.4015	X8 Cr17 X8 Cr18	5632-61	Kh17

APPLICATIONS

En 60 has a corrosion resistance higher than En 56A, approaching that of the austenitic stainless steels of En 58. The stee also possesses good resistance to scaling at temperatures up to 800°C. Typical applications are: Motor car trim, domestic and catering equipment, spoons, forks and domestic and industrial heater parts, furnace linings, nitric acid plant and equipment.

WELDING

The welding properties are fair, but there is a tendency for grain growth near the weld, with consequent brittleness.

MACHINABILITY

The machinability is approximately 50% of that for mild steel (En 3).

HOT WORKING AND HEAT TREATMENT TEMPERATURES

Forging, rolling and stamping	1100°C, finish above 900°C.
Softening	700—780°C, air cool.

In general En 60 is not hardenable by heat treatment, but steel with a carbon content above 0·1% may be hardened by air cooling or oil quenching from 980 to 1020°C to a tensile strength of approximately 64 tons/sq.in. (300 HV) but accompanied with low ductility.

PHYSICAL PROPERTIES

SPECIFIC GRAVITY (d)

Chemical Composition (%)				Heat Treatment °C	Specific Gravity at 20°C	Ref.
C	Si	Mn	Cr			
0·06	0·40	0·60	16·5	Softened 750	7·70	1
0·10	0·50	0·30	16·0	Softened	7·73	2
Within specification				"	7·72	3
Within AISI 430 Specification				"	7·70	4 & 5

SPECIFIC HEAT (c)

Chemical Composition	Heat Treatment °C	Mean Specific Heat cal/g.deg. C					Ref.
		20 to 100°C	20 to 200°C	20 to 400°C	20 to 600°C	20 to 800°C	
Within specification	Softened		0·126	0·137	0·157	0·164	3
	"	0·115					6
	"	0·12					2

MEAN COEFFICIENT OF THERMAL EXPANSION (α)

Chemical Composition (%)					Heat Treatment °C	10^6 × Mean Coefficient of Thermal Expansion per deg. C								Ref.
C	Si	Mn	Ni	Cr		20 to 100°C	20 to 200°C	20 to 300°C	20 to 400°C	20 to 500°C	20 to 600°C	20 to 700°C	20 to 800°C	
0·06	0·40	0·60		16·5	Softened 750°C	10·0		11·0		12·0				1
0·07	0·20	0·25		17·0	Softened	10·7	10·9		11·3		12·0	12·6		7
0·08	0·49	0·44	0·44	17·2	Softened 750°C	10·4	10·7	11·0	11·3	11·5	11·8	12·0		1
Within specification							10·6		11·4		12·0		12·7	3
Within AISI 430 Specification					Softened	9·2							11·9	5
" " "					"	10·4 (0—100°C)					11·9 (0—650°C)			4
" " "					"		10·2	11·0	11·2	11·6	11·9			8

ELECTRICAL RESISTIVITY (ρ)

Chemical Composition (%)					Heat Treatment °C	Electrical Resistivity microhm-cm									Ref.
C	Si	Mn	Ni	Cr		20°C	100°C	200°C	300°C	400°C	500°C	600°C	700°C	800°C	
0·06	0·30	0·96	0·36	16·7		59·1	66·0	74·9	83·0	91·0	99·0	106·8	114·0	118·0	9
0·06	0·40	0·60		16·5	Softened 750	62·0									1
0·06	0·51		0·19	17·6	WQ780	62·4									"
0·07	0·28		0·19	16·2	WQ780	55·1									"
0·07	0·35		0·19	17·3	WQ780	58·5									"
Within specification					Softened	58·0	65·4	73·6							3
Within AISI 430 Specification					"	60·0									4 & 5

THERMAL CONDUCTIVITY (k)

Chemical Composition (%)					Heat Treatment °C	Thermal Conductivity cal/cm sec. deg. C										Ref.
C	Si	Mn	Ni	Cr		20°C	100°C	200°C	300°C	400°C	500°C	600°C	700°C	800°C	850°C	
0·06	0·30	0·96	0·36	16·7		0·058	0·057	0·056	0·056	0·055	0·055	0·055	0·055	0·058	0·060	9
0·06	0·40	0·60		16·5	Softened 750	0·052										1
0·10	0·50	0·30		16·0	Softened	0·055										2
Within specification					"		0·056	0·057	0·058	0·059	0·061	0·062	0·063			3
Within AISI 430 Specification					"	0·062					0·063					4

YOUNG'S MODULUS (E)

Chemical Composition (%)	Young's Modulus tons/sq.in.	Ref.
Within specification	13,400	6
Within AISI 430 Specification	12,900	4 & 5

MAGNETIC PROPERTIES

Chemical Composition	Heat Treatment °C	Magnetic Induction B (Gauss) for magnetizing field H Oersted =						Remanence Brem	Coercive Force H_c	Permea-bility max.	Ref.
		5	10	25	50	100	250				
Within specification	Softened 750°C AC	680	4,050	8,900	12,100	13,600	15,600	6,900	10		3
"	"						16,500	5,000	9	290	1

MECHANICAL PROPERTIES AT ROOM TEMPERATURE

Chemical Composition (%)						Section Size dia. (in.)	Heat Treatment °C	Tensile				Izod ft.lb.	Ref.
C	Si	Mn	Ni	Cr	N_2			Y.P. (Rs) tons/sq.in.	T.S. (Rm) tons/sq.in.	El % (A)	R.A. % (Z)		
0·06	0·40	0·60		16·5			AC750	22·0	35·0	32	50		1
0·06				17·2	0·11	1	3hrs. 750 AC	28·1	42·2	27	61	62	10
0·07				16·4	0·12	1	" " "	27·2	41·1	27	66	86	10
0·07				17·9	0·32	1	" " "	19·9	32·1	29	60	42	"
0·08				18·5	0·17	1	" " "	30·4	46·0	25	57	61	10
0·10				16·2	0·03	1	" " "	18·7	32·1	30	56	60	10
0·10				17·2	0·031	1	" " "	21·1	34·4	28	56	48	"
0·10	0·72	0·22	0·18	17·5		$\frac{7}{8}$	As rolled	28·5	38·1	24	41		1
"	"	"	"	"		"	Muffle cool 800	22·4	34·4	33½	68		"
"	"	"	"	"		"	OQ1100 T600	32·0	42·4	24	50		"
"	"	"	"	"		"	" T700	27·6	37·4	28½	53		"
"	"	"	"	"		"	" T800	22·1	33·5	30	58		"
0·10	0·17	0·63	0·14	16·1			As rolled	21·0	41·6	12	25	7	"
"	"	"	"	"			Muffle cool 950	19·0	33·0	39	67	44	"
"	"	"	"	"			OQ975 T3hrs. 750	21·2	35·6	36	68½	90	"
0·10	0·67	0·89	0·14	16·2			As rolled	24·8	35·0	16	20	1	"
"	"	"	"	"			Muffle cool 950	22·2	34·4	33	61	8	"
"	"	"	"	"			OQ950 T3hrs. 750	22·2	33·8	31½	63	40	"
Within AISI 430 Specification							Annealed	20·1	33·5	22	65		4
" " " "							Annealed	15·6	26·8	20	40	3 to 85	5
						Sheet				on 2'' / on 4'' / on 8''			
0·09	0·40	0·49		16·5		0·18 thick	Hot rolled	46·6	59·2	12 / 7½ / 4½			11
"	"	"		"		"	Annealed	22·0	31·2	31 / 23 / 20			"
"	"	"		"		0·075	Hard temper	50·0	52·0	6 / 3½ / 2			"
"	"	"		"		"	Soft "	24·2	34·9	25 / 22 / 16½			"
"	"	"		"		0·047	Hard "	47·6	49·5	4 / 2½ / 1			"
"	"	"		"		"	Soft "	20·7	33·0	30 / 24 / 20			"
"	"	"		"		0·031	Hard "	47·8	50·0	3 / 1½ / 1			"
"	"	"		"		"	Soft "	21·1	31·7	30 / 25 / 20			"

MECHANICAL PROPERTIES AT LOW TEMPERATURE

Chemical Composition (%) C	Si	Mn	Ni	Cr	Section Size dia. (in.)	Heat Treatment °C	Test Temperature °C	Tensile Y.P. (Rs) tons/sq.in.	Tensile T.S. (Rm) tons/sq.in.	El % (A)	R.A. % (Z)	Izod ft.lb.	Ref.
Within AISI 430 Specification						Softened	+21	17·0	29·0	37	73	35	5
" " " "						"	0	17·9	30·8	37	72	20	"
" " " "						"	—40	18·3	33·9	36	72	10	"
" " " "						"	—62	19·6	36·2	36	70	8	"
" " " "						"	—196	38·8	40·2	0	4	2	"

MECHANICAL PROPERTIES AT HIGH TEMPERATURE

Chemical Composition (%) C	Si	Mn	Ni	Cr	Section Size dia. (in.)	Heat Treatment °C	Test Temperature °C	Tensile L.P. tons/sq.in.	Tensile T.S. (Rm) tons/sq.in.	El % (A)	R.A. % (Z)	Ref.
0·09	0·84	0·34	0·19	16·7	Tube	Softened	21	17·0	34·5	38		12
"	"	"	"	"	"	"	482	9·1	25·1	23		"
"	"	"	"	"	"	"	538	7·6	19·7	25½		"
"	"	"	"	"	"	"	593	5·9	15·3	31½		"
"	"	"	"	"	"	"	677	2·2	7·2	49½		"
"	"	"	"	"	"	"	816	0·6	3·2	71		"
Within AISI 430 Specification						Softened	RT		33·5			4
" " " "						"	93		31·7			"
" " " "						"	204		29·5			"
" " " "						"	316		27·2			"
" " " "						"	427		24·6			"
" " " "						"	482		21·7			"
" " " "						"	593		12·9			"
" " " "						"	704		6·2			"
" " " "						"	816		2·9			"
" " " "							871		1·6			"

CREEP AND RUPTURE PROPERTIES

Chemical Composition	Heat Treatment	Test Temperature °C	Stress (tons/sq.in.) for 1% Creep Strain		Stress (tons/sq.in.) for Rupture		Ref.
			in 10,000 hrs.	100,000 hrs.	in 1,000 hrs.	10,000 hrs.	
Within AISI 430 Specification	Softened	482	6·8		13·4	10·7	4
" " " "	"	538	3·7	3·1	7·8	6·0	"
" " " "	"	593	2·2	1·8	4·1	2·9	"
" " " "	"	649	1·0	0·7	2·2	1·5	"
" " " "	"	704	0·6	0·4			"
" " " "	"	760	0·4	0·3			"

LIST OF REFERENCES

1. Private correspondence with Firth-Vickers Stainless Steel Co. Ltd.
2. Private correspondence with Richard Thomas and Baldwins Ltd.
3. Private correspondence with The United Steel Companies Ltd.
4. Data published by Atlas Steel Ltd. U.S.A.
5. Data published by Allegheny Ludlum Steel Corporation.
6. Private correspondence with Brown Bayley Steels Ltd.
7. Private correspondence with Samuel Osborn and Co. Ltd.
8. Data published by the United States Steel Corporation.
9. Private correspondence with the De Havilland Aircraft Co. Ltd.
10. R. Franks, Chromium Steels of High Nitrogen Content. *Trans. Amer. Soc. Metals*, vol. 23, December, 1935, 968—994.
11. *The Book of Stainless Steels*. Edited by E. E. Thum. Amer. Soc. Metals, 2nd Edition, 1935, 328—334.
12. C. L. Clarke and A. E. White. The Stability of Metals at Elevated Temperatures. University of Michigan, Dept. of Eng. Research *Bulletin*, No. 11, 1928.

FERRITIC CHROMIUM RUST-RESISTING STEEL 20 PER CENT CHROMIUM

SPECIFICATION

CHEMICAL COMPOSITION (%)

C	Si	Mn	S	P	Ni	Cr
0·12 max.	1·00 max.	1·00 max.	0·045 max.	0·045 max.	0·50 max.	20·0—22·0

MECHANICAL PROPERTIES

No mechanical properties are specified for this steel.

RELATED SPECIFICATIONS

United Kingdom	America	
En No.	SAE	AISI
61	51442	442

APPLICATIONS

En 61 has similar characteristics to those of En 60, and is used for applications requiring greater oxidation and corrosion resistance at temperatures up to 900°C where strength is not of considerable importance. Typical applications are similar to En 60, e.g. oil burner combustion chambers, furnace and stack dampers, boiler baffles.

WELDING

The welding properties are poor, the welds tending to be brittle.

MACHINABILITY

The machinability is approximately 50% of that for mild steel (En 3).

HOT WORKING AND HEAT TREATMENT TEMPERATURES

Forging, rolling and stamping	1100°C, finish above 900°C.
Softening	700°C—780°C, air cool or oil quench (or water quench for thick sections.)

This steel cannot be hardened by heat treatment.

PHYSICAL PROPERTIES

SPECIFIC GRAVITY (d)

Chemical Composition (%)					Heat Treatment °C	Specific Gravity at 20°C	Ref.
C	Si	Mn	Ni	Cr			
0·07	0·40	0·80		21·0	Softened 750	7·66	1
0·09	0·37	0·26	0·14	22·7	AC 950	7·66	2
0·12	1·00	0·90		21·0	Softened	7·66	3

SPECIFIC HEAT (c)

Chemical Composition (%)					Heat Treatment °C	Mean Specific Heat cal/g. deg. C								Ref.
C	Si	Mn	Ni	Cr		20 to 100°C	20 to 200°C	20 to 300°C	20 to 400°C	20 to 500°C	20 to 600°C	20 to 700°C	20 to 800°C	
0·07	0·40	0·80		21·0	Softened 750	0·115								1
0·09	0·37	0·26	0·14	22·7	AC950			0·132		0·143		0·160	0·160	2
Within specification					Softened		0·129		0·136		0·152		0·160	"

MEAN COEFFICIENT OF THERMAL EXPANSION (α)

Chemical Composition (%)					Heat Treatment °C	10^6 × Mean Coefficient of Thermal Expansion per deg. C										Ref.
C	Si	Mn	Ni	Cr		20 to 100°C	20 to 200°C	20 to 300°C	20 to 400°C	20 to 500°C	20 to 600°C	20 to 700°C	20 to 800°C	20 to 900°C	20 to 1000°C	
0·08	0·35	0·80	0·35	20·8	AC750	10·2	10·5	10·8	11·2	11·4	11·6	11·9	12·5	13·1		1
0·09	0·37	0·26	0·14	22·7	AC950				10·0		10·7		11·8			2
0·12	1·0	0·90		21·0	Softened	10·6	10·8		11·1		11·6		12·4		13·5	3
Within specification					Softened		10·5	10·8	11·1	11·4	11·6	11·9	12·4	13·0	13·6	2

ELECTRICAL RESISTIVITY (ϱ)

Chemical Composition (%)					Heat Treatment °C	Electrical Resistivity microhm-cm			Ref.
C	Si	Mn	Ni	Cr		20°C	100°C	200°C	
0·04	0·24		0·16	21·7	WQ820	56·6			1
0·04	0·30		0·22	21·0	"	58·5			"
0·05	0·31		0·52	20·6	"	60·8			"
0·06	0·24		0·18	20·8	"	56·6			"
0·06	0·37		0·32	20·7	"	59·1			"
0·07	0·40	0·80		21·0	AC750	62·0			"
Within specification					Softened	57·0	66·0	76·5	2

THERMAL CONDUCTIVITY (k)

Chemical Composition (%)				Heat Treatment °C	Thermal Conductivity cal/cm sec. deg. C											Ref.
C	Si	Mn	Cr		20°C	100°C	200°C	300°C	400°C	500°C	600°C	700°C	800°C	900°C	1000°C	
0·07	0·30	0·60	20·6		0·056	0·056	0·056	0·056	0·057	0·058	0·059	0·060	0·063	0·066	0·070	1
0·07	0·04	0·80	21·0	AC750	0·052											1
Within specification				Softened		0·051	0·053	0·055	0·057	0·058	0·060	0·062	0·064			2

YOUNG'S MODULUS (E)

Chemical Composition (%)				Heat Treatment °C	Young's Modulus, tons/sq.in.							Ref.
C	Si	Mn	Cr		20°C	100°C	200°C	300°C	400°C	500°C	600°C	
0·07	0·04	0·80	21·0	AC750	13,800	13,500	13,100	12,600	12,100	11,450	10,700	1

MECHANICAL PROPERTIES AT ROOM TEMPERATURE

Chemical Composition (%)							Section Size dia. (in.)	Heat Treatment °C	Tensile							Ref.
C	Si	Mn	S	P	Ni	Cr			0·1% P.S. (0·1Rp) tons/sq.in.	0·2% P.S. (0·2Rp) tons/sq.in.	0·5% P.S. (0·5Rp) tons/sq.in.	Y.P. (Rs) tons/sq.in.	T.S. (Rm) tons/sq.in.	El % (A) $e=4\sqrt{A}$	R.A. % (Z)	
0·03	0·36	0·59	0·021	0·020	0·14	20·0	CR sheet	8 min. 850 WQ	18·4		19·1		29·8	45		1
"	"	"	"	"	"	"	" "	30 " 850 "	18·6		19·5		29·2	44		"
"	"	"	"	"	"	"	" "	8 " 800 "	21·0		22·5		31·4	31		"
"	"	"	"	"	"	"	" "	30 " 800 "	18·3		18·7		29·9	45		"
0·03	0·37	0·66	0·033	0·018	0·14	20·3	CR sheet	8 min. 850 WQ	19·7		20·0		31·1	39		"
"	"	"	"	"	"	"	" "	30 " 850 "	19·3		19·7		30·4	48		"
0·03	0·16	0·65	0·024	0·014	0·23	19·7	¾	½ hr. 830 WQ	15·8	16·4	17·3		26·3	49½	77½	"
"	"	"	"	"	"	"	"	CD 20%	28·8	31·1	31·9		31·9	33½	73	"
0·06	0·25	0·73	0·017	0·016	0·21	20·8	¾	2 hrs. 800 WQ	18·9	19·8	19·9		30·7	43½	76	"
"	"	"	"	"	"	"	"	CD 37·8%	36·3	41·8	43·4		43·4	20	61	"
"	"	"	"	"	"	"	"	CD 61·1%	42·4	45·5	46·0		46·6	19	61½	"
0·06	0·25	0·70	0·017	0·015	0·21	20·7	¾	2 hrs. 800 WQ	21·4	21·9	22·1		31·4	41	73	"
0·07	0·40	0·80				21·0		AC750	25·4	25·9	26·4		35·2	34½	60	"
0·08						21·0		Softened			29·2		37·7	25	—	2
Not stated							1⅛					21·8	28·2	35½	63	5

MECHANICAL PROPERTIES AT HIGH TEMPERATURE

Chemical Composition (%)				Section Size dia. (in.)	Heat Treatment °C	Test Temperature °C	Tensile						Ref.
C	Si	Mn	Cr				0·1% P.S. (0·1Rp) tons/sq.in.	0·2% P.S. (0·2Rp) tons/sq.in.	0·5% P.S. (0·5Rp) tons/sq.in.	T.S. (Rm) tons/sq.in.	El % (A)	R.A. % (Z)	
0·07	0·40	0·80	21·0		AC750	20	25·4	25·9	26·4	35·2	34½	60	1
"	"	"	"		"	100	22·6	23·0	23·8	31·2	32	62	"
"	"	"	"		"	200	21·3	21·6	22·0	30·1	26½	58½	"
"	"	"	"		"	300	19·6	20·2	21·5	29·0	24	58½	"
"	"	"	"		"	400	18·4	19·4	20·7	28·1	24	58½	"
"	"	"	"		"	500	17·4	18·4	19·6	25·5	24	49	"
"	"	"	"		"	600	11·8	12·6	13·5	14·2	52	79½	"
"	"	"	"		"	700				6·9	59	95	"
"	"	"	"		"	800				4·5	60	96½	"

Rate of straining 0·0085 in./min. to YP then 0·125 in./min.

FATIGUE PROPERTIES

Chemical Composition (%) C	Chemical Composition (%) Si	Chemical Composition (%) Ni	Chemical Composition (%) Cr	Section Size dia. (in.)	Heat Treatment °C	Test Temperature °C	Tensile 0·1% P.S. (0·1Rp) tons/sq.in.	Tensile 0·2% P.S. (0·2Rp) tons/sq.in.	Tensile 0·5% P.S. (0·5Rp) tons/sq.in.	Tensile T.S. (Rm) tons/sq.in.	El % (A)	R.A. % (Z)	Reversed Axial Fatigue Limit (10^7 cycles) tons/sq.in.	Ref.
0·05	0·25	0·21	20·8	¾	2 hrs. 800WQ	RT	21·0	21·4	21·7	30·9	39	72	±13	1

LIST OF REFERENCES

1. Private correspondence with Firth-Vickers Stainless Steels Ltd.
2. Private correspondence with The United Steel Companies Ltd.
3. Private correspondence with Samuel Osborn and Co. Ltd.
4. E. Maures. Heat Conductivity of Steels containing Chromium at High Temperatures. *Arch. für das Eisenhüttenwesen*, vol. 10, 1936, 145—154.
5. Private correspondence with Hall and Pickles Ltd.

LOW ALLOY STEEL

SPECIFICATION

CHEMICAL COMPOSITION (%)

	C	Si	Mn	S	P	Ni	Cr	Mo
En 100	0·35—0·45	0·50max.	1·20—1·50	0·050 max.	0·050 max.	0·50—1·00	0·30—0·60	0·15—0·25
En 100A[1]	0·25—0·30	0·10—0·50	1·20—1·50	0·050 max.	0·050 max.	0·50—1·00	0·30—0·60	0·15—0·25
En 100B[1]	0·30—0·35	0·10—0·50	1·20—1·50	0·050 max.	0·050 max.	0·50—1·00	0·30—0·60	0·15—0·25
En 100C[1]	0·35—0·40	0·10—0·50	1·20—1·50	0·050 max.	0·050 max.	0·50—1·00	0·30—0·60	0·15—0·25
En 100D[1]	0·40—0·45	0·10—0·50	1·20—1·50	0·050 max.	0·050 max.	0·50—1·00	0·30—0·60	0·15—0·25
En 100E[2]	0·25—0·35	0·10—0·50	1·20—1·50	0·050 max.	0·050 max.	0·50—1·00	0·30—0·60	0·15—0·25

(1) Supplied to composition only. Mechanical properties are not part of the contractual obligations.

(2) For small ruling sections or lower tensile ranges or when the steel is to be hardened in water, by agreement between the purchaser and the manufacturers.

MECHANICAL PROPERTIES

Property	Hardened and Tempered Condition				
	R	S	T	U	V
Limiting ruling sections, in.	6	4	$2\frac{1}{2}$	$1\frac{1}{8}$	$1\frac{1}{8}$
Tensile strength, tons/sq.in. min. (Rm)	45	50	55	60	65
Yield stress, tons/sq.in. min.[1] (RS)	34	38	44	48	52
0·2% Proof stress, tons/sq.in. min.[2] (0·2Rp)	32	36	41	46	50
Elongation, %, min. (A)	22	20	18	17	16
Izod impact value, ft. lb., min.	40	40	40	35	35
Brinell hardness number,[3] HB	201/255	223/277	248/302	269/321	293/341

(1) Not to be used as acceptance values except by special arrangement between the purchaser and manufacturer.

(2) When specifically requested in the enquiry or order.

(3) For information purposes only; not contractual part of the specification.

RELATED SPECIFICATIONS

United Kingdom		Germany	
En No.	BS Air	Werkstoff No.	Name
100	2S94	1·6511	36 Cr Ni Mo4
100B		1·6511	36 Cr Ni Mo4
100C	2S94	1·6511	36 Cr Ni Mo4
100D	2S94		

APPLICATIONS

A low-alloy steel capable of being heat treated to produce tensile strengths of the order of 60 tons/sq.in. in small sections, and 50 tons/sq.in. in medium sections combined with good ductility and resistance to shock. Typical applications are for general engineering components, high tensile bolts and studs, crankshafts, gears, boring bars, lead screws, shafting, milling and boring cutter bodies, collets.

WELDING

(See Introduction for key to symbols used in this table.)

	Welding Process					
	MA	G	IG	S	F	B
En 100	d	bx	g	e	b	j

Reproduced by kind permission of H.M.S.O.

MACHINABILITY

In the softened condition the machinability is approximately 60% of that for mild steel. In the hardened and tempered condition the machinability is 45 to 55% of that for mild steel (En 3).

HOT WORKING AND HEAT TREATMENT TEMPERATURES

Forging, rolling and stamping	1200°C, finish above 900°C.
Softening (sub-critical annealing)	650°C—690°C, air cool or oil quench.
Hardening	830°C—860°C, oil quench.*
Tempering	550°C—660°C, air cool or oil quench.

* En 100E may be water quenched when stated on order.

PHYSICAL PROPERTIES

SPECIFIC GRAVITY (d)

Chemical Composition (%)								Heat Treatment °C	Specific Gravity at 20°C	Ref.
C	Si	Mn	S	P	Ni	Cr	Mo			
0·39	0·24	1·35	0·042	0·040	0·65	0·48	0·17	OQ850 T620 1hr.	7·86	1
Within specification								Softened	7·85	2
" "									7·84	3

SPECIFIC HEAT (c)

Chemical Composition	Mean Specific Heat cal/g.deg. C						Ref.
	20 to 100°C	20 to 200°C	20 to 300°C	20 to 400°C	20 to 500°C	20 to 600°C	
Within specification	0·115						1
	0·116	0·119	0·122	0·125	0·130	0·137	*

* These values have been estimated by comparison of published results for steels of near composition to En 100.

MEAN COEFFICIENT OF THERMAL EXPANSION (α)

Chemical Composition (%)								Heat Treatment °C	10^6 × Mean Coefficient of Thermal Expansion per deg. C							Ref.
C	Si	Mn	S	P	Ni	Cr	Mo		20 to 100°C	20 to 200°C	20 to 300°C	20 to 400°C	20 to 500°C	20 to 600°C	20 to 700°C	
0·39	0·24	1·35	0·042	0·042	0·65	0·48	0·17	OQ850 T620	12·0	12·7	13·4	14·0	14·3	14·7		1
Within specification								Softened	11·0		13·0		14·0		15·0	2
" "									12·4		13·4		14·2		14·7 (680°C)	3

ELECTRICAL RESISTIVITY (ρ)

Chemical Composition (%)								Heat Treatment °C	Electrical Resistivity microhm-cm							Ref.
C	Si	Mn	S	P	Ni	Cr	Mo		20°C	100°C	200°C	300°C	400°C	500°C	600°C	
0·39	0·24	1·35	0·042	0·042	0·65	0·48	0·17	OQ850 T620	24·7	28·2	34·0	41·9	52·0	60·0	74·7	1

YOUNG'S MODULUS (E)

Chemical Composition (%)								Heat Treatment °C	Young's Modulus* tons/sq.in.							Ref.
C	Si	Mn	S	P	Ni	Cr	Mo		20°C	100°C	200°C	300°C	400°C	500°C	600°C	
0·39	0·24	1·35	0·042	0·040	0·65	0·48	0·17	OQ850 T620	13,700	13,300	13,050	12,550	12,100	11,500	10,650	1

* Vibration method.

TRANSFORMATION CHARACTERISTICS

ISOTHERMAL TRANSFORMATION

Chemical Composition (%)							
C	Si	Mn	S	P	Ni	Cr	Mo
0·40	0·21	1·34	0·027	0·028	1·03	0·53	0·22

Acid Open Hearth.
McQuaid Ehn Grain Size 5 (ASTM).
As quenched Grain Size 6 (ASTM).
Bars hot rolled to ⅜ in. dia., cold drawn to 11 SWG with intermediate annealing at 600°C.
Austenitizing Temperature 845°C, 30 min. (Ref. 4)

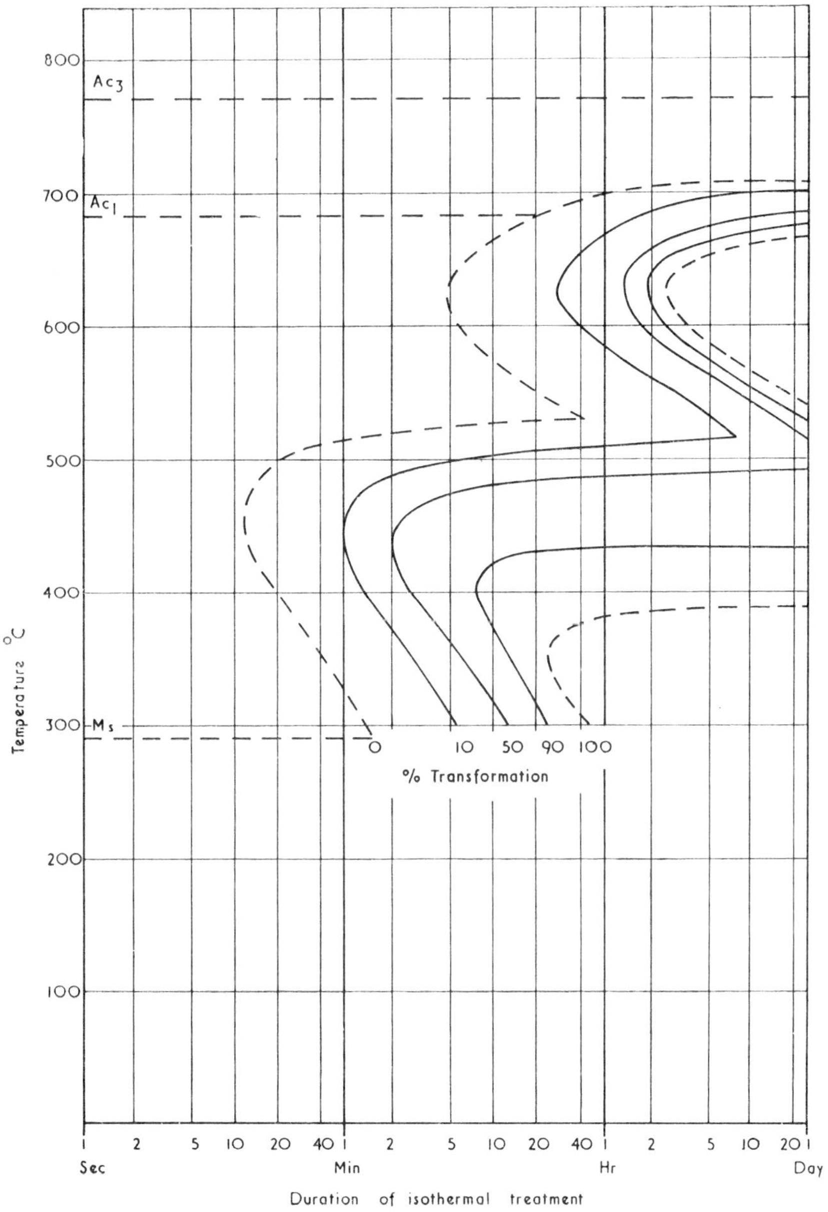

Fig. 100.1

En 100

ISOTHERMAL TRANSFORMATION (Continued)

Chemical Composition (%)							
C	Si	Mn	S	P	Ni	Cr	Mo
0·40	0·24	1·38	0·031	0·033	0·74	0·53	0·16

Acid Open Hearth.
McQuaid Ehn Grain Size 4 (ASTM).
As quenched Grain Size 8 (ASTM).
Austenitizing Temperature 860°C.

(Ref. 5)

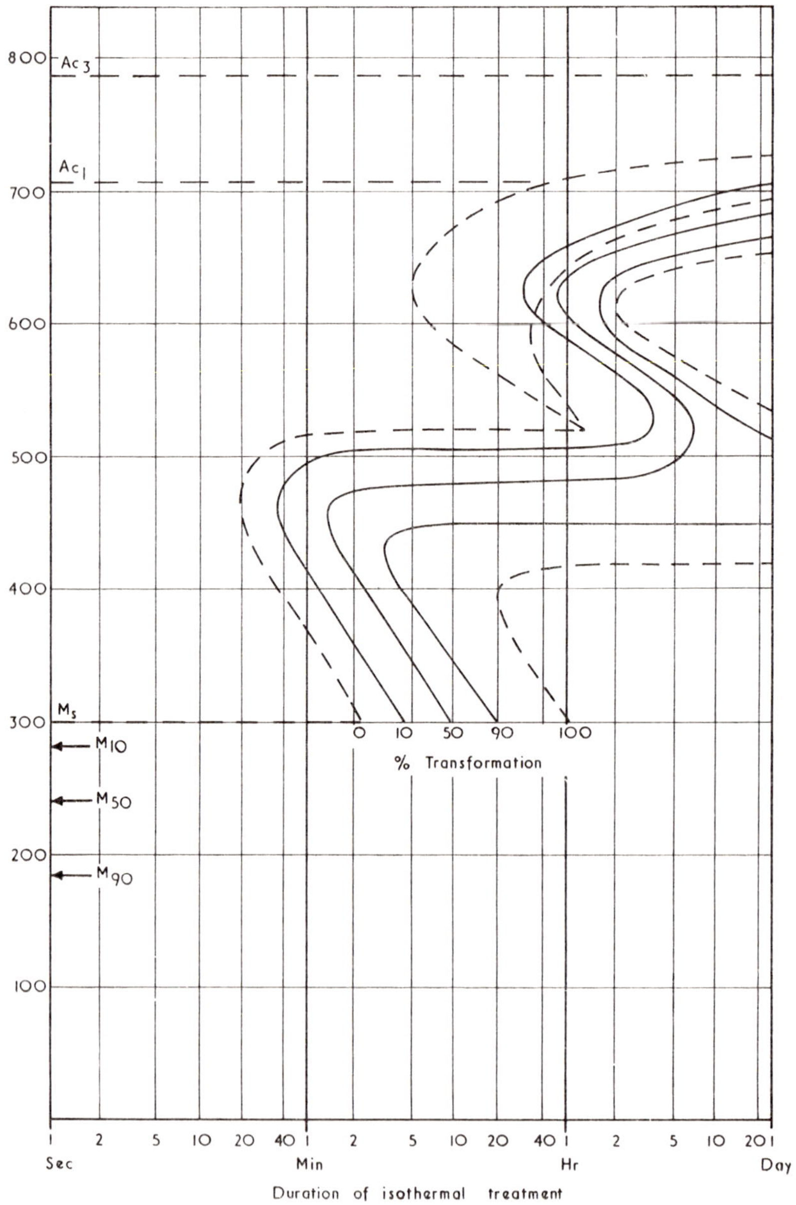

Fig. 100.2

CONTINUOUS COOLING TRANSFORMATION

Chemical Composition (%)							
C	Si	Mn	S	P	Ni	Cr	Mo
0·40	0·24	1·38	0·031	0·033	0·74	0·53	0·16

Acid Open Hearth.
McQuaid Ehn Grain Size 4 (ASTM).
As quenched Grain Size 8 (ASTM).
Austenitizing Temperature 860°C.

(Ref. 5)

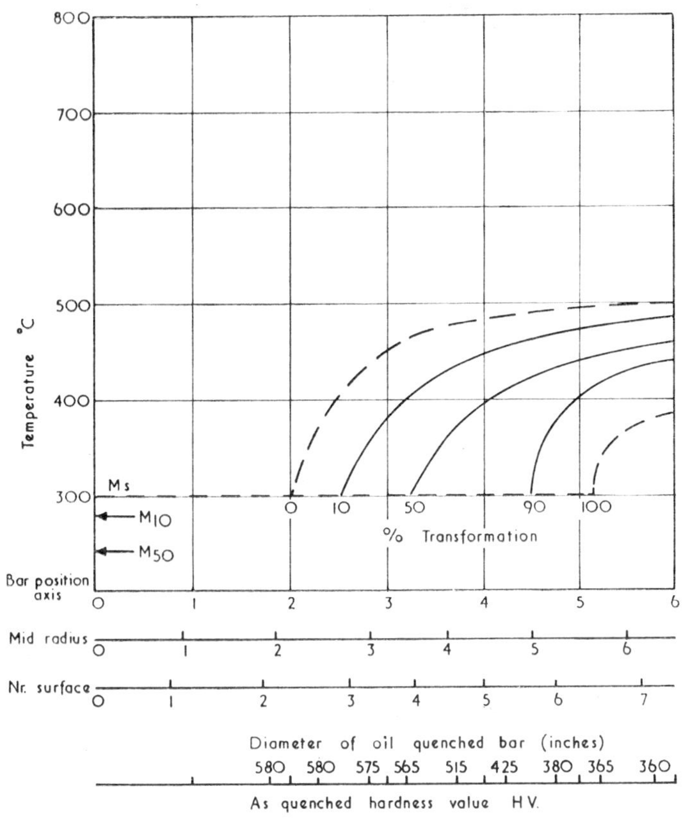

Fig. 100.3

CONTINUOUS COOLING TRANSFORMATION (Continued)

Air cooled from 850°C.

(Ref.1)

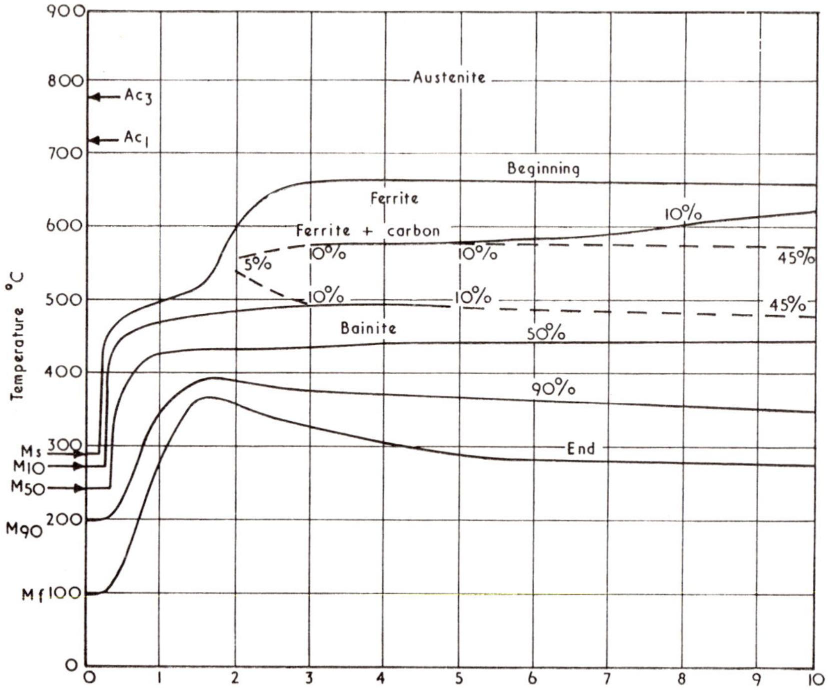

Bar diameter – equivalent cooling rate – inches

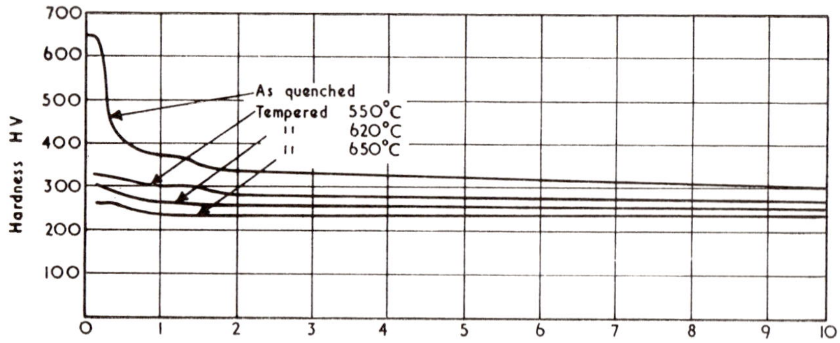

Hardness after transformation

Fig. 100.4

CONTINUOUS COOLING TRANSFORMATION (Continued)

Oil quenched from 850°C. (Ref. 1)

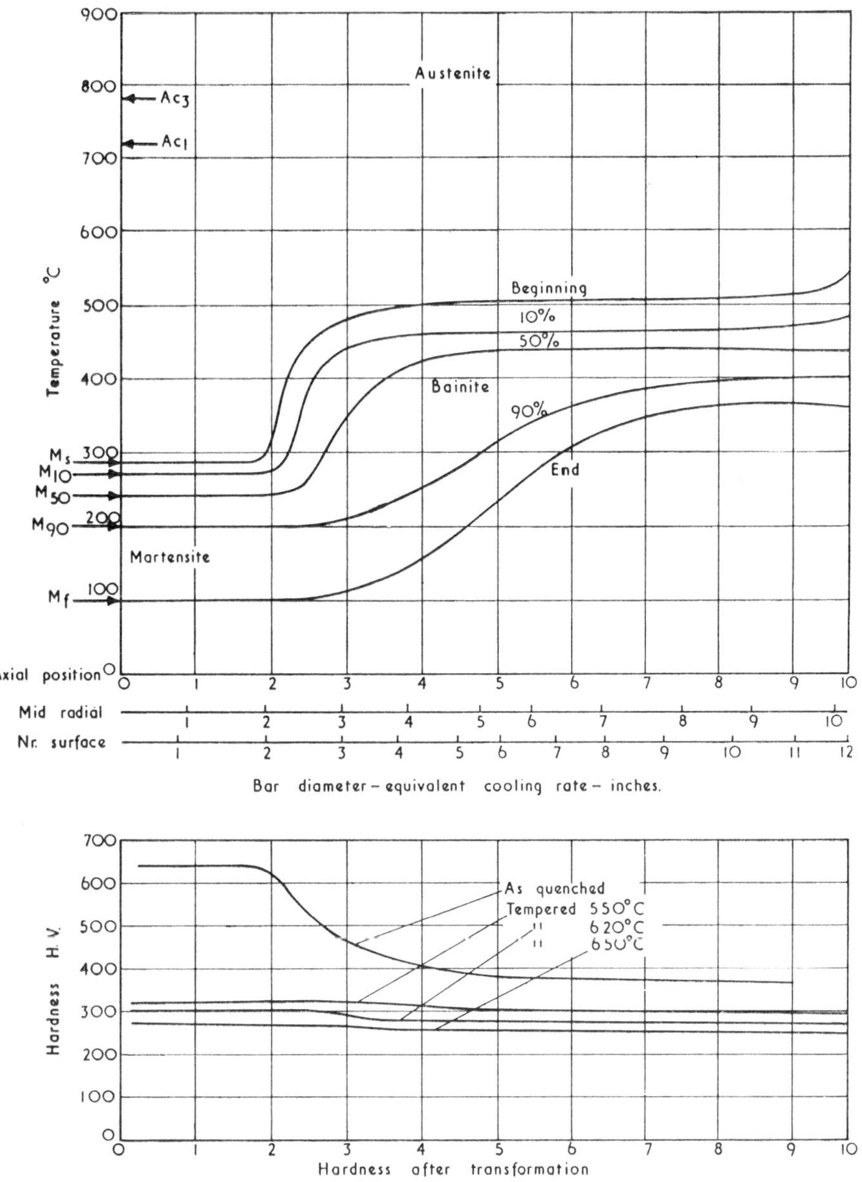

Fig. 100.5

CONTINUOUS COOLING TRANSFORMATION (Continued)

Water quenched from 850°C.

(Ref. 1)

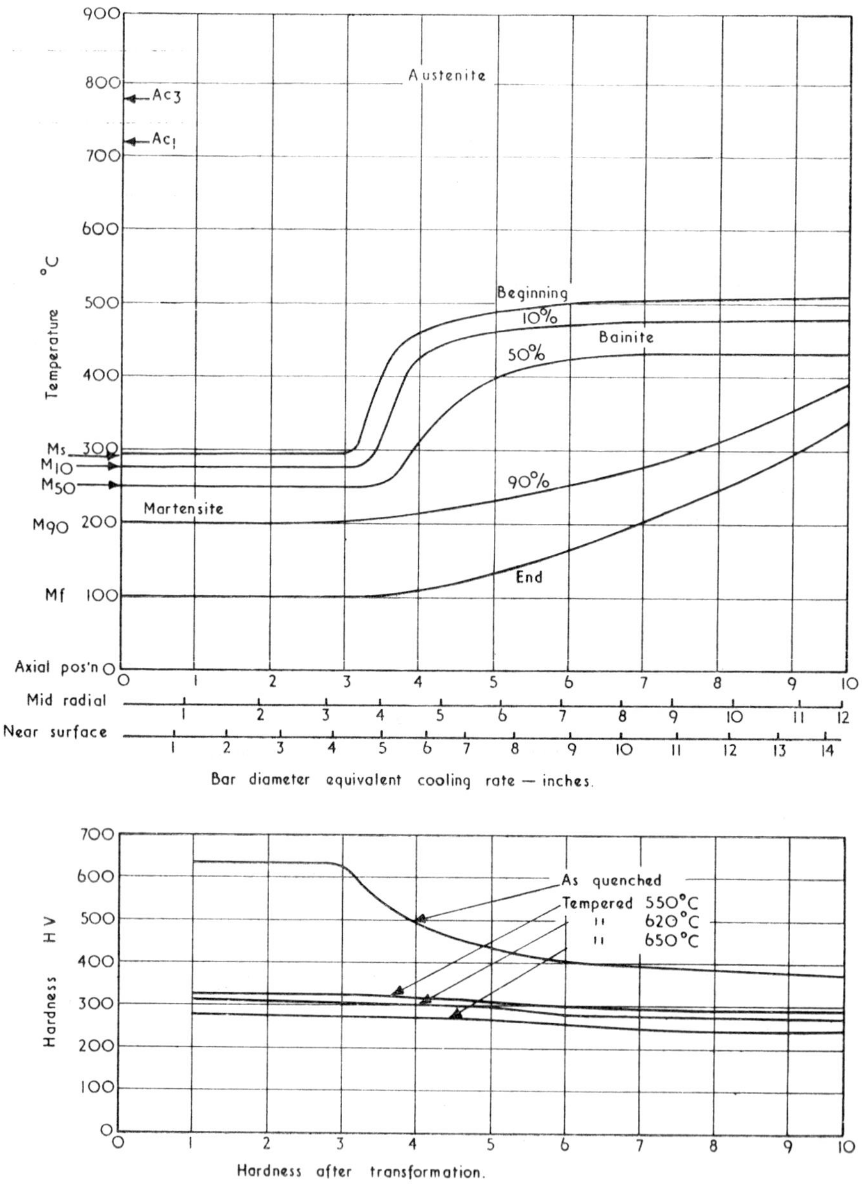

Fig. 100.6

HARDENABILITY

END QUENCH HARDENABILITY DATA

Chemical Composition (%)							
C	Si	Mn	S	P	Ni	Cr	Mo
0·40	0·21	1·34	0·027	0·028	1·03	0·53	0·22

Acid Open Hearth.
McQuaid Ehn Grain Size 5 (ASTM).
As quenched Grain Size 6 (ASTM).
End quenched from 845°C. (Ref. 4)

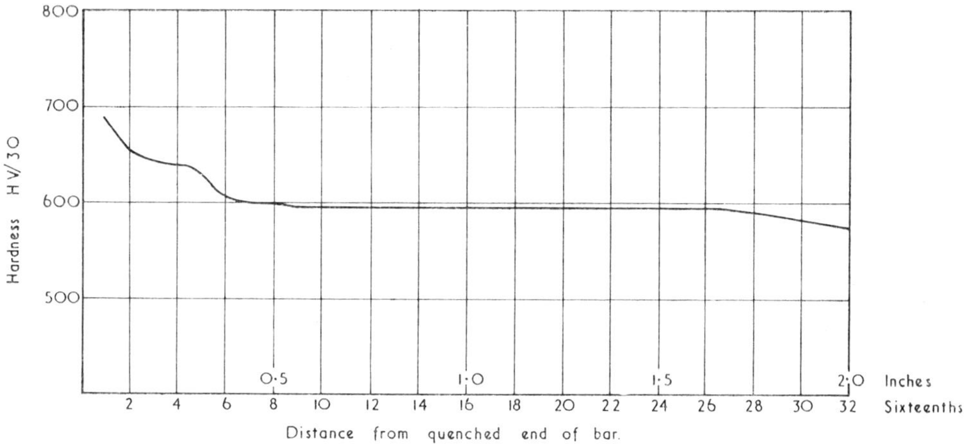

Fig. 100.7

Chemical Composition (%)							
C	Si	Mn	S	P	Ni	Cr	Mo
0·40	0·24	1·38	0·031	0·033	0·74	0·53	0·16

Acid Open Hearth.
McQuaid Ehn Grain Size 4 (ASTM).
As quenched Grain Size 8 (ASTM).
End quenched 860°C. Tempered as indicated. (Ref. 5)

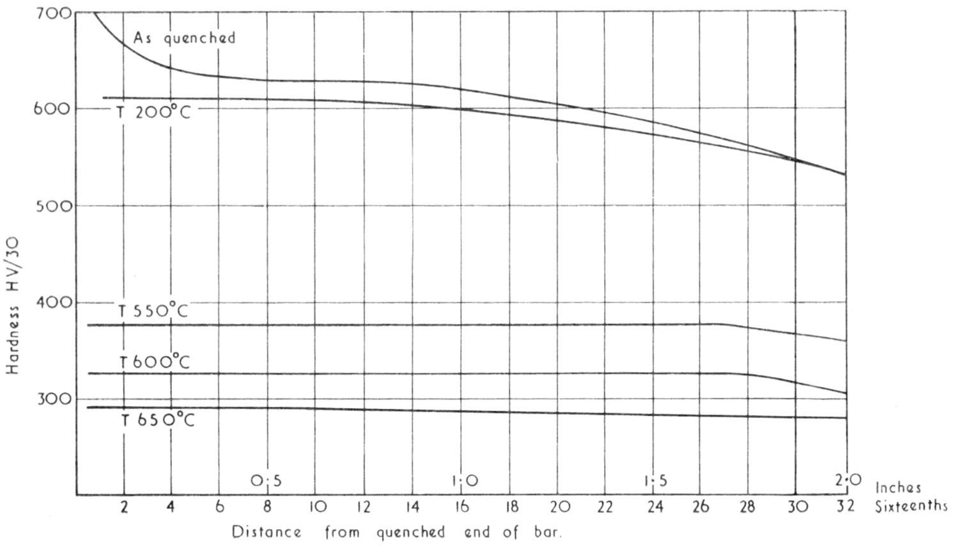

Fig. 100.8

END QUENCH HARDENABILITY DATA (Continued)

Curve No.	Chemical Composition (%)								McQuaid Ehn Grain Size (ASTM)
	C	Si	Mn	S	P	Ni	Cr	Mo	
A	0·39	0·30	1·36	0·034	0·036	1·08	0·35	0·21	6—7
B	0·40	0·30	1·43	0·035	0·034	0·88	0·38	0·20	6—7
C	0·40	0·27	1·34	0·026	0·035	0·85	0·30	0·16	3—4
D	0·39	0·29	1·45	0·031	0·035	0·87	0·34	0·17	5—6
E	0·39	0·29	1·45	0·033	0·038	0·90	0·38	0·18	6
F	0·41	0·26	1·35	0·035	0·038	1·02	0·44	0·18	3—5

End quenched 850°C. (Ref. 6)

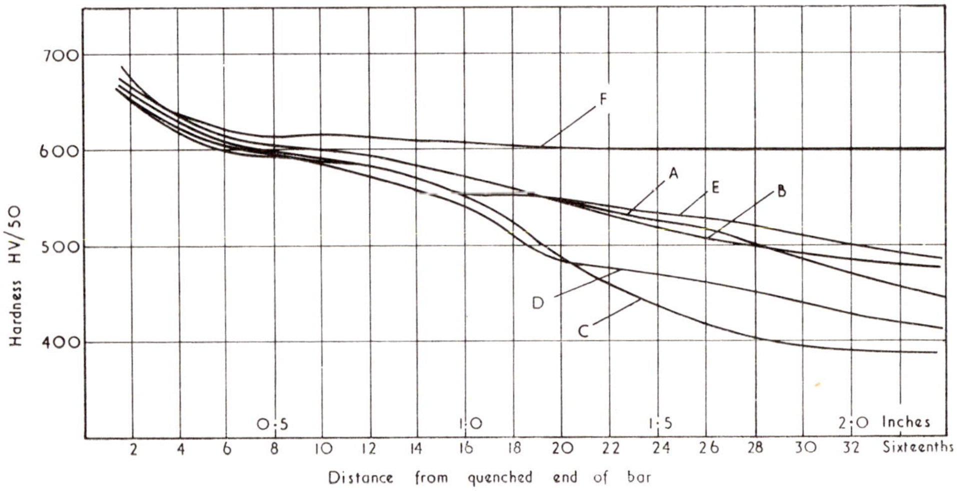

Fig. 100.9

End quenched 850°C. Tempered 640°C. (Ref. 6)

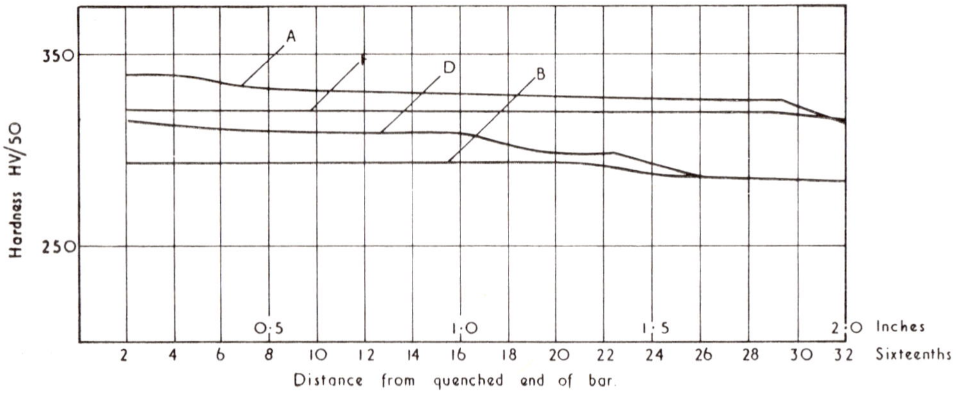

Fig. 100.10

END QUENCH HARDENABILITY DATA (Continued)

Curve No.	Chemical Composition (%)								McQuaid Ehn Grain Size (ASTM)	Pre-treatment °C	End quenched from °C
	C	Si	Mn	S	P	Ni	Cr	Mo			
G	0·40	0·22	1·33	0·024	0·018	0·65	0·51	0·20	1—6	N850	850
H	0·39	0·17	1·37	0·013	0·015	0·90	0·44	0·18	1—6	"	"
J	0·39	0·16	1·33	0·017	0·011	0·65	0·50	0·22	1—6	"	"
K	0·37	0·22	1·43	0·041	0·033	1·00	0·40	0·22	"	"	"
L	0·33	0·19	1·45	0·010	0·026	0·89	0·56	0·12	"	N900	"
M	0·32	0·23	1·28	0·034	0·028	0·85	0·46	0·16	"	"	"
N	0·35	0·22	1·18	0·027	0·027	0·91	0·36	0·19	4—6	N850	"
O	0·36	0·22	1·16	0·026	0·025	0·91	0·36	0·19	8	"	"

(Ref. 6)

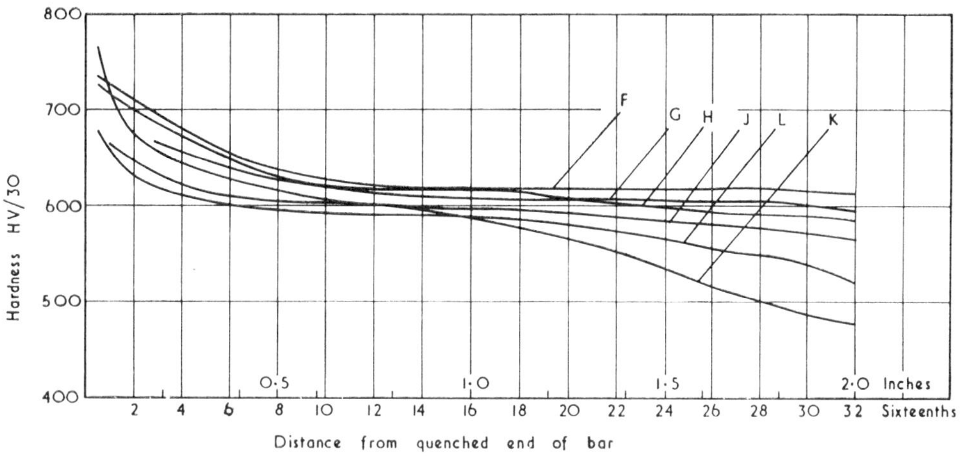

Fig. 100.11

END QUENCH HARDENABILITY DATA (Continued)

Chemical Composition (%)							
C	Si	Mn	S	P	Ni	Cr	Mo
0·39	0·27	1·29	0·026	0·011	0·70	0·44	0·18

High Frequency Melt.
McQuaid Ehn Grain Size 6—7 (5) (ASTM).
As quenched Grain Size 6—7 (ASTM).
Bars air cooled 900°C. End quenched 850°C. Tempered 600°C.

(Ref. 6)

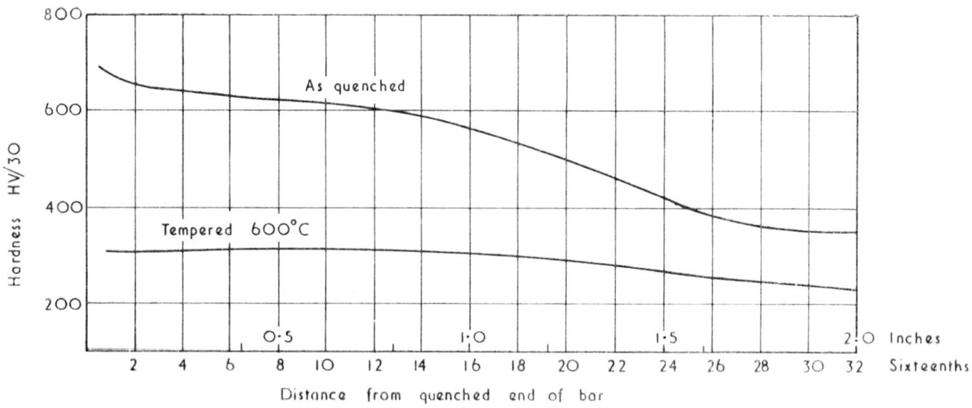

Fig. 100.12

END QUENCH HARDENABILITY BAND

(Refs. 1 and 3)

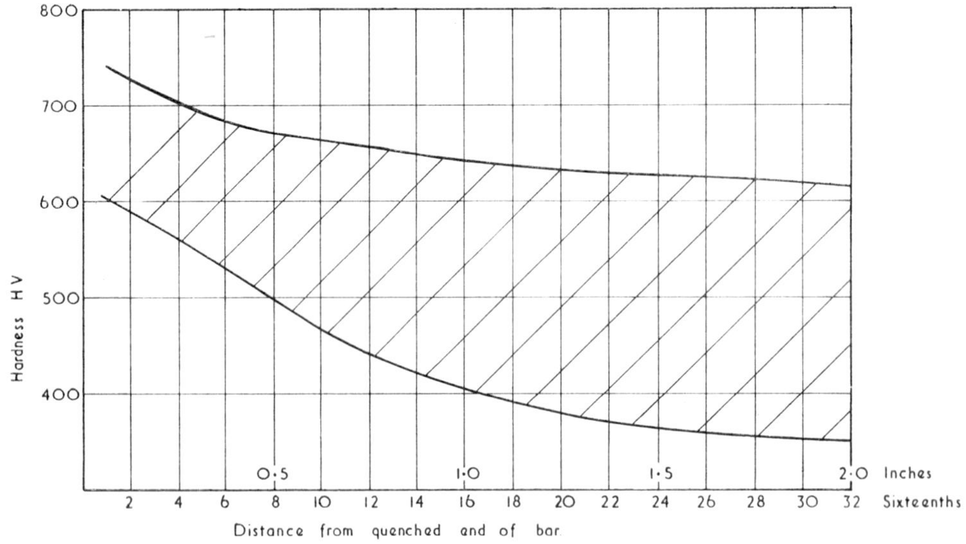

Fig. 100.13

These curves have been constructed from results obtained on several casts of En 100.

HARDNESS TRAVERSE DATA

Chemical Composition (%)							
C	Si	Mn	S	P	Ni	Cr	Mo
0·39	0·27	1·29	0·026	0·011	0·70	0·44	0·18

High Frequency Melt.
McQuaid Ehn Grain Size 6—7 (5) (ASTM).
As quenched Grain Size 6—7 (ASTM).
Pre-treatment. Air cooled 900°C.
Oil quenched 850°C. (Ref. 6)

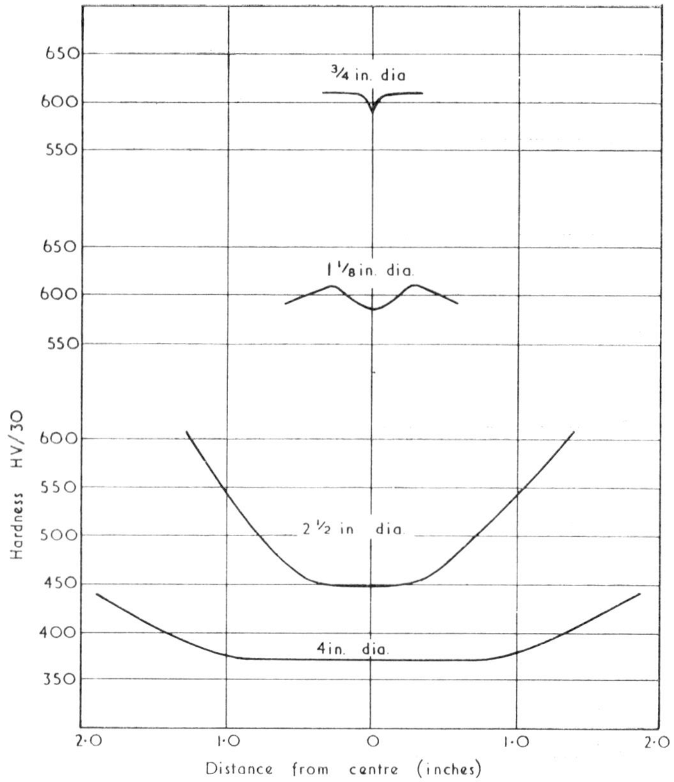

Fig. 100.14

Oil quenched 850°C. Tempered 600°C. (Ref. 6)

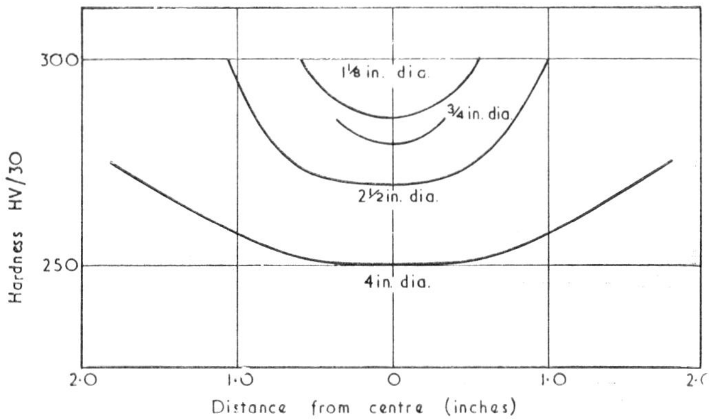

Fig. 100.15

HARDNESS TRAVERSE DATA (Continued)

Chemical Composition (%)							
C	Si	Mn	S	P	Ni	Cr	Mo
0·41	0·23	1·34	0·033	0·026	0·85	0·35	0·21

Basic Open Hearth.
Tests were carried out in two laboratories ('A' and 'B' below).

Laboratory A:
Bars T650°C, AC. Oil quenched 850°C.

(Ref. 6)

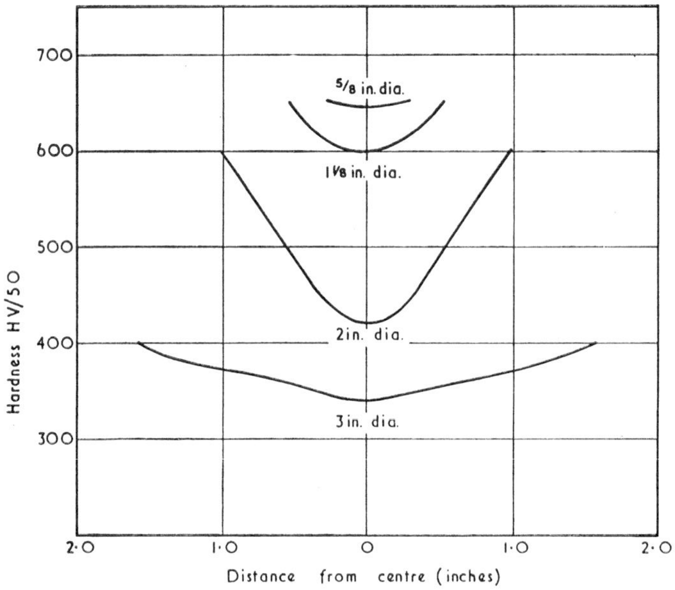

Fig. 100.16

Laboratory B:
Bars N850, T650. Oil quenched 850°C.

(Ref. 6)

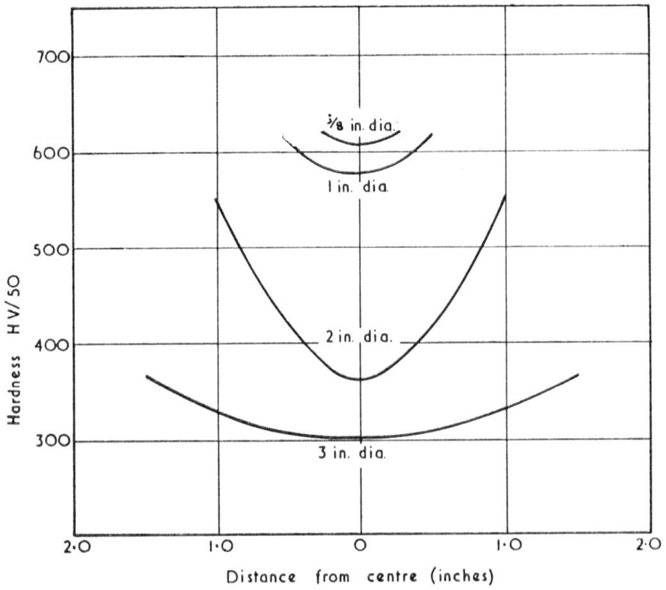

Fig. 100.17

MECHANICAL PROPERTIES AT ROOM TEMPERATURE

Process of Manufacture	ASTM Grain Size	Chemical Composition (%) C	Si	Mn	S	P	Ni	Cr	Mo	Section Size dia. (in.)	Heat Treatment °C	Tensile 0·05% P.S. (0·05Rp) tons/sq.in.	0·1% P.S. (0·1Rp) tons/sq.in.	0·2% P.S. (0·2Rp) tons/sq.in.	0·5% P.S. (0·5Rp) tons/sq.in.	Y.P. (Rs) tons/sq.in.	T S. (Rm) tons/sq.in.	El % (A)	R.A. % (Z)	Izod ft.lb.	Ref.
B.E.		0.29		1.38			0.70	0.64	0.22	6	WQ850 T640					50·2	59·0	20	50	53	2
		0.30	0.21	1.32	0.029	0.030	0.68	0.56	0.21	⅞	OQ840 T670					51·2	55·2	23	61	64	7
A.O.H.		0·32	0·23	1·28	0·034	0·028	0·85	0·46	0·16	1⅛ (C)	OQ850 T580WQ					61·4	67·0	18	54	35	2
"		"	"	"	"	"	"	"	"	2 (C)	OQ850 T620WQ					48·4	55·6	22	59	42	"
"		"	"	"	"	"	"	"	"	2½ × 1¼ (O)	WQ850 T620WQ					54·8	61·2	20½	59	70	"
"		"	"	"	"	"	"	"	"	" (C)	" "					53·0	59·4	23½	58	69	"
"		"	"	"	"	"	"	"	"	" (O)	OQ850 T620WQ					44·8	54·2	22	58	46	"
"		"	"	"	"	"	"	"	"	" (C)	" "					43·2	52·2	20½	57	30	"
"		"	"	"	"	"	"	"	"	2½ sq. (O)	" T650WQ					46·4	54·2	22	60	56	"
"		"	"	"	"	"	"	"	"	" (C)	" " "					44·4	52·2	21	58	32	"
"		"	"	"	"	"	"	"	"	3 (O)	" T625WQ					43·4	52·8	21	59	81	"
"		"	"	"	"	"	"	"	"	" (C)	" " "					39·0	48·0	24	60	65	"
"		"	"	"	"	"	"	"	"	5 sq. (O)	" T650WQ					39·2	49·2	23½	59	84	"
"		"	"	"	"	"	"	"	"	" (C)	" " "					38·8	46·0	22½	51	87	"
B.E		0·33	0·19	1·45	0·010	0·026	0·89	0·56	0·12	1⅛ (C)	OQ850 T580WQ					63·2	68·6	20	54	51	2
"		"	"	"	"	"	"	"	"	2 (C)	" T605WQ					51·6	58·4	21	60	72	"
"		"	"	"	"	"	"	"	"	2½ sq. (O)	" T620WQ					46·8	55·2	22	61	80	"
"		"	"	"	"	"	"	"	"	" (C)	" " "					44·4	52·8	20½	57	35	"
"		"	"	"	"	"	"	"	"	3 (O)	" T625WQ					50·0	58·0	20	57	55	"
"		"	"	"	"	"	"	"	"	" (C)	" " "					45·6	54·2	21½	57	43	"
"		"	"	"	"	"	"	"	"	5 sq. (O)	" T620WQ					46·0	55·2	19½	54	35	"
"		"	"	"	"	"	"	"	"	" (C)	" " "					44·0	53·4	20	46	37	"
"		"	"	"	"	"	"	"	"	" (O)	" T650WQ					40·0	50·0	23½	61	95	"
"		"	"	"	"	"	"	"	"	" (C)	" " "					37·8	47·4	23	54	100	"
		0·35		1·28			0·85	0·46	0·16	1⅛	OQ850 T580					61·4	67·0	18		35	2
		0·35		1·50			0·93	0·57	0·18	2½	OQ850 T650					44·8	55·2	21½		89	8
		"		"			"	"	"	"	" T600					52·2	62·8	19		43	"
		"		"			"	"	"	4	" T650					48·0	57·2	24		87	"
		"		"			"	"	"	"	" T600					58·5	66·8	21		64	"
B.E.		0·35		1·33			0·91	0·43	0·15	4¼	OQ850 T550					44·9	55·9	21	57	46	2
"		"		"			"	"	"	2⅛	WQ850 T600					56·1	63·0	20½	59	54	"

MECHANICAL PROPERTIES AT ROOM TEMPERATURE (Continued)

Process of Manufacture	ASTM Grain Size	Chemical Xomposition (%) C	Si	Mn	S	P	Ni	Cr	Mo	Section Size dia, (in.)	Heat Treatment °C	Tensile 0·05% P.S. (0·05Rp) tons/sq.in.	0·1% P.S. (0·1Rp)	0·2% P.S. (0·2Rp)	0·5% P.S. (0·5Rp)	Y.P. (Rs)	T.S. (Rm)	El % (A)	R.A. % (Z)	Izod ft.lb.	Ref.
	8 (some 6)	0·35	0·25	0·33	0·042	0·032	0·91	0·43	0·15	$\frac{11}{16}$ (L)	AC850	33·2	37·7	42·9	50·0		70·7	14	25½	11	3
	"	"	"	"	"	"	"	"	"	" "	" T2hrs. 200AC	38·0	42·5	48·1	56·3		69·7	17½	42	15	"
	"	"	"	"	"	"	"	"	"	" "	" " 300 "	48·0	49·1	51·5	57·4		67·1	17	49	11	"
	"	"	"	"	"	"	"	"	"	" "	" " 450 "	49·7	49·7	49·9	52·5		62·4	18	53	19	"
	"	"	"	"	"	"	"	"	"	" "	" " 500 "	46·9	47·0	47·5	49·0		58·4	19½	53	29	"
	"	"	"	"	"	"	"	"	"	" "	" " 550 "				46·3		55·9	21	55	56	"
	"	"	"	"	"	"	"	"	"	" "	" " 600 "	40·2	40·2	40·2	40·3		51·3	23½	60	66	"
	"	"	"	"	"	"	"	"	"	" "	" " 650 "	35·7	35·7	35·8	37·5		47·1	28	64	92	"
	"	"	"	"	"	"	"	"	"	" "	" " 675 "				37·7		46·6	27	64½	86	"
	"	"	"	"	"	"	"	"	"	" "	" " 700 "	29·1	29·2	29·6	31·3		46·2	25½	60½	89	"
	"	"	"	"	"	"	"	"	"	1⅛ (L)	AC850	36·2	40·3	45·1	51·6		71·6	15½	31½	12	"
	"	"	"	"	"	"	"	"	"	" "	+T2hrs. " 200AC	37·6	42·5	47·7	55·0		70·4	17	40½	15	"
	"	"	"	"	"	"	"	"	"	" "	" " 300 "	43·1	45·7	49·6	56·9		68·9	15	46½	7	"
	"	"	"	"	"	"	"	"	"	" "	" " 400 "	47·0	48·2	50·7	55·7		66·1	17	46	8	"
	"	"	"	"	"	"	"	"	"	" "	" " 500 "	46·0	46·3	47·3	49·5		58·6	20	52	24	"
	"	"	"	"	"	"	"	"	"	" "	" " 550 "	42·4	42·4	42·9	44·9		55·5	23½	55	37	"
	"	"	"	"	"	"	"	"	"	" "	" " 600 "				41·4		51·5	24	59½	65	"
	"	"	"	"	"	"	"	"	"	" "	" " 650 "				37·8		47·5	26	63	74	"
	"	"	"	"	"	"	"	"	"	" "	" " 675 "				36·2		46·5	26½	63	86	"
	"	"	"	"	"	"	"	"	"	" "	" " 700 "	30·4	30·5	30·5	32·7		46·4	26	64½	99	"
	"	"	"	"	"	"	"	"	"	2⅛ (L)	AC850	32·7	36·0	40·4	46·5		63·9	18½	40½	10	"
	"	"	"	"	"	"	"	"	"	" "	" T2hrs.400AC	42·1	43·4	46·1	50·6		61·3	24	51½	11	"
	"	"	"	"	"	"	"	"	"	" "	" " 450 "	43·4	44·2	45·7	48·9		59·1	20	54½	16	"
	"	"	"	"	"	"	"	"	"	" "	" " 500 "	43·0	43·0	43·5	45·2		55·9	22	54½	26	"
	"	"	"	"	"	"	"	"	"	" "	" " 550 "	40·5	40·6	40·9	42·0		53·5	21	54½	52	"
	"	"	"	"	"	"	"	"	"	" "	" " 600 "	38·0	38·1	38·4	39·7		50·9	22½	56½	60	"
	"	"	"	"	"	"	"	"	"	" "	" " 625 "	37·4	37·5	37·6	39·1		49·2	25½	60	72	"
	"	"	"	"	"	"	"	"	"	" "	" " 650 "				36·7		47·4	25½	63	81	"
	"	"	"	"	"	"	"	"	"	" "	" " 675 "				34·9		46·1	28	63	100	"
	"	"	"	"	"	"	"	"	"	" "	" " 700 "				32·4		43·9	29	66	88	"
	"	"	"	"	"	"	"	"	"	3 (L)	AC850	32·5	35·7	40·2	45·9		61·9	18	39	11	"
	"	"	"	"	"	"	"	"	"	" "	" T2hrs.400AC	39·0	41·0	43·4	48·1		59·0	21	53½	13	"
	"	"	"	"	"	"	"	"	"	" "	" " 450 "	40·0	41·2	43·1	46·2		56·8	21	54	19	"
	"	"	"	"	"	"	"	"	"	" "	" " 500 "	41·4	41·6	41·8	43·1		54·0	20	54½	29	"
	"	"	"	"	"	"	"	"	"	" "	" " 550 "	40·3	40·4	41·2	42·8		52·5	22	55½	36	"
	"	"	"	"	"	"	"	"	"	" "	" " 600 "				38·4		49·2	24	55½	61	"
	"	"	"	"	"	"	"	"	"	" "	" " 625 "				36·6		47·5	26½	60½	76	"
	"	"	"	"	"	"	"	"	"	" "	" " 650 "				35·4		45·8	27½	63½	78	"
	"	"	"	"	"	"	"	"	"	" "	" " 675 "				33·9		46·0	25½	63	77	"

MECHANICAL PROPERTIES AT ROOM TEMPERATURE (Continued)

Process of Manufacture	ASTM Grain Size	Chemical Composition (%) C	Si	Mn	S	P	Ni	Cr	Mo	Section Size dia. (in.)	Heat Treatment °C	Tensile 0·05% P.S. (0·05Rp) tons/sq.in.	0·1% P.S. (0·1Rp)	0·2% P.S. (0·2Rp)	0·5% P.S. (0·5Rp)	Y.P. (Rs)	T.S. (Rm)	El % (A)	R.A. % (Z)	Izod ft.lb.	Ref.
	8 (some 6)	0·35	0·26	1·35	0·042	0·032	0·91	0·43	0·15	3 (L)	AC850 T2hrs.700AC				35.6		45.2	28½	64	89	3
	"	"	"	"	"	"	"	"	"	5½ (L)	" " 600AC				39·0		51·4	22	54½	48	"
	"	"	"	"	"	"	"	"	"	" "	" " 625 "				38·3		49.2	24½	57½	55	"
	"	"	"	"	"	"	"	"	"	" "	" " 650 "				35·7		47·3	26	59½	65	"
	"	"	"	"	"	"	"	"	"	" "	" " 675 "				35·3		46·4	26½	60½	64	"
	"	"	"	"	"	"	"	"	"	" "	" " 700 "				33·1		44·2	29	63	75	"
	"	"	"	"	"	"	"	"	"	0·564 (L)	OQ850 T2hrs. 100AC	70·5	77·9	86·3	100·3		122·0	14	31½	9	"
	"	"	"	"	"	"	"	"	"	" "	" " 200 "	78·8	84·7	90·8	100·5		116·4	12½	35	11	"
	"	"	"	"	"	"	"	"	"	" "	" " 300 "	91·0	93·1	96·3	103·6		112·7	13	45½	6	"
	"	"	"	"	"	"	"	"	"	" "	" " 400 "	91·0	91·7	93·1	98·2		106·6	12	44½	5	"
	"	"	"	"	"	"	"	"	"	11/16 (L)	OQ850	72·8	80·7	89·7	101·8		126·6	11½	37½	13	"
	"	"	"	"	"	"	"	"	"	" "	" T2hrs. 100AC	62·3	71·1	81·6	—		—	—	—	10	"
	"	"	"	"	"	"	"	"	"	" "	" " 200 "	81·6	87·9	94·4	103·4		118·2	15	39½	12	"
	"	"	"	"	"	"	"	"	"	" "	" " 250 "	95·4	96·0	97·5	—		114·8	13	44½	8	"
	"	"	"	"	"	"	"	"	"	" "	" " 300 "	94·5	94·8	96·0	99·7		108·6	13	43½	5	"
	"	"	"	"	"	"	"	"	"	" "	" " 400 "				86·9		92·6	14	49½	9	"
	"	"	"	"	"	"	"	"	"	" "	" " 450 "	80·0	80·2	80·2	81·0		84·6	14½	50½	15	"
	"	"	"	"	"	"	"	"	"	" "	" " 500 "				71·2		74·7	17½	55	26	"
	"	"	"	"	"	"	"	"	"	" "	" " 550 "	59·6	59·7	59·7	59·7		65·4	20½	59½	53	"
	"	"	"	"	"	"	"	"	"	" "	" " 600 "	55·0	55·0	55·0	55·7		61·2	22½	61½	63	"
	"	"	"	"	"	"	"	"	"	" "	" " 650 "				47·4		52·8	25½	64½	76	"
	"	"	"	"	"	"	"	"	"	" "	" " 675 "				47·8		51·8	27	65	95	"
	"	"	"	"	"	"	"	"	"	" "	" " 700 "	25·4	27·5	29·2	32·5		54·2	20	48	55	"
	"	"	"	"	"	"	"	"	"	1⅛ (L)	OQ850	78·0	85·4	92·7	103·7		117·1	10½	39½	10	"
	"	"	"	"	"	"	"	"	"	" "	" T2hrs.100AC	78·8	85·8	92·1	100·6		108·9	13	44	10	"
	"	"	"	"	"	"	"	"	"	" "	" " 200 "	71·8	79·8	87·2	97·4		112·1	12	38	12	"
	"	"	"	"	"	"	"	"	"	" "	" " 250 "	84·3	89·2	93·8	100·9		109·2	11½	42	6	"
	"	"	"	"	"	"	"	"	"	" "	" " 300 "	87·5	91·4	94·1	101·9		109·0	12½	43	5	"
	"	"	"	"	"	"	"	"	"	" "	" " 350 "	77·7	83·1	89·0	100·1		107·1	11	42½	5	"
	"	"	"	"	"	"	"	"	"	" "	" " 400 "	75·6	79·2	81·5	84·8		89·2	12½	45½	9	"
	"	"	"	"	"	"	"	"	"	" "	" " 450 "	71·6	73·5	74·6	76·5		80·4	15	50	15	"
	"	"	"	"	"	"	"	"	"	" "	" " 500 "	69·2	69·4	69·3	70·3		74·4	18½	55	26	"
	"	"	"	"	"	"	"	"	"	" "	" " 550 "	58·4	58·9	59·1	59·6		65·3	19	55½	49	"
	"	"	"	"	"	"	"	"	"	" "	" " 600 "	51·5	51·5	51·6	51·7		59·6	24	61½	75	"
	"	"	"	"	"	"	"	"	"	" "	" " 650 "				47·2		54·0	25½	66	93	"
	"	"	"	"	"	"	"	"	"	" "	" " 675 "				45·5		50·2	27	64½	96	"
	"	"	"	"	"	"	"	"	"	" "	" " 700 "				37·8		49·5	27½	64½	90	"
	"	"	"	"	"	"	"	"	"	2⅛ (L)	OQ850	34·0	40·0	46·9	53·8		72·2	13½	37	6	"

MECHANICAL PROPERTIES AT ROOM TEMPERATURE (Continued)

Process of Manufacture	ASTM Grain Size	Chemical Composition (%)								Section Size dia. (in.)	Heat Treatment °C	Tensile								Izod ft.lb.	Ref.
		C	Si	Mn	S	P	Ni	Cr	Mo			0·05% P.S. (0·05Rp)	0·1% P.S. (0·1Rp)	0·2% P.S. (0·2Rp)	0·5% P.S. (0·5Rp)	Y.P. (Rs)	T.S. (Rm)	El % (A)	R.A. % (Z)		
												tons/sq.in.									
	8 (some 6)	0·35	0·26	1·35	0·042	0·032	0·91	0·43	0·15	2⅛ (L)	OQ850 T2hrs.400AC	44·9	48·9	52·8	58·4		67·5	15½	49½	6	3
	”	”	”	”	”	”	”	”	”	” ”	” ” 450 ”	48·7	50·0	52·1	55·6		64·2	18	53	9	”
	”	”	”	”	”	”	”	”	”	” ”	” ” 500 ”	47·7	48·2	49·2	51·1		61·6	20	54½	23	”
	”	”	”	”	”	”	”	”	”	” ”	” ” 550 ”	46·6	46·7	46·8	47·8		58·4	21	56½	52	”
	”	”	”	”	”	”	”	”	”	” ”	” ” 600 ”	42·1	42·2	42·2	42·4		54·0	24	62½	70	”
	”	”	”	”	”	”	”	”	”	” ”	” ” 625 ”				40·9		51·5	26	63	91	”
	”	”	”	”	”	”	”	”	”	” ”	” ” 650 ”				38·1		49·8	27	66	80	”
	”	”	”	”	”	”	”	”	”	” ”	” ” 675 ”				38·3		48·5	27½	61	111	”
	”	”	”	”	”	”	”	”	”	” ”	” ” 700 ”				36·7		47·0	27	67	105	”
	”	”	”	”	”	”	”	”	”	3 (L)	OQ850	34·6	39·2	44·0	52·0		65·7	18	42½	8	”
	”	”	”	”	”	”	”	”	”	” ”	” T2hrs.400AC	41·8	44·6	47·7	52·8		62·7	17½	52½	6	”
	”	”	”	”	”	”	”	”	”	” ”	” ” 450 ”	49·0	49·9	51·6	55·5		63·5	18	54½	13	”
	”	”	”	”	”	”	”	”	”	” ”	” ” 500 ”	47·2	47·6	47·6	48·7		59·3	21½	55	30	”
	”	”	”	”	”	”	”	”	”	” ”	” ” 550 ”	43·6	44·0	44·1	45·5		55·5	21½	57½	36	”
	”	”	”	”	”	”	”	”	”	” ”	” ” 600 ”	41·4	41·5	41·5	42·0		52·6	23	58	53	”
	”	”	”	”	”	”	”	”	”	” ”	” ” 625 ”				40·4		51·1	25	60	70	”
	”	”	”	”	”	”	”	”	”	” ”	” ” 650 ”				41·0		51·1	24	63	79	”
	”	”	”	”	”	”	”	”	”	” ”	” ” 675 ”				35·7		46·5	27	64½	87	”
	”	”	”	”	”	”	”	”	”	” ”	” ” 700 ”				35·2		45·5	29	66½	93	”
	”	”	”	”	”	”	”	”	”	4¼ (L)	OQ850	47·2	48·9	51·8	57·0		66·2	17	51	8	”
	”	”	”	”	”	”	”	”	”	” ”	” T2hrs.400AC	44·5	47·8	51·7	57·2		66·3	14	46½	5	”
	”	”	”	”	”	”	”	”	”	” ”	” ” 500 ”	46·4	47·2	48·5	50·7		60·6	19	50½	19	”
	”	”	”	”	”	”	”	”	”	” ”	” ” 550 ”				44·9		55·9	21	57	47	”
	”	”	”	”	”	”	”	”	”	” ”	” ” 575 ”				43·5		54·4	23	57½	60	”
	”	”	”	”	”	”	”	”	”	” ”	” ” 600 ”	42·1	42·4	42·4	42·4		54·4	22	56½	74	”
	”	”	”	”	”	”	”	”	”	” ”	” ” 625 ”				40·8		50·8	24	60	61	”
	”	”	”	”	”	”	”	”	”	” ”	” ” 650 ”				38·7		49·6	25	61½	75	”
	”	”	”	”	”	”	”	”	”	” ”	” ” 675 ”				36·7		46·9	27	64½	87	”
	”	”	”	”	”	”	”	”	”	” ”	” ” 700 ”				35·6		45·8	29	65	82	”
	”	”	”	”	”	”	”	”	”	5½ (L)	OQ850 T2hrs.500AC	45·4	46·2	47·4	49·3		59·4	18½	52	20	”
	”	”	”	”	”	”	”	”	”	” ”	” ” 550 ”				43·9		54·9	22	56½	35	”
	”	”	”	”	”	”	”	”	”	” ”	” ” 575 ”				42·2		53·1	23½	57	53	”
	”	”	”	”	”	”	”	”	”	” ”	” ” 600 ”	41·0	41·8	42·8	44·4		55·4	20	50	51	”
	”	”	”	”	”	”	”	”	”	” ”	” ” 625 ”				39·2		50·5	24½	60	61	”
	”	”	”	”	”	”	”	”	”	” ”	” ” 650 ”				37·5		49·6	24½	60	64	”
	”	”	”	”	”	”	”	”	”	” ”	” ” 675 ”				34·8		46·4	27½	63½	72	”
	”	”	”	”	”	”	”	”	”	” ”	” ” 700 ”				34·8		45·5	25½	64	80	”
	”	”	”	”	”	”	”	”	”	$\frac{11}{16}$ (L)	WQ850 T2hrs.300AC				91·8		104·7	14	47	5	”
	”	”	”	”	”	”	”	”	”	” ”	” ” 400 ”				84·5		89·1	10	30½	10	”
	”	”	”	”	”	”	”	”	”	” ”	” ” 450 ”	76·0	76·0	76·0	77·7		81·3	15	52½	24	”

MECHANICAL PROPERTIES AT ROOM TEMPERATURE (Continued)

Process of Manufacture	ASTM Grain Size	Chemical Composition (%) C	Si	Mn	S	P	Ni	Cr	Mo	Section Size dia. (in.)	Heat Treatment °C	Tensile 0·05% P.S. (0·05Rp) tons/sq.in.	0·1% P.S. (0·1Rp)	0·2% P.S. (0·2Rp)	0·5% P.S. (0·5Rp)	Y.P. (Rs)	T.S. (Rm)	El % (A)	R.A. % (Z)	Izod ft.lb.	Ref.
	8 to 6	0·35	0·26	1·35	0·042	0·032	0·91	0·43	0·15	11/16 (L)	WQ850 T2hrs.500AC	66·7	66·7	67·0	69·3		72·0	18½	55	36	3
	"	"	"	"	"	"	"	"	"	" "	" " 550 "	57·6	58·0	58·2	59·7		64·5	16	45½	49	"
	"	"	"	"	"	"	"	"	"	" "	" " 600 "				53·3		60·8	23½	60	58	"
	"	"	"	"	"	"	"	"	"	" "	" " 650 "				48·7		52·7	26½	64½	76	"
	"	"	"	"	"	"	"	"	"	" "	" " 675 "				45·9		50·5	27	65	89	"
	"	"	"	"	"	"	"	"	"	" "	" " 700 "				33·3		50·8	28	54	83	"
	"	"	"	"	"	"	"	"	"	1⅛ (L)	WQ850 T2hrs.300AC				91·8		104·7	14	47½	5	"
	"	"	"	"	"	"	"	"	"	" "	" " 400 "				84·5		89·1	10	30½	10	"
	"	"	"	"	"	"	"	"	"	" "	" " 500 "				77·7		81·3	15	52	24	"
	"	"	"	"	"	"	"	"	"	" "	" " 550 "				61·9		65·9	20	56	42	"
	"	"	"	"	"	"	"	"	"	" "	" " 600 "				56·2		60·0	22½	61	79	"
	"	"	"	"	"	"	"	"	"	" "	" " 650 "				49·0		53·3	24½	63½	84	"
	"	"	"	"	"	"	"	"	"	" "	" " 675 "				46·7		51·2	26	64	96	"
	"	"	"	"	"	"	"	"	"	" "	" " 700 "	29·6	30·4	31·3	34·3		50·5	24½	53	70	"
	"	"	"	"	"	"	"	"	"	2⅛ (L)	WQ850 T2hrs.400 "	76·1	77·6	79·4	82·5		87·4	11	41½	7	"
	"	"	"	"	"	"	"	"	"	" "	" " 450 "	71·4	73·0	73·1	74·7		80·2	15	49½	21	"
	"	"	"	"	"	"	"	"	"	" "	" " 500 "	63·0	63·4	63·8	64·8		70·9	17	52½	46	"
	"	"	"	"	"	"	"	"	"	" "	" " 550 "				60·1		67·1	19	55	45	"
	"	"	"	"	"	"	"	"	"	" "	" " 600 "				53·5		60·5	21	61	55	"
	"	"	"	"	"	"	"	"	"	" "	" " 625 "				48·8		56·1	24	63	81	"
	"	"	"	"	"	"	"	"	"	" "	" " 650 "				43·6		52·6	25	65	80	"
	"	"	"	"	"	"	"	"	"	" "	" " 675 "						50·4	26	65½	101	"
	"	"	"	"	"	"	"	"	"	" "	" " 700 "				40·9		47·9	28½	66	113	"
	"	"	"	"	"	"	"	"	"	3 (L)	WQ850 T2hrs.400AC	59·2	63·5	68·3	73·7		81·1	13	44	5	"
	"	"	"	"	"	"	"	"	"	" "	" " 450 "	59·3	62·4	65·5	69·0		74·7	15	48½	11	"
	"	"	"	"	"	"	"	"	"	" "	" " 500 "	60·2	61·2	61·8	63·3		69·7	18	52	35	"
	"	"	"	"	"	"	"	"	"	" "	" " 550 "	52·9	53·9	54·7	55·9		64·0	18½	52½	36	"
	"	"	"	"	"	"	"	"	"	" "	" " 600 "	48·6	49·0	49·2	50·1		59·1	20	58	52	"
	"	"	"	"	"	"	"	"	"	" "	" " 625 "	45·4	45·5	45·5	45·5		54·1	25	63½	60	"
	"	"	"	"	"	"	"	"	"	" "	" " 650 "				42·3		51·1	26	64½	70	"
	"	"	"	"	"	"	"	"	"	" "	" " 675 "				40·2		48·2	28	67½	79	"
	"	"	"	"	"	"	"	"	"	" "	" " 700 "				39·1		47·9	28½	66½	93	"
	"	"	"	"	"	"	"	"	"	4¼ (L)	WQ850	72·0	74·6	75·5	78·5		84·5	13	46	9	"
	"	"	"	"	"	"	"	"	"	" "	" T2hrs.400AC	51·2	56·2	61·0	65·0		72·7	13½	39	7	"
	"	"	"	"	"	"	"	"	"	" "	" " 500 "	57·9	59·1	60·0	61·7		68·8	17	51½	29	"
	"	"	"	"	"	"	"	"	"	" "	" " 550 "				59·6		66·2	19	57	43	"
	"	"	"	"	"	"	"	"	"	" "	" " 575 "	54·3	54·4	54·4	54·6		62·3	20	59	58	"
	"	"	"	"	"	"	"	"	"	" "	" " 600 "	47·5	48·5	49·2	50·0		58·8	21	55½	66	"
	"	"	"	"	"	"	"	"	"	" "	" " 625 "				48·9		56·2	23	63	70	"

MECHANICAL PROPERTIES AT ROOM TEMPERATURE (Continued)

Process of Manufacture	ASTM Grain Size	Chemical Composition (%)								Section Size dia. (in.)	Heat Treatment °C	Tensile								Izod ft.lb.	Ref.
		C	Si	Mn	S	P	Ni	Cr	Mo			0·05% P.S. (0·05Rp)	0·1% P.S. (0·1Rp)	0·2% P.S. (0·2Rp)	0·5% P.S. (0·5Rp)	Y.P. (Rs)	T.S. (Rm)	El % (A)	R.A. % (Z)		
												tons/sq.in.									
	8 (some 6)	0·35	0·26	1·35	0·042	0·032	0·91	0·43	0·15	4¼	WQ850 T2hrs.650AC	41·2	41·2	41·2	41·2		52·2	24	61	81	3
	"	"	"	"	"	"	"	"	"	"	" " 675 "				41·7		51·6	25	63½	86	"
	"	"	"	"	"	"	"	"	"	"	" " 700 "				38·7		48·0	28	64½	82	"
	"	"	"	"	"	"	"	"	"	5½ (L)	" " 500 "	53·0	53·9	57·0	59·4		67·5	17½	48	20	"
	"	"	"	"	"	"	"	"	"	" "	" " 550 "	53·0	54·0	54·9	56·4		64·2	19	54	41	"
	"	"	"	"	"	"	"	"	"	" "	" " 575 "	50·0	51·4	51·8	52·8		60·9	21½	57	47	"
	"	"	"	"	"	"	"	"	"	" "	" " 600 "	45·5	46·4	47·4	48·8		58·6	21½	55½	51	"
	"	"	"	"	"	"	"	"	"	" "	" " 625 "				45·2		55·1	23½	62½	58	"
	"	"	"	"	"	"	"	"	"	" "	" " 650 "				43·8		54·1	23	60½	61	"
	"	"	"	"	"	"	"	"	"	" "	" " 675 "				39·2		51·0	25½	62½	72	"
	"	"	"	"	"	"	"	"	"	" "	" " 700 "				37·0		47·4	28	65½	87	"
	"	"	"	"	"	"	"	"	"	4¼ (T)	AC850 T2hrs.500AC					43·0	55·0	6	6½	6	"
	"	"	"	"	"	"	"	"	"	" "	" " 600 "					39·0	50·0	7	8½	9	"
	"	"	"	"	"	"	"	"	"	" "	" " 650 "					35·0	47·0	11	15	12	"
	"	"	"	"	"	"	"	"	"	5½ (T)	AC850 T2hrs.600AC					39·0	51·0	12½	15	11	"
	"	"	"	"	"	"	"	"	"	" "	" " 650 "					36·0	48·0	15	20	17	"
	"	"	"	"	"	"	"	"	"	4¼ (T)	OQ850 T2hrs.500AC					48·0	59·5	6	6½	5	"
	"	"	"	"	"	"	"	"	"	" "	" " 600 "					42·0	53·0	6	6½	7	"
	"	"	"	"	"	"	"	"	"	" "	" " 650 "					38·0	49·0	9	10	13	"
	"	"	"	"	"	"	"	"	"	5½ (T)	OQ850 T2hrs.500AC					48·0	59·0	8	8½	8	"
	"	"	"	"	"	"	"	"	"	" "	" " 600 "					42·0	54·0	10	11½	10	"
	"	"	"	"	"	"	"	"	"	" "	" " 650 "					38·0	50·0	16	18	17	"
	"	"	"	"	"	"	"	"	"	4¼ (T)	WQ850 T2hrs.500AC					50·0	62·0	5	5	5	"
	"	"	"	"	"	"	"	"	"	" "	" " 600 "					48·0	59·0	6	6½	7	"
	"	"	"	"	"	"	"	"	"	" "	" " 650 "					43·0	52·0	7	6½	9	"
	"	"	"	"	"	"	"	"	"	5½ (T)	WQ850 T2hrs.500AC					57·0	67·0	6	8½	6	"
	"	"	"	"	"	"	"	"	"	" "	" " 600 "					48·0	58·8	12	16½	12	"
	"	"	"	"	"	"	"	"	"	" "	" " 650 "					42·0	53·0	14	23	14	"
		0·36		1·28			0·74	0·46	0·26	6 sq.	OQ850 T650					46·0	56·2	21½		75	8
		0·36		1·36			0·79	0·46	0·19	2¼	OQ850 T610					53·6	59·0	21		69	2
		0·36	0·20	1·28	0·026	0·036	1·03	0·55	0·20	$\frac{21}{32}$	OQ830 T650					59·2	65·6	20	57	60	7
		0·36		1·28			0·74	0·46	0·25	1⅛	OQ850 T550 OQ					73·0	76·6	16½	52	36	1
		"		"			"	"	"	"	" 600 "					60·0	66·6	20½	59	69	"
		"		"			"	"	"	"	" 650 "					54·0	60·9	21½	64	78	"
		"		"			"	"	"	2½	OQ850 T550 OQ					68·2	72·8	16½	52	26	"
		"		"			"	"	"	"	" 600 "					60·0	66·7	20½	59	65	"
		"		"			"	"	"	"	" 650 "					50·0	58·3	22	64	76	"
		"		"			"	"	"	4 sq.	OQ850 T550 OQ					62·0	68·3	16½	55	21	"
		"		"			"	"	"	"	" 600 "					52·4	61·4	21	57	57	"

MECHANICAL PROPERTIES AT ROOM TEMPERATURE (Continued)

Process of Manufacture	ASTM Grain Size	Chemical Composition (%) C	Si	Mn	S	P	Ni	Cr	Mo	Section Size dia. (in.)	Heat Treatment °C	Tensile 0·05% P.S. (0·05Rp) tons/sq.in.	0·1% P.S. (0·1Rp) tons/sq.in.	0·2% P.S. (0·2Rp) tons/sq.in.	0·5% P.S. (0·5Rp) tons/sq.in.	Y.P. (Rs) tons/sq.in.	T.S. (Rm) tons/sq.in.	El % (A)	R.A. % (Z)	Izod ft.lb.	Ref.
		0·36		1·28			0·74	0·46	0·25	4 sq.	OQ850 T650 OQ					47·6	56·3	21½	64	79	1
		"		"			"	"	"	6 sq.	OQ850 T550 "					65·2	71·4	16	47	29	"
		"		"			"	"	"	"	" 600 "					55·6	63·3	19	50	64	"
		"		"			"	"	"	"	" 650 "					46·0	56·2	22½	57	75	"
		"		"			"	"	"	6 sq. × 6	OQ850 T650 OQ(L)					42·4	52·0	24	57	78	"
		"		"			"	"	"	"	" " " (T)					42·7	51·5	16	28	23	"
		0·37		1·30			0·84	0·43	0·11	1⅛	OQ850 T600					58·4	65·8	20		60	8
		0·37	0·19	1·24			0·83	0·43	0·19	1¾	OQ850 T620		48·9				58·5	24		72	3
		0·37	0·29	1·32			0·79	0·59	0·22	2	OQ850 T610					47·0	56·0	25	63½	86	"
		0·37	0·20	1·25			0·76	0·47	0·21	3	OQ850 T650					43·0	50·0	25	61	85	"
B.E.	—	0·38		1·26			0·56	0·50	0·21	6	OQ850 T650					45·0	57·6	21	57	50	2
A.O.H.	7 (5—8)	0·38	0·25	1·44	0·036	0·025	0·57	0·42	0·15	⅞	OQ850 T625					55·2	62·2	23	56		"
		0·38	0·33	1·26	0·016	0·020	0·56	0·50	0·21	1⅛	WQ850 T400		83·6				94·0	14	42	9	9
		"	"	"	"	"	"	"	"	"	" 450		79·0				88·0	15½	45	15	"
		"	"	"	"	"	"	"	"	"	" 500		74·8				82·2	15½	45	17	"
		"	"	"	"	"	"	"	"	"	" 550		71·0				80·0	18	47	18	"
		"	"	"	"	"	"	"	"	"	" 600		62·2				68·8	19	55	53	"
		"	"	"	"	"	"	"	"	"	" 650		52·7				60·0	23	57	82	"
		"	"	"	"	"	"	"	"	2½	WQ850 T450		76·4				85·2	14	45	15	"
		"	"	"	"	"	"	"	"	"	" 550		71·6				78·4	15	45	15	"
		"	"	"	"	"	"	"	"	"	" 600		62·0				70·8	18	52½	44	"
		"	"	"	"	"	"	"	"	"	" 620		54·0				63·6	20	59½	64	"
		"	"	"	"	"	"	"	"	4	WQ850 T570		66·0				74·0	19	55	40	9
		"	"	"	"	"	"	"	"	"	" 580		63·6				70·4	18	52½	45	"
		"	"	"	"	"	"	"	"	"	" 600		60·8				68·0	20	57	60	"
		"	"	"	"	"	"	"	"	"	" 650		52·0				60·8	22	59½	74	"
		"	"	"	"	"	"	"	"	"	OQ850 T580		57·3				69·0	19	55	38	"
		"	"	"	"	"	"	"	"	"	" 600		54·2				64·8	22	61½	65	"
		"	"	"	"	"	"	"	"	"	" 620		51·8				62·0	22	61½	69	"
		"	"	"	"	"	"	"	"	6	" 650		45·0				57·6	21	57	50	"
		0·39		1·37			0·73	0·48	0·18	2½	OQ850 T650					51·2	58·2	22		81	8
		"		"			"	"	"	5 sq.	" "					43·1	53·4	17½		68	"
		0·39		1·47			0·73	0·49	0·17	4	OQ850 T650					45·1	55·7	24		81	"
		"		"			"	"	"	6 sq.	" "					46·8	56·9	21½		80	"
		0·39		1·41			0·90	0·38	0·13	2½ sq.	OQ850 T650					51·2	57·6	24		94	"

MECHANICAL PROPERTIES AT ROOM TEMPERATURE (Continued)

Process of Manufacture	ASTM Grain Size	Chemical Composition (%) C	Si	Mn	S	P	Ni	Cr	Mo	Section Size dia. (in.)	Heat Treatment °C	Tensile 0·05% P.S. (0·05Rp)	0·1% P.S. (0·1Rp)	0·2% P.S. (0·2Rp)	0·5% P.S. (0·5Rp)	Y.P. (Rs)	T.S. (Rm)	El % (A)	R.A. % (Z)	Izod ft.lb.	Ref.
												tons/sq.in.									
		0·39		1·41			0·90	0·38	0·18	4 sq.	OQ850 T650					46·8	57·0	22		69	8
		0·39		1·40			0·81	0·43	0·24	0·86	OQ850 T590					72·0	74·4	$17\frac{1}{2}$		42	2
		0·39		1·41			0·77	0·46	0·20	$4\frac{1}{4}$	OQ850 T650					50·6	57·6	$21\frac{1}{2}$		63	"
	6—7	0·39	0·30	1·36	0·034	0·036	1·08	0·35	0·21	$1\frac{1}{8}$	OQ850 T640						58·8	21	57	60	6
	"	"	"	"	"	"	"	"	"	$2\frac{1}{2}$	" T630						59·8	22	56	72	"
	5—6	0·39	0·29	1·45	0·031	0·035	0·87	0·34	0·17	$1\frac{1}{8}$	" T640						56·9	23	58	74	"
	"	"	"	"	"	"	"	"	"	$2\frac{1}{2}$	" T630						60·4	20	59	43	"
	6	0·39	0·50	1·41	0·033	0·038	0·90	0·38	0·18	$1\frac{1}{8}$	" T640						59·6	24	$59\frac{1}{2}$	86	"
	"	"	"	"	"	"	"	"	"	$2\frac{1}{2}$	" T650						57·6	24	59	94	"
	"	"	"	"	"	"	"	"	"	4	" T650						57·0	22	57	69	"
		0·39	0·22	1·38	0·016	0·019	0·70	0·64	0·22	6	WQ850 T640		50·2				59·0	20	50	53	9
	McQuaid Actual Ehn																				
H.F.	6—7(5) 6—7	0·39	0·27	1·29	0·026	0·011	0·70	0·44	0·18	$\frac{3}{4}$	N900 OQ850 T600					54·0	60·0	$23\frac{1}{2}$	64	73	6
"	" "	"	"	"	"	"	"	"	"	$1\frac{1}{8}$	" " "					54·4	60·1	$23\frac{1}{2}$	$64\frac{1}{2}$	75	"
"	" "	"	"	"	"	"	"	"	"	$2\frac{1}{2}$ (C)	" " "					47·6	56·2	24	67	80	"
"	" "	"	"	"	"	"	"	"	"	" (M)	" " "					51·0	58·0	$23\frac{1}{2}$	66	84	"
"	" "	"	"	"	"	"	"	"	"	" (O)	" " "					51·2	58·5	24	67	77	"
"	" "	"	"	"	"	"	"	"	"	4 (C)	" " "					42·6	52·6	24	65	72	"
"	" "	"	"	"	"	"	"	"	"	" (M)	" " "					43·2	54·6	24	67	83	"
"	" "	"	"	"	"	"	"	"	"	" (O)	" " "					46·1	56·0	$24\frac{1}{2}$	66	82	"
B.E.		0·39	0·30	1·33	0·015	0·024	0·63	0·43	0·21	$\frac{5}{8}$	OQ850 T640					52·0	58·8	24	$60\frac{1}{2}$	72	9
"		"	"	"	"	"	"	"	"	$1\frac{5}{8}$	" "					50·0	57·0	24	64	76	"
"		0·39		1·35			0·65	0·48	0·17	$2\frac{1}{2}$ sq.	OQ850 T640 OQ					51·0	59·1	$21\frac{1}{2}$	59	62	1
		0·40		1·36			1·00	0·53	0·23	$\frac{7}{8}$	OQ850 T580					69·2	73·6	19		38	2
		"		"			"	"	"	$1\frac{1}{8}$	" T600					62·0	67·4	$20\frac{1}{2}$		65	"
		0·40		1·41			0·79	0·40	0·20	$1\frac{1}{2}$	" T620					52·0	59·2	22		80	"
		0·40	0·20	1·23	0·023	0·033	0·87	0·49	0·16	$1\frac{1}{16}$	" T670					52·8	60·2	22	57	75	7
	6—7	0·40	0·30	1·43	0·035	0·034	0·88	0·38	0·20	$1\frac{1}{8}$	OQ850 T640						61·0	20	56	60	6
	"	"	"	"	"	"	"	"	"	$2\frac{3}{8}$	" T650						54·8	$25\frac{1}{2}$	$60\frac{1}{2}$	92	"
	3—4	0·40	0·27	1·34	0·026	0·035	0·85	0·30	0·16	$1\frac{1}{8}$	OQ850 T640						62·5	20	56	60	"
	"	"	"	"	"	"	"	"	"	$2\frac{3}{8}$	" T640						57·6	20	57	72	"
		0·40	0·32	1·46			0·77	0·45	0·20	$\frac{7}{8}$	OQ850 T600					62·8	68·0	20	$52\frac{1}{2}$	42	3

MECHANICAL PROPERTIES AT ROOM TEMPERATURE (Continued)

Process of Manufacture	ASTM Grain Size	Chemical Composition (%) C	Si	Mn	S	P	Ni	Cr	Mo	Section Size dia. (in.)	Heat Treatment °C	Tensile 0·05% P.R. (0.05 Rp) tons/sq.in.	0·1% P.S. (0·1Rp) tons/sq.in.	0·2% P.S. (0·2Rp) tons/sq.in.	0·5% P.S. (0·5Rp) tons/sq.in.	Y.P. (Rs) tons/sq.in.	T.S. (Rm) tons/sq.in.	El % (A)	R.A. % (Z)	Izod ft.lb.	Ref.
A.O.H.	8	0·40	0·30	1·51	0·036	0·034	0·89	0·21	0·16	1⅛	OQ850 T400 AC		84·8	84·9	86·0		92·4	13	43	6	10
"	"	"	"	"	"	"	"	"	"	"	" 450 "		76·8	77·6	78·0		81·6	14½	44½	13	"
"	"	"	"	"	"	"	"	"	"	"	" 500 "		70·0	70·8	72·0		76·0	16	46	14	"
"	"	"	"	"	"	"	"	"	"	"	" 550 "		66·8	66·8	66·8		70·4	17½	47	44	"
"	"	"	"	"	"	"	"	"	"	"	" 600 "		63·4	63·4	63·4		66·8	19½	51	66	"
"	"	"	"	"	"	"	"	"	"	"	" 650 "		52·4	52·4	52·4		55·2	26	61½	93	"
"	"	"	"	"	"	"	"	"	"	"	" 700 "		27·6	30·8	36·0		60·0	19	40½	63	"
"	"	"	"	"	"	"	"	"	"	2½	OQ850 T400 AC		69·4	74·6	79·6		86·8	14	46	3	"
"	"	"	"	"	"	"	"	"	"	"	" 450 "		68·8	71·4	75·6		80·0	14½	47	9	"
"	"	"	"	"	"	"	"	"	"	"	" 500 "		64·8	66·0	66·8		74·0	16½	47	15	"
"	"	"	"	"	"	"	"	"	"	"	" 550 "		61·2	62·0	62·8		70·0	17	48½	29	"
"	"	"	"	"	"	"	"	"	"	"	" 600 "		56·4	56·4	56·4		63·0	19½	57	63	"
"	"	"	"	"	"	"	"	"	"	"	" 650 "		48·6	48·6	48·6		55·0	24½	62½	83	"
"	"	"	"	"	"	"	"	"	"	"	" 700 "		26·4	28·8	33·2		56·0	21½	49½	52	"
"	"	"	"	"	"	"	"	"	"	4	OQ850 T400 AC		53·2	55·6	60·0		70·6	15	43	6	"
"	"	"	"	"	"	"	"	"	"	"	" 450 "		52·0	55·2	58·8		68·6	16	47	7	"
"	"	"	"	"	"	"	"	"	"	"	" 500 "		52·4	54·0	56·4		66·4	17½	47	10	"
"	"	"	"	"	"	"	"	"	"	"	" 550 "		47·6	48·6	50·0		61·2	19½	49½	20	"
"	"	"	"	"	"	"	"	"	"	"	" 600 "		46·4	47·0	48·0		58·6	20	53½	32	"
"	"	"	"	"	"	"	"	"	"	"	" 650 "		42·2	42·2	42·2		53·0	23½	60½	74	"
"	"	"	"	"	"	"	"	"	"	"	" 700 "		37·2	37·2	37·2		48·2	29	63½	93	"
A.O.H.	8	"	"	"	"	"	"	"	"	6	OQ850 T400 AC		53·6	58·0	62·8		70·8	4	5	3	10
"	"	"	"	"	"	"	"	"	"	"	" 450 "		49·6	52·8	56·8		66·0	14	35	4	"
"	"	"	"	"	"	"	"	"	"	"	" 500 "		48·6	51·2	54·1		64·6	16½	44½	8	"
"	"	"	"	"	"	"	"	"	"	"	" 550 "		47·2	48·8	50·4		61·2	19	46	19	"
"	"	"	"	"	"	"	"	"	"	"	" 600 "		46·0	47·2	48·2		58·6	20	49½	35	"
"	"	"	"	"	"	"	"	"	"	"	" 650 "		43·6	44·2	44·8		54·4	21½	54½	54	"
"	"	"	"	"	"	"	"	"	"	"	" 700		26·8	28·4	30·8		52·0	21½	49½	69	"
		0·41		1·43			0·65	0·48	0·23	3½	OQ850 T640					53·6	61·2	22		56	2
	3—5	0·41	0·26	1·35	0·035	0·038	1·02	0·44	0·11	1⅛	OQ850 T640						62·8	21	57	68	6
	"	"	"	"	"	"	"	"	"	2½	" T650						60·8	22½	58	71	"
	"	"	"	"	"	"	"	"	"	4	" T650						60·0	21½	57	37	"
		0·41	0·19	1·32			0·71	0·60	0·21	1¾ sq.	OQ850 T640		47·0				61·6	24		72	3
		0·41	0·20	1·37			0·80	0·50	0·20	2	" T630					54·0	62·0	21	57	71	"
		0·41	0·23	1·41			0·79	0·43	0·22	8½	" T620					39·0	51·0	21½	56	54	"
B.E.		0·41	0·29	1·47	0·012	0·016	0·66	0·48	0·22	1⅝	OQ850 T630					47·9	56·0	24	61½	76	9
		0·42	0·19	1·48	0·032	0·028	0·64	0·42	0·19	59/64	OQ830 T670					52·0	59·0	21	52	67	7
		"	"	"	"	"	"	"	"	2 7/16	" T650					48·8	57·2	21	50	83	"

MECHANICAL PROPERTIES AT ROOM TEMPERATURE (Continued)

Process of Manufacture	ASTM Grain Size	Chemical Composition (%)								Section Size dia. (in.)	Heat Treatment °C	Tensile								Izod ft.lb.	Ref.
		C	Si	Mn	S	P	Ni	Cr	Mo			0.05% P.S. (0·05 Rp) tons/sq.in.	0·1% P.S. (0·1Rp) tons/sq.in.	0·2% P.S. (0·2Rp) tons/sq.in.	0·5% P.S. 0·5Rp) tons/sq.in.	Y.P. (Rs) tons/sq.in.	T.S. (Rm) tons/sq.in.	El % (A)	R.A. % (Z)		
H.F.	McQuaid Actual Ehn 7—8 8—9	0·42	0·37	1·29	0·038	0·035	0·65	0·43	0·14	1	AC875 OQ850 T550					66·5	71·9	17½	53½	30	6
		0·42	0·27	1·36			0·87	0·49	0·21	2½	OQ850 T660					45·0	54·0	27	61½	84	3
		0·42	0·26	1·38			0·82	0·43	0·24	6¾	OQ850 T610					45·0	56·0	21	56	56	"
	8	0·42		1·40			0·77	0·49	0·18	1⅛	OQ850 T200					111·3	121·7	3	21	6	1
	"	"		"			"	"	"	"	" T560					64·5	68·8	15	47	31	"
	"	"		"			"	"	"	"	" T590					61·3	67·9	17	47	47	"
	"	"		"			"	"	"	"	" T620					58·0	64·4	18½	55	56	"
	"	"		"			"	"	"	"	" T650					50·2	57·5	20½	55	81	"
	"	"		"			"	"	"	2½	OQ850 T560					61·5	68·8	15½	44	42	"
	"	"		"			"	"	"	"	" T590					57·9	64·8	18½	50	58	"
	"	"		"			"	"	"	"	" T620					52·8	61·3	18½	55	68	"
	"	"		"			"	"	"	"	" T650					48·3	58·0	18½	59	77	"
	8	"		"			"	"	"	4 (C)	OQ850 T560					56·4	65·4	15	31	30	1
	"	"		"			"	"	"	" (O)	" "					58·3	65·8	18½	55	42	"
	"	"		"			"	"	"	" (C)	" T590					53·1	67·0	17½	47	57	"
	"	"		"			"	"	"	" (O)	" "					57·5	66·0	20	56	56	"
	"	"		"			"	"	"	" (C)	" T620					49·7	62·0	20½	50	66	"
	"	"		"			"	"	"	" (O)	" "					52·8	61·9	21½	59	64	"
	"	"		"			"	"	"	" (C)	" T650					45·6	56·4	23	62	66	"
	"	"		"			"	"	"	" (O)	" "					47·8	58·3	23½	64	76	"
		0·43		1·34			0·97	0·56	0·21	2¼	OQ850 T630					51·2	58·6	23½		69	2
H.F.	McQuaid Actual Ehn 8 (7) 8	0·43	0·35	1·31	0·036	0·038	0·64	0·42	0·14	1	AC875 OQ850 T550					67·0	72·1	16½	49	33	6
		0·43	0·27	1·50			0·74	0·55	0·20	$1\frac{1}{32}$	OQ850 T630					58·5	66·0	23	54½	53	3
		0·43	0·21	1·40			0·95	0·54	0·22	17¾×3¾	OQ850 T620					49·2	55·7	20	52	52	"
		0·44		1·43			0·68	0·50	0·17	1⅛	OQ850 T550					70·0	76·8	16½		27	8
		"		"			"	"	"	"	" T600					64·2	70·5	18½		48	"
		"		"			"	"	"	"	" T650					54·7	61·7	22½		80	"
		"		"			"	"	"	2½	OQ850 T550					68·8	74·7	17½		27	"
		"		"			"	"	"	"	" T600					60·1	68·2	17½		57	"
		"		"			"	"	"	"	" T650					51·0	59·2	23		86	"
		"		"			"	"	"	4 sq.	OQ850 T550					58·8	67·8	16		21	"
		"		"			"	"	"	"	" T600					53·6	64·2	19½		57	"
		"		"			"	"	"	"	" T650					46·2	58·2	21		75	"
		"		"			"	"	"	6 sq.	OQ850 T550					59·9	68·1	15½		22	"
		"		"			"	"	"	"	" T600					54·5	63·3	17		43	"
		"		"			"	"	"	"	" T650					49·7	59·2	20		59	"

EFFECT OF TEMPERING ON THE MECHANICAL PROPERTIES AT ROOM TEMPERATURE

Chemical Composition (%)							
C	Si	Mn	S	P	Ni	Cr	Mo
0·34	0·23	1·20	0·016	0·027	0·63	0·35	0·24

Basic Electric Arc.
Test pieces oil quenched 860°C. Tempered as indicated.

(Ref. 9)

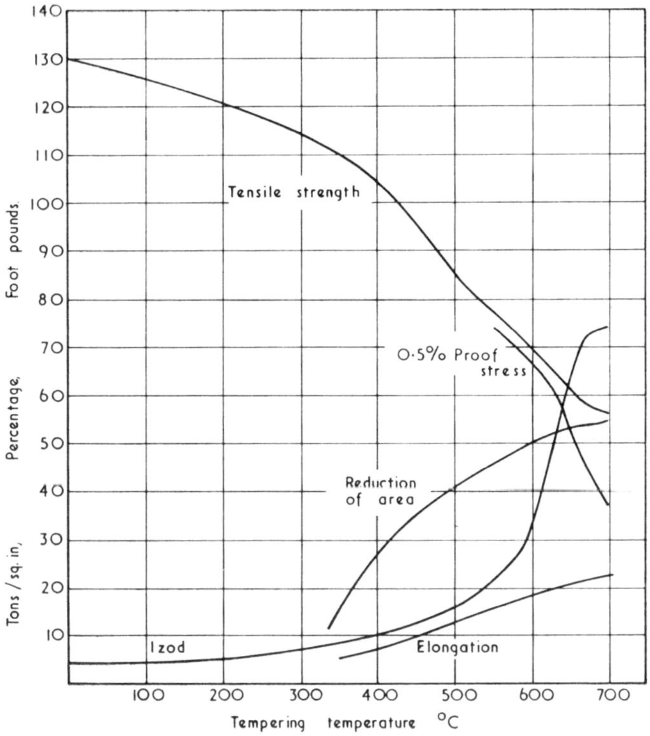

Fig. 100.18

EFFECT OF TEMPERING ON THE MECHANICAL PROPERTIES AT ROOM TEMPERATURE

(Continued)

Chemical Composition (%)							
C	Si	Mn	S	P	Ni	Cr	Mo
0·42	0·37	1·30	0·029	0·022	0·89	0·42	0·25

Basic Electric Arc.
1⅛ in. dia. bars. Oil quenched 840°C. Tempered as indicated.

(Ref. 9)

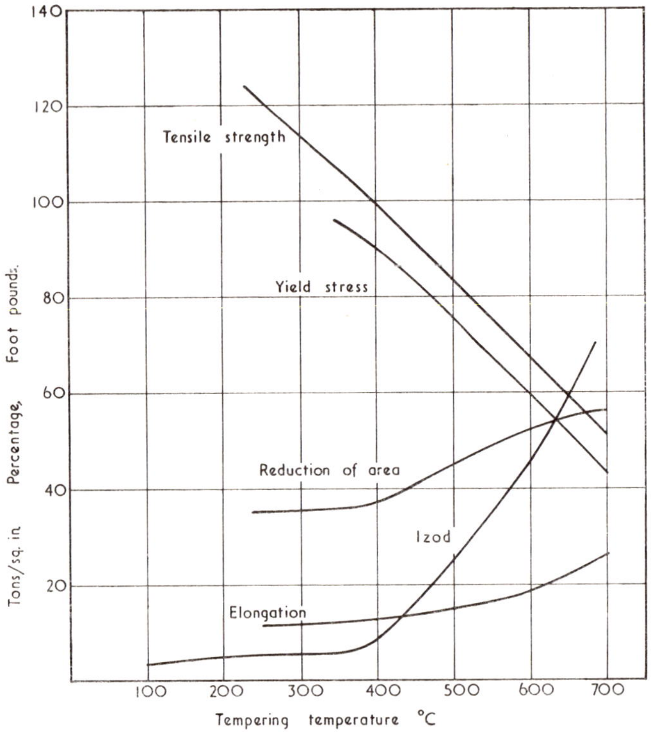

Fig. 100.19

EFFECT OF TEMPERING ON THE MECHANICAL PROPERTIES AT ROOM TEMPERATURE

(Continued)

Chemical Composition (%)							
C	Si	Mn	S	P	Ni	Cr	Mo
0·45	0·28	1·53	0·015	0·025	1·14	0·55	0·25

Basic Electric Arc.
Test piece. Oil quenched 830°C. Tempered as indicated.

(Ref. 9)

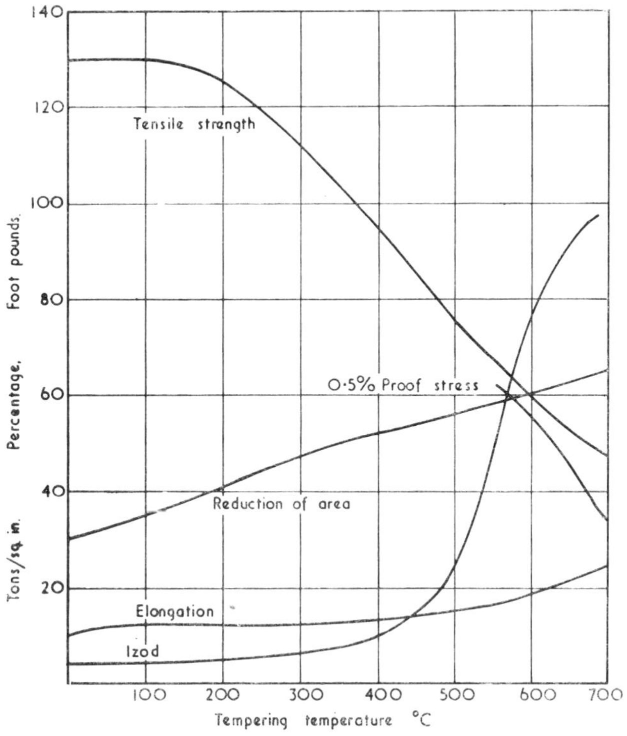

Fig. 100.20

EFFECT OF TEMPERING ON THE MECHANICAL PROPERTIES AT ROOM TEMPERATURE

(Continued)

Nominal Chemical Composition (%)				
C	Mn	Ni	Cr	Mo
0·40	1·35	0·75	0·45	0·20

1¼ in. dia. bars. Oil quenched 850°C. Tempered as indicated.

(Ref. 3)

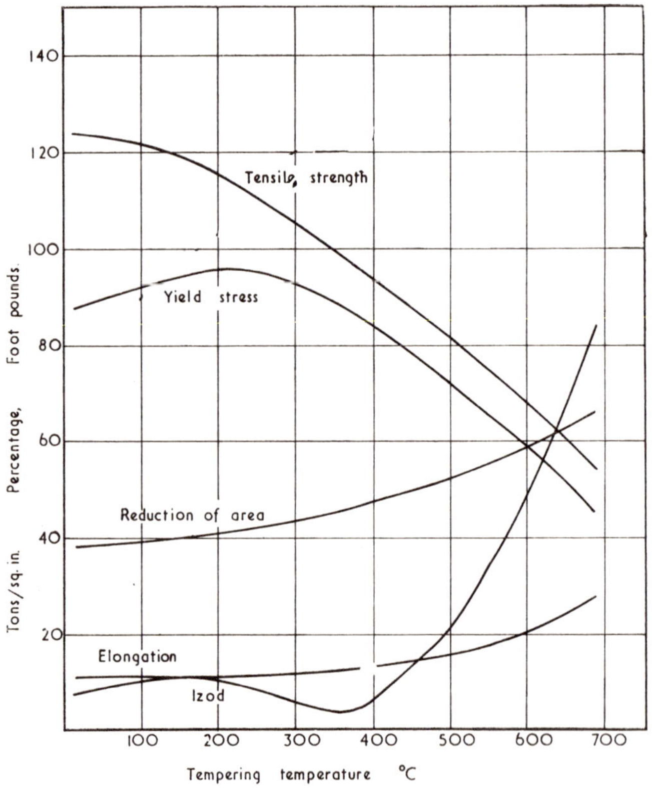

Fig. 100.21

EFFECT OF SECTION SIZE ON THE MECHANICAL PROPERTIES AT ROOM TEMPERATURE
(Typical curves)

Oil quenched 850°C. Tempered 600°C.

(Ref. 3)

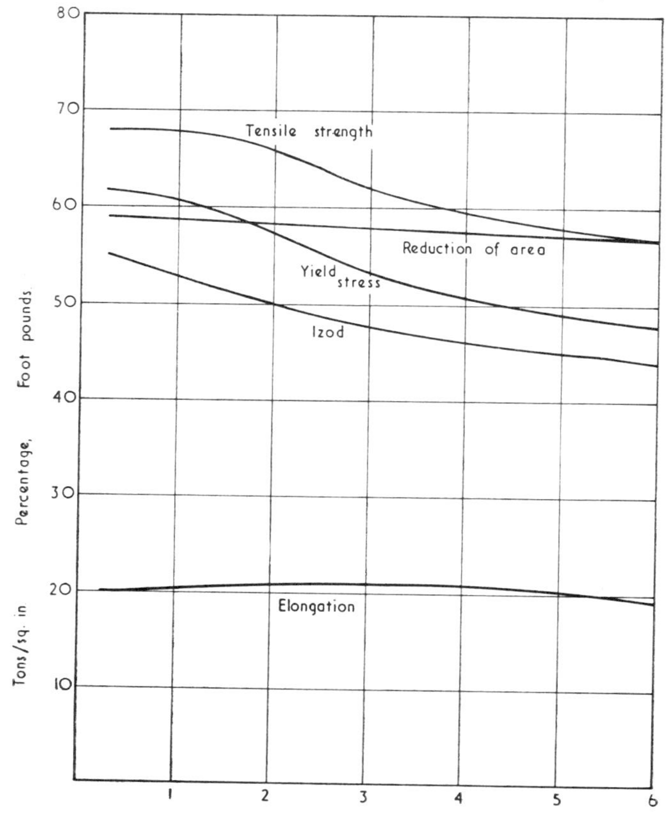

Fig. 100.22

Oil quenched 850°C. Tempered 625°C.

(Ref. 3)

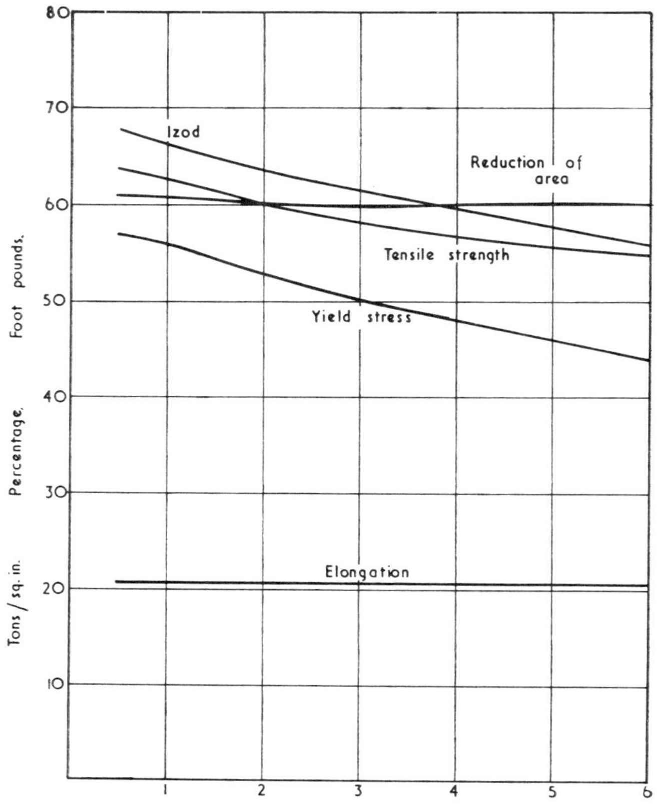

Fig. 100.23

EFFECT OF SECTION SIZE ON THE MECHANICAL PROPERTIES AT ROOM TEMPERATURE

(Typical curves)

(Continued)

Oil quenched 850°C. Tempered 650°C.

(Ref. 3)

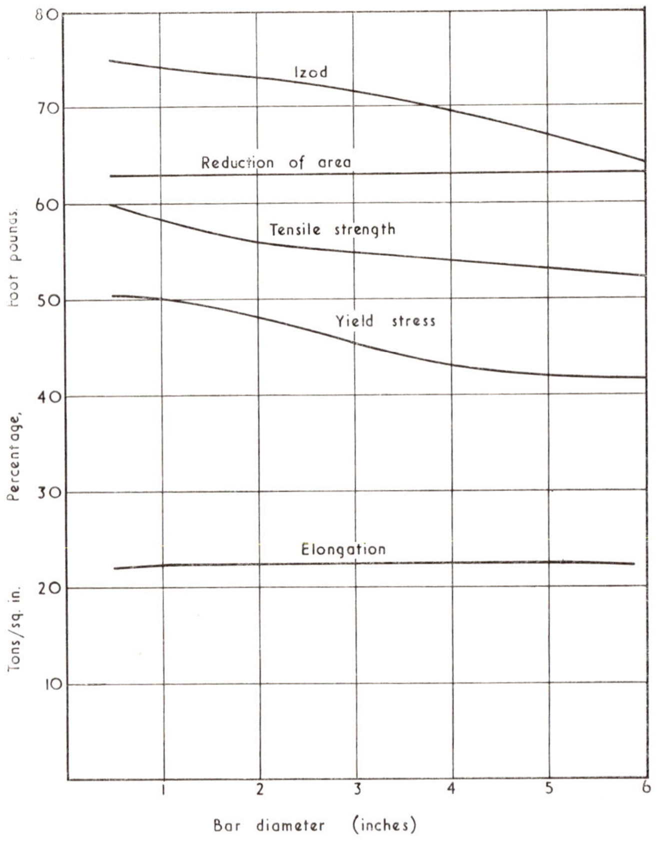

Fig. 100.24

IMPACT TRANSITION TEMPERATURE DATA

Chemical Composition (%)							
C	Si	Mn	S	P	Ni	Cr	Mo
0·41	0·23	1·31	0·038	0·038	0·80	0·50	0·20

McQuaid Ehn Grain Size 6—7 (ASTM).
As treated Grain Size 0.001 in.
1⅜ in. dia. bars. Oil quenched 850°C. Tempered 620/630°C, air cool.

Tensile Test Results

0·05% P.S. (0·05Rp)	0·1% P.S. (0·1Rp)	0·2% P.S. (0·2Rp)	0·5% P.S. (0·5Rp)	T.S. (Rm)	El (A) %	R.A (Z) %
tons/sq.in.						
51·3	53·3	55·2	56·5	64·3	21·8	55·1

(Ref. 3)

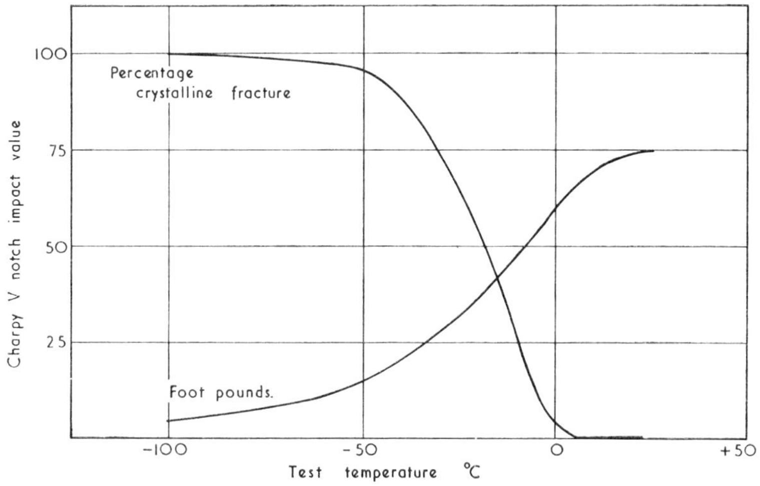

Fig. 100.25

MECHANICAL PROPERTIES AT LOW TEMPERATURE

Chemical Composition (%) C	Ni	Cr	Mo	Heat Treatment °C	Test Temperature °C	Tensile Y.P. (Rm) tons/sq.in.	T.S. (Rs) tons/sq.in.	El % (A)	R.A. % (Z)	Charpy V ft.lb.	Ref.
0·30	0·55	0·52	0·18	Normalized	+20	35·6	39·7	15½	58	29	11
"	"	"	"	"	—78	40·2	45·5	17½	55	4	"
"	"	"	"	"	—192		67·0	13	32	3	"
"	"	"	"	"	—253					3	"
"	"	"	"	OQ815 T455	+20	53·5	62·5	15½	59½	39	"
"	"	"	"	" "	—78	60·3	66·0	18½	59	21	"
"	"	"	"	" "	—127					16	"
"	"	"	"	" "	—198					5	"
"	"	"	"	" "	—253					5	"

MECHANICAL PROPERTIES IN TORSION

Chemical Composition (%)								Heat Treatment	Test Temperature	Tensile				Torsion			Ref.
C	Si	Mn	S	P	Ni	Cr	Mo			0·1% P.S. (0·1Rp)	T.S. (Rm)	El % (A)	R.A. % (Z)	Y.P.	Apparent Shear Strength	Probable max. Shear Strength	
										tons/sq.in.				tons/sq.in.			
0·41	0·25	1·42	0·041	0·033	0·66	0·46	0·16	H&T	RT	43·9	59·4	25		32·8	52·4	39·3	12

FATIGUE PROPERTIES

ASTM Grain Size	Chemical Composition (%)								Section Size dia. (in.)	Heat Treatment °C	Test Temperature °C	Tensile				Fatigue Limit			Ref.
	C	Si	Mn	S	P	Ni	Cr	Mo				Y.P. (Rs)	T.S. (Rm)	El %	R.A. % (Z)	Rotating Bending (10^7 cycles)	Fluctuating Axial (5×10^7 cycles)	Reverse Bending (10^7 cycles)	
												tons/sq.in.							
5—8	0·38	0·25	0·44	0·036	0·025	0·57	0·42	0·15	$\frac{7}{8}$	OQ850 T625	RT	55·2	62·2	23	56	±33·5			2
	Within specification								$\frac{1}{2}$			64·2	65·9	17	—		26±22		13
	Within specification										+25	55·5	62·5	15½	59½			±29	11
											—78	60·3	66·0	18½	59			±36	"
											—196							±54	"

LIST OF REFERENCES

1. Private correspondence with The United Steel Companies Ltd.
2. Private correspondence with Firth Brown Ltd.
3. Private correspondence with The English Steel Corporation Ltd.
4. Atlas of Isothermal Transformation Diagrams of the B. S. En Steels, Iron and Steel Institute Special Report No. 56, 2nd Edition, 1956.
5. Transformation Characteristics of Direct Hardening Nickel-alloy Steels. International Nickel Ltd. 3rd. Edition, 1958.
6. Symposium on the Hardenability of Steel organised by the Technical Advisory Committee of the Special and Alloy Steels Committee (Ministry of Supply) and the Iron and Steel Institute. Iron and Steel Institute Special Report No 36, 1946.
7. Private correspondence with The Park Gate Iron and Steel Co. Ltd.
8. British Standard 971: Commentary on the British Standard Wrought Steels, En Series. British Standards Institution, London, 1950.
9. Private correspondence with Jessop–Saville Ltd.
10. Private correspondence with Brown Bayley Steels Ltd.
11. J. L. Zambrow and M. G. Fontana. Mechanical Properties including Fatigue of Aircraft Alloys at very low temperature. Am. Soc. Metal. Preprint No. 19, Oct., 1948.
12. Private correspondence with Dowty Rotol Ltd.
13. Private correspondence with the Royal Aircraft Establishment, Farnborough.

LOW NICKEL–CHROMIUM–MOLYBDENUM STEEL

SPECIFICATION

CHEMICAL COMPOSITION (%)

C	Si	Mn	S	P	Ni	Cr	Mo
0·35—0·45	0·10—0·35	0·40—0·80	0·050max.	0·050max.	1·20—1·60	0·90—1·40	0·10—0·20

MECHANICAL PROPERTIES

	Hardened and Tempered Condition				
	S	T	U	V	W
Limiting ruling section in.	6	4	$2\frac{1}{2}$	$1\frac{1}{8}$	$1\frac{1}{8}$
Tensile strength, tons/sq.in. min. (Rm)	50	55	60	65	70
Yield stress(1), tons/sq.in. min. (Rs)	38	44	48	52	58
0·2% Proof stress(2), tons/sq.in. min. (0·2Rp)	36	41	46	50	55
Elongation % min. (A)	20	18	17	16	15
Izod impact value, ft.lb. min.	40	40	35	35	30
Brinell hardness(3) number, HB	223/277	248/302	269/321	293/341	311/375

When supplied in the softened condition the material shall have a Brinell hardness number not exceeding 269.

(1) Not to be used as acceptance tests except by special arrangement between purchaser and manufacturer.
(2) When specifically requested in enquiry or order.
(3) For information purposes only. Not contractual part of the specification.

RELATED SPECIFICATIONS

United Kingdom			America		France	Germany			Sweden	USSR	
En No.	BS	Air	SAE	AISI	AFNOR	Werkstoff No.	Name	DIN	SIS	GOST	Mark
110		S118	4340	4340	35NOD6	1.6511 1.6582	36Cr Ni Mo4 34Cr Ni Mo6	17200 17200	14.2541	4543-61	40KhMA

APPLICATIONS

An alloy steel of similar composition to En 24 but with lower molybdenum and nickel contents and consequently having a lower hardenability and a more limited range of mechanical properties. It can be heat treated to give tensile strengths of the order of 55 tons/sq.in. in fairly large sections and 65 tons/sq.in. in small sections combined with good ductility and resistance to shock. Applications are similar to those of En 24, e.g. general machine parts, bolts, nuts, gears, axles, connecting rods, shafts.

WELDING

(See Introduction for key to symbols used in this table.)

	Welding Process					
	MA	G	IG	S	F	B
En 110	d	bx	g	e	b	j

Reproduced by kind permission of H.M.S.O.

MACHINABILITY

In the softened condition the machinability is approximately 55% of that for mild steel. When hardened to 55 to 60 tons/sq. in., the machinability is approximately 40% of that for mild steel (En 3).

HOT WORKING AND HEAT TREATMENT TEMPERATURES

Forging, rolling and stamping	1200°C, finish above 900°C.
Retarded cooling is advisable for large or intricate sections.	
Annealing	820°C—850°C, furnace cool.
Softening (sub-critical annealing)	630°C—670°C, air or furnace cool.
Hardening	820°C—850°C, oil quench.
Tempering	550°C—660°C, air cool or oil quench.

PHYSICAL PROPERTIES

(See also En 24)

SPECIFIC GRAVITY (d)

Chemical Composition (%)								Heat Treatment °C	Specific Gravity at 20°C	Ref.
C	Si	Mn	S	P	Ni	Cr	Mo			
0·38	0·25	0·65			1·40	1·20	0·15	Softened	7·86	1
0·39	0·22	0·65	0·044	0·028	1·39	1·02	0·14	OQ840, T6650 OQ	7·84	2
		(Within specification)							7·85	3

SPECIFIC HEAT (c)

Chemical Composition (%)						Heat Treatment °C	Specific Heat cal/g. deg. C	Ref.
C	Si	Mn	Ni	Cr	Mo		20 to 100°C	
0·38	0·25	0·65	1·40	1·20	0·15	Softened	0·118	1

MEAN COEFFICIENT OF THERMAL EXPANSION (α)

Chemical Composition (%)						Heat Treatment °C	$10^6\times$ Mean Coefficient of Thermal Expansion per deg. C								Ref.
C	Si	Mn	Ni	Cr	Mo		20 to 100°C	20 to 200°C	20 to 300°C	20 to 400°C	20 to 500°C	20 to 600°C	20 to 650°C	20 to 700°C	
0·38	0·25	0·65	1·40	1·20	0·15	Softened	11		13		14			15	1
0·39	0·22	0·65	1·39	1·02	0·14	OQ840, T650 OQ	12·0	12·6	13·2	13·6	14·0	14·3	14·4		2
													20 to 680°C		
	(Within specification)					Softened	12·5		13·3		14·0			14·5	3

ELECTRICAL RESISTIVITY (ϱ)

Chemical Composition (%)						Heat Treatment °C	Electrical Resistivity microhm-cm								Ref.
C	Si	Mn	Ni	Cr	Mo		20°C	100°C	200°C	300°C	400°C	500°C	600°C	650°C	
0·39	0·22	0·65	1·39	1·02	0·14	OQ840, T650 OQ	24·8	29·2	35·6	42·8	54·0	65·2	78·0	85·6	2

YOUNG'S MODULUS (E)

Chemical Composition (%)						Heat Treatment °C	Young's Modulus tons/sq.in.								Method of Determi-nation	Ref.
C	Si	Mn	Ni	Cr	Mo		20°C	100°C	200	300°C	400°C	500°C	600°C	650°C		
0·39	0·22	0·65	1·39	1·02	0·14	OQ840, T640 OQ	13,600	13,300	12,900	12,500	12,000	11,400	10,600	10,100	Vibration	2
0·39	0·25	0·61	1·45	1·06	0·16	H & T (70 tons/sq.in.)	13,500	12,950	12,850	12,200					Static	4

TRANSFORMATION CHARACTERISTICS

ISOTHERMAL TRANSFORMATION

Chemical Composition (%)							
C	Si	Mn	S	P	Ni	Cr	Mo
0·39	0·23	0·62	0·018	0·021	1·44	1·11	0·18

Basic Electric Arc.
McQuaid Ehn Grain Size 2—3 ASTM.
As quenched Grain Size 7 ASTM.

Bars hot rolled to ⅜ in. dia., cold drawn to 11 SWG, with intermediate annealing at 600°C.

Austenitizing Temperature 845°C, 30 min.

(Ref. 5)

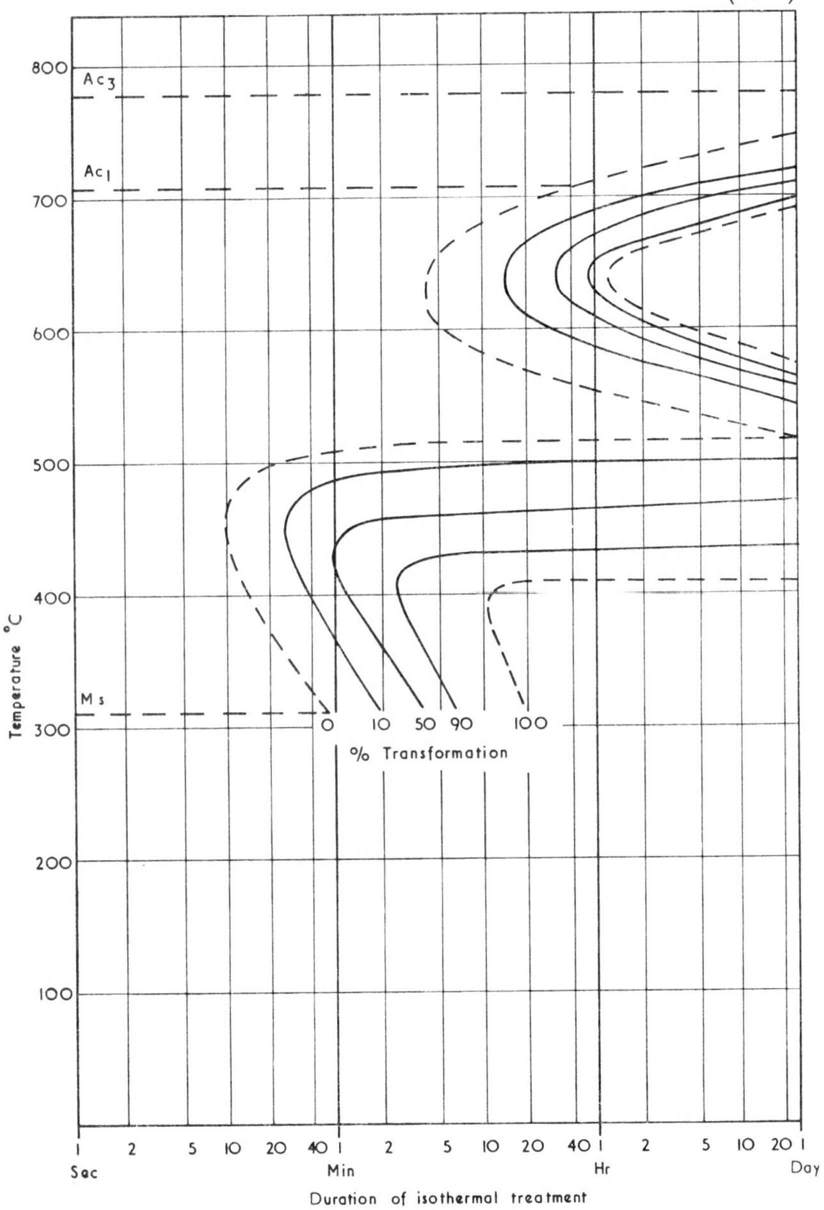

Fig. 110.1

En 110

ISOTHERMAL TRANSFORMATION (Continued)

Chemical Composition (%)							
C	Si	Mn	S	P	Ni	Cr	Mo
0·44	0·23	0·58	0·004	0·029	1·40	1·26	0·11

Basic Electric Arc.
McQuaid Ehn Grain Size 6 (ASTM).
As quenched Grain Size 7—8 (ASTM).
Austenitizing Temperature 860°C.

(Ref. 6)

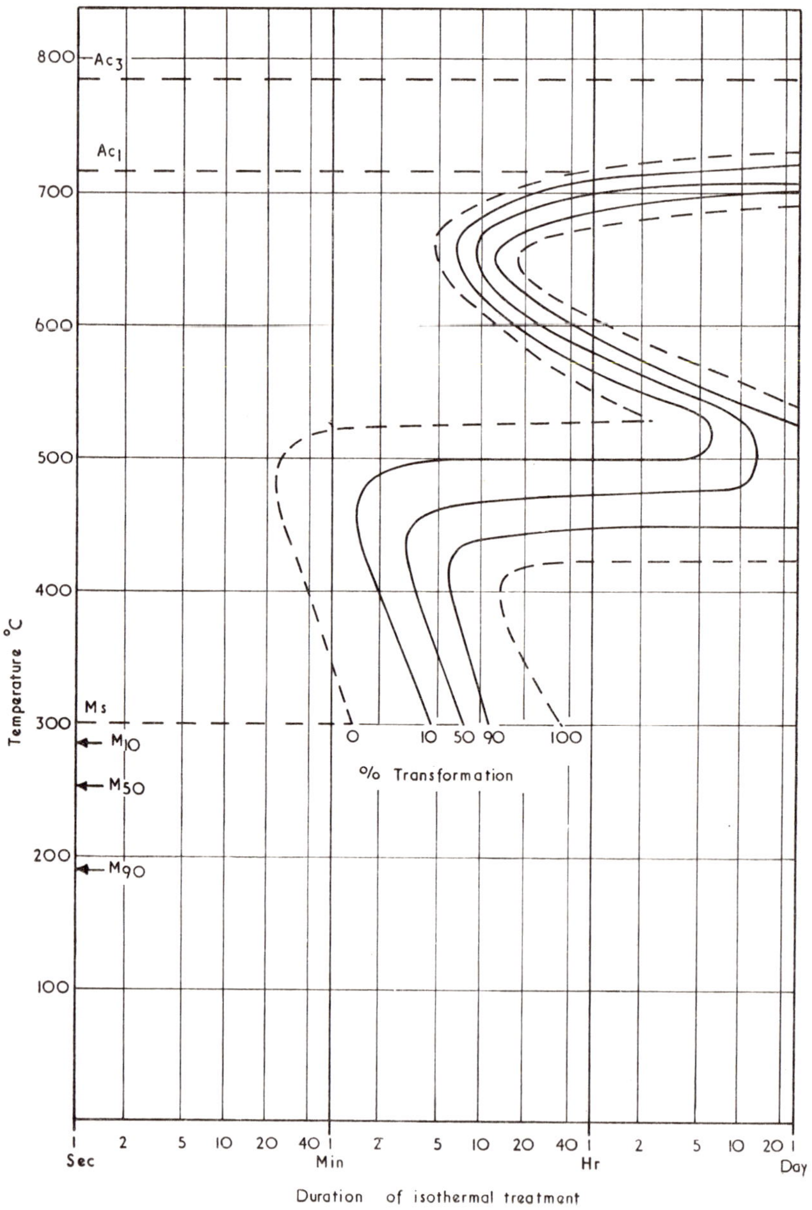

Fig. 110.2

CONTINUOUS COOLING TRANSFORMATION DIAGRAMS

Chemical Composition (%)							
C	Si	Mn	S	P	Ni	Cr	Mo
0·44	0·23	0·58	0·004	0·029	1·40	1·26	0·11

Basic Electric Arc.
McQuaid Ehn Grain Size 6 (ASTM).
As quenched Grain Size 7—8 (ASTM).
Austenitizing Temperature 860°C.

(Ref. 6)

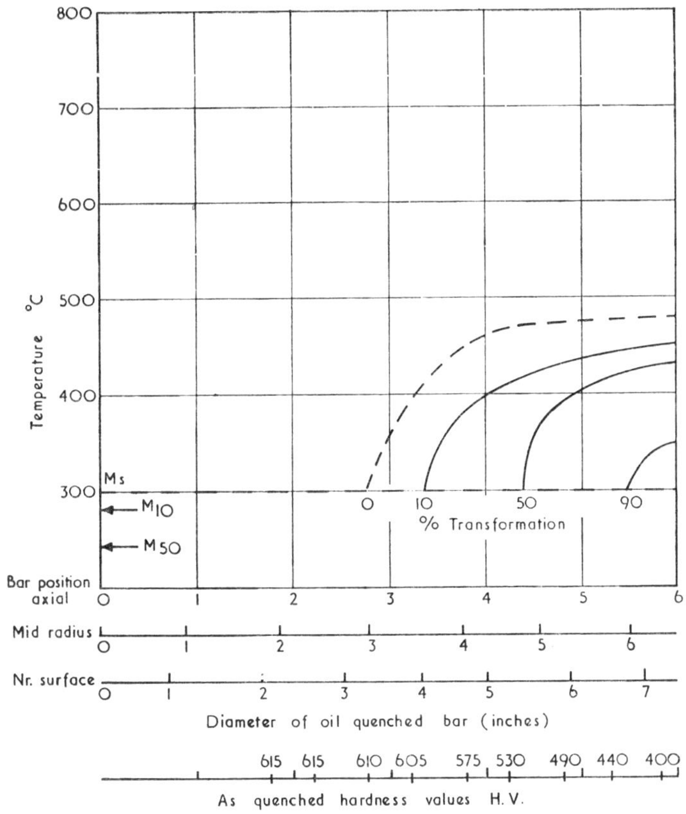

Fig. 110.3

CONTINUOUS COOLING TRANSFORMATION DIAGRAMS (Continued)

Air cooled from 840°C.

(Ref. 2)

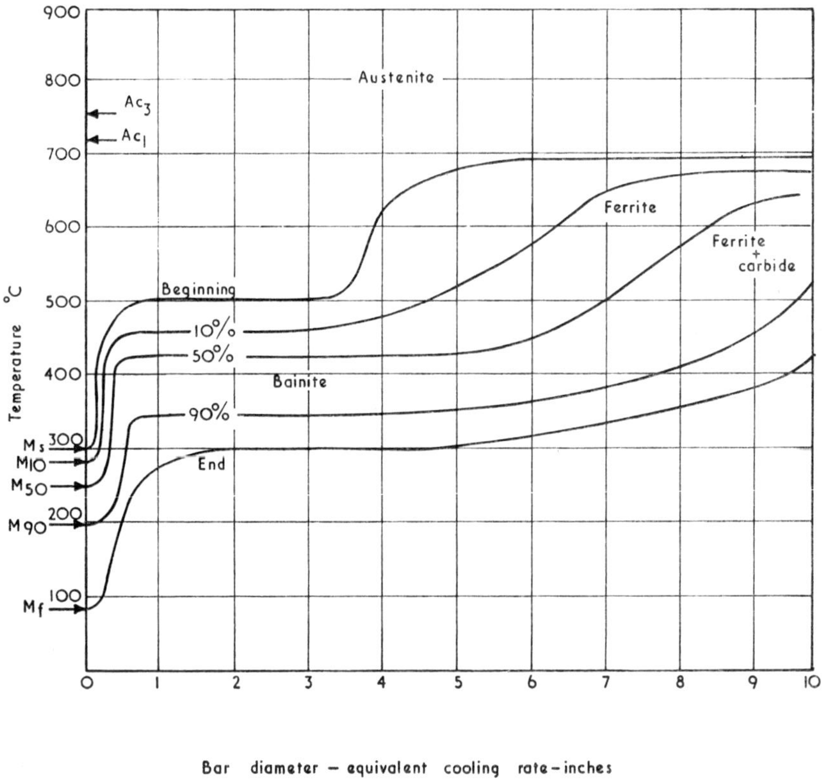

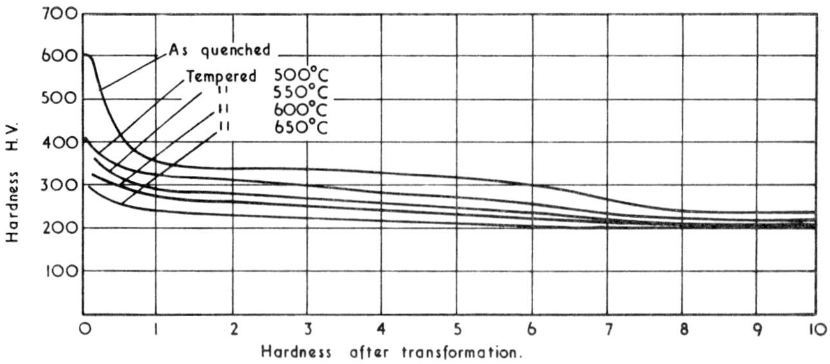

Fig. 110.4

CONTINUOUS COOLING TRANSFORMATION DIAGRAMS (Continued)

Oil quenched from 840°C.

(Ref. 2)

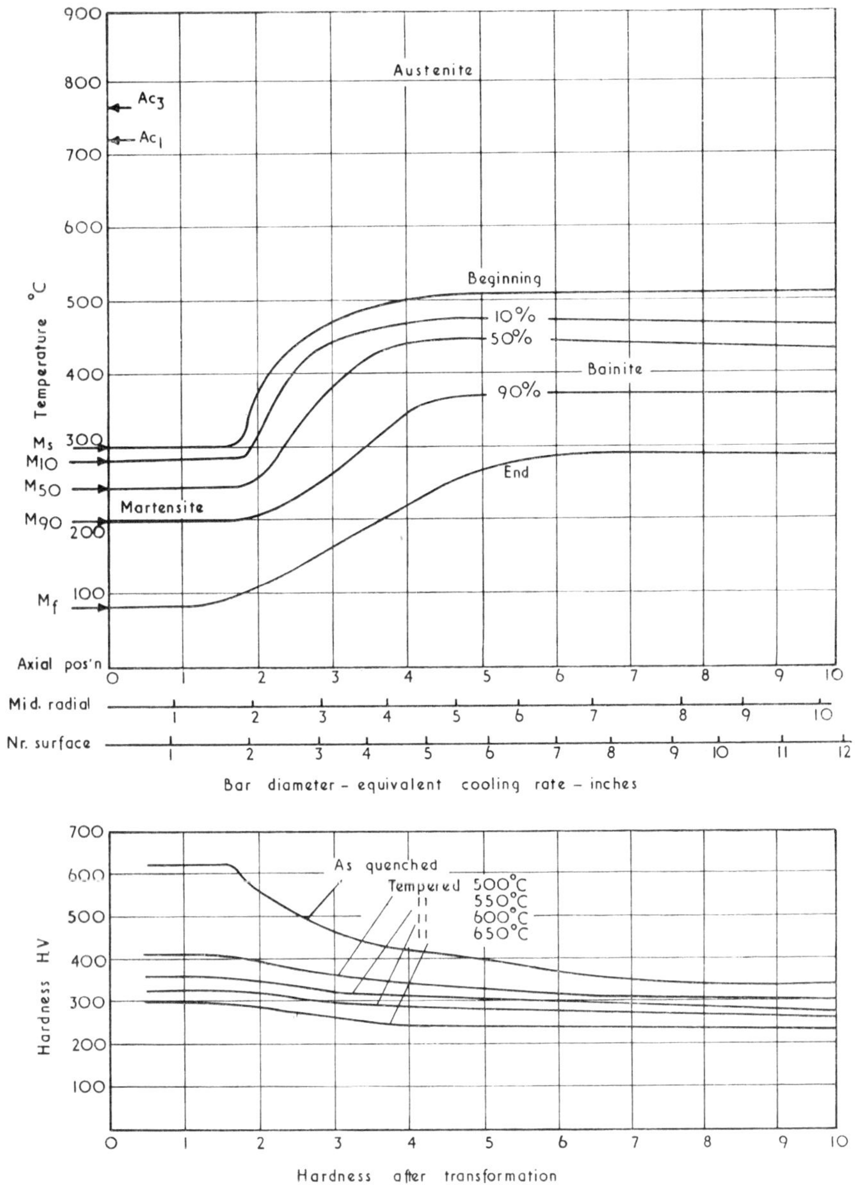

Fig. 110.5

En 110

CONTINUOUS COOLING TRANSFORMATION DIAGRAMS (Continued)

Water quenched from 840°C.

(Ref. 2)

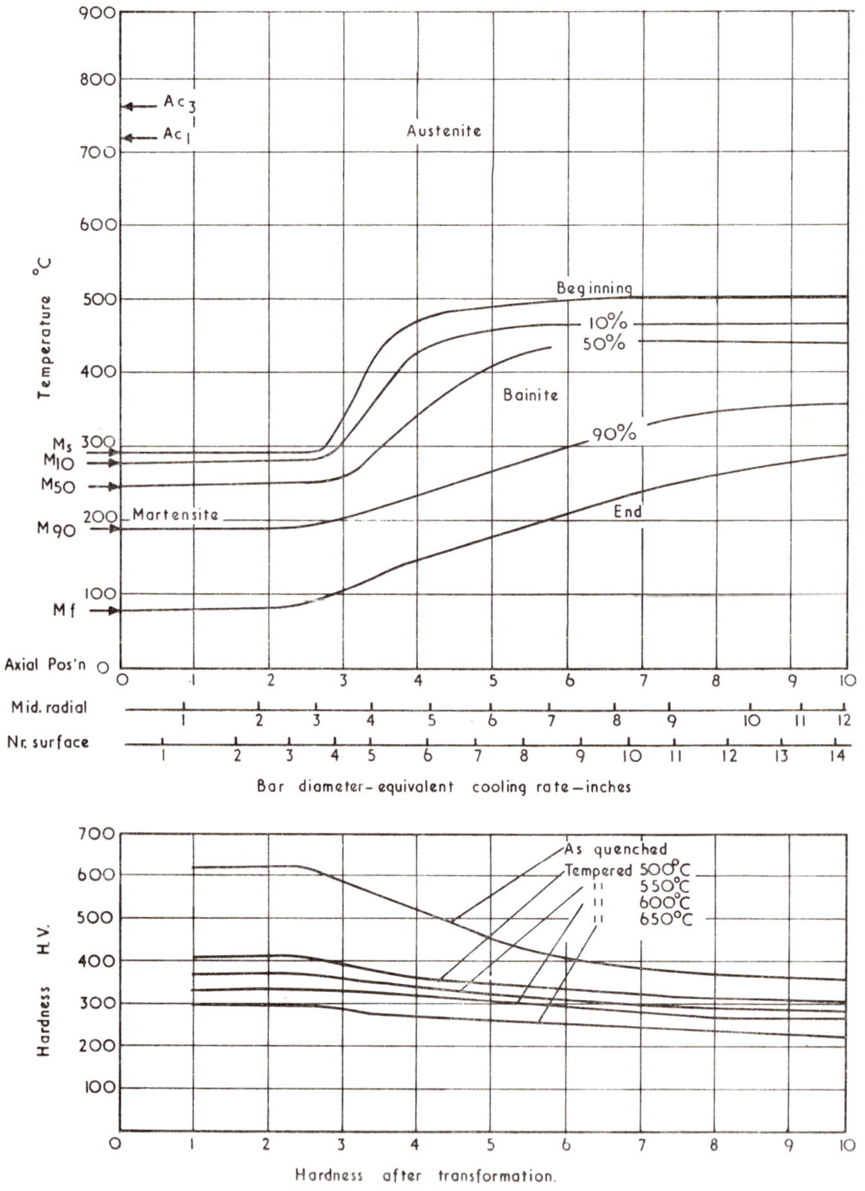

Fig. 110.6

HARDENABILITY

END QUENCH HARDENABILITY DATA

Chemical Composition (%)							
C	Si	Mn	S	P	Ni	Cr	Mo
0·39	0·23	0·62	0·018	0·021	1·44	1·11	0·18

Basic Electric Arc.
McQuaid Ehn Grain Size 2—3 (ASTM).
As quenched Grain Size 7 (ASTM).
End quenched from 845°C. (Ref. 5)

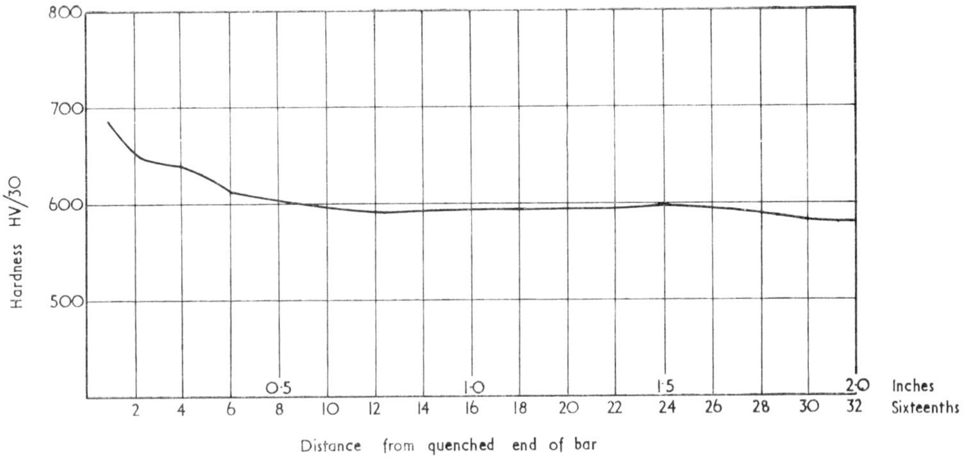

Fig. 110.7

Chemical Composition (%)							
C	Si	Mn	S	P	Ni	Cr	Mo
0·44	0·23	0·58	0·004	0·029	1·40	1·26	0·11

Basic Electric Arc.
McQuaid Ehn Grain Size 6 (ASTM).
As quenched Grain Size 7—8 (ASTM).
End quenched from 860°C. Tempered as indicated. (Ref. 6)

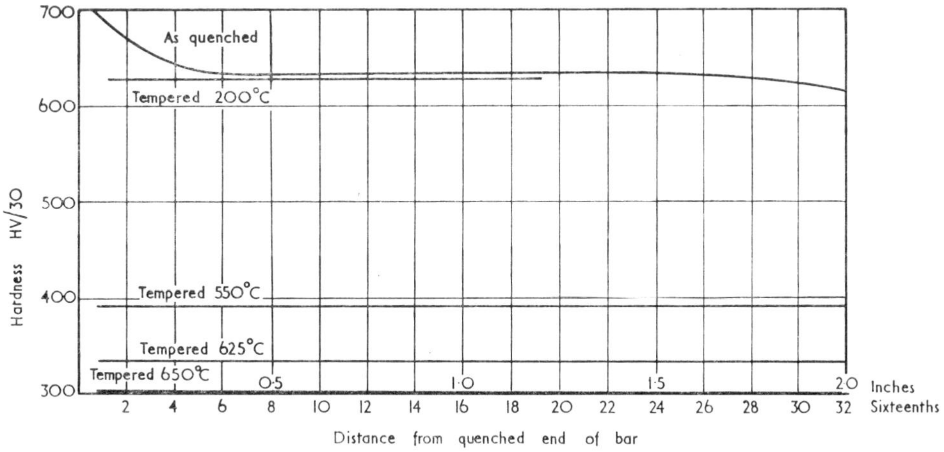

Fig. 110.8

En 110

END QUENCH HARDENABILITY DATA (Continued)

Curve No.	Chemical Composition (%)								McQuaid Ehn Grain Size (ASTM)
	C	Si	Mn	S	P	Ni	Cr	Mo	
A	0·41	0·24	0·62	0·027	0·036	1·28	0·81	0·15	6—7
B	0·36	0·25	0·54	0·035	0·026	1·38	0·74	0·24	3—5
C	0·38	0·27	0·60	0·033	0·034	1·48	0·83	0·26	6
D	0·43	0·28	0·63	0·019	0·030	1·41	1·10	0·26	5—7
E	0·42	0·29	0·60	0·039	0·030	1·36	0·96	0·27	3—5

End quenched 850°C.

(Ref. 7)

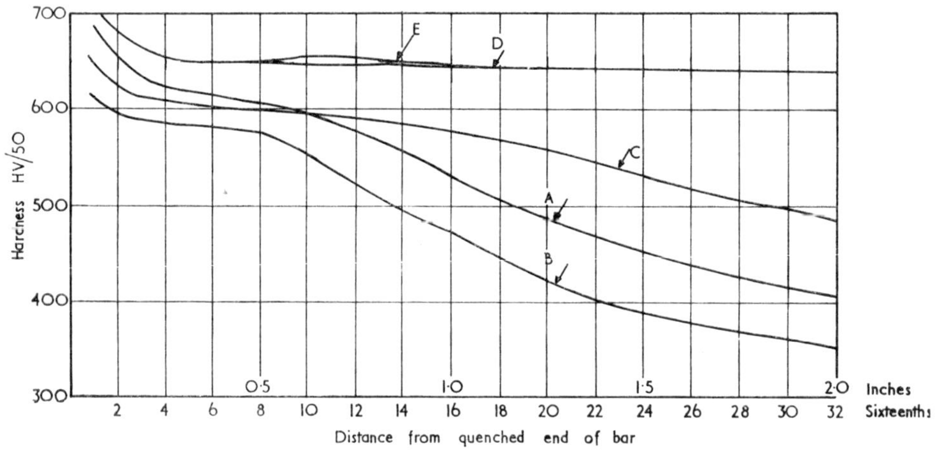

Fig. 110.9

End quenched 850. Tempered 640°C.

(Ref. 7)

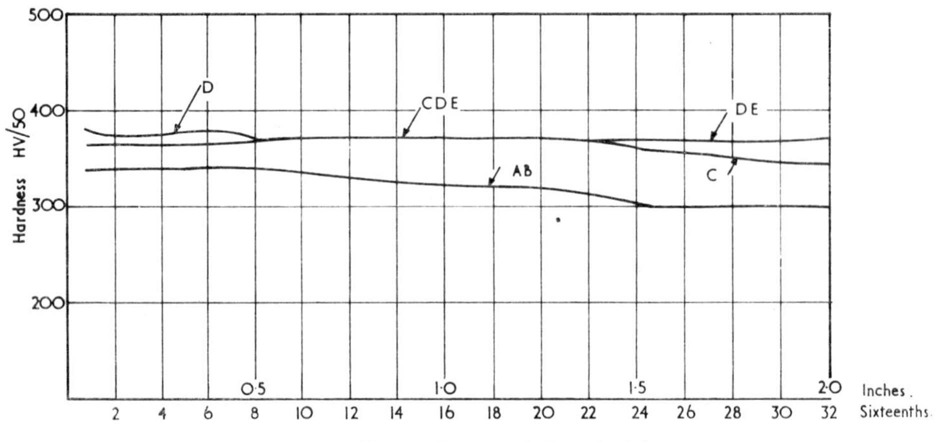

Fig. 110.10

END QUENCH HARDENABILITY DATA (Continued)

Curve No.	Chemical Composition (%)								McQuaid Ehn Grain Size (ASTM)	Pretreatment °C	Quenching Temperature °C
	C	Si	Mn	S	P	Ni	Cr	Mo			
F	0·42	0·21	0·62	0·012	0·012	1·45	1·20	0·18	3—7	N850	850
G	0·40	0·23	0·63	0·012	0·018	1·47	1·00	0·25	2—7	"	"
H	0·40	0·22	0·59	0·010	0·021	1·60	1·09	0·17	6—8	"	"
J	0·40	0·25	0·60	0·017	0·028	1·45	1·13	0·24	2—7	"	"
K	0·39	0·17	0·61	0·019	0·021	1·55	1·10	0·18	1—7	"	"
L	0·37	0·29	0·63	0·021	0·029	1·33	0·97	0·19	4—5	N850, T650	"

(Ref. 7)

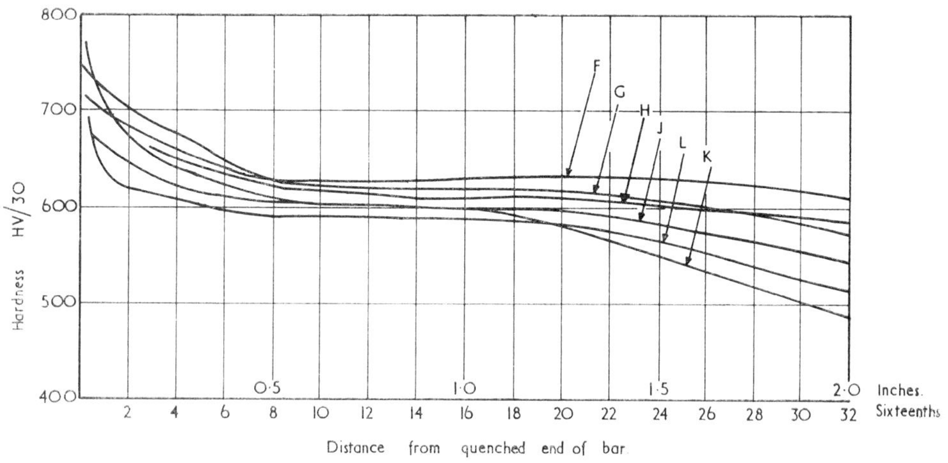

Fig. 110.11

END QUENCH HARDENABILITY BAND

(Ref. 2 and 3)

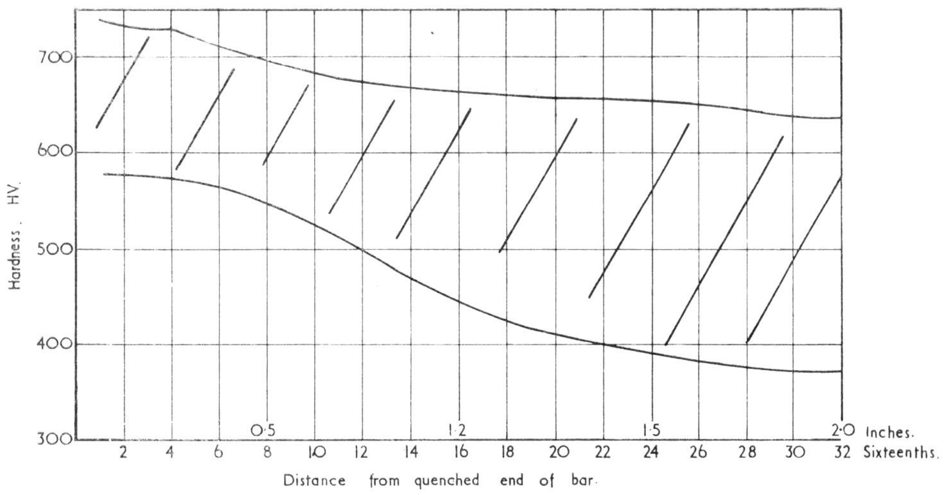

Fig. 110.12

These curves have been constructed from results obtained on several casts of En 110.

HARDNESS TRAVERSE DATA

Chemical Composition (%)							
C	Si	Mn	S	P	Ni	Cr	Mo
0·37	0·29	0·63	0·021	0·029	1·33	0·97	0·19

Basic Electric Arc.
Tests carried out in two laboratories ("A" and "B" below).

Laboratory A:
Bars T650°C, air cooled. Oil quenched 850°C.

(Ref. 7)

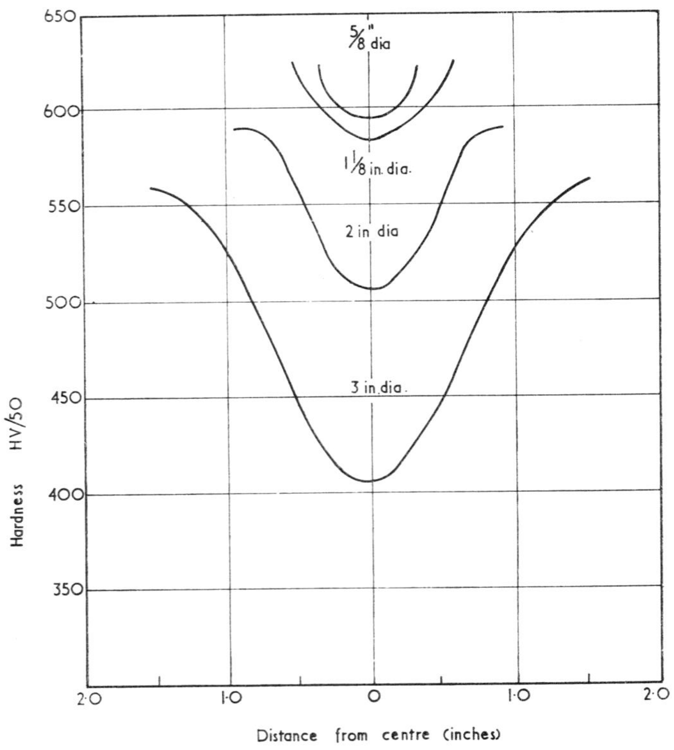

Fig. 110.13

Laboratory B:
Bars N850°C, T650°C. Oil quenched 850°C.

(Ref. 7)

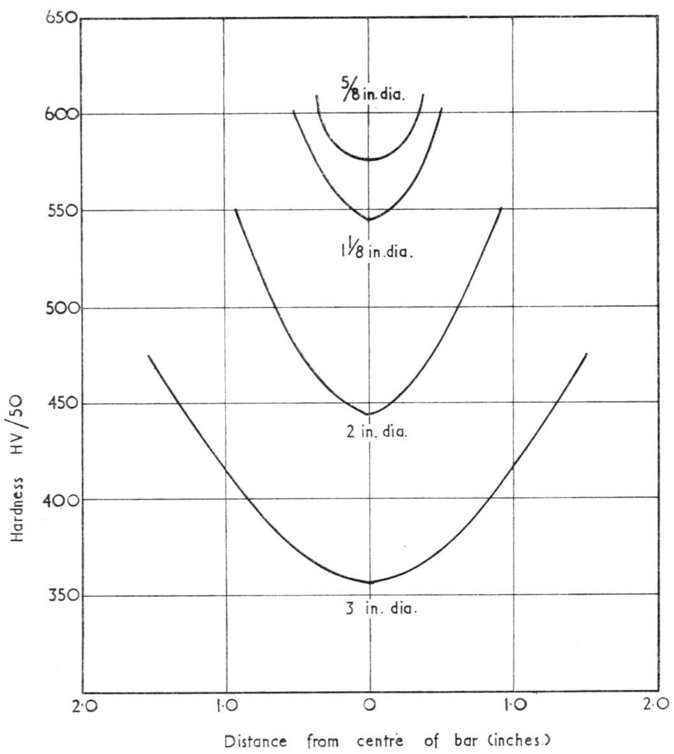

Fig. 110.14

MECHANICAL PROPERTIES AT ROOM TEMPERATURE

Process of Manufacture	ASTM Grain Size	Chemical Composition (%) C	Si	Mn	S	P	Ni	Cr	Mo	Section Size dia (in.)	Heat Treatment °C	Tensile 0·1% P.S. (0·1Rp) tons/sq. in.	0·2% P.S. (0·2Rp)	Y.P. (Rs)	T.S. (Rm)	El % (A)	R.A. % (Z)	Izod ft.lb.	Ref.
B.E.		0·34		0·58			1·40	1·46	0·21	$1\frac{1}{8}$	OQ850 T200			98·0	113·5	$11\frac{1}{2}$	$44\frac{1}{2}$		1
	3—5	0·36	0·25	0·54	0·035	0·026	1·38	0·74	0·24	$1\frac{1}{8}$	OQ850 T640				62·7	$20\frac{1}{2}$	$60\frac{1}{2}$	60	7
		"	"	"	"	"	"	"	"	4	" T600				57·2	19	42	16	"
		"	"	"	"	"	"	"	"	6	" T600				55·2	$15\frac{1}{2}$	35	18	"
		0·36					1·40	1·05	0·16	$1\frac{1}{8}$	OQ850 T590				67·4	$20\frac{1}{2}$		59	1
		0·36	0·23	0·64	0·038	0·031	1·44	0·96	0·16	$\frac{3}{4}$	OH T200		101·8		125·0	$15\frac{1}{2}$	48		2
	7	0·37		0·67			1·38	1·15	0·15	$1\frac{1}{8}$	OQ840 T560			70·3	74·3	17	55	34	2
	"	"		"			"	"	"	"	" T590			65·2	69·8	$20\frac{1}{2}$	62	56	"
	"	"		"			"	"	"	"	" T620			58·8	66·3	21	62	64	"
	"	"		"			"	"	"	"	" T650			52·8	60·2	$23\frac{1}{2}$	63	72	"
	"	"		"			"	"	"	$2\frac{1}{2}$ sq.	OQ840 T560			65·1	71·5	19	57	44	"
	"	"		"			"	"	"	"	" T590			61·5	68·0	21	62	64	"
	"	"		"			"	"	"	"	" T620			55·4	64·0	20	62	72	"
	"	"		"			"	"	"	"	" T650			50·1	58·2	$23\frac{1}{2}$	66	88	"
	"	"		"			"	"	"	4 sq.	OQ840 T560			61·7	68·8	19	57	33	"
	"	"		"			"	"	"	"	" T590			55·0	62·9	20	59	65	"
	"	"		"			"	"	"	"	" T620			51·5	60·2	21	62	74	"
	"	"		"			"	"	"	"	" T650			48·6	56·8	24	64	86	"
	"	"		"			"	"	"	6 sq.	OQ840 T560			62·0	68·0	$15\frac{1}{2}$	39	19	"
	"	"		"			"	"	"	"	" T590			56·2	63·4	$17\frac{1}{2}$	50	45	"
	"	"		"			"	"	"	"	" T620			50·6	58·5	$19\frac{1}{2}$	52	66	"
	"	"		"			"	"	"	"	" T650			45·7	56·4	$21\frac{1}{2}$	55	72	"
B.E.	0·37						1·36	1·08	0·15	$3\frac{3}{4}$ (M)	OQ870 T615			46·0	55·4	$21\frac{1}{2}$	61	74	1
"	"						"	"	"	$3\frac{5}{8}$ (M)	OQ850 T600			51·6	59·8	21	58	36	"
B.E.	0·37						1·33	1·05	0·19	$1\frac{3}{8}$ (C)	OQ850 T640			57·0	63·0	22	61	70	"
"	"						"	"	"	6 (M)	OQ870 T670			41·4	51·4	24	64	65	"
"	0·37			0·60			1·40	1·06	0·12	6	OQ840 T620			50·8	58·7	$19\frac{1}{2}$		79	8
	"			"			"	"	"	"	" T650			47·6	56·4	22		81	"
	0·37	0·27	0·67				1·47	1·09	0·18	$2\frac{5}{8} \times 11\frac{1}{2}$	OQ850 T640			48·0	58·0	$23\frac{1}{2}$	62	83	3

En 110

MECHANICAL PROPERTIES AT ROOM TEMPERATURE (Continued)

Process of Manufacture	ASTM Grain Size	Chemical Composition (%)								Section Size dia. (in.)	Heat Treatment °C	Tensile						Izod ft. lb	Ref.
		C	Si	Mn	S	P	Ni	Cr	Mo			0·1% P.S. (0·1Rp) tons/sq. in.	0·2% P.S. (0·2Rp) tons/sq. in.	Y.P. (Rs) tons/sq. in.	T.S. (Rm) tons/sq. in.	El % (A)	R.A. % (Z)		
	6	0·38	0·27	0·60	0·033	0·034	1·48	0·83	0·26	1⅛	OQ850 T640				62·6	21½	57	68	7
	"	"	"	"	"	"	"	"	"	4	" T600				61·6	18	45	32	"
	"	"	"	"	"	"	"	"	"	6	" T600				60·0	13½	28	22	"
		0·38					1·36	1·14	0·16	1	OQ850 T580				68·4	20		64	1
		"					"	"		2⅞ (M)	" T615			51·0	60·8	22	62	71	"
		0·38	0·22	0·63			1·30	1·20	0·15	4	OQ850 T630			46·0	56·0	23	63	85	3
		0·39		0·56			1·50	0·99	0·10	4 sq.	OQ840 T650			52·0	58·5	21		66	8
		"		"			"	"	"	6 sq.	" "			47·6	57·0	20		62	"
		0·39					1·31	1·15	0·20	½	OQ850 T570				74·0	19		44	1
		0·39					1·42	1·12	0·19	1⅞	OQ850 T630			56·0	63·0	21		56	"
		0·39					1·37	1·11	0·16	6	OQ850 T660			44·6	53·0	24		58	"
		0·39	0·18	0·67			1·22	1·16	0·17	$2\frac{3}{32}$	OQ850 T610			60·0	70·0	21	59	59	3
		0·39	0·22	0·62			1·29	1·17	0·16	2¼	OQ850 T620			58·0	68·0	21	56	45	"
		0·39		0·65			1·39	1·02	0·14	1⅛	OQ830 T650			59·7	65·6	20½	59	50	2
		0·40					1·30	1·05	0·17	⅞	OQ850 T600			61·6	69·2	19½		44	1
		"					"	"	"	1⅛	" "			52·3	62·8	23		63	8
		"					"	"	"	2½	" "			49·3	60·4	24		70	"
		"		0·63			1·45	1·10	0·2	4	OQ850 T650			46·2	56·8	23		68	"
		0·40	0·25	0·65			1·37	1·08	0·18	2¼	OQ830 T630			52·0	64·0	23	57	60	3
		0·40	0·26	0·64			1·33	1·16	0·18	4	OQ850 T650	48·0			59·0	23		72	"
		0·40	0·23	0·69			1·37	1·13	0·17	5½	OQ850 T640			49·4	59·4	20½	59	60	"
		0·40	0·19	0·70			1·35	1·22	0·17	7	OQ850 T640			48·4	58·2	21	61	59	"
	6—7	0·41	0·24	0·62	0·027	0·036	1·28	0·81	0·15	1⅛	OQ850 T650				60·0	26	66	87	7
	"	"	"	"	"	"	"	"	"	4	" T600				57·8	18	45	41	"
	"	"	"	"	"	"	"	"	"	6	" T600				56·0	19	44½	33	"
		0·41					1·41	1·09	0·18	$1\frac{1}{16}$	OQ850 T610			60·2	71·2	21		44	1
		0·41					1·42	1·16	0·14	2⅞	OQ850 T640			55·6	65·2	21½		64	"
		0·41		0·58			1·48	1·12	0·16	1⅛	OQ850 T550			72·5	78·8	17		23	8
		"		"			"	"	"	"	" T600			64·3	70·8	20		56	"
		"		"			"	"	"	"	" T650			54·4	62·0	23		73	"
		"		"			"	"	"	2½	OQ850 T550			68·4	75·6	15		16	"
		"		"			"	"	"	"	" T600			57·9	66·8	20		63	"
		"		"			"	"	"	"	" T650			47·2	58·0	23		82	"
		"		"			"	"	"	4	OQ850 T550			59·0	68·8	15		14	"

MECHANICAL PROPERTIES AT ROOM TEMPERATURE (Continued)

Process of Manufacture	ASTM Grain Size	Chemical Composition (%) C	Si	Mn	S	P	Ni	Cr	Mo	Section Size dia. (in.)	Heat Treatment °C	Tensile 0·1% P.S. (0·1Rp) tons/sq.in.	0·2% P.S. (0·2Rp) tons/sq.in.	Y.P. (Rs) tons/sq.in.	T.S. (Rm) tons/sq.in.	El % (A)	R.A. % (Z)	Izod ft.lb.	Ref.
		0·41		0·58			1·48	1·12	0·16	4	OQ850 T600			50·0	60·8	19		36	8
		"		"			"	"	"	"	" T650			42·3	54·4	22		81	"
	3—5	0·42	0·29	0·60	0·039	0·030	1·36	0·96	0·27	$1\frac{1}{8}$	OQ850 T640				64·6	$20\frac{1}{2}$	56	52	7
	"	"	"	"	"	"	"	"	"	4	" T600				66·0	15	64	53	"
	"	"	"	"	"	"	"	"	"	6	" T600				61·6	13	26	25	"
		0·42					1·37	1·00	0·16	$3\frac{7}{8}$	OQ850 T640			51·6	59·6	$21\frac{1}{2}$		60	1
		0·42	0·23	0·64			1·36	1·06	0·12	$1\frac{3}{4}$	OQ850 T620			56·0	66·0	20	$49\frac{1}{2}$	52	3
		0·42	0·28	0·70			1·30	1·19	0·18	$3\frac{1}{4}$	OQ850 T640			53·0	64·0	24	59	65	"
	5—7	0·43	0·28	0·63	0·019	0·030	1·41	1·10	0·26	$1\frac{1}{8}$	OQ850 T650				66·6	21	$54\frac{1}{2}$	60	7
	"	"	"	"	"	"	"	"	"	4	" T600				64·0	$18\frac{1}{2}$	46	51	"
	"	"	"	"	"	"	"	"	"	6	" T600				63·2	$15\frac{1}{2}$	32	26	"
		0·43					1·38	1·15	0·15	$3\frac{3}{8}$	OQ850 T650			53·2	59·8	$22\frac{1}{2}$		67	1
		0·43		0·64			1·53	1·07	0·08	$1\frac{1}{8}$	OQ840 T550			71·9	75·8	$15\frac{1}{2}$	50	23	2
		"		"			"	"	"	"	" T600			67·6	72·6	17	54	45	"
		"		"			"	"	"	"	" T650			56·8	63·9	20	59	63	"
		"		"			"	"	"	$2\frac{1}{2}$	OQ840 T550			71·0	76·2	$16\frac{1}{2}$	50	17	"
		"		"			"	"	"	"	" T600			66·6	72·0	$17\frac{1}{2}$	52	39	"
		"		"			"	"	"	"	" T650			59·0	65·0	$21\frac{1}{2}$	59	59	"
		"		"			"	"	"	4	OQ840 T550			69·9	75·7	$16\frac{1}{2}$	50	16	"
		"		"			"	"	"	"	" T600			66·0	72·3	$17\frac{1}{2}$	54	30	"
		"		"			"	"	"	"	" T650			55·7	63·9	21	59	61	"
		"		"			"	"	"	6 sq.	OQ840 T550			69·5	74·0	17	47	16	"
		"		"			"	"	"		" T600			65·6	72·0	$16\frac{1}{2}$	52	31	"
		"		"			"	"	"		" T650			51·8	63·2	20	57	58	"
		0·45		0·65			1·53	1·04	0·20	$1\frac{1}{8}$	OQ840 T650			58·0	65·7	$21\frac{1}{2}$	57	55	2
		"		"			"	"	"	$2\frac{1}{2}$	" "			60·2	66·8	$21\frac{1}{2}$	57	62	"
		"		"			"	"	"	$3\frac{3}{4}$	" "			57·8	65·8	20	55	59	"

EFFECT OF TEMPERING ON THE MECHANICAL PROPERTIES AT ROOM TEMPERATURE

Chemical Composition (%)				
C	Mn	Ni	Cr	Mo
0·39	0·60	1·4	1·15	0·15

1⅛ in. dia. bars. Oil quenched 830°C. Tempered as indicated.

(Ref. 9)

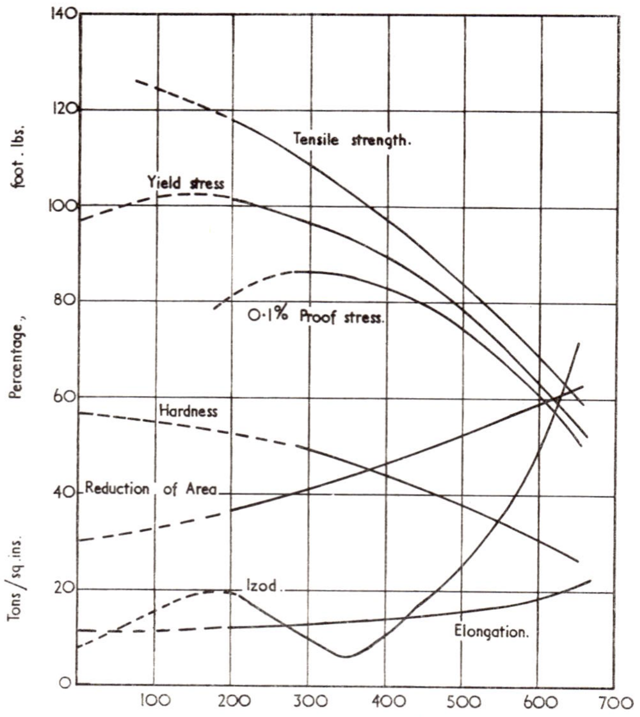

Fig. 110.15

Chemical Composition (%)				
C	Mn	Ni	Cr	Mo
0·40	0·60	1·30	1·20	0·15

1⅛ in. dia. bars. Oil quenched 850°C. Tempered as indicated.

(Ref. 3)

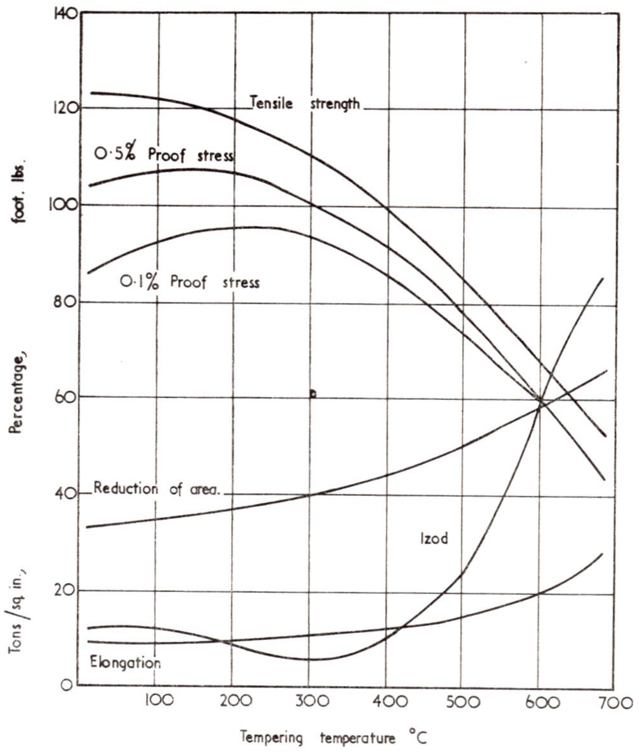

Fig. 110.16

EFFECT OF SECTION SIZE ON THE MECHANICAL PROPERTIES AT ROOM TEMPERATURE

(Typical Curves)

Oil quenched 850°C. Tempered 600°C.

(Ref. 3)

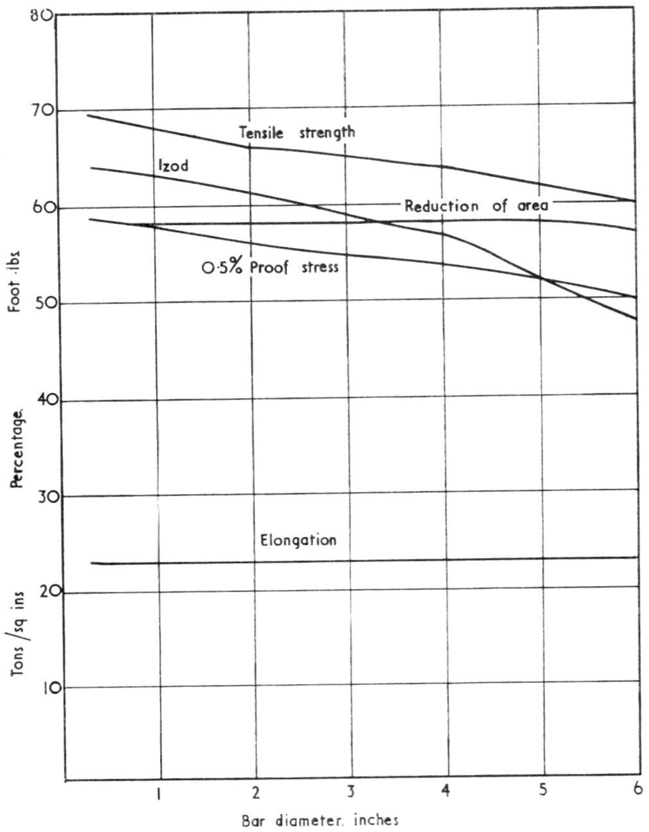

Fig. 110.17

Oil quenched 850°C. Tempered 625°C.

(Ref. 3)

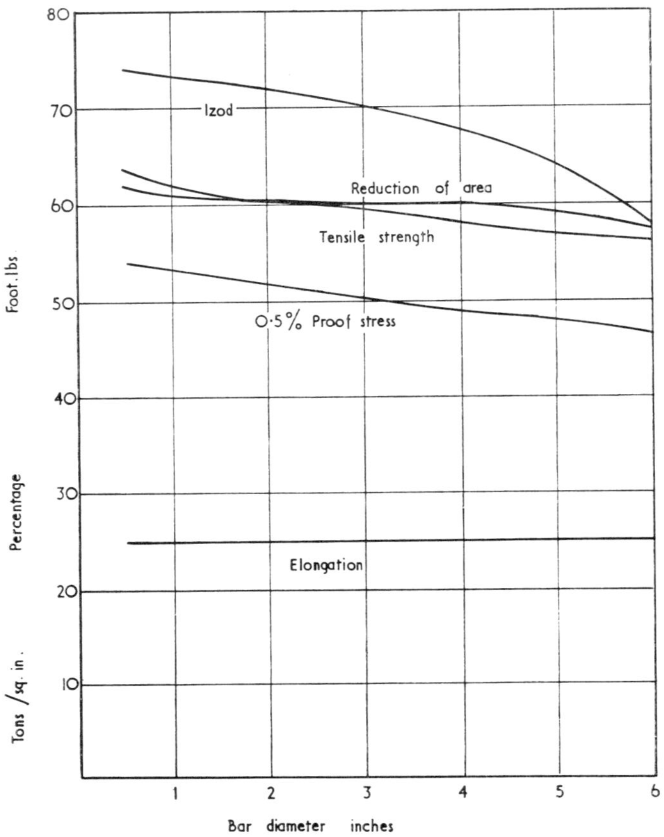

Fig. 110.18

EFFECT OF SECTION SIZE ON THE MECHANICAL PROPERTIES AT ROOM TEMPERATURE

(Continued)

Oil quenched 850°C. Tempered 650°C.

(Ref. 3)

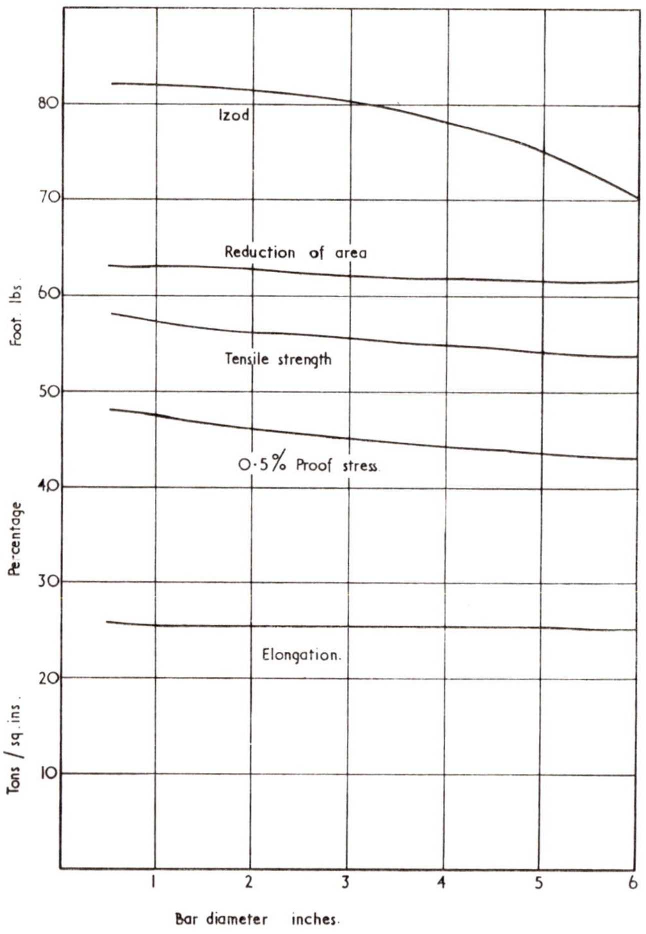

Fig. 110.19

MECHANICAL PROPERTIES AT HIGH TEMPERATURE

Chemical Composition (%)								Heat Treatment	Test Temperature °C	Time at Temperature hrs.	Tensile				Izod ft.lb.	Ref.
C	Si	Mn	S	P	Ni	Cr	Mo				0·1% P.S. (0·1Rp) tons/sq. in.	T.S. (Rm) tons/sq. in.	El % (A)	R.A. % (Z)		
0·39	0·25	0·61	0·016	0·014	1·45	1·06	0·16	H & T (70 tons/sq.in.)	RT	20	56·0	72·2	19		43	4
"	"	"	"	"	"	"	"		100	20	57·8	69·2	17		—	"
"	"	"	"	"	"	"	"		150	20	55·7	68·8	18		—	"
"	"	"	"	"	"	"	"		200	20	54·4	67·5	19½		—	"
"	"	"	"	"	"	"	"		250	1	53·2	69·0	16½		41	"
"	"	"	"	"	"	"	"		"	20	55·6	68·3	15		39	"
"	"	"	"	"	"	"	"		"	200	52·4	68·0	15		39	"
"	"	"	"	"	"	"	"		"	1000	56·8	68·8	17½		44	"
"	"	"	"	"	"	"	"		300	1	48·9	67·8	15		45	"
"	"	"	"	"	"	"	"		"	20	52·5	66·4	18		39	"
"	"	"	"	"	"	"	"		"	200	52·7	67·1	17½		38	"
"	"	"	"	"	"	"	"		"	1000	52·4	65·0	17		39	"
"	"	"	"	"	"	"	"		350	1	49·4	63·7	19		41	"
"	"	"	"	"	"	"	"		"	20	47·8	61·7	20½		38	"
"	"	"	"	"	"	"	"		"	200	50·6	63·3	21½		44	"
"	"	"	"	"	"	"	"		"	1000	53·0	64·4	19		39	"

FATIGUE PROPERTIES

Chemical Composition (%)								Section Size dia (in.)	Heat Treatment °C	Test Temperature °C	Tensile				Rotating Bending Fatigue Limit		Ref.
C	Si	Mn	S	P	Ni	Cr	Mo				0·2% P.S. (0·2Rp) tons/sq.in.	T.S. (Rm) tons/sq.in.	El % (A)	R.A. % (Z)	Plain tons/sq.in.	Notched* tons/sq.in.	
0·36	0·23	0·64	0·038	0·031	1·44	0·96	0·16	¾	OH T200	RT	101·8	125·0	15½	48	50·0	24·0	2

* Details of Notch: 55°V, 0·008 in. root radius, 0·020 in. deep, test dia. 0·335 in., root diameter 0·295. Neuber Stress Concentration Factor 2·85.

LIST OF REFERENCES

1. Private correspondence with Firth Brown Ltd.
2. Private correspondence with The United Steel Companies Ltd.
3. Private correspondence with The English Steel Corporation Ltd.
4. Private correspondence with the Royal Aircraft Establishment, Farnborough.
5. Atlas of Isothermal Transformation Diagrams of the B. S. En Steels. 2nd Edition, Iron and Steel Institute Special Repor. No. 56, 1956.
6. *Transformation Characteristics of Direct-hardening Nickel Alloy Steels*. 3rd Edition. International Nickel Ltd., 1958.
7. Symposium on the Hardenability of Steel organised by the Technical Advisory Committee of the Special and Alloy Steels Committee (Ministry of Supply) and the Iron and Steel Institute. Iron and Steel Institute Special Report No. 36, 1946.
8. British Standard 971: Commentary on the British Standard Wrought Steels, En Series. British Standards Institution, 1950
9. Private correspondence with Jessop—Saville Ltd.

LOW NICKEL–CHROMIUM STEEL

SPECIFICATION

CHEMICAL COMPOSITION (%)

	C	Si	Mn	S	P	Ni	Cr
En 111	0·30—0·40	0·10—0·35	0·60—0·90	0·050 max.	0·050 max.	1·00—1·50	0·45—0·75
En 111A*	0·33—0·38	0·10—0·35	0·60—0·90	0·050 max.	0·050 max.	1·00—1·50	0·45—0·75

* Supplied to composition only. Mechanical properties are not a contractual part of the specification.

MECHANICAL PROPERTIES

Property	Hardened and Tempered Condition			
	R	S	T	U
Limiting ruling section, in.	6	4	$2\frac{1}{2}$	$1\frac{1}{8}$
Tensile strength, tons/sq.in min. (Rm)	45	50	55	60
Yield stress[1], tons/sq.in. min. (Rs)	34	38	44	48
0·2% Proof stress[2], tons/sq.in. min. (0·2 Rp)	32	36	41	46
Elongation, %, min. (A)	22	20	18	17
Izod impact value, ft. lb., min.	40	40	40	35
Brinell hardness number,[3] H. B.	201/255	223/277	248/302	269/321

(1) Not to be used as acceptance tests except by special arrangement between purchaser and manufacturer.
(2) When specifically requested in the enquiry or order.
(3) For information purposes only. Not contractual part of the specification.

RELATED SPECIFICATIONS

United Kingdom		America		Germany			Sweden	USSR	
En No.	BS Air	SAE	AISI	Werkstoff No.	Name	DIN	SIS	GOST	Mark
En 111	S103 S122	3135	3135	1.5710	28(36)Ni Cr 6	1662	14.2530	4543—61	40KhN
En 111	S103 S122	3135	3135	1.5710	28(36)Ni Cr 6	1662	14.2530	4543—61	40KhN

APPLICATIONS

This low-alloy steel can be heat treated to give tensile strengths of 45 to 70 tons/sq. in., depending on the section size. Due to the low alloy content the hardenability is shallow and water quenching is sometimes recommended for sections greater than 1½ in. in order to obtain the higher tensile strengths. Typical uses are high tensile bolts for cylinder heads and crankshaft main bearings, motor-car wheel studs, connecting rods, gun parts, and vehicle components.

WELDING

(See Introduction for key to symbols used in this table.)

	Welding Process					
	MA	G	IG	S	F	B
En 111	d	bx	g	e	b	j

Reproduced by kind permission of H.M.S.O.

MACHINABILITY

In the annealed condition the machinability is approximately 60% of that for mild steel. When hardened and tempered to 45 to 55 tons/sq. in. tensile strength the machinability is 50%, and to 60 to 70 tons/sq. in. tensile strength 40% of that for mild steel (En 3).

HOT WORKING AND HEAT TREATMENT TEMPERATURES

Forging, rolling and stamping	1200°C, finish above 900°C.
Annealing	820°C—850°C, furnace cool.
Softening (sub-critical annealing)	630°C—670°C, air or furnace cool.
Hardening	820°C—850°C, oil or water quench.
Tempering	550°C—660°C.

PHYSICAL PROPERTIES

SPECIFIC GRAVITY (d)

Chemical Composition (%)							Heat Treatment °C	Specific Gravity at 20°C	Ref.
C	Si	Mn	S	P	Ni	Cr			
0·35	0·25	0·80			1·25	0·60	Softened	7·80	1
0·35	0·23	0·66	0·025	0·017	1·27	0·59	OQ850 T600	7·84	2
(Within specification)							Softened	7·85	3

SPECIFIC HEAT (c)

Chemical Composition (%)					Heat Treatment °C	Specific Heat cal/g. deg. C					Ref.
C	Si	Mn	Ni	Cr		20 to 100°C	20 to 200°C	20 to 300°C	20 to 400°C	20 to 500°C	
0·35	0·16	0·62	1·17	0·63	Softened	0·117	0·120	0·124	0·128	0·133	*
0·35	0·25	0·80	1·25	0·60	Softened	0·115					1

Chemical Composition (%)						Specific eat at stated temperature cal/g. deg. C					Ref.
C	Si	Mn	Ni	Cr		100°C	200°C	300°C	400°C	500°C	
0·30	0·35	0·60	1·50	0·50		0·094	0·111	0·126	0·144	0·164	4

* Values estimated by comparison of published results for steels of near composition.

MEAN COEFFICIENT OF THERMAL EXPANSION (α)

Chemical Composition (%)							Heat Treatment °C	$10^6 \times$ Mean Coefficient of Thermal Expansion per deg. C							Ref.
C	Si	Mn	S	P	Ni	Cr		20 to 100°C	20 to 200°C	20 to 300°C	20 to 400°C	20 to 500°C	20 to 600°C	20 to 700°C	
0·35	0·23	0·66	0·025	0·017	1·27	0·59	OQ850 T600	11·5	12·4	13·0	13·5	14·0	14·4		2
0·35	0·25	0·80			1·25	0·60	Softened	11		13		14		15 (20 to 680)	1
(Within specification)							Softened	12·7		13·4		14·0		14·5 (20 to 680)	3
0·35	0·16	0·62			1·17	0·63	Softened	11·7	12·5	13·1	13·6	14·0	14·4		*

* Values estimated by comparison of published results for steels of near composition.

ELECTRICAL RESISTIVITY (ρ)

Chemical Composition (%)							Heat Treatment °C	Electrical Resistivity microhm—cm									Ref.
C	Si	Mn	S	P	Ni	Cr		0°C	20°C	100°C	200°C	300°C	400°C	500°C	600°C	700°C	
0·35	0·23	0·66	0·025	0·017	1·27	0·59	OQ850 T600		23·9	28·5	35·5	43·3	53·0	64·0	78·0		2
0·35	0·16	0·62			1·17	0·63	Softened	21·1	22·2	27·1	34·3	42·8	52·9	64·8	78·8	95·0	*

* Values estimated by comparison of published results for steels of near composition.

THERMAL CONDUCTIVITY (k)

Chemical Composition (%)					Heat Treatment °C	Thermal Conductivity (cal/cm sec. deg. C)									Ref.
C	Si	Mn	Ni	Cr		0°C	20°C	100°C	200°C	300°C	400°C	500°C	600°C	700°C	
0·35	0·16	0·62	1·17	0·63	Softened	0·104	0·104	0·104	0·102	0·098	0·092	0·087	0·081	0·071	*
0·35			1·37	0·46		0·110		0·105	0·101	0·097	0·093	0·089	0·085		5

* Values estimated by comparison of published results for steels of near composition.

YOUNG'S MODULUS (E)

Chemical Composition (%)							Heat Treatment °C	Young's Modulus* tons/sq. in.							Ref.
C	Si	Mn	S	P	Ni	Cr		20°C	100°C	200°C	300°C	400°C	500°C	600°C	
0·35	0·23	0·66	0·025	0·017	0·027	0·59	OQ850 T600	13,700	13,350	13,050	12,700	12,250	11,600	10,450	2

* Vibration method.

TRANSFORMATION CHARACTERISTICS

ISOTHERMAL TRANSFORMATION

Chemical Composition (%)						
C	Si	Mn	S	P	Ni	Cr
0·35	0·13	0·65	0·032	0·035	1·27	0·55

McQuaid Ehn Grain Size 2—3 (ASTM).
As quenched Grain Size 7 (ASTM).
Austenitizing Temperature 845°C. (Ref. 6)

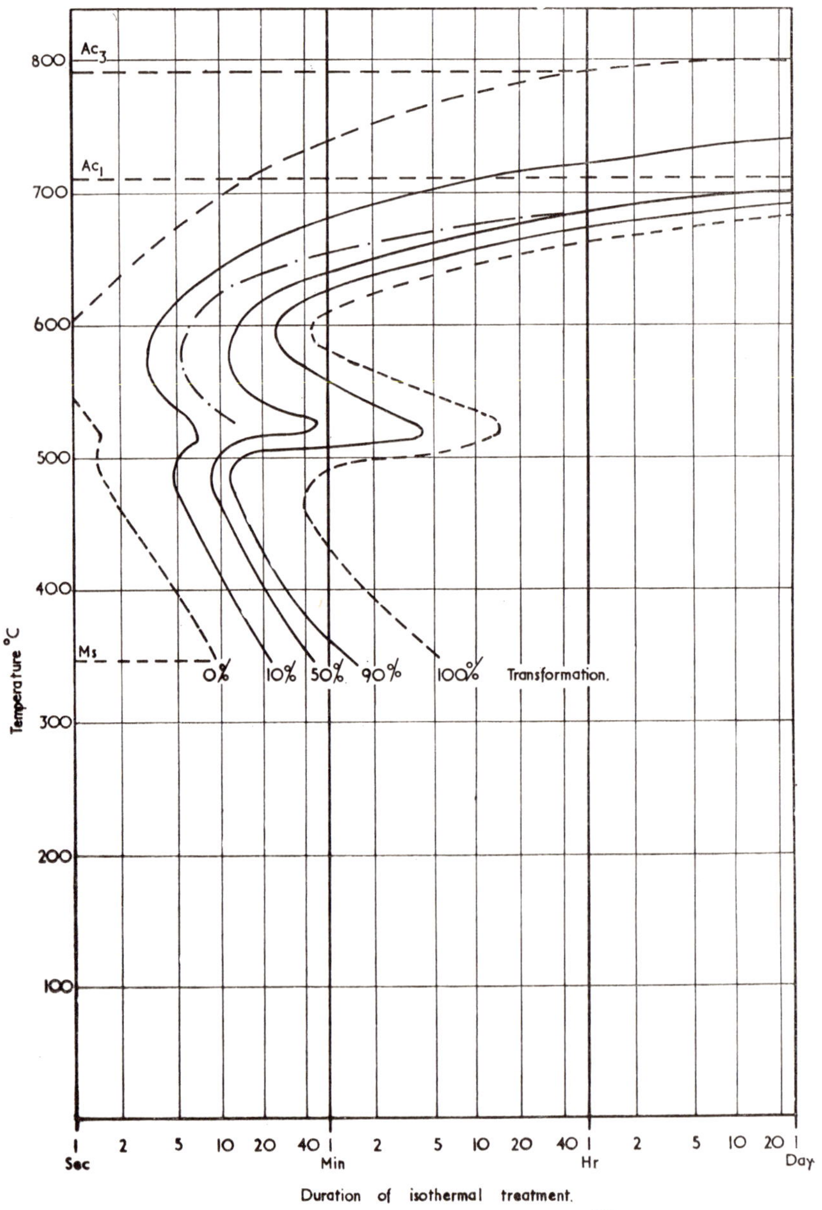

Fig. 111.1

ISOTHERMAL TRANSFORMATION (Continued)

Chemical Composition (%)							
C	Si	Mn	S	P	Ni	Cr	Mo
0·37	0·28	0·89	0·035	0·025	1·24	0·63	0·05

Acid Open Hearth.
McQuaid Ehn Grain Size 5 (ASTM).
As quenched Grain Size 8 (ASTM).

Bars hot rolled to $\frac{3}{8}$ in. dia. rod, cold drawn to 0·144 in. dia. wire, with intermediate annealing at 600°C, ground to 0·119 in. dia., rolled to tape 0·03 in. × $\frac{3}{16}$ in.

Austenitizing Temperature 845°C, 30 min. (Ref. 7)

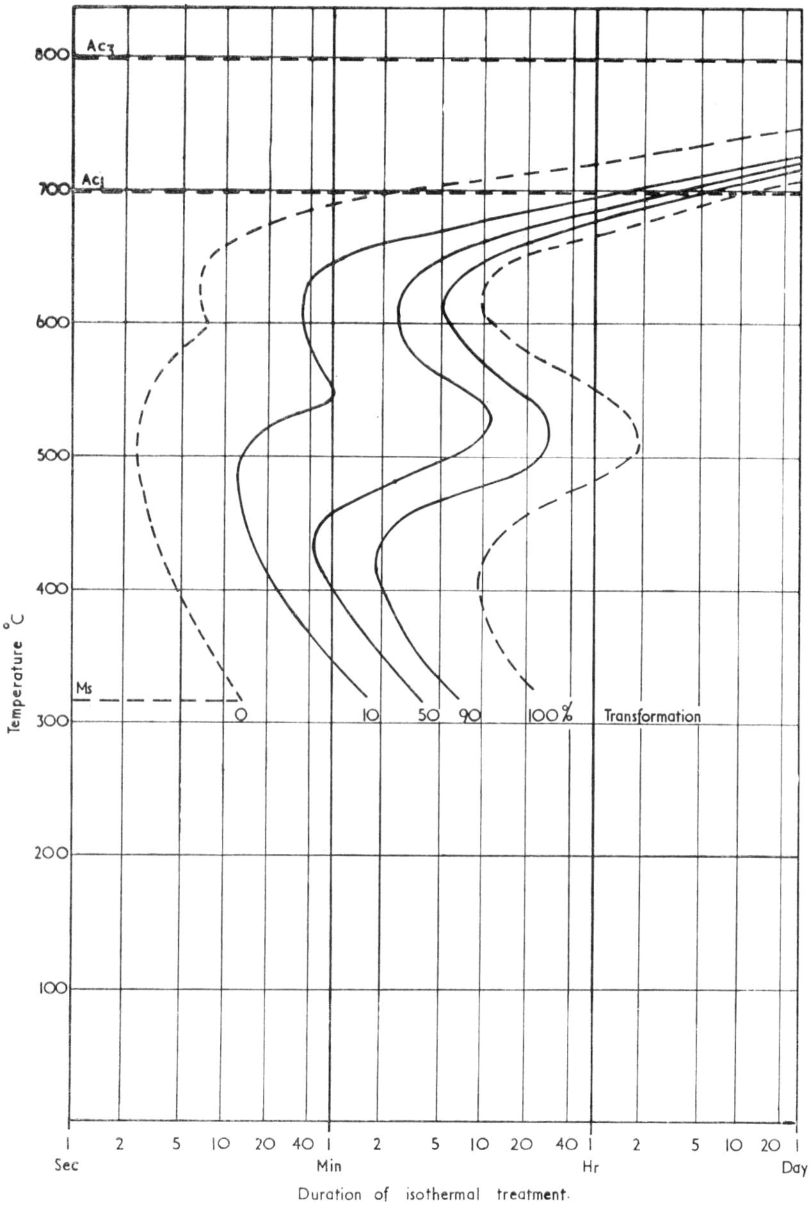

Fig. 111.2

En 111

ISOTHERMAL TRANSFORMATION (Continued)

Chemical Composition (%)			
C	Mn	Ni	Cr
0·38	0·72	1·32	0·49

Grain Size 7—8 (ASTM).
Austenitizing Temperature 844°C.

(Ref. 8)

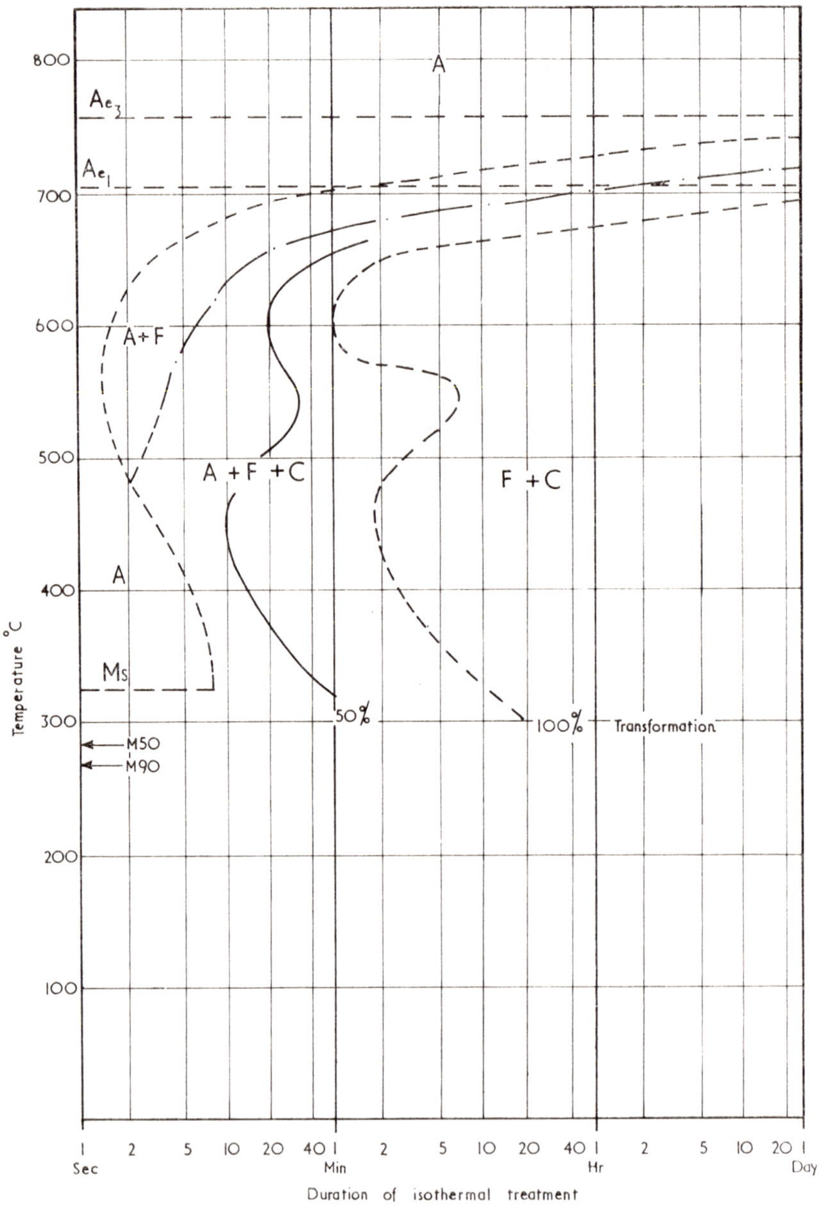

Fig. 111.3

CONTINUOUS COOLING TRANSFORMATION

Chemical Composition (%)						
C	Si	Mn	S	P	Ni	Cr
0·35	0·13	0·65	0·032	0·035	1·27	0·55

McQuaid Ehn Grain Size 2—3 (ASTM).
As quenched Grain Size 7 (ASTM).
Austenitizing Temperature 845°C.

(Ref. 6)

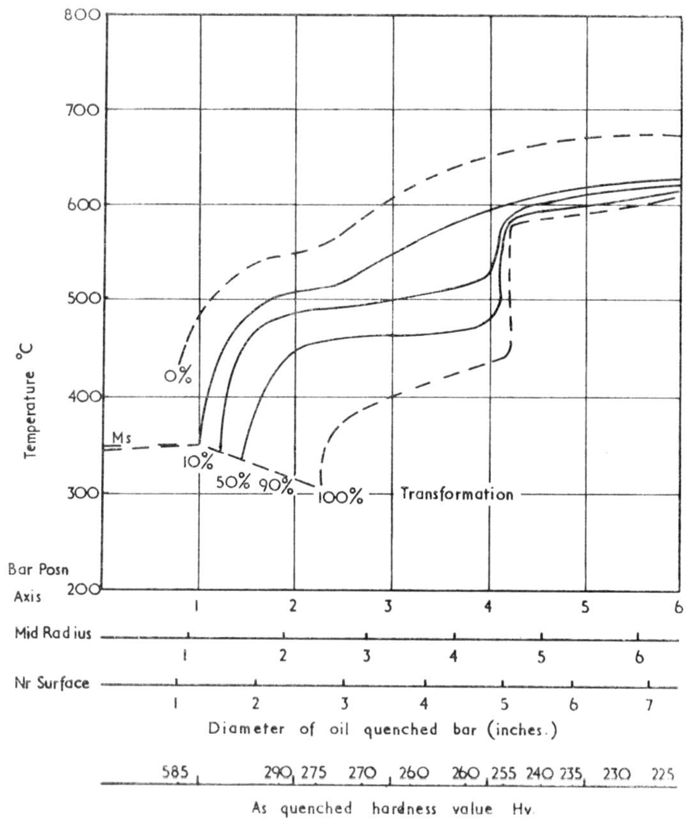

Fig. 111.4

CONTINUOUS COOLING TRANSFORMATION (Continued)

Air cooled from 850°C.

(Ref. 2)

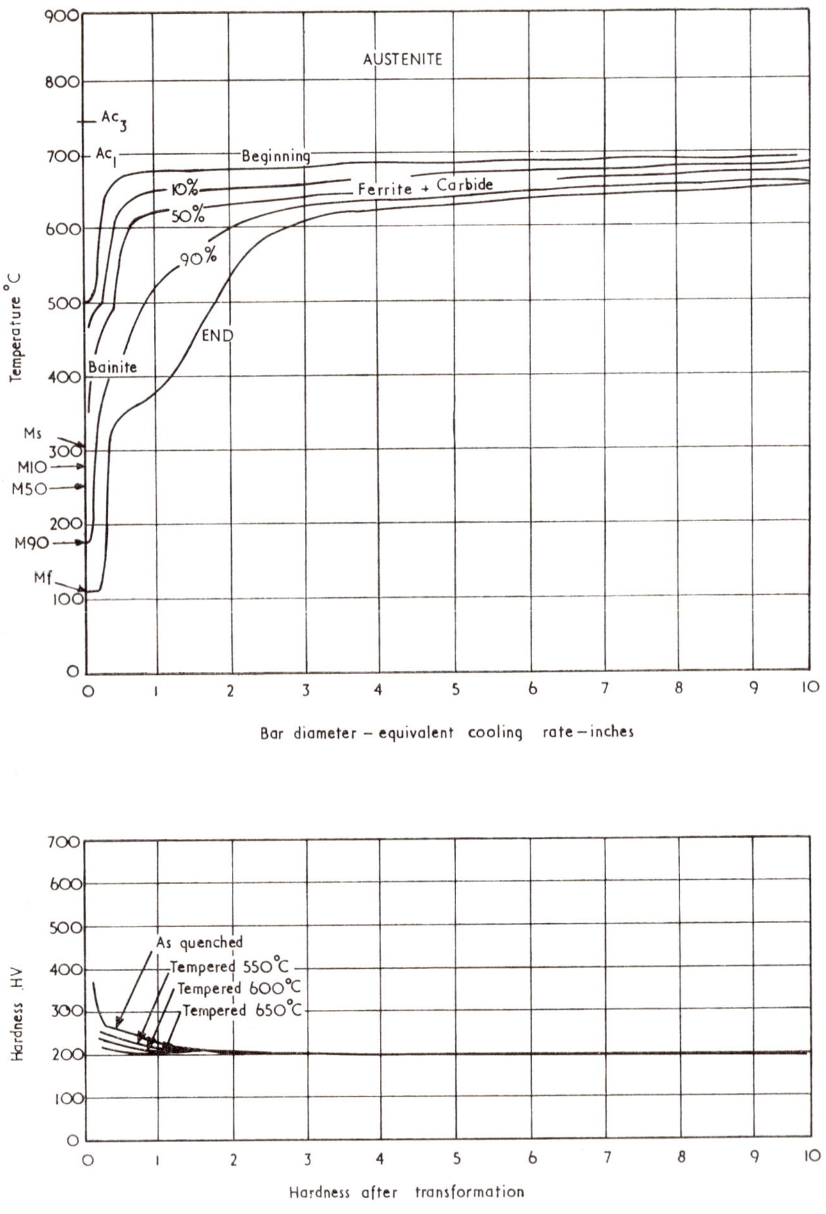

Fig. 111.5

CONTINUOUS COOLING TRANSFORMATION (Continued)

Oil quenched 850°C.

(Ref. 2)

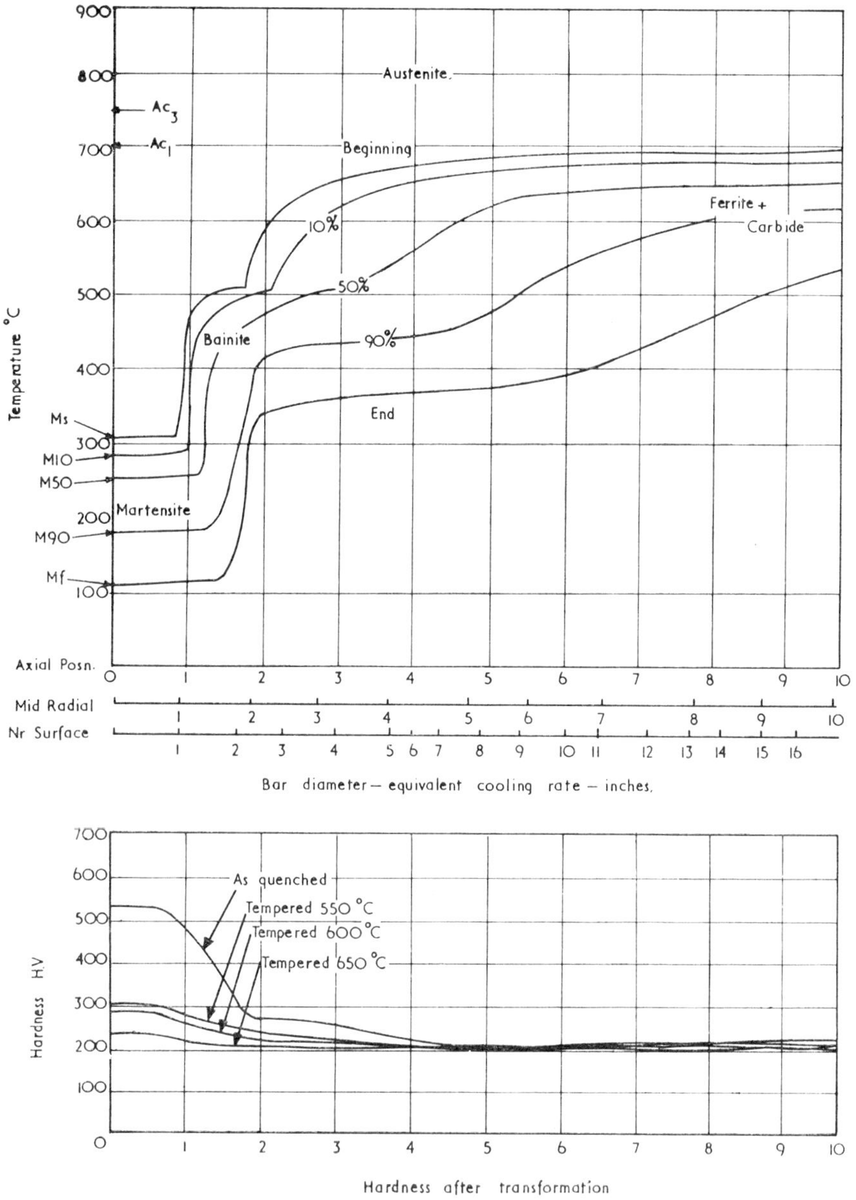

Fig. 111.6

En 111

CONTINUOUS COOLING TRANSFORMATION (Continued)

Water quenched 850°C.

(Ref. 2)

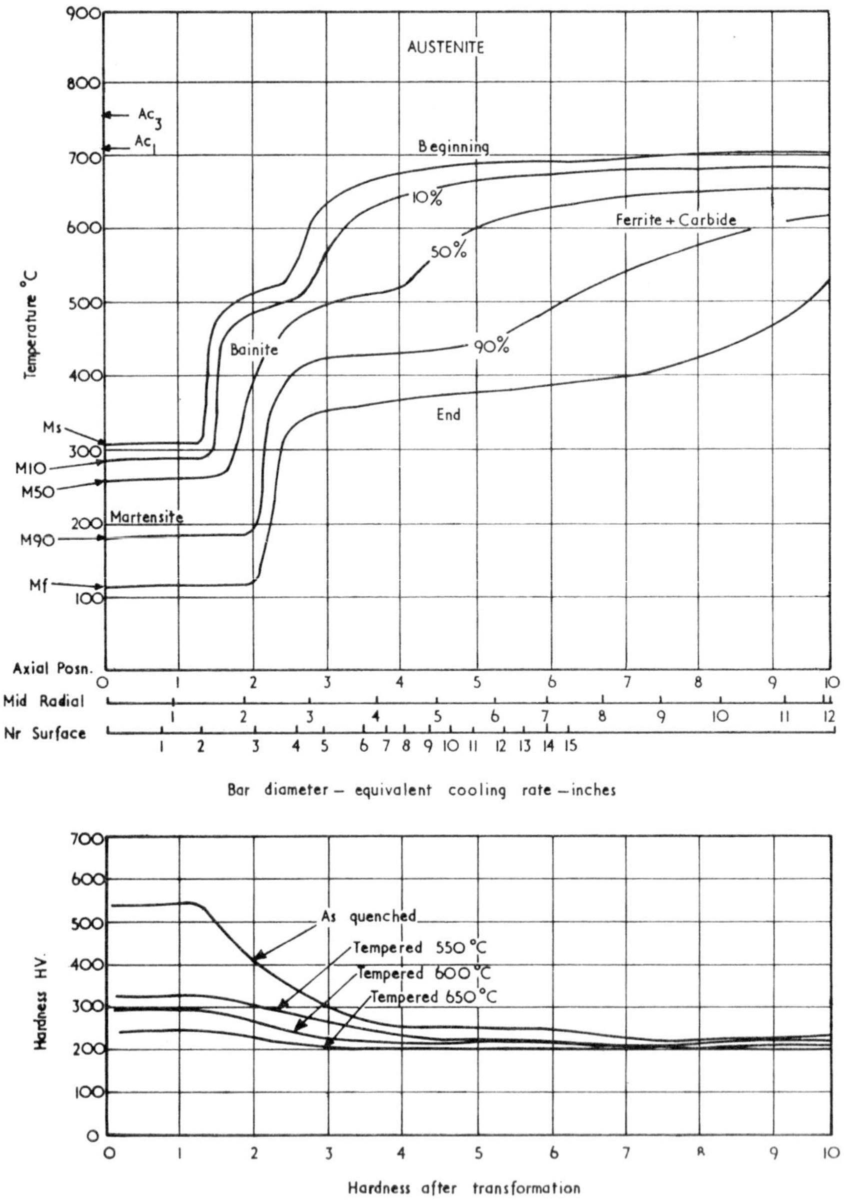

Fig. 111.7

HARDENABILITY

END QUENCH HARDENABILITY DATA

Chemical Composition (%)						
C	Si	Mn	S	P	Ni	Cr
0·35	0·13	0·65	0·032	0·035	1·27	0·55

McQuaid Ehn Grain Size 2—3 (ASTM).
As quenched Grain Size 7 (ASTM).
End quenched 845°C. Tempered as indicated. (Ref. 6)

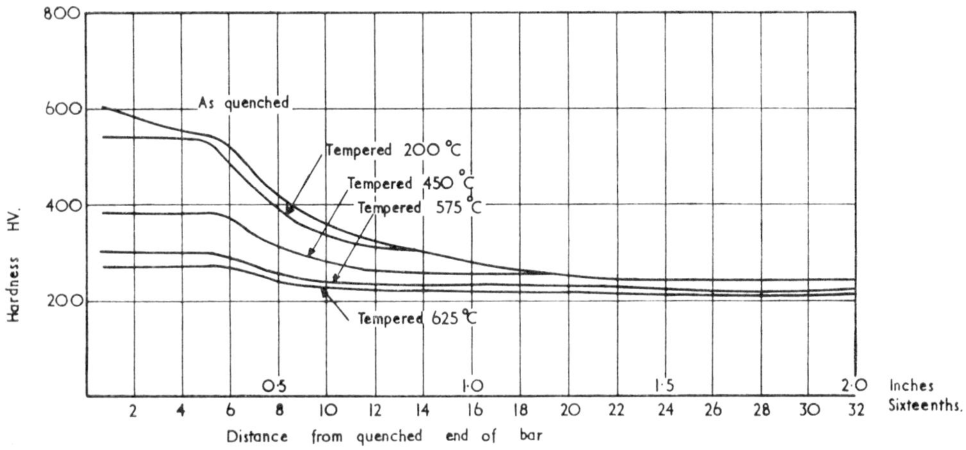

Fig. 111.8

Chemical Composition (%)							
C	Si	Mn	S	P	Ni	Cr	Mo
0·37	0·28	0·89	0·035	0·025	1·24	0·63	0·05

Acid Open Hearth.
McQuaid Grain Size 5 (ASTM).
As quenched Grain Size 8 (ASTM).
End quenched from 845°C. (Ref. 7)

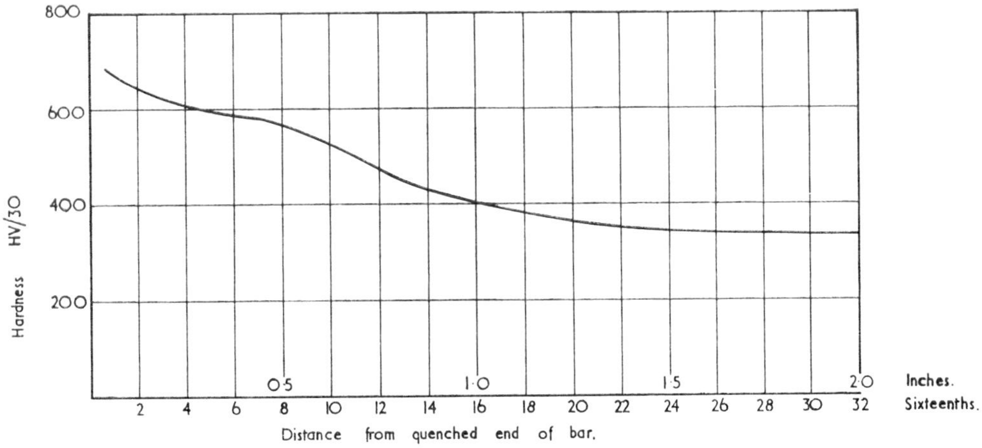

Fig. 111.9

END QUENCH HARDENABILITY DATA (Continued)

Chemical Composition (%)			
C	Mn	Ni	Cr
0·38	0·72	1·32	0·49

Grain Size 7—8 (ASTM).
End quenched from 844°C.

(Ref. 8)

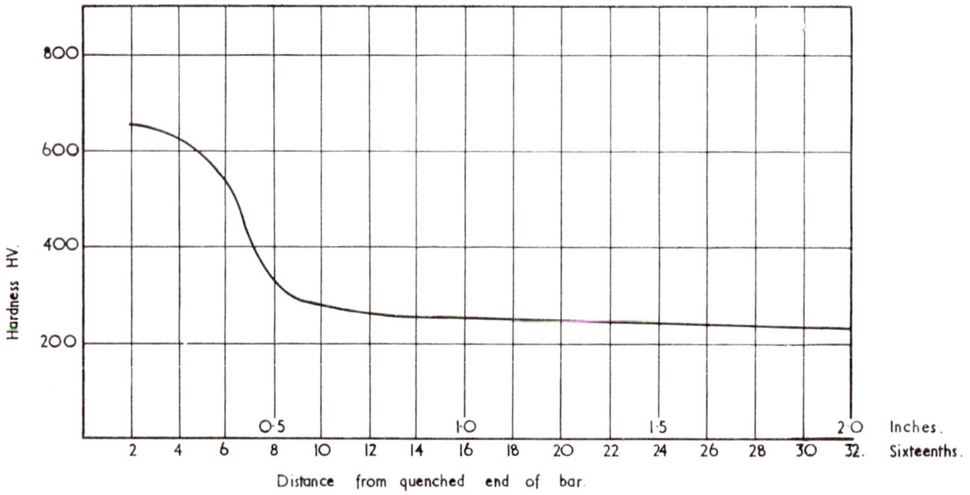

Fig. 111.10

MECHANICAL PROPERTIES AT ROOM TEMPERATURE

Process of Manufacture	ASTM Grain Size	Chemical Composition (%) C	Si	Mn	S	P	Ni	Cr	Section Size dia (in.)	Heat Treatment °C	Tensile 0·2% P.S. (0·2Rp) tons/sq.in.	Y.P. (Rs) tons/sq.in.	T.S. (Rm) tons/sq.in.	El % (A)	R.A. % (Z)	Izod. ft.lb.	Ref.
		0·31		0·65			1·23	0·63	1⅛	OQ850 T550		54·3	58·2	21	59	64	2
		″		″			″	″	″	″ T600		45·8	53·5	22½	64	74	″
		″		″			″	″	″	WQ850 T550		59·0	63·8	19½	59	57	″
		″		″			″	″	″	″ T600		52·4	58·8	22	66	70	″
		″		″			″	″	2½	OQ850 T550		44·3	52·8	21	57	62	″
		″		″			″	″	″	WQ850 T550		52·6	59·0	17½	55	53	″
		0·33		0·71			1·12	0·61	1⅛	OQ850 T550		45·9	56·0	22	62	71	2
		″		″			″	″	″	″ T600		43·8	50·9	24	66	79	″
		″		″			″	″	″	WQ850 T550		54·8	60·6	21½	64	65	″
		″		″			″	″	2	OQ850 T550		42·0	52·5	23	59	76	″
		″		″			″	″	″	″ T600		39·9	50·1	26	64	88	″
		″		″			″	″	2½	OQ850 T550		42·1	52·0	23	59	75	″
		″		″			″	″	″	″ T600		38·1	49·1	25	64	84	″
		″		″			″	″	″	WQ850 T550		48·0	57·0	22	62	72	″
B.O.H.		0·33	0·20	0·74	0·034	0·023	1·19	0·50	5/16	OQ840 T620		46·8	54·8	24		69	9
		0·34					1·27	0·63	1⅜	WQ850 T590		55·2	60·2	23		77	1
B.E.		0·34	0·26	0·71	0·009	0·012	1·16	0·61	1	OQ850 T590WQ		52·8	58·8	23	61	77	″
″		″	″	″	″	″	″	″	1⅞	OQ850 T550WQ		51·2	56·8	20½	55	40	″
″		″	″	″	″	″	″	″	″	WQ850 T600WQ		53·6	58·6	23	62	70	″
B.O.H.		0·35	0·23	0·79	0·030	0·026	1·24	0·55	19/32	OQ850 T630		48·5	56·1	26		78	9
B.O.H.		0·35	0·24	0·81	0·030	0·027	1·20	0·62	19/32	OQ850 T640		47·2	58·4	24		65	″
B.O.H.		0·35	0·22	0·78	0·027	0·030	1·16	0·53	21/32	OQ850 T630		48·0	55·6	24		81	″
″		″	″	″	″	″	″	″	19/32	″ ″		47·4	54·6	24		77	″
		0·35					1·25	0·65	7/8	OQ850 T570		57·2	61·4	18½		48	1
		0·35					1·26	0·72	1⅜	OQ850 T570		57·2	62·2	22		67	″
B.O.H.		0·36	0·22	0·77	0·030	0·038	1·19	0·60	19/32	OQ850 T620		53·6	60·2	21		58	9
B.O.H.		0·36	0·19	0·75	0·040	0·036	1·20	0·58	19/32	OQ850 T615		54·6	60·2	22		71	″
B.O.H.		0·36	0·20	0·78	0·027	0·031	1·22	0·55	19/32	OQ850 T630		50·0	57·4	25		74	″
B.O.H.		0·36	0·26	0·78	0·029	0·028	1·18	0·54	19/32	OQ840 T640		51·2	58·8	25		71	″
B.O.H.		0·36	0·23	0·77	0·031	0·025	1·15	0·51	19/32	OQ840 T640		50·4	57·2	24		76	″

MECHANICAL PROPERTIES AT ROOM TEMPERATURE (Continued)

Process of Manufacture	ASTM Grain Size	Chemical Composition (%)							Section Size dia (in.)	Heat Treatment °C	Tensile					Izod ft.lb.	Ref.
		C	Si	Mn	S	P	Ni	Cr			0·2% P.S. (0·2Rp) tons/sq.in.	Y.P. (Rs) tons/sq.in.	T.S. (Rm) tons/sq.in.	El % (A)	R.A. % (Z)		
B.O.H.	1—3	0·36	0·21	0·78	0·036	0·043	1·20	0·47	$\frac{19}{32}$	OQ840 T640		46·4	57·2	22		59	9
B.O.H.		0·36	0·22	0·78	0·026	0·027	1·16	0·57	$\frac{15}{16}$	OQ840 T620		50·4	58·0	23		74	"
		0·36					1·25	0·71	$1\frac{1}{2}$	WQ850 T580		59·0	63·8	21		57	1
B.O.H.		0·36	0·25	0·72	0·026	0·017	1·23	0·60	$1\frac{1}{8}$	OQ850 T200AC		88·0	110·0	12	32	9	"
"		"	"	"	"	"	"	"	"	" T350AC		94·0	101·0	9	35	3	"
B.E.		0·36	0·14	0·67	0·026	0·028	1·15	0·65	$1\frac{1}{8}$	OQ850 T225AC		98·0	117·0	11	31	6	"
"		"	"	"	"	"	"	"	"	" T275AC		"	"	"	"	4	"
"		"	"	"	"	"	"	"	"	" T325WQ		"	104·8	9	$33\frac{1}{2}$	4	"
"		"	"	"	"	"	"	"	"	" T375WQ		"	94·0	11	38	5	"
B.O.H.	6—7	0·37	0·23	0·81	0·026	0·035	1·20	0·60	$\frac{19}{32}$	OQ850 T630		53·8	58·2	25		79	9
B.O.H.		0·37	0·21	0·73	0·024	0·032	1·21	0·55	$\frac{19}{32}$	OQ840 T640		52·0	58·4	23		69	"
		0·37					1·40	0·64	$1\frac{3}{8}$	OQ850 T590		54·4	60·0	$21\frac{1}{2}$		55	1
		0·37					1·30	0·73	$1\frac{1}{8}$	OQ850 T610		53·4	60·0	22		78	"
		0·37					1·39	0·65	$1\frac{9}{16}$	OQ850 T630		47·8	55·0	$23\frac{1}{2}$		60	"
		0·37					1·40	0·67	$2\frac{3}{4}$	WQ850 T600		55·4	60·0	22		73	"
		0·37					1·37	0·70	3	WQ850 T620		51·0	56·8	23		75	"
		0·37					1·28	0·55	2	OQ850 T620		45·8	55·0	23		44	"
A.O.H.		0·37	0·25	0·75	0·042	0·033	1·28	0·55	$2\frac{1}{2}$	OQ850 T620WQ		47·4	55·0	$22\frac{1}{2}$	56	20	"
		"	"	"	"	"	"	"	2	" "		46·0	53·6	$24\frac{1}{2}$	58	33	"
		"	"	"	"	"	"	"	"	WQ850 T620WQ		52·2	58·0	23	59	60	"
B.E.		0·37	0·23	0·73	0·006	0·015	1·30	0·73	$1\frac{5}{8}$	OQ850 T580WQ	46·2		59·9	20	57	37	"
		0·37		0·70			1·35	0·65	$2\frac{1}{2}$	OQ830 T600		47·4	55·6	22		75	10
		0·37		0·68			1·04	0·63	$1\frac{1}{8}$	OQ830 T600		50·8	59·2	$21\frac{1}{2}$		59	"
		0·37		0·70			1·23	0·62	$1\frac{1}{8}$	OQ830 T580		61·1	66·5	$20\frac{1}{2}$		65	"
B.O.H.		0·38	0·22	0·75	0·040	0·036	1·20	0·58	$\frac{19}{32}$	OQ850 T615		53·2	60·3	20		89	9
B.O.H.		0·38	0·22	0·76	0·027	0·032	1·26	0·59	$\frac{19}{32}$	OQ840 T640		53·0	60·3	22		67	"
B.O.H.		0·38	0·22	0·82	0·027	0·027	1·18	0·65	$1\frac{5}{16}$	OQ840 T620		53·6	60·6	20		60	"
		0·38					1·37	0·63	$1\frac{5}{16}$	OQ850 T620		53·4	58·6	21		65	1
		0·38					1·29	0·58	$2\frac{1}{2}$	WQ850 T620		53·0	57·8	$22\frac{1}{2}$		63	"

MECHANICAL PROPERTIES AT ROOM TEMPERATURE (Continued)

Process of Manufacture	ASTM Grain Size	Chemical Composition (%)							Section Size dia (in.)	Heat Treatment °C	Tensile					Izod ft.lb.	Ref.
		C	Si	Mn	S	P	Ni	Cr			0·2% P.S. (0·2Rp)	Y.P. (Rs)	T.S (Rm)	El % (A)	R.A. % (Z)		
											tons/sq.in.						
B.E.		0·38	0·27	0·69	0·005	0·015	1·51	0·73	2½	OQ850 T580WQ		50·8	58·6	20½	58	42	1
		0·38		0·65			1·10	0·64	3¾	OQ830 T600		36·8	50·9	26		75	10
		0·38		0·69			1·02	0·65	1⅛	OQ850 T600		51·4	59·3	20	55	50	2
		"		"			"	"	"	" T650		45·5	54·8	22	59	60	"
		"		"			"	"	2½	OQ850 T600		43·9	53·7	21½	59	61	"
		"		"			"	"	"	" T650		40·9	51·2	23½	62	66	"
		"		"			"	"	4sq.	OQ850 T600		39·6	52·6	23	59	57	"
		"		"			"	"	"	" T650		39·5	50·2	25	59	64	"
B.O.H.		0·39	0·23	0·80	0·026	0·025	1·24	0·55	15/16	OQ840 T640		50·8	58·4	23		71	9
B.E.		0·39	0·21	0·78	0·010	0·017	1·23	0·71	2¼	OQ850 T580WQ		46·2	55·0	22½	59	55	1
B.E.		0·39	0·28	0·65	0·007	0·016	1·27	0·73	1½	OQ850 T620WQ		48·4	55·6	25	61	80	"
		0·39					1·27	0·64	2½	OQ850 T620		46·8	54·6	21½		32	"
	4—5	0·39		0·67			1·35	0·61	0·564	OQ830 T200		109·6	125·3	13	36	8	2
	"	"		"			"	"	"	" T300		107·6	111·8	12	44	3	"
	"	"		"			"	"	"	" T400		90·4	92·0	14	50	7	"
	"	"		"			"	"	"	" T450		83·2	84·6	14	52	25	"
	"	"		"			"	"	1⅛	OQ830 T500		69·6	72·6	15½	52	26	"
	"	"		"			"	"	"	" T550		63·2	68·3	17½	55	42	"
	"	"		"			"	"	"	" T600		59·1	66·0	19	57	47	"
	"	"		"			"	"	"	" T650		52·4	59·3	21½	62	61	"
	"	"		"			"	"	2½	OQ830 T550		51·2	58·6	19	55	53	"
	"	"		"			"	"	"	" T600		47·0	56·3	21½	59	62	"
	"	"		"			"	"	"	" T650		45·1	54·6	22	62	71	"
	"	"		"			"	"	3(C)	OQ830 T550		51·0	57·8	19	55	52	"
	"	"		"			"	"	"(O)	" "		51·7	59·3	19½	55	53	"
	"	"		"			"	"	"(C)	" T600		50·4	56·8	19½	55	56	"
	"	"		"			"	"	"(O)	" "		47·2	57·3	19	55	59	"
	"	"		"			"	"	"(C)	" T650		44·0	52·4	23	59	72	"
	"	"		"			"	"	"(O)	" "		43·8	54·3	22½	62	70	"
		0·40		0·67			1·27	0·61	3½	OQ830 T600		39·6	50·8	23		49	11
B.E.		0·40	0·25	0·68	0·008	0·013	1·48	0·62	1¼	OQ850 T600WQ		55·2	62·5	22½	58	65	1

MECHANICAL PROPERTIES AT ROOM TEMPERATURE (Continued)

Process of Manufacture	ASTM Grain Size	Chemical Composition (%) C	Si	Mn	S	P	Ni	Cr	Section Size dia (in.)	Heat Treatment °C	Tensile 0·2% P.S. (0·2Rp) tons/sq.in.	Y.P. (Rs) tons/sq.in.	T.S. (Rm) tons/sq.in.	El % (A)	R.A. % (Z)	Izod ft.lb.	Ref.
	5—8	0·40	0·27	0·90	0·025	0·018	1·21	0·62	1	A816		27·3	44·6	24½	51	34	12
	"	"	"	"	"	"	"	"	½	AC871		42·3	66·2	16½	43	21	"
	"	"	"	"	"	"	"	"	1	"		38·8	57·7	19½	57½	39	"
	"	"	"	"	"	"	"	"	2	"		32·8	53·2	21½	59	45	"
	"	"	"	"	"	"	"	"	4	"		31·7	52·7	21	56½	30	"
	"	"	"	"	"	"	"	"	½	OQ830 T538		62·7	67·2	17½	55	46	"
	"	"	"	"	"	"	"	"	"	" T593		56·1	61·6	20	59	57	"
	"	"	"	"	"	"	"	"	"	" T649		52·3	57·3	21½	62	72	"
	"	"	"	"	"	"	"	"	1	OQ830 T538		58·9	65·5	17½	57	47	"
	"	"	"	"	"	"	"	"	"	" T593		53·1	58·8	21	61	65	"
	"	"	"	"	"	"	"	"	"	" T649		48·7	55·6	21½	60	66	"
	"	"	"	"	"	"	"	"	2	OQ830 T538		45·9	58·1	19½	60½	64	"
	"	"	"	"	"	"	"	"	"	" T593		42·6	55·0	21½	63½	69	"
	"	"	"	"	"	"	"	"	"	" T649		37·9	51·1	23½	61½	92	"
	"	"	"	"	"	"	"	"	4	OQ830 T538		40·7	54·1	20	57	50	"
	"	"	"	"	"	"	"	"	"	" T593		34·9	48·8	23½	65	85	"
	"	"	"	"	"	"	"	"	"	" T649		35·3	48·2	24	65½	91	"

EFFECT OF TEMPERING ON THE MECHANICAL PROPERTIES AT ROOM TEMPERATURE

Chemical Composition (%)			
C	Mn	Ni	Cr
0·35	0·75	1·25	0·60

$1\frac{1}{8}$ in. dia. bars. Hardened in oil from 845°C.

(Ref. 10)

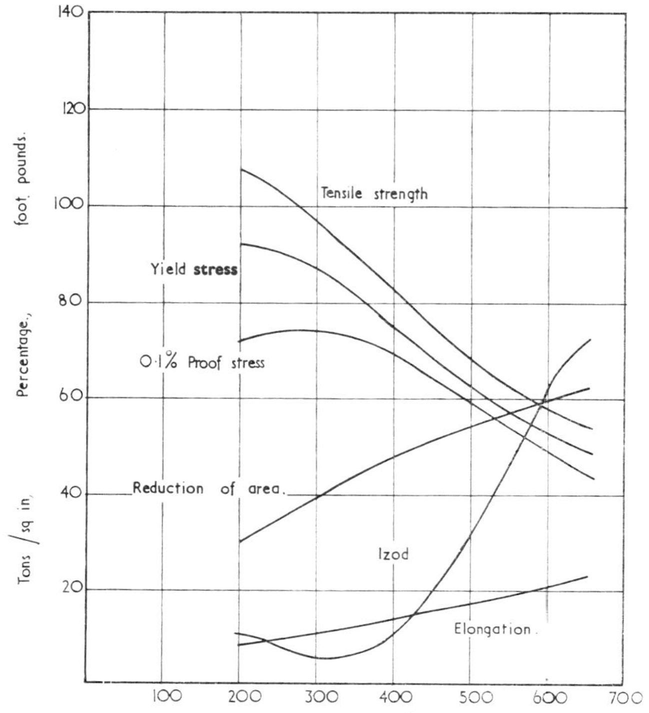

Fig. 111.11

Chemical Composition (%)							
C	Si	Mn	S	P	Ni	Cr	Mo
0·39	0·25	0·76	0·026	0·013	1·20	0·65	0·08

Grain Size 6—7 (ASTM).
0·530 in. dia. bars. Normalized 870°C. Oil quenched 830°C. Tempered as indicated.

(Ref. 12)

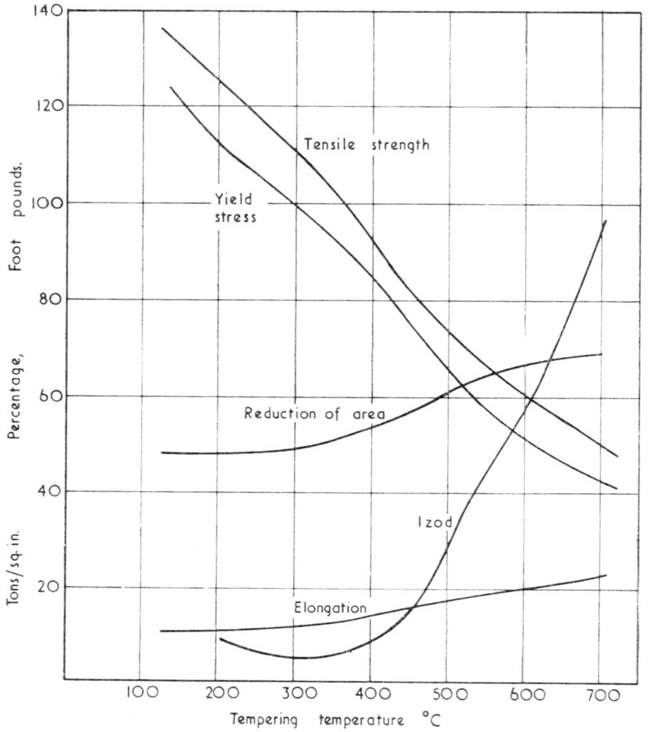

Fig. 111.12

EFFECT OF SECTION SIZE ON THE MECHANICAL PROPERTIES AT ROOM TEMPERATURE
(Typical curves)

Oil quenched from 830°C.
Tempered 550°C.

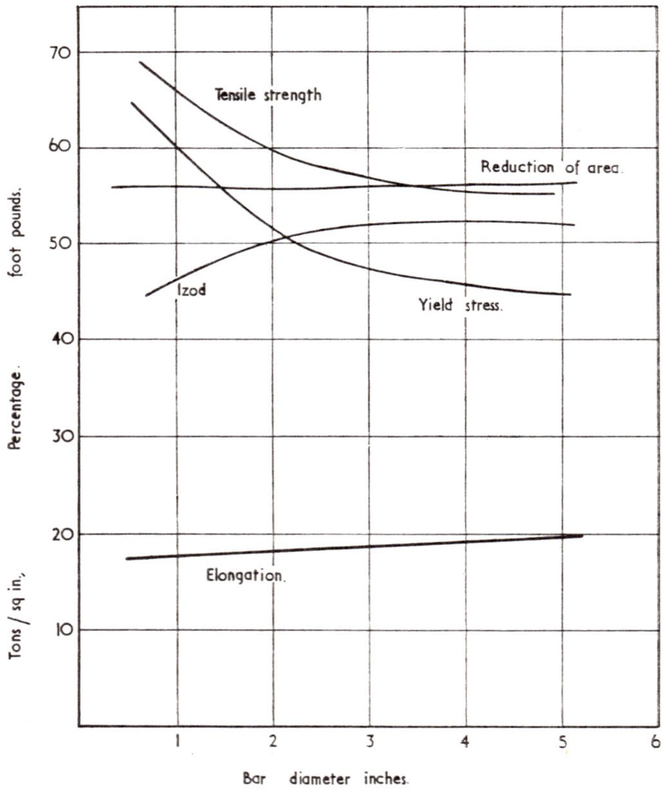

Fig. 111.13

Oil quenched from 830°C.
Tempered 600°C.

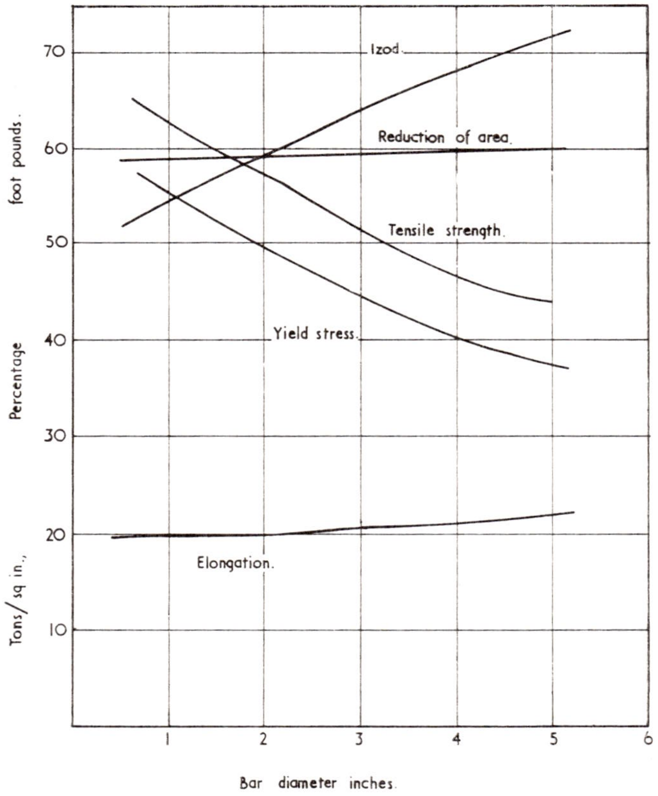

Fig. 111.14

EFFECT OF SECTION SIZE ON THE MECHANICAL PROPERTIES AT ROOM TEMPERATURE

(Continued)

(Typical curves)

Oil quenched from 830°C. Tempered 650°C.

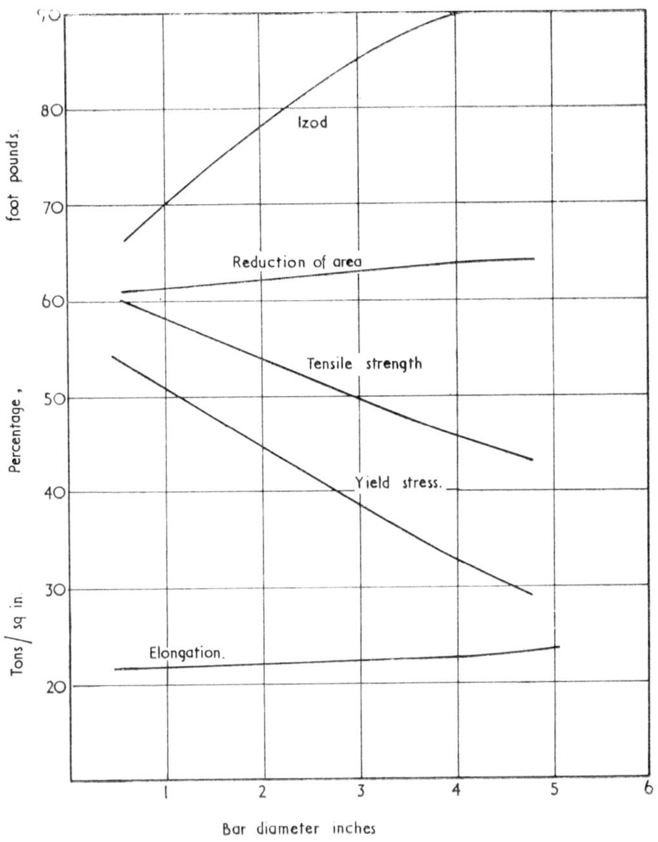

Fig. 111.15

IMPACT TRANSITION TEMPERATURE DATA

ASTM Grain Size	Chemical Composition (%) C	Si	Mn	S	P	Ni	Cr	Mo	Section Size dia. (in.)	Heat Treatment °C	Charpy V Notch impact ft.lb. RT	—40°C	—70°C	Ref.
	0·32	0·21	0·71			1·16	0·60			N927	89	30	12	13
	"	"	"			"	"			N927 T350Q	120+	120+	108	"
											Charpy V Notch 80% Max. Energy Transition Temperature °C			
8—9	0·38	0·19	0·77	0·031	0·013	1·26	0·64		$\frac{1}{2}\times2\frac{1}{2}$	OQ845 T177	—90			14
"	"	"	"	"	"	"	"		"	" T204	—62			"
"	"	"	"	"	"	"	"		"	" T316	+4			"
"	"	"	"	"	"	"	"		"	" T427	—34			"
"	"	"	"	"	"	"	"		"	" T538	—73			"
"	"	"	"	"	"	"	"		"	" T649	—73			"

MECHANICAL PROPERTIES AT HIGH TEMPERATURE

Chemical Composition (%)				
C	Si	Mn	Ni	Cr
0·36	0·26	0·60	1·33	0·56

¾ in. dia. bar. Oil quenched from 840°C. Tempered 540°C.

(Ref. 15)

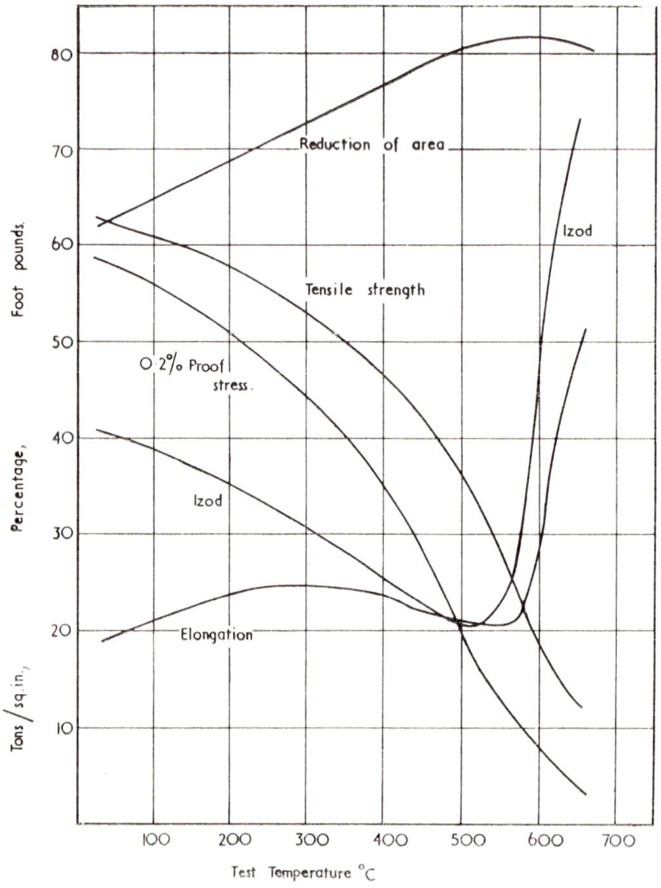

Fig. 111.16

FATIGUE PROPERTIES

Chemical Composition (%)							Section Size dia. (in.)	Heat Treatment °C	Test Temperature °C	Tensile				Izod ft.lb.	Rotating Bending Fatigue Limit	Ref.
C	Si	Mn	S	P	Ni	Cr				Y.P. (Rs)	T.S. (Rm)	El % (A)	R.A. % (Z)			
										tons/sq.in.					tons/sq.in.	
0·34	0·23	0·71	0·042	0·034	1·15	0·52	1⅛	OQ840 T650	RT	37·6	47·4	26	66½	27	±27·7	16
0·36	0·18	0·63	0·038	0·032	1·11	0·55	1⅛	OQ840 T650	RT	34·6	44·4	29	67	94	±26·0	"

LIST OF REFERENCES

1. Private correspondence with Firth Brown Ltd.
2. Private correspondence with The United Steel Companies Ltd.
3. Private correspondence with The English Steel Corporation Ltd.
4. H. Esser, W. Eilender and E. Putz. The Thermal Conductivity of Commercially Pure Iron and Various Steels. *Archiv für das Eisenhüttenwesen*, vol. 11, 1937—1938 (1938), 619—622.
5. S. M. Shelton and W. H. Swanger. Thermal Conductivity of Irons and Steels and Some Other Metals in the Temperature Range 0—600°C. *Trans. Am. Soc. Steel Treating*, vol. 21, 1933, 1061—1078.
6. Transformation Characteristics of Direct-Hardening Nickel Alloy Steels. International Nickel Ltd., 1958.
7. Atlas of Isothermal Transformation Diagrams of the B. S. En Steels. Iron and Steel Institute Special Report No. 56, 1956.
8. Atlas of Isothermal Transformation Diagrams. United States Steel Corporation, 1957.
9. Private correspondence with The Park Gate Iron and Steel Co. Ltd.
10. Private correspondence with International Nickel Ltd.
11. British Standard 971: Commentary on the British Standard Wrought Steels, En Series. British Standards Institution, London, 1950.
12. *Modern Steels and their Properties, Carbon and Alloy Steel Bars*. Bethlehem Steel Company, 6th Edition, 1961.
13. H. W. Gillett. Impact Resistance and Tensile Properties of Metals at Sub-atmospheric Temperatures. Project No. 13 of the joint ASME—ASTM Research Committee on the Effect of Temperature on the Properties of Metals.
14. J. P. Sheehan and H. Schwartzbart. Impact Properties of Quenched and Tempered Alloy Steels, *Proceedings*, ASTM, vol. 56, 1956, 483—520.
15. War Office, unpublished report.
16. Private correspondence with Edgar Allen and Co. Ltd.

2 PER CENT NICKEL–MOLYBDENUM STEEL

SPECIFICATION

CHEMICAL COMPOSITION (%)

	C	Si	Mn	S	P	Ni	Mo
En 160	0·35—0·45	0·10—0·35	0·30—0·60	0·050 max.	0·050 max.	1·50—2·00	0·20—0·35
En 160A*	0·38—0·43	0·10—0·35	0·30—0·60	0·050 max.	0·050 max.	1·50—2·00	0·20—0·35

* Supplied to composition only Mechanical properties are not part of the contractual specification.

MECHANICAL PROPERTIES

Property	Hardened and Tempered Condition			
	R	S	T	U
Limiting ruling section, in.	6	4	$2\frac{1}{2}$	$1\frac{1}{8}$
Tensile strength, tons/sq. in. min. (Rm)	45	50	55	60
Yield stress (1), tons/sq. in. min. (Rs)	34	38	44	48
0·2% Proof stress(2), tons/sq. in. min. (0·2Rp)	32	36	41	46
Elongation % min. (A)	22	20	18	17
Izod impact value ft. lb. min.	40	40	40	35
Brinell hardness number,(3) (HB)	201/255	223/227	248/302	269/321

(1) Not to be used as acceptance tests except by special arrangement between the purchaser and the manufacturer.
(2) When specifically requested in the enquiry or order.
(3) For information purposes only; not contractual part of the specification.

APPLICATIONS

This steel has similar mechanical properties to En 19 and En 22 and is used in similar applications. It is capable of being heat treated to 45 to 55 tons/sq. in. sizes up to 6 in. diameter and to 60 to 70 tons/sq. in. in sizes up to $1\frac{1}{8}$ in. diameter.

Typical applications are highly stressed shafts, spindles and gears.

WELDING

The welding characteristics will be similar to those of En 22.

MACHINABILITY

The machinability of this steel when hardened to 50—55 tons/sq. in. is approximately 50% of that for mild steel (En 3). In the softened condition the machinability is approximately 60% of that for mild steel.

HOT WORKING AND HEAT TREATMENT TEMPERATURES

Forging, rolling and stamping	1180°C, finish above 900°C.
Annealing	830°C—860°C, furnace cool.
Softening	630°C—650°C, air or furnace cool.
Hardening	830°C—860°C, oil or water quench.
Tempering	550°C—660°C, air cool.

PHYSICAL PROPERTIES

No data are available for the physical properties of En 160. The values quoted below have been estimated as outlined in the Introduction for a steel of nominal composition 0·40% C, 0·25% Si, 0·45% Mn, 1·75% Ni, 0·25% Mo.

SPECIFIC GRAVITY (d)

7·85 at 20°C.

SPECIFIC HEAT (c)

0·116 cal/g. deg. C at 20°C to 100°C.

MEAN COEFFICIENT OF THERMAL EXPANSION (α)

$11{\cdot}5 \times 10^{-6}$ deg. C at 20°C to 100°C.

ELECTRICAL RESISTIVITY (ϱ)

22 microhm-cms at 20°C.

THERMAL CONDUCTIVITY (k)

0·11 cal/cm sec. deg. C at 20°C.

YOUNG'S MODULUS (E)

13,500 tons/sq. in.

En 160

TRANSFORMATION CHARACTERISTICS

ISOTHERMAL TRANSFORMATION

Chemical Composition (%)							
C	Si	Mn	S	P	Ni	Cr	Mo
0·41	0·13	0·48	0·043	0·016	1·75	0·17	0·22

McQuaid Ehn Grain Size 2—3 (ASTM).
As Quenched Grain Size 6—7 (ASTM).
Austenitizing Temperature 845°C.

(Ref. 1)

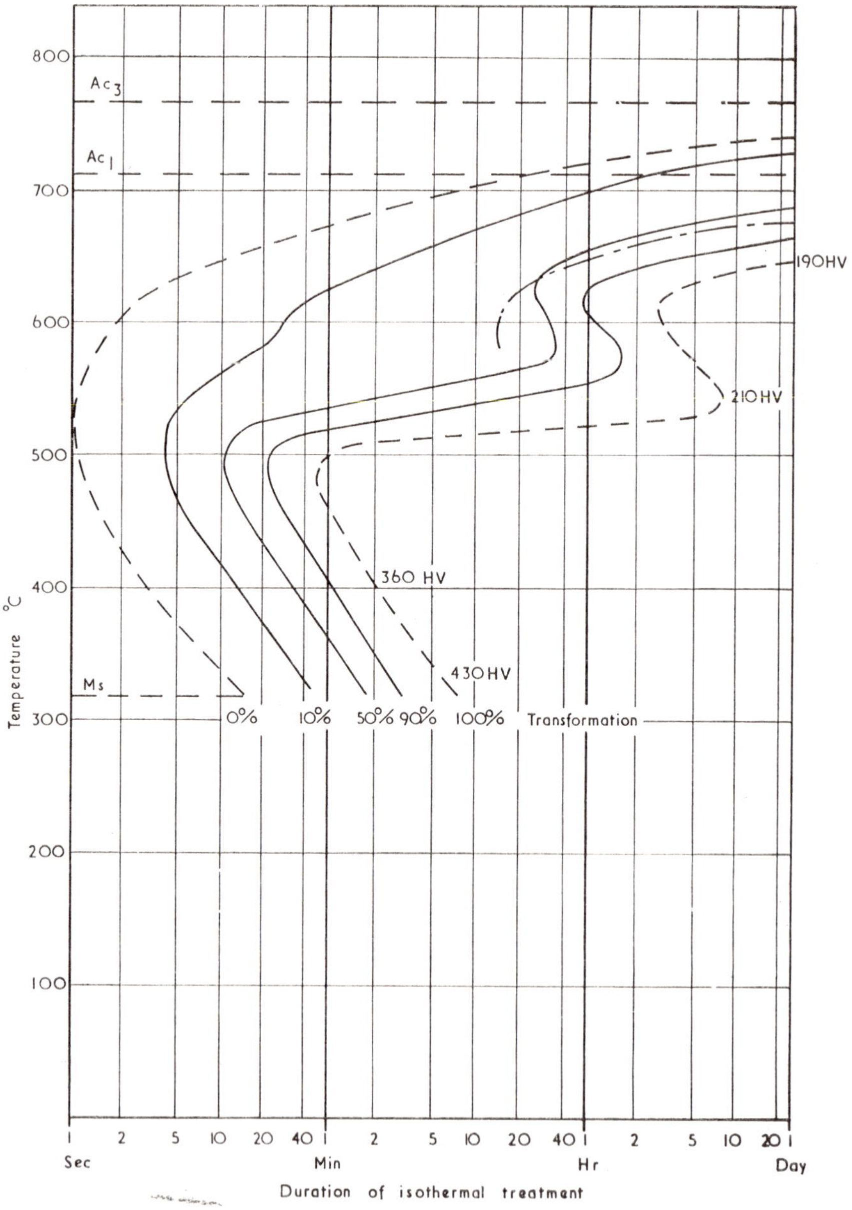

Fig. 160.1

The hardness values are those of the fully transformed steel.

ISOTHERMAL TRANSFORMATION (Continued)

Chemical Composition (%)			
C	Mn	Ni	Mo
0·36	0·63	1·84	0·23

Grain Size 7—8 ASTM.
Austenitizing Temperature 843°C.

(Ref. 2)

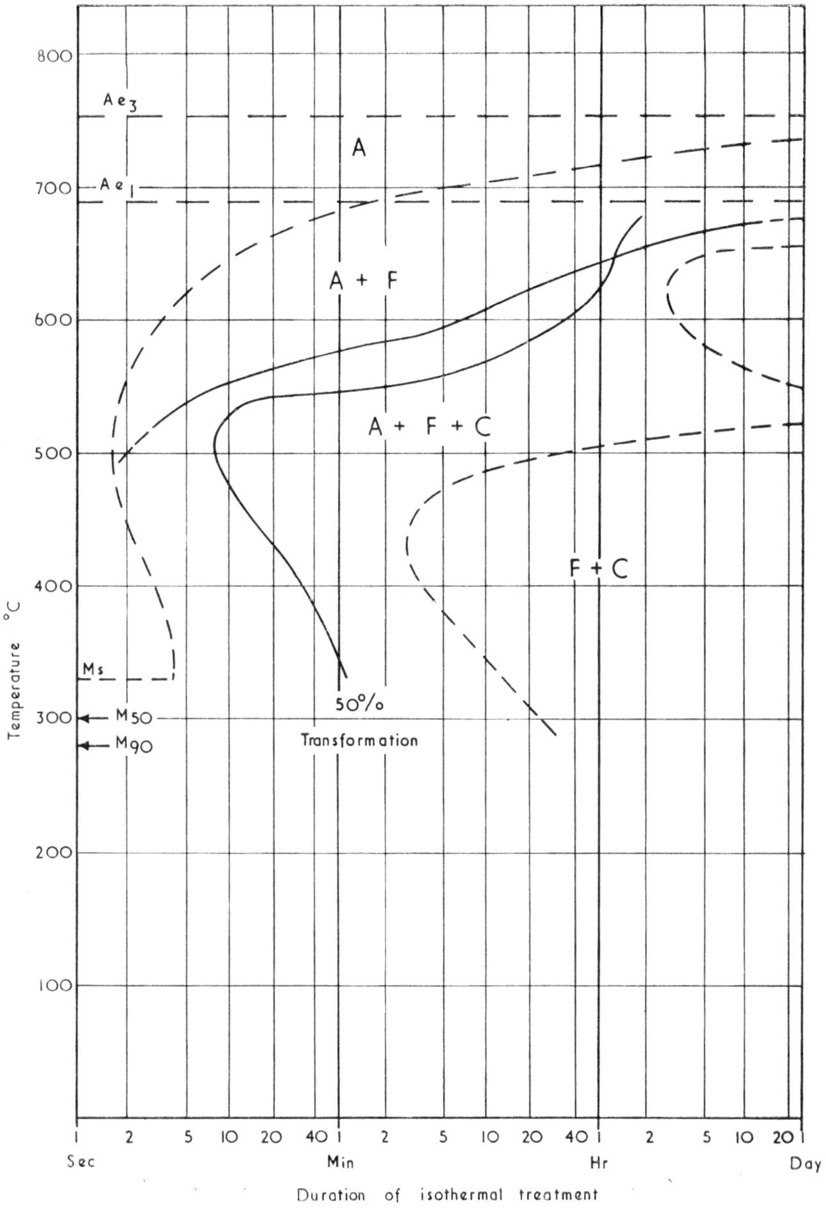

Fig. 160.2

En 160

CONTINUOUS COOLING TRANSFORMATION

Chemical Composition (%)							
C	Si	Mn	S	P	Ni	Cr	Mo
0·41	0·13	0·48	0·043	0·016	1·75	0·17	0·22

McQuaid Ehn Grain Size 2—3 (ASTM).
As quenched Grain Size 6—7 (ASTM).
Austenitizing Temperature 845°C.

(Ref. 1)

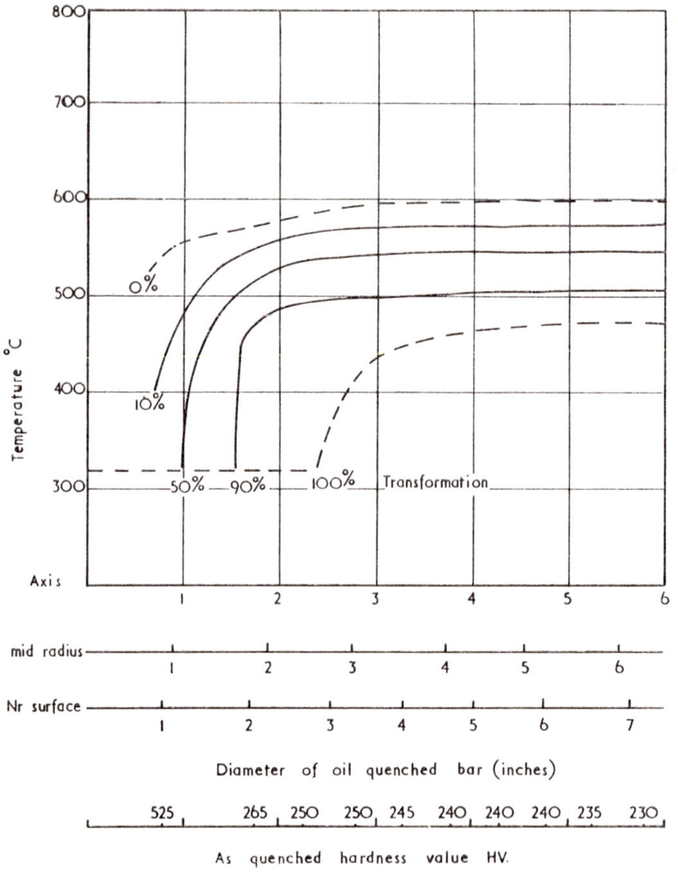

Fig. 160.3

HARDENABILITY

END QUENCH HARDENABILITY

Chemical Composition (%)							
C	Si	Mn	S	P	Ni	Cr	Mo
0·41	0·13	0·48	0·043	0·016	1·75	0·17	0·22

McQuaid Ehn Grain Size 2—3 (ASTM).
As quenched Grain Size 6—7 (ASTM).
End quenched from 845°C. Tempered as indicated. (Ref. 1)

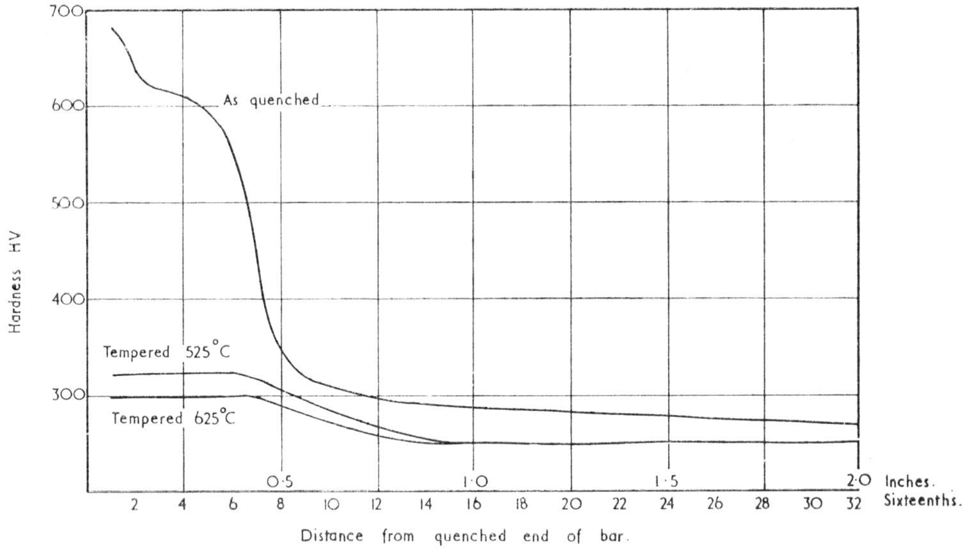

Fig. 160.4

Chemical Composition (%)			
C	Mn	Ni	Mo
0·36	0·63	1·84	0·23

Grain Size 7—8 (ASTM). End quenched from 843°C. (Ref. 2)

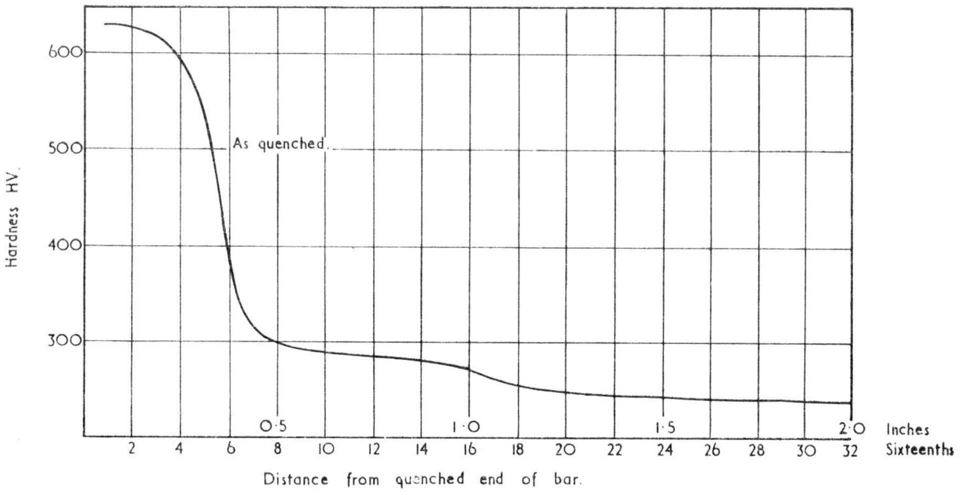

Fig. 160.5

En 160

MECHANICAL PROPERTIES AT ROOM TEMPERATURE

Process of Manufacture	ASTM Grain Size	Chemical Composition (%)								Section Size dia. (in.)	Heat Treatment °C	Tensile							Izod ft.lb.	Ref.
		C	Si	Mn	S	P	Ni	Cr	Mo			0·1% P.S. (0·1Rp)	0·2% P.S. (0·2Rp)	0·5% P.S. (0·5Rp)	Y.P. (Rs)	T.S. (Rm)	El % (A)	R.A. % (Z)		
												tons/sq.in.								
B.E.A.	3—6	0·38	0·19	0·73	0·012	0·011	1·96	0·21	0·32	0·424	OQ 850	70·0	80·7	89·9		130·3	12	29	19	3
"	"	"	"	"	"	"	"	"	"	"	" T2hrs. 200 AC	90·5	99·6	110·0		117·2	13	38	19	"
"	"	"	"	"	"	"	"	"	"	"	" 300 "	89·1	92·8	99·2		108·2	12	45	14	"
"	"	"	"	"	"	"	"	"	"	"	" 400 "	79·4	80·4	83·6		88·3	14½	47½	21	"
"	"	"	"	"	"	"	"	"	"	"	" 500 "	70·8	71·4	74·4		76·7	16	50½	35	"
"	"	"	"	"	"	"	"	"	"	"	" 550 "				67·0	73·5	18½	53½	46	"
"	"	"	"	"	"	"	"	"	"	"	" 600 "				59·2	63·9	21	53½	60	"
"	"	"	"	"	"	"	"	"	"	"	" 625 "				53·9	59·5	23½	59½	71	"
"	"	"	"	"	"	"	"	"	"	"	" 650 "				47·3	53·2	25	61	88	"
"	"	"	"	"	"	"	"	"	"	"	" 675 "	34·1	34·8	36·1		51·8	24	59½	77	"
"	"	"	"	"	"	"	"	"	"	¾	OQ 850	63·8	75·5	94·7		119·0			12	"
"	"	"	"	"	"	"	"	"	"	"	" T2hrs. 200 AC	89·2	95·9	105·7		119·7	11	28½	13	"
"	"	"	"	"	"	"	"	"	"	"	" " 300 "	86·8	89·1	94·0		101·7	11	38	9	"
"	"	"	"	"	"	"	"	"	"	"	" " 400 "	78·5	80·2	84·0		86·2	11½	41	15	"
"	"	"	"	"	"	"	"	"	"	"	" " 500 "	70·5	71·4	73·3		77·9	16	47	29	"
"	"	"	"	"	"	"	"	"	"	"	" " 550 "	67·0	67·4	68·6		73·4	18	50	37	"
"	"	"	"	"	"	"	"	"	"	"	" " 600 "	58·5	58·6	59·4		64·4	19½	56½	57	"
"	"	"	"	"	"	"	"	"	"	"	" " 625 "				54·4	60·2	21	58½	67	"
"	"	"	"	"	"	"	"	"	"	"	" " 650 "				47·8	54·3	25	62	78	"
"	"	"	"	"	"	"	"	"	"	"	" " 675 "	36·9	37·0	37·8		51·8	26	59½	74	"
"	"	"	"	"	"	"	"	"	"	1⅛	OQ 850	70·3	79·2	89·9		108·9	11	28½	11	"
"	"	"	"	"	"	"	"	"	"	"	" T2hrs. 100 AC	79·1	87·9	98·0		113·0	8½	25	11	"
"	"	"	"	"	"	"	"	"	"	"	" " 200 "	74·8	84·4	95·2		107·1	10	29	12	"
"	"	"	"	"	"	"	"	"	"	"	" " 300 "	80·0	85·7	92·2		99·8	9½	32	8	"
"	"	"	"	"	"	"	"	"	"	"	" " 400 "	76·6	79·3	83·5		87·8	11½	39	14	"
"	"	"	"	"	"	"	"	"	"	"	" " 500 "	68·7	70·2	72·5		77·6	16½	44½	29	"
"	"	"	"	"	"	"	"	"	"	"	" " 550 "	63·4	65·0	66·9		71·8	18	52½	45	"
"	"	"	"	"	"	"	"	"	"	"	" " 600 "	53·4	54·1	55·1		61·7	21½	59½	64	"
"	"	"	"	"	"	"	"	"	"	"	" " 625 "	50·3	50·4	50·5		58·3	23½	63½	81	"
"	"	"	"	"	"	"	"	"	"	"	" " 650 "	46·8	47·0	47·2		55·4	24	63½	86	"
"	"	"	"	"	"	"	"	"	"	"	" " 675 "	37·7	38·1	39·0		51·3	26	51	79	"
"	"	"	"	"	"	"	"	"	"	1½	OQ 850	56·1	65·5	76·3		89·6	11	33½	9	3
"	"	"	"	"	"	"	"	"	"	"	" T2hrs. 200 AC	57·8	65·9	75·2		85·1	11½	33	11	"
"	"	"	"	"	"	"	"	"	"	"	" " 300 "	48·4	55·3	64·2		74·1	14	44	10	"
"	"	"	"	"	"	"	"	"	"	"	" " 400 "	49·9	54·7	59·3		68·2	15	47	13	"
"	"	"	"	"	"	"	"	"	"	"	" " 500 "	51·3	54·3	57·5		66·2	17	50	29	"
"	"	"	"	"	"	"	"	"	"	"	" " 550 "	52·9	55·0	57·4		65·5	19	56½	38	"
"	"	"	"	"	"	"	"	"	"	"	" " 600 "	54·1	55·1	56·2		63·5	22	60	70	"

MECHANICAL PROPERTIES AT ROOM TEMPERATURE (Continued)

Process of Manufacture	ASTM Grain Size	Chemical Composition (%) C	Si	Mn	S	P	Ni	Cr	Mo	Section Size dia. (in.)	Heat Treatment °C	Tensile 0·1% P.S. (0·1Rp) tons/sq.in.	0.2% P.S. (0·2Rp)	0.5% P.S. (0·5Rp)	Y.P. (Rs)	T.S. (Rm)	El % (A)	R.A. % (Z)	Izod ft.lb.	Ref.
B.E.A.	3—6	0·38	0·19	0·73	0·012	0·011	1·96	0·21	0·32	1½	OQ 850									
"	"	"	"	"	"	"	"	"	"	"	T2hrs. 625 AC	45·7	46·3	47·0		56·3	23	60	87	3
"	"	"	"	"	"	"	"	"	"	"	" " 650 "			43·7		54·0	25	65½	90	"
"	"	"	"	"	"	"	"	"	"	"	" " 675 "	34·3	34·7	35·4		49·9	26	60½	87	"
"	"	"	"	"	"	"	"	"	"	2½	OQ 850	43·6	49·8	53·2		72·4	12½	30½	8	"
"	"	"	"	"	"	"	"	"	"	"	" T2hrs. 200 AC	45·1	52·0	58·6		71·0	13	37½	11	"
"	"	"	"	"	"	"	"	"	"	"	" " 300 "	48·5	54·8	61·8		72·2	15	48½	8	"
"	"	"	"	"	"	"	"	"	"	"	" " 400 "	51·8	53·9	59·3		69·0	16	48½	13	"
"	"	"	"	"	"	"	"	"	"	"	" " 500 "	50·6	52·7	55·9		64·5	19	53	22	"
"	"	"	"	"	"	"	"	"	"	"	" " 550 "	49·2	51·0	53·7		62·5	18½	53½	25	"
"	"	"	"	"	"	"	"	"	"	"	" " 600 "	46·1	47·2	50·0		57·6	23	59½	42	"
"	"	"	"	"	"	"	"	"	"	"	" " 625 "	44·6	45·6	48·3		55·9	22½	62	57	"
"	"	"	"	"	"	"	"	"	"	"	" " 650 "	39·5	40·0	41·1		52·2	26	63½	86	"
"	"	"	"	"	"	"	"	"	"	"	" " 675 "			37·6		50·7	26	64½	78	"
"	"	"	"	"	"	"	"	"	"	0·424	WQ 850	88·2	99·4	115·5		127·1	1	7½	6	"
"	"	"	"	"	"	"	"	"	"	"	" T2hrs. 100 AC	88·9	98·7	111·3		138·3	—	—	5	"
"	"	"	"	"	"	"	"	"	"	"	" " 200 "	91·0	97·3	105·0		136·3	13	44	25	"
"	"	"	"	"	"	"	"	"	"	"	" " 500 "	68·9	70·0	70·7		84·0	17½	50½	42	"
"	"	"	"	"	"	"	"	"	"	"	" " 550 "	66·5	67·4	68·6		72·8	18½	54	46	"
"	"	"	"	"	"	"	"	"	"	"	" " 600 "	55·3	57·4	59·5		68·1	19½	56	60	"
"	"	"	"	"	"	"	"	"	"	"	" " 625 "			54·1		59·2	23	59½	70	"
"	"	"	"	"	"	"	"	"	"	"	" " 650 "	48·3	49·0	49·4		57·1	23	67	77	"
"	"	"	"	"	"	"	"	"	"	"	" " 675 "	42·7	44·1	45·5		52·5	27½	65	90	"
"	"	"	"	"	"	"	"	"	"	¾	WQ 850									
"	"	"	"	"	"	"	"	"	"	"	T2hrs. 500 AC	71·2	71·3	73·1		77·4	17½	49½	34	3
"	"	"	"	"	"	"	"	"	"	"	" " 550 "	66·6	66·7	67·1		73·0	14	36	44	"
"	"	"	"	"	"	"	"	"	"	"	" " 600 "			59·6		65·8	21½	57	58	"
"	"	"	"	"	"	"	"	"	"	"	" " 625 "			53·1		59·6	22½	59	70	"
"	"	"	"	"	"	"	"	"	"	"	" " 650 "			47·1		53·0	26½	61½	81	"
"	"	"	"	"	"	"	"	"	"	"	" " 675 "	33·7	34·9	37·5		53·7	24½	54½	77	"
"	"	"	"	"	"	"	"	"	"	1⅛	WQ 850									
"	"	"	"	"	"	"	"	"	"	"	T2hrs. 450 AC			77·2		80·2	15	49	25	"
"	"	"	"	"	"	"	"	"	"	"	" " 500 "			72·3		78·3	17	50½	31	"
"	"	"	"	"	"	"	"	"	"	"	" " 550 "	67·0	67·4	68·5		73·3	17	40½	44	"
"	"	"	"	"	"	"	"	"	"	"	" " 600 "			59·0		64·9	21	59½	60	"
"	"	"	"	"	"	"	"	"	"	"	" " 625 "			52·9		59·8	22	49½	74	"
"	"	"	"	"	"	"	"	"	"	"	" " 650 "			46·9		53·9	25	62	83	"
"	"	"	"	"	"	"	"	"	"	"	" " 675 "	31·1	32·6	35·4		55·0	25	51	71	"
"	"	"	"	"	"	"	"	"	"	2⅛	WQ 850									
"	"	"	"	"	"	"	"	"	"	"	T2hrs. 450 AC	74·0	74·0	76·8		81·8	14½	47½	23	"
"	"	"	"	"	"	"	"	"	"	"	" " 500 "	69·7	70·5	72·5		77·4	17½	49	33	"

En 160

MECHANICAL PROPERTIES AT ROOM TEMPERATURE (Continued)

Process of Manufacture	ASTM Grain Size	Chemical Composition (%) C	Si	Mn	S	P	Ni	Cr	Mo	Section Size dia. (in.)	Heat Treatment °C	Tensile 0·1% P.S. (0·1Rp) tons/sq.in.	0·2% P.S. (0·2Rp) tons/sq.in.	0·5% P.S. (0·5Rp) tons/sq.in.	Y.P. (Rs) tons/sq.in.	T.S. (Rm) tons/sq.in.	El % (A)	R.A. % (Z)	Izod ft.lb.	Ref.
B.E.A.	3—6	0·38	0·19	0·73	0·012	0·011	1·96	0·21	0·32	2⅛	WQ 850									
"	"	"	"	"	"	"	"	"	"	"	T2hrs. 550 AC	65·7	66·0	67·0		72·3	20	53½	46	3
"	"	"	"	"	"	"	"	"	"	"	" " 600 "	58·5	58·5	58·7		64·8	21½	58	63	"
"	"	"	"	"	"	"	"	"	"	"	" " 625 "			50·7		58·0	24	62½	77	"
"	"	"	"	"	"	"	"	"	"	"	" " 650 "			46·8		53·9	25	63	88	"
"	"	"	"	"	"	"	"	"	"	"	" " 675 "	32·5	33·0	33·4		51·3	27½	56	83	"
"	"	"	"	"	"	"	"	"	"	0·424	AC 850	33·4	36·4	43·8		57·4	18½	47	22	"
"	"	"	"	"	"	"	"	"	"	"	" T2hrs. 200 AC	35·5	39·4	44·6		57·2	19½	48	24	"
"	"	"	"	"	"	"	"	"	"	"	" " 400 "	42·9	46·2	50·6		56·2	20½	52	24	"
"	"	"	"	"	"	"	"	"	"	"	" " 500 "	41·0	42·8	44·6		55·5	20	52	32	"
"	"	"	"	"	"	"	"	"	"	"	" " 550 "			42·0		53·7	23	54	33	"
"	"	"	"	"	"	"	"	"	"	"	" " 600 "			41·0		51·6	24	56½	43	"
"	"	"	"	"	"	"	"	"	"	"	" " 625 "			38·7		50·6	22½	57	65	"
"	"	"	"	"	"	"	"	"	"	"	" " 650 "			36·9		48·1	26	60½	67	"
"	"	"	"	"	"	"	"	"	"	"	" " 675 "			32·9		46·2	25½	61	80	"
"	"	"	"	"	"	"	"	"	"	¾	AC 850	33·7	37·9	43·4		56·5	19	47	12	"
"	"	"	"	"	"	"	"	"	"	"	" T2hrs. 200 AC	35·1	39·2	44·4		57·2	18	46	13	"
"	"	"	"	"	"	"	"	"	"	"	" " 400 "	39·0	40·8	44·9		55·9	18	48	16	"
"	"	"	"	"	"	"	"	"	"	"	" " 500 "	40·8	41·4	43·6		54·8	21	51½	19	"
"	"	"	"	"	"	"	"	"	"	"	" " 550 "	41·9	42·0	43·8		54·5	22	51½	23	"
"	"	"	"	"	"	"	"	"	"	"	" " 600 "			42·0		52·3	22½	53	32	"
"	"	"	"	"	"	"	"	"	"	"	" " 625 "			40·2		51·3	23	56½	34	"
"	"	"	"	"	"	"	"	"	"	"	" " 650 "			38·0		48·6	24½	56½	45	"
"	"	"	"	"	"	"	"	"	"	"	" " 675 "			34·2		46·3	26½	61½	58	"
"	"	"	"	"	"	"	"	"	"	1⅛	AC 850	33·0	37·2	42·4		55·7	18	42	13	"
"	"	"	"	"	"	"	"	"	"	"	" T2hrs. 100 AC	33·0	37·2	42·4		55·7	18½	42	13	"
"	"	"	"	"	"	"	"	"	"	"	" " 200 "	33·9	37·7	42·6		55·4	18½	42	13	"
"	"	"	"	"	"	"	"	"	"	"	" " 300 "	35·2	38·7	43·8		55·7	19	48½	11	"
"	"	"	"	"	"	"	"	"	"	"	" " 400 "	36·4	39·6	44·0		55·3	19	46½	12	"
"	"	"	"	"	"	"	"	"	"	"	" " 500 "	39·8	41·3	44·1		54·5	20	49	19	"
"	"	"	"	"	"	"	"	"	"	"	" " 550 "	40·5	41·0	42·6		53·6	22	51½	21	"
"	"	"	"	"	"	"	"	"	"	"	" " 600 "		39·9	40·4		51·5	22	56	26	"
"	"	"	"	"	"	"	"	"	"	"	" " 625 "	38·9	39·0	39·6		50·4	23	55	34	"
"	"	"	"	"	"	"	"	"	"	"	" " 650 "			36·7		48·4	24½	57	44	"
"	"	"	"	"	"	"	"	"	"	"	" " 675 "			33·7		45·6	25	57	67	"
"	"	"	"	"	"	"	"	"	"	1½	AC 850	33·2	39·4	42·0		55·2	19	44	12	"
"	"	"	"	"	"	"	"	"	"	"	" T2hrs. 200 AC	32·3	36·3	42·0		54·9	19	42	12	"
"	"	"	"	"	"	"	"	"	"	"	" " 500 "	37·3	39·1	41·9		53·1	22	40	20	"
"	"	"	"	"	"	"	"	"	"	"	" " 550 "	37·7	39·1	41·3		52·1	22	52½	21	"
"	"	"	"	"	"	"	"	"	"	"	" " 600 "	39·4	39·5	40·5		51·7	23	54½	24	"
"	"	"	"	"	"	"	"	"	"	"	" " 625 "			38·1		49·7	24	56	30	"

MECHANICAL PROPERTIES AT ROOM TEMPERATURE (Continued)

Process of Manufacture	ASTM Grain Size	Chemical Composition (%) C	Si	Mn	S	P	Ni	Cr	Mo	Section Size dia. (in.)	Heat Treatment °C	Tensile 0·1% P.S. (0·1Rp) tons/sq.in.	0·2% P.S. (0·2Rp) tons/sq.in.	0·5% P.S. (0·5Rp) tons/sq.in.	Y.P. (Rs) tons/sq.in.	T.S. (Rm) tons/sq.in.	El % (A)	R.A. % (Z)	Izod ft.lb.	Ref.
B.E.A.	3—6	0·38	0·19	0·73	0·012	0·011	1·96	0·21	0·32	1½	AC 850									
”	”	”	”	”	”	”	”	”	”	”	T2hrs. 650 AC			36·8		48·4	24½	55	37	3
”	”	”	”	”	”	”	”	”	”	2⅛	” ” 675 ”			32·1		45·9	26	58½	49	”
”	”	”	”	”	”	”	”	”	”	”	AC 850	34·2	38·0	43·4		54·4	18½	49½	11	”
”	”	”	”	”	”	”	”	”	”	”	” T2hrs. 200 AC	35·0	39·0	43·6		55·1	20½	47	11	”
”	”	”	”	”	”	”	”	”	”	”	” ” 400 ”	36·1	39·9	45·3		53·4	20	49½	13	”
”	”	”	”	”	”	”	”	”	”	”	” ” 500 ”	38·9	40·5	43·4		54·1	22½	51	16	”
”	”	”	”	”	”	”	”	”	”	”	” ” 550 ”	41·2	43·1	47·4		53·2	22	52½	19	”
”	”	”	”	”	”	”	”	”	”	”	” ” 600 ”	39·5	40·5	44·6		51·2	23	53½	27	”
”	”	”	”	”	”	”	”	”	”	”	” ” 625 ”	38·3	39·5	43·0		50·2	24½	54½	34	”
”	”	”	”	”	”	”	”	”	”	”	” ” 650 ”			36·3		47·8	26	57½	38	”
”	”	”	”	”	”	”	”	”	”	”	” ” 675 ”			33·8		46·1	27	60	44	”
		0·40					1·87		0·22	1⅛	AC 850				54·0	73·2	11	22		4
		”					”		”	”	OQ 850T650				53·1	61·7	21½	58½		”
B.E.A.		0·40	0·26	0·46	0·011	0·015	1·76	0·15	0·27	1⅛	OQ 870 T550					69·6	20	51	44	5
”		”	”	”	”	”	”	”	”	”	” T600				58·6	64·8	21	56	57	”
”		”	”	”	”	”	”	”	”	5sq.	” T580				41·2	53·6	23	50	19	”
”		”	”	”	”	”	”	”	”	”	” T610				38·4	50·6	25	59	47	”
		0·41		0·73			1·96	0·21	0·31	1¼	OQ850 T650				51·0	57·1	25		82	6
		”		”			”	”	”	1·65	” T550				68·0	72·3	18		44	”
B.E.A.		0·42	0·36	0·56	0·004	0·010	1·89	0·32	0·30	13×12 ring	OQ860 T635 SC				40·6	55·2	18	42	31	5
		0·45		0·82			1·90	0·16	0·27	1⅛	OQ840 T600				64·9	69·8	19½		48	6
		”		”			”	”	”	2½	” T650				52·3	57·5	24		79	”
B.E.A.		0·45	0·31	0·55	0·007	0·013	2·02	0·31	0·34	13×12 ring	OQ860 T635 SC				39·0	52·4	20	37	46	5

En 160

EFFECT OF TEMPERING ON THE MECHANICAL PROPERTIES AT ROOM TEMPERATURE

Chemical Composition (%)				
C	Mn	Ni	Cr	Mo
0·40	0·70	1·85	0·20	0·30

$1\frac{1}{8}$ in. dia. bars. Oil quenched from 830°C.

(Ref. 6)

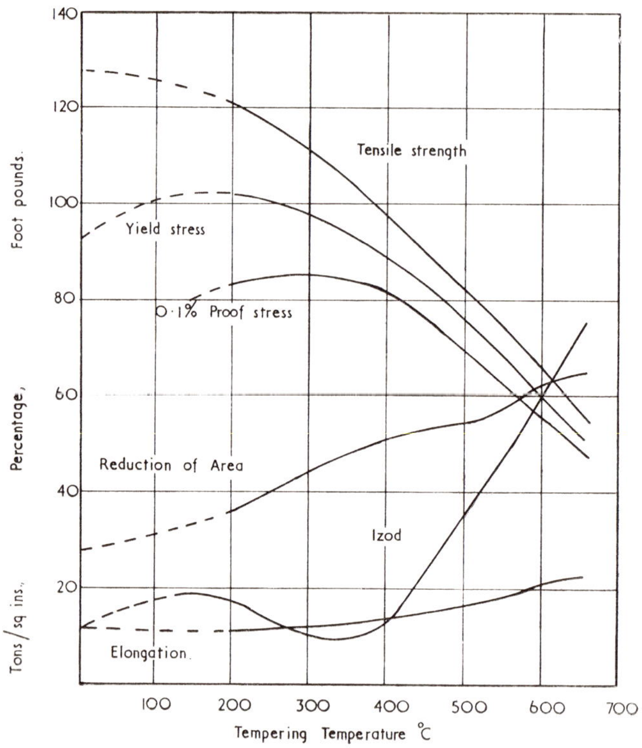

Fig. 160.6

EFFECT OF TEMPERING ON THE MECHANICAL PROPERTIES AT ROOM TEMPERATURE (Continued)

Chemical Composition (%)							
C	Si	Mn	S	P	Ni	Cr	Mo
0·38	0·19	0·73	0·012	0·011	1·96	0·21	0·32

Basic Electric Arc.
McQuaid Ehn Grain Size 3—6 (ASTM).
$1\frac{1}{8}$ in. dia. bar. Quenched and tempered as indicated.

Oil quenched from 850°C. (Ref. 3)

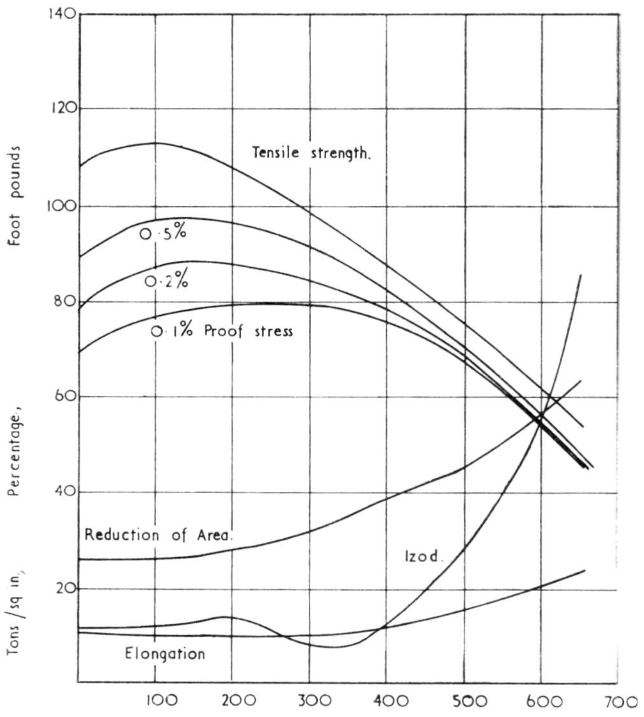

Fig. 160.7

Water quenched from 850°C. (Ref. 3)

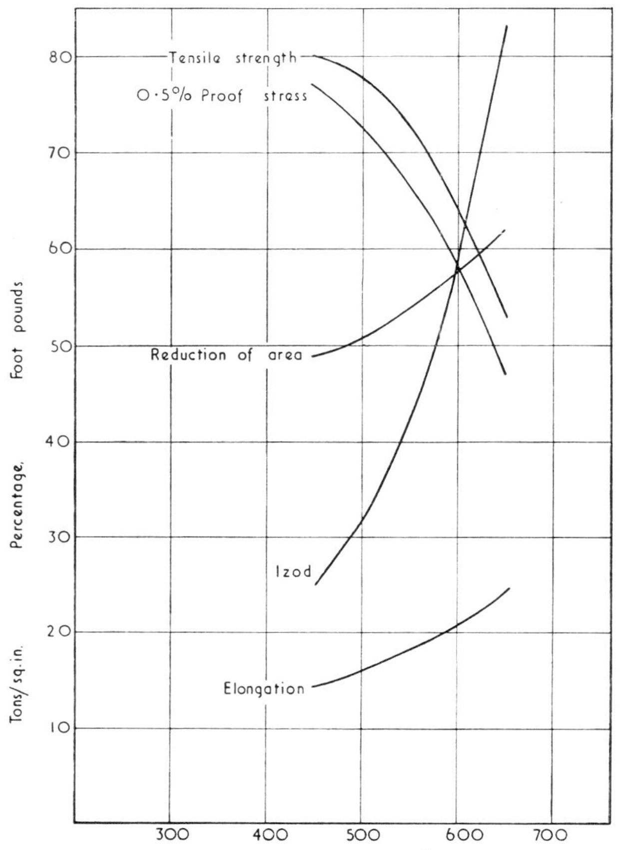

Fig. 160.8

TORSIONAL PROPERTIES AT ROOM TEMPERATURE

Chemical Composition (%)					Section Size dia. (in.)	Heat Treatment °C	Tensile		Torsion		Ref.
C	Mn	Ni	Cr	Mo			0·1% P.S. (0·1Rp)	0·5% P.S. (0·5Rp)	Apparent max. Shear Stress	Probable max. Shear Stress	
							tons/sq.in.		tons/sq.in.		
0·42	0·73	1·96	0·21	0·31	0·6	OQ830 T300	63·1	69·4	79·9	59·9	6
"	"	"	"	"	"	" T400	49·9	56·0	64·8	48·6	"
"	"	"	"	"	"	" T450	47·3	53·3	63·1	47·3	"
"	"	"	"	"	2	" T500	44·0	48·7	58·7	44·0	"
"	"	"	"	"	2	" T300	52·1	63·1	78·3	58·7	"
"	"	"	"	"	"	" T400	51·3	58·8	66·1	49·6	"
"	"	"	"	"	"	" T450	49·9	55·4	65·6	49·2	"
"	"	"	"	"	"	" T500	44·1	49·3	62·0	46·5	"

LIST OF REFERENCES

1. Transformation Characteristics of Direct-Hardening Nickel Alloy Steels. International Nickel Ltd. 3rd Edition, 1958.
2. Atlas of Isothermal Transformation Diagrams. United States Steel Corporation, 1957.
3. Private correspondence with The English Steel Corporation Ltd.
4. Private correspondence with Edgar Allen and Co. Ltd.
5. Private correspondence with Firth Brown Ltd.
6. Private correspondence with International Nickel Ltd.

CARBON–MANGANESE CASE-HARDENING STEEL

SPECIFICATION

CHEMICAL COMPOSITION (%)

C	Si	Mn	S	P
0·18max.	0·05—0·35	1·10—1·50	0·050max.	0·050max.

MECHANICAL PROPERTIES

Tensile strength, tons/sq.in. min. (Rm)	40
Elongation, % min. (A)	20
Izod impact value, ft. lb. min.	40

The heat treatment given to 1⅛ in. dia. test bars shall be blank carburize at a temperature between 880°C and 930°C,(1) (2) refine at a temperature of 870°C—900°C, cool in air, oil or water, harden in oil or water from a temperature of 770—790°C

(1) By arrangement with the purchaser the test may be carried out on samples from which the blank carburizing treatment has been omitted or modified to a short heating period.

(2) A single quench from 810°C to 830°C may be used in certain cases when the refining treatment after carburizing is omitted by special arrangement between the purchaser and the manufacturer.

RELATED SPECIFICATIONS

United Kingdom		Sweden	USSR	
En No.	BS Air	SIS	GOST	Mark
201	2T6	14.1431	1050—52	14G
		14.1432		
		14.2101	1050—62	15G
			5058—57	14G

APPLICATIONS

This steel is suitable for case-hardened parts requiring a slightly higher core strength than can be obtained with En 32. In the normalized condition it has excellent notched impact properties and may be used for parts requiring a greater degree of toughness than mild steel. Typical applications are gudgeon pins, small components for motor cycles, rifles, general constructional parts in textile and agricultural machinery, bolts subject to shock loading.

WELDING

(See Introduction for key to symbols used in this table.)

	Welding Process					
	MA	G	IG	S	F	B
En 201	a	ax	a	e	a	h

Reproduced by kind permission of H.M.S.O.

MACHINABILITY

In the normalized condition the machinability is approximately 80% of that for mild steel (En 3).

HOT WORKING AND HEAT TREATMENT TEMPERATURES

Forging, rolling and stamping	1200°C, finish above 900°C.
Normalizing	870°C to 930°C, air cool.
Softening	650°C to 670°C, air cool.
Carburizing	880°C to 930°C.
Refining*	870°C to 900°C, air cool, oil or water quench.
Hardening*	770°C to 790°C, oil and water quench.
Tempering (for stress relief)	150°C to 220°C.

* A single oil quench from 810°C to 830°C may be adopted when steel is specifically requested for this treatment ,

PHYSICAL PROPERTIES

No data are available on the physical properties of En 201. Data given for En 14 (Volume 1), which is of similar composition, will be comparable with those obtainable from En 201.

The following data have been estimated for a steel of nominal composition 0·16% C, 0·25% Si, 1·30% Mn, as outlined in the Introduction.

SPECIFIC GRAVITY (d)

7·85 at 20°C.

MEAN SPECIFIC HEAT (c)

0·116 cal/g. deg. C at 20°C to 100°C.

MEAN COEFFICIENT OF THERMAL EXPANSION (α)

$12{\cdot}0 \times 10^{-6}$ per deg. C at 20°C to 100°C.

ELECTRICAL RESISTIVITY (ϱ)

21 microhm-cm at 20°C.

THERMAL CONDUCTIVITY (k)

0·11 cal/cm sec. deg. C at 20°C.

YOUNG'S MODULUS (E)

13,500 tons/sq. in. at 20°C.

MECHANICAL PROPERTIES AT ROOM TEMPERATURE

Process of Manufacture	ASTM Grain Size	Chemical Composition (%)							Section Size dia. (in.)	Heat Treatment °C	Tensile							Izod. ft.lb.	Ref.
		C	Si	Mn	S	P	Ni	Cr			0·1% P.S. (0·1Rp)	0·2% P.S. (0·2Rp)	0·5% P.S. (0·5Rp)	Y.P. (Rm)	T.S. (Rs)	El % (A)	R.A. % (Z)		
											tons/sq.in.								
B.O.H.	1 to 3	0·11	0·20	1·28	0·038	0·020	0·25		$1\frac{1}{8}$	WQ900, WQ780					43·0	32		101	1
"	"	0·11		1·47					$\frac{7}{8}$	BC,OQ900 OQ780				32·8	45·4	29	63		2
"	"	"		"					$1\frac{1}{8}$	" " "				30·5	44·0	32	47		"
"	"	"		"					2	" " "				31·0	42·1	32	48		"
"	"	"		"					$\frac{7}{8}$	BC,WQ900 WQ780				37·0	50·9	28	63		"
"	"	"		"					$1\frac{1}{8}$	" " "				33·5	48·4	31	54		"
"	"	"		"					2	" " "				31·4	44·5	33	61		"
B.O.H.	7 to 8	0·13	0·16	1·30	0·047	0·032	0·25		$1\frac{1}{16}$	WQ900 WQ770					54·4	27		76	1
B.O.H.	"	0·13	0·14	1·32	0·043	0·030	0·18	0·07	$1\frac{1}{16}$	WQ900 WQ770					56·0	27		51	"
"	"	"	"	"	"	"	"	"	"	" OQ780					43·4	34		94	"
B.O.H.	7 to 8	0·13	0·17	1·27	0·029	0·018	0·16		$1\frac{1}{16}$	WQ900 WQ770					54·4	30		68	"
B.O.H.	6 to 7	0·13	0·16	1·25	0·045	0·028	0·22		$1\frac{1}{8}$	WQ900 WQ770					48·4	30		57	"
B.O.H.	6 to 7	0·14	0·15	1·37	0·020	0·033	0·17	0·10	$1\frac{1}{8}$	WQ900 WQ770					46·8	31		72	1
B O.H.	5 to 6	0·14	0·11	1·28	0·047	0·027	0·21		$1\frac{1}{8}$	WQ900 WQ770					49·6	29		58	"
B.O.H.	7 to 8	0·15	0·16	1·30	0·037	0·021	0·21		$1\frac{1}{8}$	WQ900 WQ770					49·4	28		60	"
B.O.H.	"	0·15	0·15	1·35	0·042	0·025	0·20	0·07	$1\frac{1}{8}$	WQ900 WQ770					46·6	30		75	"
B.O.H.	7	0,15	0·15	1·38	0·034	0·022	0·23		$1\frac{1}{8}$	WQ900 WQ770					51·8	26		59	"
B.O.H.	7	0·15	0·14	1·22	0·031	0·020	0·22		$1\frac{1}{8}$	WQ900 WQ770					46·8	30		77	"
B.O.H.	6 to 8	0·15	0·16	1·34	0·036	0·034	0·20	0·08	$1\frac{1}{8}$	WQ900 WQ770					51·6	27		54	"
B.O.H.		0·15	0·18	1·13	0·043	0·025	0·21		$1\frac{1}{8}$	WQ900 WQ770					50·6	29		86	"
B.O.H.	7	0·16	0·15	1·17	0·037	0·022	0·19		$1\frac{1}{16}$	WQ900 WQ780					44·6	34		93	1
		0·16	0·15	1·13						A788 FC				22·5	30·8	$34\frac{1}{2}$	67		3
		"	"	"						" T649				22·8	30·4	$32\frac{1}{2}$	68		"
B.E.A.		0·16	0·27	1·61	0·011	0·015	0·17	0·16	$\frac{3}{4}$	AC880	19·9	21·7	24·2		42·1	32	60	96	4
"		"	"	"	"	"	"	"	"	" T2hrs. 500AC				29·5	38·7	$34\frac{1}{2}$	72	116	"
"		"	"	"	"	"	"	"	"	" " 550 "				28·0	36·6	38	$75\frac{1}{2}$	119	"
"		"	"	"	"	"	"	"	"	" " 600 "				27·5	36·3	38	77	119	"
"		"	"	"	"	"	"	"	"	" " 650 "				26·0	34·8	39	$78\frac{1}{2}$	119	"
"		"	"	"	"	"	"	"	"	" " 675 "				26·1	34·6	$39\frac{1}{2}$	$78\frac{1}{2}$	119	"
"		"	"	"	"	"	"	"	$1\frac{1}{8}$	AC880				24·1	39·9	$32\frac{1}{2}$	$61\frac{1}{2}$	104	"
"		"	"	"	"	"	"	"	"	" T2hrs. 500AC				28·2	37·6	$34\frac{1}{2}$	72	114	"
"		"	"	"	"	"	"	"	"	" T2hrs. 550 "				27·3	36·5	36	73	115	"
"		"	"	"	"	"	"	"	"	" T2hrs. 600 "				25·3	34·4	38	75	116	"
"		"	"	"	"	"	"	"	"	" T2hrs. 650 "				24·8	33·4	39	77	116	"
"		"	"	"	"	"	"	"	"	" T2hrs. 675 "				24·5	33·1	39	78	119	"

MECHANICAL PROPERTIES AT ROOM TEMPERATURE (Continued)

Process of Manufacture	ASTM Grain Size	Chemical Composition (%)							Section Size dia. (in.)	Heat Treatment °C	Tensile							Izod. ft.lb.	Ref.
		C	Si	Mn	S	P	Ni	Cr			0·1% P.S. (0·1Rp)	0·2% P.S. (0·2Rp)	0·5% P.S. (0·5Rp)	Y.P. (Rm)	T.S. (Rs)	El % (A)	R.A. % (Z)		
											tons/sq.in.								
B.E.		0·16	0·27	1·61	0·011	0·015	0·17	0·16	2½	AC880				24·8	38·6	34	65½	117	4
"		"	"	"	"	"	"	"	"	" T2hrs. 500AC				27·1	37·4	35½	73½	117	"
"		"	"	"	"	"	"	"	"	" " 550 "				26·4	36·5	37	73	101	"
"		"	"	"	"	"	"	"	"	" " 600 "				25·8	35·6	37	75½	117	"
"		"	"	"	"	"	"	"	"	" " 650 "				24·3	33·9	39	77½	119	"
"		"	"	"	"	"	"	"	"	" " 675 "	21·8	23·3	24·0		33·6	36	77	118	"
"		"	"	"	"	"	"	"	4¼	AC880				23·5	37·8	34	71½	118	"
"		"	"	"	"	"	"	"	"	" T2hrs. 500AC				25·7	37·6	35½	75	118	"
"		"	"	"	"	"	"	"	"	" " 550 "				26·0	37·6	37	75	118	"
"		"	"	"	"	"	"	"	"	" " 600 "				24·2	35·8	38	76	118	"
"		"	"	"	"	"	"	"	"	" " 650 "				24·4	34·6	39	78	118	"
"		"	"	"	"	"	"	"	"	" " 675 "				23·7	33·8	39	76½	118	"
"		"	"	"	"	"	"	"	6	AC880				24·6	38·2	36	73½	114	"
"		"	"	"	"	"	"	"	"	" T2hrs. 500AC				25·8	37·5	38½	76	117	"
"		"	"	"	"	"	"	"	"	" " 550 "				25·1	37·5	38	76	113	"
"		"	"	"	"	"	"	"	"	" " 600 "				24·1	36·6	38	76½	114	"
"		"	"	"	"	"	"	"	"	" " 650 "				24·2	34·7	39½	77	118	"
"		"	"	"	"	"	"	"	"	" " 675 "				23·8	34·0	40	77	118	"
"		"	"	"	"	"	"	"	¾	OQ880	29·6	32·2	38·7		53·6	24½	56½	76	"
"		"	"	"	"	"	"	"	"	" T2hrs. 500AC				36·7	45·8	25½	71	118	"
"		"	"	"	"	"	"	"	"	" " 550 "				34·5	43·5	28	74	119	"
"		"	"	"	"	"	"	"	"	" " 600 "				29·2	39·1	32½	76½	119	"
"		"	"	"	"	"	"	"	"	" " 650 "				29·2	38·3	33	77½	119	"
"		"	"	"	"	"	"	"	"	" " 675 "				28·7	36·1	37½	80	119	"
"		"	"	"	"	"	"	"	1⅛	OQ880	24·9	27·9	33·5		50·3	27	56	77	4
"		"	"	"	"	"	"	"	"	" T2hrs. 500AC				34·0	42·9	29	70½	116	"
"		"	"	"	"	"	"	"	"	" " 550 "				31·5	40·2	31½	75½	117	"
"		"	"	"	"	"	"	"	"	" " 600 "				29·3	37·9	33	75½	119	"
"		"	"	"	"	"	"	"	"	" " 650 "				27·3	35·4	36	77½	119	"
"		"	"	"	"	"	"	"	"	" " 675 "				27·5	35·1	36	78	119	"
"		"	"	"	"	"	"	"	2½	OQ880	22·3	24·5	28·3		45·6	30	61	118	"
"		"	"	"	"	"	"	"	"	" T2hrs. 500AC				31·4	40·6	31	71½	118	"
"		"	"	"	"	"	"	"	"	" " 550 "				30·2	39·2	33	69	117	"
"		"	"	"	"	"	"	"	"	" " 600 "				29·5	37·8	33½	75	117	"
"		"	"	"	"	"	"	"	"	" " 650 "				27·5	35·4	37	77½	118	"
"		"	"	"	"	"	"	"	"	" " 675 "				27·3	35·8	34½	80	118	"
"		"	"	"	"	"	"	"	4¼	OQ880	28·0	28·6	29·4		48·8	33	75½	83	"
"		"	"	"	"	"	"	"	"	" T2hrs. 500AC				29·5	39·8	33½	72	118	"
"		"	"	"	"	"	"	"	"	" " 550 "				28·5	38·7	33	73	118	"

MECHANICAL PROPERTIES AT ROOM TEMPERATURE (Continued)

Process of Manufacture	ASTM Grain Size	Chemical Composition (%) C	Si	Mn	S	P	Ni	Cr	Section Size dia. (in.)	Heat Treatment °C	Tensile 0·1% P.S. (0·1Rp) tons/sq.in.	0·2% P.S. (0·2Rp)	0·5% P.S. (0·5Rp)	Y.P. (Rm)	T.S. (Rs)	El % (A)	R.A. % (Z)	Izod ft.lb.	Ref.
B.E.		0·16	0·27	1·61	0·011	0·015	0·17	0·16	4¼	OQ880 T2hrs. 600AC				27·2	37·4	36½	75	118	4
"		"	"	"	"	"	"	"	"	" " 650 "				26·1	35·3	37	76½	118	"
"		"	"	"	"	"	"	"	"	" " 675 "				25·4	34·6	38	77½	118	"
"		"	"	"	"	"	"	"	6	OQ880	21·0	23·3	26·1		42·7	33	65	93	"
"		"	"	"	"	"	"	"	"	" T2hrs. 500AC				28·7	39·1	34½	73	118	"
"		"	"	"	"	"	"	"	"	" " 550 "				27·9	38·8	34	76½	113	"
"		"	"	"	"	"	"	"	"	" " 600 "				27·0	36·9	35½	76	115	"
"		"	"	"	"	"	"	"	"	" " 650 "				25·6	35·2	38	77½	118	"
"		"	"	"	"	"	"	"	"	" " 675 "				27·9	33·2	33	73	118	"
"		"	"	"	"	"	"	"	¾	WQ880	44·2	52·6	63·7		77·3	14½	42½	30	"
"		"	"	"	"	"	"	"	"	" T2hrs. 500AC				49·6	50·0	21½	67½	108	"
"		"	"	"	"	"	"	"	"	" " 550 "				43·1	49·1	26	73	119	"
"		"	"	"	"	"	"	"	"	" " 600 "				36·8	45·2	29	77½	119	"
"		"	"	"	"	"	"	"	"	" " 650 "				32·5	41·2	32	77½	119	"
"		"	"	"	"	"	"	"	"	" " 675 "				29·8	39·0	33	78½	119	"
"		"	"	"	"	"	"	"	1⅛	WQ880	40·5	45·4	52·7		66·6	19	49½	39	4
"		"	"	"	"	"	"	"	"	" T2hrs. 550AC	40·8	41·0	41·5		48·3	25	69	104	"
"		"	"	"	"	"	"	"	"	" " 550 "				38·3	45·3	28½	74	113	"
"		"	"	"	"	"	"	"	"	" " 600 "				33·8	41·4	32	76½	119	"
"		"	"	"	"	"	"	"	"	" " 650 "				31·5	39·0	35	79	119	"
"		"	"	"	"	"	"	"	"	" " 675 "				29·9	37·3	35	79	119	"
"		"	"	"	"	"	"	"	2½	WQ880	32·3	35·5	40·5		54·8	24	57½	72	"
"		"	"	"	"	"	"	"	"	" T2hrs. 500AC				36·5	45·5	28	71½	118	"
"		"	"	"	"	"	"	"	"	" " 550 "				32·9	41·4	30	74½	118	"
"		"	"	"	"	"	"	"	"	" " 600 "				32·5	41·0	30	75½	118	"
"		"	"	"	"	"	"	"	"	" " 650 "				29·2	37·2	35	78½	118	"
"		"	"	"	"	"	"	"	"	" " 675 "				26·3	34·7	38	78½	119	"
"		"	"	"	"	"	"	"	4¼	WQ880	27·8	29·5	32·7		49·0	26½	61	59	"
"		"	"	"	"	"	"	"	"	" T2hrs. 500AC				32·5	42·7	29	70½	118	"
"		"	"	"	"	"	"	"	"	" " 550 "				32·4	42·3	29	71½	118	"
"		"	"	"	"	"	"	"	"	" " 600 "	23·1	24·6	27·7		44·0	32	63	118	"
"		"	"	"	"	"	"	"	"	" " 650 "				29·2	37·2	35	78½	118	"
"		"	"	"	"	"	"	"	"	" " 675 "				26·3	34·7	38	78½	119	"
"		"	"	"	"	"	"	"	6	WQ880	24·5	27·5	32·1		49·3	27	59½	56	"
"		"	"	"	"	"	"	"	"	" T2hrs. 500AC				34·7	49·3	32½	72½	118	"

MECHANICAL PROPERTIES AT ROOM TEMPERATURE (Continued)

Process of Manufacture	ASTM Grain Size	Chemical Composition (%) C	Si	Mn	S	P	Ni	Cr	Section Size dia. (in.)	Heat Treatment °C	Tensile 0·1% P.S. (0·1Rp) tons/sq.in.	0·2% P.S. (0·2Rp) tons/sq.in.	0·5% P.S. (0·5Rp) tons/sq.in.	Y.P. (Rm) tons/sq.in.	T.S. (Rs) tons/sq.in.	El % (A)	R.A. % (Z)	Izod. ft.lb.	Ref.
B.E.		0·16	0·27	1·61	0·011	0·015	0·17	0·16	6	WQ880 T2hrs. 550AC				31·4	41·0	30	$71\frac{1}{2}$	118	4
"		"	"	"	"	"	"	"	"	" " 600 "				28·9	39·1	$32\frac{1}{2}$	$76\frac{1}{2}$	114	"
"		"	"	"	"	"	"	"	"	" " 650 "				27·2	36·7	$35\frac{1}{2}$	$76\frac{1}{2}$	118	"
"		"	"	"	"	"	"	"	"	" " 675 "				26·8	35·6	$35\frac{1}{2}$	77	118	"
"		"	"	"	"	"	"	"	$\frac{3}{4}$	OQ880 OQ780	21·6	24·4	29·4		53·3	28	$53\frac{1}{2}$	61	"
"		"	"	"	"	"	"	"	$1\frac{1}{8}$	" "	21·9	24·1	31·0		50·4	$31\frac{1}{2}$	57	81	"
"		"	"	"	"	"	"	"	$2\frac{1}{2}$	" "	20·0	21·5	25·0		47·0	31	59	81	"
"		"	"	"	"	"	"	"	$4\frac{1}{4}$	" "	20·8	22·4	25·2		47·0	$31\frac{1}{2}$	63	63	"
"		"	"	"	"	"	"	"	6	" "	21·0	22·4	24·6		45·7	$32\frac{1}{2}$	63	54	"
"		"	"	"	"	"	"	"	$\frac{3}{4}$	OQ880 WQ780	28·8	33·7	41·9		64·6	20	$39\frac{1}{2}$	34	"
"		"	"	"	"	"	"	"	$1\frac{1}{8}$	" "	27·7	31·1	37·0		58·7	$22\frac{1}{2}$	47	47	"
"		"	"	"	"	"	"	"	$2\frac{1}{2}$	" "	25·5	28·3	34·0		54·7	25	47	—	"
"		"	"	"	"	"	"	"	$4\frac{1}{4}$	" "	20·6	22·5	27·1		50·8	30	59	46	"
"		"	"	"	"	"	"	"	6	" "	21·5	23·4	27·3		50·6	27	$57\frac{1}{2}$	43	"
B.E.		"	"	"	"	"	"	"	$\frac{3}{4}$	WQ880 OQ780	23·1	25·6	30·3		51·2	31	58	89	4
"		"	"	"	"	"	"	"	$1\frac{1}{8}$	" "	22·9	24·4	26·8		48·9	33	$60\frac{1}{2}$	101	"
"		"	"	"	"	"	"	"	$2\frac{1}{2}$	" "	23·6	24·5	26·7		47·7	34	62	100	"
"		"	"	"	"	"	"	"	$4\frac{1}{4}$	" "	22·9	24·0	26·2		46·6	33	$64\frac{1}{2}$	80	"
"		"	"	"	"	"	"	"	6	" "	23·6	24·5	26·7		46·4	$30\frac{1}{2}$	$64\frac{1}{2}$	73	"
"		"	"	"	"	"	"	"	$\frac{3}{4}$	WQ880 WQ780	28·3	32·6	40·3		63·9	22	42	48	"
"		"	"	"	"	"	"	"	$1\frac{1}{8}$	" "	28·9	31·4	36·4		57·1	$25\frac{1}{2}$	55	61	"
"		"	"	"	"	"	"	"	$2\frac{1}{2}$	" "	24·9	26·4	30·0		51·8	$28\frac{1}{2}$	$56\frac{1}{2}$	63	"
"		"	"	"	"	"	"	"	$4\frac{1}{4}$	" "	23·0	25·2	29·3		49·7	31	$61\frac{1}{2}$	70	"
"		"	"	"	"	"	"	"	6	" "	22·9	24·4	27·6		49·6	31	$61\frac{1}{2}$	61	"

EFFECT OF TEMPERING ON THE MECHANICAL PROPERTIES AT ROOM TEMPERATURE

Chemical Composition (%)						
C	Si	Mn	S	P	Ni	Cr
0·16	0·27	1·61	0·011	0·015	0·17	0·16

Basic Electric Arc.
$1\frac{1}{8}$ in. dia. bar, quenched and tempered as indicated.

Oil quenched from 880°C. (Ref. 4)

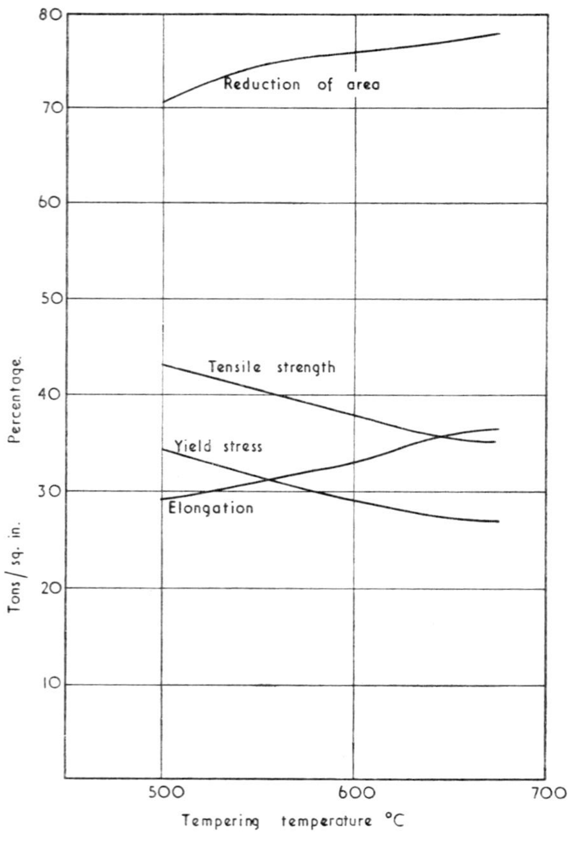

Fig. 201.1

Water quenched from 880°C. (Ref. 4)

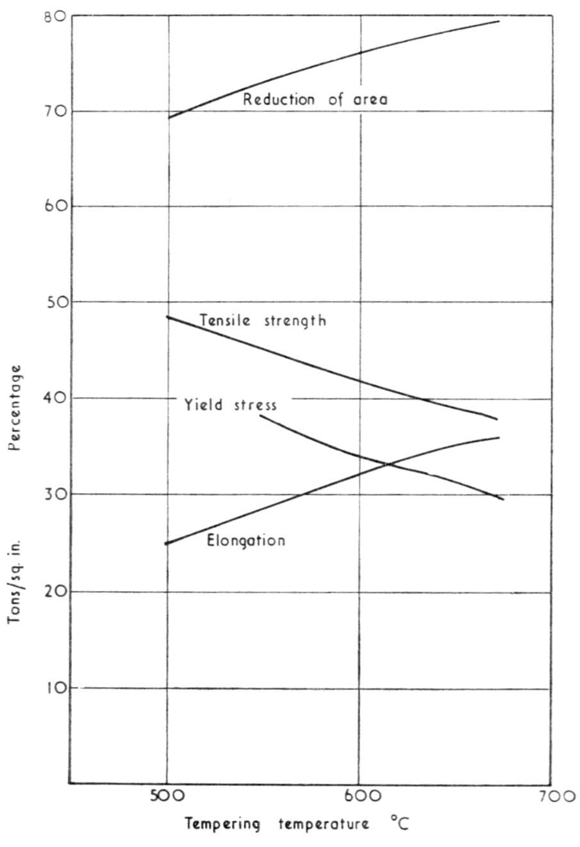

Fig. 201.2

EFFECT OF SECTION SIZE ON THE CORE PROPERTIES AT ROOM TEMPERATURE

Chemical Composition (%)						
C	Si	Mn	S	P	Ni	Cr
0·16	0·27	1·61	0·011	0·015	0·17	0·16

Basic Electric Arc.

Oil quenched 880°C. Oil quenched 780°C.

(Ref. 4)

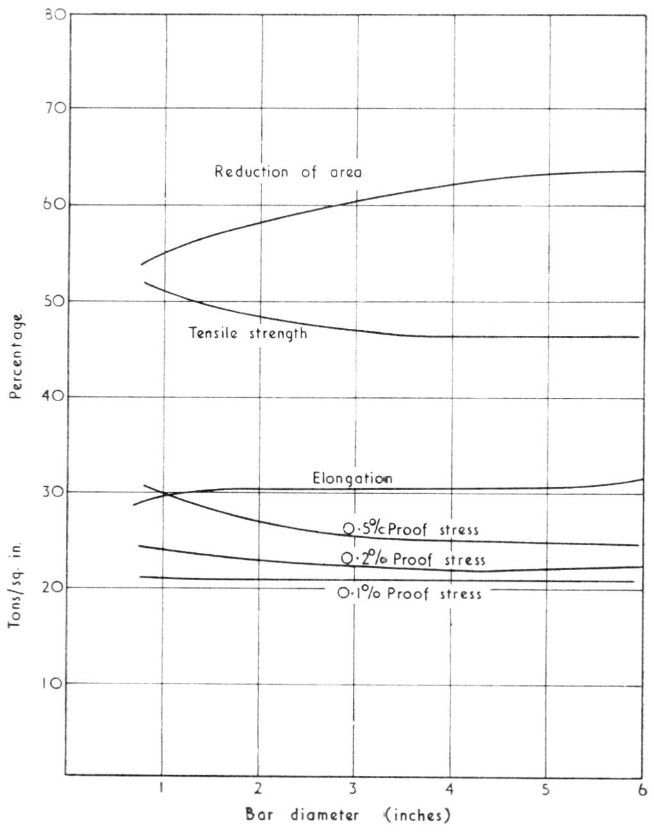

Fig. 201.3

Oil quenched 880°C. Water quenched 780°C.

(Ref. 4)

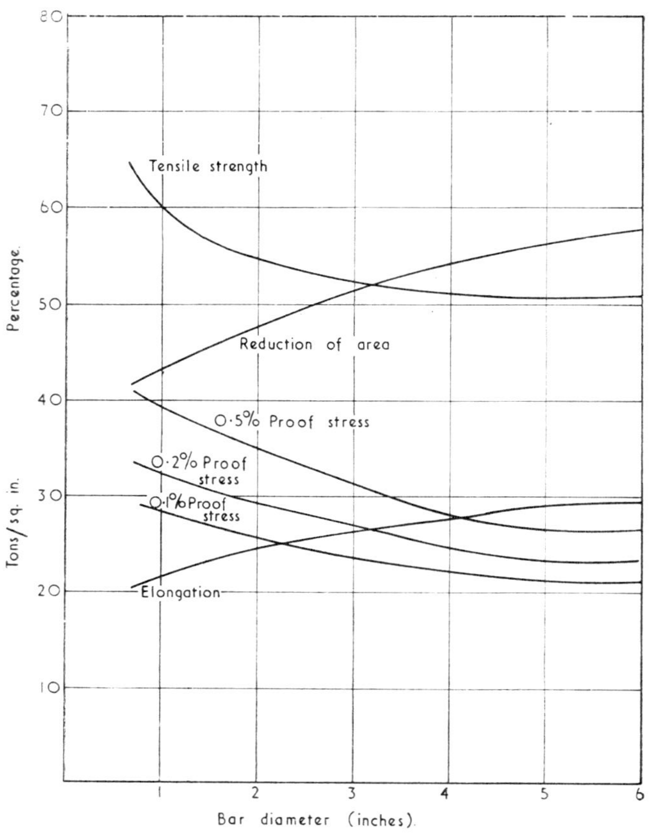

Fig. 201.4

EFFECT OF SECTION SIZE ON THE CORE PROPERTIES AT ROOM TEMPERATURE (Continued)

Water quenched 880°C. Oil quenched 780°C.

(Ref. 4)

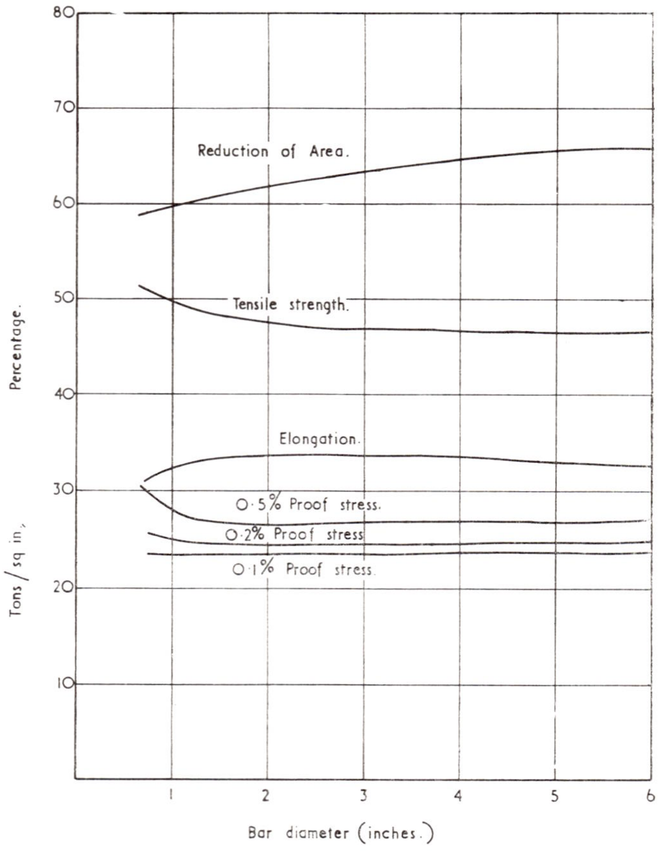

Fig. 201.5

Water quenched 880°C. Water quenched 780°C.

(Ref. 4

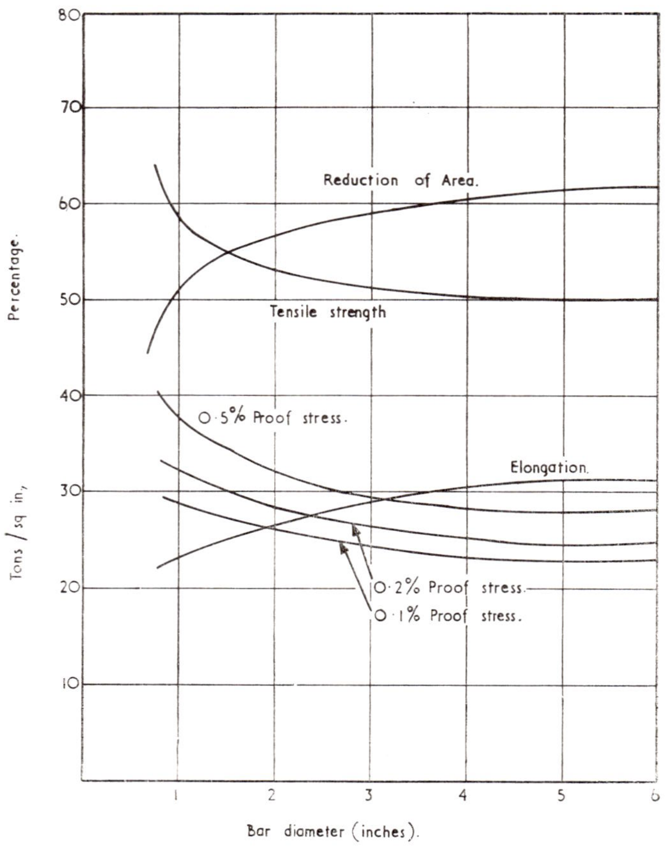

Fig. 201.6

LOW TEMPERATURE IMPACT DATA

Process of Manufacture	Chemical Composition (%)				Section Size dia. (in.)	Heat Treatment °C	Tensile at RT				Charpy V Impact ft.lb. at °C								Ref.
	C	Si	Mn	Al			Y.P. (Rm) tons/sq.in.	T.S. (Rs) tons/sq.in.	El % (A)	R.A. % (Z)	RT	—18	—40	—65	—78	—96	—120	—150	
	0·14	0·11	1·20	0·041		N900	21·4	30.4	41	75½	120+		120+		87	19	8½	2½	3
	"	"	"	"		N1000	19·9	29.7	40½	75	120+		120+		75	6½		1½	"
	"	"	"	"		N1100	19·0	30.4	37	72½	120+		89	48	2				"
	"	"	"	"		WQ900 T677	30·8	33.9	36	79	120+	97	120+		118	88		3	"
H.F. (Basic)	0·15	0·13	1·37	0·035		N900	26·6	35.3	37	74½	120+		119		89	66		3	3
"	"	"	"	"		N1000	25·4	34.6	35	73	120+		119	77	36	40	38		"
"	"	"	"	"		N1100	23·2	34.8	31½	70	107		49	31	18				"
"	"	"	"	"		WQ900 T677	36·2	37.7	32	73½	120+		113		99	110	94		"

Process of Manufacture	Chemical Composition (%)			Section Size dia. (in.)	Heat Treatment °C	Tensile at RT				Charpy Keyhole ft.lb. at °C				Ref.
	C	Si	Mn			Y.P. (Rm) tons/sq.in.	T.S. (Rs) tons/sq.in.	El % (A)	R.A. % (Z)	RT	—19	—65	—184	
H.F.	0·16	0·15	1·13		FC788	22·5	30·8	34½	67	53	49	12	½	3
"	"	"	"		" T649	22·8	30·4	32½	68	54	47	26	1	"

Process of Manufacture	Chemical Composition (%)						Section Size dia. (in.)	Heat Treatment °C	Tensile at RT				Charpy Keyhole ft.lb. at °C						Ref.
	C	Si	Mn	S	P	V			Y.P. (Rm) tons/sq.in.	T.S. (Rs) tons/sq.in.	El % (A)	R.A. % (Z)	RT	—4	—18	—32	—46	—60	
B.O.H.	0·17	0·23	1·38	0·023	0·023	0·1	½ plate	As rolled	28·6	37·2	35	55	45	43	42	31	27	25	3
"	"	"	"	"	"	"	"	AC927					54	66	42	39	32	30	"
"	"	"	"	"	"	"	"	" T649	27·5	35·7	42	54½	40	40	42	37	32	24	"

LIST OF REFERENCES

1. Private correspondence with The Park Gate Iron and Steel Co. Ltd.
2. British Standard 971: Commentary on the British Standard Wrought Steels, En Series. British Standards Institution, London, 1960.
3. H. W. Gillett. Impact Resistance and Tensile Properties of Metals at Sub-atmospheric Temperatures. Project No. 13 of the Joint ASME-ASTM Research Committee on the Effect of Temperatures on the Properties of Metals.
4. Private correspondence with The English Steel Corporation Ltd.

CARBON–MANGANESE CASEHARDENING STEEL
(Semi-free cutting)

SPECIFICATION

CHEMICAL COMPOSITION (%)

C	Si	Mn	S	P
0·18 max.	0·05—0·35	1·20—1·50	0·10—0·18	0·050 max.

MECHANICAL PROPERTIES

Tensile strength, tons/sq. in. min. (Rm)	38
Elongation, % min. (A)	20
Izod impact, ft. lb. min.	30

The heat treatment given to 1⅛ in. dia. test bars shall be: Blank carburize at a temperature between 880°C and 930°C[1], [2]. Refine at a temperature of 870/900°C, cool in air, oil or water. Harden in oil or water from a temperature of 770/790°C.

(1) By arrangement with the purchaser the test may be carried out on samples from which the blank carburizing treatment has been omitted or modified to a short heating period.
(2) A single quench from 810°C to 830°C may be used in certain cases when the refining treatment after carburizing is omitted by special arrangement between the purchaser and the manufacturers.

RELATED SPECIFICATIONS

United Kingdom	America		France	Sweden
En No.	SAE	AISI	AFNOR	SIS
En 202	1118	C1118	20MF4	14.1922

APPLICATIONS

This steel is suitable for case-hardened parts where a core strength of approximately 40 tons/sq. in. is required and where the hardened case is not subjected to shock loading. The addition of sulphur to the steel increases its machinability. Typical uses are low-duty pinions, motor cycle and bicycle parts and case-hardened components requiring automatic or difficult machining operations.

The steel specification does not differ greatly from that of En 7A.

WELDING

(See Introduction for key to symbols used in this table.)

	Welding Process					
	MA*	G	IG	S	F	B
En 202	c	f	f	f	f	h

* Basic coated electrodes are required.

Reproduced by kind permission of H.M.S.O.

MACHINABILITY

In the normalized condition the machinability is approximately 140% of that for mild steel (En 3).

HOT WORKING AND HEAT TREATMENT TEMPERATURES

Forging, rolling and stamping	1200°C, finish above 900°C.
Normalizing	870°C—930°C, air cool.
Softening	630°C—680°C, air cool.
Carburizing	880°C—930°C.
Refining*	870°C—900°C, cool in air, oil or water.
Hardening*	770°C—790°C, oil or water quench.
Tempering (stress relief)	150°C—200°C.

* A single oil quench from 810°C—830*C may be adopted when steel is specifically requested for this treatment.

PHYSICAL PROPERTIES

No data are available for the physical properties of this steel. The data given for En 7A (Volume 1) and En 201 will be very similar to those obtained on this steel.

MECHANICAL PROPERTIES AT ROOM TEMPERATURE

(See also En 7A)

Process of Manufacture	Chemical Composition (%)							Section Size dia. (in.)	Heat Treatment °C	Tensile			Izod ft.lb.	Ref.
	C	Si	Mn	S	P	Ni	Cr			T.S. (Rm) tons/sq.in.	El % (A)	R.A. % (Z)		
B.O.H.	0·12	0·15	1·22	0·158	0·027	0·21		$1\frac{1}{8}$	WQ900 WQ770	37·6	32		77	1
"	"	"	"	"	"	"		"	" "	39·6	31		65	"
"	0·12	0·24	1·20	0·140	0·026	0·16		$1\frac{1}{16}$	WQ900 WQ770	40·6	31		67	"
B.O.H.	0·13	0·17	1·28	0·160	0·027	0·18		$1\frac{1}{8}$	WQ900 WQ770	40·6	33		62	"
"	"	"	"	"	"	"		$1\frac{1}{8}$	" "	42·0	28		57	"
"	0·13	0·15	1·28	0·150	0·026	0·16		$1\frac{1}{8}$	WQ900 WQ770	44·0	30		65	"
"	"	"	"	"	"	"		"	" "	40·2	32		78	"
"	0·13	0·17	1·32	0·146	0·024	0·23	0·05	1	WQ900 WQ770	43·0	29		53	"
B.O.H.	0·14	0·14	1·25	0·150	0·021	0·18		$1\frac{1}{8}$	WQ900 WQ770	41·8	30		63	"
"	0·14	0·14	1·32	0·140	0·027	0·16		$1\frac{1}{8}$	WQ900 WQ770	46·0	25		75	"
"	"	"	"	"	"	"		$\frac{5}{8}$	WQ900 OQ770	46·4	30		47	"
B.O.H.	0·15	0·14	1·26	0·150	0·030	0·20		$1\frac{1}{8}$	WQ900 WQ770	50·0	20		42	"
"	0·15	0·09	1·34	0·150	0·026	0·19		$1\frac{1}{8}$	WQ900 WQ770	44·0	27		58	"
"	"	"	"	"	"	"		$\frac{7}{8}$	" "	49·2	28		61	"
"	0·15	0·16	1·39	0·140	0·025	0·23		$1\frac{1}{8}$	WQ900 WQ770	47·0	27		62	"
"	0·15	0·14	1·26	0·135	0·028	0·18		$1\frac{1}{8}$	WQ900 WQ770	42·8	30		83	"
"	0·15	0·10	1·30	0·150	0·024	0·22		$1\frac{1}{8}$	WQ900 WQ770	40·8	29		67	"
"	"	"	"	"	"	"		$\frac{5}{8}$	" "	58·4	23		50	"
"	0·15	0·19	1·40	0·112	0·021	0·15		$1\frac{1}{8}$	WQ900 WQ770	46·1	$27\frac{1}{2}$		62	"
"	0·15	0·15	1·37	0·134	0·029	0·23		$\frac{3}{4}$	WQ900 WQ780	41·0	31		70	"
B.O.H.	0·16	0·08	1·43	0·142	0·026	0·18		$1\frac{1}{8}$	WQ900 WQ770	46·0	28		59	"
"	0·16	0·10	1·34	0·135	0·030	0·17		$1\frac{1}{16}$	WQ900 WQ770	39·8	32		68	"
"	"	"	"	"	"	"		$1\frac{1}{8}$	" "	46·0	26		53	"

LIST OF REFERENCES

1. Private correspondence with The Park Gate Iron and Steel Co. Ltd.

LOW CHROMIUM CASE-HARDENING STEEL

SPECIFICATION

CHEMICAL COMPOSITION (%)

C	Si	Mn	S	P	Cr
0·12—0·17	0·10—0·35	0·30—0·50	0·050 max.	0·050 max.	0·30—0·50

MECHANICAL PROPERTIES

No mechanical properties are specified for this steel, which is supplied to composition only.

RELATED SPECIFICATIONS

United Kingdom	America		Germany		
En No.	SAE	AISI	Werkstoff No.	Name	DIN
En 206	5015	5015	1.7015	15Cr3	17210

APPLICATIONS

En 206 is a water-hardening case-carburizing steel used to a limited extent for special applications, mainly in the motor-car industry. The steel is capable of producing a hard case with a core strength slightly greater than that of En 32 but not so great as the alloy case-hardening steels. Typical uses are gudgeon pins, etc., for petrol engines in the automobile industries.

WELDING

The welding characteristics will be similar to those of En 32 (Volume 2).

MACHINABILITY

The machinability will be similar to that of En 32 (Volume 2).

HOT WORKING AND HEAT TREATMENT TEMPERATURE

Forging, rolling and stamping	1200°C, finish above 900°C.
Normalizing	880°C—930°C, cool in air.
Softening	630°C—670°C, cool in air.
Carburizing	880°C—930°C.
Refining*	870°C—900°C, air cool, oil or water quench.
Hardening*	760°C—780°C, water quench.

* A single water quench from 810°C—830°C may be adopted when the steel is specifically requested for this treatment.

The carburizing characteristics of the steel will be similar to those of En 32 (Volume 2).

PHYSICAL PROPERTIES

No data are available on the physical properties of En 206. Data given for En 32 (Volume 2) and En 207 (Volume 3) will be comparable with those obtained from this steel.

The following data have been estimated for a steel of nominal composition 0·15% C, 0·25% Si, 0·40% Mn, 0·4% Cr, in the softened condition as outlined in the Introduction.

SPECIFIC GRAVITY (d)

7·86 at 20°C.

MEAN SPECIFIC HEAT (c)

0·115 cal/g. deg. C at 20°C to 100°C.

MEAN COEFFICIENT OF THERMAL EXPANSION (α)

12×10^{-6}/deg. C at 20°C to 100°C.

ELECTRICAL RESISTIVITY (ρ)

16·5 microhm-cm at 20°C.

THERMAL CONDUCTIVITY (k)

0·13 cal/cm sec. deg. C at 20°C.

YOUNG'S MODULUS (E)

13,500 tons/sq.in.

MECHANICAL PROPERTIES

No worthwhile data have been found for this steel, but quoted for En 32 (Volume 2), and En 207 in the present volume, are comparable.

LOW CHROMIUM CASE-HARDENING STEEL

SPECIFICATION

CHEMICAL COMPOSITION (%)

C	Si	Mn	S	P	Cr
0·16—0·21	0·10—0·35	0·60—0·80	0·050 max.	0·050 max.	0·60—0·80

MECHANICAL PROPERTIES

No mechanical properties are specified for this steel, which is supplied to composition only.

RELATED SPECIFICATIONS

United Kingdom	America		France
En No.	SAE	AISI	AFNOR
En 207	5120	5120	18C3

APPLICATIONS

This is a water-hardening case-carburizing steel similar to En 206 and is used to a limited extent in the motor car and motor cycle industry. In the water-hardened condition the core strength is slightly greater than that of En 206. The steel may also be satisfactorily oil-hardened to give a core strength of 40 to 50 tons/sq.in. Typical applications are similar to En 206, i.e. gudgeon pins of commercial-vehicle diesel and passenger-car petrol engines.

WELDING

The welding characteristics are similar to those of En 32.

MACHINABILITY

The machinability in the softened condition is approximately 80% of that for mild steel (En 3).

HOT WORKING AND HEAT TREATMENT TEMPERATURE

Forging, rolling and stamping	1150°C, finish above 900°C.
Normalizing	870°C—920°C, air cool.
Softening	640°C—680°C, air cool.
Carburizing	880°C—930°C.
Refining*	870°C—900°C air cool, oil or water quench.
Hardening*	760°C—780°C, water quench.
Tempering	150°C—200°C, air cool.

*A single water quench from 810°C to 830°C may be adopted when the steel is specifically requested for this treatment.

PHYSICAL PROPERTIES

SPECIFIC GRAVITY (d)

Chemical Composition (%)						Heat Treatment °C	Specific Gravity at 20°C	Ref.
C	Si	Mn	S	P	Cr			
0·18	0·28	0·72	0·023	0·033	0·73	N880	7·82	1
"	"	"	"	"	"	BC880 OQ820	7·80	"

SPECIFIC HEAT (c)

Chemical Composition (%)						Heat Treatment °C	Mean Specific Heat cal/g.deg.C						Ref.
C	Si	Mn	S	P	Cr		20 to 200°C	20 to 300°C	20 to 400°C	20 to 500°C	20 to 600°C	20 to 700°C	
0·18	0·28	0·72	0·023	0·033	0·73	N880	0·112	0·120	0·129	0·137	0·145	0·154	1
"	"	"	"	"	"	BC880 OQ820	0·113	0·121	0·128	0·136	0·144	0·153	"

MEAN COEFFICIENT OF THERMAL EXPANSION (α)

Chemical Composition (%)						Heat Treatment °C	$10^6\times$ Mean Coefficient of Thermal Expansion/deg.C							Ref.
C	Si	Mn	S	P	Cr		20 to 100°C	20 to 200°C	20 to 300°C	20 to 400°C	20 to 500°C	20 to 600°C	20 to 700°C	
0·18	0·28	0·72	0·023	0·033	0·73	N880	11·65	12·60	13·20	13·55	14·00	14·35	14·60	1
"	"	"	"	"	"	BC880 OQ820	12·05	12·90	13·85	13·75	14·00	14·40	14·80	"

En 207

ELECTRICAL RESISTIVITY (ϱ)

Chemical Composition (%)						Heat Treatment °C	Electrical Resistivity, microhm-cm								Ref.
C	Si	Mn	S	P	Cr		20°C	100°C	200°C	300°C	400°C	500°C	600°C	700°C	
0·18	0·28	0·72	0·023	0·033	0·73	N880	25·6	30·7	38·4	47·5	58·5	68·5	83·7	101·2	1
"	"	"	"	"	"	BC880 OQ820	26·0	31·2	38·5	46·0	55·5	67·0	80·5	96·5	"

THERMAL CONDUCTIVITY (k)

Chemical Composition (%)						Heat Treatment °C	Thermal Conductivity cal/cm sec. deg.C							Ref.
C	Si	Mn	S	P	Cr		100°C	200°C	300°C	400°C	500°C	600°C	700°C	
0·18	0·28	0·72	0·023	0·033	0·73	N880	0·123	0·117	0·107	0·100	0·094	0·089	0·085	1
"	"	"	"	"	"	BC880 OQ820	0·123	0·117	0·107	0·100	0·094	0·089	0·085	"

YOUNG'S MODULUS (E)

Chemical Composition (%)						Heat Treatment °C	Young's Modulus* tons/sq.in.								Ref.
C	Si	Mn	S	P	Cr		20°C	100°C	200°C	300°C	400°C	500°C	600°C	700°C	
0·18	0·28	0·72	0·023	0·033	0·73	N880	13,700	13,400	12,960	12,480	11,950	11,350	10,660	9,630	1
"	"	"	"	"	"	BC880 OQ820	13,550	13,220	12,780	12,330	11,830	11,270	10,570	9,500	"

* Vibration method.

MAGNETIC PROPERTIES

Chemical Composition (%)						Heat Treatment °C	Magnetic Induction B (Gauss) for Magnetizing field of H (Oersted)=								Ref.
C	Si	Mn	S	P	Cr		2·5	5	8	10	25	50	100	250	
0·18	0·28	0·72	0·023	0·033	0·73	N880	750	5,290	8,330	9,670	13,800	15,600	16,880	18,980	1
"	"	"	"	"	"	BC880 OQ820	360	990	2,535	3,640	9,600	13,220	15,220	17,700	"

TRANSFORMATION CHARACTERISTICS

ISOTHERMAL TRANSFORMATION

CORE CHARACTERISTICS

Chemical Composition (%)							
C	Si	Mn	S	P	Ni	Cr	Mo
0·20	0·30	0·72	0·010	0·010	0·27	0·79	0·02

Basic Electric Arc.
McQuaid Ehn Grain Size 6 (ASTM).
Actual Grain Size 6 (ASTM).
Austenitizing Temperature 850°C 30 min. (Ref. 2)

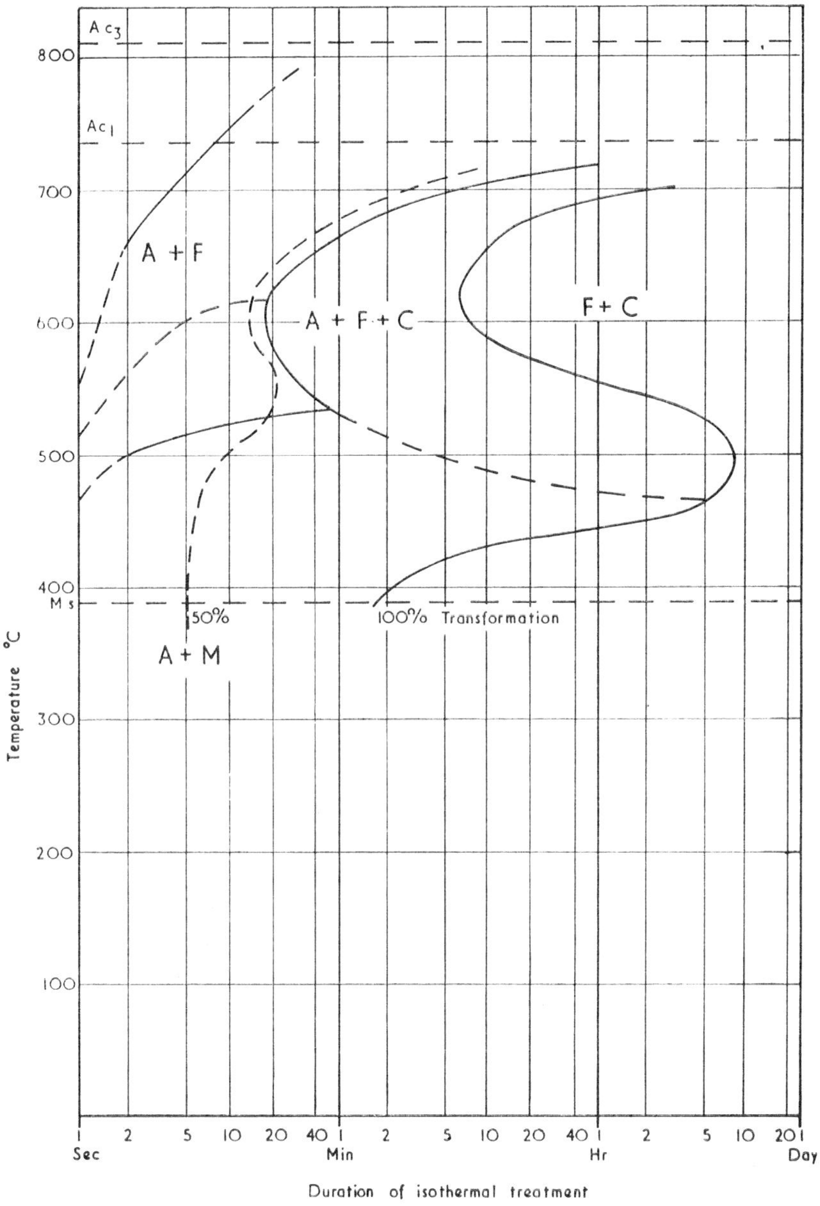

Fig. 207.1

ISOTHERMAL TRANSFORMATION (Continued)

CASE CHARACTERISTICS

Chemical Composition (%)						
C	Si	Mn	S	P	Ni	Cr
0·88	0·24	0·41	0·010	0·010	0·10	0·78

Basic Electric Arc.
Austenitizing Temperature 850°C, 30 min.

(Ref. 2)

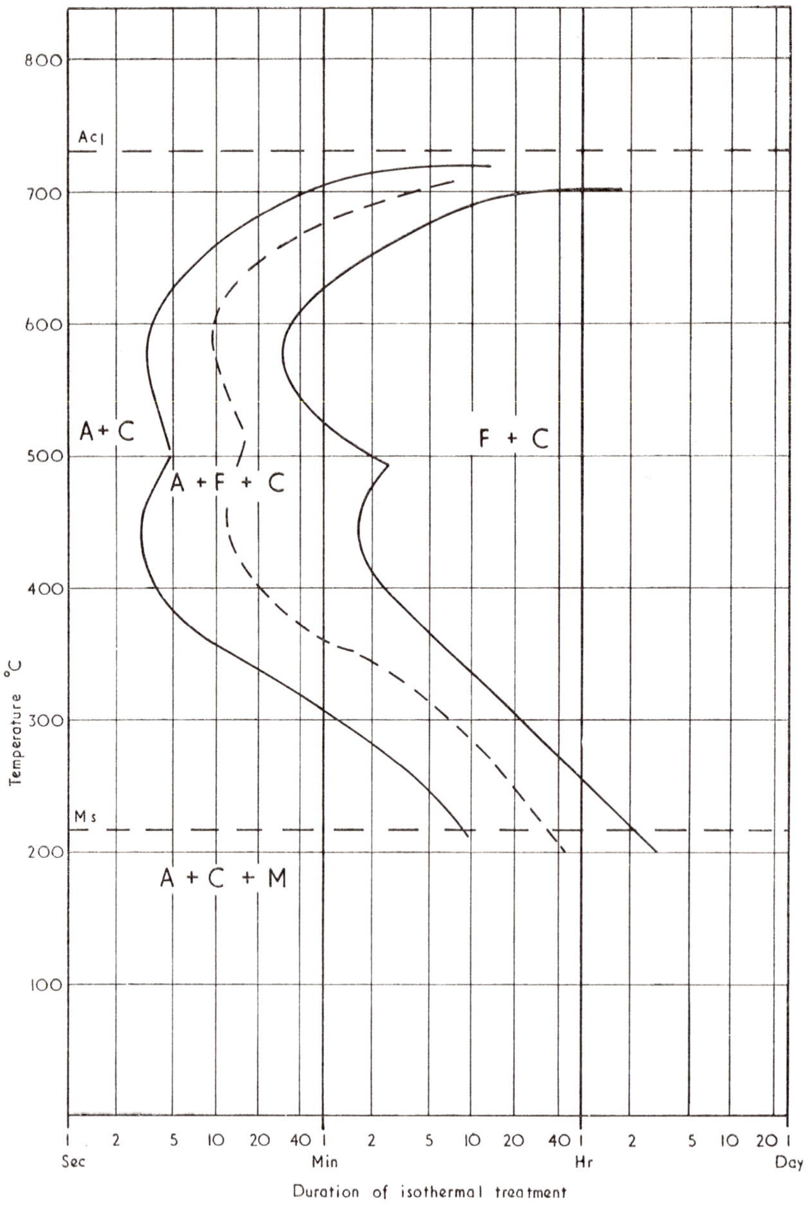

Fig. 207.2

HARDENABILITY

END QUENCH HARDENABILITY

Chemical Composition (%)							
C	Si	Mn	S	P	Ni	Cr	Mo
0·20	0·30	0·72	0·010	0·010	0·27	0·79	0·02

Basic Elec tric Arc. Actual Grain Size 6 (ASTM).
McQuaid Ehn Grain Size 6 (ASTM). End quenched Grain Size 850°C. (Ref. 2)

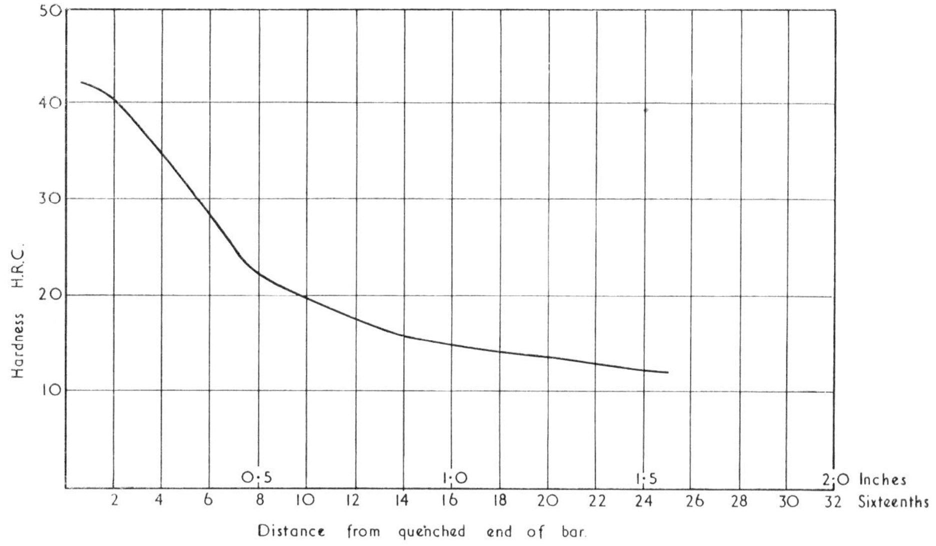

Fig. 207.3

Curve	Chemical Composition (%)				McQuaid Ehn Grain Size ASTM	Pretreatment	End Quench Temperature °C
	C	Si	Mn	Cr			
A	0·22	0·24	0·65	0·60	8	N927	927
B	0·21	0·27	0·72	0·62	8	"	"
C	0·22	0·25	0·72	0·61	7	"	"
D	0·17	0·26	0·71	0·77	7	"	"
E	0·20	0·29	0·77	0·64	6—7	"	"

(Ref. 3)

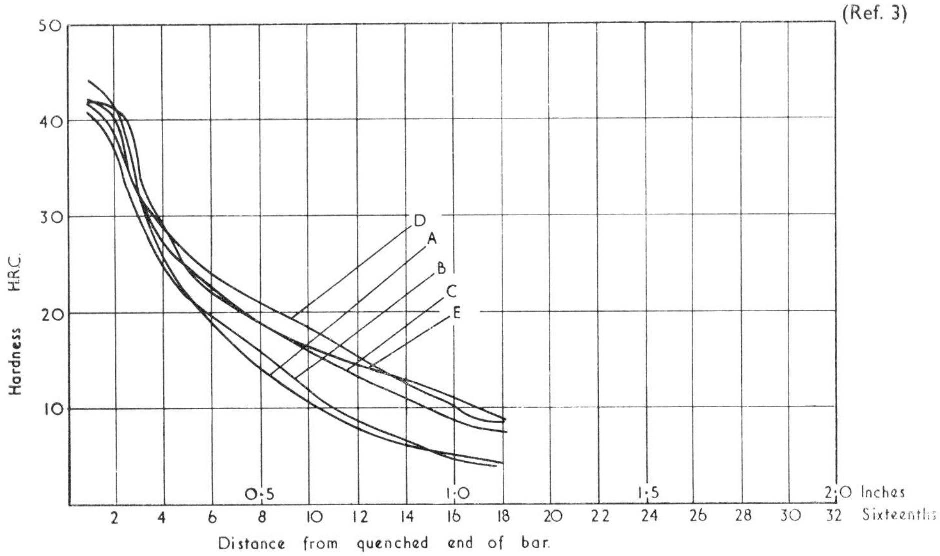

Fig. 207.4

MECHANICAL PROPERTIES AT ROOM TEMPERATURE

ASTM Grain Size	Chemical Composition (%)							Section Size dia. (in.)	Heat Treatment °C	Tensile					Izod ft.lb.	Ref.
	C	Si	Mn	S	P	Ni	Cr			Y.P. (Rm) tons/sq.in.	T.S. (Rs) tons/sq.in.	El % (A) $4\sqrt{A}$	El % (A) 5d	R.A. % (Z)		
	0·17	0·31	0·32	0·028	0·040	0·13	0·69	$1\frac{1}{8}$	BC920 OQ840		49·5	24		47	12	4
7	0·17	0·26	0·71				0·77	$1\frac{1}{8}$	Treatment A	32·2	50·4	25	$20\frac{1}{2}$	51		3
"	"	"	"				"	$\frac{9}{16}$	" A	50·7	54·0	$20\frac{1}{2}$	$15\frac{1}{2}$	55	44	"
"	"	"	"				"	$1\frac{1}{8}$	" B	38·2	53·4	18	15	31	70	"
"	"	"	"				"	$\frac{9}{16}$	" B	51·2	64·1	15	12	36	44	"
6	0·18	0·28	0·72	0·023	0·033		0·73	$1\frac{1}{8}$	BC910 OQ880 OQ780	20·0	40·8	28		49	33	5
"	"	"	"	"	"		"	$1\frac{1}{2}$	" " "	25·1	43·7	27		51	50	"
"	"	"	"	"	"		"	$1\frac{1}{8}$	BC950 OQ820	25·2	44·4	28		45	29	"
"	"	"	"	"	"		"	$1\frac{1}{2}$	" "	23·8	38·9	30		49	30	"
6	0·19	0·31	0·79	0·021	0·032		0·74	$1\frac{1}{8}$	BC910 OQ880 OQ780	28·0	50·0	24		48	36	"
"	"	"	"	"	"		"	$1\frac{1}{2}$	" " "	23·1	42·3	26		54	57	"
"	"	"	"	"	"		"	$1\frac{1}{8}$	BC950 OQ820	32·4	52·8	23		63	18	"
"	"	"	"	"	"		"	$1\frac{1}{2}$	" "	23·0	41·4	26		56	44	"
	0·20	0·26	0·70	0·042	0·040	0·12	0·65	$1\frac{1}{8}$	BC920 OQ840		49·5	24		47	12	4
6—7	0·20	0·22	0·72	0·037	0·024		0·71	$1\frac{1}{8}$	WQ880 WQ760		57·7	22			13	6
5—6	0·20	0·15	0·65	0·028	0·030		0·65	$1\frac{1}{8}$	WQ900 WQ770		55·2	24			32	"
6—7	0·20		0·72				0·74	$1\frac{1}{2}$	BC910 OQ880 OQ780	20·3	42·3	30		58	66	5
"	"		"				"	$1\frac{1}{8}$	BC950 OQ820	32·4	51·2	23		42	18	"
"	"		"				"	$1\frac{1}{2}$	" "	20·1	37·8	31		62	43	"
6—7	0·20	0·29	0·77				0·64	$1\frac{1}{8}$	Treatment A	37·5	52·7	25	$20\frac{1}{2}$	53		3
"	"	"	"				"	$\frac{9}{16}$	" A	49·8	62·5	19	$14\frac{1}{2}$	51	53	"
"	"	"	"				"	$1\frac{1}{8}$	" B	43·7	54·7	$18\frac{1}{2}$	15	43		"
"	"	"	"				"	$\frac{9}{16}$	" B	59·0	70·1	$16\frac{1}{2}$	13	51	38	"
	0·21	0·28	0·69	0·033	0·039	0·13	0·73	$1\frac{1}{8}$	BC920 OQ840		61·0	20		42	22	4
8	0·21	0·27	0·72				0·62	$1\frac{1}{8}$	Treatment A	32·7	51·2	$24\frac{1}{2}$	20			3
"	"	"	"					$\frac{9}{16}$	" A	44·5	59·4	$22\frac{1}{2}$	17			"
"	"	"	"				"	$1\frac{1}{8}$	" B	36·0	52·4	$18\frac{1}{2}$	$15\frac{1}{2}$			"
"	"	"	"				"	$\frac{9}{16}$	" B	50·4	63·8	14	11			"

MECHANICAL PROPERTIES AT ROOM TEMPERATURE (Continued)

ASTM Grain Size	Chemical Composition (%) C	Si	Mn	S	P	Ni	Cr	Section Size dia. (in.)	Heat Treatment °C	Tensile Y.P. (Rm) tons/sq.in.	T.S. (Rs) tons/sq.in.	El % (A) $4\sqrt{A}$	El % (A) 5d	R.A. % (Z)	Izod ft.lb.	Ref.
7	0·21	0·28	0·71	0·030	0·031		0·73	1⅛	BC910 OQ880 OQ700	31·1	49·7	26		44	34	5
"	"	"	"	"	"		"	1½	" " "	25·7	45·9	24		52	44	"
"	"	"	"	"	"		"	1⅛	BC950 OQ820	27·8	47·0	22		51	26	"
"	"	"	"	"	"		"	1½	" "	23·0	40·6	29		62	41	"
6	0·22	0·24	0·65				0·60	1⅛	Treatment A	31·0	48·3	25½	20½	54		3
"	"	"	"				"	9/16	" A	46·2	62·0	20½	15½	56	45	"
"	"	"	"				"	1⅛	" B	34·9	50·5	20	16½	35		"
"	"	"	"				"	9/16	" B	50·7	65·3	15	11	35	28	"
7	0·22	0·25	0·72				0·61	1⅛	" A	36·0	52·3	24½	20	52		"
"	"	"	"				"	9/16	" A	47·7	62·6	22½	17½	53	55	"
"	"	"	"				"	1⅛	" B	38·0	53·0	18½	15½	34		"
"	"	"	"				"	9/16	" B	51·5	65·1	16½	13	40	70	"

Treatment A: Test bars treated at 915°C for 2 hrs, FC to 836°C, held at 836°C and then OQ, T190°C.
Treatment B: Heated in continuous gas carburizing furnace with full load at 915°C to 927°C for 9 to 16 hours, then carbon diffused for 2 hrs at 836°C, OQ from 836°C, T190°C.

LIST OF REFERENCES

1. Private correspondence with The United Steel Companies Ltd.
2. G. Delbart, A. Constant, and A. Clerc. Courbes de Transformation des Aciers de Fabrication Française, vol II. L'Institute de Rercherches de la Sidérurgie (IRSID).
3. Private correspondence with the Ford Motor Co. Ltd., Dagenham.
4. Private correspondence with the Clyde Alloy Steel Co. Ltd.
5. Private correspondence with Forgings and Presswork Ltd.
6. Private correspondence with The Park Gate Iron and Steel Co. Ltd.

LOW NICKEL–CHROMIUM–MOLYBDENUM CASE-HARDENING STEEL

SPECIFICATION

CHEMICAL COMPOSITION (%)

C	Si	Mn	S	P	Ni	Cr	Mo
0·22 max.	0·10—0·35	0·45—0·65	0·050 max.	0·050 max.	1·50—2·00	0·40—0·60	0·20—0·30

MECHANICAL PROPERTIES

Tensile strength, tons/sq. in. min. (Rm)	55
Elongation, %, min. (A)	15
Izod impact value, ft. lb. min.	30

The heat treatment given to 1⅛ in. dia. test bars shall be blank carburize at a temperature between 880°C and 930°C.[1],[2]. Refine at a temperature of 850°C to 880°C, cool in air, oil or water, harden in oil from a temperature of 770°C to 800°C.

[1] By special arrangement with the purchaser the test may be carried out on samples from which the blank carburizing treatment has been omitted or modified to a short heating period.

[2] A single quench from 810°C to 830°C may be used in certain cases when the refining treatment after carburizing is omitted by special arrangement between the purchaser and the manufacturer.

RELATED SPECIFICATIONS

United Kingdom	America	
En No.	AISI	SAE
325	4317 4320	4317 4320

APPLICATIONS

A low-alloy case-hardening steel which may be used as an alternative for En 35A and En 36A. It can be carburized and hardened to produce a hard wear-resistant case with a tough medium-strength core. The steel is suitable for the manufacture of components of large section for motor car transmission and differential gears.

WELDINGS

(See Introduction for key to symbols used in this table.)

	Welding Process					
	MA	G	IG	S	F	B
En 325	a	ax	g	e	a	h

Reproduced by kind permission of H.M.S.O.

MACHINABILITY

In the softened condition the machinability is approximately 70% of that for mild steel (En 3).

HOT WORKING AND HEAT TREATMENT TEMPERATURES

Forging, rolling and stamping	1150°C, finish above 900°C.
Normalizing	850°C—880°C, air cool.
Softening (sub-critical annealing)	630°C—670°C, air or furnace cool.
Carburizing	880°C—930°C.
Refining*	850°C—880°C, air cool, oil or water quench.
Hardening*	770°C—800°C, oil quench.
Tempering (for stress relief)	150°C—200°C.

* A single oil quench from 800°C to 830°C may be adopted when steel is specifically requested for this treatment.

PHYSICAL PROPERTIES

No data are available on the physical properties of this steel. The following data have been estimated for a steel of nominal composition 0·19% C, 0·25% Si, 0·55% Mn, 1·75% Ni, 0·50% Cr, 0·25% Mo, in the softened condition as outlined in the Introduction.

SPECIFIC GRAVITY (d)

7·86 at 20°C.

MEAN SPECIFIC HEAT (c)

0·116 cal/g. deg. C at 20°C to 100°C.

MEAN COEFFICIENT OF THERMAL EXPANSION (α)

12×10^{-6} per deg. C at 20°C to 100°C.

ELECTRICAL RESISTIVITY (ρ)

23 microhm-cm at 20°C.

THERMAL CONDUCTIVITY (k)

0·095 cal/cm sec. deg. C at 20°C.

YOUNG'S MODULUS (E)

13,500 tons/sq. in. at 20°C.

TRANSFORMATION CHARACTERISTICS

ISOTHERMAL TRANSFORMATION

Chemical Composition (%)				
C	Mn	Ni	Cr	Mo
0·17	0·57	1·87	0·45	0·24

Grain Size 7(ASTM).
Austenitizing Temperature 927°C. (Ref. 1)

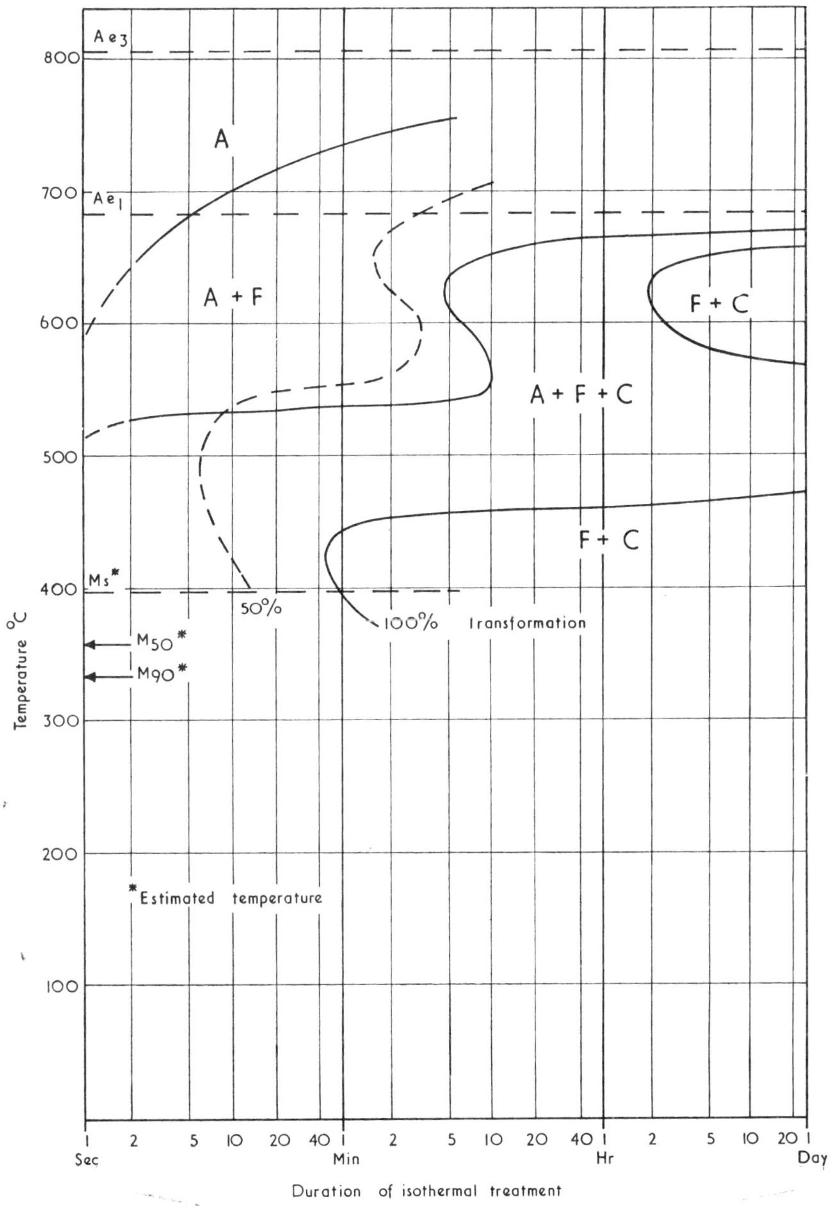

Fig. 325.1

HARDENABILITY

END QUENCH HARDENABILITY DATA

Chemical Composition (%)				
C	Mn	Ni	Cr	Mo
0·17	0·57	1·87	0·45	0·24

Grain Size 7(ASTM).
End quenched from 927°C.

(Ref. 1)

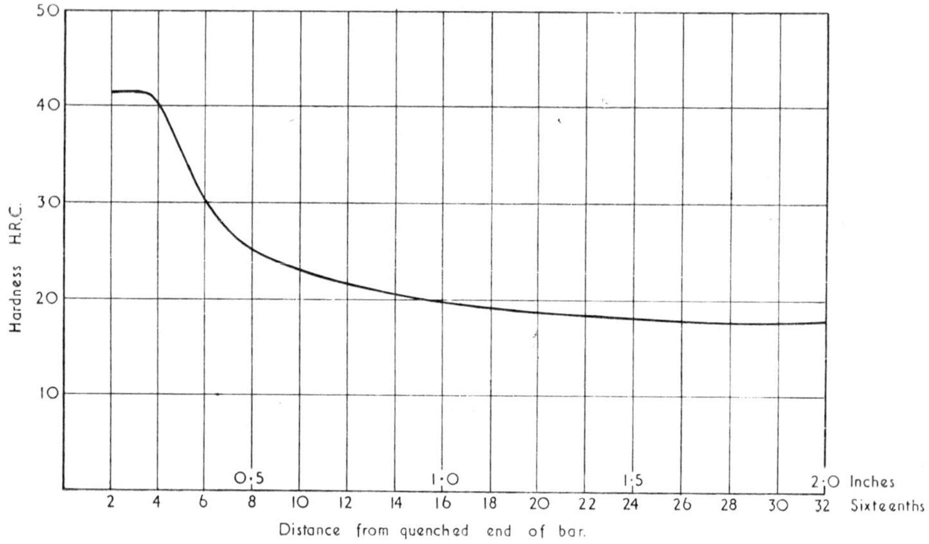

Fig. 325.2

MECHANICAL PROPERTIES AT ROOM TEMPERATURE

ASTM Grain Size	Chemical Composition (%)								Section Size dia. (in.)	Heat Treatment °C	Tensile							Izod ft.lb.	Ref.
	C	Si	Mn	S	P	Ni	Cr	Mo			0·1% P.S. (0·1Rp)	0·2% P.S. (0·2Rp)	0·5% P.S. (0·5Rp)	Y.P. (Rm)	T.S. (Rs)	El % (A)	R.A. % (Z)		
											tons/sq.in.								
	0·17		0·50			1·75	0·45	0·25	1⅛	OQ830 OQ770				36·0	58·0	21		60	2
	"		"			"	"	"	"	" OQ820				42·0	60·0	19		57	"
	0·17		0·50			1·85	0·42	0·23	1⅛	OQ870 OQ770				42·5	57·4	27		52	"
8	0·19	0·15	0·59	0·014	0·013	1·70	0·50	0·28	1⅛	OQ700			33·2		40·3	33	70	118	3
"	"	"	"	"	"	"	"	"	"	OQ720	24·0	26·5	30·0		47·7	25	55½	89	"
"	"	"	"	"	"	"	"	"	"	OQ740	26·8	31·2	38·0		54·3	26	50	58	"
"	"	"	"	"	"	"	"	"	"	OQ760	25·0	28·8	34·2		54·0	27	50½	62	"
"	"	"	"	"	"	"	"	"	"	OQ780	38·2	42·6	48·3		54·2	23	54	75	"
"	"	"	"	"	"	"	"	"	"	OQ800	30·7	34·9	40·2		53·7	23	51½	81	"
"	"	"	"	"	"	"	"	"	"	OQ820	30·3	34·5	40·2		54·3	23	56	72	"
"	"	"	"	"	"	"	"	"	"	OQ840	29·3	33·9	39·1		52·8	24	57	74	"
"	"	"	"	"	"	"	"	"	0·424	AC900	23·5	26·6	31·0		45·7	28	59½	87	"
"	"	"	"	"	"	"	"	"	¾	"	23·8	27·8	33·5		44·6	28	59	83	"
"	"	"	"	"	"	"	"	"	1	"	21·5	24·5	28·6		42·6	27	55½	75	"
"	"	"	"	"	"	"	"	"	1⅛	"	21·0	24·1	28·2		43·2	27	57	80	"
"	"	"	"	"	"	"	"	"	2⅛	"	22·1	25·4	29·4		43·3	29	61½	92	"
"	"	"	"	"	"	"	"	"	3	"	20·5	22·9	26·1		39·8	32	65	100	"
"	"	"	"	"	"	"	"	"	4(O)	"	22·0	23·6	26·8		39·0	32½	63½	102	"
"	"	"	"	"	"	"	"	"	4(C)	"	22·0	23·8	26·3		38·9	32	63½	105	"
"	"	"	"	"	"	"	"	"	0·424	OQ820	38·5	45·3	54·6		68·3	18	55½	60	"
"	"	"	"	"	"	"	"	"	¾	"	33·8	39·0	47·0		56·7	25	57½	69	"
"	"	"	"	"	"	"	"	"	1	"	30·3	35·0	40·8		53·7	25½	58½	74	"
"	"	"	"	"	"	"	"	"	1⅛	"	30·3	34·5	40·2		54·3	23	56	72	"
"	"	"	"	"	"	"	"	"	2⅛	"	29·2	33·0	36·6		50·0	27	62½	90	"
"	"	"	"	"	"	"	"	"	3	"	27·6	30·6	34·8		48·0	26½	63½	99	"
"	"	"	"	"	"	"	"	"	4(O)	"	30·0	33·0	36·0		47·0	27	64½	98	"
"	"	"	"	"	"	"	"	"	4(C)	"	25·2	28·3	32·5		46·3	26½	59	96	"
"	"	"	"	"	"	"	"	"	0·424	OQ860 OQ780	38·2	44·1	51·8		65·3	18	58½	58	3
"	"	"	"	"	"	"	"	"	¾	" "	32·3	37·6	46·0		57·5	24	55½	64	"
"	"	"	"	"	"	"	"	"	1	" "	27·4	31·4	37·3		54·6	25½	52½	60	"
"	"	"	"	"	"	"	"	"	1⅛	" "	26·1	30·3	36·1		53·9	25	54	68	"
"	"	"	"	"	"	"	"	"	2⅛	" "	24·4	27·2	31·8		49·2	28½	61½	88	"
"	"	"	"	"	"	"	"	"	3	" "	24·2	27·4	32·5		48·1	29	61½	87	"
"	"	"	"	"	"	"	"	"	4(O)	" "	24·6	27·0	31·1		47·0	29½	63	92	"
"	"	"	"	"	"	"	"	"	4(C)	" "	22·0	25·0	29·0		45·1	29	61	96	"
"	"	"	"	"	"	"	"	"	0·424	OQ860 OQ800	36·8	43·1	51·6		65·8	20	56½	59	"
"	"	"	"	"	"	"	"	"	¾	" "	37·2	43·8	51·0		57·2	23½	59	75	"
"	"	"	"	"	"	"	"	"	1	" "	32·4	36·7	42·3		55·6	22	51	74	"

MECHANICAL PROPERTIES AT ROOM TEMPERATURE (Continued)

ASTM Grain Size	Chemical Composition (%) C	Si	Mn	S	P	Ni	Cr	Mo	Section Size dia. (in.)	Heat Treatment C°	Tensile 0·1% P.S. (0·1Rp) tons/sq.in.	0·2% P.S. (0·2Rp)	0·5% P.S. (0·5Rp)	Y.P. (Rm)	T.S. (Rs)	El % (A)	R.A. % (Z)	Izod ft.lb.	Ref.
8	0·19	0·15	0·59	0·014	0·013	1·70	0·50	0·28	1⅛	OQ860 OQ800	31·0	35·2	40·9		55·2	24½	53	74	3
,,	,,	,,	,,	,,	,,	,,	,,	,,	2⅛	,, ,,	26·6	29·9	34·5		50·0	28	61½	88	,,
,,	,,	,,	,,	,,	,,	,,	,,	,,	3	,, ,,	24·2	27·5	31·8		47·8	29	61½	86	,,
,,	,,	,,	,,	,,	,,	,,	,,	,,	4(O)	,, ,,	23·4	26·4	30·5		46·2	29	61½	95	,,
,,	,,	,,	,,	,,	,,	,,	,,	,,	4(C)	,, ,,	26·6	29·3	32·8		48·0	30	63½	88	,,
	0·20		0·56			1·82	0·44	0·22	1⅛	OQ870 OQ770				46·0	61·1	22		40	2
6—8	0·20	0·25	0·59	0·021	0·018	1·77	0·47	0·23	0·565	BC927 OQ927 T149				71·2	96·9	13	50	32	4
,,	,,	,,	,,	,,	,,	,,	,,	,,	,,	,, ,, T232				70·9	96·2	12½	49½	26	,,
,,	,,	,,	,,	,,	,,	,,	,,	,,	,,	BC927 OQ816 T149				79·5	97·4	13½	48	27	,,
,,	,,	,,	,,	,,	,,	,,	,,	,,	,,	,, ,, T232				77·2	94·4	12½	51	28	,,
,,	,,	,,	,,	,,	,,	,,	,,	,,	,,	BC927 OQ816 CQ774 T149				43·3	67·7	19½	49½	49	,,
,,	,,	,,	,,	,,	,,	,,	,,	,,	,,	,, ,, T232				42·2	65·1	22	56½	48	,,
,,	,,	,,	,,	,,	,,	,,	,,	,,	1	A849 FC				27·4	37·5	29	58½	81	4
,,	,,	,,	,,	,,	,,	,,	,,	,,	½	N893				33·2	54·2	24	54½	54	,,
,,	,,	,,	,,	,,	,,	,,	,,	,,	1	,,				30·0	51·3	21	51	54	,,
,,	,,	,,	,,	,,	,,	,,	,,	,,	2	,,				26·3	45·7	23½	59	67	,,
,,	,,	,,	,,	,,	,,	,,	,,	,,	4	,,				25·4	45·7	22½	55	40	,,
,,	,,	,,	,,	,,	,,	,,	,,	,,	½	BC927 OQ816 T149				72·8	94·6	12	45½	28	,,
,,	,,	,,	,,	,,	,,	,,	,,	,,	1	,, ,, ,,				47·8	68·0	17	51	44	,,
,,	,,	,,	,,	,,	,,	,,	,,	,,	2	,, ,, ,,				38·4	59·2	22½	56½	58	,,
,,	,,	,,	,,	,,	,,	,,	,,	,,	4	,, ,, ,,				33·5	53·5	24	57	53	,,
,,	,,	,,	,,	,,	,,	,,	,,	,,	½	BC927 OQ816 T149				66·7	83·4	14	53	28	,,
,,	,,	,,	,,	,,	,,	,,	,,	,,	1	,, ,, ,,				46·8	66·5	18	55	44	,,
,,	,,	,,	,,	,,	,,	,,	,,	,,	2	,, ,, ,,				37·9	58·0	21	64	49	,,
,,	,,	,,	,,	,,	,,	,,	,,	,,	4	,, ,, ,,				33·5	52·7	22½	52	60	,,

LIST OF REFERENCES

1. Atlas of Isothermal Transformation Diagrams. United States Steel Corporation, 1957.
2. Private correspondence with International Nickel Ltd.
3. Private correspondence with The English Steel Corporation Ltd.
4. *Modern Steels and Their Properties, Carbon and Alloy Steel Bars*. The Bethlehem Steel Co. Ltd. 6th Edition, 1961.

$\frac{3}{4}$ PER CENT NICKEL–CHROMIUM CASE-HARDENING STEEL

SPECIFICATION

CHEMICAL COMPOSITION

C	Si	Mn	S	P	Ni	Cr	Mo
0·20 max.	0·35 max.	0·60—1·00	0·050 max.	0·050 max.	0·60—1·00	0·40—0·80	0·10 max.

MECHANICAL PROPERTIES

Tensile strength, tons/sq.in.min. (Rm)	45
Elongation, % min.	18
Izod impact value, ft lb. min.	30

The heat treatment given to $1\frac{1}{8}$ in. dia. test bars shall be blank carburize at a temperature between 880°C and 930°C.[1],[2]. Refine at a temperature of 850°C to 880°C, cool in air, oil or water, harden in oil from a temperature of 780°C to 820°C.

(1) By special arrangement with the purchaser the test may be carried out on samples from which the blank carburizing treatment has been omitted or modified to a short heating period

(2) A single quench from 810°C to 830°C may be used in certain cases when the refining treatment after carburizing is omitted by special arrangement between the purchaser and the manufacturer.

RELATED SPECIFICATIONS

United Kingdom	
En No.	DTD
351	5002

APPLICATIONS

A low-alloy case-hardening steel which may be used as an alternative to En 33 and En 34. It can be carburized and hardened to produce a hard wear-resistant case developing a core strength of the order of 45 tons/sq. in. with good resistance to shock. Typical applications are similar to those given for En 33 and En 34, e.g. gears, gudgeon pins, shafts, levers, camshafts, magnets, drive wheels, clutch plates for the motor car and aircraft industries as well as components for small arms.

WELDING

(See Introduction for key to symbols used in this table.)

	Welding Process					
	MA	G	IG	S	F	B
En 351	c	ax	g	e	a	h

Reproduced by kind permission of H.M.S.O.

MACHINABILITY

In the softened condition the machinability is approximately 80% of that for mild steel (En 3).

HOT WORKING AND HEAT TREATMENT TEMPERATURES

Forging, rolling and stamping	1150°C, finish above 900°C.
Normalizing	850°C—880°C, air cool.
Softening (subcritical annealing)	630°C—670°C, air or furnace cool.
Carburizing	880°C—930°C.
Refining*	850°C—880°C, cool in air, oil or water.
Hardening*	780°C—820°C, oil quench.
Tempering (for stress relief)	150°C—200°C.
(for maximum core toughness)	200°C—250°C.

* A single oil quench from 810°C to 830°C may be adopted when steel is specifically requested for this treatment.

CARBURIZING DATA

Carburizing Temperature °C	Depth of case produced in stated time (solid pack carburizing)			Ref.
	4 hrs.	8 hrs.	16 hrs.	
	0·03 in.	0·05 in.		1
	0·03 in.	0·045 in.	0·06 in.	2
880	0·028 in.	0·038 in.		3
900	0·034 in.	0·048 in.		"
925	0·041 in.	0·059 in.		"

PHYSICAL PROPERTIES

SPECIFIC GRAVITY (d)

Chemical Composition (%)						Heat Treatment °C	Specific Gravity at 20°C	Ref.
C	Si	Mn	Ni	Cr	Mo			
0·15	0·25	0·70	0·80	0·60		Softened	7·85	2
0·17	0·23	0·82	0·86	0·71	0·06	BC910 FC OQ820	7·85	5
(Within specification)						Softened	7·86, 7·87	4

SPECIFIC HEAT (c)

Chemical Composition (%)						Heat Treatment °C	Specific Heat cal/g. deg. C	Ref.
C	Si	Mn	Ni	Cr	Mo		20 to 100°C	
0·15	0·25	0·70	0·80	0·60		Softened	0·118	2

MEAN COEFFICIENT OF THERMAL EXPANSION (α)

Chemical Composition (%)						Heat Treatment °C	$10^6 \times$ Mean Coefficient of Thermal Expansion per deg.C					Ref.
C	Si	Mn	Ni	Cr	Mo		20 to 100°C	20 to 200°C	20 to 300°C	20 to 500°C	20 to 700°C	
0·15	0·25	0·70	0·80	0·60		Softened	12		13	14	14·5	2
0·17	0·23	0·82	0·86	0·71	0·06	BC910 FC OQ820	12·8	13·1				5
(Within Specification)						Softened	12·0		13·6	14·4	(680) 14·8	4
"	"					"	12·2		13·3	14·0	(700) 14·6	"

ELECTRICAL RESISTIVITY (ϱ)

Chemical Composition (%)						Heat Treatment °C	Electrical Resistivity microhm-cm			Ref.
C	Si	Mn	Ni	Cr	Mo		20°C	100°C	200°C	
0·15	0·25	0·70	0·80	0·60		Softened	25			2
"	"	"	"	"		Hardened	27			"
0·17	0·23	0·82	0·86	0·71	0·06	BC910 FC OQ820	29·1	34·2	41·1	5

THERMAL CONDUCTIVITY (k)

No data are available. The estimated value at room temperature based on an electrical resistivity of 25 microhm-cm is approximately 0·085 cals/cm sec. deg. C.

YOUNG'S MODULUS (E)

Chemical Composition (%)						Heat Treatment °C	Young's Modulus tons/sq. in.			Ref.
C	Si	Mn	Ni	Cr	Mo		20°C	100°C	200°C	
0·17	0·23	0·82	0·86	0·71	0·06	BC910 FC OQ820	13,500	13,200	12,800	5

TRANSFORMATION CHARACTERISTICS

ISOTHERMAL TRANSFORMATION

CORE CHARACTERISTICS

Chemical Composition (%)							
C	Si	Mn	S	P	Ni	Cr	Mo
0·17	0·22	0·88	0·016	0·019	0·86	0·59	0·05

Basic Electric Arc.
McQuaid Ehn Grain Size 6 (ASTM).
As quenched Grain Size 6 (ASTM).
Bars hot rolled to $\frac{3}{8}$ in. dia., followed by 900°C (2hrs.) furnace cool.
Austenitizing Temperature 870°C, 30 min. (Ref. 6)

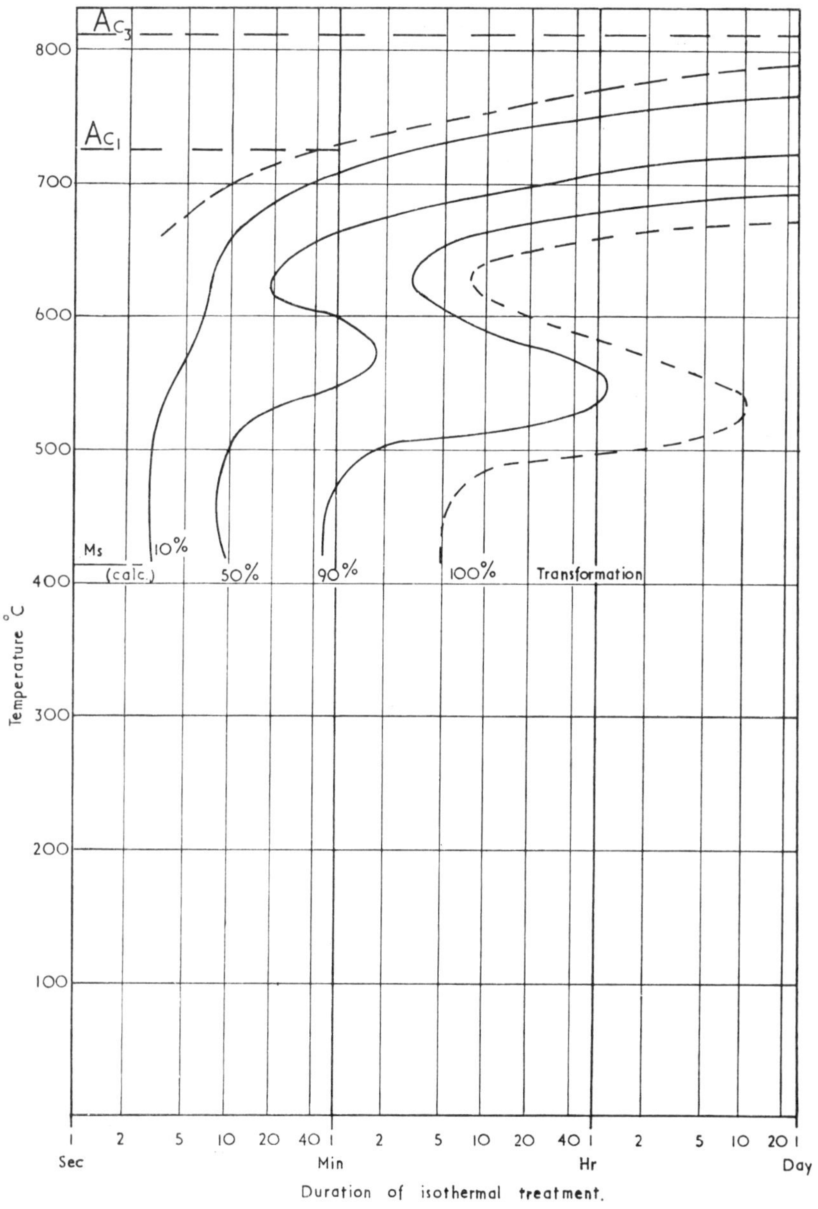

Fig. 351.1

En 351

ISOTHERMAL TRANSFORMATION (Continued)

CORE CHARACTERISTICS

Chemical Composition (%)							
C	Si	Mn	S	P	Ni	Cr	Mo
0·17	0·22	0·88	0·016	0·019	0·86	0·59	0·05

Basic Electric Arc.
McQuaid Ehn Grain Size 6 (ASTM).
As quenched Grain Size 6 (ASTM).
Bars hot rolled to ⅜ in. dia., followed by 900°C air cool.
Austenitizing Temperature 830°C, 30 min.

(Ref. 6)

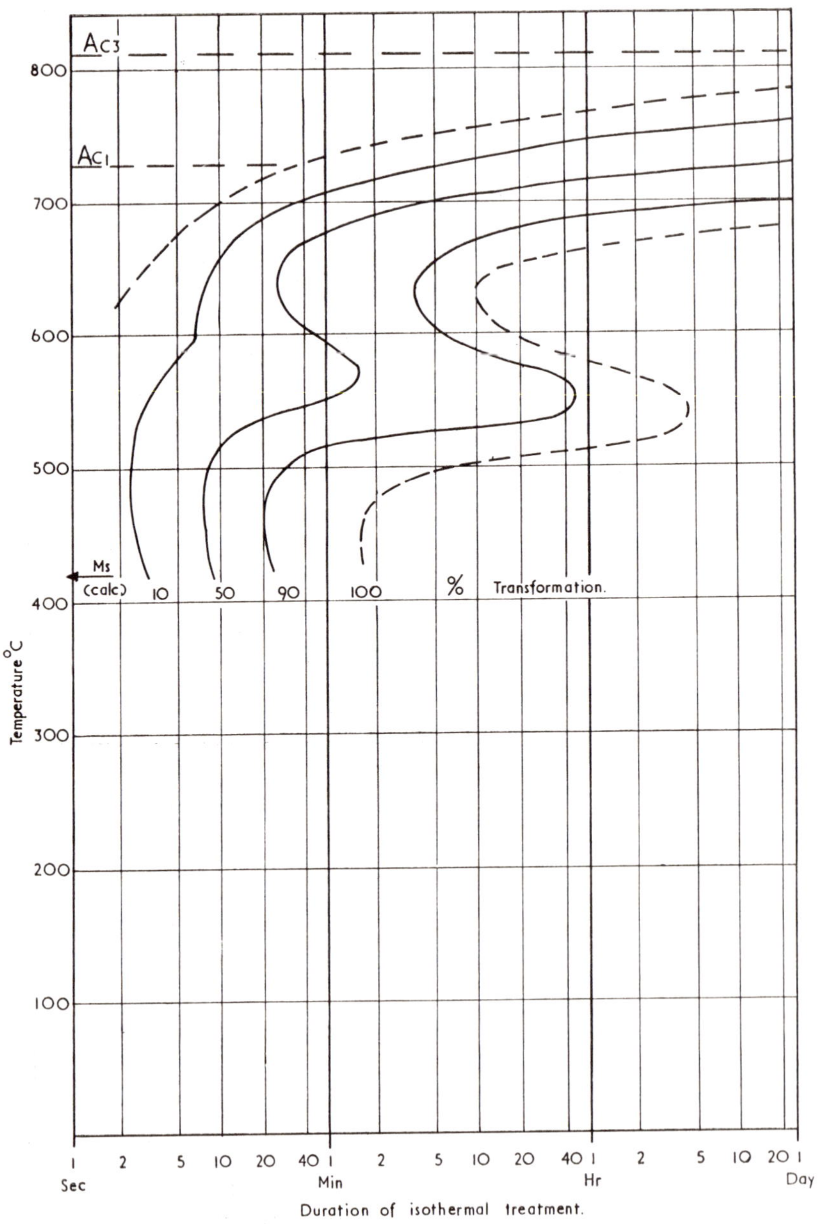

Fig. 351.2

ISOTHERMAL TRANSFORMATION (Continued)

CASE CHARACTERISTICS

Chemical Composition (%)							
C	Si	Mn	S	P	Ni	Cr	Mo
0·92	0·30	0·93	0·019	0·028	0·90	0·57	0·03

High Frequency Melt.
McQuaid Ehn Grain Size 7—8 (ASTM).
As quenched Grain Size finer than 8 (ASTM).
Bars hot rolled to $\frac{3}{8}$ in. dia., followed by 900°C (2hrs.) furnace cool.
Austenitizing Temperature 870°C, 30 min.

(Ref. 6)

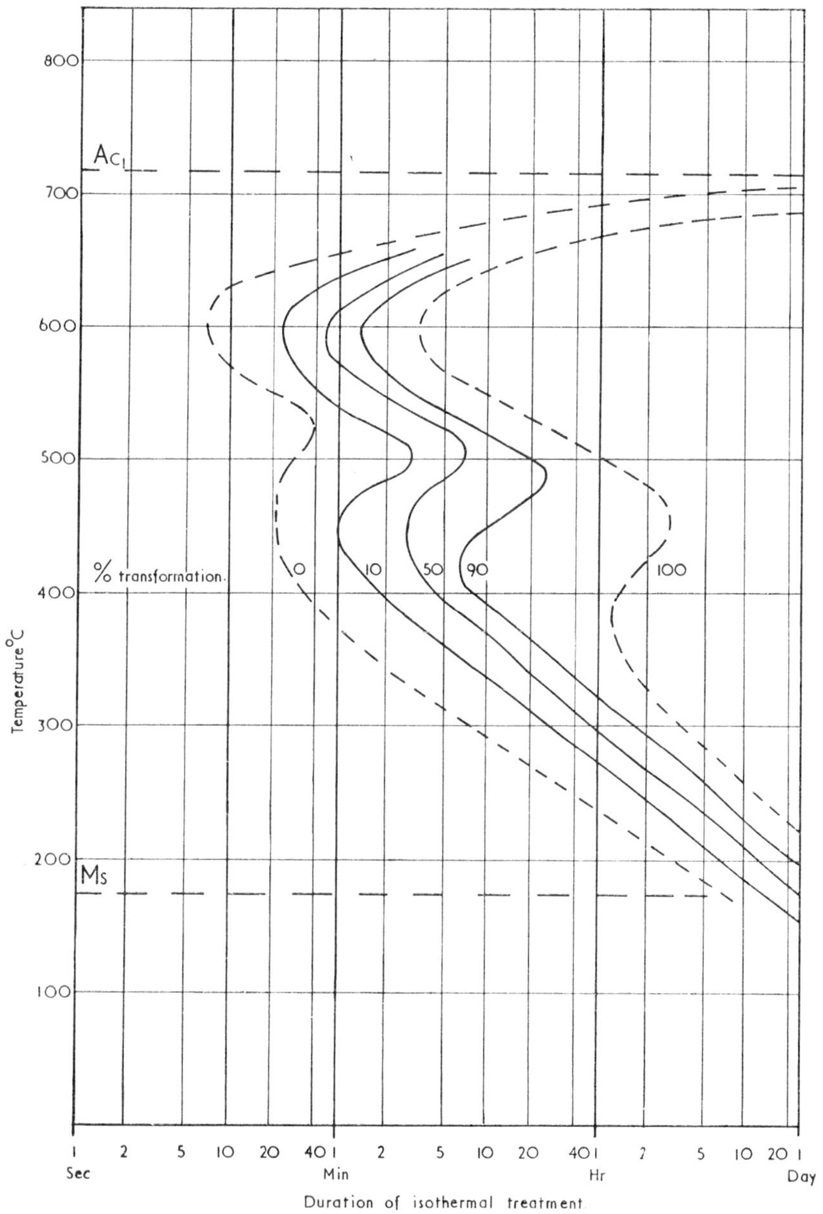

Fig. 351.3

ISOTHERMAL TRANSFORMATION (Continued)

CASE CHARACTERISTICS

Chemical Composition (%)							
C	Si	Mn	S	P	Ni	Cr	Mo
0·92	0·30	0·93	0·019	0·028	0·90	0·57	0·03

High Frequency Melt.
McQuaid Ehn Grain Size 7—8(ASTM).
As quenched Grain Size finer than 8(ASTM).
Bars hot rolled to ⅜ in. dia., followed by 900°C (2hrs.) furnace cool then 870°C (½ hr.) oil quench.
Austenitizing Temperature 830°C, 30 min.

(Ref. 6)

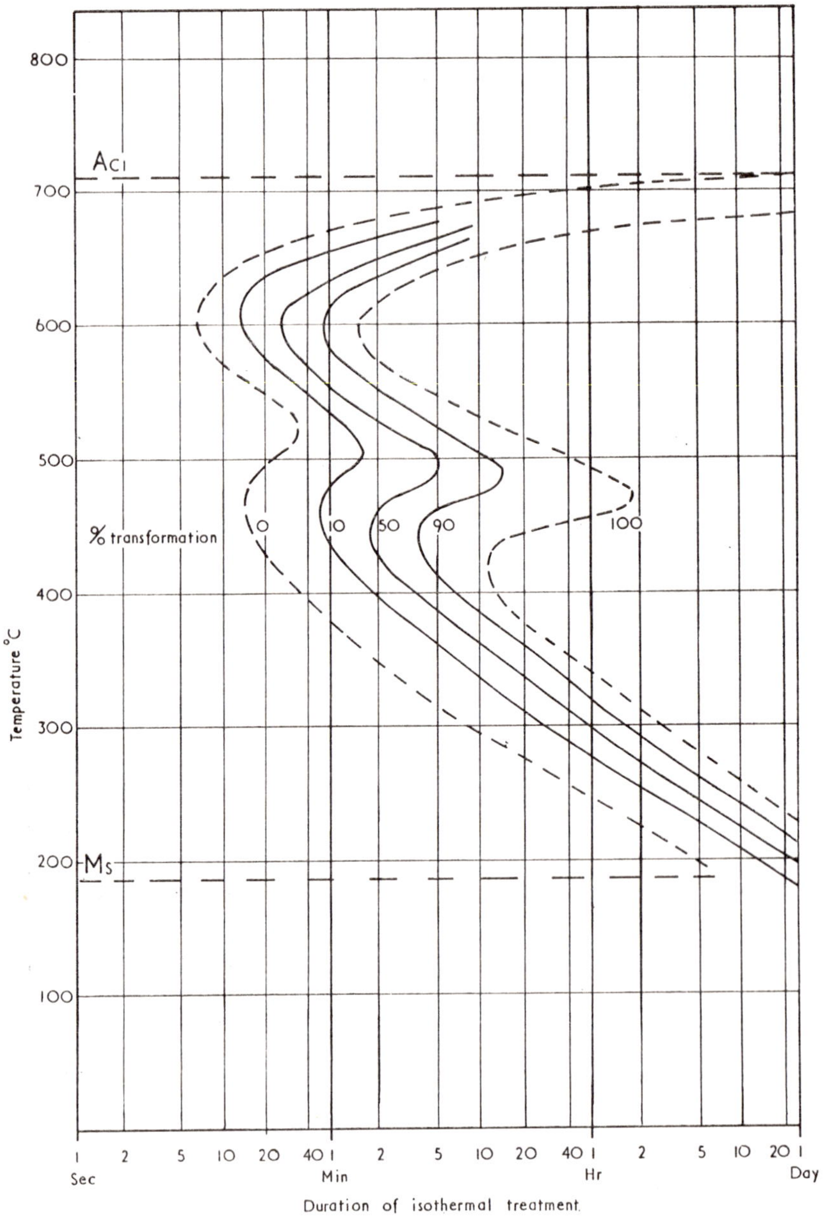

Fig. 351.4

CONTINUOUS COOLING TRANSFORMATION

Air cooled from 820°C.

(Ref. 5)

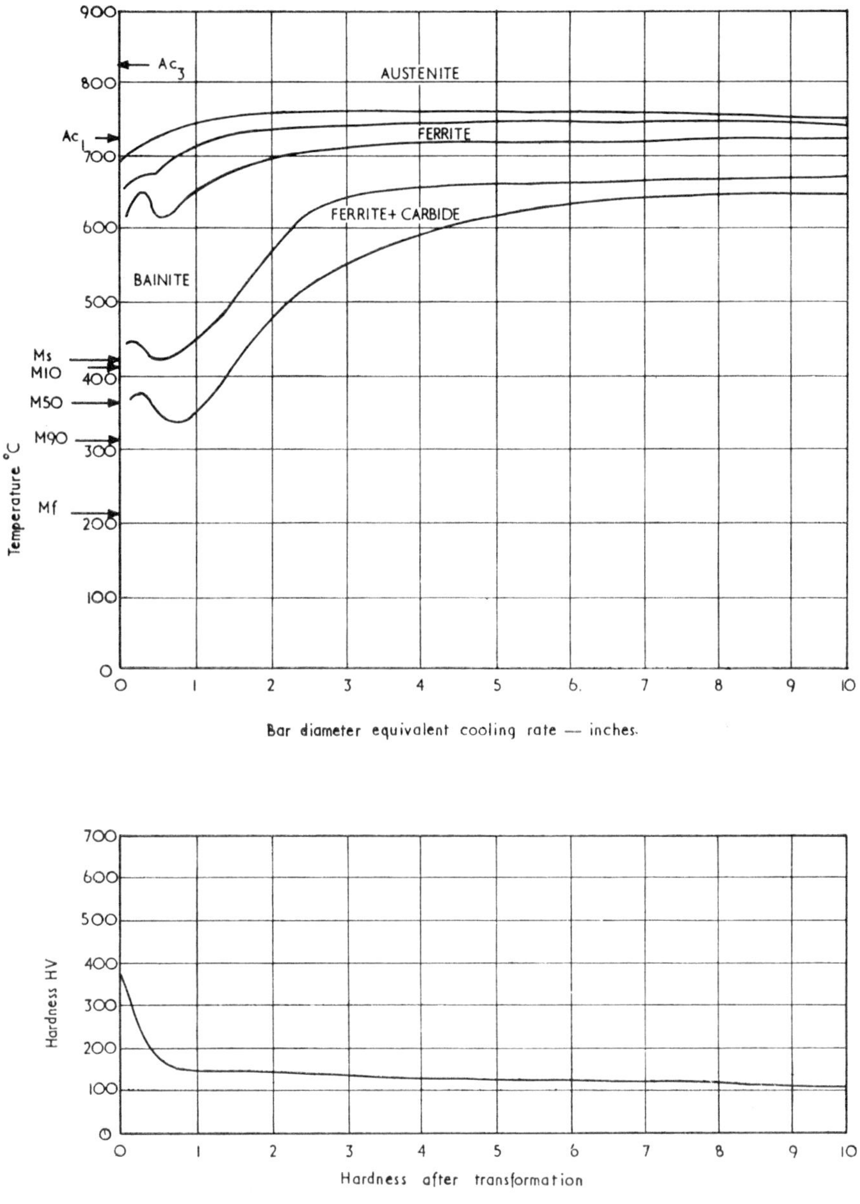

Fig. 351.5

CONTINUOUS COOLING TRANSFORMATION (Continued)

Oil quenched 820°C.

(Ref. 5)

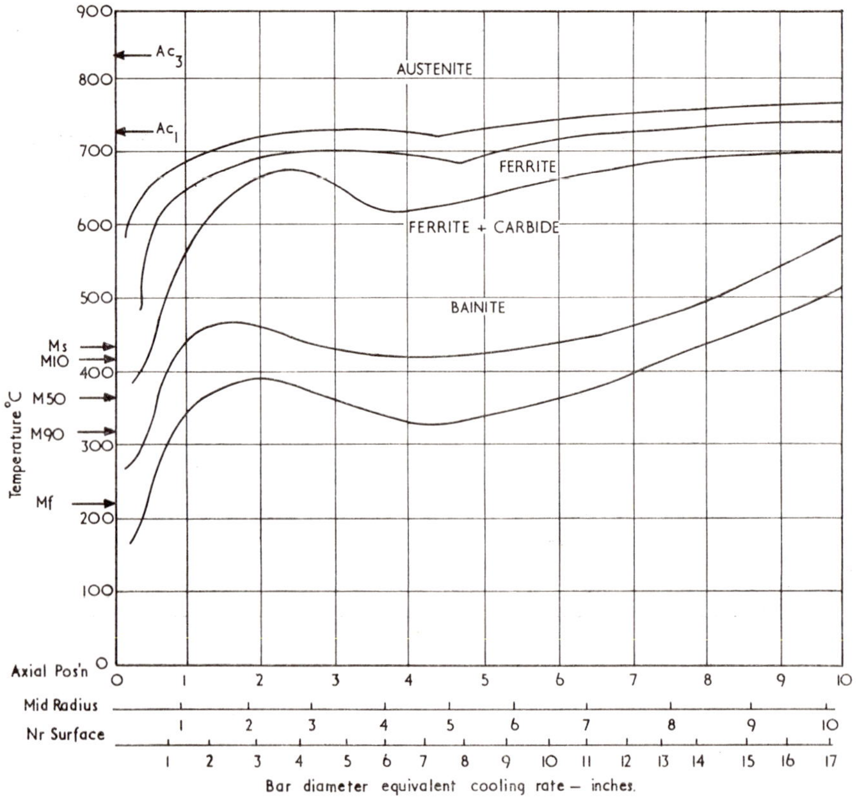

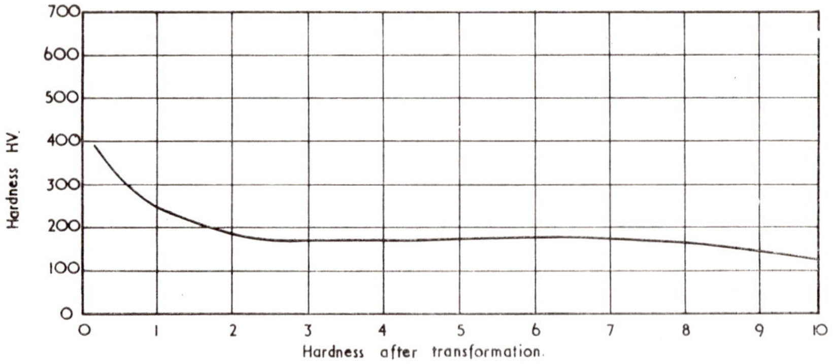

Fig. 351.6

CONTINUOUS COOLING TRANSFORMATION (Continued)

Water quenched 820°C.

(Ref. 5)

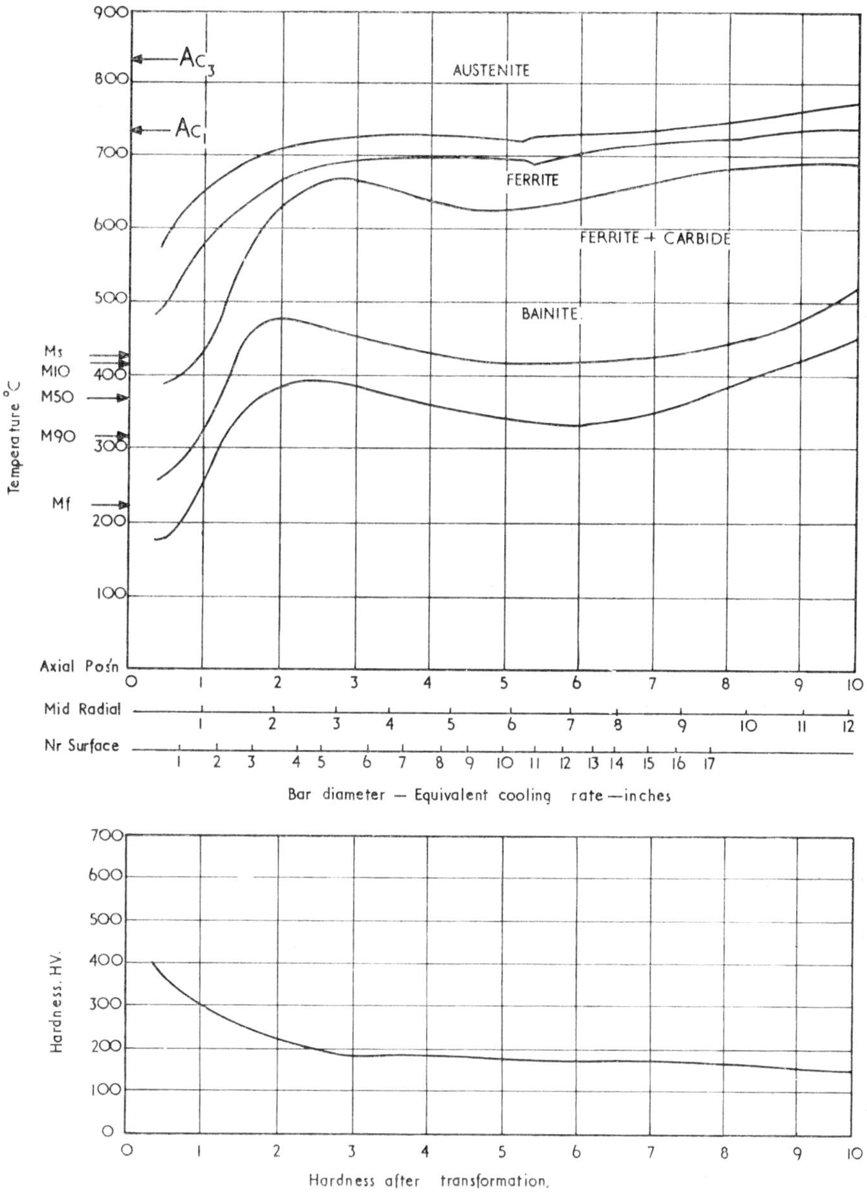

Fig. 351.7

HARDENABILITY

END QUENCH HARDENABILITY DATA

Chemical Composition (%)							
C	Si	Mn	S	P	Ni	Cr	Mo
0·17	0·22	0·88	0·016	0·019	0·86	0·59	0·05

Basic Electric Arc.
McQuaid Ehn Grain Size 6(ASTM).
As quenched Grain Size 6(ASTM).
Curve A: End quenched from 870°C.
Curve B: End quenched from 830°C.

(Ref. 6)

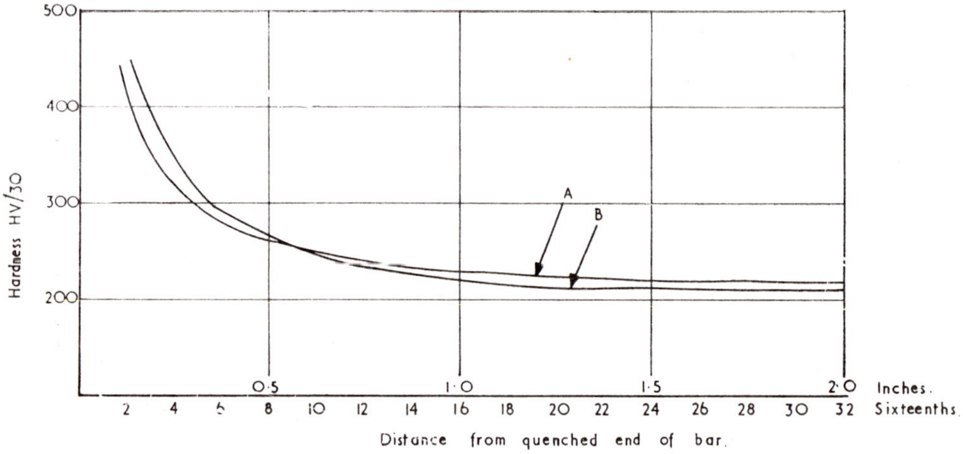

Fig. 351.8

END QUENCH HARDENABILITY BAND

These curves have been constructed from the results obtained on several casts of En 351.

(Refs. 4 and 5)

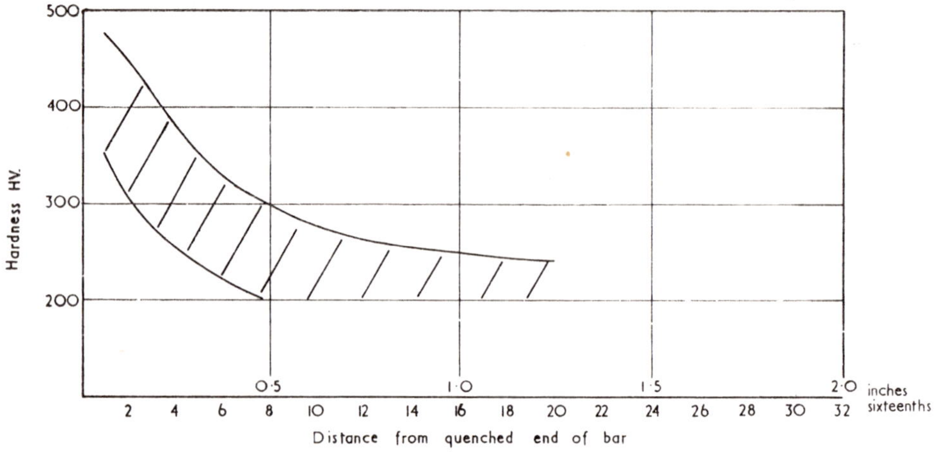

Fig. 351.9

MECHANICAL PROPERTIES AT ROOM TEMPERATURE

Process of Manufacture	ASTM Grain Size	Chemical Composition (%)								Section Size dia. (in.)	Heat Treatment °C	Tensile							Izod ft.lb.	Ref.
		C	Si	Mn	S	P	Ni	Cr	Mo			0·1% P.S. (0·1Rp) tons/sq.in.	0·2% P.S. (0·2Rp) tons/sq.in.	0·5% P.S. (0·5Rp) tons/sq.in.	Y.P. (Rm) tons/sq.in.	T.S. (Rs) tons/sq.in.	El % (A)	R.A. % (Z)		
	7	0·13	0·24	0·91	0·025	0·025	0·75	0·68	0·09	1¼	BC900 OQ860 OQ800					48·0	28	54½	65	2
		0·13	0·17	0·95			0·84	0·69	0·10	1⅛	BC900 OQ860 OQ800					47·8	24½	46	53	4
		0·13		0·65			0·95	0·44	0·06	⅞	WQ820				46·8	58·8	18	50	34	5
		"		"			"	"	"	"	OQ820				33·4	47·4	26	57	60	"
		"		"			"	"	"	"	OQ860 OQ800				33·1	48·8	26	52	47	"
		"		"			"	"	"	1⅛	BC920 FC OQ800				35·4	45·0	26	55	28	"
		"		"			"	"	"	"	" OQ820				39·8	46·7	24	57	58	"
		"		"			"	"	"	"	" OQ840				39·0	46·5	25½	59	56	"
		"		"			"	"	"	"	OQ860 OQ800				31·9	43·3	28	57	51	"
		"		"			"	"	"	"	" OQ840				40·2	47·0	25	57	60	"
A.O.H.		0·14	0·27	0·72	0·038	0·029	0·94	0·55	0·05	1⅛	OQ800					52·4	23½	43	48	2
"		"	"	"	"	"	"	"	"	"	OQ820					51·4	24½	49	44	"
	7	0·14	0·18	0·92	0·027	0·022	0·74	0·70	0·06	1¼	BC900 OQ860 OQ800					48·0	28	52	69	"
	5—8	0·14	0·19	0·84	0·021	0·014	0·75	0·67	0·06	1¼	BC900 OQ860 OQ800					45·5	30	60½	78	"
B.E.		0·14	0·23	0·94	0·020	0·011	0·81	0·69	0·06	¼	BC AC870 OQ810 T140	26·3				57·2	28			7
"		"	"	"	"	"	"	"	"	½	BC AC870 OQ810 T140	19·6				51·4	24		48	"
"		"	"	"	"	"	"	"	"	¾	BC OQ870 OQ810 T140	19·3				50·8	25		47	"
"		"	"	"	"	"	"	"	"	1⅛	BC OQ 870 OQ810 T140	19·3				47·1	30		81	"
"		"	"	"	"	"	"	"	"	1¼	BC OQ870 OQ810 T140	27·6			41·4	53·8	24½	53½	66	"
		0·14	0·19	0·82			0·77	0·72	0·05	1⅛	BC900 OQ860 OQ800					47·8	24½	46	53	4
		0·14	0·28	0·91			0·89	0·71	0·10	1⅛	BC900 OQ860 OQ800					49·8	22½	47	66	"
		0·14	0·17	0·90			0·73	0·65	0·10	1⅛	BC900 OQ860 OQ800					51·0	24	50	57	"
		0·14	0·21	0·94			0·84	0·73	0·10	1⅛	BC900 OQ860 OQ800					54·2	22	42	54	"
		0·14	0·21	0·87			0·85	0·69	0·10	1⅛	BC900 OQ880 OQ780					55·0	21	42	44	"
A.O.H.		0·14	0·33	0·88	0·036	0·029	0·69	0·52	0·10	1⅛	BC900 OQ880 OQ780					45·0	31	64	51	1
"		"	"	"	"	"	"	"	"	"	" WQ880 OQ820					47·4	25	50	32	"
"		"	"	"	"	"	"	"	"	"	" OQ780					47·2	25	50	22	"
B.E.		0·14	0·29	0·82	0·014	0·018	0·73	0·57	0·10	1⅛	BC900 WQ880 OQ780					46·8	32	61½	82	"
"		"	"		"	"	"	"	"	"	" " OQ820					48·0	28	57	67	"
		0·14		0·82			0·83	0·77	0·07	1⅛	OQ850 OQ800					49·2	28		46	8
A.O.H.		0·15	0·33	0·81	0·037	0·031	0·86	0·66	0·06	1⅛	OQ810				40·0	53·2	23	46½	25	2
"		"	"	"	"	"	"	"	"	"	OQ820					56·4	22½	49	36	"
B.E.	7	0·15	0·34	0·95	0·017	0·019	0·75	0·71	0·07	1¼	AC900 OQ860 OQ800					53·0	29	54½	56	"
		0·15	0·21	0·80			0·77	0·68	0·10	1⅛	BC900 OQ860 OQ800					50·2	25½	50	77	4

MECHANICAL PROPERTIES AT ROOM TEMPERATURE (Continued)

Process of Manufacture	ASTM Grain Size	Chemical Composition (%)								Section Size dia. (in.)	Heat Treatment °C	Tensile							Izod ft.lb.	Ref.
		C	Si	Mn	S	P	Ni	Cr	Mo			0·1% P.S. (0·1Rp)	0·2% P.S. (0·2Rp)	0·5% P.S. 0·5Rp)	Y.P. (Rm)	T.S. (Rs)	El % (A)	R.A. % (Z)		
												tons/sq.in.								
		0·15		0·78			0·84	0·40	0·05	1⅛	OQ880 OQ820					47·0	25		59	8
N.O.H.	6—8	0·15	0·22	0·76	0·030	0·030	0·78	0·50	0·11	½	BC920 OC880 OQ800				41·2	51·0	28	52	53	7
"	"	"	"	"	"	"	"	"	"	⅞	" " "				37·6	49·0	27½	48½	44	"
"	"	"	"	"	"	"	"	"	"	1⅛	" " "				32·0	47·5	26	48½	56	"
"	"	"	"	"	"	"	"	"	"	"	OQ880 OQ800				34·2	46·6	27	47	57	"
"	"	"	"	"	"	"	"	"	"	1¼	BC900 OQ870 OQ810 T140	23·3			33·6	47·5	26½	52½	54	"
"	"	"	"	"	"	"	"	"	"	1½	BC920 OQ880 OQ800				34·6	44·8	28½	54½	59	"
"	"	"	"	"	"	"	"	"	"	2½	" " "				29·6	42·5	28	52	49	"
B.O.H.	5—6	0·15	0·24	0·79	0·041	0·032	0·81	0·49	0·08	1¼	BC900 OQ870 OQ810 T140	24·3			38·1	52·4	23	48½	36	"
"	"	"	"	"	"	"	"	"	"	1⅛	OQ960 OQ800				40·6	53·1	25	52½	43	"
A.O.H.	7	0·15	0·24	0·83	0·032	0·024	0·76	0·67	0·06	1⅛	OQ740	21·2	24·6	31·7		51·8	22	38½	16	4
"	"	"	"	"	"	"	"	"	"	"	OQ760	22·2	25·6	30·8		51·9	24	44	26	"
"	"	"	"	"	"	"	"	"	"	"	OQ780	21·0	24·3	29·3		49·4	25	48	27	"
"	"	"	"	"	"	"	"	"	"	"	OQ860	21·6	24·4	29·3		50·0	26	48	32	"
"	"	"	"	"	"	"	"	"	"	"	OQ820	23·6	27·2	32·5		51·2	26	50½	45	"
"	"	"	"	"	"	"	"	"	"	"	OQ840	22·9	26·6	31·8		51·2	25	40	45	"
"	"	"	"	"	"	"	"	"	"	0·424	AC900	15·8	18·0	21·7		42·0	27	55	40	"
"	"	"	"	"	"	"	"	"	"	¾	"	16·6	18·7	22·6		42·0	32	62	60	"
"	"	"	"	"	"	"	"	"	"	1	"	20·4	20·9	22·1		38·2	32	63½	75	"
"	"	"	"	"	"	"	"	"	"	1⅛	"	21·3	21·5	21·5		38·0	35	69½	83	"
"	"	"	"	"	"	"	"	"	"	2⅛	"			24·5		35·5	38	73½	89	"
"	"	"	"	"	"	"	"	"	"	3	"			24·1		35·1	36	71½	88	"
"	"	"	"	"	"	"	"	"	"	4¼ (O)				23·4		34·2	39	73	87	"
"	"	"	"	"	"	"	"	"	"	4¼ (C)				24·5		35·3	39	71½	90	"
"	"	"	"	"	"	"	"	"	"	0·424	AC900 OQ820	27·5	32·9	40·1		59·0	21	47	35	"
"	"	"	"	"	"	"	"	"	"	¾	" "	22·5	26·1	32·5		52·8	24½	46	44	"
"	"	"	"	"	"	"	"	"	"	1	" "	23·2	26·4	31·6		51·4	25½	51	39	"
"	"	"	"	"	"	"	"	"	"	1⅛	" "	22·2	25·2	30·1		50·9	29	50½	53	"
"	"	"	"	"	"	"	"	"	"	2⅛	" "	20·6	23·6	28·0		45·4	29	59	54	"
"	"	"	"	"	"	"	"	"	"	3	" "	20·6	23·6	26·2		43·6	30	59	48	4
"	"	"	"	"	"	"	"	"	"	4¼ (O)	" "	18·2	20·4	24·0		42·0	31	59	60	"
"	"	"	"	"	"	"	"	"	"	4¼ (C)	" "	20·6	22·9	26·7		43·2	31	57	53	"
"	"	"	"	"	"	"	"	"	"	0·424	AC900 OQ860 OQ800	22·9	26·3	32·4		55·7	22	50	47	"
"	"	"	"	"	"	"	"	"	"	¾	" " "	21·5	24·3	29·8		51·3	28	52	57	"
"	"	"	"	"	"	"	"	"	"	1	" " "	21·9	24·0	28·0		49·4	28½	54½	59	"
"	"	"	"	"	"	"	"	"	"	1⅛	" " "	21·9	23·8	28·2		49·5	29	55½	62	"

MECHANICAL PROPERTIES AT ROOM TEMPERATURE (Continued)

Process of Manufacture	ASTM Grain Size	Chemical Composition (%) C	Si	Mn	S	P	Ni	Cr	Mo	Section Size dia. (in.)	Heat Treatment °C	Tensile 0·1% P.S. (0·1Rp) tons/sq.in.	0·2% P.S. (0·2Rp)	0·5% P.S. (0·5Rp)	Y.P. (Rm)	T.S. (Rs)	El % (A) 4√A	R.A. % (Z)	Izod ft.lb.	Ref.
A.O.H.	7	0·15	0·24	0·83	0·032	0·024	0·76	0·67	0·06	2⅛	AC900 OQ860 OQ800	20·4	22·6	26·4		46·6	30	61½	73	4
"	"	"	"	"	"	"	"	"	"	3	" " "	20·8	24·0	28·7		45·5	30½	61½	80	"
"	"	"	"	"	"	"	"	"	"	4¼ (O)	" " "	19·0	21·2	25·8		44·1	30	61½	75	"
"	"	"	"	"	"	"	"	"	"	4¼ (C)	" " "	18·6	21·1	25·5		45·1	30	58½	63	"
B.O.H.		0·15	0·23	0·74	0·047	0·028	0·74	0·53	"	1⅛	WQ880 OQ780					41·6	32		83	9
B.E.	7	0·16	0·21	0·76	0·006	0·015	0·95	0·64	0·10	1⅛	OQ870 OQ810				39·2	56·4	22½	50	43	2
"	"	"	"	"	"	"	"	"	"	"	" OQ830					54·6	23½	58	62	"
"	8	0·16	0·14	0·84	0·018	0·018	0·78	0·78	0·08	1⅛	BC900 OQ860 OQ800					50·0	26	52	69	"
B.O.H.		0·16	0·23	0·63	0·030	0·020	0·75	0·48	0·10	¼	BC AC870 OQ810 T140	29·1				63·3	18	39		7
"		"	"	"	"	"	"	"	"	½	BC AC870 OQ810 T140	20·7				55·5	20	41	15	"
"		"	"	"	"	"	"	"	"	"	BC875 OQ870 OQ780 T140	23·6				52·0	26	54½	53	"
"		"	"	"	"	"	"	"	"	"	" " OQ810 "	28·2				57·6	20	44		"
"		"	"	"	"	"	"	"	"	"	" " " T180	35·5				64·6	18	41½		"
"		"	"	"	"	"	"	"	"	"	BC900 OQ870 OQ760 T140	22·9				52·2	30	58	53	"
"		"	"	"	"	"	"	"	"	"	" " OQ780 "	24·5				53·0	27	56	58	"
"		"	"	"	"	"	"	"	"	"	" " OQ810 "	31·2				64·6	18	43	23	"
"		"	"	"	"	"	"	"	"	"	" " " T180	29·5				60·7	18	45	31	"
"		"	"	"	"	"	"	"	"	¾	BC OQ870 OQ810 T140	19·7				51·3	23	47	21	"
"		"	"	"	"	"	"	"	"	1⅛	" " " "	17·7				46·8	29	50	39	"
"		"	"	"	"	"	"	"	"	1¼	" " " "	23·6			34·5	47·8	27	57	46	
A.O.H.		0·16	0·22	0·70	0·031	0·022	0·85	0·66	0·07	1⅛	BC900 OQ870 OQ800				28·0	45·8	33 5d 27	60		2
A.O.H.	1—5	0·16	0·28	0·74	0·028	0·029	0·85	0·59	0·07	1⅛	OQ870 OQ810				39·2	57·6	23	50	23	"
"	"	"	"	"	"	"	"	"	"	"	" OQ840				48·0	58·0	20	49	32	"
A.O.H.	7	0·16	0·30	0·82	0·028	0·024	0·78	0·65	0·05	1⅛	OQ870 OQ810				39·2	60·8	19	42	20	"
"	"	"	"	"	"	"	"	"	"	"	" OQ840				51·2	62·0	20½	42	25	"
B.E.	6—8	0·16	0·25	0·94	0·008	0·020	0·79	0·71	0·06	¼	BC900 AC870 OQ810 T140	37·8				72·8	20	55	—	7
"	"	"	"	"	"	"	"	"	"	½	" " " "	22·9				61·8	20	41	30	"
"	"	"	"	"	"	"	"	"	"	¾	" OQ870 " "	23·2				59·4	20	39	33	"
"	"	"	"	"	"	"	"	"	"	1⅛	" " " "	20·9				53·8	27	39	59	"
"	"	"	"	"	"	"	"	"	"	1¼	" " " "	27·2			41·4	54·3	24½	54	59	"

MECHANICAL PROPERTIES AT ROOM TEMPERATURE (Continued)

Process of Manufacture	ASTM Grain Size	Chemical composition (%)								Section Size dia. (in.)	Heat Treatment °C	Tensile							Izod ft.lb.	Ref.
		C	Si	Mn	S	P	Ni	Cr	Mo			0·1% P.S. (0·1Rp)	0·2% P.S. (0·2Rp)	0·5% P.S. (0·5Rp)	Y.P. (Rm)	T.S. (Rs)	El % (A)	R.A. % (Z)		
												tons/sq.in.								
B.O.H.		0·16	0·23	0·76	0·037	0·026	0·80	0·56		$\frac{11}{16}$	WQ880 OQ780					46·4	32		69	9
B.O.H.	6	0·16	0·16	0·77	0·041	0·031	0·70	0·62		$1\frac{1}{8}$	WQ880 OQ780					45·9	$31\frac{1}{2}$		84	"
B.O.H.		0·16	0·22	0·76	0·036	0·022	0·95	0·53		$1\frac{1}{8}$	WQ880 OQ780					47·8	28		61	"
B.O.H.		0·16	0·22	0·78	0·048	0·017	0·77	0·50		$1\frac{1}{8}$	WQ880 OQ780					44·2	33		107	"
B.O.H.		0·16	0·20	0·78	0·036	0·020	0·67	0·58		$1\frac{1}{8}$	WQ880 OQ780					44·0	32		68	"
A.O.H.	5—6	0·16	0·24	0·75	0·041	0·032	0·84	0·50	0·09	$1\frac{1}{8}$	BC OQ800				42·0	52·5	20	39	13	"
"	"	"	"	"	"	"	"	"	"	"	" OQ820				41·2	54·4	22	$44\frac{1}{2}$	17	"
"	"	"	"	"	"	"	"	"	"	"	" OQ840				43·6	54·5	19	$49\frac{1}{2}$	38	"
"	"	"	"	"	"	"	"	"	"	"	" OQ860				49·1	57·2	21	$52\frac{1}{2}$	40	"
"	"	"	"	"	"	"	"	"	"	"	BC OQ860 OQ800				36·0	50·1	29	55	40	"
"	"	"	"	"	"	"	"	"	"	"	" " OQ840				45·6	56·1	$21\frac{1}{2}$	$52\frac{1}{2}$	47	"
"	"	"	"	"	"	"	"	"	"	"	OQ860 OQ800				36·2	50·7	29	$52\frac{1}{2}$	40	"
"	"	"	"	"	"	"	"	"	"	"	" OQ840				45·0	55·5	21	$52\frac{1}{2}$	46	"
B.O.H.		0·16	0·23	0·62	0·030	0·020	0·77	0·46	0·10	0·236	BC AC870 OQ810 T140	39·5				73·9	14	35		7
"		"	"	"	"	"	"	"	"	0·367	" " " " "	41·8				70·5	12	40		"
"		"	"	"	"	"	"	"	"	0·574	" " " " "	31·2				61·2	20	43	28	"
"		"	"	"	"	"	"	"	"	$\frac{7}{8}$	" " " "	26·0				52·8	21	46		"
"		"	"	"	"	"	"	"	"	$1\frac{1}{4}$	" " " "	22·4				48·8	21	50	23	"
		0·17		0·80			0·88	0·62	0·05	$1\frac{1}{8}$	BC920 FC OQ800				46·5	58·0	$19\frac{1}{2}$	39	14	5
		"		"			"	"	"	"	" OQ820				48·4	58·2	21	42	24	"
		"		"			"	"	"	"	" OQ840				50·8	56·9	$20\frac{1}{2}$	52	35	"
		"		"			"	"	"	"	OQ860 OQ800				44.6	56·9	$24\frac{1}{2}$	44	18	"
		"		"			"	"	"	"	" OQ840				53·4	61·1	19	50	35	"
		"		"			"	"	"	"	BC920 OQ				52·3	55·5	20	44	44	"
		"		"			"	"	"	"	BC920 Cool to 810 $\frac{1}{2}$hr. OQ				47·3	53·8	$22\frac{1}{2}$	47	45	"
H.F.		0·17	0·33	0·65	0·020	0·016	0·70	0·83	0·06	$1\frac{1}{8}$	BC OQ880 OQ780					51·2	23	39	24	"
"		"	"	"	"	"	"	"	"	"	" " OQ800					51·6	22	42	28	1
B.E.	6	0·17	0·22	0·88	0·016	0·019	0·86	0·59		$\frac{1}{2}$	OQ870 OQ810		61·3			77·5	11	41	28	2
"	"	"	"	"	"	"	"	"		$\frac{7}{8}$	" "		41·0			56·6	24	54	52	"
"	"	"	"	"	"	"	"	"		$1\frac{1}{8}$	" "		36·7			54·2	24	$60\frac{1}{2}$	51	"
"	"	"	"	"	"	"	"	"		$2\frac{1}{2}$	" "		32·6			49·4	25	54	51	"
"	"	"	"	"	"	"	"	"		$1\frac{1}{8}$	" OQ830					59·0	22	51	41	"

MECHANICAL PROPERTIES AT ROOM TEMPERATURE (Continued)

Process of Manufacture	ASTM Grain Size	Chemical Composition (%) C	Si	Mn	S	P	Ni	Cr	Mo	Section Size dia. (in.)	Heat Treatment °C	Tensile 0·1% P.S. (0·1Rp) tons/sq.in.	0·2% P.S. (0·2Rp)	0·5% P.S. (0·5Rp)	Y.P. (Rm)	T.S. (Rs)	El % (A)	R.A. % (Z)	Izod ft.lb.	Ref.
N.O.H.	6—8	0·17	0·24	0·80	0·033	0·028	0·78	0·50	0·12	¾	BC900 AC870 OQ810 T140	25·9				61·6	18	30		7
"	"	"	"	"	"	"	"	"	"	½	" " " "	21·7				57·6	21	41	21	"
"	"	"	"	"	"	"	"	"	"	¾	" OQ870 " "	20·4				55·0	21	36	26	"
"	"	"	"	"	"	"	"	"	"	1⅛	" " " "	19·0				49·0	27	47	44	"
"	"	"	"	"	"	"	"	"	"	1¼	" AC870 " "	28·8			40·2	52·6	24	51	52	"
B.E.	6	0·17	0·21	0·78	0·006	0·015	0·92	0·68	0·11	1⅛	OQ870 OQ800					55·4	25½	47	36	"
"	"	"	"	"	"	"	"	"	"	"	BC900 OQ870 OQ810 T140	28·6			43·2	57·1	23½	54½	55	"
"	"	"	"	"	"	"	"	"	"	1¼	" " " "	30·6			44·2	57·8	24½	54	60	"
"	"	"	"	"	"	"	"	"	"	1½	" " " "	30·0			43·2	56·6	24½	55	60	"
B.O.H.		0·17	0·23	0·78	0·037	0·025	0·84	0·64	"	1⅛	WQ880 OQ780					49·4	28		60	9
		0·17	0·20	0·90	0·013	0·025	0·75	0·65	0·09	1¼	BC900 OQ860 OQ800					55·0	26	49½	58	2
B.O.H.	6—7	0·18	0·23	0·82	0·037	0·029	0·84	0·61	0·08	1¼	BC900 OQ870 OQ760 T140	23·1				53·9	24	34½	38	7
"	"	"	"	"	"	"	"	"	"	"	" " OQ780 "	22·6				51·8	23½	32	31	"
"	"	"	"	"	"	"	"	"	"	"	" " OQ810 "	30·0				57·5	18½	35½		"
"	"	"	"	"	"	"	"	"	"	"	" " " T180	28·9				55·5	19½	33½	39	"
"	"	"	"	"	"	"	"	"	"	"	BC925 " OQ760 T140	21·5				52·9	30	53½	33	"
"	"	"	"	"	"	"	"	"	"	"	" " OQ780 "	22·8				52·0	29½	52½	23	"
"	"	"	"	"	"	"	"	"	"	"	" " OQ810 "	30·8				58·3	22½	46½		"
"	"	"	"	"	"	"	"	"	"	"	" " " T180	31·3				56·0	24	51	43	"
A.O.H.	4—5	0·18	0·20	0·79	0·033	0·030	0·93	0·60		½	OQ870 OQ810				61·5	84·0	15	46	20	2
"	"	"	"	"	"	"	"	"		⅞	" "				46·5	59·9	20	56	28	"
"	"	"	"	"	"	"	"	"		1⅛	" "				39·5	54·0	24	54	31	"
"	"	"	"	"	"	"	"	"		"	" OQ840				47·6	59·6	19	49½	23	"
A.O.H.	1—5	0·18	0·31	0·78	0·035	0·034	0·94	0·59	0·10	1⅛	OQ870 OQ810				44·4	61·2	20½	44	25	"
"	"	"	"	"	"	"	"	"	"	"	" OQ830					56·6	20½	48	31	"
B.O.H.		0·18	0·24	0·84	0·034	0·017	0·75	0·64	0·04	1⅛	BC OQ860 OQ780				31·2	57·6	21	40½	18	10
"		"	"	"	"	"	"	"	"	"	" " OQ800				33·2	58·0	19	38	23	"
"		"	"	"	"	"	"	"	"	"	" " OQ820				34·8	56·4	21	44½	20	"
"		"	"	"	"	"	"	"	"	"	" " OQ840				36·4	56·4	19	43	31	"
A.O.H.	5—6	0·18	0·34	0·76	0·022	0·030	0·79	0·68	0·06	0·424	AC900	19·8	24·7	27·7		45·7	27½	56½	31	4
"	"	"	"	"	"	"	"	"	"	¾	"	19·4	22·4	26·7		43·8	26½	51½	29	"
"	"	"	"	"	"	"	"	"	"	1	"	18·6	22·1	26·4		41·1	27½	56	34	"
"	"	"	"	"	"	"	"	"	"	1⅛	"	19·6	22·0	25·0		41·8	28½	57	38	"
"	"	"	"	"	"	"	"	"	"	2⅛	"		21·5	24·0		38·5	30½	69½	54	"

MECHANICAL PROPERTIES AT ROOM TEMPERATURE (Continued)

Process of Manufacture	ASTM Grain Size	Chemical Composition (%) C	Si	Mn	S	P	Ni	Cr	Mo	Section Size dia. (in.)	Heat Treatment °C	Tensile 0·1% P.S. (0·1Rp) tons/sq. in.	0·2% P.S. (0·2Rp) tons/sq. in.	0·5% P.S. (0·5Rp) tons/sq. in.	Y.P. (Rm) tons/sq. in.	T.S. (Rs) tons/sq. in.	El % (A)	R.A. % (Z)	Izod ft.lb.	Ref.
A.O.H.	5—6	0·18	0·34	0·76	0·022	0·030	0·79	0·68	0·06	3	AC900			24·0		36·4	33½	70½	102	4
”	”	”	”	”	”	”	”	”	”	4¼ (O)	”			25·1		36·5	35	73½	100	”
”	”	”	”	”	”	”	”	”	”	4¼ (C)	”			23·4		35·7	36	71½	102	”
”	”	”	”	”	”	”	”	”	”	0·424	AC900 OQ820	34·1	40·1	46·9		63·4	19	56½	33	”
”	”	”	”	”	”	”	”	”	”	¾	” ”	31·4	36·8	44·3		58·7	21½	48	41	”
”	”	”	”	”	”	”	”	”	”	1	” ”	31·4	35·3	40·8		55·2	23	54	42	”
”	”	”	”	”	”	”	”	”	”	1⅛	” ”	29·6	33·4	39·2		55·1	23½	53½	38	”
”	”	”	”	”	”	”	”	”	”	2⅛	” ”	21·7	25·0	30·1		49·4	25	52½	26	”
”	”	”	”	”	”	”	”	”	”	3	” ”	21·3	24·0	28·8		47·8	26½	54½	25	”
”	”	”	”	”	”	”	”	”	”	4¼ (O)	” ”	20·8	24·7	30·8		47·0	29	56	27	”
”	”	”	”	”	”	”	”	”	”	4¼ (C)	” ”	20·3	22·7	27·3		44·7	29	58	31	”
”	”	”	”	”	”	”	”	”	”	0·424	AC 900 OQ860 OQ800	27·8	33·6	40·8		61·4	20	48½	25	”
”	”	”	”	”	”	”	”	”	”	¾	” ” ”	25·8	29·0	35·2		56·3	23	48	28	”
”	”	”	”	”	”	”	”	”	”	1	” ” ”	25·4	28·3	32·4		52·2	27	53	39	”
”	”	”	”	”	”	”	”	”	”	1⅛	” ” ”	24·0	26·6	31·7		52·7	27	53½	45	”
”	”	”	”	”	”	”	”	”	”	2⅛	” ” ”	23·7	25·2	27·6		48·6	29	58½	40	”
”	”	”	”	”	”	”	”	”	”	3	” ” ”	21·3	22·9	26·7		47·2	29½	58½	44	”
”	”	”	”	”	”	”	”	”	”	4¼ (O)	” ” ”	21·7	25·3	31·3		46·4	31	59	47	”
”	”	”	”	”	”	”	”	”	”	4¼ (C)	” ” ”	21·0	23·8	29·2		44·2	30	63½	46	”
B.O.H.	6—7	0·18	0·24	0·82	0·035	0·028	0·84	0·60	0·07	1⅛	BC OQ800				49·7	59·8	21	36½	12	5
”	”	”	”	”	”	”	”	”	”	”	” OQ820				50·1	59·7	20	42	22	”
”	”	”	”	”	”	”	”	”	”	”	” OQ840				52·0	58·9	22½	49½	39	”
”	”	”	”	”	”	”	”	”	”	”	” OQ860				50·9	58·9	21	47	36	”
”	”	”	”	”	”	”	”	”	”	”	BC OQ860 OQ800				46·8	58·3	22	42	24	”
”	”	”	”	”	”	”	”	”	”	”	BC OQ860 OQ840				55·7	62·8	18½	46½	33	”
”	”	”	”	”	”	”	”	”	”	1¼	BC900 OQ870 OQ760 T140	23·1				53·9	24	34½	38	7
											” ” CQ780 T140	22·6				51·8	23½	32	31	”
”	”	”	”	”	”	”	”	”	”	”	” ” OQ810 ”	30·0				57·5	18½	35½	”	”
”	”	”	”	”	”	”	”	”	”	”	” ” OQ810 T180	28·9				55·5	19½	33½	39	”

MECHANICAL PROPERTIES AT ROOM TEMPERATURE (Continued)

Process of Manufacture	ASTM Grain Size	Chemical Composition (%) C	Si	Mn	S	P	Ni	Cr	Mo	Section Size dia. (in.)	Heat Treatment °C	Tensile 0·1% P.S. (0·1Rp) tons/sq.in.	0·2% P.S. (0·2Rp) tons/sq.in.	0·5% P.S. (0·5Rp) tons/sq.in.	Y.P. (Rm) tons/sq.in.	T.S. (Rs) tons/sq.in.	El % (A)	R.A. % (Z)	Izod ft.lb.	Ref.
B.O.H.	6—7	0·18	0·24	0·82	0·035	0·028	0·84	0·60	0·07	1¼	BC900 OQ870 OQ760 T140	21·5				52·9	30	53½	33	7
"	"	"	"	"	"	"	"	"	"	"	" " OQ780 "	22·8				52·0	29½	52½	23	"
"	"	"	"	"	"	"	"	"	"	"	" " OQ810 "	30·8				58·3	22½	46½	—	"
"	"	"	"	"	"	"	"	"	"	"	" " OQ810 T180	31·3				56·0	24	51	43	"
B.O.H.	"	0·18	0·25	0·80	0·036	0·028	0·78	0·61		1⅛	WQ880 OQ780					48·7	26½		51	9
B.O.H.		0·18	0·22	0·88	0·026	0·020	0·79	0·61		1⅛	WQ880 OQ780					57·2	21		23	"
"		"	"	"	"	"	"	"		15/16	" "					50·0	26		60	"
B.O.H.	6—7	0·18	0·18	0·84	0·039	0·023	0·73	0·56		1⅛	WQ880 OQ780					49·2	29½		86	"
"		0·18	0·23	0·75	0·034	0·023	0·79	0·48		15/16	WQ880 OQ780					59·2	16		50	"
"		"	"	"	"	"	"	"		1⅛	" "					46·2	31		63	"
A.O.H.	6—7	0·19	0·29	0·75	0·030	0·023	0·90	0·57	0·07	1⅛	OQ870 OQ810				42·0	60·4	20½	41½	24	2
"	"	"	"	"	"	"	"	"	"	"	" OQ830					58·8	21½	45	35	"
"	"	"	"	"	"	"	"	"	"	"	" OQ840				42·0	62·6	18	44	23	"
A.O.H.	7 (some 2 & 3)	0·19	0·29	0·76	0·042	0·032	0·83	0·59	0·08	1⅛	OQ870 OQ810				38·0	58·4	20	40	24	"
"	"	"	"	"	"	"	"	"	"	"	" OQ840				47·2	60·2	20½	46	28	"
B.O.H.		0·19	0·23	0·83	0·030	0·036	0·77	0·60		1⅛	WQ880 WQ780					49·2	26		54	9
B.O.H.		0·19	0·23	0·80	0·029	0·023	0·75	0·57		1⅛	WQ880 WQ780					48·8	29		46	"
B.O.H.		0·19	0·23	0·83	0·046	0·018	0·76	0·66		1⅛	WQ880 WQ780					51·4	25		57	"
B.O.H.		0·19	0·23	0·83	0·038	0·018	0·69	0·57		21/32	WQ880 WQ780					55·2	25		40	"
"		"	"	"	"	"	"	"		1⅛	" "					56·6	22		21	"
B.E.	6—7	0·20	0·31	0·84	0·027	0·012	0·91	0·63		1⅛	OQ870 OQ810				41·0	60·8	21½	45	29	2
		0·20		0·76			1·03	0·56	0·07	1⅛	BC920 FC OQ800				51·2	60·2	19	34	11	5
		"		"			"	"	"	"	" OQ820				50·8	63·8	19	39	15	"
		"		"			"	"	"	"	" OQ840				55·2	63·8	21	44	24	"
		"		"			"	"	"	"	" OQ860				57·2	64·7	19	44	31	"
B.O.H.		0·20	0·21	0·78	0·028	0·024	0·76	0·61		1⅛	WQ880 OQ780					50·6	29		40	9
B.O.H.		0·20	0·24	0·82	0·039	0·027	0·81	0·76		1⅛	WQ880 OQ780					61·6	16		18	9
B.O.H.		0·20	0·26	0·88	0·044	0·022	0·71	0·70		1⅛	WQ880 OQ780					53·2	19		35	"
B.E.	7	0·20	0·27	0·88	0·021	0·011	0·84	0·58		½	OQ870 OQ810				58·2	75·0	15½	40	28	2
"	"	"	"	"	"	"	"	"		⅞	" "				39·3	56·3	26	49½	51	"
"	"	"	"	"	"	"	"	"		1⅛	" "				35·0	54·2	24	45	44	"
"	"	"	"	"	"	"	"	"		2½	" "				32·2	52·8	25	44	38	"
"	"	"	"	"	"	"	"	"		4	" "				24·8	43·4	27½		72	"
"	"	"	"	"	"	"	"	"		1⅛	" OQ830					56·0	23½	51	42	"

EFFECT OF TEMPERING TEMPERATURE ON CASE HARDNESS

(Typical Curve)

Carburized 900°C. Oil quenched 860°C. Oil quenched 800°C.

(Ref. 4)

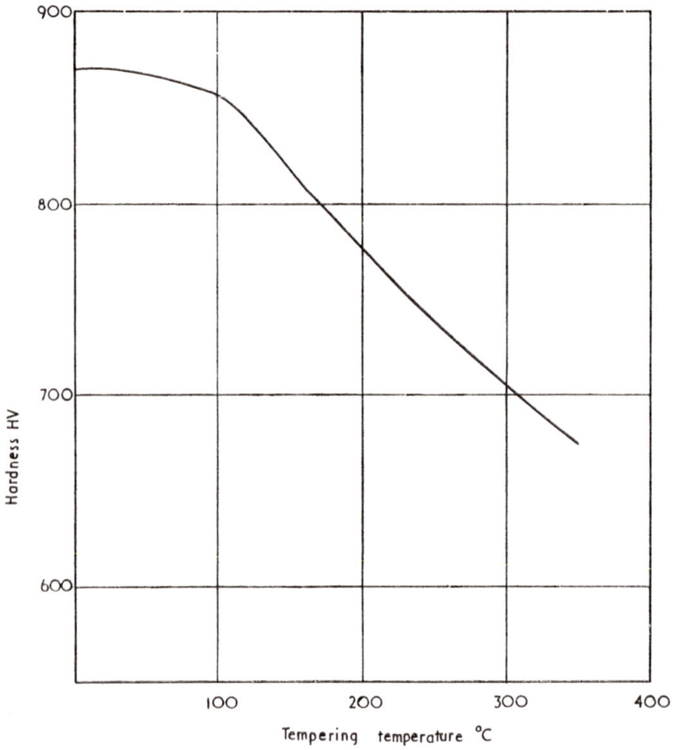

Fig. 351.10

EFFECT OF SECTION SIZE ON CORE PROPERTIES

(Typical Curves)

Blank carburized 900°C.
Oil refined 860°C.
Oil hardened 800°C.

(Ref. 4)

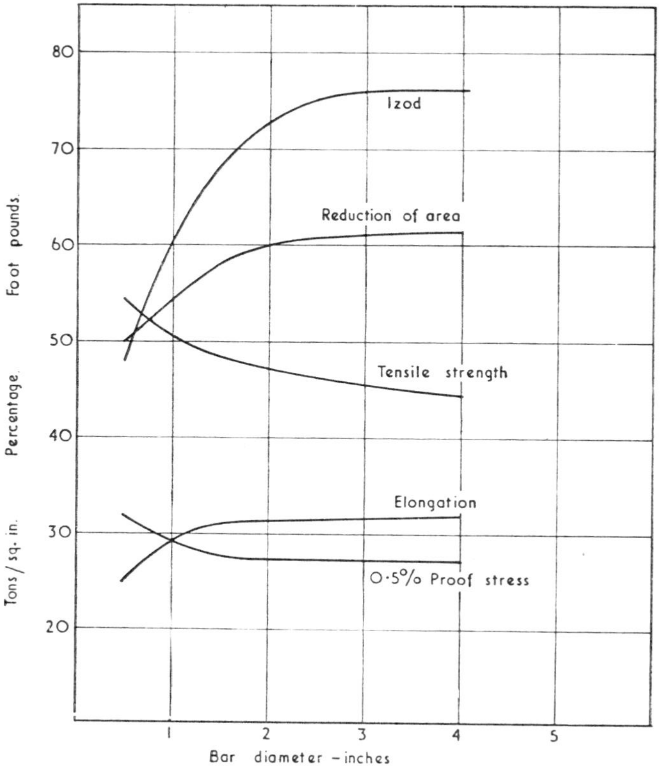

Fig. 351.11

Blank carburized 900°C.
Oil hardened 820°C.
(Single quenched quality).

(Ref. 4)

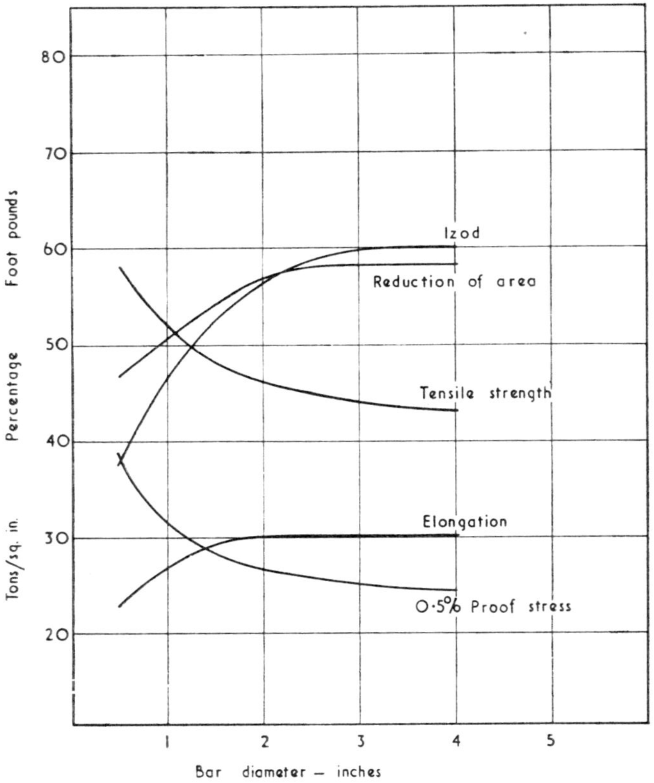

Fig. 351.12

MECHANICAL PROPERTIES AT LOW TEMPERATURE

Chemical Composition (%)								Section Size	Heat Treatment °C	Test Temperature °C	Tensile					Ref.
C	Si	Mn	S	P	Ni	Cr	Mo				0·1% P.S. (0·1Rp) tons/sq.in.	0·2% P.S. (0·2Rp) tons/sq.in.	T.S. (Rs) tons/sq.in.	El % (A)	R.A. % (Z)	
0·17	0·21	0·88	0·014	0·018	0·87	0·62	0·11	T.P.S.	BC930 OQ880 OQ780	+ 20	52·4	57·2	67·7	2	3½	11
"	"	"	"	"	"	"	"	"	" " "	—196	58·4	66·0	88·0	1½	2	"
"	"	"	"	"	"	"	"	"	Carb.* 930 OQ880 OQ780	+ 20	51·7	56·0	67·0	½	1	"
"	"	"	"	"	"	"	"	"	" " "	—196	42·8	60·4	74·3	½	1½	"
"	"	"	"	"	"	"	"	"	Carb.* 930 OQ780	+ 20	50·6	56·4	62·5	1	2½	"
"	"	"	"	"	"	"	"	"	" " "	—196	60·0	70·4	82·2	½	1½	"

* Test piece carburized to case depth of 0·045 in., finish ground to case depth of 0·02 in.

IMPACT TRANSITION TEMPERATURE DATA

Curve	Chemical Composition (%) C	Si	Mn	S	P	Ni	Cr	Mo	Section Size dia. (in.)	Heat Treatment °C	Charpy Mesnager Notch 50% max. energy Transition Temperature °C	Ref.
A	0·17	0·21	0·88	0·014	0·018	0·87	0·62	0·11	TPS*	BC930 OQ880 OQ780	—184	11
B	"	"	"	"	"	"	"	"		Carb. 930 OQ880 OQ780	+147	"
C	"	"	"	"	"	"	"	"		Carb. 930 OQ780	+153	"

* Test piece machined to size before carburizing or heat treatment. Notch 2 mm deep, root radius 2·00 mm, case depth 0·02 in.

Curve A, BC 930°C OQ 880°C OQ 780°C.
Curve B, carburized 930°C OQ 880°C OQ 780°C.
Curve C, carburized 930°C OQ 780°C.

(Ref. 11)

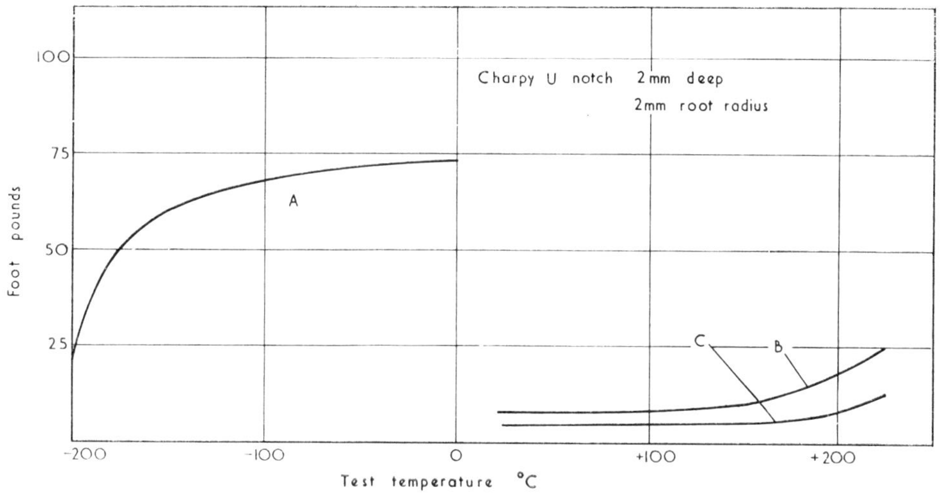

Fig. 351.13

LIST OF REFERENCES

1. Private correspondence with Jessop–Saville Ltd.
2. Private correspondence with Firth Brown Ltd.
3. Private correspondence with the Royal Aircraft Establishment, Farnborough.
4. Private correspondence with The English Steel Corporation Ltd.
5. Private correspondence with The United Steel Companies Ltd.
6. Atlas of Isothermal Transformation Diagrams of the B.S. En Steels. 2nd Edition. Iron and Steel Institute Special Report No. 56, 1956.
7. ASRTC S & T memo., No. 3/54.
8. Private correspondence with International Nickel Ltd.
9. Private correspondence with The Park Gate Iron and Steel Co. Ltd.
10. Private correspondence with Handfields Ltd.
11. War Office, unpublished report.

1 PER CENT NICKEL–CHROMIUM CASE-HARDENING STEEL

SPECIFICATION

CHEMICAL COMPOSITION (%)

C	Si	Mn	S	P	Ni	Cr	Mo
0·20 max.	0·35 max.	0·50—1·00	0·050 max.	0·050 max.	0·85—1·25	0·60—1·00	0·10 max.

MECHANICAL PROPERTIES

Tensile strength, tons/sq. in min. (Rm)	55
Elongation, %, min. (A)	15
Izod impact value, ft. lb. min.	20

The heat treatment given to 1⅛ in. dia. test bars shall be blank carburize at a temperature between 880°C and 830°C. (1),(2) Refine at a temperature of 850°C to 880°C, cool in air, oil or water. Harden in oil from a temperature of 780°C to 820°C.

(1) By special arrangement with the purchaser the test may be carried out on samples from which the blank carburizing treatment has been omitted or modified to a short heating period.

(2) A single quench from 810°C to 830°C may be used in certain cases when the refining treatment after carburizing is omitted by special arrangement between the purchaser and the manufacturer.

RELATED SPECIFICATIONS

United Kingdom	France	USSR	
En No.	AFNOR	GOST	Mark
352	16NCD5	4543—61	20KhN

APPLICATIONS

A low-alloy case-hardening steel developed as an alternative to En 35 and En 36A, which may be carburized and hardened to produce a hard wear-resistant case. It develops a core strength of the order of 55 tons/sq. in. in relatively small section sizes with moderate resistance to shock. Typical applications are gears, pinions, layshafts, cams, camshafts, gudgeon pins and rifle components.

WELDING

(See Introduction for key to symbols used in this table.)

	Welding Process					
	MA	G	IG	S	F	B
En 352	c	ax	g	e	a	h

Reproduced by kind permission of H.M.S.O.

MACHINABILITY

In the softened condition the machinability of this steel is approximately 75% of that for mild steel (En 3).

HOT WORKING AND HEAT TREATMENT TEMPERATURES

Forging, rolling and stamping	1150°C, finish above 900°C.
Softening	630°C—670°C, air or furnace cool.
Carburizing	880°C—930°C.
Refining*	850°C—880°C, cool in air, oil or water.
Hardening*	780°C—820°C, oil quench.
Tempering (for stress relief)	150°C—200°C.
(for maximum case toughness)	200°C—250°C.

* A single oil quench from 810°C —830°C may be adopted when steel is specifically requested for this treatment.

PHYSICAL PROPERTIES

SPECIFIC GRAVITY (d)

Chemical Composition (%)								Heat Treatment °C	Specific Gravity at 20°C	Ref.
C	Si	Mn	S	P	Ni	Cr	Mo			
0·18	0·25	0·70			1·00	0·80		Softened	7·85	1
0·18	0·24	0·76	0·018	0·024	1·05	0·83	0·07	BC910 OQ820	7·81	2
		(Within specification)						Softened	7·86—7·87	3

SPECIFIC HEAT (c)

Chemical Composition (%)					Heat Treatment °C	Specific Heat cal/g.deg.C	Ref.
C	Si	Mn	Ni	Cr		20 to 100°C	
0·18	0·25	0·70	1·00	0·80	Softened	0·118	1

MEAN COEFFICIENT OF THERMAL EXPANSION (α)

Chemical Composition (%)						Heat Treatment °C	$10^6 \times$ Mean Coefficient of Thermal Expansion per deg.C					Ref.
C	Si	Mn	Ni	Cr	Mo		20 to 100°C	20 to 200°C	20 to 300°C	20 to 500°C	20 to 700°C	
0·18	0·25	0·70	1·00	0·80	—	Softened	12		13	14	14·5	1
		(Within specification)				Softened	12·3		13·4	14·2	14·7	3
0·18	0·24	0·76	1·05	0·83	0·07	BC 910 FC OQ820	12·2	12·3				2

ELECTRICAL RESISTIVITY (ϱ)

Chemical Composition (%)						Heat Treatment °C	Electrical Resistivity microhm-cm			Ref.
C	Si	Mn	Ni	Cr	Mo		20°C	100°C	200°C	
0·18	0·25	0·70	1·00	0·80	—	Softened	26			1
0·18	0·24	0·76	1·05	0·83	0·07	BC910 FC OQ820	30·2	35·8	42·2	2

YOUNG'S MODULUS (E)

Chemical Composition (%)						Heat Treatment °C	Young's Modulus* tons/sq.in.			Ref.
C	Si	Mn	Ni	Cr	Mo		RT	100°C	200°C	
0·18	0·24	0·76	1·05	0·83	0·07	BC910 FC OQ820	13,400	13,100	12,800	2

*Vibration method.

TRANSFORMATION CHARACTERISTICS

ISOTHERMAL TRANSFORMATION

CORE CHARACTERISTISCS

Chemical Composition (%)							
C	Si	Mn	S	P	Ni	Cr	Mo
0·20	0·15	0·71	0·018	0·032	1·13	0·80	0·05

Basic Electric Arc.
McQuaid Ehn Grain Size 5—6 (ASTM).
As quenched Grain Size 6—7 (ASTM).
Bars hot rolled to ⅜ in. dia., annealed and cold drawn, and rolled to 0·030 in. thick tape, followed by 900°C (2 hrs.) furnace cool.
Austenitizing Temperature 865°C, 30 min. (Ref. 4)

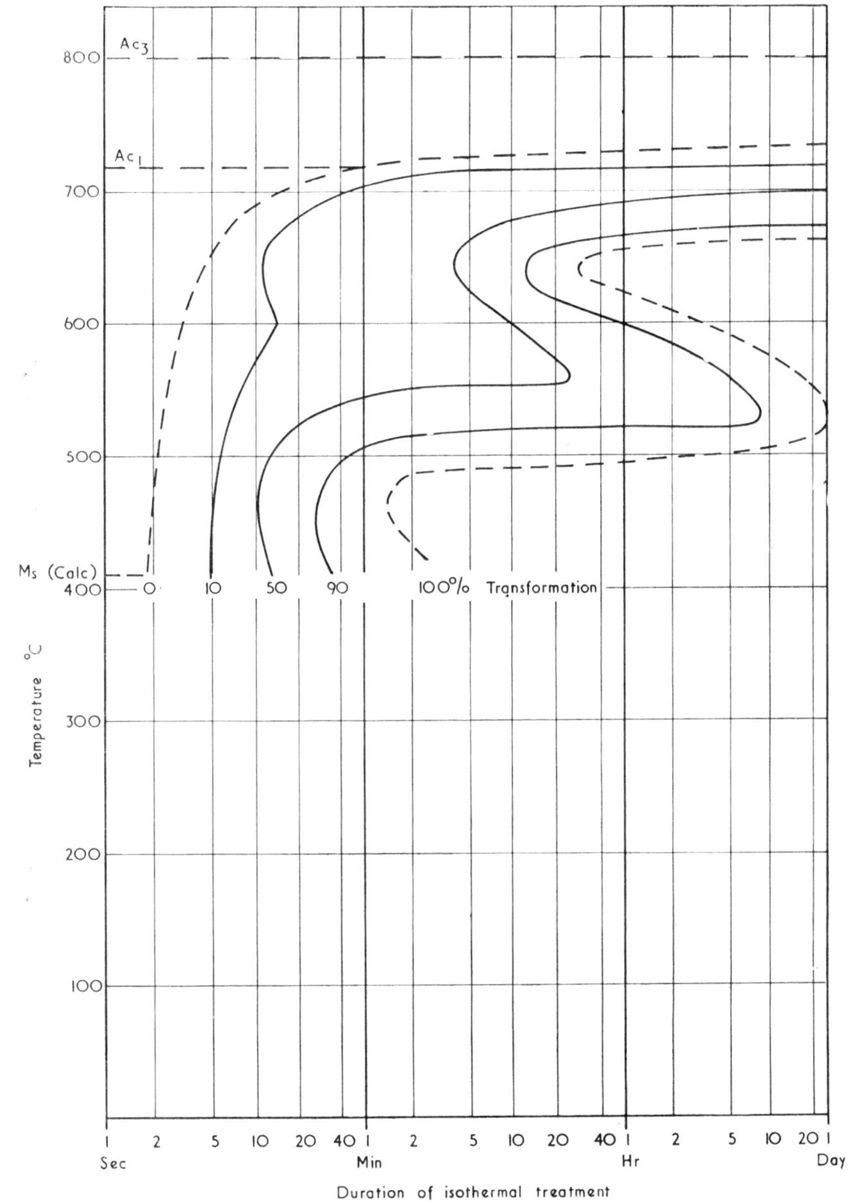

Fig. 352.1

En 352

ISOTHERMAL TRANSFORMATION (Continued)

CORE CHARACTERISTICS

Chemical Composition (%)							
C	Si	Mn	S	P	Ni	Cr	Mo
0·20	0·15	0·71	0·018	0·032	1·13	0·80	0·05

Basic Electric Arc.
McQuaid Ehn Grain Size 5—6 (ASTM).
As quenched Grain Size 6—7 (ASTM).
Bars hot rolled to ⅜ in. dia., annealed and cold drawn, and rolled to 0·030 in. thick tape, followed by 900°C (2 hrs.) furnace cool, then 865°C (½ hr.) oil quench.
Austenitizing Temperature 800°C, 30 min. (Ref. 4)

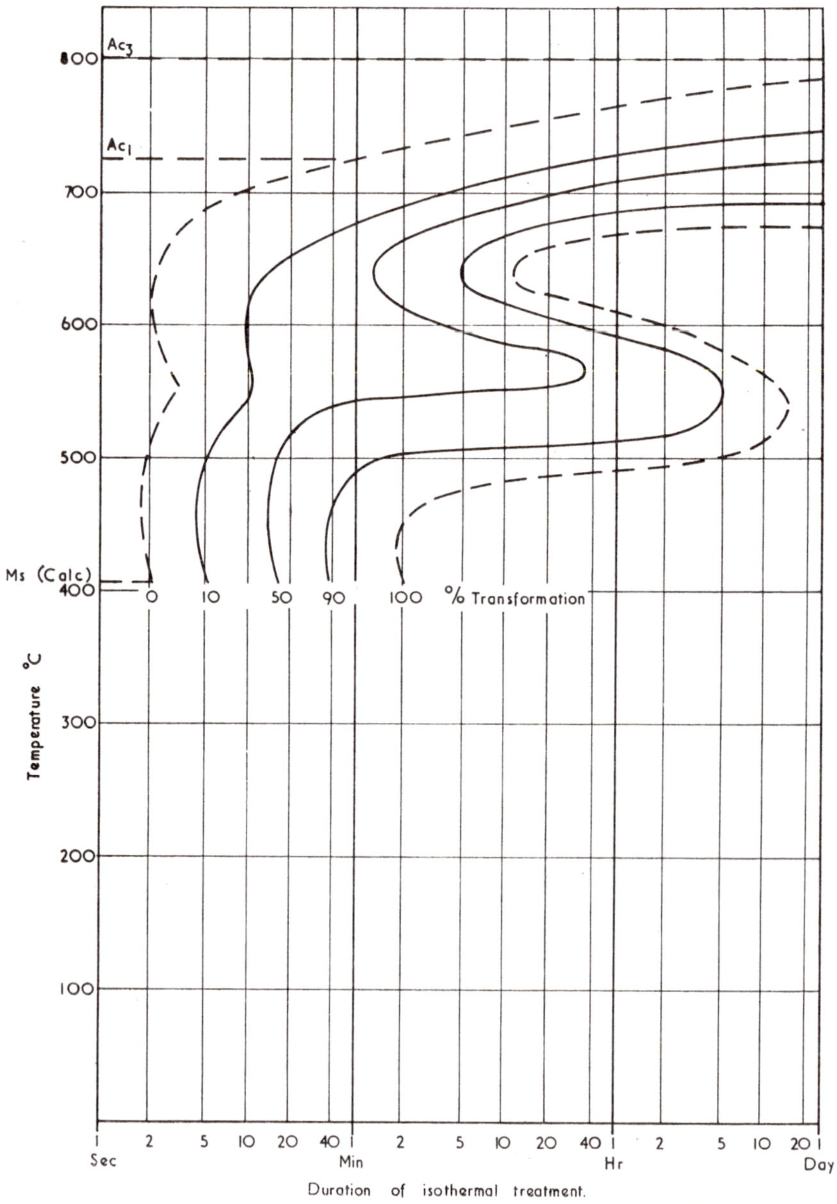

Fig. 352.2

ISOTHERMAL TRANSFORMATION (Continued)

CASE CHARACTERISTICS

Chemical Composition (%)							
C	Si	Mn	S	P	Ni	Cr	Mo
0·96	0·26	0·74	0·016	0·029	1·19	0·84	0·09

High Frequency Melt.
McQuaid Ehn Grain Size finer than 8 (ASTM).
As quenched Grain Size finer than 8 (ASTM).
Bars hot rolled to ⅜ in. dia., annealed and cold drawn to 11 SWG, followed by 900°C (2 hrs). furnace cool.
Austenitizing Temperature 865°C, 30 min. (Ref. 4)

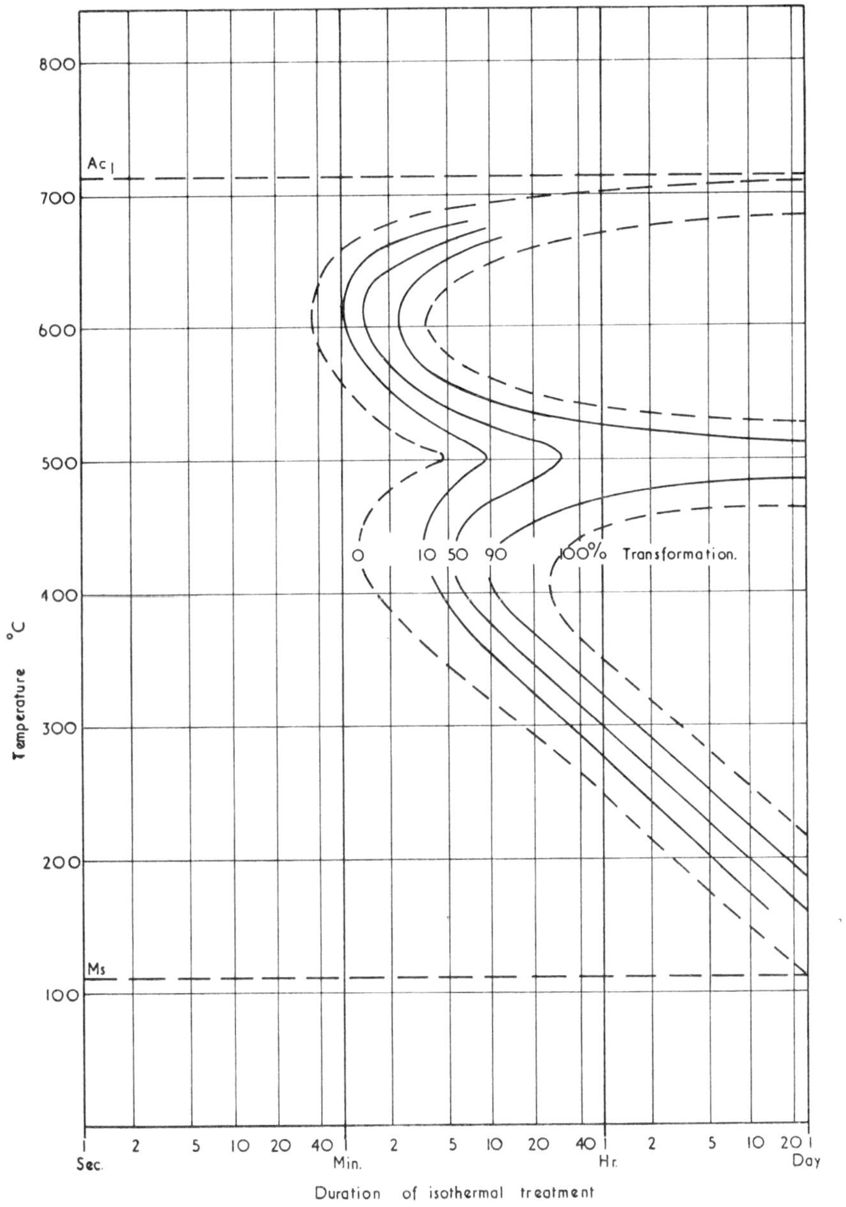

Fig. 352.3

ISOTHERMAL TRANSFORMATION (Continued)

CASE CHARACTERISTICS

Chemical Composition (%)							
C	Si	Mn	S	P	Ni	Cr	Mo
0·96	0·26	0·74	0·016	0·029	1·19	0·84	0·09

High Frequency Melt.
McQuaid Ehn Grain Size finer than 8 (ASTM).
As quenched Grain Size finer than 8 (ASTM).
Bars hot rolled to ⅜ in. dia., annealed and cold drawn to 11 SWG, followed by water quenching from 865°C (½ hr.) sealed in 1 in. dia. tube.
Austenitizing Temperature 800°C, 30 min.

(Ref. 4)

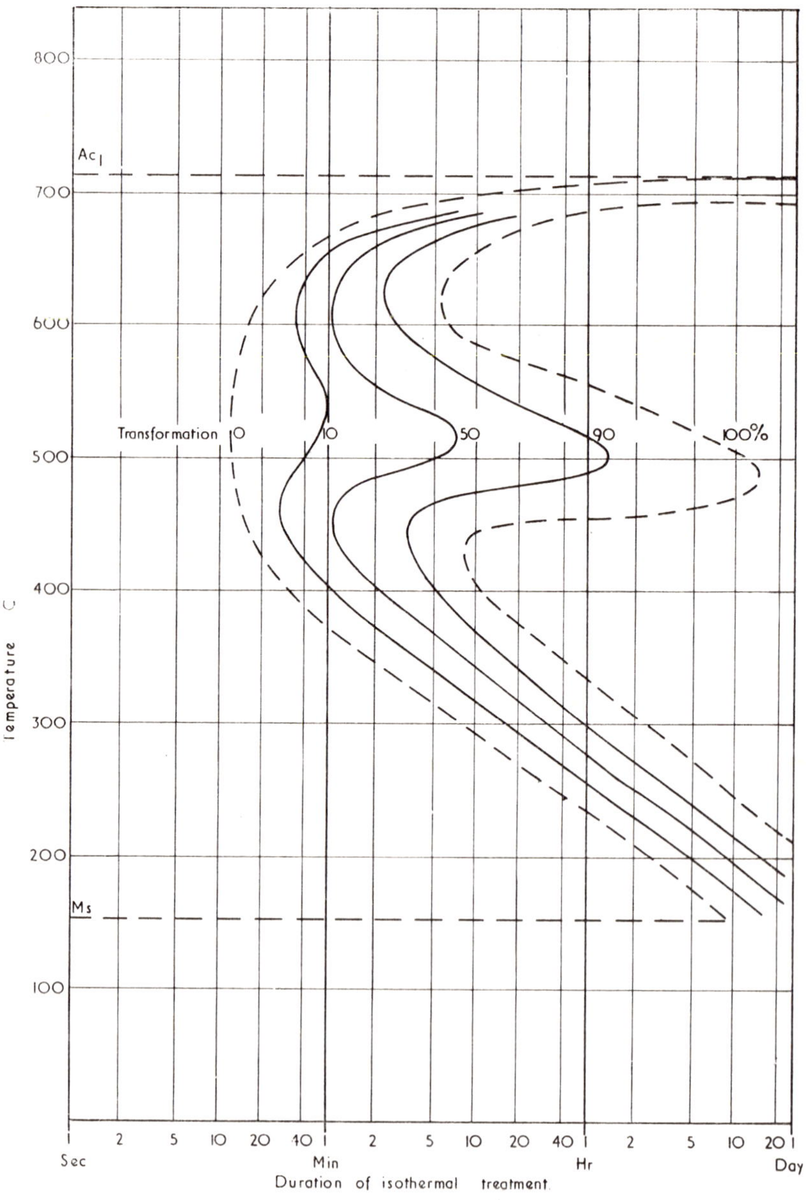

Fig. 352.4

CONTINUOUS COOLING TRANSFORMATION

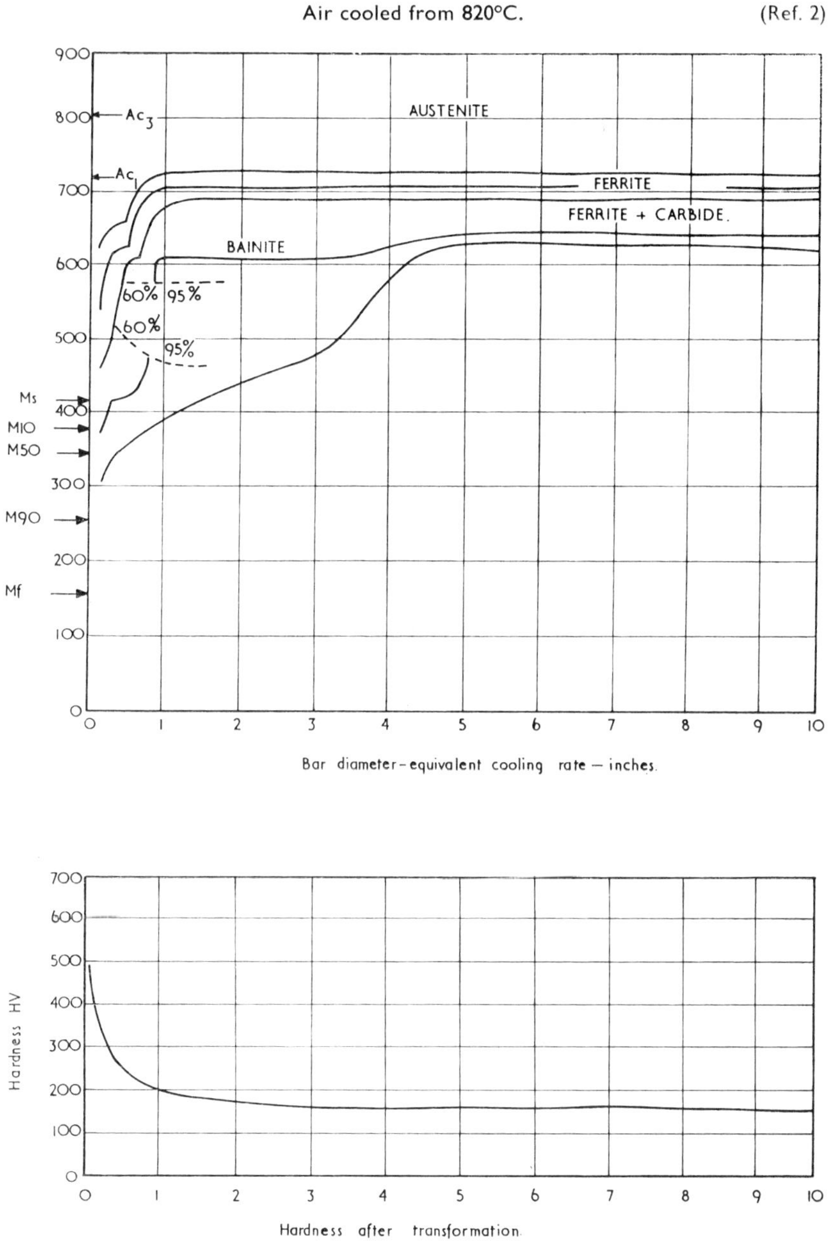

Fig. 352.5

En 352

CONTINUOUS COOLING TRANSFORMATION (Continued)

Oil quenched from 820°C. (Ref. 2)

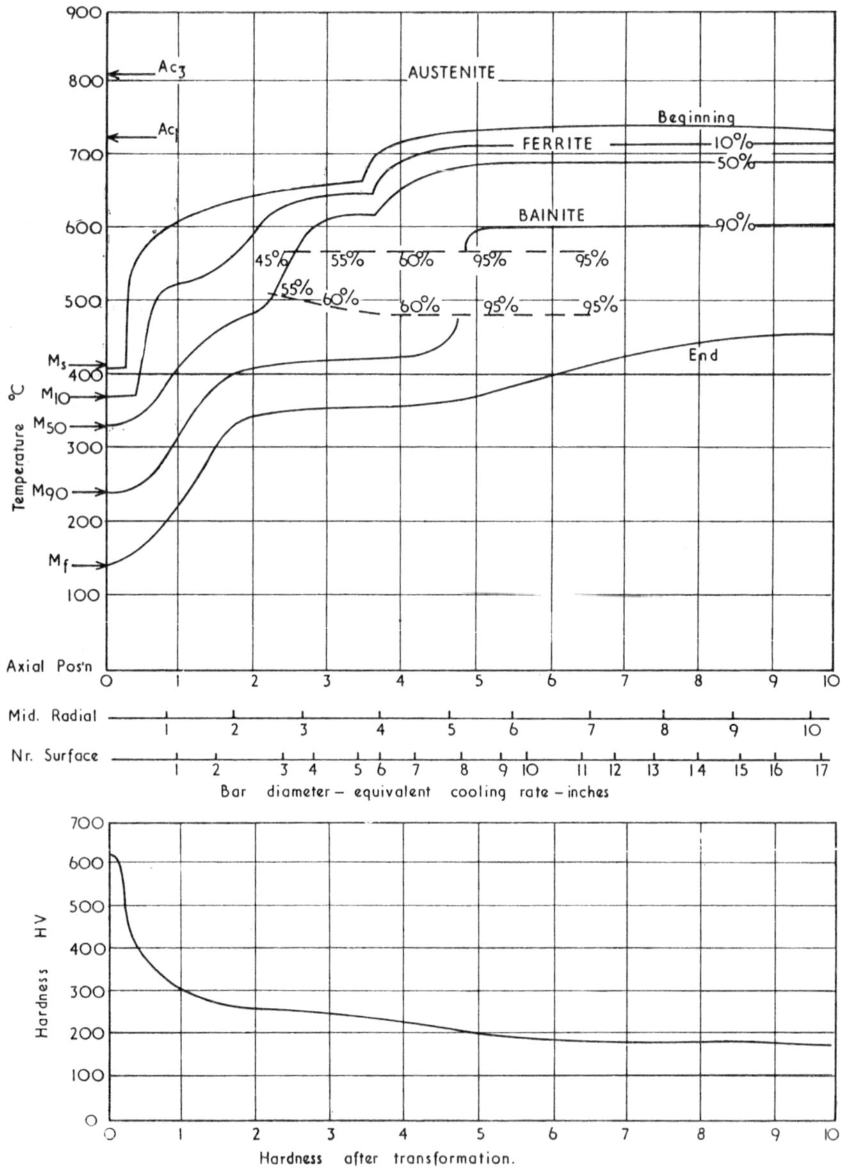

Fig.352.6

CONTINUOUS COOLING TRANSFORMATION (Continued)

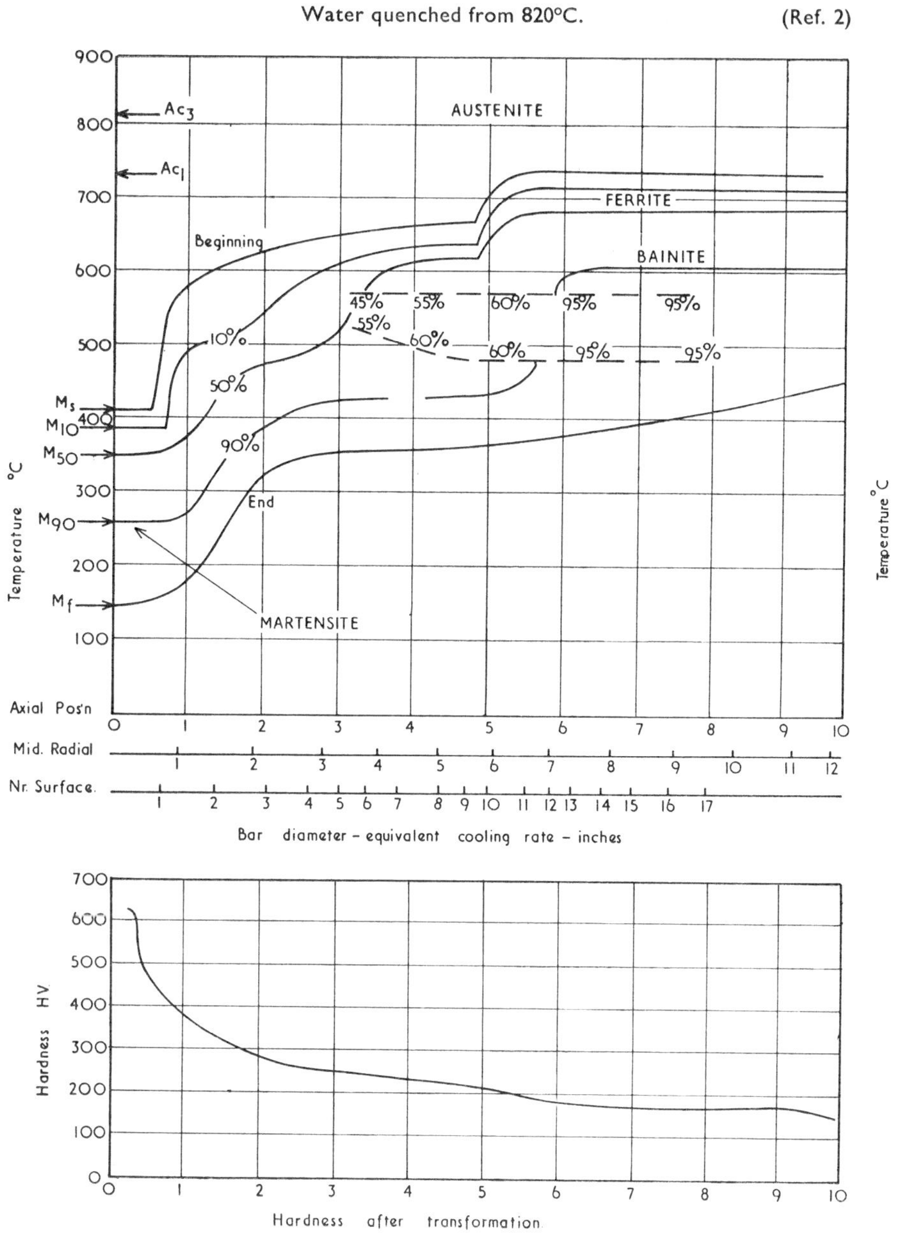

Fig. 352.7

En 352

HARDENABILITY

END QUENCH HARDENABILITY DATA

Chemical Composition (%)							
C	Si	Mn	S	P	Ni	Cr	Mo
0·20	0·15	0·71	0·018	0·032	1·13	0·80	0·05

Basic Electric Arc.
Curve A: Bars end quenched from 865°C.
Curve B: Bars end quenched from 800°C.

(Ref. 4)

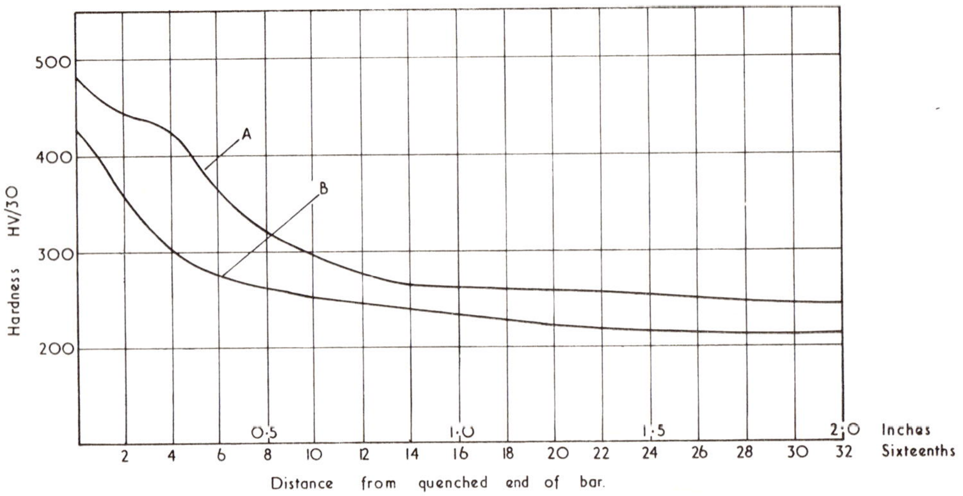

Fig. 352.8

Curve	Chemical Composition (%)								McQuaid Ehn Grain Size ASTM	Quenching Temperature °C
	C	Si	Mn	S	P	Ni	Cr	Mo		
C	0·16	0·31	0·75	0·024	0·017	0·99	0·78	0·05	6—7	860
D	0·18	0·25	0·80	0·033	0·020	0·96	0·79	0·09	6	"
E	0·19	0·30	0·78	0·015	0·018	1·01	0·80	0·07	5—6	"

(Ref. 5)

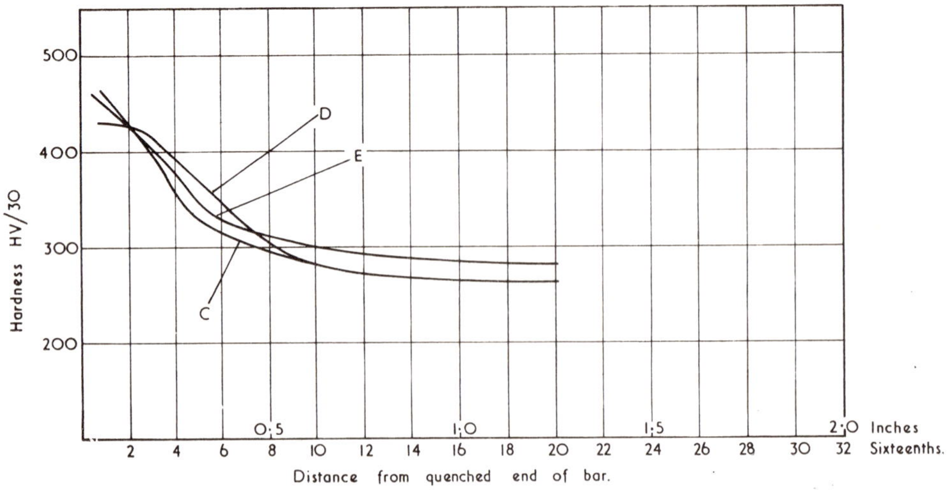

Fig. 352.9

END QUENCH HARDENABILITY BAND

These curves have been constructed from the results obtained on several casts of En 352.

(Refs. 2 and 3)

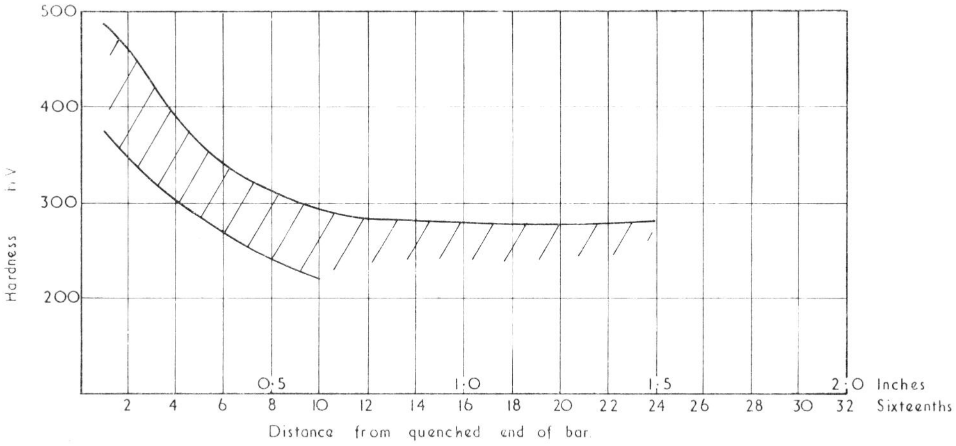

Fig. 352.10

MECHANICAL PROPERTIES AT ROOM TEMPERATURE

Process of Manufacture	ASTM Grain Size	Chemical Composition (%) C	Si	Mn	S	P	Ni	Cr	Mo	Section Size dia. (in.)	Heat Treatment °C	Tensile 0·1% PS. (0·1Rp) tons/sq.in.	0·2% P.S. (0·2Rp)	0·5% P.S. (0·5Rp)	Y.P. (Rm)	T.S. (Rs)	El % (A)	R.A. % (Z)	Izod ft.lb.	Ref.
B.E.	8 (some 4)	0·13	0·23	0·90	0·017	0·019	1·08	0·90	0·11	1⅛	AC900 OQ740	21·9	25·2	30·5		50·7	20½	43½	27	3
”	”	”	”	”	”	”	”	”	”	”	” OQ760	24·9	28·9	34·6		54·9	25	47½	33	”
”	”	”	”	”	”	”	”	”	”	”	” OQ780	26·0	30·0	36·0		55·8	24½	52	36	”
”	”	”	”	”	”	”	”	”	”	”	” OQ800	30·5	35·3	41·6		57·4	22	55	56	”
”	”	”	”	”	”	”	”	”	”	”	” OQ820	31·4	36·5	43·1		58·0	22	56	66	”
”	”	”	”	”	”	”	”	”	”	”	” OQ840	31·3	36·3	43·1		58·2	24½	56	66	”
”	”	”	”	”	”	”	”	”	”	0·424	AC900	20·8	25·4	32·6		49·0	26	49½	46	”
”	”	”	”	”	”	”	”	”	”	¾	”	21·2	25·8	33·2		48·7	24	47	37	”
”	”	”	”	”	”	”	”	”	”	1	”	21·2	25·2	31·3		49·2	21½	40½	39	”
”	”	”	”	”	”	”	”	”	”	1⅛	”	19·4	23·5	30·2		47·4	25	45	50	”
”	”	”	”	”	”	”	”	”	”	2⅛	”	20·1	23·2	28·6		46·7	25½	44½	53	”
”	”	”	”	”	”	”	”	”	”	4¼ (O)	”	18·6	21·1	25·5		42·6	28	50	78	”
”	”	”	”	”	”	”	”	”	”	4¼ (C)	”	18·6	20·7	24·6		41·9	26½	51½	77	”
”	”	”	”	”	”	”	”	”	”	0·424	AC900 OQ820	36·8	43·1	51·6		63·7	20½	59½	60	”
”	”	”	”	”	”	”	”	”	”	¾	” ”	33·6	38·7	45·4		58·3	22	58	65	”
”	”	”	”	”	”	”	”	”	”	1	” ”	32·7	36·8	42·6		56·5	22	40	69	”
”	”	”	”	”	”	”	”	”	”	1⅛	” ”	30·8	35·0	40·6		55·0	24	56½	78	”
”	”	”	”	”	”	”	”	”	”	2⅛	” ”	27·4	30·8	36·1		52·4	25	54½	68	”
”	”	”	”	”	”	”	”	”	”	3	” ”	24·3	27·8	33·7		50·6	25½	54	68	”
”	”	”	”	”	”	”	”	”	”	4¼ (O)	” ”	23·6	26·8	31·5		49·2	28	52	63	”
”	”	”	”	”	”	”	”	”	”	4¼ (C)	” ”	22·0	25·4	30·5		48·4	27	53½	73	”
”	”	”	”	”	”	”	”	”	”	0·424	AC900 OQ860 OQ800	36·6	43·2	52·5		64·1	17½	51½	52	”
”	”	”	”	”	”	”	”	”	”	¾	” ” ”	31·7	37·4	45·0		60·5	21	52½	55	”
”	”	”	”	”	”	”	”	”	”	1	” ” ”	29·1	34·0	40·6		57·1	24	54	62	”
”	”	”	”	”	”	”	”	”	”	1⅛	” ” ”	28·3	32·1	37·9		54·7	24½	55	74	”
”	”	”	”	”	”	”	”	”	”	2⅛	” ” ”	24·7	28·0	33·9		52·5	26	55	69	”
”	”	”	”	”	”	”	”	”	”	3	” ” ”	23·3	26·5	32·2		50·6	27	54½	64	”
”	”	”	”	”	”	”	”	”	”	4¼ (O)	” ” ”	21·5	24·2	28·6		49·4	28½	54½	68	”
”	”	”	”	”	”	”	”	”	”	4¼ (C)	” ” ”	21·1	24·0	28·8		48·7	27½	54½	76	”
A.O.H.		0·14	0·29	0·74	0·036	0·032	1·02	0·81	0·09	1⅛	BC925 OQ870 OQ800			29·1		47·8	4√A 5d 32 26½	57		1
		0·14	0·27	0·85			1·19	0·87	0·06	1⅛	BC900 OQ860 OQ800					56·0	22	54½	73	3

MECHANICAL PROPERTIES AT ROOM TEMPERATURE (Continued)

Process of Manufacture	ASTM Grain Size	Chemical Composition (%)								Section Size dia. (in.)	Heat Treatment °C	Tensile							Izod ft.lb.	Ref.
		C	Si	Mn	S	P	Ni	Cr	Mo			0·1% P.S. (0·1Rp)	0·2% P.S. (0·2Rp)	0·5% P.S. (0·5Rp)	Y.P. (Rm)	T.S. (Rs)	El % (A)	R.A. % (Z)		
												tons/sq.in.								
B.E.	7	0·15	0·18	0·77	0·012	0·019	1·06	0·81	0·08	$1\frac{1}{4}$	BC900 OQ860 OQ800					48·5	29	$60\frac{1}{2}$	75	1
"	6—7	0·15	0·22	0·67	0·007	0·020	1·24	0·74	0·07	$1\frac{1}{8}$	OQ870				41·6	58·2	22	51	35	6
B.E.		0·15	0·16	0·64	0·008	0·014	1·13	0·93	0·12	$1\frac{1}{8}$	OQ870 OQ810				44·4	63·4	19	$50\frac{1}{2}$	33	1
B.O.H.	7	0·15	0·25	0·76	0·039	0·031	1·17	0·72	0·08	$1\frac{1}{8}$	BC OQ800				46·2	57·7	21	42	19	2
"	"	"	"	"	"	"	"	"	"	"	" OQ820				50·2	58·4	21	$44\frac{1}{2}$	21	"
"	"	"	"	"	"	"	"	"	"	"	" OQ840				51·2	60·4	22	$49\frac{1}{2}$	31	"
"	"	"	"	"	"	"	"	"	"	"	" OQ860				52·3	59·8	22	$52\frac{1}{2}$	40	"
"	"	"	"	"	"	"	"	"	"	"	BC OQ860 OQ800				37·7	56·0	26	$49\frac{1}{2}$	50	"
"	"	"	"	"	"	"	"	"	"	"	" " OQ840				51·2	61·2	$17\frac{1}{2}$	47	53	"
		0·15	0·22	0·73			1·23	0·85	0·14	$1\frac{1}{8}$	BC OQ800				47·5	60·2	$18\frac{1}{2}$	44	17	2
		"	"	"			"	"	"	"	" OQ820				49·6	63·2	20	47	25	"
		"	"	"			"	"	"	"	" OQ840				54·0	61·5	20	57	39	"
		"	"	"			"	"	"	"	" OQ860				54·1	62·0	20	55	39	"
		"	"	"			"	"	"	"	BC OQ860 OQ800				45·6	57·7	22	50	33	"
		"	"	"					"	"	" " OQ840				55·1	61·6	20	55	46	"
		"	"	"			"	"	"	"	OQ860 OQ800				44·0	58·8	22	50	37	"
		"	"	"			"	"	"	"	" OQ840				52·8	63·7	19	55	35	"
H.F.		0·16	0·28	0·64	0·006	0·014	1·04	1·07	0·04	$1\frac{1}{8}$	BC927 OQ900 OQ780	22·3		33·0		55·2	$21\frac{1}{2}$	41	33	1
"		"	"	"	"	"	"	"	"	"	" " OQ810	31·8		44·3		58·2	$21\frac{1}{2}$	54	50	"
"		"	"	"	"	"	"	"	"	"	" " OQ840	32·8		45·0		58·2	21	51	41	"
"		"	"	"	"	"	"	"	"	"	" " WQ780	23·4		35·6		60·0	21	35	17	"
"		"	"	"	"	"	"	"	"	"	" " WQ810	43·6		63·6		81·2	15	$39\frac{1}{2}$	18	"
"		"	"	"	"	"	"	"	"	$\frac{1}{2}$	BC927 OQ870 OQ810			65·2		84·0	$15\frac{1}{2}$	$50\frac{1}{2}$	34	"
"		"	"	"	"	"	"	"	"	$\frac{7}{8}$	" " "			54·9		71·0	$18\frac{1}{2}$	$51\frac{1}{2}$	30	"
"		"	"	"	"	"	"	"	"	$1\frac{1}{8}$	" " "			44·3		58·2	$21\frac{1}{2}$	54	50	"
"		"	"	"	"	"	"	"	"	$2\frac{1}{2}$	" " "			34·7		51·2	24	$50\frac{1}{2}$	49	"
A.O.H.	4—6	0·16	0·28	0·62	0·023	0·031	1·11	0·78	0·08	$1\frac{1}{8}$	BC OQ850 OQ780				38·5	54·8	$23\frac{1}{2}$	50	34	7
		0·16	0·27	0·84	0·013	0·025	1·07	0·99	0·06	$1\frac{1}{4}$	BC900 OQ860 OQ800					58·0	24	$49\frac{1}{2}$	47	1
		0·16	0·34	0·72	0·011	0·013	0·99	0·70	0·10	$1\frac{1}{8}$	BC900 WQ880 OQ800					53·6	24	37	65	8
		"	"	"	"	"	"	"	"	$1\frac{1}{8}$	" OQ880 OQ780					58·8	21	45	33	"
B.E.		0·16	0·21	0·76	0·006	0·015	0·95	0·64	0·10	$\frac{1}{2}$	OQ810				60·5	76·7	$16\frac{1}{2}$	49	32	1
"		"	"	"	"	"	"	"	"	$\frac{7}{8}$	"				48·5	64·8	22	48	35	"
"		"	"	"	"	"	"	"	"	$1\frac{1}{8}$	"				42·9	60·4	21	$49\frac{1}{2}$	37	"
"		"	"	"	"	"	"	"	"	$1\frac{1}{8}$	OQ830					54·6	$23\frac{1}{2}$	58	62	"
"		"	"	"	"	"	"	"	"	$2\frac{1}{2}$	OQ810				34·4	51·8	$27\frac{1}{2}$	$56\frac{1}{2}$	74	"
"		"	"	"	"	"	"	"	"	4	"				26·0	44·0	29	61	66	"

MECHANICAL PROPERTIES AT ROOM TEMPERATURE (Continued)

Process of Manufacture	ASTM Grain Size	Chemical Composition (%) C	Si	Mn	S	P	Ni	Cr	Mo	Section Size dia. (in.)	Heat Treatment °C	Tensile 0·1% P.S. (0·1Rp) tons/sq.in.	0·2% P.S. (0·2Rp)	0·5% P.S. (0·5Rp)	Y.P. (Rm)	T.S. (Rs)	El % (A)	R.A. % (Z)	Izod ft.lb.	Ref.
A.O.H.		0·16	0·28	0·72	0·023	0·031	1·11	0·78	0·08	½	OQ810				59·6	73·9	16½	53	34	1
"		"	"	"	"	"	"	"	"	⅞	"				45·6	59·9	24½	56	50	"
"		"	"	"	"	"	"	"	"	1⅛	"				42·2	57·8	22	49½	40	"
"		"	"	"	"	"	"	"	"	2½	"				34·6	52·2	24	48½	47	"
"		"	"	"	"	"	"	"	"	4	"				27·5	45·0	27	53	68	"
A.O.H.		0·16	0·28	0·74	0·028	0·029	0·85	0·60	0·07	½	OQ810				67·5	85·0	15½	50½	21	"
"		"	"	"	"	"	"	"	"	⅞	"				52·6	67·0	19½	48	21	"
"		"	"	"	"	"	"	"	"	1⅛	"				46·2	62·0	20½	47½	18	"
"		"	"	"	"	"	"	"	"	2½	"				32·2	48·6	26½	54½	39	"
"		"	"	"	"	"	"	"	"	4	"				26·6	43·2	28	57	38	"
B.O.H.	4—6	0·16	0·24	0·64	0·044	0·020	1·09	0·87	0·06	0·564	OQ860 OQ810				53·1	62·1	19½	44½	33	2
"	"	"	"	"	"	"	"	"	"	1⅛	" OQ780				45·1	56·0	24½	49½	31	"
"	"	"	"	"	"	"	"	"	"	"	" OQ800				45·2	56·4	25	49½	35	"
"	"	"	"	"	"	"	"	"	"	"	" OQ820				45·4	57·8	20½	47	30	"
"	"	"	"	"	"	"	"	"	"	"	" OQ860				50·2	60·0	23	55	43	"
"	"	"	"	"	"	"	"	"	"	"	BC920 FC OQ820				52·2	58·9	19½	42	27	"
"	"	"	"	"	"	"	"	"	"	"	" OQ840				52·7	58·7	21	52	44	"
"	"	"	"	"	"	"	"	"	"	"	" OQ860				53·9	59·2	20½	55	42	"
"	"	"	"	"	"	"	"	"	"	2½	BC920 FC OQ850				34·6	49·6	25	52½	42	"
B.E.	7—8	0·16	0·18	0·72	0·023	0·017	1·10	0·82	0·03	1⅛	BC OQ800				46·8	58·1	21½	42	25	"
"	"	"	"	"	"	"	"	"	"	"	" OQ820				46·1	58·0	22	47	39	"
"	"	"	"	"	"	"	"	"	"	"	" OQ840				48·1	59·0	22	47	42	"
"	"	"	"	"	"	"	"	"	"	"	" OQ860				50·0	59·8	22½	47	41	"
"	"	"	"	"	"	"	"	"	"	"	BC OQ860 OQ800				44·8	57·7	24½	44½	36	"
"	"	"	"	"	"	"	"	"	"	"	" " OQ840				49·2	60·6	20½	47	44	"
"	"	"	"	"	"	"	"	"	"	"	OQ860 OQ800				45·2	58·8	24½	44½	39	"
"	"	"	"	"	"	"	"	"	"	"	" OQ840				53·2	63·2	21½	47	43	"
B.E.		0·17	0·12	0·71	0·009	0·012	1·02	0·63	0·05	1⅛	OQ810				44·0	57·6	21	47	31	1
"	7	0·17	0·24	0·85	0·017	0·014	0·98	0·80	0·10	1¼	BC900 OQ860 OQ800					57·5	24	51	52	"
A.O.H.		0·17	0·27	0·70	0·037	0·031	0·88	0·62	0·09	1⅛	OQ810				38·0	56·8	22½	44	27	"
"		"	"	"	"	"	"	"	"	"	OQ820					54·4	22	44	25	"
		0·17	0·20	0·95			1·12	0·99	0·03	1⅛	OQ870 OQ780				32·8	58·0	16½		37	9
B.E.	7—8	0·17	0·19	0·70	0·026	0·035	1·12	0·76	0·05	1⅛	BC OQ800				47·9	60·3	18½	36½	15	2
"	"	"	"	"	"	"	"	"	"	"	" OQ820				50·4	61·4	19	42	29	"
"	"	"	"	"	"	"	"	"	"	"	" OQ840				50·8	62·2	19	42	31	"
"	"	"	"	"	"	"	"	"	"	"	" OQ860				52·0	62·6	18½	44½	32	"

MECHANICAL PROPERTIES AT ROOM TEMPERATURE (Continued)

Process of Manufacture	ASTM Grain Size	Chemical Composition (%) C	Si	Mn	S	P	Ni	Cr	Mo	Section Size dia. (in.)	Heat Treatment °C	Tensile 0·1% P.S. (0·1Rp) tons/sq.in.	0·2% P.S. (0·2Rp) tons/sq.in.	0·5% P.S. (0·5Rp) tons/sq.in.	Y.P. (Rm) tons/sq.in.	T.S. (Rs) tons/sq.in.	El % (A)	R.A. % (Z)	Izod ft.lb.	Ref.
B.E.	7—8	0·17	0·19	0·70	0·026	0·035	1·12	0·76	0·05	1⅛	BC OQ860 OQ800				44·4	60·1	19½	42	23	2
"	"	"	"	"	"	"	"	"	"	"	" " OQ840				50·4	62·2	19	44½	30	"
"	"	"	"	"	"	"	"	"	"	"	OQ860 OQ800				45·2	60·6	21	39	26	"
"	"	"	"	"	"	"	"	"	"	"	" OQ840				57·9	65·6	19½	44½	31	"
B.E.	7—8	0·17	0·13	0·75	0·016	0·029	1·12	0·82	0·04	1⅛	BC OQ800				61·2	66·9	17½	42	12	"
"	"	"	"	"	"	"	"	"	"	"	" OQ820				60·2	67·0	19	49½	21	"
"	"	"	"	"	"	"	"	"	"	"	" OQ840				59·9	69·0	16½	47	20	"
"	"	"	"	"	"	"	"	"	"	"	" OQ860				59·8	66·4	17	47	22	"
"	"	"	"	"	"	"	"	"	"	"	BC OQ860 OQ800				62·0	70·8	17½	42	18	"
"	"	"	"	"	"	"	"	"	"	"	" " OQ840				59·8	65·9	17½	49½	25	"
"	"	"	"	"	"	"	"	"	"	"	OQ860 OQ800				66·8	71·8	15	42	20	"
"	"	"	"	"	"	"	"	"	"	"	" OQ840				64·8	69·3	17½	44½	23	"
"		0·17	0·21	0·85			1·06	0·89	0·05	1⅛	BC900 OQ860 OQ800					59·0	22	52	70	3
		0·17	0·24	0·82			1·09	0·85	0·07	1⅛	BC900 OQ860 OQ800					60·0	22	49½	47	"
		0·17	0·27	0·89			1·07	0·92	0·10	1⅛	BC900 OQ860 OQ800					61·0	23	47	45	"
		0·17	0·29	0·86			1·17	0·87	0·05	1⅛	BC900 OQ860 OQ800					63·0	22	48½	50	"
		0·17	0·25	0·94			1·08	0·85	0·10	1⅛	BC900 OQ860 OQ800					63·0	24	49½	54	"
		0·17	0·28	0·95			1·18	0·92	0·07	1⅛	BC900 OQ860 OQ800					65·0	20	48½	43	"
A.O.H.	6	0·17	0·27	0·68	0·043	0·028	1·07	0·85	0·09	1⅛	CQ870 OQ810				44·0	62·2	19	48½	24	1
"	"	"	"	"	"	"	"	"	"	"	" OQ840				48·4	63·0	19	43	28	"
B.E.	7	0·17	0·28	0·93	0·012	0·022	1·11	0·90	0·09	1⅛	BC OQ860 OQ800	27·8	31·7	38·1		59·8	25	52½	64	"
A.O.H.	6	0·18	0·32	0·79	0·033	0·034	1·23	0·81		1⅛	OQ870 OQ810				45·6	63·4	20	47	25	1
"	"	"	"	"	"	"	"	"		"	" OQ840				51·2	72·8	17	46	21	"
B.O.H.		0·18	0·25	0·83	0·023	0·028	1·09	0·68	0·09	⅜	BC OQ850 OQ780					63·8	25	44½		7
"		"	"	"	"	"	"	"	"	¾	" " "				35·4	59·8	25	52	36	"
"		"	"	"	"	"	"	"	"	1⅛	" " "				25·6	54·8	27	53½	48	"
"		"	"	"	"	"	"	"	"	1¾	" " "				28·0	52·8	27	53½	46	"
"		"	"	"	"	"	"	"	"	2	" " "				28·8	53·2	27	52½	45	"
	6—7	0·18	0·18	0·81	0·019	0·025	1·11	0·95	0·09	1¼	BC900 OQ860 OQ800					58·5	20	51	41	1
B.E.	7—8	0·18	0·16	0·70	0·030	0·022	1·06	0·83	0·04	1⅛	BC OQ800				48·2	60·2	20	39	18	2
"	"	"	"	"	"	"	"	"	"	"	" OQ820				54·9	62·4	19½	44½	36	"
"	"	"	"	"	"	"	"	"	"	"	" OQ840				52·4	63·3	20½	49½	37	"
"	"	"	"	"	"	"	"	"	"	"	" OQ860				53·6	62·9	22	47	38	"
"	"	"	"	"	"	"	"	"	"	"	" OQ860 OQ800				52·8	63·8	18	39	22	"
"	"	"	"	"	"	"	"	"	"	"	" " OQ840				53·6	64·1	19	44½	35	"
"	"	"	"	"	"	"	"	"	"	"	OQ860 OQ800				54·8	65·4	17½	36½	20	"
"	"	"	"	"	"	"	"	"	"	"	" OQ840				56·2	65·8	18½	47	36	"

MECHANICAL PROPERTIES AT ROOM TEMPERATURE (Continued)

Process of Manufacture	ASTM Grain Size	Chemical Composition (%) C	Si	Mn	S	P	Ni	Cr	Mo	Section Size dia. (in.)	Heat Treatment °C	Tensile 0·1% P.S. (0·1Rp)	0·2% P.S. (0·2Rp)	0·5% P.S. (0·5Rp)	Y.P. (Rm)	T.S. (Rs)	El % (A)	R.A % (Z)	Izod ft.lb.	Ref.
												tons/sq. in.								
		0·18	0·29	0·95			1·10	0·95	0·10	$1\frac{1}{8}$	BC900 OQ860 OQ800					64·0	22	$54\frac{1}{2}$	55	3
B.O.H.	7	0·18	0·25	0·72	0·037	0·026	1·12	0·69	0·06	$1\frac{1}{8}$	BC OQ800				46·3	58·9	$19\frac{1}{2}$	39	17	2
”	”	”	”	”	”	”	”	”	”	”	” OQ820				45·3	59·6	$20\frac{1}{2}$	42	21	”
”	”	”	”	”	”	”	”	”	”	”	” OQ840				42·4	58·2	21	47	37	”
”	”	”	”	”	”	”	”	”	”	”	” OQ860				49·8	62·2	20	47	35	”
B.E.		0·18	0·34	0·86	0·006	0·010	1·09	0·91		$1\frac{1}{8}$	BC OQ870 OQ810			50·0		69·2	$20\frac{1}{2}$	46	32	1
A.O.H.		0·18	0·32	0·79	0·033	0·034	1·23	0·81		$1\frac{1}{8}$	BC OQ870 OQ810			45·6		63·4	20	47	25	”
B.E.		0·19	0·20	0·63	0·011	0·010	1·21	0·91	0·14	$\frac{1}{2}$	OQ810				67·7	85·5	16	$54\frac{1}{2}$	39	1
”		”	”	”	”	”	”	”	”	$\frac{7}{8}$	”				55·6	70·9	19	$57\frac{1}{2}$	34	”
”		”	”	”	”	”	”	”	”	$1\frac{1}{8}$	”				45·5	61·6	$21\frac{1}{2}$	52	43	”
”		”	”	”	”	”	”	”	”	$2\frac{1}{2}$	”				41·0	56·0	23	$57\frac{1}{2}$	53	”
”		”	”	”	”	”	”	”	”	4	”				38·8	54·0	$24\frac{1}{2}$	$56\frac{1}{2}$	51	”
		0·19		0·80			1·10	0·87		$1\frac{1}{8}$	OQ870 OQ810				50·0	67·2	$20\frac{1}{2}$		28	”
	6—7	0·19	0·24	0·91	0·010	0·014	1·17	0·95	0·08	$1\frac{1}{4}$	” ”					67·0	19	47	50	”
		0·19	0·29	0·89			1·15	0·85	0·10	$1\frac{1}{8}$	BC900 OQ860 OQ800					64·5	20	$48\frac{1}{2}$	51	3
B.O.H.		0·19	0·23	0·85	0·037	0·022	1·02	0·60		$1\frac{1}{8}$	OQ880 OQ780					54·9	19		30	10
A.O.H.		0·19	0·25	0·63	0·039	0·026	1·20	0·66		$1\frac{1}{8}$	OQ870 OQ810				47·2	63·2	17	$46\frac{1}{2}$	19	1
”		”	”	”	”	”	”	”		”	” OQ840				53·2	64·2	17	43	23	”
B.E.		0·19	0·24	0·80	0·009	0·012	1·10	0·87		$1\frac{1}{8}$	BC OQ870 OQ810			50·0		67·2	$20\frac{1}{2}$	51	28	”
B.E.		0·20	0·31	0·84	0·027	0·012	0·91	0·63		$1\frac{1}{8}$	OQ810				41·0	60·8	$21\frac{1}{2}$	45	29	1
”		”	”	”	”	”	”	”		”	OQ820					66·2	$20\frac{1}{2}$	38	21	”
B.O.H.	5—8	0·20	0·25	0·94	0·037	0·023	1·03	0·77	0·13	$\frac{3}{8}$	BC OQ850 OQ780					68·7	$13\frac{1}{2}$			7
”	”	”	”	”	”	”	”	”	”	$\frac{3}{4}$	” ” ”				46·0	66·0	22	$48\frac{1}{2}$	28	”
”	”	”	”	”	”	”	”	”	”	$1\frac{1}{8}$	” ” ”				30·8	60·8	24	$48\frac{1}{2}$	35	”
”	”	”	”	”	”	”	”	”	”	$1\frac{3}{4}$	” ” ”				28·8	59·2	23	46	29	”
”	”	”	”	”	”	”	”	”	”	2	” ” ”				30·4	57·2	23	45	31	”
B.E.	1—7																			
	mixed	0·20	0·19	0·75	0·015	0·018	1·05	0·80	0·04	$\frac{3}{8}$	BC OQ850 OQ780					75·6	16	42		
”	”	”	”	”	”	”	”	”	”	$\frac{3}{4}$	” ” ”				49·5	69·3	17	$37\frac{1}{2}$	16	”
”	”	”	”	”	”	”	”	”	”	$1\frac{1}{8}$	” ” ”				31·2	61·2	21	39	20	”
B.E.		0·20	0·23	0·73	0·017	0·030	1·09	0·75	0·08	$\frac{3}{8}$	BC OQ850 OQ780					71·0	18	47	—	”
”		”	”	”	”	”	”	”	”	$\frac{3}{4}$	” ” ”				42·5	67·2	20	41	23	”
”		”	”	”	”	”	”	”	”	$1\frac{1}{8}$	” ” ”				35·2	58·2	23	47	33	”
”		”	”	”	”	”	”	”	”	$1\frac{3}{4}$	” ” ”				28·4	56·0	21	43	30	”
”		”	”	”	”	”	”	”	”	$2\frac{1}{2}$	” ” ”				26·0	54·4	22	43	38	”

MECHANICAL PROPERTIES AT ROOM TEMPERATURE (Continued)

Process of Manufacture	ASTM Grain Size	Chemical Composition (%)								Section Size dia. (in.)	Heat Treatment °C	Tensile							Izod ft.lb.	Ref.
		C	Si	Mn	S	P	Ni	Cr	Mo			0·1% P.S. (0·1Rp)	0·2% P.S. (0·2Rp)	0.5% P.S. (0·5Rp)	Y.P. (Rm)	T.S. (Rs)	El % (A)	R.A. % (Z)		
												tons/sq. in.								
B.O.H.	7—8	0·20	0·27	0·75	0·038	0·034	1·04	0·77	0·05	1⅛	BC OQ800				53·2	64·4	17½	33½	13	2
"	"	"	"	"	"	"	"	"	"	"	" OQ820				49·2	66·8	16½	33½	13	"
"	"	"	"	"	"	"	"	"	"	"	" OQ840				55·2	66·2	19½	44½	26	"
"	"	"	"	"	"	"	"	"	"	"	" OQ860				56·8	68·2	19½	44½	30	"
"	"	"	"	"	"	"	"	"	"	"	BC OQ860 OQ800				47·6	63·4	20½	42	23	"
"	"	"	"	"	"	"	"	"	"	"	" " OQ840				60·0	70·1	19	44½	28	"
B.O.H.	7	0·20	0·25	0·75	0·041	0·029	1·24	0·83	0·06	1⅛	BC OQ800				59·0	71·9	15½	33½	14	"
"	"	"	"	"	"	"	"	"	"	"	" OQ820				61·0	70·1	17½	42	18	"
"	"	"	"	"	"	"	"	"	"	"	" OQ840				62·9	72·0	14½	47	27	"
"	"	"	"	"	"	"	"	"	"	"	" OQ860				67·4	73·3	17½	47	30	"
B.E.	"	0·20	0·25	0·86	0·006	0·009	1·12	0·86		2½	BC OQ870 OQ810				42·5	60·8	21½	45	31	1
		0·20		0·87			1·12	0·92	0·08	1⅛	BC920 FC OQ800				64·4	75·8	16	39	15	2
		"		"			"	"	"	"	" OQ820				70·4	77·0	18	47	27	"
		"		"			"	"	"	"	" OQ840				70·0	76·7	16	50	25	"
		"		"			"	"	"	"	" OQ860				75·2	80·4	17	52	28	"
		"		"			"	"	"	"	OQ860 OQ800				73·6	78·4	16½	52	31	"
		"		"			"	"	"	"	" OQ840				77·6	83·6	15	44	25	"
		"		"			"	"	"	"	BC920 FC OQ860 OQ800				71·6	78·3	16½	42	19	"
		"		"			"	"	"	"	BC920 OQ860 OQ800				80·4	82·8	15	50	30	"

En 352

EFFECT OF TEMPERING TEMPERATURE ON CASE HARDNESS

(TYPICAL CURVE)

Carburized 900°C.
Oil refined 860°C.
Oil hardened 800°C.
Tempered as indicated.

(Ref. 3)

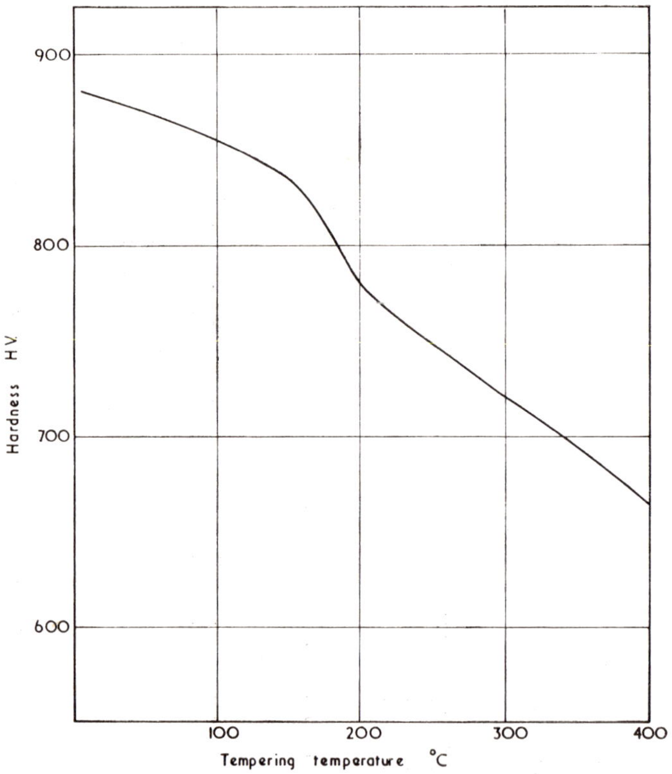

Fig. 352.11

EFFECT OF SECTION SIZE ON THE MECHANICAL PROPERTIES AT ROOM TEMPERATURE

(TYPICAL CURVES)

Blank carburized 900°C.
Oil refined 860°C.
Oil hardened 800°C.

(Ref. 3)

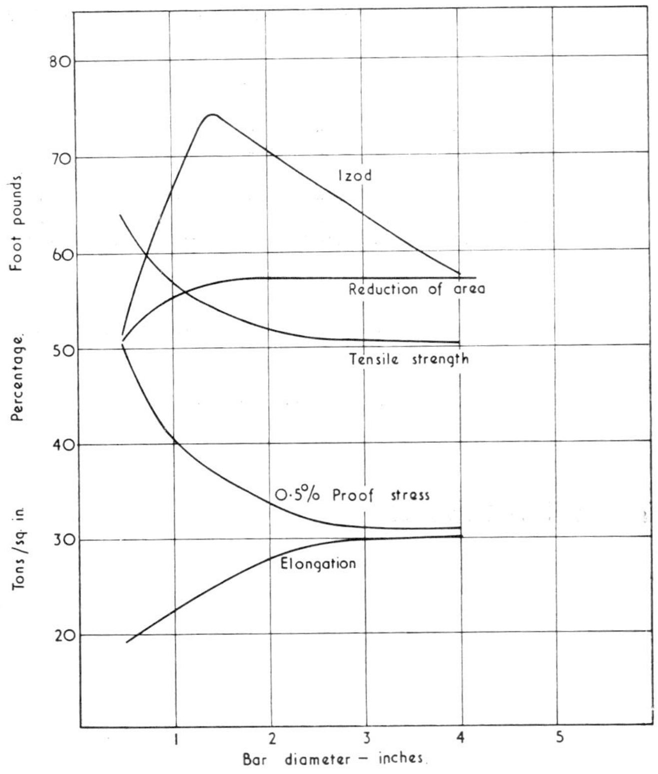

Fig. 352.12

Blank carburized 900°C.
Oil hardened 820°C.
(Single quench quality).

(Ref. 3)

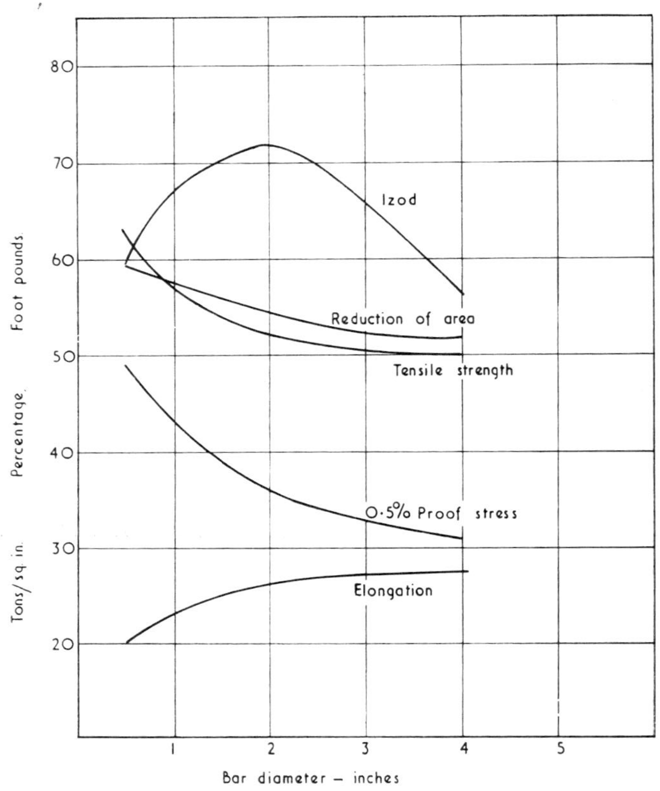

Fig. 352.13

En 352

MECHANICAL PROPERTIES AT LOW TEMPERATURE

Chemical Composition (%)								Section Size	Heat Treatment °C	Test Temperature °C	Tensile					Ref.
C	Si	Mn	S	P	Ni	Cr	Mo				0·1% P.S. (0·1Rp) tons/sq.in.	0·2% P.S. (0·2Rp) tons/sq.in.	T.S. (Rs) tons/sq.in.	El % (A)	R.A. % (Z)	
0·20	0·24	0·85	0·006	0·013	1·10	0·91	0·14	T.P.S.	BC930 OQ880 OQ780	+ 20	49·6	57·6	80·7	1½	4½	11
"	"	"	"	"	"	"	"		" " "	—196	65·6	78·8	105·2	1½	2	"
"	"	"	"	"	"	"	"		Carb.* 930 OQ880 OQ780	+ 20	44·0	56·0	69·4	½	1	"
"	"	"	"	"	"	"	"									
"	"	"	"	"	"	"	"		" " "	—196	60·8	70·0	91·2	½	2½	"
"	"	"	"	"	"	"	"		" OQ780	+ 20	49·2	57·2	72·1	1	2½	"
"	"	"	"	"	"	"	"		" "	—196	63·6	76·0	97·5	1	2	"

* Test pieces carburized to case depth of 0·045 in., finish ground to case depth of 0·02 in.

IMPACT TRANSITION TEMPERATURE DATA

Curve No.	Chemical Composition (%)								Heat Treatment (1) °C	Charpy Mesnager 50% Mean Energy Impact Transition Temperature °C	Ref.
	C	Si	Mn	S	P	Ni	Cr	Mo			
A	0·20	0·24	0·85	0·006	0·013	1·10	0·91	0·14	BC930 OQ880OQ790	—189	11
B	"	"	"	"	"	"	"	"	Carb. 930 OQ880 OQ780	+107	"
C	"	"	"	"	"	"	"	"	Carb. 930 OQ780	+107	"

(1) Test piece machined to size before carburizing or heat treatment. Notch 2 mm deep, root radius 2·0 mm, case depth 0·02 in.

(Ref. 11)

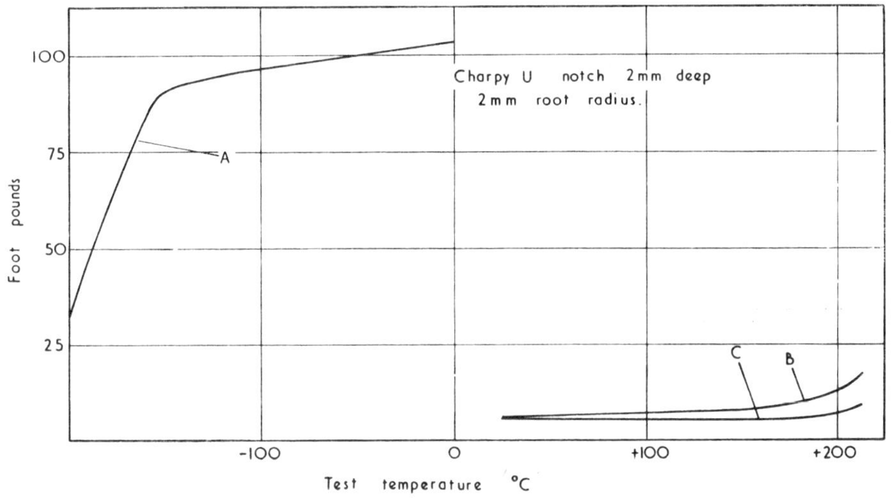

Fig. 352.14

LIST OF REFERENCES

1. Private correspondence with Firth Brown Ltd.
2. Private correspondence with The United Steel Companies Ltd.
3. Private correspondence with The English Steel Corporation Ltd.
4. Atlas of Isothermal Transformation Diagrams of the British Standard En Steels. Iron and Steel Institute Special Report No. 56, 1956.
5. Private correspondence with Garringtons Ltd.
6. Private correspondence with International Nickel Ltd.
7. Private correspondence with Brown Bayley Steels Ltd.
8. Private correspondence with Jessop–Saville Ltd.
9. Private correspondence with Kayser Ellison and Co. Ltd.
10. Private correspondence with The Park Gate Iron and Steel Co. Ltd.
11. War Office unpublished report.

$1\frac{1}{4}$ PER CENT NICKEL–CHROMIUM CASE-HARDENING STEEL

SPECIFICATION

CHEMICAL COMPOSITION (%)

C	Si	Mn	S	P	Ni	Cr	Mo
0·20 max.	0·35 max.	0·50—1·00	0·050 max.	0·050 max.	1·00—1·50	0·75—1·25	0·08—0·15

MECHANICAL PROPERTIES

Tensile strength, tons/sq.in. min. (Rm)	65
Elongation % min.	12
Izod impact value, ft. lb., min.	20

The heat treatment given to $1\frac{1}{8}$ in. dia. test bars shall be blank carburize at a temperature between 880°C and 930°C. Refine at a temperature of 850°C to 880°C, cool in air or oil.[(1), (2)] Harden in oil from a temperature of 780°C to 820°C.

(1) By special arrangement with the purchaser the test may be carried out on samples from which the blank carburizing treatment has been omitted or modified to a short heating period.
(2) A single quench from 810°C to 830°C may be used in certain cases when the refining treatment after carburizing is omitted by special arrangement between the purchaser and the manufacturer.

RELATED SPECIFICATIONS

United Kingdom	France	Sweden	USSR	
En No.	AFNOR	SIS	GOST	Mark
352	16NCD	14.2511	4543—61	20KhN

APPLICATIONS

A low-alloy case-hardening steel, developed as an alternative to En 36C, which may be carburized and hardened and tempered to produce a hard wear-resistant case. It develops a core strength of the order of 65 tons/sq. in. in relatively small section sizes with moderate resistance to shock. Typical applications are heavy duty gears, shafts, pinions, rocker arms, camshafts, gudgeon pins, clutch plates and weights, push rod leads, levers, bushes and small arms components.

WELDING

(See Introduction for key to symbols used in this table.)

	Welding Process					
	MA	G	IG	S	F	B
En 353	cx	ax	g	e	a	h

Reproduced by kind permission of H.M.S.O.

MACHINABILITY

In the softened condition the machinability of this steel is approximately 75% of that for mild steel (En 3).

HOT WORKING AND HEAT TREATMENT TEMPERATURES

Forging, rolling and stamping	1150°C, finish above 900°C.
Softening	630°C—670°C, air or furnace cool.
Carburizing	880°C—930°C.
Refining*	850°C—880°C, cool in air or oil.
Hardening*	780°C—820°C, oil quench.
Tempering (for stress relief)	150°C—200°C.
(for maximum toughness of case)	200°C—250°C.

* A single oil quench from 810°C to 830°C may be adopted when steel is specifically requested for this treatment.

CARBURIZING DATA

Carburizing Temperature °C	Depth of case produced in stated time (solid pack carburizing)			Ref.
	4 hrs.	8 hrs.	16 hrs.	
880	0·030 in.	0·041 in.		1
900	0·037 in.	0·050 in.		"
"	0·027 in.	0·042 in.	0·059 in.	2
925	0·042 in.	0·062 in.		1

PHYSICAL PROPERTIES

SPECIFIC GRAVITY (d)

Chemical Composition (%)								Heat Treatment °C	Specific Gravity at 20°C	Ref.
C	Si	Mn	S	P	Ni	Cr	Mo			
0·17	0·23	0·85	0·015	0·011	1·25	1·02	0·15	BC910 FC, OQ810	7·87	3
0·18	0·25	0·70			1·30	1·00	0·10	Softened	7·85	2
Within specification								Softened	7·87	4

SPECIFIC HEAT (c)

Chemical Composition (%)						Heat Treatment °C	Specific Heat cals/g.deg.C	Ref.
C	Si	Mn	Ni	Cr	Mo		20 to 100°C	
0·18	0·25	0·70	1·30	1·00	0·10	Softened	0·118	2

MEAN COEFFICIENT OF THERMAL EXPANSION (α)

Chemical Composition (%)						Heat Treatment °C	$10^6 \times$ Mean Coefficient of Thermal Expansion per deg.C						Ref.
C	Si	Mn	Ni	Cr	Mo		20 to 100°C	20 to 200°C	20 to 250°C	20 to 300°C	20 to 500°C	20 to 700°C	
0·17	0·23	0·85	1·25	1·02	0·15	BC910 FC OQ810	11·3	12·4	13·0				3
0·18	0·25	0·70	1·30	1·00	0·10	Softened	12			13	14	14·5	2
Within specification						Softened	12·2			13·2	13·8	14·4	4

ELECTRICAL RESISTIVITY (ρ)

Chemical Composition (%)						Heat Treatment °C	Electrical Resistivity microhm-cm.			Ref.
C	Si	Mn	Ni	Cr	Mo		20°C	100°C	200°C	
0·17	0·23	0·85	1·25	1·02	0·15	BC910 FC OQ810	31·8	36·6	43·2	3
0·18	0·25	0·70	1·30	1·00	0·10	Softened	26			2

YOUNG'S MODULUS (E)

Chemical Composition (%)						Heat Treatment °C	Young's Modulus* tons/sq.in.			Ref.
C	Si	Mn	Ni	Cr	Mo		20°C	100°C	200°C	
0·17	0·23	0·85	1·25	1·02	0·15	BC910 FC OQ810	13,200	13,000	12,600	3

*Vibration method.

TRANSFORMATION CHARACTERISTICS

ISOTHERMAL TRANSFORMATION

CORE CHARACTERISTICS

Chemical Composition (%)							
C	Si	Mn	S	P	Ni	Cr	Mn
0·18	0·26	0·93	0·008	0·016	1·34	1·11	0·11

High Frequency Melt.
McQuaid Ehn Grain Size 7(ASTM).
As quenched Grain Size finer than 8 (ASTM).
Bars hot rolled to 5 SWG, annealed, cold drawn and rolled to 0·030 in. thick tape, followed by 900°C (2 hrs.) furnace cool.
Austenitizing Temperature 870°C, 30 min. (Ref. 5)

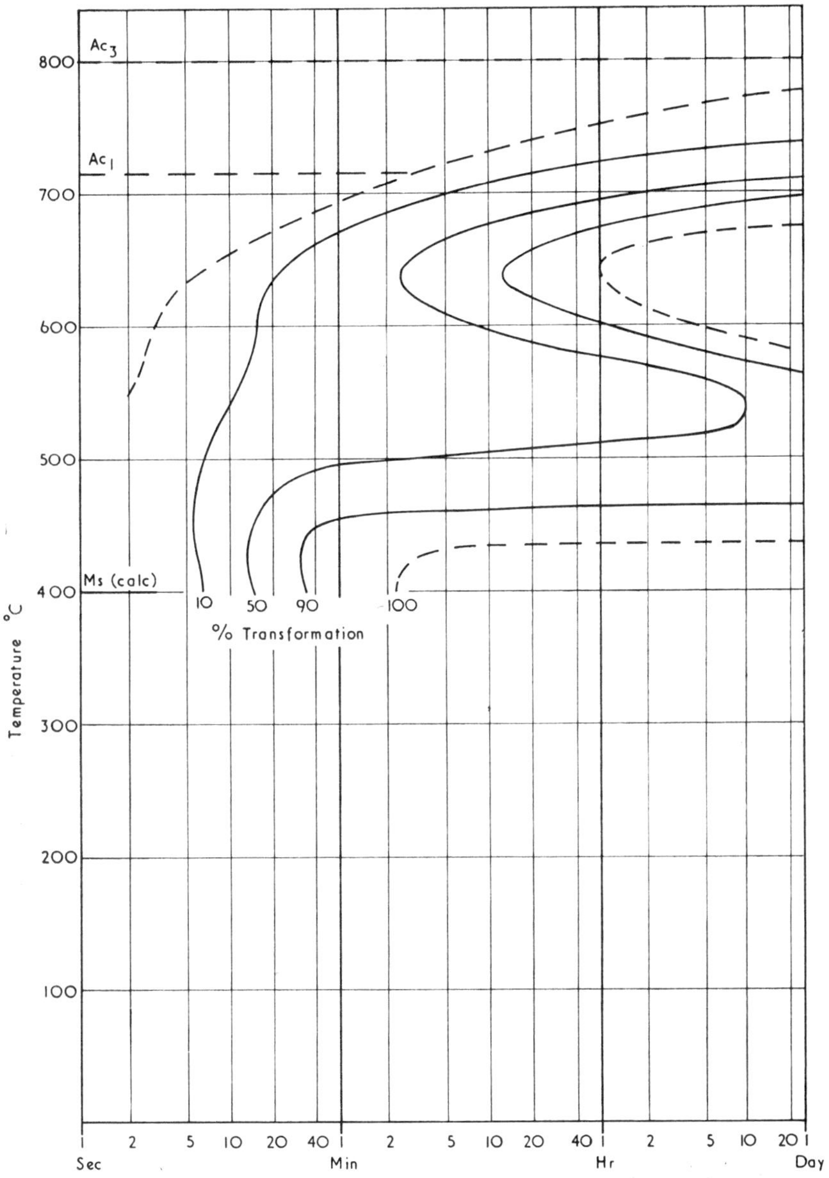

Fig. 353.1

ISOTHERMAL TRANSFORMATION (Continued)

CORE CHARACTERISTICS

Chemical Composition (%)							
C	Si	Mn	S	P	Ni	Cr	Mo
0·18	0·26	0·93	0·008	0·016	1·34	1·11	0·11

High Frequency Melt.
McQuaid Ehn Grain Size 7 (ASTM).
As quenched Grain Size finer than 8 (ASTM).
Bars hot rolled to 5 SWG, annealed, cold drawn and rolled to 0·030 in. thick tape, followed by 880°C (½ hr.) air cool.
Austenitizing Temperature 800°C, 30 min.

(Ref. 5)

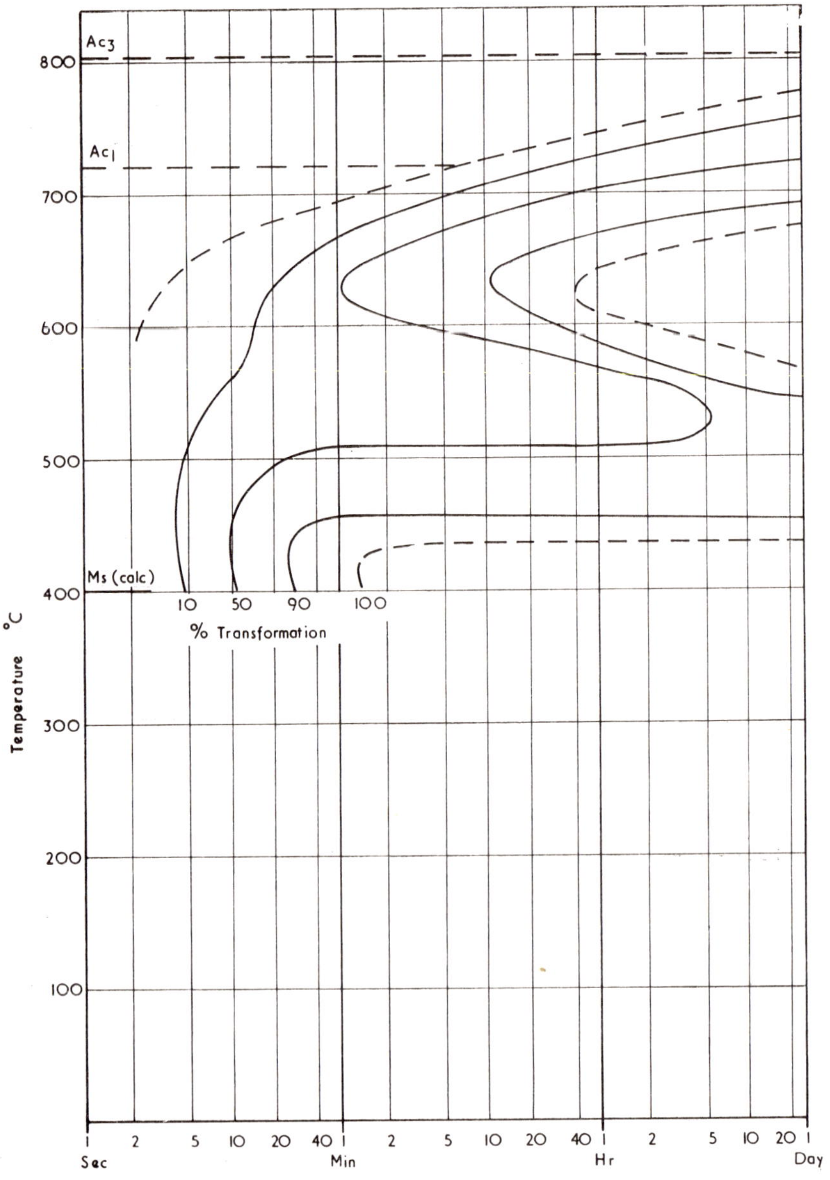

Fig. 353.2

ISOTHERMAL TRANSFORMATION (Continued)

CASE CHARACTERISTICS

Chemical Composition (%)							
C	Si	Mn	S	P	Ni	Cr	Mo
1·00	0·28	0·99	0·012	0·023	1·42	1·12	0·11

High Frequency Melt.
McQuaid Ehn Grain Size 6—8 (ASTM).
As quenched Grain Size finer than 8 (ASTM).
Bars hot rolled to ⅜ in. dia. followed by 775°C (2hrs.) furnace cool, then 900°C (2hrs.) furnace cool.
Austenitizing Temperature 880°C, 30 min. (Ref. 5)

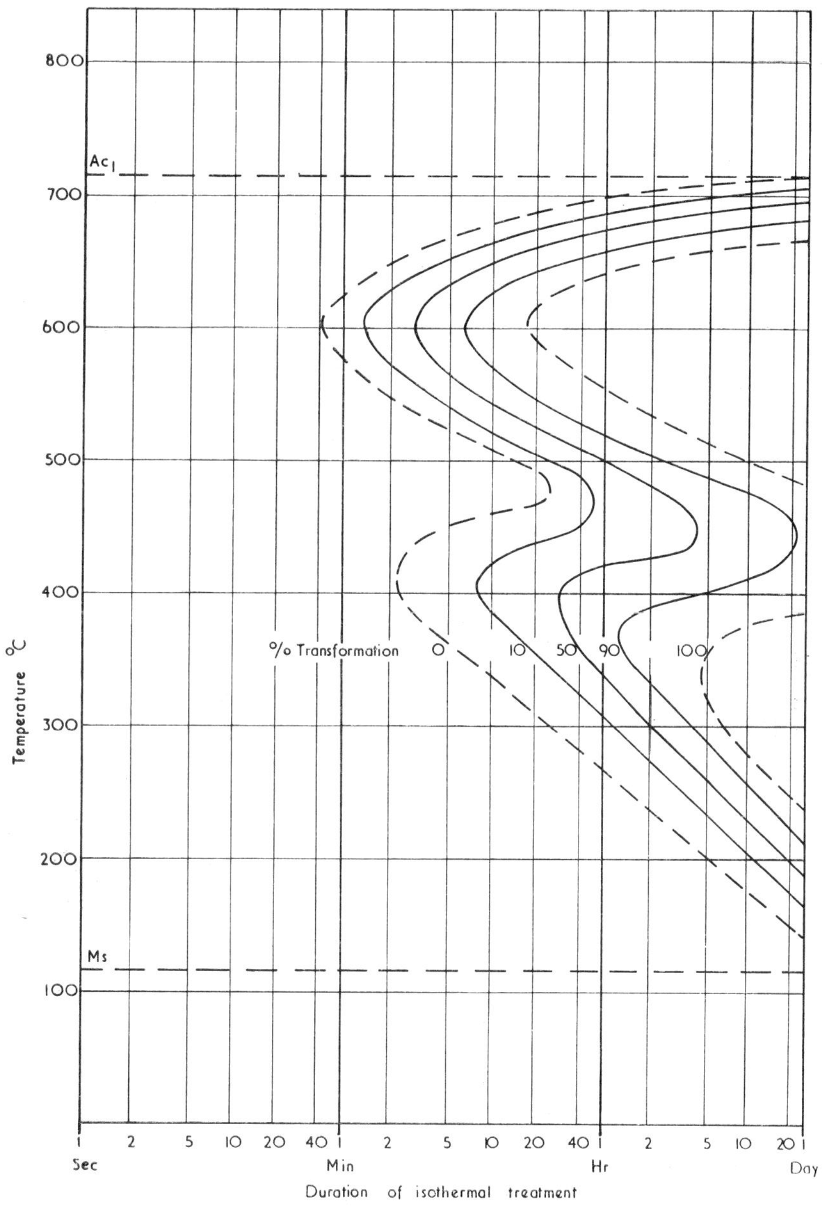

Fig. 353.3

ISOTHERMAL TRANSFORMATION (Continued)

CASE CHARACTERISTICS

Chemical Composition (%)							
C	Si	Mn	S	P	Ni	Cr	Mo
1·0	0·28	0·99	0·012	0·023	1·42	1·12	0·11

High Frequency Melt.
McQuaid Ehn Grain Size 6—8 (ASTM).
As quenched Grain Size finer than 8 (ASTM).
Bars hot rolled to ⅜ in. dia, followed by 775°C (2hrs.) furnace cool, then 880°C (½ hr.) air cool.
Austenitizing Temperature 800°C, 30 min. (Ref. 5)

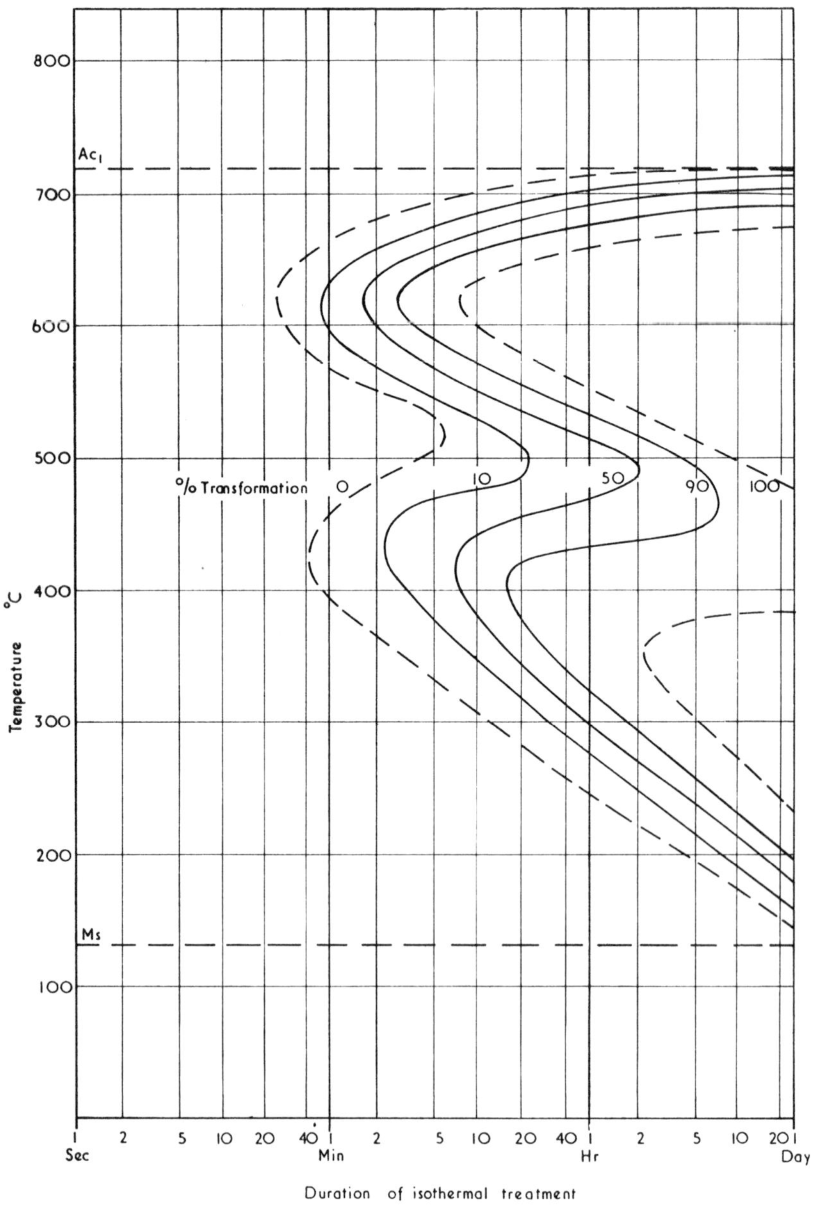

Fig. 353.4

ISOTHERMAL TRANSFORMATION (Continued)

CORE CHARACTERISTICS

Chemical Composition (%)							
C	Si	Mn	S	P	Ni	Cr	Mo
0·15	0·30	0·55	0·001	0·013	1·30	0·82	0·09

Basic Electric Arc.
McQuaid Ehn Grain Size 4 (ASTM).
Actual Grain Size 8—11 (ASTM).
Austenitizing Temperature 900°C, 30 min. (Ref. 6)

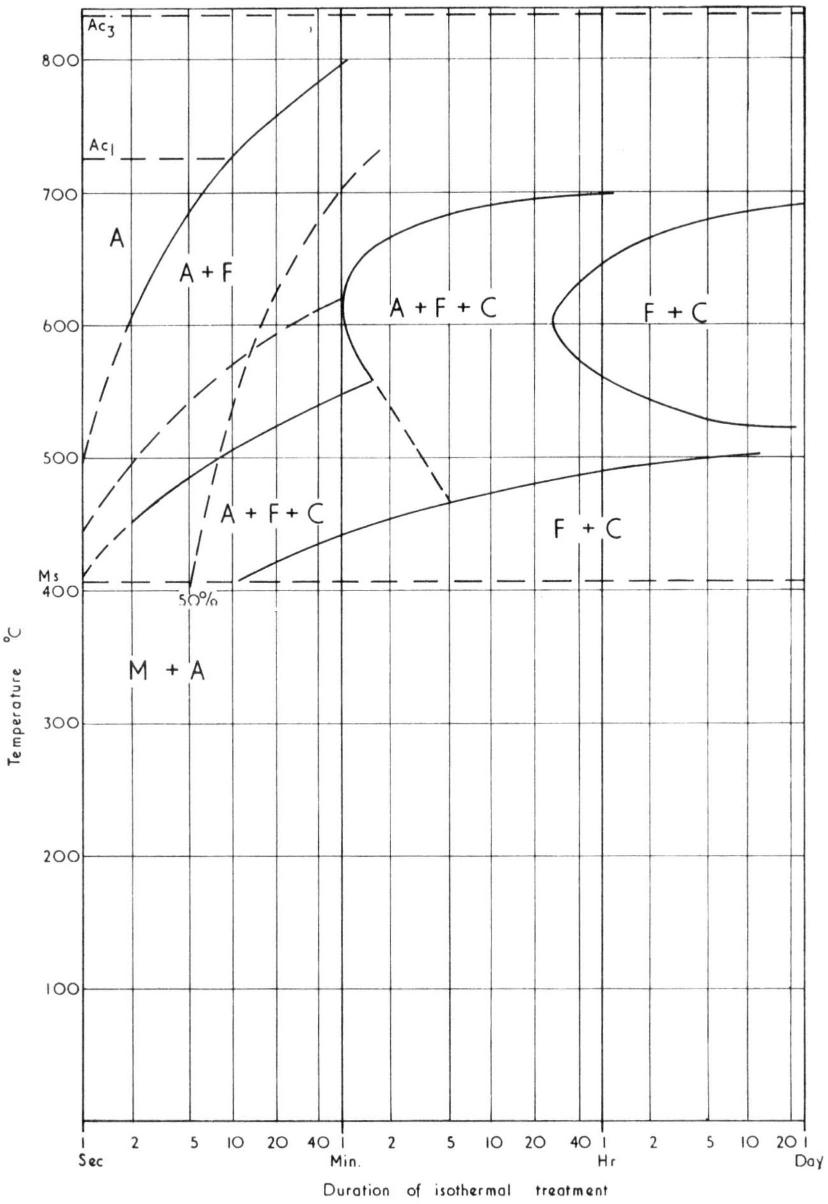

Fig. 353.5

En 353

CONTINUOUS COOLING TRANSFORMATION

Air cooled from 810°C.

(Ref. 3)

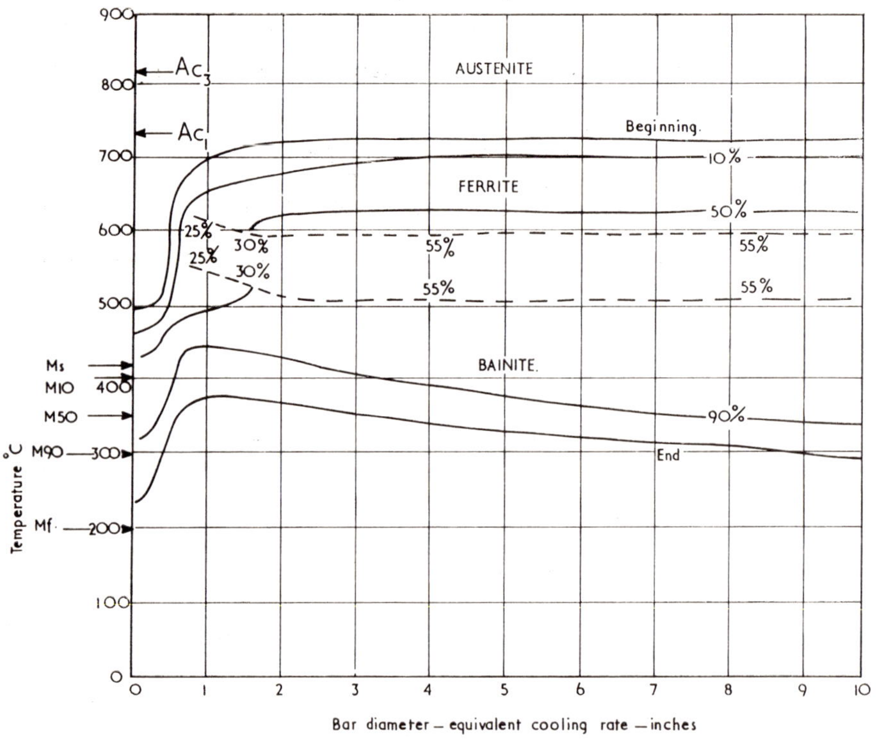

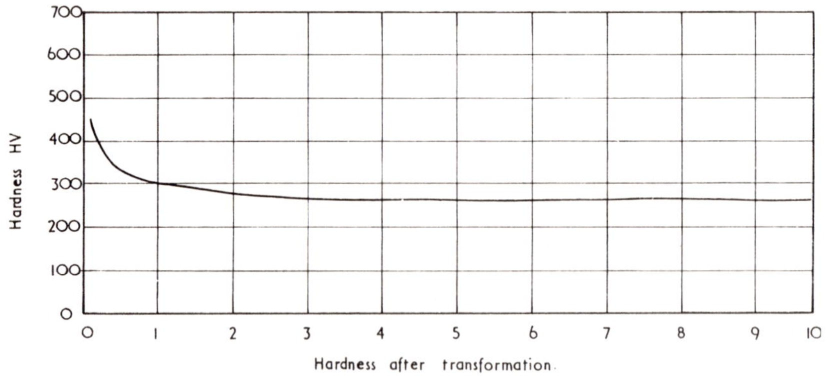

Fig. 353.6

CONTINUOUS COOLING TRANSFORMATION (Continued)

Oil quenched from 810°C.

(Ref. 3)

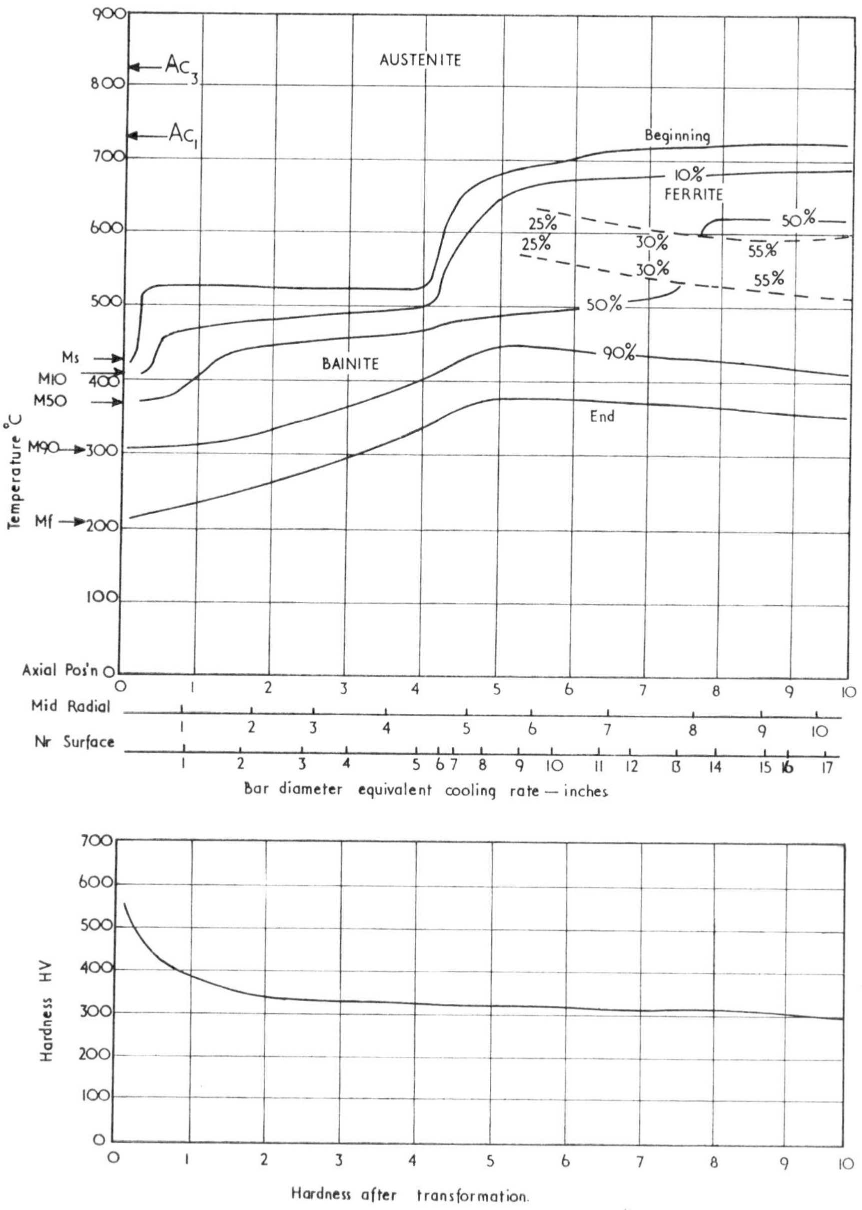

Fig. 353.7

En 353

CONTINUOUS COOLING TRANSFORMATION (Continued)

Water quenched from 810°C.

(Ref. 3)

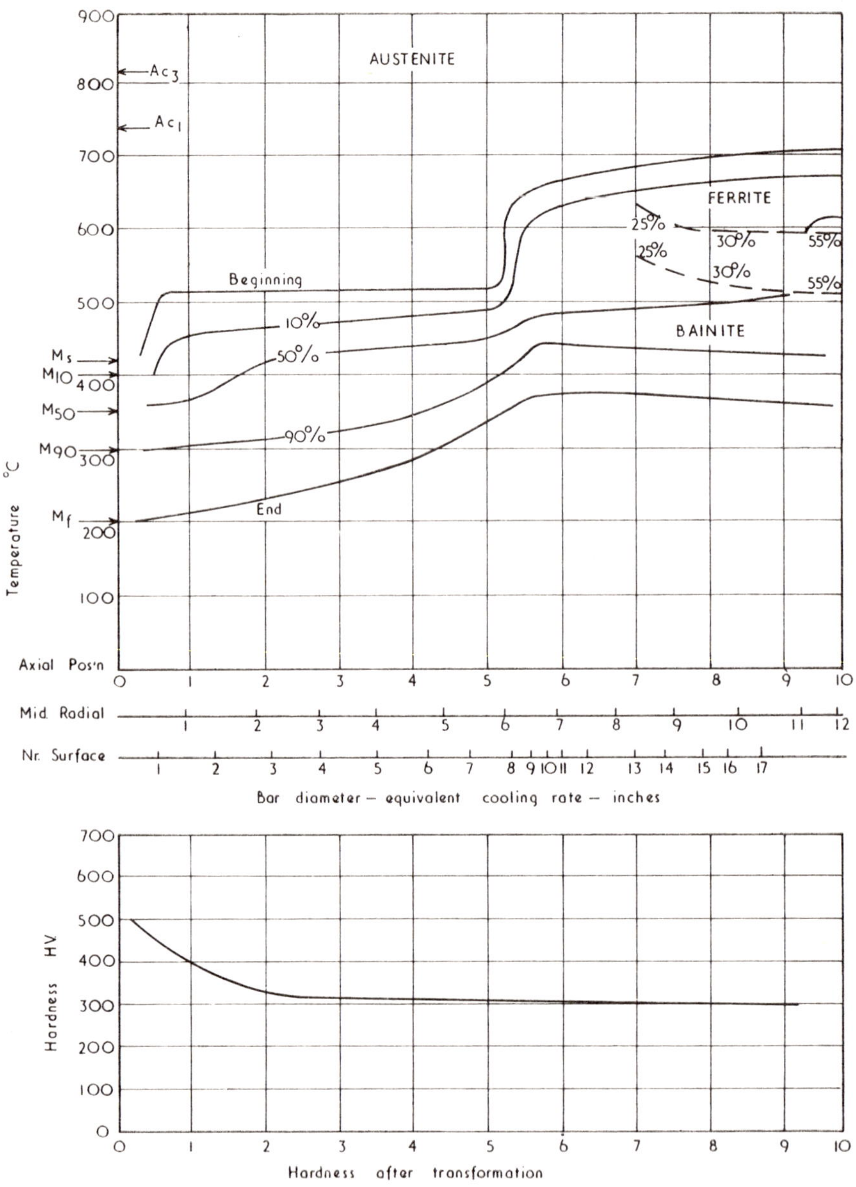

Fig. 353.8

HARDENABILITY

END QUENCH HARDENABILITY DATA

Chemical Composition (%)							
C	Si	Mn	S	P	Ni	Cr	Mo
0·18	0·26	0·93	0·008	0·016	1·34	1·11	0·11

High Frequency Melt.
McQuaid Ehn Grain Size 7(ASTM).
As quenched Grain Size finer than 8 (ASTM).
Curve A: End quenched from 870°C.
Curve B: End quenched from 800°C.
(Ref. 5)

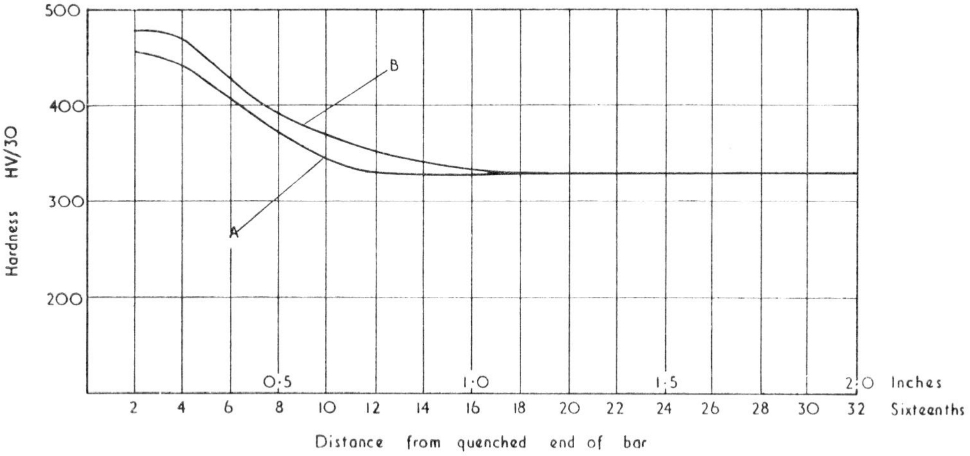

Fig. 353.9

Chemical Composition (%)							
C	Si	Mn	S	P	Ni	Cr	Mo
0·15	0·30	0·55	0·001	0·013	1·30	0·82	0·09

Basic Electric Arc.
McQuaid Ehn Grain Size 4 (ASTM).
Actual Grain Size 8—11 (ASTM).
End quenched from 900°C.
(Ref. 6)

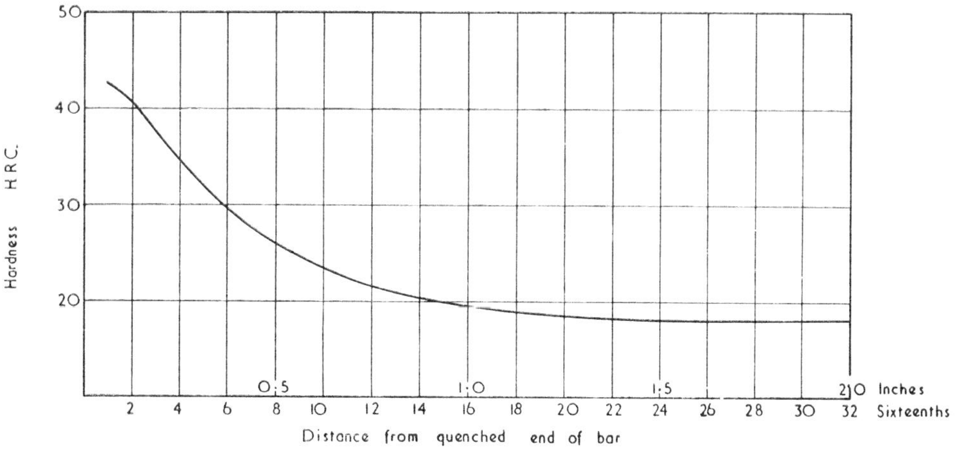

Fig. 353.10

En 353

END QUENCH HARDENABILITY BAND

These curves have been constructed from results obtained on several casts of En 353.

(Refs. 3 and 4)

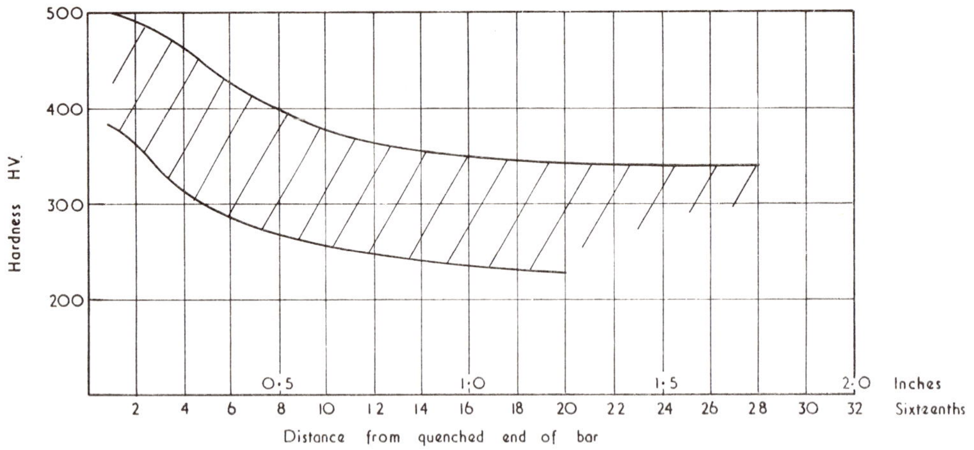

Fig. 353.11

MECHANICAL PROPERTIES AT ROOM TEMPERATURE

Process of Manufacture	ASTM Grain Size	Chemical Composition (%) C	Si	Mn	S	P	Ni	Cr	Mo	Section Size dia. (in.)	Heat Treatment °C	Tensile 0·1% P.S. (0·1Rp) tons/sq.in.	0·2% P.S. (0·2Rp)	0·5% P.S. (0·5Rp)	Y.P. (Rm)	T.S. (Rs)	El % (A)	R.A. % (Z)	Izod ft.lb.	Ref.
B.O.H.	7	0·15	0·22	0·73	0·037	0·036	1·23	0·85	0·14	1⅛	BC OQ800				47·5	60·2	18½	44½	17	3
"	"	"	"	"	"	"	"	"	"	"	" OQ820				49·6	63·2	20	47	25	"
"	"	"	"	"	"	"	"	"	"	"	" OQ840				54·0	61·5	20	57	39	"
"	"	"	"	"	"	"	"	"	"	"	" OQ860				54·1	62·0	20	55	39	"
"	"	"	"	"	"	"	"	"	"	"	" OQ860 OQ800				45·6	57·7	22	49½	33	"
"	"	"	"	"	"	"	"	"	"	"	" " OQ840				55·1	61·6	20	55	46	"
"	"	"	"	"	"	"	"	"	"	"	OQ860 OQ800				44·0	58·8	22	49½	37	"
"	"	"	"	"	"	"	"	"	"	"	" OQ840				52·8	63·7	19	55	35	"
B.E.		0·15	0·25	0·58	0·006	0·012	1·36	0·84	0·21	1⅛	OQ810				50·0	62·8	20½	52½	30	2
B.E.		0·16	0·19	0·65	0·018	0·015	1·27	0·77	0·04	1⅛	OQ810				42·4	56·2	21	44½	32	2
B.E.		0·16	0·22	0·63	0·010	0·015	1·25	0·88	0·13	1⅛	OQ810					66·8	20½	45	30	"
A.O.H.		0·16	0·27	0·76	0·024	0·026	1·36	0·97	0·14	1⅛	BC925 FC OQ870 OQ800			42·8		59·3	$4\sqrt{A}$ 25, 5d 20	58	—	"
		0·16	0·24	0·90			1·23	1·02	0·08	1⅛	BC900 OQ860 OQ800					66·0	18½	36½	46	4
		0·16	0·31	0·91			1·41	0·95	0·09	1⅛	BC900 OQ860 OQ800					67·0	16	39	49	"
		0·16	0·22	0·88			1·35	0·98	0·10	1⅛	BC900 OQ860 OQ800					68·6	15	35	34	"
		0·16	0·25	0·95			1·36	0·95	0·10	1⅛	BC900 OQ860 OQ800					71·6	12½	33½	34	"
A.O.H.		0·16	0·30	0·84	0·044	0·036	1·25	1·04	0·17	1⅛	BC900 OQ850 OQ800					70·4	17	48	43	7
"		"	"	"	"	"	"	"	"	"	" OQ880 OQ780					64·8	20	42	46	"
B.E.		0·17	0·17	0·60	0·008	0·014	1·29	0·89	0·09	1⅛	OQ810				48·0	61·6	20	53½	32	2
B.O.H.		0·17	0·25	0·89	0·022	0·032	1·19	0·82	0·13	⅜	BC OQ870 OQ810					90·6	16	54½		8
"		"	"	"	"	"	"	"	"	¾	" " "				55·2	69·4	17	59½	31	"
"		"	"	"	"	"	"	"	"	1⅛	" " "				43·2	64·5	21	52½	34	"
"		"	"	"	"	"	"	"	"	1¾	" " "				35·2	55·6	22	59	49	"
"		"	"	"	"	"	"	"	"	2½	" " "				35·6	54·0	24	58	49	"
		0·17	0·31	0·90			1·42	0·90	0·08	1⅛	BC900 OQ860 OQ800					66·8	19½	33½	57	4
		0·17	0·22	0·87			1·35	0·89	0·10	1⅛	BC900 OQ860 OQ800					68·0	14	28	36	"
		0·17	0·20	0·90			1·29	1·06	0·15	1⅛	BC900 OQ860 OQ800					69·6	15	43½	53	"
		0·17	0·26	0·89			1·35	0·92	0·08	1⅛	BC900 OQ860 OQ800					70·5	12½	36½	36	"
B.E.		0·17	0·27	0·79	0·010	0·020	1·46	1·01	0·14	1⅛	BC900 OQ850 OQ800					71·2	17	42	33	7
"		"	"	"	"	"	"	"	"	"	" OQ880 OQ780					71·2	15	33½	36	"
"		"	"	"	"	"	"	"	"	"	" OQ780					70·0	13	25	28	"

MECHANICAL PROPERTIES AT ROOM TEMPERATURE (Continued)

Process of Manufacture	ASTM Grain Size	Chemical Composition (%)								Section Size dia. (in.)	Heat Treatment °C	Tensile							Izod ft.lb.	Ref.
		C	Si	Mn	S	P	Ni	Cr	Mo			0·1% P.S. (0·1Rp)	0·2% P.S. (0·2Rp)	0·5% P.S. (0·5Rp)	Y.P. (Rm)	T.S. (Rs)	El % (A)	R.A. % (Z)		
												tons/sq.in.								
	8	0·17	0·35	0·85	0·019	0·020	1·24	0·98	0·13	1¼	BC900 OQ860 OQ800					73·0	14	36½	41	2
B.E.		0·17	0·21	0·62	0·009	0·018	1·24	0·90	0·10	1⅛	OQ870 OQ810				52·0	65·2	20	48½	25	9
		0·17		0·80			1·21	0·78	0·09	1⅛	" OQ820				55·4	67·9	18½		23	"
		"		"			"	"	"	"	OQ850 OQ800				59·0	70·6	17½		30	"
A.O.H.		0·18	0·31	0·74	0·036	0·031	1·20	0·94	0·14	1⅛	OQ810					73·4	17½	48	22	2
A.O.H.		0·18	0·30	0·76	0·034	0·035	1·26	0·94	0·17	1⅛	OQ810				50·6	65·2	21	53	29	"
H.F.		0·18	0·26	0·93	0·008	0·016	1·34	1·11	0·11	½	OQ870 OQ810			65·3		83·1	13½	45	36	"
"		"	"	"	"	"	"	"	"	⅞	" "			67·3		85·5	16½	49½	25	"
"		"	"	"	"	"	"	"	"	1⅛	" "			56·0		73·8	16	44½	35	"
"		"	"	"	"	"	"	"	"	2¼	" "			47·0		66·8	16	35½	36	"
B.E.		0·18	0·17	0·67	0·012	0·013	1·32	0·91	0·12	1⅛	OQ370 OQ810					70·2	18½	43	25	"
"		"	"	"	"	"	"	"	"	0·236	BC AC870 OQ810	61·5				100·0	15	57		1
"		"	"	"	"	"	"	"	"	0·367	" " "	58·4				98·4	15	50		"
"		"	"	"	"	"	"	"	"	0·574	" " "	53·1				92·8	13	45	19	"
"		"	"	"	"	"	"	"	"	⅞	" " "	45·0				85·5	16	45		"
"		"	"	"	"	"	"	"	"	1¼	" " "	36·1				71·0	18	45	29	"
B.E.		0·18	0·21	0·67	0·006	0·017	1·26	0·88	0·14	0·159	BC AC870 OQ810	59·8				90·0	16½	62		"
"		"	"	"	"	"	"	"	"	0·236	" " "	61·0				92·5	15	62		"
"		"	"	"	"	"	"	"	"	0·367	" " "	58·8				92·0	18½	57		"
"		"	"	"	"	"	"	"	"	0·574	" " "	54·4				86·8	15	57	19	"
"		"	"	"	"	"	"	"	"	⅞	" " "	43·8				74·4	18	57		"
"		"	"	"	"	"	"	"	"	1¼	" " "	35·2				62·4	21	57	45	"
"		"	"	"	"	"	"	"	"	1⅛	OQ870 OQ800					66·6	19	45	30	4
B.E.	7	0·18	0·21	0·87	0·020	0·020	1·07	0·85	0·09	1⅛	AC900 OQ740	23·9	28·0	34·3		56·7	19	36	18	"
"	"	"	"	"	"	"	"	"	"	"	" OQ760	26·5	30·8	37·4		59·8	22	42½	26	"
"	"	"	"	"	"	"	"	"	"	"	" OQ780	31·2	36·4	44·2		58·4	22½	46	26	"
"	"	"	"	"	"	"	"	"	"	"	" OQ800	28·5	33·4	40·7		60·3	21	44½	33	"
"	"	"	"	"	"	"	"	"	"	"	" OQ820	31·0	35·9	42·9		60·2	22	50	38	"
"	"	"	"	"	"	"	"	"	"	"	" OQ840	30·2	35·3	42·2		59·8	21½	47½	40	"

MECHANICAL PROPERTIES AT ROOM TEMPERATURE (Continued)

Process of Manufacture	ASTM Grain Size	Chemical Composition (%)								Section Size dia. (in.)	Heat Treatment °C	Tensile							Izod ft.lb.	Ref.
		C	Si	Mn	S	P	Ni	Cr	Mo			0·1% P.S. (0·1Rp)	0·2% P.S. (0·2Rp)	0·5% P.S. (0·5Rp)	Y.P. (Rm)	T.S. (Rs)	El % (A)	R.A. % (Z)		
												tons/sq.in.								
B.E.	7	0·18	0·21	0·87	0·020	0·020	1·07	0·85	0·09	0·424	AC900	21·7	25·6	30·3		51·8	22	50½	34	4
,,	,,	,,	,,	,,	,,	,,	,,	,,	,,	¾	,,	21·6	25·0	31·5		49·7	23	47½	36	,,
,,	,,	,,	,,	,,	,,	,,	,,	,,	,,	1	,,	20·2	23·2	28·2		47·1	24	49½	38	,,
,,	,,	,,	,,	,,	,,	,,	,,	,,	,,	1⅛	,,	20·2	24·1	29·1		47·2	24	48	47	,,
,,	,,	,,	,,	,,	,,	,,	,,	,,	,,	2⅛	,,	19·8	22·3	25·8		44·4	23	47	70	,,
,,	,,	,,	,,	,,	,,	,,	,,	,,	,,	3	,,	21·0	22·9	27·2		40·4	26	60	76	,,
,,	,,	,,	,,	,,	,,	,,	,,	,,	,,	4¼ (O)	,,			24·9		38·4	35	71½	105	,,
,,	,,	,,	,,	,,	,,	,,	,,	,,	,,	4¼ (C)	,,			22·8		41·9	27	54½	75	,,
,,	,,	,,	,,	,,	,,	,,	,,	,,	,,	0·424	AC900 OQ820	40·1	46·2	52·9		70·2	17½	55	34	,,
,,	,,	,,	,,	,,	,,	,,	,,	,,	,,	¾	,, ,,	37·6	42·0	48·6		63·5	22	54	40	,,
,,	,,	,,	,,	,,	,,	,,	,,	,,	,,	1	,, ,,	33·2	38·6	46·6		61·4	21	53	49	,,
,,	,,	,,	,,	,,	,,	,,	,,	,,	,,	1⅛	,, ,,	38·2	40·5	45·2		59·0	20	62½	53	,,
,,	,,	,,	,,	,,	,,	,,	,,	,,	,,	2⅛	,, ,,	26·2	29·6	35·2		54·3	25	52	53	,,
,,	,,	,,	,,	,,	,,	,,	,,	,,	,,	3	,, ,,	25·3	29·2	33·5		52·2	25	54½	55	,,
,,	,,	,,	,,	,,	,,	,,	,,	,,	,,	4¼ (O)	,, ,,	23·6	26·9	31·4		50·4	27½	55	52	,,
,,	,,	,,	,,	,,	,,	,,	,,	,,	,,	4¼ (C)	,, ,,	24·8	29·6	36·4		51·0	27	51	41	,,
,,	,,	,,	,,	,,	,,	,,	,,	,,	,,	0·424	AC900 OQ860 OQ800	36·1	42·5	49·4		70·2	20	52	38	,,
,,	,,	,,	,,	,,	,,	,,	,,	,,	,,	¾	,, ,, ,,	38·7	45·3	55·1		69·5	19	51½	33	,,
,,	,,	,,	,,	,,	,,	,,	,,	,,	,,	1	,, ,, ,,	28·6	33·9	41·5		61·2	22	57½	45	,,
,,	,,	,,	,,	,,	,,	,,	,,	,,	,,	1⅛	,, ,, ,,	28·4	33·1	40·4		60·4	23	51½	48	,,
,,	,,	,,	,,	,,	,,	,,	,,	,,	,,	2⅛	,, ,, ,,	24·4	27·9	34·6		55·0	24	50½	55	,,
,,	,,	,,	,,	,,	,,	,,	,,	,,	,,	3	,, ,, ,,	22·4	25·4	30·9		53·1	27	54½	49	,,
,,	,,	,,	,,	,,	,,	,,	,,	,,	,,	4¼ (O)	,, ,, ,,	21·6	25·3	32·5		51·4	28½	53½	65	,,
,,	,,	,,	,,	,,	,,	,,	,,	,,	,,	4¼ (C)	,, ,, ,,	22·6	26·0	31·4		53·4	23	50	61	,,
B.O.H.	6—7	0·19	0·26	0·80	0·035	0·027	1·30	1·05	0·11	1⅛	BC910 OQ800				64·6	77·6	17	39	16	3
,,	,,	,,	,,	,,	,,	,,	,,	,,	,,	,,	,, OQ820				74·2	78·6	16½	47	22	,,
,,	,,	,,	,,	,,	,,	,,	,,	,,	,,	,,	,, OQ840				74·2	78·5	17	49½	25	,,
,,	,,	,,	,,	,,	,,	,,	,,	,,	,,	,,	,, OQ860				74·4	79·3	19	49½	21	,,
,,	,,	,,	,,	,,	,,	,,	,,	,,	,,	,,	OQ860 OQ800				73·2	81·8	16½	44½	22	,,
,,	,,	,,	,,	,,	,,	,,	,,	,,	,,	,,	,, OQ840				79·2	82·4	17	49½	28	,,
B.E.		0·19	0·24	0·80	0·009	0·012	1·10	0·87		1⅛	OQ810				50·0	69·2	20½	46	32	2

MECHANICAL PROPERTIES AT ROOM TEMPERATURE (Continued)

Process of Manufacture	ASTM Grain Size	Chemical Composition (%) C	Si	Mn	S	P	Ni	Cr	Mo	Section Size dia. (in.)	Heat Treatment °C	Tensile 0·1% P.S. (0·1Rp) tons/sq.in.	0·2% P.S. (0·2Rp)	0·5% P.S. (0·5Rp)	Y.P. (Rm)	T.S. Rs)	El % (A)	R.A. % (Z)	Izod ft.lb.	Ref.
A.O.H.		0·19	0·29	0·72	0·031	0·029	1·57	0·92	0·13	1⅛	OQ810					72·2	16½	41	32	2
		0·19	0·34	0·85			1·20	1·03	0·18	1⅛	BC900 OQ860 OQ800					72·2	15½	35	41	4
		0·19	0·26	0·88			1·38	0·95	0·13	1⅛	BC900 OQ860 OQ800					74·0	13½	35	35	"
B.E.		0·19	0·20	0·63	0·011	0·010	1·21	0·91	0·14	1⅛	OQ870 OQ810					67·4	20	50	43	2
B.E.		"	"	"	"	"	"	"	"	0·236	BC AC870 OQ860	62·4				96·5	19	63		"
"		"	"	"	"	"	"	"	"	0·367	" " "	62·4				98·5	9	57		1
"		"	"	"	"	"	"	"	"	0·574	" " "	51·2				89·6	14	50	21	"
"		"	"	"	"	"	"	"	"	⅞	" " "	40·0				74·8	18	50		"
"		"	"	"	"	"	"	"	"	1¼	" " "	36·8				67·2	21	53	37	"
"		"	"	"	"	"	"	"	"	½	OQ870 OQ810		67·7			85·5	16	54½	39	2
"		"	"	"	"	"	"	"	"	⅞	" "		55·6			70·9	19	57½	34	"
"		"	"	"	"	"	"	"	"	1⅛	" "		45·0			61·6	12½	52	43	"
"		"	"	"	"	"	"	"	"	2½	" "		41·0			56·0	13	57½	53	"
"		"	"	"	"	"	"	"	"	4	" "		38·8			54·0	24½	56½	51	"
B.E.		0·19	0·13	0·93	0·018	0·025	1·34	0·97	0·15	0·236	BC AC870 OQ800	59·7				91·0	19	62		"
"		"	"	"	"	"	"	"	"	0·367	" " "	61·2				91·5	9	57		"
"		"	"	"	"	"	"	"	"	0·574	" " "	50·8				86·8	15	50	11	"
"		"	"	"	"	"	"	"	"	⅞	" " "	42·8				76·8	15	53		"
"		"	"	"	"	"	"	"	"	1¼	" " "	38·0				68·0	18	52	25	"
B.E.		0·19	0·27	0·90	0·012	0·015	1·32	0·91	0·14	0·159	BC AC870 OQ800	65·5				100·0	16½	57		"
"		"	"	"	"	"	"	"	"	0·236	" " "	65·5				96·0	16	53		"
"		"	"	"	"	"	"	"	"	0·367	" " "	63·6				96·0	15	53		"
"		"	"	"	"	"	"	"	"	0·574	" " "	62·1				95·2	16	53	21	"
"		"	"	"	"	"	"	"	"	⅞	" " "	54·0				88·4	17	53		"
"		"	"	"	"	"	"	"	"	1¼	" " "	42·2				74·4	18	55	33	"
B.E.		0·20	0·25	0·86	0·006	0·009	1·12	0·86		1⅛	OQ810					73·0	17½	45	27	2
"		"	"	"	"	"	"	"		2½	"				42·5	60·8	21½	45	31	"
"		"	"	"	"	"	"	"		4	"				38·1	57·2	23½	51½	31	"
B.E.		0·20	0·35	0·88	0·013	0·024	1·31	0·96	0·10	1⅛	BC900 OQ850 OQ880					76·8	15	39	34	7
											" OQ880 OQ780					68·0	20	39	32	"
		0·20		0·97			1·34	1·00	0·10	0·424	OQ820					94·1	16½		29	9
		"		"			"	"	"	1⅛	OQ860 OQ800					74·2	16½		43	"
		"		"			"	"	"	"	OQ820					75·4	16½		39	"
		0·20		0·97			1·34	1·00	0·10	2⅛	OQ860 OQ800					67·8	21½		43	9
		"		"			"	"	"	"	OQ820					66·9	19½		50	"
		"		"			"	"	"	3	OQ860 OQ800					66·3	20		51	"
		"		"			"	"	"	4½	" "					63·8	23½		35	"

EFFECT OF TEMPERING ON CASE HARDNESS

(TYPICAL CURVE)

Carburized 900°C.
Oil refined 860°C.
Oil hardened 800°C.

(Ref. 4)

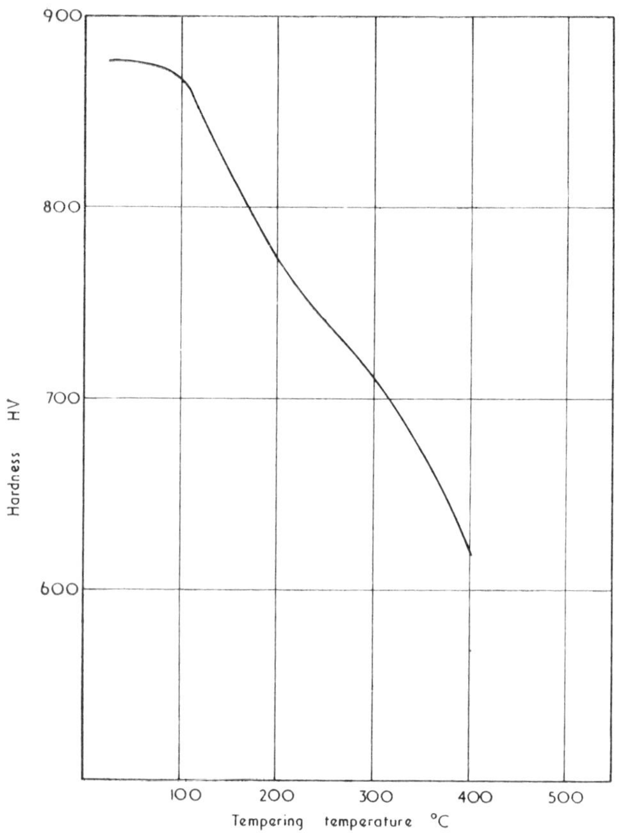

Fig. 353.12

EFFECT OF SECTION SIZE ON THE CORE MECHANICAL PROPERTIES AT ROOM TEMPERATURE

(TYPICAL CURVES)

Blank carburized 900°C.
Oil refined 860°C.
Oil hardened 800°C.

(Ref. 4)

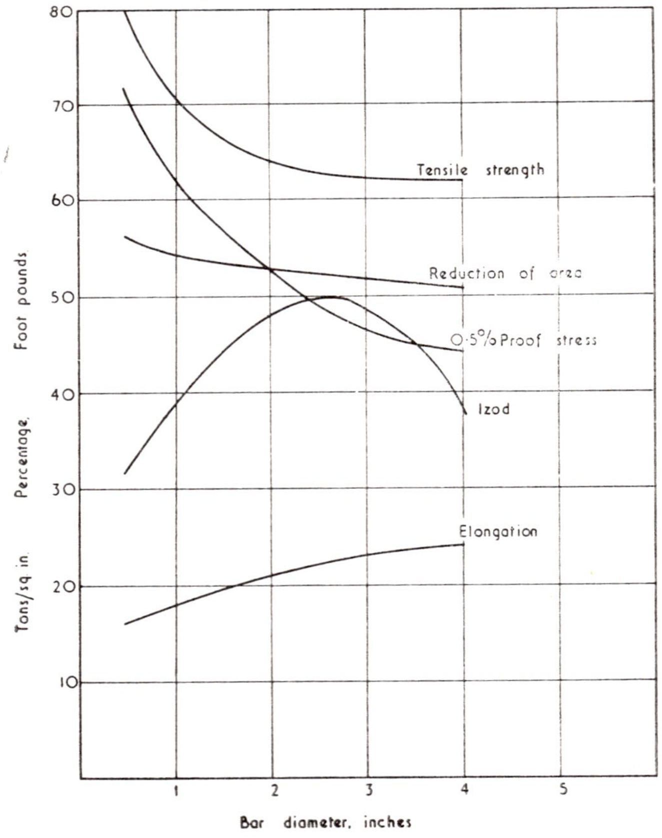

Fig. 353.13

Blank carburized 900°C.
Oil hardened 820°C.
(Single quench quality)

(Ref. 4)

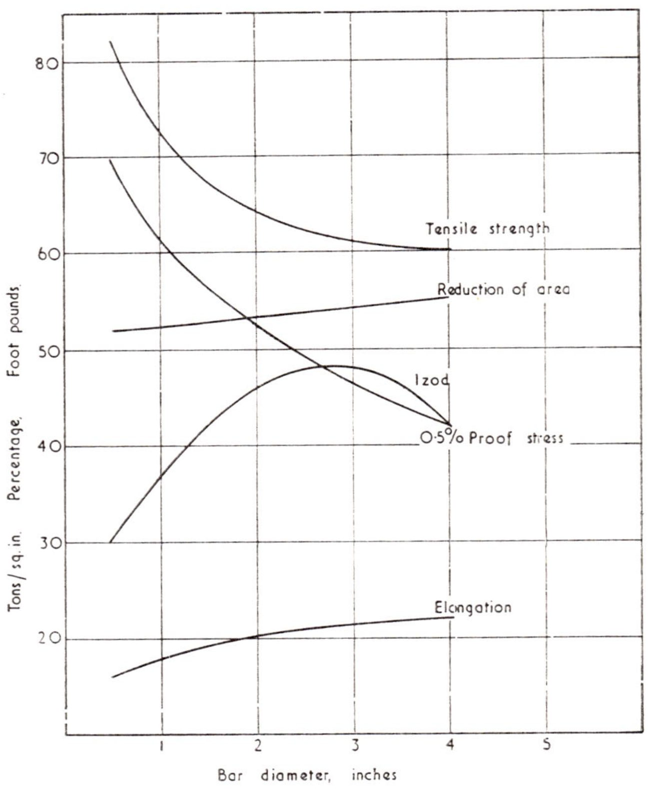

Fig. 353.14

LIST OF REFERENCES

1. Private correspondence with the Royal Aircraft Establishment, Farnborough.
2. Private correspondence with Firth Brown Ltd.
3. Private correspondence with The United Steel Companies Ltd.
4. Private correspondence with The English Steel Corporation Ltd.
5. Atlas of Isothermal Transformation Diagrams of the B. S. En Steels, Iron and Steel Institute Special Report No. 36, 1956.
6. G. Delbart and A. Constant. Courbes de Transformation des Aciers de Fabrication Française, vol. 1. L'Institute de Recherches de la Sidérurgie (IRSID).
7. Private correspondence with Jessop–Saville Ltd.
8. Private correspondence with Brown Bayley Steels Ltd.
9. Private correspondence with International Nickel Ltd.

$1\frac{3}{4}$ PER CENT NICKEL–CHROMIUM–MOLYBDENUM CASE-HARDENING STEEL

SPECIFICATION

CHEMICAL COMPOSITION (%)

C	Si	Mn	S	P	Ni	Cr	Mo
0·20 max.	0·35 max.	0·50—1·00	0·050 max.	0·050 max.	1·50—2·00	0·75—1·25	0·10—0·20

MECHANICAL PROPERTIES

Tensile strength, tons/sq.in. min. (Rm)	75
Elongation, % min. (A)	12
Izod impact, value, ft.lb., min.	20

In the softened condition the material shall have a Brinell hardness number not exceeding 277.

The heat treatment given to $1\frac{1}{8}$ in. dia. test bars shall be blank carburize at a temperature between 880°C and 930°C.[(1), (2)] Refine at a temperature of 850°C to 880°C, cool in air or oil. Harden in oil from a temperature of 780°C to 820°C.

(1) By special arrangement with the purchaser the test may be carried out on samples from which the blank carburizing treatment has been omitted or modified to a short heating period.

(2) A single oil quench from 810°C to 830°C may be used in certain cases when the refining treatment after carburizing is omitted by special arrangement between the purchaser and the manufacturer.

RELATED SPECIFICATIONS

United Kingdom En No.	France AFNOR
354	18NCD6

APPLICATIONS

A medium-alloy case-hardening steel which may be carburized and hardened to produce a hard wear-resistant case. It develops a core strength of the order of 75 tons/sq. in. with a moderate resistance to shock. Typical applications are heavy duty gears, heavy vehicle and motor car transmission components, cams and camshafts.

WELDING

(See Introduction for key to symbols used in this table.)

	Welding Process					
	MA	G	IG	S	F	B
En 354	cx	ax	g	e	a	h

Reproduced by kind permission of H.M.S.O.

MACHINABILITY

In the softened condition the machinability is approximately 70% of that for mild steel (En 3).

HOT WORKING AND HEAT TREATMENT TEMPERATURES

Forging, rolling and stamping	1150°C, finish above 900°C.
Softening (sub-critical annealing)	630°C—670°C, air or furnace cool.
Carburizing	880°C—930°C.
Refining*	850°C—880°C, air or oil cool.
Hardening*	780°C—820°C, oil quench.
Tempering (for stress relief)	150°C—200°C.
(for maximum case toughness)	200°C—260°C.

* A single oil quench from 810°C to 830°C may be adopted when steel is specifically requested for this treatment.

CARBURIZING DATA

Carburizing Temperature °C	Depth of case produced in stated time (solid pack carburizing)		Ref.
	4 hrs.	8 hrs.	
880	0·035 in.	0·043 in.	1
900	0·037 in.	0·050 in.	"
925	0·041 in.	0·054 in.	"

PHYSICAL PROPERTIES

SPECIFIC GRAVITY (d)

Chemical Composition (%)						Heat Treatment °C	Specific Gravity at 20°C	Ref.
C	Si	Mn	Ni	Cr	Mo			
0·18	0·25	0·70	1·75	1·00	0·15	Softened	7·85	2
not stated						"	7·87	3

SPECIFIC HEAT (c)

Chemical Composition (%)						Heat Treatment °C	Specific Heat cal/g. deg.C 20 to 100°C	Ref.
C	Si	Mn	Ni	Cr	Mo			
0·18	0·25	0·70	1·75	1·00	0·15	Softened	0·118	2

TRANSFORMATION CHARACTERISTICS

ISOTHERMAL TRANSFORMATION

CORE CHARACTERISTICS

Chemical Composition (%)							
C	Si	Mn	S	P	Ni	Cr	Mo
0·19	0·21	0·90	0·015	0·017	1·87	1·08	0·18

Basic Electric Arc.
McQuaid Ehn Grain Size 7—8 (ASTM).
As quenched Grain Size finer than 8 (ASTM).
Steel hammered to 1 in. dia. bars, annealed, specimen cut axially, then 920°C (2hrs.) furnace cool.
Austenitizing Temperature 870°C, 30 min. (Ref. 4)

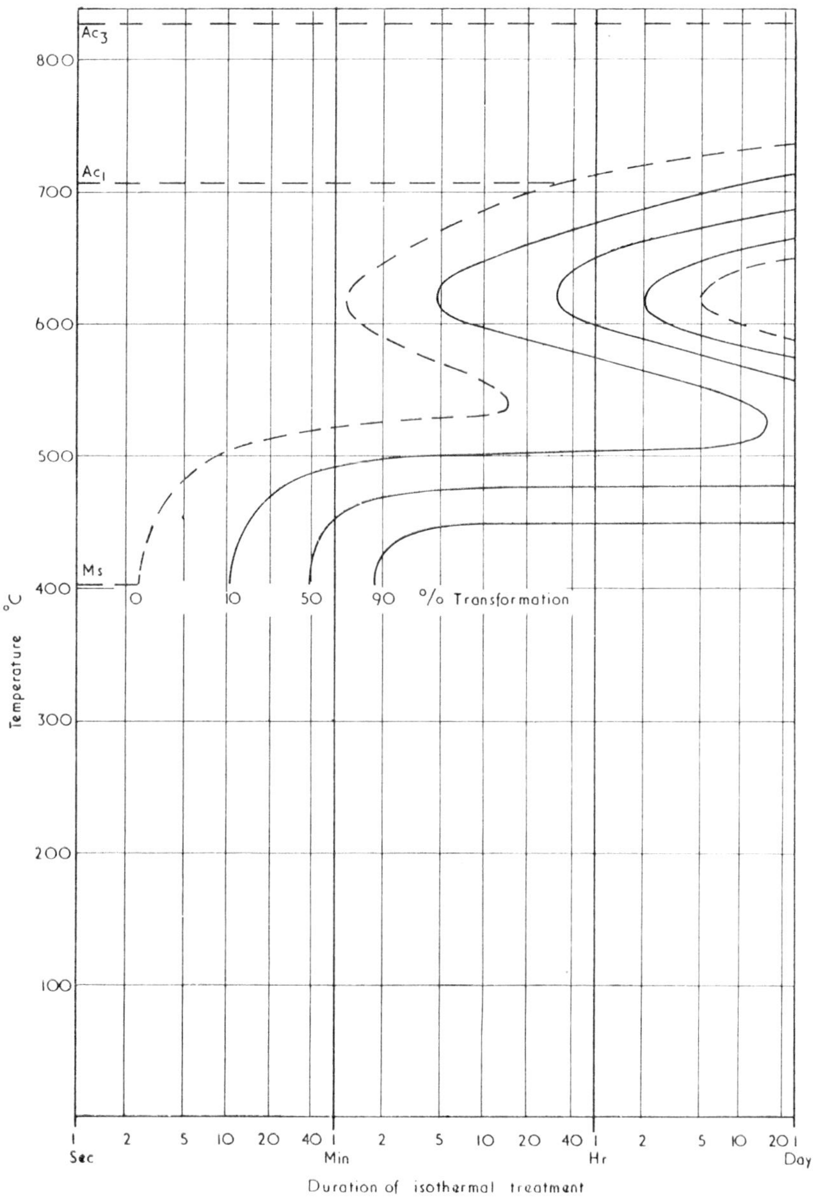

Fig. 354.1

ISOTHERMAL TRANSFORMATION (Continued)

CORE CHARACTERISTICS

Chemical Composition (%)							
C	Si	Mn	S	P	Ni	Cr	Mo
0·19	0·21	0·90	0·015	0·017	1·87	1·08	0·18

Basic Electric Arc.
McQuaid Ehn Grain Size 7—8 (ASTM).
As quenched Grain Size finer than 8 (ASTM).
Steel hammered to 1 in. dia. bars, annealed, specimens cut axially and water quenched from 860°C (1hr.) sealed in 1 in. dia. tube.
Austenitizing Temperature 820°C, 30 min. (Ref. 4)

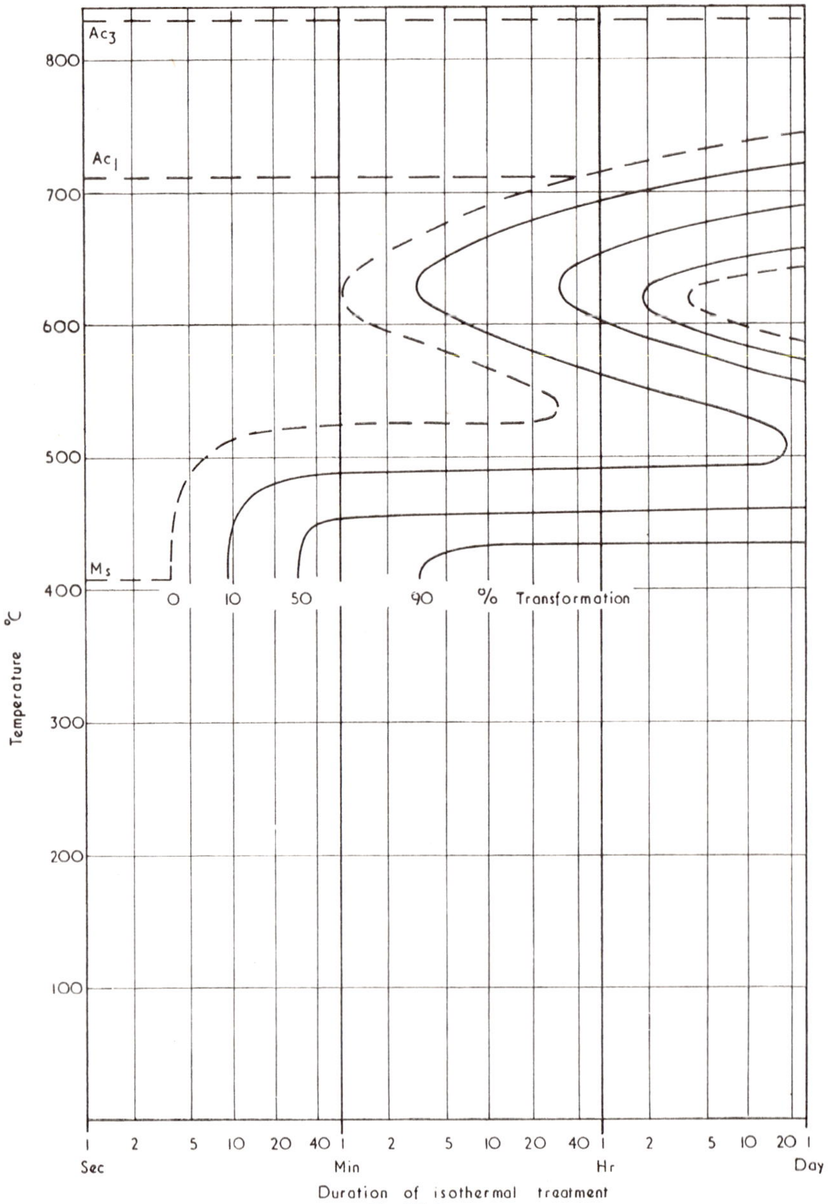

Fig. 354.2

ISOTHERMAL TRANSFORMATION (Continued)

CASE CHARACTERISTICS

Chemical Composition (%)							
C	Si	Mn	S	P	Ni	Cr	Mo
0·97	0·33	1·00	0·018	0·029	1·93	1·13	0·23

High Frequency Melt.
McQuaid Ehn Grain Size 8 (ASTM).
As quenched Grain Size 8 (ASTM).
Steel hammered to $\frac{5}{16}$ in. dia. bar, annealed, specimens cut axially, then 900°C (2hrs.) furnace cool
Austenitizing Temperature 870°C, 30 min. (Ref. 4)

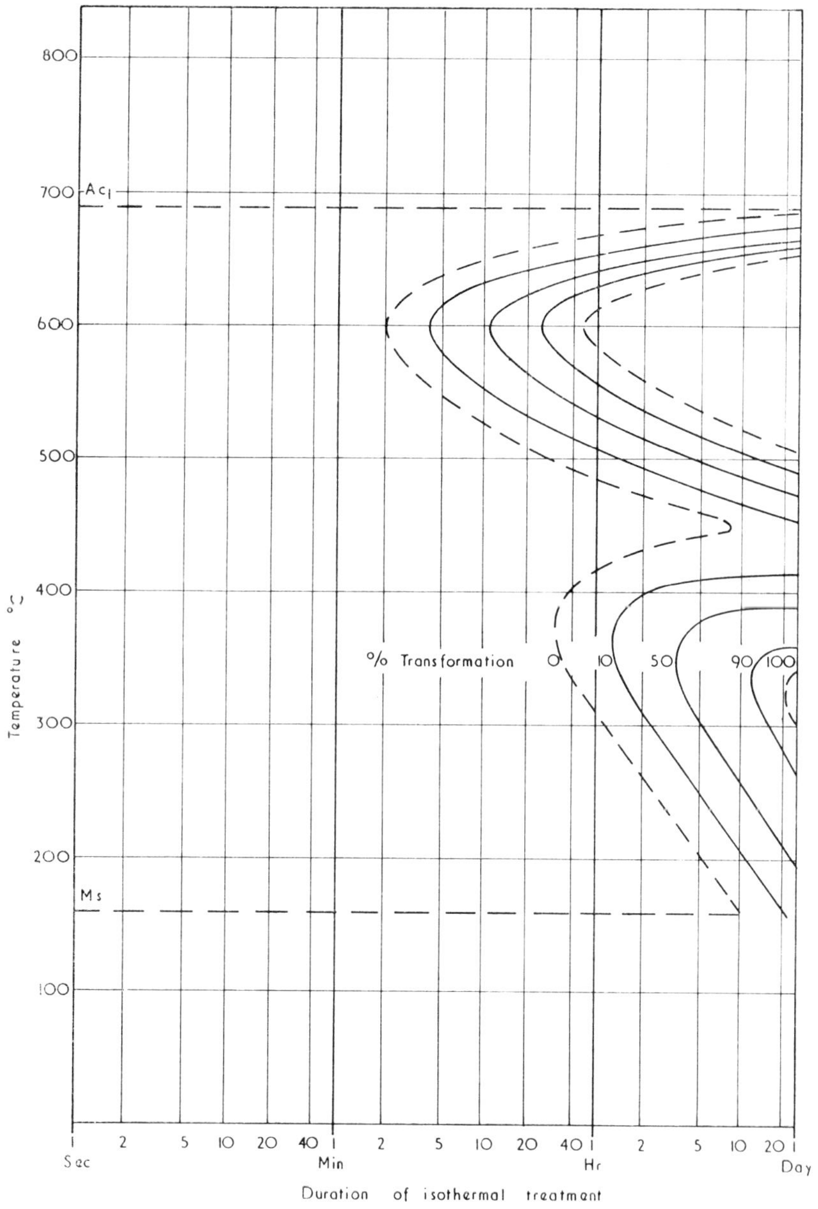

Fig. 354.3

ISOTHERMAL TRANSFORMATION (Continued)

CASE CHARACTERISTICS

Chemical Composition (%)							
C	Si	Mn	S	P	Ni	Cr	Mo
0·97	0·33	1·00	0·018	0·029	1·93	1·13	0·23

High Frequency Melt.
McQuaid Ehn Grain Size 8 (ASTM).
As quenched Grain Size 8 (ASTM).
Steel hammered to $\frac{5}{16}$ in. dia. bar, annealed, specimens cut axially, then 900°C (2hrs.) furnace cool, followed by 870°C ($\frac{1}{2}$ hr.) oil quench.
Austenitizing Temperature 820°C, 30 min. (Ref. 4)

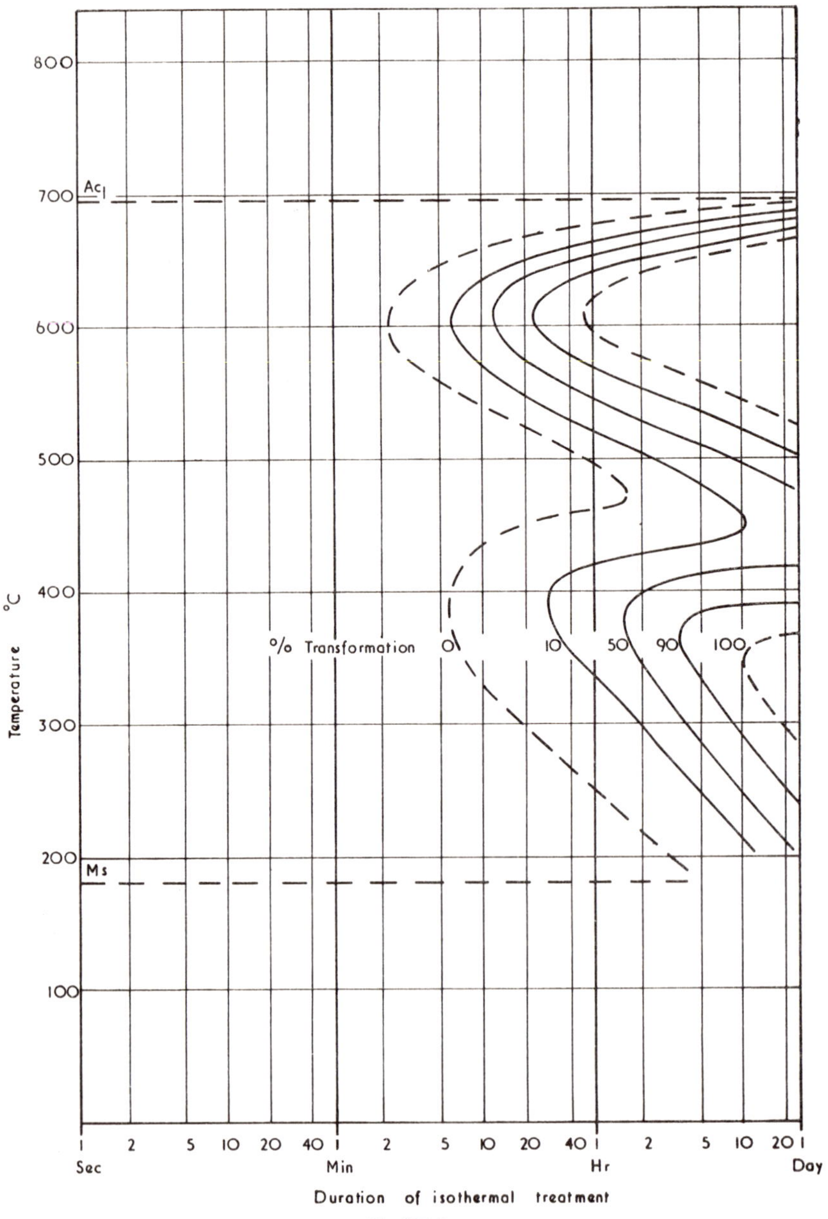

Fig. 354.4

HARDENABILITY

END QUENCH HARDENABILITY DATA

Chemical Composition (%)							
C	Si	Mn	S	P	Ni	Cr	Mo
0·19	0·21	0·90	0·015	0·017	1·87	1·08	0·18

Basic Electric Arc.
McQuaid Ehn Grain Size 7—8 (ASTM).
As quenched Grain Size finer than 8 (ASTM).
Curve A: End quenched 870°C.
Curve B: End quenched 820°C. (Ref. 4)

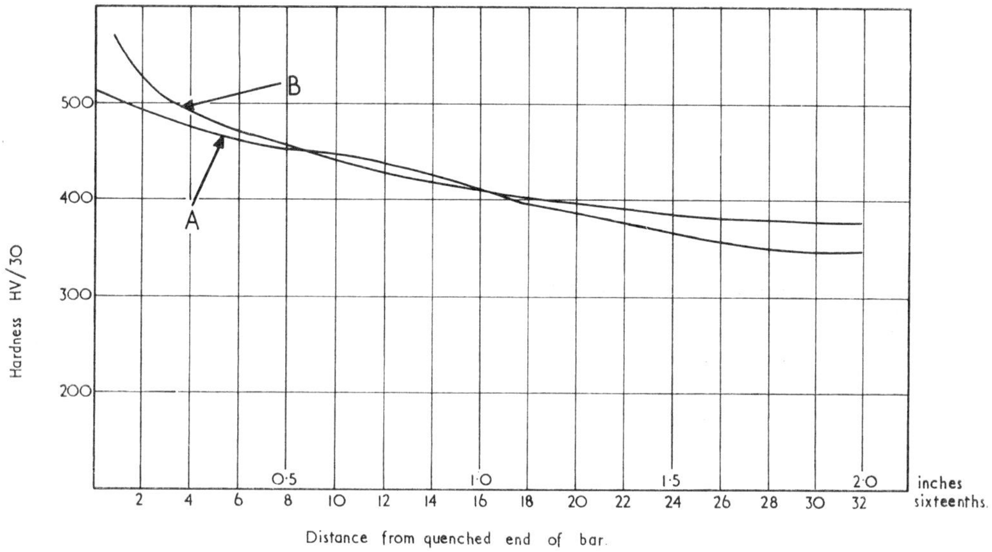

Fig. 354.5

END QUENCH HARDENABILITY BAND

These curves have been compiled from the results obtained on several casts of En 354.

(Ref. 3)

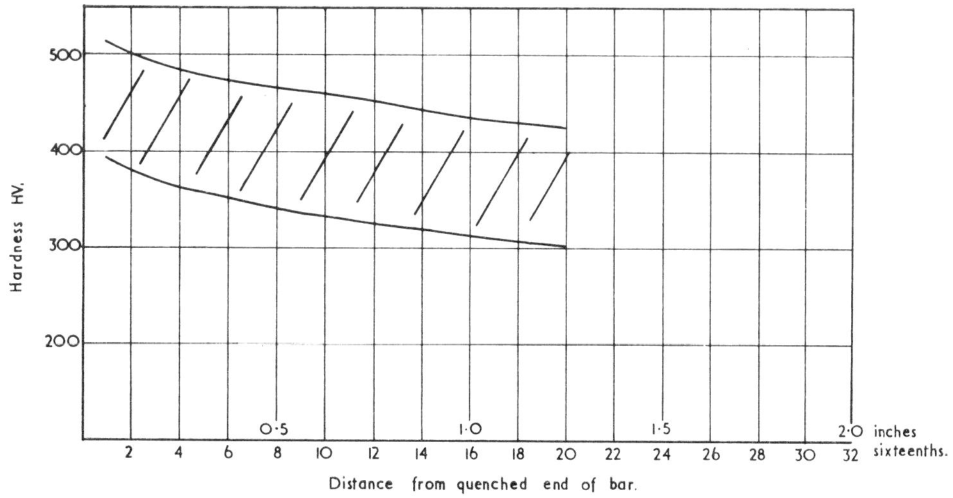

Fig. 354.6

En 354

MECHANICAL PROPERTIES AT ROOM TEMPERATURE

Process of Manufacture	ASTM Grain Size	Chemical Composition (%) C	Si	Mn	S	P	Ni	Cr	Mo	Section Size dia. (in.)	Heat Treatment °C	Tensile 0·1% P.S. (0·1Rp) tons/sq.in.	0·2% P.S. (0·2Rp)	0·5% P.S. (0·5Rp)	Y.P. (Rm)	T.S. (Rs)	El % (A)	R.A. % (Z)	Izod ft.lb.	Ref.
																	4√A 5d			
B.E.		0·15	0·22	0·82	0·007	0·008	1·88	1·11	0·17	1⅛	BC925 FC OQ870 OQ800				53·0	69·0	20½ 15½	60		2
B.E.	7	0·15	0·24	0·68	0·007	0·014	1·69	0·87	0·24	1⅛	OQ370 OQ810				52·0	69·8	20	58	38	"
		0·16		0·82			1·58	1·15	0·16	1⅛	OQ820					78·7	17		34	5
		0·16		0·75			1·80	0·90	0·17	1⅛	OQ850 OQ800					74·6	16		31	"
		0·16	0·29	0·86			1·71	1·07	0·18	1⅛	BC900 OQ860 OQ800					75·5	18	51	34	3
		0·16	0·09	0·60			1·71	1·24	0·14	⅞	BC927 OQ810	41·3		58·4		70·8	20	60½	30	2
		"	"	"			"	"	"	"	" OQ840	"		58·5		70·5	"	59	28	"
		"	"	"			"	"	"	½	" OQ810	49·5		68·0		83·5	16½	54½	42	"
B.E.		0·17	0·29	0·66	0·008	0·012	1·85	0·93	0·15	1⅛	OQ870 OQ810				54·0	68·4	19½	53½	51	2
"		"	0·23	0·63	0·008	0·014	1·76	0·89	0·14	1⅛	OQ870 OQ810				60·0	71·5	18	51½	26	"
"		"	0·21	0·60	0·015	0·013	1·73	0·87	0·12	½	OQ870 OQ810				72·1	90·0	14½	51	36	"
"		"	"	"	"	"	"	"	"	⅜	" "				59·9	75·9	17½	49½	25	"
"		"	"	"	"	"	"	"	"	1⅛	" "				49·6	66·0	19½	51	32	"
"		"	"	"	"	"	"	"	"	2½	" "				40·8	56·8	24	55½	60	"
"		"	"	"	"	"	"	"	"	4	" "				35·2	51·8	25½	55	53	"
"		"	0·31	0·71	0·009	0·013	1·74	1·20	0·19	½	OQ870 OQ810	50·0			74·5	96·7	15½	49	32	"
"		"	"	"	"	"	"	"	"	⅞	" "				80·7	100·5	18	51½	30	"
"		"	"	"	"	"	"	"	"	1⅛	" "	45·0			65·0	84·0	15½	46½	23	"
"		"	"	"	"	"	"	"	"	2½	" "				52·8	71·2	16	40	40	"
"		"	"	"	"	"	"	"	"	5	" "				44·6	67·4	10½	16½	26	"
"		"	0·36	0·91			1·72	1·01	0·19	1⅛	BC900 OQ860 OQ800					76·7	14	31	48	3
"		"	0·23	0·92			1·80	1·16	0·20	1⅛	BC900 OQ860 OQ800					77·0	20	52	65	"
H.F.		"		0·59			1·63	0·91	0·19	0·504	OQ860 OQ810				79·4	85·2	15½	49½	40	6
"		"		"			"	"	"	1⅛	" OQ780				44·8	70·5	18	47	31	"
"		"		"			"	"	"	1⅛	" OQ800				53·0	69·4	15½	52½	44	"
"		"		"			"	"	"	1⅛	" OQ820				59·2	71·9	16½	49½	36	"
"		"		"			"	"	"	1⅛	" OQ860				63·2	71·0	17	49½	31	"
"		"		"			"	"	"	1⅛	BC OQ820				63·2	68·4	17	52½	36	"
"		"		"			"	"	"	1⅛	" OQ850				60·0	66·4	17½	52½	33	"
"		"		"			"	"	"	2½	OQ860 OQ810				48·8	58·2	22	49½	58	"
"		"		"			"	"	"	4	" "				48·0	55·4	21	52½	60	"

MECHANICAL PROPERTIES AT ROOM TEMPERATURE (Continued)

Process of Manufacture	ASTM Grain Size	Chemical Composition (%)								Section Size dia. (in.)	Heat Treatment °C	Tensile							Izod ft.lb.	Ref.
		C	Si	Mn	S	P	Ni	Cr	Mo			0·1% P.S. (0·1Rp)	0·2% P.S. (0·2Rp)	0·5% P.S. (0·5Rp)	Y.P. (Rm)	T.S. (Rs)	El % (A)	R.A. % (Z)		
												tons/sq.in.								
B.E.	7	0·18	0·26	0·66	0·006	0·011	1·76	0·85	0·16	½	OQ870 OQ810				68·0	87·2	16½	54½	44	2
"	"	"	"	"	"	"	"	"	"	⅞	" "				58·1	74·0	19	56	38	"
"	"	"	"	"	"	"	"	"	"	1⅛	" "				47·4	64·8	21	51	47	"
"	"	"	"	"	"	"	"	"	"	2½	" "				40·6	68·8	24	54½	48	"
"	"	"	"	"	"	"	"	"	"	4	" "				37·6	55·6	24½	53	54	"
"		"	0·25	0·65	0·010	0·013	1·72	0·92	0·14	1⅛	OQ870 OQ810				68·0	79·6	16	53	27	"
"	"	"	0·30	0·44	0·008	0·016	1·93	1·25	0·16	1⅛	" "				68·0	84·0	16	54	26	"
		"	0·27	0·68	0·022	0·025	1·80	0·94	0·21	1⅛	" "				58·0	78·5	16	51½	22	"
		"	0·30	0·92			1·78	1·12	0·19	1⅛	BC900 OQ860 OQ800					78·0	18	52	55	3
B.E.		0·19	0·22	0·63	0·015	0·019	1·79	0·93	0·15	1⅛	OQ870 OQ810				68·0	82·4	15	47	20	2
"	8	"	0·24	0·81	0·008	0·015	1·81	1·09	0·14	½	" "	49·0		74·5		95·0	12½	45	34	"
"	"	"	"	"	"	"	"	"	"	⅞	" "	53·4		76·2		95·4	15	48	29	"
"	"	"	"	"	"	"	"	"	"	1⅛	" "	48·0		68·4		89·2	14½	44½	28	"
"	"	"	"	"	"	"	"	"	"	2½	" "			56·0		76·4	16	40	30	"
"	"	"	"	"	"	"	"	"	"	⅞	" OQ850	54·5		76·5		96·1	16	48	28	"
B.E.	6—7	"	0·21	0·90	0·015	0·017	1·87	1·08	0·18	1¼	OQ860 OQ800					80·8	17	53½	38	"
		"	0·19	0·91			1·85	1·13	0·21	1⅛	BC900 OQ860 OQ800					78·0	18	56	60	3
		"	0·23	0·89			1·89	1 09	026	1⅛	" " "					80·8	20	54½	44	"
		"	0·33	0·95			1·80	1·06	010	1⅛	" " "					82·0	19	49½	45	"
"		"	0·25	0·67	0·011	0·013	1·59	0·86	0·20	0·169	AC870 OQ810	65·5				102·0	18	60		2
"		"	"	"	"	"	"	"	"	0·236	" "	69·5				104·0	17½	58		"
"		"	"	"	"	"	"	"	"	0·367	" "	68·0				102·0	16	53		"
"		"	"	"	"	"	"	"	"	0·574	" "	62·4				102·4	15	52	20	"
"		"	"	"	"	"	"	"	"	1⅛	OQ870 "					78·8	17	44	25	"
"		"	"	"	"	"	"	"	"	1¼	AC870 "	46·7				79·6	16	50		"
"		"	0·26	0·95	0·014	0·011	1·89	1 04	0·19	0·236	AC870 OQ810	66·4				93·7	17½	56		1
"		"	"	"	"	"	"	"	"	0·367	" "	60·6				94·3	10	50		"
"		"	"	"	"	"	"	"	"	0·574	" "	56·4				92·8	15	50		"
"		"	"	"	"	"	"	"	"	⅞	" "	53·5				87·6	16	50	23	"
"		"	"	"	"	"	"	"	"	$1\frac{9}{16}$	" "	40·5				75·0	17	53		"
"		"	"	"	"	"	"	"	"	1½	" "								27	"

MECHANICAL PROPERTIES AT ROOM TEMPERATURE (Continued)

Process of Manufacture	ASTM Grain Size	Chemical Composition (%) C	Si	Mn	S	P	Ni	Cr	Mo	Section Size dia. (in.)	Heat Treatment °C	Tensile 0·1% P.S. (0·1Rp) tons/sq.in.	0·2% P.S. (0·2Rp)	0·5% P.S. (0·5Rp)	Y.P. (Rm)	T.S. (Rs)	El % (A)	R.A. % (Z)	Izod ft.lb.	Ref.
B.E.		0·20	0·24	0·70	0·009	0·019	1·61	0·91	0·16	1⅛	OQ870 OQ810				70·0	86·0	14	47½	17	2
"		"	0·20	0·64	0·010	0·012	1·66	0·90	0·21	1⅛	" "				68·0	79·6	16	47½	24	"
"		"	0·20	0·70	0·008	0·016	1·67	1·06	0·17	1⅛	" "				61·6	92·0	14½	48	22	"
"	7	"	0·30	0·89	0·015	0·018	1·79	1·07	0·18	1¼	" "					83·0	16	47	32	"
"		"	0·21	0·93	0·015	0·017	1·91	1·09	0·19	0·169	AC870 OQ810	64·6				100·0	16½	64		1
"		"	"	"	"	"	"	"	"	0·236	" "	63·5				93·8	16	58		"
"		"	"	"	"	"	"	"	"	0·367	" "	64·4				96·0	17	55		"
"		"	"	"	"	"	"	"	"	0·574	" "	60·8				94·8	17	55	19	"
"		"	"	"	"	"	"	"	"	⅞	" "	56·8				93·6	16	55		"
"		"	"	"	"	"	"	"	"	1¼	" "	49·9				84·0	16	58	28	"
"		"	0·25	0·66	0·010	0·013	1·66	0·96	0·15	0·236	AC870 OQ810	62·5				95·5	16	56		"
"		"	"	"	"	"	"	"	"	0·367	" "	60·8				96·6	10	55		"
"		"	"	"	"	"	"	"	"	0·574	" "	56·8				92·0	17	53	23	"
"		"	"	"	"	"	"	"	"	⅞	" "	47·2				81·0	16	50		"
"		"	"	"	"	"	"	"	"	1⅛	OQ870					77·0	17½	44	22	2
"		"	"	"	"	"	"	"	"	1¼	AC870	39·2				76·4	19	57	34	1
"		"	0·20	0·67	0·009	0·018	1·73	0·92	0·17	0·169	AC870 OQ810	56·5				100·0	15	45		"
"		"	"	"	"	"	"	"	"	0·236	" "	66·0				102·3	17½	56		"
"		"	"	"	"	"	"	"	"	0·367	"	69·2				103·0	16	55		"
"		"	"	"	"	"	"	"	"	0·574	"	60·8				95·2	17	55	22	"
"		"	"	"	"	"	"	"	"	⅞	" "	50·8				87·6	17	54		"
"		"	"	"	"	"	"	"	"	1⅛	OQ870 "					77·4	15½	42	24	2
"		"	"	"	"	"	"	"	"	1¼	AC870 "	42·6				74·4	18	55	35	1
"		"	0·30	0·90	0·017	0·020	1·83	1·02	0·19	0·236	" "	58·0				93·0	17½	60		"
"		"	"	"	"	"	"	"	"	0·367	" "	56·8				94·5	16½	56		"
"		"	"	"	"	"	"	"	"	0·574	" "	54·4				91·2	17	54	33	"
"		"	"	"	"	"	"	"	"	⅞	" "	49·2				86·5	16	55		"
"		"	"	"	"	"	"	"	"	1¼	" "	42·0				78·0	18	55	44	"

MECHANICAL PROPERTIES AT ROOM TEMPERATURE (Continued)

Process of Manufacture	ASTM Grain Size	Chemical Composition (%) C	Si	Mn	S	P	Ni	Cr	Mo	Section Size dia. (in.)	Heat Treatment °C	Tensile 0·1% P.S. (0·1Rp) tons/sq.in.	0·2% P.S. (0·2Rp)	0·5% P.S. (0·5Rp)	Y.P. (Rm)	T.S. (Rs)	El % (A)	R.A. % (Z)	Izod ft.lb.	Ref.
B.E.	7	0·21	0·19	0·81	0·015	0·017	1·96	1·07	0·19	0·424	AC900	41·1	49·5	61·8		70·5	20	51½	20	3
"	"	"	"	"	"	"	"	"	"	¾	"	40·4	46·2	54·0		71·6	20	48½	18	"
"	"	"	"	"	"	"	"	"	"	1	"	37·3	43·5	50·0		67·5	19	48½	19	"
"	"	"	"	"	"	"	"	"	"	1⅛	"	35·3	42·4	51·3		70·0	18½	47	19	"
"	"	"	"	"	"	"	"	"	"	2⅛	"	45·5	51·4	57·5		66·2	19	49	14	"
"	"	"	"	"	"	"	"	"	"	3	"	39·4	44·0	50·5		64·7	20	49	17	"
"	"	"	"	"	"	"	"	"	"	4¼(O)	"	37·6	42·7	49·6		63·3	20	51½	14	"
"	"	"	"	"	"	"	"	"	"	4¼(C)	"	37·8	42·8	49·1		63·2	19	50½	15	"
"	"	"	"	"	"	"	"	"	"	0·424	AC900 OQ820	57·4	66·9	80·2		98·9	16½	52½	30	"
"	"	"	"	"	"	"	"	"	"	¾	" "	60·5	69·4	82·1		98·4	16½	51½	24	"
"	"	"	"	"	"	"	"	"	"	1	" "	69·3	75·7	84·5		94·9	15½	55½	20	"
"	"	"	"	"	"	"	"	"	"	1⅛	" "	69·0	75·1	83·0		91·7	16	56	29	"
"	"	"	"	"	"	"	"	"	"	2⅛	" "	48·3	55·5	65·0		78·8	16	52	33	"
"	"	"	"	"	"	"	"	"	"	3	" "	44·2	50·7	59·5		73·8	19½	51	45	"
"	"	"	"	"	"	"	"	"	"	4¼(O)	" "	46·4	53·6	62·6		70·7	19	61½	36	"
"	"	"	"	"	"	"	"	"	"	4¼(C)	" "	48·3	55·3	63·0		73·9	17½	54½	27	"
"	"	"	"	"	"	"	"	"	"	0·424	AC900 OQ860 OQ800	51·1	62·1	77·7		99·4	16½	52½	34	"
"	"	"	"	"	"	"	"	"	"	¾	" " "	60·0	68·8	81·7		97·4	16	52½	23	"
"	"	"	"	"	"	"	"	"	"	1	" " "	65·2	72·2	82·2		94·6	16	54	30	"
"	"	"	"	"	"	"	"	"	"	1⅛	" " "	59·5	67·3	78·5		90·5	16	56	28	"
"	"	"	"	"	"	"	"	"	"	2⅛	" " "	49·7	56·2	64·8		78·7	26½	53	35	"
"	"	"	"	"	"	"	"	"	"	3	" " "	47·8	51·9	59·8		76·1	18½	49	46	"
"	"	"	"	"	"	"	"	"	"	4¼(O)	" " "	41·1	47·4	55·2		67·8	21	57½	30	"
"	"	"	"	"	"	"	"	"	"	4¼(C)	" " "	42·6	49·5	58·3		72·9	16	51	23	"
B.E.	"	0·21		0·80			1·98	1·07	0·17	1¼	OQ860 OQ800					84·0	15	49½	32	2

En 354

EFFECT OF TEMPERING ON CASE HARDNESS

(TYPICAL CURVE)

Carburized 900°C.
Oil refined 860°C.
Oil hardened 800°C.

(Ref. 3)

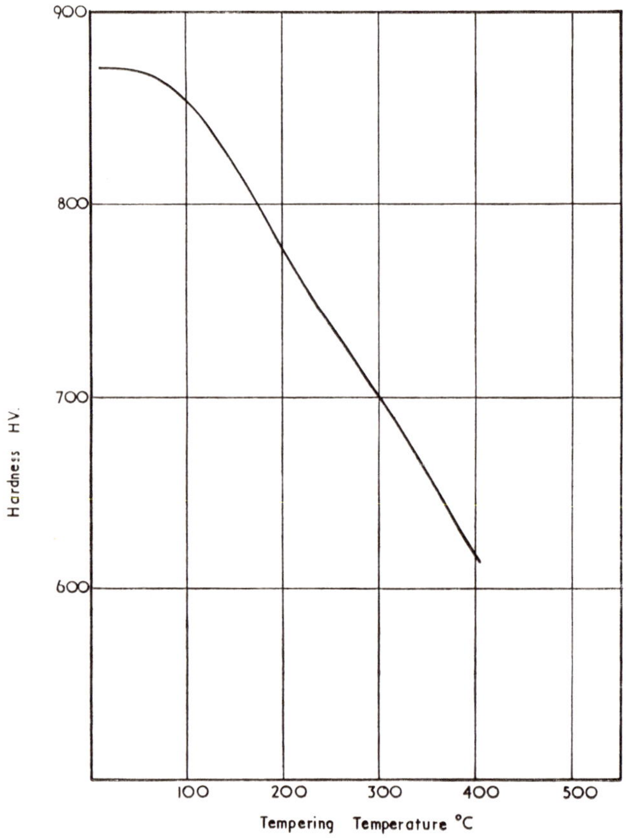

Fig. 354.7

EFFECT OF SECTION SIZE ON THE CORE PROPERTIES AT ROOM TEMPERATURE

(TYPICAL CURVES)

Blank carburized 900°C.
Oil refined 860°C.
Oil hardened 800°C.

(Ref. 3)

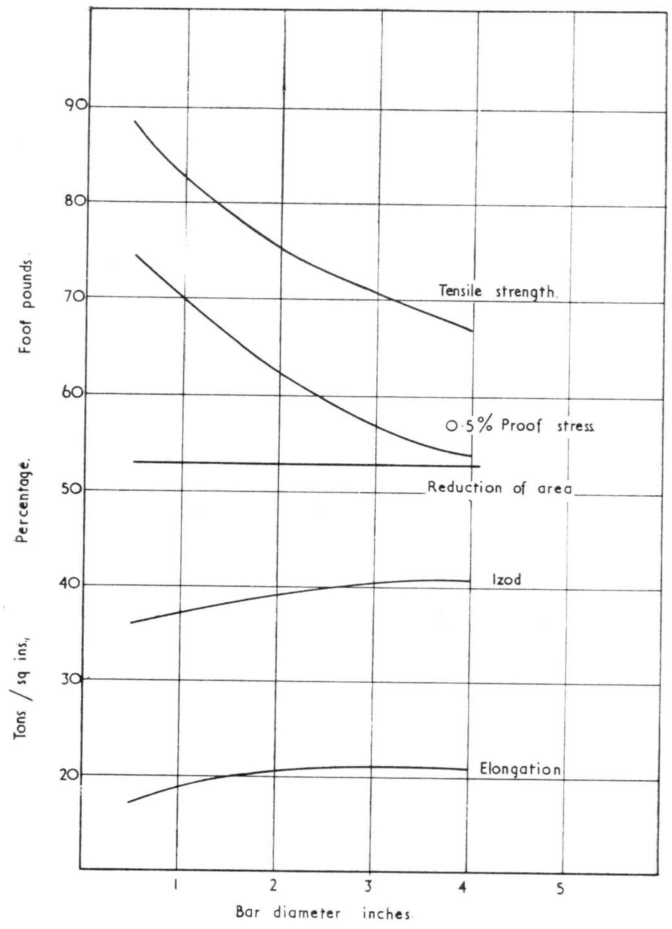

Fig. 354.8

Blank carburized 900°C.
Oil hardened 820°C.
(Single quenched quality).

(Ref. 3)

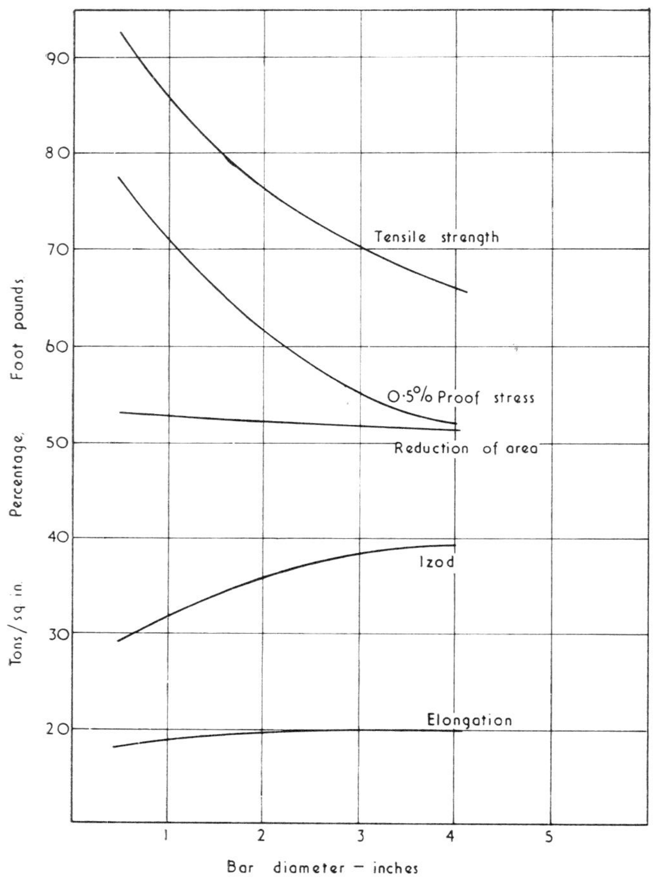

Fig. 354.9

MECHANICAL PROPERTIES AT LOW TEMPERATURE

Chemical Composition (%)								Section Size	Heat Treatment °C	Test Temperature °C	Tensile					Ref.
C	Si	Mn	S	P	Ni	Cr	Mo				0·1% P.S. (0·1Rp) tons/sq.in.	0·2% P.S. (0·2Rp) tons/sq.in.	T.S. (Rs) tons/sq.in.	El % (A)	R.A. % (Z)	
0·20	0·21	0·70	0·010	0·015	1·64	1·14	0·28	TPS	BC930 OQ880 OQ780	RT	66·0	76·4	93·6	1½	3	7
"	"	"	"	"	"	"	"	"	BC OQ880 OQ780	—196	83·0	95·6	121·0	1½	2½	"
"	"	"	"	"	"	"	"	"	Carb.* 930 OQ880 OQ780	"	64·0	72·8	92·3	1	1½	"
"	"	"	"	"	"	"	"	"	Carb.* 930 " "	"	64·8	76·0	97·7	1½	2	"
"	"	"	"	"	"	"	"	"	Carb.* 930 OQ780	"	64·6	73·2	93·8	1½	3	"
"	"	"	"	"	"	"	"	"	Carb.* 930 OQ780	"	67·2	80·8	102·1	1	1½	"

* Test pieces carburized to case depth of 0·045 in., finish ground to case depth of 0·02 in.

IMPACT TRANSITION TEMPERATURE DATA

Curve	Chemical Composition (%)								Heat Treatment(1) °C	Charpy Mesnager 50% Mean Energy Impact Transition Temperature °C
	C	Si	Mn	S	P	Ni	Cr	Mo		
A	0·20	0·21	0·70	0·010	0·015	1·64	1·14	0·28	BC930 OQ880 OQ780	—193
B	"	"	"	"	"	"	"	"	Carb.930 OQ880 OQ780	+100
C	"	"	"	"	"	"	"	"	Carb. 930 OQ780	+100

(1) Test piece machined to size before carburizing and heat treatment. Notch 2 mm deep, root radius 2·00 mm, case depth 0·02 in.

(Ref. 7)

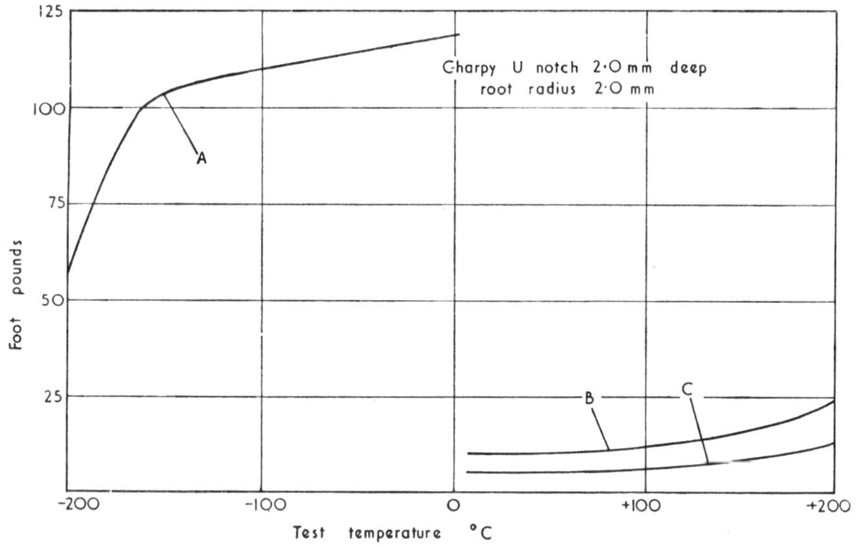

Fig. 354.10

LIST OF REFERENCES

1. Private correspondence with the Royal Aircraft Establishment, Farnborough.
2. Private correspondence with Firth Brown Ltd.
3. Private correspondence with The English Steel Corporation Ltd.
4. Atlas of Isothermal Transformation Diagrams of the British Standard En Steels. Iron and Steel Institute Special Report No. 56, 1956.
5. Private correspondence with International Nickel Ltd.
6. Private correspondence with The United Steel Companies Ltd.
7. War Office unpublished report.

En 355

2 PER CENT NICKEL–CHROMIUM–MOLYBDENUM CASE-HARDENING STEEL

SPECIFICATION

CHEMICAL COMPOSITION (%)

C	Si	Mn	S	P	Ni	Cr	Mo
0·20 max.	0·35 max.	0·40—0·70	0·050 max.	0·050 max.	1·80—2·20	1·40—1·70	0·15—0·25

MECHANICAL PROPERTIES

Tensile strength, tons/sq. in. min. (Rm)	85
Elongation, % min. (A)	12
Izod impact value, ft. lb., min.	25

In the softened condition the material shall have a Brinell hardness number not exceeding 277.

The heat treatment given to 1⅛ in. dia. test bars shall be blank carburize at a temperature between 880°C and 930°C.[1],[2] Refine at a temperature of 850°C to 880°C, cool in air or oil. Harden in oil from a temperature of 780°C to 820°C.

[1] By special arrangement with the purchaser the test may be carried out on samples from which the blank carburizing treatment has been omitted or modified to a short heating period.
[2] A single oil quench from 810°C to 830°C may be used in certain cases when the refining treatment after carburizing is omitted by special arrangement between the purchaser and the manufacturer.

APPLICATIONS

An alloy case-hardening steel developed as an alternative to En 39B which may be carburized and hardened to produce a hard wear-resistant case. It develops a core strength of the order of 85 tons/sq. in. with good resistance to shock. Typical applications are heavy duty gears for all types of machinery, pinions, gudgeon pins, valve rockers, camshafts, clutch plates, collets and small arms parts.

WELDING

(See Introduction for key to symbols used in this table.)

	Welding Process					
	MA	G	IG	S	F	B
En 355	cx	ax	g	e	a	h

Reproduced by kind permission of H.M.S.O.

MACHINABILITY

In the softened condition the machinability is approximately 70% of mild steel (En 3).

HOT WORKING AND HEAT TREATMENT TEMPERATURES

Forging, rolling and stamping	1150°C, finish above 900°C.
Softening	630°C—660°C, cool in air or furnace.
Carburizing	880°C—930°C.
Refining*	850°C—880°C, cool in air or oil.
Hardening	780°C—820°C, oil quench.
Tempering (for stress relief)	150°C—200°C.
(for maximum case toughness)	200°C—250°C.

* A single oil quench from 810°C to 830°C may be adopted when steel is specifically requested for this treatment.

CARBURIZING DATA

The carburizing characteristics of this steel will be similar to those given for En 39 (volume 2) and En 354.

Carburizing Temperature C	Depth of case produced in stated time (solid pack carburizing)		Ref.
	4 hrs.	8 hrs.	
880	0·035 in.	0·048 in.	1
900	0·041 in.	0·056 in.	"
925	0·048 in.	0·061 in.	"

En 355

PHYSICAL PROPERTIES

SPECIFIC GRAVITY (d)

Chemical Composition (%)								Heat Treatment °C	Specific Gravity at 20°C	Ref.
C	Si	Mn	S	P	Ni	Cr	Mo			
0·18	0·25	0·70			2·00	1·60	0·20	Softened	7·85	2
0·19	0·26	0·57	0·009	0·011	1·97	1·73	0·17	DQ* and T140	7·83	"
Within specification								Softened	7·85	3
0·16	0·27	0·56	0·011	0·018	2·00	1·50	0·20	FC910 OQ810	7·84	4

*DQ = double quench.

SPECIFIC HEAT (c)

Chemical Composition (%)						Heat Treatment °C	Mean Specific Heat cal/g.deg.C	Ref.
C	Si	Mn	Ni	Cr	Mo		20 to 100°C	
0·18	0·25	0·70	2·00	1·60	0·20	Softened	0·118	2

MEAN COEFFICIENT OF THERMAL EXPANSION (α)

Chemical Composition (%)						Heat Treatment °C	$10^6 \times$ Mean Coefficient of Thermal Expansion per deg.C					Ref.
C	Si	Mn	Ni	Cr	Mo		20 to 100°C	20 to 200°C	20 to 300°C	20 to 500°C	20 to 700°C	
0·18	0·25	0·70	2·00	1·60	0·20	Softened	12		13	14	14·5	2
0·19	0·26	0·57	1·97	1·73	0·17	DQ*+T140	11·4	11·9	12·3			"
Within specification						Softened	11·8		13·3	13·7	14·0	3
0·16	0·27	0·56	2·00	1·50	0·20	FC910 OQ810	11·8	12·3				4

* DQ = double quench.

ELECTRICAL RESISTIVITY (ρ)

Chemical Composition (%)						Heat Treatment °C	Electrical Resistivity microhm-cm			Ref.
C	Si	Mn	Ni	Cr	Mo		20°C	100°C	200°C	
0·16	0·27	0·57	1·97	1·73	0·17	FC910 OQ810	34·5	39·2	45·7	4

YOUNG'S MODULUS (E)

Chemical Composition (%)						Heat Treatment °C	Young's Modulus tons/sq. in.			Ref.
C	Si	Mn	Ni	Cr	Mo		20°C	100°C	200°C	
Within specification						Softened	13,250[1]			2
0·16	0·27	0·56	2·0	1·5	0·20	FC910 OQ810	13,300[2]	13,000[2]	12,700[2]	4

(1) Static method.
(2) Vibration method.

TRANSFORMATION CHARACTERISTICS

ISOTHERMAL TRANSFORMATION

CORE CHARACTERISTICS

Chemical Composition (%)							
C	Si	Mn	S	P	Ni	Cr	Mo
0·20	0·23	0·61	0·011	0·015	2·00	1·65	0·19

Basic Electric Arc.
McQuaid Ehn Grain Size 3—5 (ASTM).
As quenched Grain Size 8 (ASTM).
Bars hammered to ¼ in. dia., annealed, specimens cut axially, then 900°C (2hrs.) air cool.
Austenitizing Temperature 870°C, 30 min. (Ref. 5)

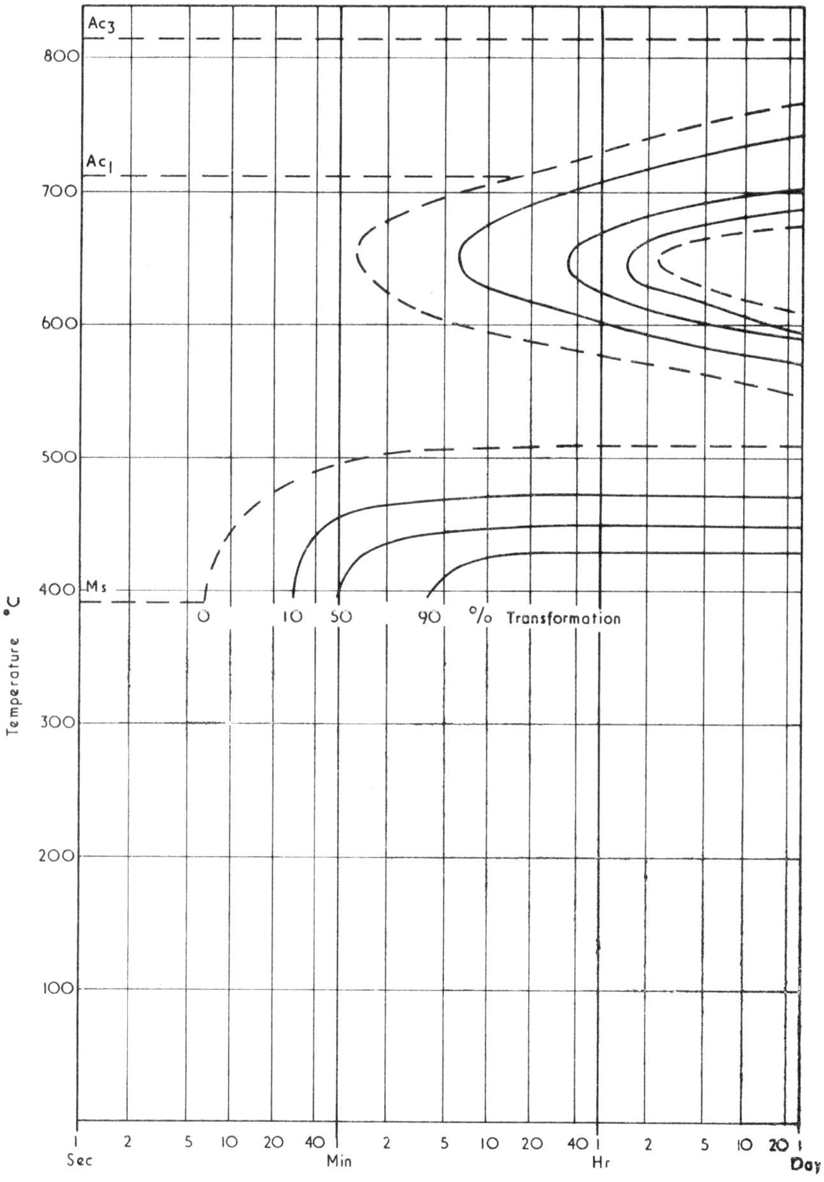

Fig. 355.1

En 355

ISOTHERMAL TRANSFORMATION (Continued)

CORE CHARACTERISTICS

Chemical Composition (%)							
C	Si	Mn	S	P	Ni	Cr	Mo
0·20	0·23	0·61	0·011	0·015	2·00	1·65	0·19

Basic Electric Arc.
McQuaid Ehn Grain Size 3—5 (ASTM).
As quenched Grain Size 8 (ASTM).
Bars hammered to ¼ in. dia. bar, annealed, specimens cut axially, then 880°C (½ hr.) air cool.
Austenitizing Temperature 800°C, 30 min.

(Ref. 5)

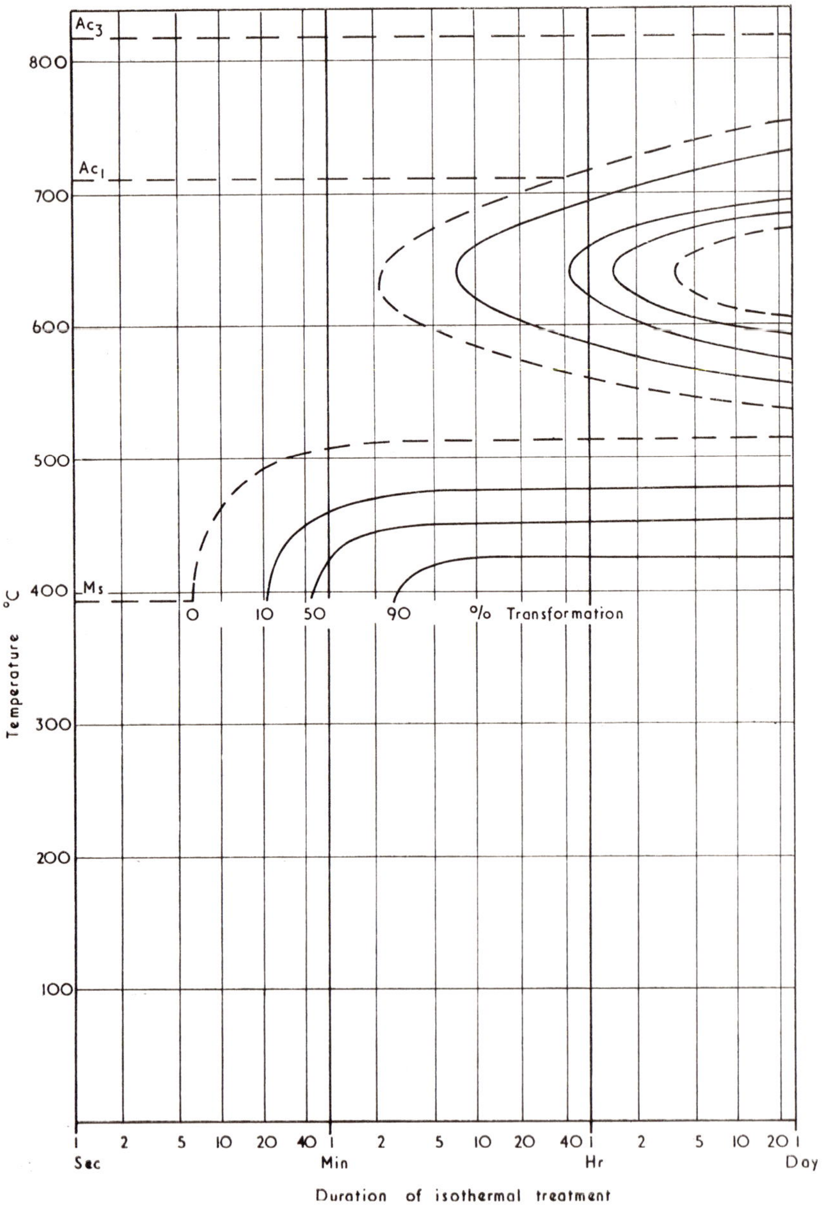

Fig. 355.2

ISOTHERMAL TRANSFORMATION (Continued)

CASE CHARACTERISTICS

Chemical Composition (%)							
C	Si	Mn	S	P	Ni	Cr	Mo
0·93	0·38	0·71	0·017	0·029	2·10	1·70	0·20

High Frequency Melt.
McQuaid Ehn Grain Size 7 (ASTM).
As quenched Grain Size 7 (ASTM).
Bars hammered to $\frac{5}{16}$ in. dia., annealed, specimens cut axially, then 900°C (2hrs.) air cool.
Austenitizing Temperature 870°C, 30 min. (Ref. 5)

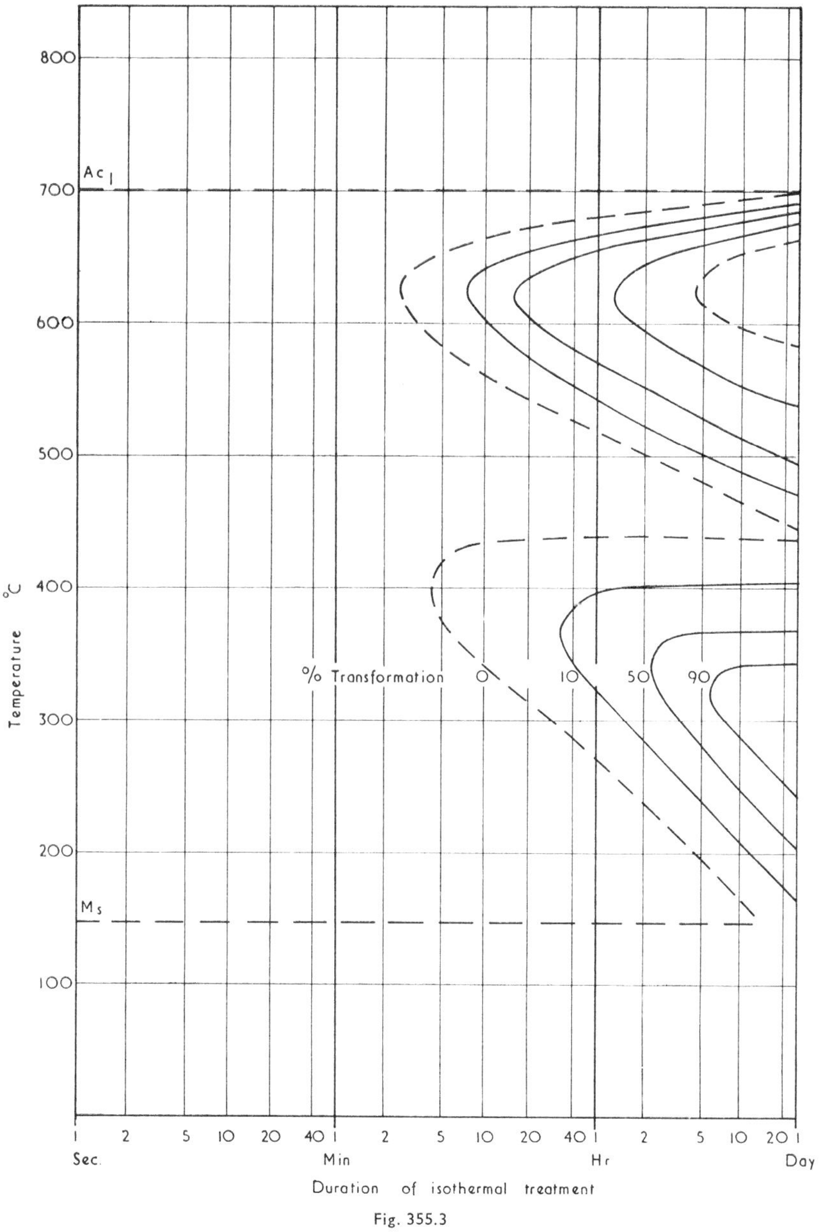

Fig. 355.3

ISOTHERMAL TRANSFORMATION (Continued)

CASE CHARACTERISTICS

Chemical Composition (%)							
C	Si	Mn	S	P	Ni	Cr	Mo
0·93	0·38	0·71	0·017	0·029	2·10	1·70	0·20

High Frequency Melt.
McQuaid Ehn Grain Size 7 (ASTM).
As quenched Grain Size 7 (ASTM).
Bars hammered to $\frac{5}{16}$ in. dia., annealed, specimens cut axially, then 880°C (2 hrs.) furnace cool.
Austenitizing Temperature 800°C, 30 min.

(Ref. 5)

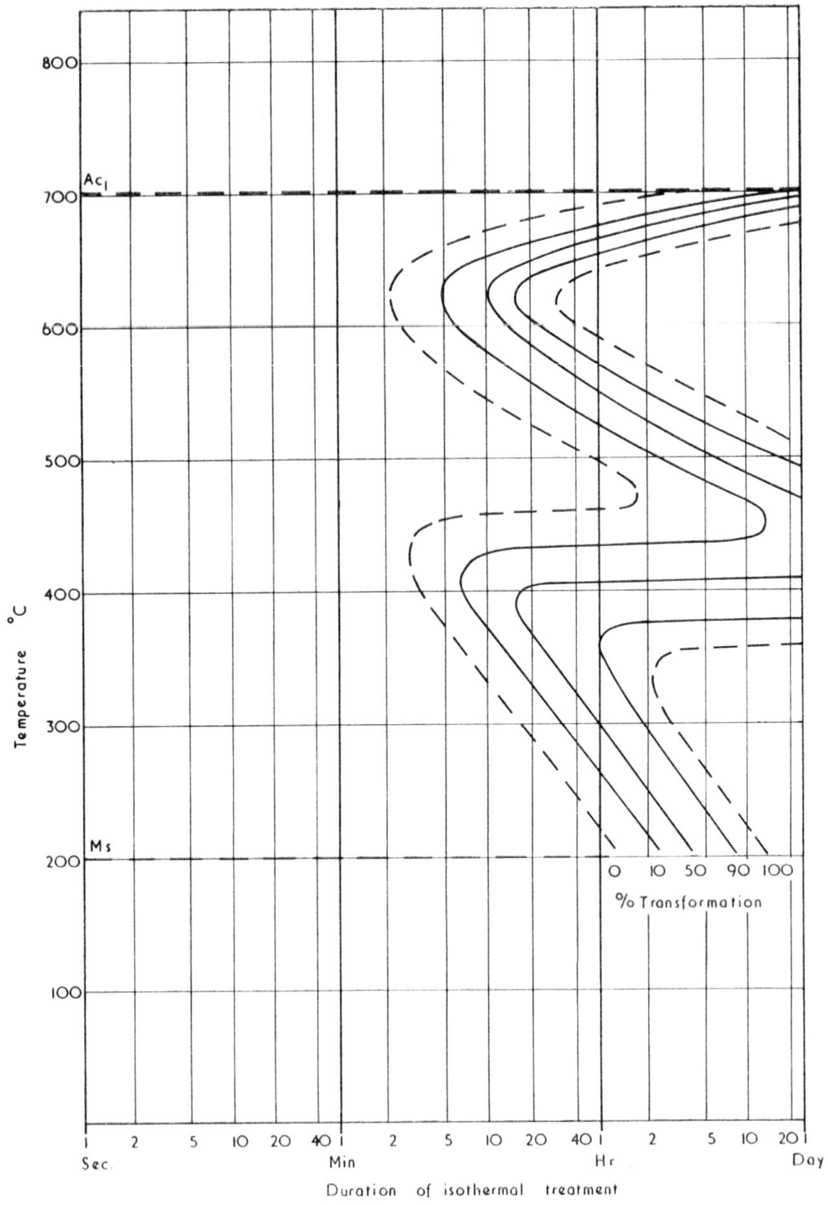

Fig. 355.4

CONTINUOUS COOLING TRANSFORMATION

Air cooled from 810°C.

(Ref. 4)

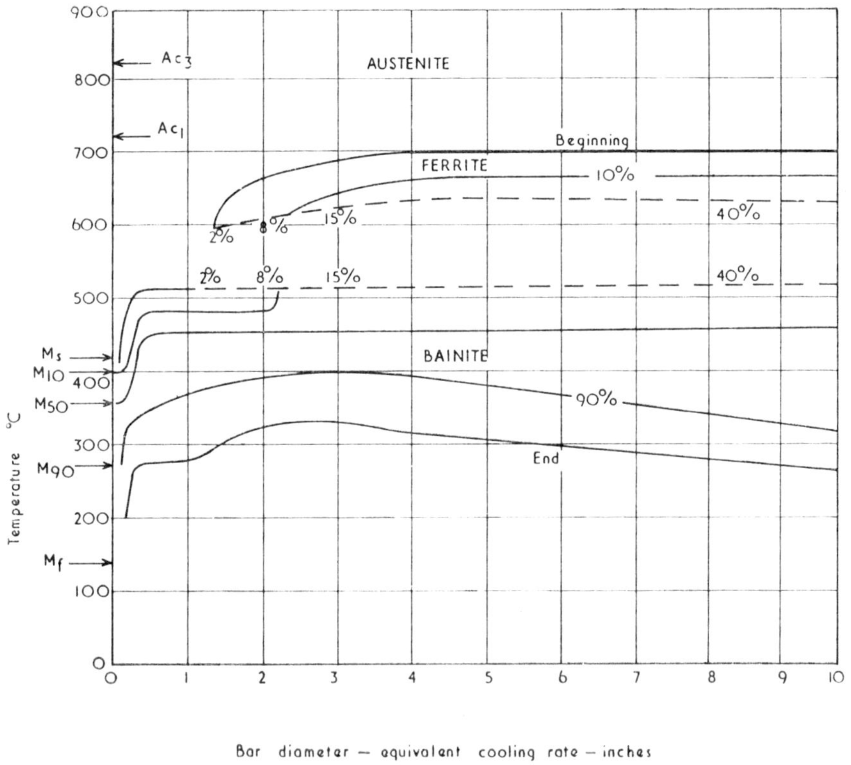

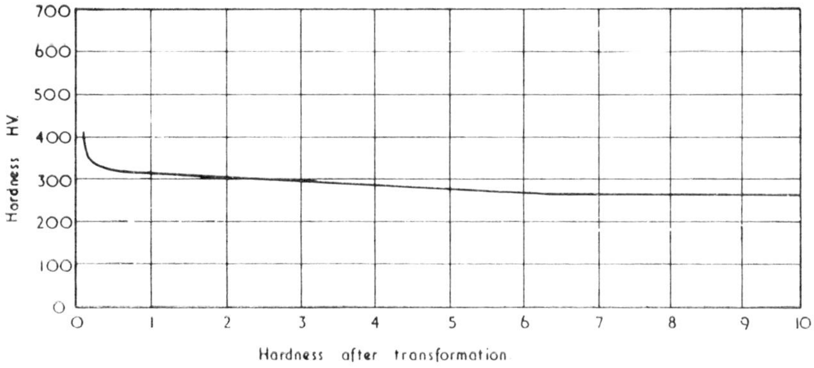

Fig. 355.5

En 355

CONTINUOUS COOLING TRANSFORMATION (Continued)

Oil quenched from 810°C.

(Ref. 4)

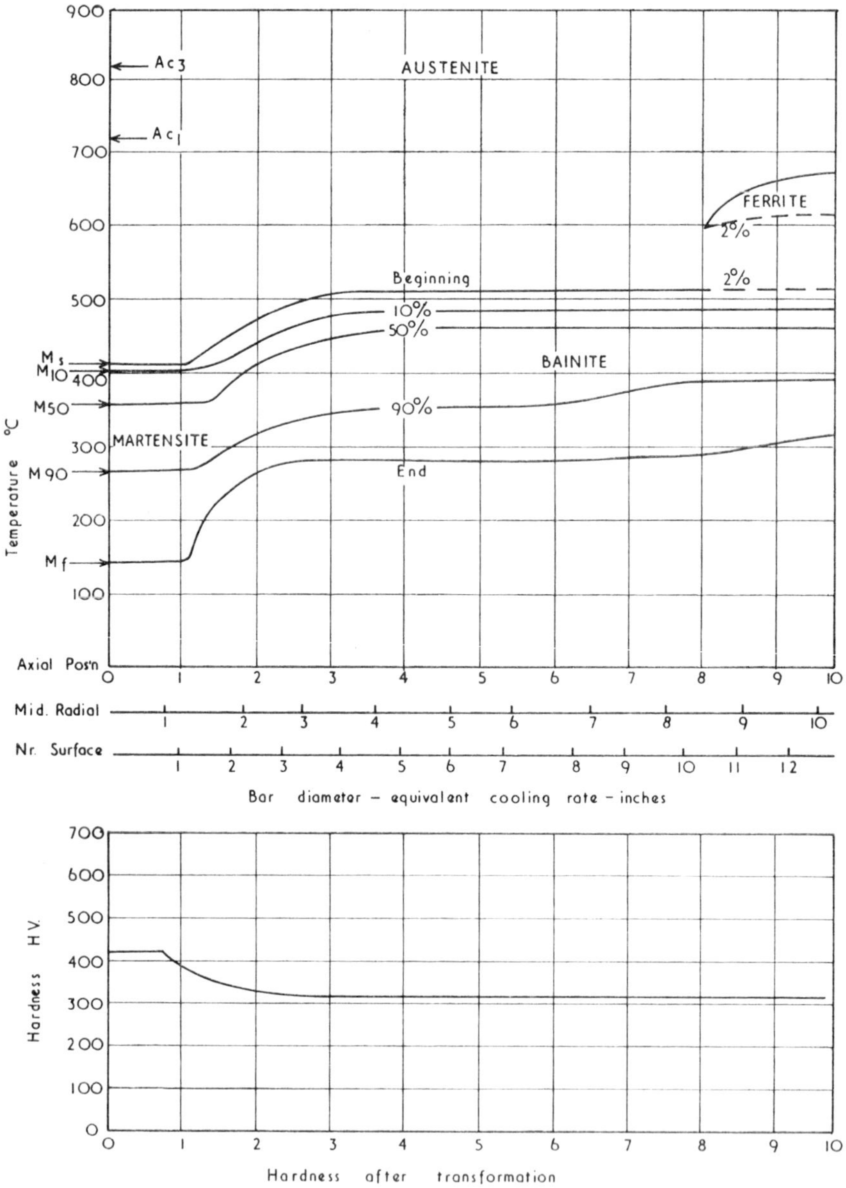

Fig. 355.6

CONTINUOUS COOLING TRANSFORMATION (Continued)

Water quenched from 810°C.

(Ref. 4)

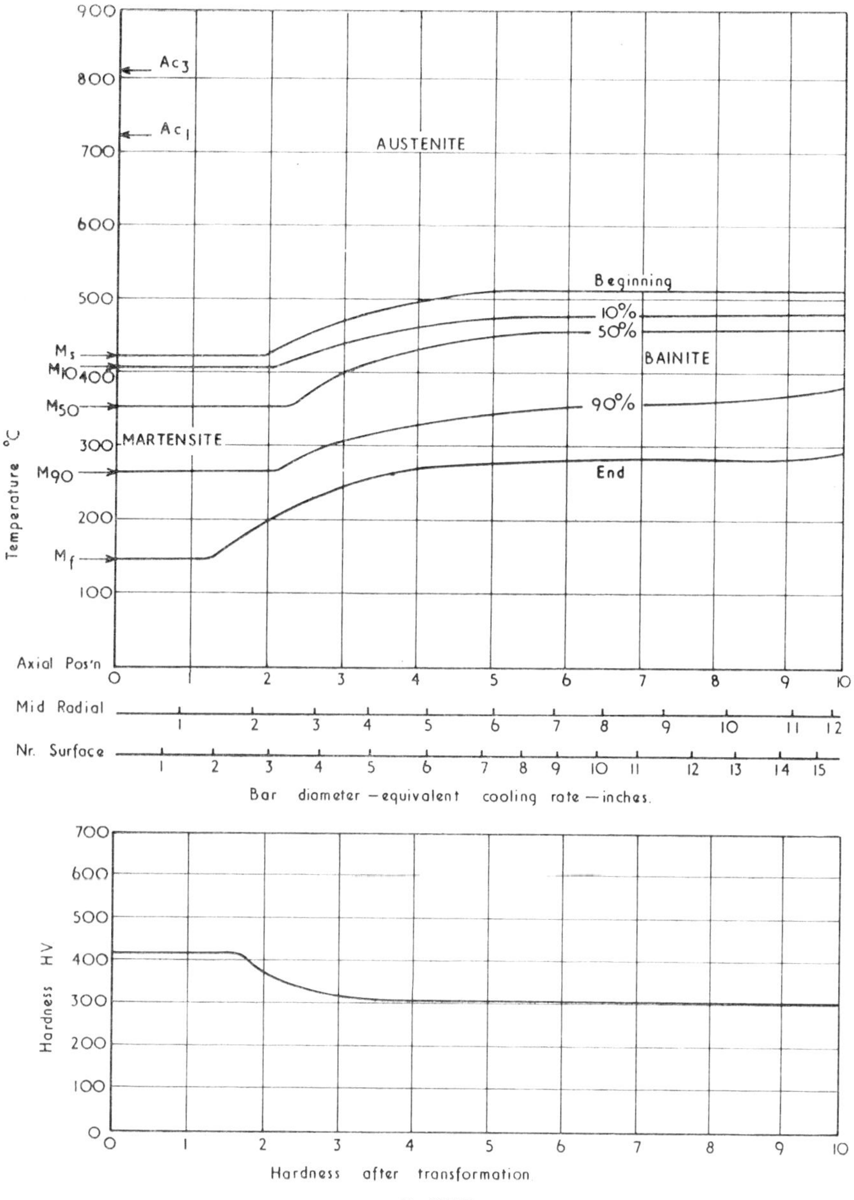

Fig. 355.7

En 355

HARDENABILITY

END QUENCH HARDENABILITY DATA

Chemical Composition (%)							
C	Si	Mn	S	P	Ni	Cr	Mo
0·20	0·23	0·61	0·011	0·015	2·00	1·65	0·19

Basic Electric Arc.
McQuaid Ehn Grain Size 3—5 (ASTM).
As quenched Grain Size 8(ASTM).
Curve A: End quenched from 870°C.
Curve B: End quenched from 800°C. (Ref. 5)

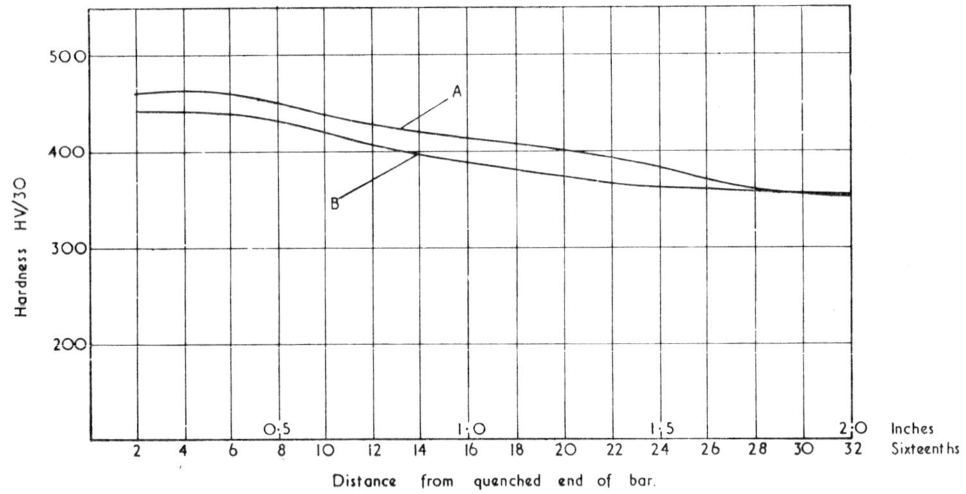

Fig. 355.8

END QUENCH HARDENABILITY BAND

These curves have been constructed from the results obtained on several casts of En 355.

(Refs. 3 and 4)

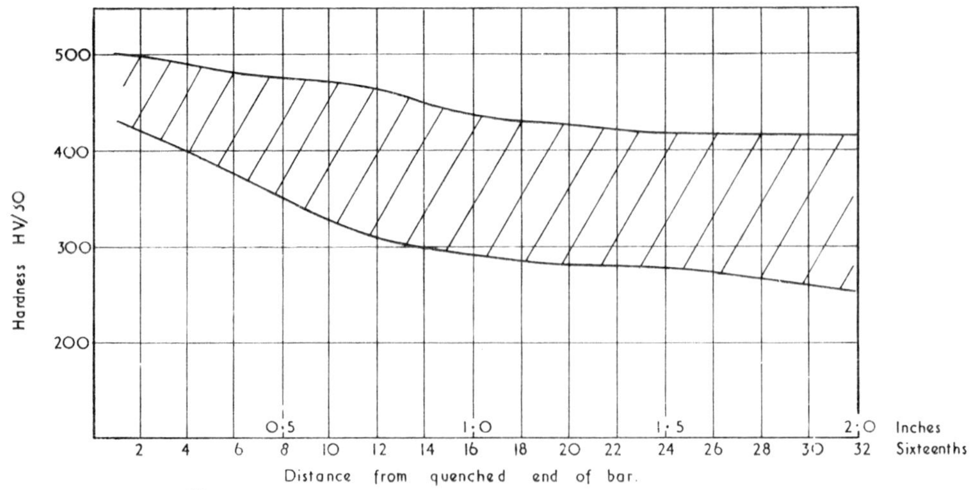

Fig. 355.9

MECHANICAL PROPERTIES AT ROOM TEMPERATURE

Process of Manufacture	ASTM Grain Size	Chemical Composition (%)								Section Size dia. (in.)	Heat Treatment °C	Tensile								Izod ft.lb.	Ref.
		C	Si	Mn	S	P	Ni	Cr	Mo			0·1% P.S. (0·1Rp)	0·2% P.S. (0·2Rp)	0·5% P.S. (0·5Rp)	Y.P. (Rm)	T.S. (Rs)	El % (A) 4√A	El % (A) 5d	R.A. % (Z)		
												tons/sq.in.									
B.E.A.		0·13	0·31	0·54	0·012	0·011	2·04	1·46	0·19	0·57	FC925 OQ830			55·6		69·0	20	15	62		2
		0·13	0·26	0·58	0·032	0·020	1·93	1·45	0·22	0·564	BC OQ810				71·6	83·2	16		49½	36	4
		"	"	"	"	"	"	"	"	"	" OQ860 OQ810				73·4	83·4	17		52½	45	"
		"	"	"	"	"	"	"	"	1⅛	" OQ810				68·0	74·3	18		49½	40	"
		"	"	"	"	"	"	"	"	"	" OQ860 OQ810				68·0	75·5	18		49½	42	"
		"	"	"	"	"	"	"	"	2 sq.	" " "				59·6	66·7	16		39	49	"
		"	"	"	"	"	"	"	"	3 sq.	" " "				57·6	66·5	17		39	48	"
		0·14		0·47			1·96	1·29	0·19	0·564	BC OQ860 OQ810				83·4	86·1	16		52	28	4
		"		"			"	"	"	1⅛	" " "				73·6	77·2	17		49	27	"
		"		"			"	"	"	2½	" " "				59·4	66·6	16		36	35	"
		0·15		0·55			1·93	1·57	0·20	0·564	BC OQ860 OQ810				86·4	88·9	16		50	33	4
		"		"			"	"	"	1⅛	" " "				80·0	86·0	16		50	30	"
		"		"			"	"	"	2½	" " "				66·0	71·0	17½		50	37	"
B.E.A.	7	0·15	0·19	0·52	0·006	0·012	1·97	1·57	0·20	½	BC OQ870 OQ810				72·7	91·8	17		53	37	2
B.E.A.	6—7	0·15	0·32	0·53	0·013	0·009	1·79	1·62	0·24	½	BC OQ870 OQ810				72·2	89·1	17		53	37	"
		0·15		0·52			1·95	1·59	0·18	1⅛	OQ800					86·0	18			34	6
		0·15	0·18	0·68			2·07	1·54	0·19	0·564	BC900 OQ860 OQ800 T200					85·0	15½		55	42	3
B.O.H.	5—6	0·16	0·28	0·52	0·032	0·025	2·02	1·67	0·19	0·564	BC OQ860 OQ810				88·4	96·2	16½		57	44	4
"	"	"	"	"	"	"	"	"	"	1⅛	" " "				87·1	90·1	17		55	40	"
"	"	"	"	"	"	"	"	"	"	2½	" " "				71·4	74·9	14		36½	48	"
B.O.H.	5—6	0·16	0·28	0·52	0·032	0·025	2·02	1·67	0·19	0·574	BC OQ870 OQ810	54·4				88·8	16		56	28	"
"	"	"	"	"	"	"	"	"	"	0·367	" " "	55·0				91·2	17		51		"
"	"	"	"	"	"	"	"	"	"	0·236	" " "	68·2				89·4	16		55		"
B.O.H.	5—7	0·16	0·25	0·53	0·028	0·023	1·96	1·58	0·20	0·564	OQ860 OQ800				84·0	91·2	16		47	31	"
"	"	"	"	"	"	"	"	"	"	"	" OQ820				83·6	89·7	16		49½	29	"
"	"	"	"	"	"	"	"	"	"	"	" OQ840				83·0	89·0	16½		49½	31	"
"	"	"	"	"	"	"	"	"	"	1⅛	" OQ810				81·9	85·9	14		43½	29	"
"	"	"	"	"	"	"	"	"	"	2½	" "				63·3	72·1	13½		33½	43	"
"	"	"	"	"	"	"	"	"	"	4	" "				52·2	65·2	16		36½	51	"
B.E.A.	6	0·16	0·22	0·51	0·008	0·011	1·97	1·55	0·20	1⅛	BC OQ870 OQ810				61·2	92·8	17		55	33	7

MECHANICAL PROPERTIES AT ROOM TEMPERATURE (Continued)

Process of Manufacture	ASTM Grain Size	Chemical Composition (%) C	Si	Mn	S	P	Ni	Cr	Mo	Section Size dia. (in.)	Heat Treatment °C	Tensile 0·1% P.S. (0·1Rp)	0·2% P.S. (0·2Rp)	0·5% P.S. (0·5Rp)	Y.P. (Rm)	T.S. (Rs)	El % (A)	R.A. % (Z)	Izod ft.lb.	Ref.
												tons/sq.in.								
B.O.H.	5—8	0·16	0·28	0·52	0·032	0·025	1·97	1·79	0·20	¼	BC AC870 OQ810 T140	58·0	67·5	77·5		89·4	20	62	—	7
"	"	"	"	"	"	"	"	"	"	½	" " " "	54·3	62·4	73·1		87·0	19	57	40	"
"	"	"	"	"	"	"	"	"	"	¾	" " " "	50·6	58·3	70·0		83·7	16	56	56	"
"	"	"	"	"	"	"	"	"	"	1⅛	" " " "	45·6	54·4	65·0		79·1	19	56	53	"
"	"	"	"	"	"	"	"	"	"	1¼	" OQ870 " "	51·8	60·6	71·8	77·0	87·0	16½	56½	44	"
B.E.A.	6—7	0·17	0·31	0·56	0·014	0·012	2·03	1·58	0·21	0·564	OQ360 OQ810				89·2	95·4	16½	52½	46	4
"	"	"	"	"	"	"	"	"	"	1⅛	" "				87·9	95·8	16½	55	38	"
"	"	"	"	"	"	"	"	"	"	2½	" "				72·8	81·1	15	36½	42	"
B.E.A.	7	0·17	0·22	0·53	0·008	0·019	1·99	1·55	0·23	1⅛	BC OQ870 OQ810				72·0	92·0	16½	56	35	2
B.E.A.	5—8	0·17	0·29	0·66	0·013	0·010	1·99	1·78	0·23	¼	BC AC870 OQ810 T140	62·0	68·3	82·5		89·4	22	65		7
"	"	"	"	"	"	"	"	"	"	½	" " " "	56·5	63·9	75·0		89·8	19	61	56	"
"	"	"	"	"	"	"	"	"	"	¾	" " " "	53·6	62·3	73·6		90·0	17	56	54	"
"	"	"	"	"	"	"	"	"	"	1⅛	" " " "	51·1	58·9	69·1		84·4	18	56	64	"
		"	"	"	"	"	"	"	"	1¼	" OQ870 " "	52·0	61·2	72·0	76·2	86·8	16½	55½	58	"
		0·17	0·22	0·58			2·10	1·43	0·20	0·564	BC900 OQ860 OQ800 T200					87·4	16½	55	50	3
B.E.A.	5	0·17	0·30	0·59	0·013	0·010	2·02	1·78	0·19	1⅛	N900 OQ740	32·0	37·3	46·3		69·4	21	47	47	"
"	"	"	"	"	"	"	"	"	"	"	" OQ760	36·7	43·6	54·0		74·8	19	61½	41	"
"	"	"	"	"	"	"	"	"	"	"	" OQ780	48·6	56·7	68·0		81·7	19	47½	41	"
"	"	"	"	"	"	"	"	"	"	"	" OQ800	51·1	60·3	73·8		91·0	17	54½	41	"
"	"	"	"	"	"	"	"	"	"	"	" OQ820	50·6	60·0	72·5		87·9	17	58	42	"
"	"	"	"	"	"	"	"	"	"	"	" OQ840	51·4	61·2	74·9		92·3	17	55	40	"
"	"	"	"	"	"	"	"	"	"	0·424	N900 OQ860 OQ800	52·5	62·3	75·3		93·8	17	53	38	"
"	"	"	"	"	"	"	"	"	"	¾	" " "	52·2	61·6	74·9		93·4	16½	52½	37	"
"	"	"	"	"	"	"	"	"	"	1	" " "	57·7	65·5	77·0		92·7	16	55	33	"
"	"	"	"	"	"	"	"	"	"	1⅛	" " "	54·9	64·6	76·9		89·2	17	53	41	"
"	"	"	"	"	"	"	"	"	"	2⅛	" " "	51·7	58·8	66·9		77·4	18	60	49	"
"	"	"	"	"	"	"	"	"	"	3	" " "	50·6	56·4	64·1		76·5	20	57	54	"
"	"	"	"	"	"	"	"	"	"	4¼ (C)	" " "	47·0	52·8	60·7		73·0	19	56	55	"
"	"	"	"	"	"	"	"	"	"	4¼ (O)	" " "	51·2	57·5	65·7		79·4	17	55½	36	"
"	"	"	"	"	"	"	"	"	"	0·424	N900 OQ820	52·7	62·0	78·2		91·7	17	53½	41	"
"	"	"	"	"	"	"	"	"	"	¾	" "	51·9	61·1	73·8		91·7	17	47½	35	"
"	"	"	"	"	"	"	"	"	"	1	" "	51·0	60·7	74·2		90·2	17	56½	36	"
"	"	"	"	"	"	"	"	"	"	1⅛	" "	50·6	60·0	72·5		87·9	17	58	42	"
"	"	"	"	"	"	"	"	"	"	2⅛	" "	52·4	59·1	68·0		80·0	17	58	56	"
"	"	"	"	"	"	"	"	"	"	3	" "	46·6	52·9	61·9		76·9	19	57	60	"
"	"	"	"	"	"	"	"	"	"	4¼ (C)	" "	46·9	52·6	60·6		74·0	19½	58	62	"
										4¼ (O)	" "	48·7	55·1	64·1		79·3	16½	52	40	"

MECHANICAL PROPERTIES AT ROOM TEMPERATURE (Continued)

Process of Manufacture	ASTM Grain Size	Chemical Composition (%) C	Si	Mn	S	P	Ni	Cr	Mo	Section Size dia. (in.)	Heat Treatment °C	Tensile 0·1% P.S. (0·1Rp) tons/sq.in.	0·2% P.S. (0·2Rp) tons/sq.in.	0·5% P.S. (0·5Rp) tons/sq.in.	Y.P. (Rm) tons/sq.in.	T.S. (Rs) tons/sq.in.	El % (A)	R.A. % (Z)	Izod ft.lb.	Ref.
B.E.A.	5	0·17	0·30	0·59	0·013	0·010	2·02	1·78	0·19	0·424	N900	44·8	51·0	59·6		71·7	19½	58	41	3
"	"	"	"	"	"	"	"	"	"	¾	"	41·5	49·2	59·4		74·6	19	54½	36	"
"	"	"	"	"	"	"	"	"	"	1	"	42·6	49·7	59·2		74·2	20	56	39	"
"	"	"	"	"	"	"	"	"	"	1⅛	"	45·5	51·4	59·6		73·2	20½	53	37	"
"	"	"	"	"	"	"	"	"	"	2⅛	"	45·1	49·8	56·2		70·4	19	55	36	"
"	"	"	"	"	"	"	"	"	"	3	"	40·3	46·3	55·4		71·0	18½	49½	35	"
"	"	"	"	"	"	"	"	"	"	4¼ (O)	"	41·4	47·0	54·4		69·4	20½	53½	33	"
										4¼ (C)	"	40·4	46·5	55·0		71·2	19	49½	20	"
	5—8	0·17	0·30	0·75	0·022	0·035	2·01	1·46	0·21	¼	BC AC870 OQ810 T140	63·5	69·8	79·4		91·6	18	62	20	7
	"	"	"	"	"	"	"	"	"	½	" " " "	55·1	63·5	73·8		91·4	18	58	27	"
	"	"	"	"	"	"	"	"	"	¾	" " " "	53·0	61·3	71·8		88·7	17	59	31	"
	"	"	"	"	"	"	"	"	"	1⅛	" " " "	48·3	56·9	56·9		84·8	19	63	34	"
	"	"	"	"	"	"	"	"	"	1¼	" OQ870 " "	58·0	65·6	76·0	82·6	90·9	16½	54	34	"
B.E.A.	5—6	0·18	0·22	0·56	0·007	0·011	1·99	1·58	0·25	½	BC OQ870 OQ810				74·5	93·5	17	54½	49	2
"	"	"	"	"	"	"	"	"	"	⅞	" " "				74·8	94·0	19	59	43	"
"	"	"	"	"	"	"	"	"	"	1⅛	" " "				74·8	94·0	17	56½	42	"
"	"	"	"	"	"	"	"	"	"	2½	" " "				69·8	86·0	18	58½	42	"
"	"	"	"	"	"	"	"	"	"	4	" " "				52·8	71·6	20	56½	43	"
B.E.A.	6—7	0·18	0·21	0·55	0·010	0·012	2·01	1·66	0·26	½	BC OQ870 OQ810				73·2	96·9	16	52	33	"
B.E.A.	6—7	0·18	0·23	0·54	0·006	0·012	1·76	1·58	0·18	1⅛	BC OQ870 OQ810				75·8	97·6	17	48	23	"
B.E.A.	5—8	0·18	0·22	0·64	0·011	0·010	2·16	1·52	0·25	0·236	BC AC870 OQ810 T140	62·5				90·8	16	60		7
"	"	"	"	"	"	"	"	"	"	0·367	" " " "	60·5				91·1	18	55		"
"	"	"	"	"	"	"	"	"	"	0·574	" " " "	55·8				90·4	16½	55		"
"	"	"	"	"	"	"	"	"	"	1¼	" OQ870 " "	51·9	63·0	73·9	77·2	90·0	17½	57½	47	"
		0·18	0·24	0·63			2·10	1·52	0·20	0·564	BC900 OQ860 OQ800 T200					93·0	13	46	30	3
		0·18	0·20	0·61			2·01	1·55	0·22	0·564	BC900 OQ860 OQ800 T200					90·6	13	51	35	"
		0·18	0·32	0·59			2·24	1·48	0·24	0·564	BC900 OQ860 OQ800 T200					87·4	16½	55	50	"
		0·18		0·55			1·80	1·64	0·22	1⅛	OQ840					88·5	16		49	6
		0·18		0·49			1·96	1·50	0·17	0·564	OQ870 OQ810				76·0	97·6	17		43	2
		0·18		0·60			2·07	1·76	0·18	1¼	OQ850 OQ810 T180				79·0	88·6	14½	52	30	4
		"		"			"	"	"	"	OQ810 T180				79·4	90·1	13	48	29	"
		"		"			"	"	"	3½ sq.	OQ850 OQ810 T180				70·5	77·3	12½	34	27	"

MECHANICAL PROPERTIES AT ROOM TEMPERATURE (Continued)

Process of Manufacture	ASTM Grain Size	Chemical Composition (%) C	Si	Mn	S	P	Ni	Cr	Mo	Section Size dia. (in.)	Heat Treatment °C	Tensile 0·1% P.S. (0·1Rp) tons/sq.in.	0·2% P.S. (0·2Rp)	0·5% P.S. (0·5Rp)	Y.P. (Rm)	T.S. (Rs)	El % (A)	R.A. % (Z)	Izod ft.lb.	Ref.
B.E.A.	6	0·19	0·25	0·54	0·009	0·010	1·88	1·70	0·25	1⅛	BC OQ870 OQ810				74·0	96·6	15½	49	24	2
B.E.A.		0·19	0·27	0·55	0·013	0·012	1·97	1·66	0·23	1⅛	BC AC870 OQ810 T140			70·8		92·0	17	52½	40	7
B.E.A.	5—8	0·19	0·26	0·57	0·009	0·011	1·97	1·73	0·17	0·219	BC OQ870 OQ810 T140	64·4	72·3	82·5		96·3	14½	64½		7
"	"	"	"	"	"	"	"	"	"	0·344	" " " "	65·3	73·1	84·0		99·8	15	57½		"
"	"	"	"	"	"	"	"	"	"	0·562	" " " "	67·8	74·8	85·2		101·8	12½	58½	40	"
"	"	"	"	"	"	"	"	"	"	0·875	" " " "	63·6	71·8	84·6		101·7	12	57½	38	"
"	"	"	"	"	"	"	"	"	"	1¼	" " " "	59·6	67·6	79·8		98·4	11½	59	42	"
"	"	"	"	"	"	"	"	"	"	"	" " " T200	69·8	77·0	85·2		98·8	17	62	42	"
"	"	"	"	"	"	"	"	"	"	"	" " " T250	73·8	80·4	89·4		92·4	16	62	38	"
"	"	"	"	"	"	"	"	"	"	"	" " " T300	76·3	81·0	88·4		95·3	16	62	25	"
"	"	"	"	"	"	"	"	"	"	"	" OQ870 OQ810	54·2	65·8	81·0		100·2	16½	58	37	"
"	"	"	"	"	"	"	"	"	"	"	" " " T140	57·8	66·1	80·2	84·6	97·3	16½	54	38	"
B.O.H.		0·20	0·22	0·57	0·017	0·027	1·95	1·64	0·21	0·564	AC860 OQ810				82·0	100·1	15½	49½	25	7
"		"	"	"	"	"	"	"	"	1⅛	" "				89·6	98·0	15½	52½	17	"
"		"	"	"	"	"	"	"	"	0·564	" "				91·8	98·5	15½	49½	29	"
"		"	"	"	"	"	"	"	"	1⅛	BC OQ800				89·2	98·3	15	49½	22	"
"		"	"	"	"	"	"	"	"	2⅞ sq.	" "				71·4	78·8	13½	31	30	"
B.E.A.	6	0·20	0·20	0·51	0·006	0·013	2·03	1·65	0·18	1⅛	BC OQ870 OQ810				77·5	93·2	17	54½	34	2
B.E.A.	6	0·20	0·24	0·52	0·020	0·012	2·11	1·55	0·22	1⅛	BC OQ870 OQ810				77·6	96·8	16	52	25	"
	5—8	0·20	0·24	0·72	0·019	0·044	1·95	1·45	0·22	0·424	BC OQ870 OQ760 T140				63·8	95·9	4½	5	22	7
	"	"	"	"	"	"	"	"	"	"	" " OQ780 "				63·2	96·4	6	17	18	"
	"	"	"	"	"	"	"	"	"	"	" " OQ810 "				60·6	96·1	10½	40½	22	"
	"	"	"	"	"	"	"	"	"	"	" " OQ810 T180				65·7	95·8	13½	59	22	"
	"	"	"	"	"	"	"	"	"	1¼	" " OQ800 T140	56·4	66·3	80·8	84·8	96·2	16½	53	33	"
B.E.A.	5—8	0·20	0·25	0·58	0·009	0·013	1·96	1·64	0·23	¼	BC AC870 OQ810 T140	67·5	73·8	83·8		97·2	20	62		7
"	"	"	"	"	"	"	"	"	"	½	" " " "	60·9	68·8	80·1		96·1	18	55	38	"
"	"	"	"	"	"	"	"	"	"	¾	" " " "	57·1	66·1	77·9		96·1	17	55	41	"
"	"	"	"	"	"	"	"	"	"	1⅛	" OQ870 " "	54·4	63·8	76·6		93·4	16	53	45	"
"	"	"	"	"	"	"	"	"	"	1¼	" " " "	54·4	64·4	77·5	82·6	95·1	16½	54½	42	"

EFFECT OF TEMPERING ON CASE HARDNESS

TYPICAL CURVE

Carburized 900°C.
Oil refined 860°C.
Oil hardened 800°C.

(Ref. 3)

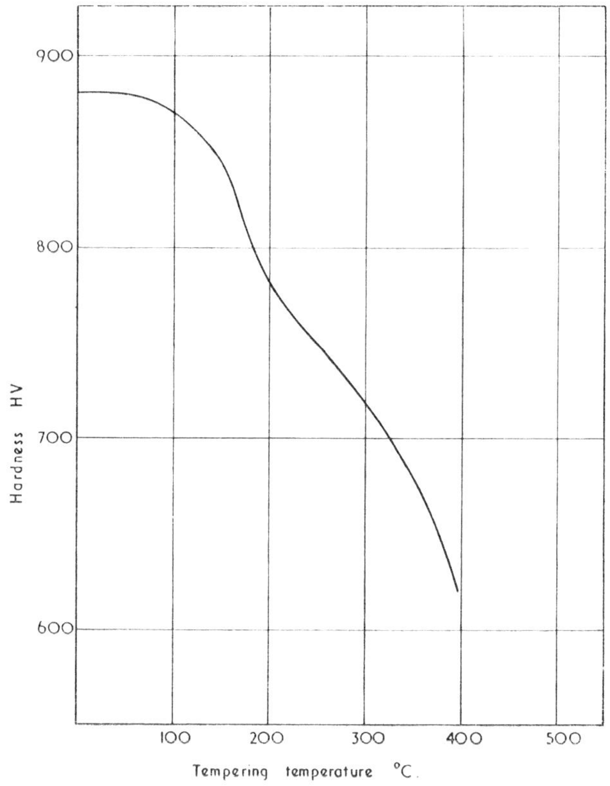

Fig. 355.10

EFFECT OF SECTION SIZE ON THE CORE PROPERTIES AT ROOM TEMPERATURE

TYPYCAL CURVE

Blank carburized 900°C.
Oil refined 860°C.
Oil hardened 800°C.
Tempered 200°C.

(Ref. 3)

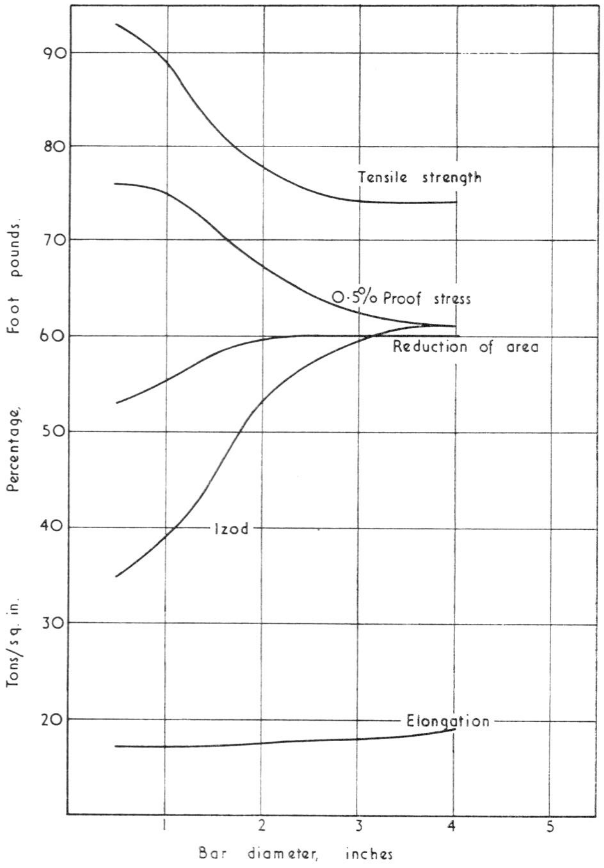

Fig. 355 11

Blank carburized 900°C.
Oil hardened 820°C.
Tempered 200°C.

(Ref. 3)

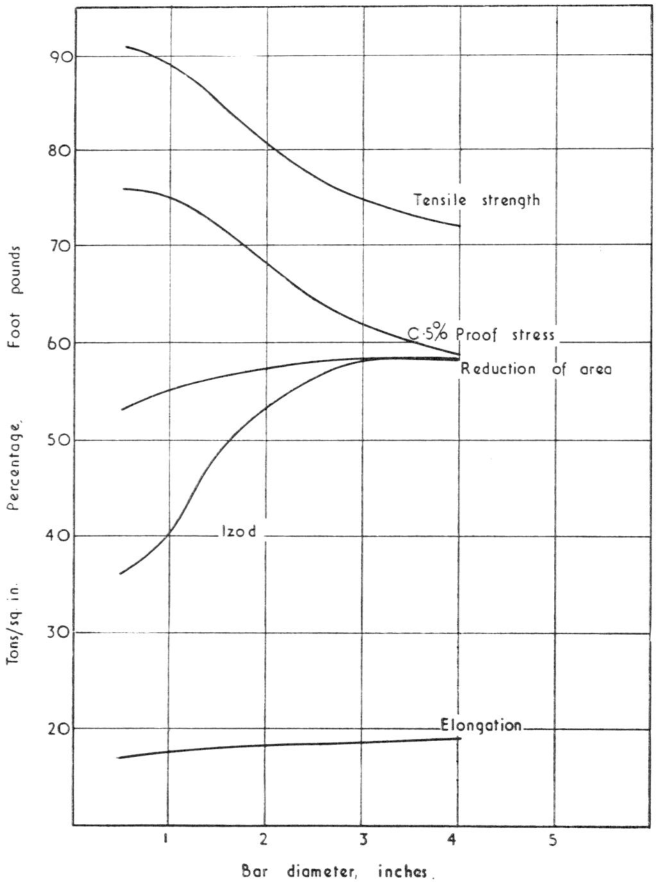

Fig. 355.12

MECHANICAL PROPERTIES AT LOW TEMPERATURE

Chemical Composition (%)								Section Size	Heat Treatment* °C	Test Temperature °C	Tensile					Ref.
C	Si	Mn	S	P	Ni	Cr	Mo				0·1% P.S. (0·1Rp)	0·2% P.S. (0·2Rp)	T.S. (Rs)	El % (A)	R.A. % (Z)	
											tons/sq.in.					
0·20	0·27	0·57	0·007	0·011	1·88	1·81	0·22	TPS	BC930 OQ880 OQ780	+20	59·5	67·6	96·3	17½	52½	8
"	"	"	"	"	"	"	"	"	" " "	—196	75·6	88·0	119·7	3	5½	"
"	"	"	"	"	"	"	"	"	Carb. 930 " "	+20	48·0	57·6	73·4	1	2	"
"	"	"	"	"	"	"	"	"	" " "	—196	61·2	74·4	105·2	1	2	"
"	"	"	"	"	"	"	"	"	" OQ780	+20	49·2	61·6	81·2	1½	3	"
"	"	"	"	"	"	"	"	"	" "	—196	68·0	78·8	107·2	1½	3½	"

* Test pieces carburized to case depth of 0·045 in., finish ground after heat treatment to case depth of 0·02 in.

IMPACT TRANSITION TEMPERATURE DATA

(for tensile test results see page 557)

Curve	Chemical Composition (%)								Section Size	Heat Treatment[1] °C	Charpy Mesnager 50% Mean Energy Impact Transition Temperature °C	Ref.
	C	Si	Mn	S	P	Ni	Cr	Mo				
A	0·20	0·27	0·57	0·007	0·011	1·88	1·81	0·22	TPS	BC930 OQ880 OQ780	—196	8
B	"	"	"	"	"	"	"	"		Carb. 930 OQ880 OQ780	0	"
C	"	"	"	"	"	"	"	"		Carb. 930 OQ780	+40	"

[1] Test piece machined to size before carburizing and heat treatment. Notch 2mm deep, root radius 2·00 mm, case depth 0·02 in.

(Ref. 8)

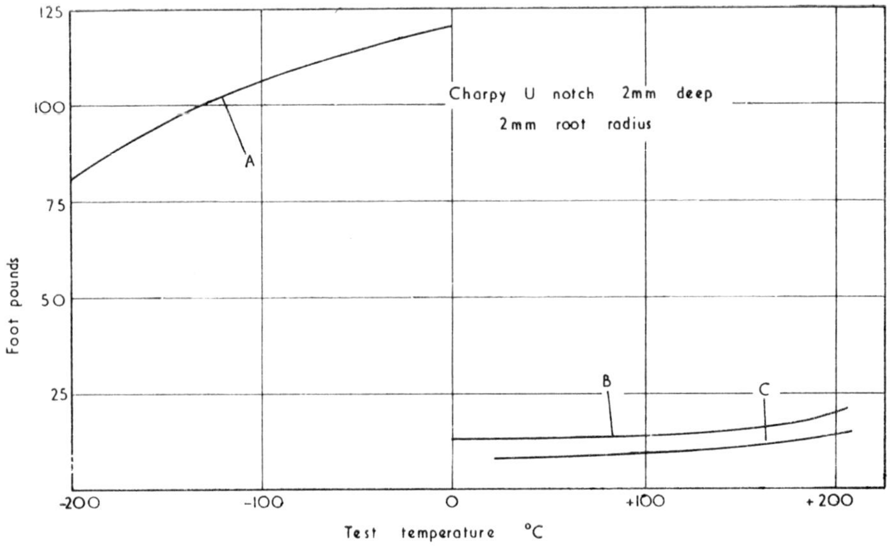

Fig. 355.13

FATIGUE PROPERTIES

Chemical Composition (%)								Section Size dia. (in.)	Heat Treatment °C	Tensile				Rotating Bending Fatigue Limit tons/sq.in.		Ref.
C	Si	Mn	S	P	Ni	Cr	Mo			0·2% P.S. (0·2Rp) tons/sq.in.	T.S. (Rs) tons/sq.in.	El % (A)	R.A. % (Z)	Unnotched	Notched*	
0·20	0·23	0·58	0·025	0·034	1·90	1·51	0·20	$\frac{3}{4}$	OH T200	74·5	94·7	16½	58	42	19	4

* Details of test piece: 0·335 in. dia. Notch 55 °V, root radius 0·008 in., depth of notch 0·020 in., diameter at root 0·295. Neuber Stress Concentration Factor 2·85.

LIST OF REFERENCES

1. Private correspondence with the Royal Aircraft Establishment, Farnborough.
2. Private correspondence with Firth Brown Ltd.
3. Private correspondence with The English Steel Corporation Ltd.
4. Private correspondence with The United Steel Companies Ltd.
5. Atlas of Isothermal Transformation Diagrams of the B. S. En Steels. Iron and Steel Institute Special Report No. 56, 1956.
6. Private correspondence with International Nickel Ltd.
7. ASRTC.
8. War Office unpublished report.

"15" CARBON LOW-ALLOY CASE-HARDENING STEEL

SPECIFICATION

CHEMICAL COMPOSITION (%)

C	Si	Mn	S	P	Ni	Cr	Mo
0·13—0·17	0·35 max.	0·70—1·00	0·050 max.	0·050 max.	0·40—0·70	0·55—0·80	0·08—0·15

MECHANICAL PROPERTIES

Tensile strength, tons/sq. in. min.(Rm)	45
Elongation, %, min.(A)	10
Izod impact value, ft. lb. min.	25

The heat treatment given to 1⅛ in. dia. test bars shall be blank carburize at a temperature between 880°C and 930°C.[1],[2] Refine at a temperature of 850°C to 880°C, cool in air, oil or water. Harden in oil from a temperature of 780°C to 820°C.

(1) By special arrangement with the purchaser the test may be carried out on samples from which the blank carburizing treatment has been omitted or modified to a short heating period.
(2) A single oil quench from 810°C to 830°C may be used in certain cases when the refining treatment after carburizing is omitted by special arrangement between the purchaser and the manufacturer.

RELATED SPECIFICATIONS

United Kingdom	America		France
En No.	AISI	SAE	AFNOR
361	8615	8615	15NCD2

APPLICATIONS

A low-alloy steel which may be carburized and hardened to produce a hard wear-resistant case and developing a core strength of the order of 45 tons/sq. in. with good resistance to shock. The steel is suitable as an alternative to En 33 and is similar to En 351. Typical applications are lightly stressed components, camshafts, gears, steering mechanisms, automobile gearbox and transmission components.

WELDING

(See Introduction for key to symbols used in this table.)

	Welding Process					
	MA	G	IG	S	F	B
En 361	c	ax	g	e	a	h

Reproduced by kind permission of H.M.S.O.

MACHINABILITY

In the softened condition the machinability is approximately 80% of that for mild steel (En 3).

HOT WORKING AND HEAT TREATMENT TEMPERATURES

Forging, rolling and stamping	1150°C, finish above 900°C.
Softening	640°C—680°C, cool in air or furnace.
Carburizing	880°C—930°C.
Refining*	850°C—880°C, cool in air, oil or water.
Hardening*	780°C—820°C, oil quench.
Tempering (for stress relief)	150°C—200°C.

* A single oil quench from 810°C to 830°C may be adopted when steel is specifically requested for this treatment.

CARBURIZING DATA

Carburizing to Temperature °C	Depth of case produced in stated time (solid pack carburizing)		Ref.
	4 hrs.	8 hrs.	
880	0·028 in.	0·038 in.	1
900	0·031 in.	0·048 in.	"
925	0·039 in.	0·052 in.	"

PHYSICAL PROPERTIES

(See also En 351)

SPECIFIC GRAVITY (d)

Chemical Composition (%)								Heat Treatment °C	Specific Gravity at 20°C	Ref.
C	Si	Mn	S	P	Ni	Cr	Mo			
Within specification								Softened	7·87	2
" "								Softened	7·85	3
0·17	0·25	0·77	0·042	0·025	0·55	0·65	0·13	BC910 OQ810	7·82	4

SPECIFIC HEAT (c)

Mean Specific Heat 20 to 100°C 0·118 cal/g. deg. C (Ref. 3).

MEAN COEFFICIENT OF THERMAL EXPANSION (α)

Chemical Composition (%)								Heat Treatment °C	$10^6 \times$ Mean Coefficient of Thermal Expansion per deg.C					Ref.
C	Si	Mn	S	P	Ni	Cr	Mo		20 to 100°C	20 to 250°C	20 to 300°C	20 to 500°C	20 to 700°C	
Within specification								Softened	12	—	13	14	14·5	3
" "								Softened	12·3	—	13·4	14·2	14·7	2
0·17	0·25	0·77	0·042	0·025	0·55	0·65	0·13	BC910 OQ810	12·8	13·5				4

ELECTRICAL RESISTIVITY (ϱ)

Chemical Composition (%)								Heat Treatment °C	Electrical Resistivity microhm—cm.		Ref.
C	Si	Mo	S	P	Ni	Cr	Mo		20°C	100°C	
Within specification								Softened	25		3
0·17	0·25	0·77	0·042	0·025	0·55	0·65	0·13	BC910 OQ810	29·1	31·7	4

YOUNG'S MODULUS (E)

Chemical Composition (%)								Heat Treatment °C	Young's Modulus* tons/sq. in.			Ref.
C	Si	Mn	S	P	Ni	Cr	Mo		20°C	100°C	200°C	
0·17	0·25	0·77	0·042	0·025	0·55	0·65	0·13	BC910 OQ810	13,400	13,200	12,800	4

* Vibration method.

TRANSFORMATION CHARACTERISTICS

ISOTHERMAL TRANSFORMATION

CORE CHARACTERISTICS

Chemical Composition (%)					
C	Si	Mn	Ni	Cr	Mo
0·15	0·19	0·81	0·55	0·63	0·11

McQuaid Ehn Grain Size 8 (ASTM).
Austenitizing Temperature 820°C, 30 min. (Ref. 2)

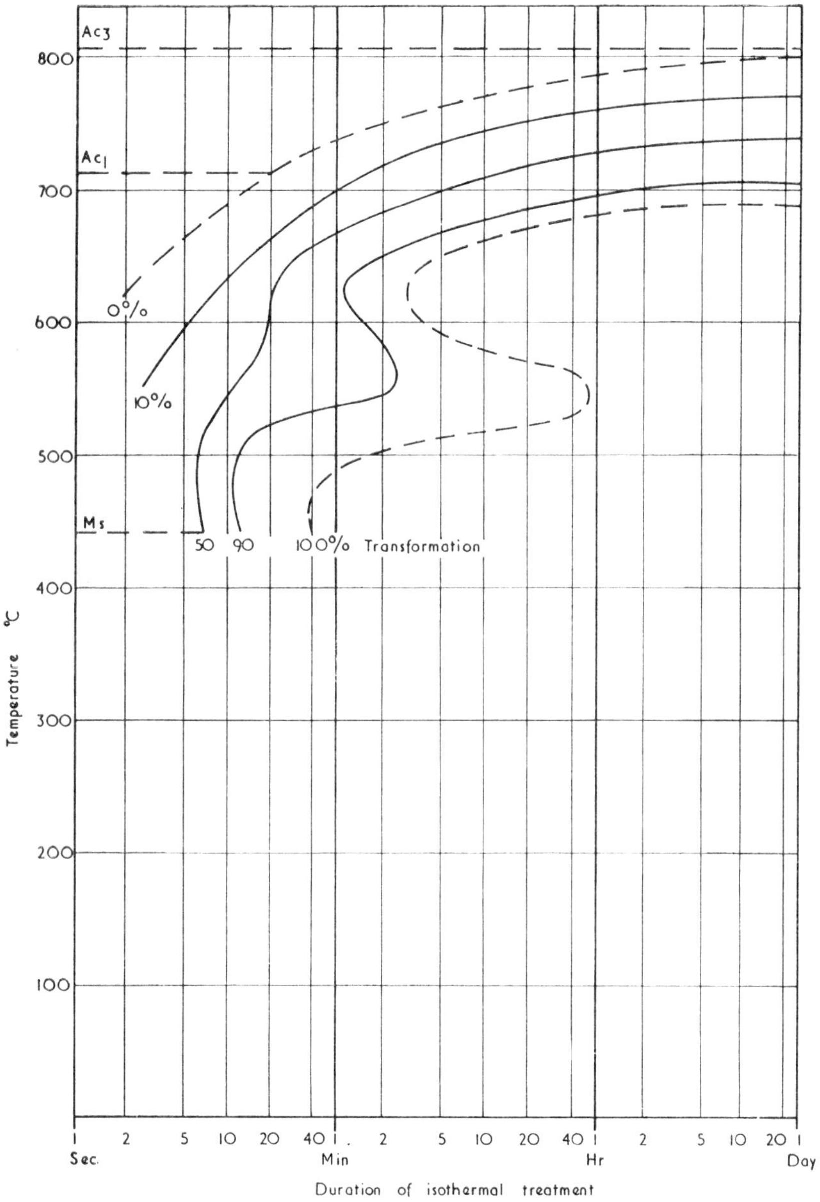

Fig. 361.1

ISOTHERMAL TRANSFORMATION (Continued)

CASE CHARACTERISTICS

C	Si	Mn	Ni	Cr	Mo
1·10	0·19	0·81	0·55	0·63	0·11

McQuaid Ehn Grain Size 8 (ASTM)
Austenitizing Temperature 820°C, 30 min.

(Ref. 2)

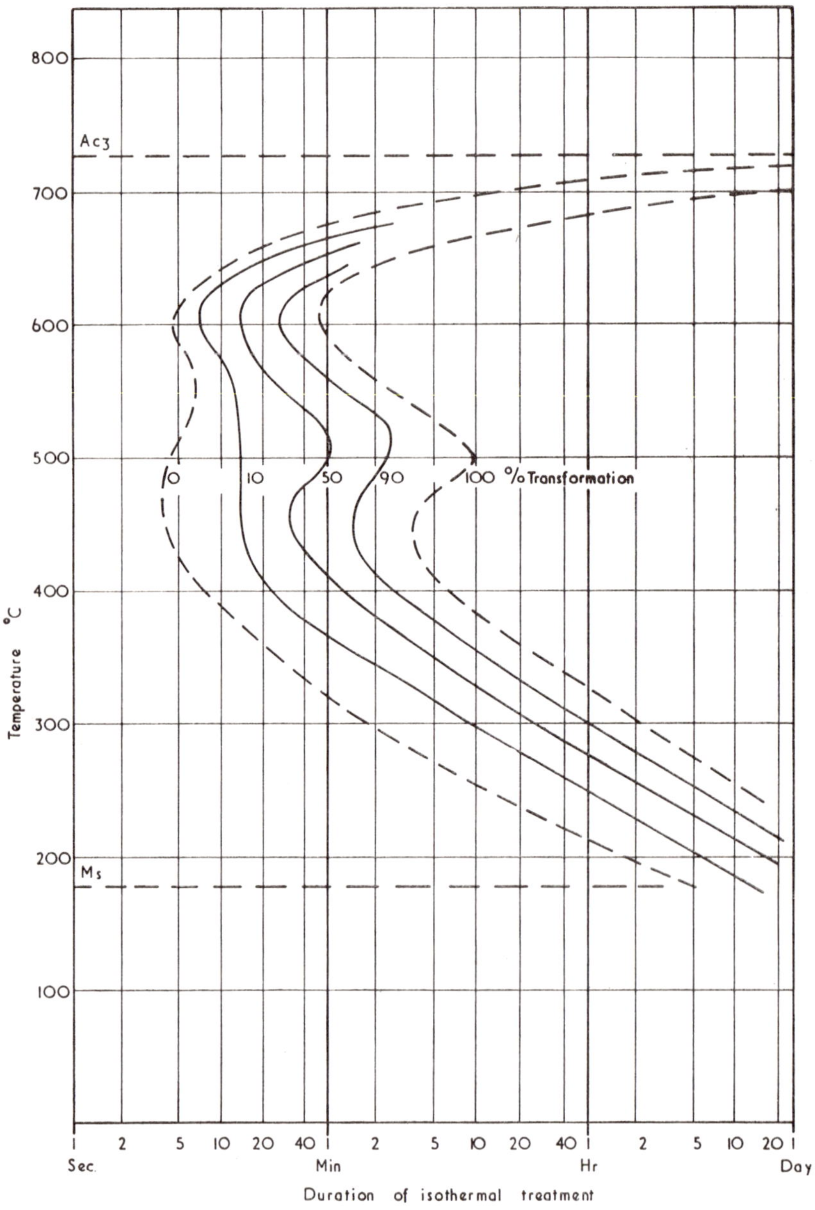

Fig. 361.2

HARDENABILITY

END QUENCH HARDENABILITY BAND

These curves have been constructed from the results obtained on several casts of En 361.

(Ref. 2)

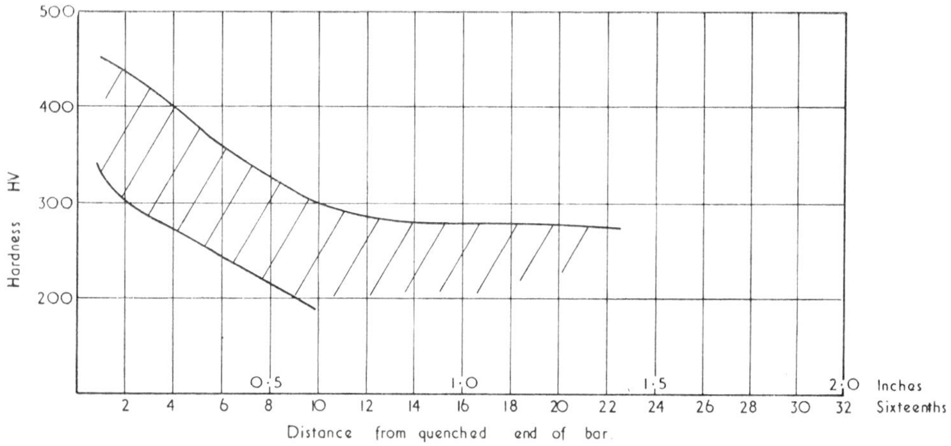

Fig. 361.3

MECHANICAL PROPERTIES AT ROOM TEMPERATURE

Process of Manufacture	ASTM Grain Size	Chemical Composition (%)								Section Size dia. (in.)	Heat Treatment °C	Tensile							Izod ft.lb.	Ref.
		C	Si	Mn	S	P	Ni	Cr	Mo			0·1% P.S. (0·1Rp)	0·2% P.S. (0·2Rp)	0·5% P.S. (0·5Rp)	Y.P. (Rm)	T.S. (Rs)	El % (A)	R.A. % (Z)		
												tons/sq.in.								
A.O.H.	7	0·13	0·27	0·81	0·016	0·030	0·60	0·60	0·10	1⅛	N900 OQ740	20·0	23·2	28·5		50·1	23	42	13	2
"	"	"	"	"	"	"	"	"	"	"	" OQ760	20·4	23·4	28·3		48·5	25	47½	24	"
"	"	"	"	"	"	"	"	"	"	"	" OQ780	20·7	23·9	28·8		48·6	21½	49½	28	"
"	"	"	"	"	"	"	"	"	"	"	" OQ800	20·6	24·0	28·9		48·2	27	50	27	"
"	"	"	"	"	"	"	"	"	"	"	" OQ820	24·4	27·9	33·5		50·9	25	50	30	"
"	"	"	"	"	"	"	"	"	"	"	" OQ840	27·0	31·1	36·5		50·6	25	57	44	"
"	"	"	"	"	"	"	"	"	"	0·424	N900 OQ860 OQ800	26·1	31·0	38·9		52·3	26½	55	65	"
"	"	"	"	"	"	"	"	"	"	¾	" " "	21·1	24·4	30·4		50·8	28½	54½	51	"
"	"	"	"	"	"	"	"	"	"	1	" " "	22·3	25·1	30·0		49·0	29	56½	62	"
"	"	"	"	"	"	"	"	"	"	1⅛	" " "	21·4	24·1	28·8		49·8	28	54½	69	"
"	"	"	"	"	"	"	"	"	"	2⅛	" " "	22·4	26·4	33·3		46·0	24½	54½	64	"
"	"	"	"	"	"	"	"	"	"	3	" " "	20·1	22·7	27·6		44·9	31	57	53	"
"	"	"	"	"	"	"	"	"	"	4¼ (O)	" " "	20·6	22·3	25·7		43·7	32½	56½	48	"
"	"	"	"	"	"	"	"	"	"	4¼ (C)	" " "	17·4	20·5	26·5		41·8	32	59	48	"
"	"	"	"	"	"	"	"	"	"	0·424	N900 OQ820	33·3	38·4	45·4		56·6	23½	60½	50	"
"	"	"	"	"	"	"	"	"	"	¾	" "	27·9	32·9	39·5		55·9	25	54	44	"
"	"	"	"	"	"	"	"	"	"	1	" "	26·1	30·0	35·9		52·5	25	53½	42	"
"	"	"	"	"	"	"	"	"	"	1⅛	" "	22·9	27·3	34·0		50·4	21½	52	41	"
"	"	"	"	"	"	"	"	"	"	2⅛	" "	20·1	24·2	30·4		45·5	26	54½	30	"
"	"	"	"	"	"	"	"	"	"	3	" "	23·4	28·2	35·0		44·9	29	55	31	"
"	"	"	"	"	"	"	"	"	"	4¼ (O)	" "	19·0	23·0	29·4		43·3	31	52½	29	"
"	"	"	"	"	"	"	"	"	"	4¼ (C)	" "	20·1	23·9	29·6		40·9	31	58½	36	"
"	"	"	"	"	"	"	"	"	"	0·424	N900	17·2	21·0	27·3		42·7	26½	53	42	"
"	"	"	"	"	"	"	"	"	"	¾	"	17·4	19·7	23·7		40·9	28	51	37	"
"	"	"	"	"	"	"	"	"	"	1	"	17·0	19·6	24·4		41·6	29	53½	45	"
"	"	"	"	"	"	"	"	"	"	1⅛	"	17·0	20·6	26·4		40·1	28	47½	40	"
"	"	"	"	"	"	"	"	"	"	2⅛	"	20·2	23·0	27·8		38·3	31	57	46	"
"	"	"	"	"	"	"	"	"	"	3	"	17·9	20·7	25·6		37·5	33	59	52	"
"	"	"	"	"	"	"	"	"	"	4¼ (O)	"	20·2	21·2	26·7		36·5	34	64	73	"
"	"	"	"	"	"	"	"	"	"	4¼ (C)	"	19·0	20·0	23·5		35·7	32	57½	58	"
		0·13	0·35	0·79			0·52	0·55	0·13	1⅛	BC900 OQ860 OQ800					45·0	30	57	70	2
		0·13	0·37	0·78			0·59	0·59	0·14	1⅛	" " "					48·0	36	57	63	"
		0·13		0·76			0·63	0·60	0·13	1⅛	OQ860 OQ810				36·0	50·5	29		34	5

MECHANICAL PROPERTIES AT ROOM TEMPERATURE (Continued)

Process of Manufacture	ASTM Grain Size	Chemical Composition (%) C	Si	Mn	S	P	Ni	Cr	Mo	Section Size dia. (in.)	Heat Treatment °C	Tensile 0·1% P.S. (0·1Rp) tons/sq.in.	0·2% P.S. (0·2Rp)	0·5% P.S. (0·5Rp)	Y.P. (Rm)	T.S. (Rs)	El % (A)	R.A. % (Z)	Izod ft.lb.	Ref.
B.E.A.	3—5	0·13	0·23	0·85	0·013	0·016	0·48	0·58	0·10	0·236	BC900 OQ870 OQ810 T140	46·0				67·5	15	60		6
"	"	"	"	"	"	"	"	"	"	0·367	" " " "	34·8				64·0	16	48		"
"	"	"	"	"	"	"	"	"	"	0·574	" " " "	32·0				56·8	20	60	18	"
"	"	"	"	"	"	"	"	"	"	$\frac{7}{8}$	" " " "	22·8				51·2	23	47		"
"	"	"	"	"	"	"	"	"	"	$1\frac{1}{4}$	" " " "	19·8				46·8	27	54	41	"
"	"	"	"	"	"	"	"	"	"	$1\frac{1}{8}$	OQ870 OQ800					50·0	25	50	54	3
		0·13	0·23	0·71			0·71	0·55	0·05	$1\frac{1}{8}$	OQ860 OQ800				22·4	40·6	$28\frac{1}{2}$	49	59	7
B.E.A.		0·14	0·17	0·82	0·007	0·012	0·51	0·65	0·11	0·236	BC900 OQ870 OQ810 T140	29·8				61·0	19	50		6
"		"	"	"	"	"	"	"	"	0·367	" " " "	30·0				63·2	12	48		"
"		"	"	"	"	"	"	"	"	0·574	" " " "	25·2				54·8	20	50	27	"
"		"	"	"	"	"	"	"	"	$\frac{7}{8}$	" " " "	21·0				50·0	21	44		"
"		"	"	"	"	"	"	"	"	$1\frac{1}{4}$	" " " "	19·8				48·8	23	46	40	"
"		"	"	"	"	"	"	"	"	$1\frac{1}{8}$	OQ870 OQ800					50·0	25	47	45	3
		0·14	0·23	0·78			0·52	0·55	0·13	$1\frac{1}{8}$	BC900 OQ860 OQ800					47·2	27	$52\frac{1}{2}$	64	2
		0·14	0·27	0·85			0·52	0·65	0·14	$1\frac{1}{8}$	" " " "					48·0	30	57	63	"
		0·15	0·19	0·81			0·55	0·63	0·11	$1\frac{1}{8}$	BC900 OQ860 OQ800					47·4	$20\frac{1}{2}$	$33\frac{1}{2}$	71	2
A.O.H.		0·15	0·23	0·72	0·015	0·021	0·63	0·71	0·11	$1\frac{1}{8}$	BC900 WQ880 OQ780					52·4	28	$52\frac{1}{2}$	42	8
"		"	"	"	"	"	"	"	"	"	" OQ880 "					52·4	26	47	35	"
B.E.A.	5	0·15	0·33	0·78	0·039	0·017	0·62	0·55	0·12	0·236	BC900 OQ870 OQ810 T140	32·0				61·5	16	40		6
"	"	"	"	"	"	"	"	"	"	0·367	" " " "	27·4				61·3	12	40		"
"	"	"	"	"	"	"	"	"	"	0·574	" " " "	23·8				56·4	20	40	15	"
"	"	"	"	"	"	"	"	"	"	$\frac{7}{8}$	" " " "	21·4				50·0	21	45		"
"	"	"	"	"	"	"	"	"	"	$1\frac{1}{4}$	" " " "	19·3				47·0	26	53	21	"
"	"	"	"	"	"	"	"	"	"	$\frac{1}{2}$	BC900 AC870 OQ810 T140	26·2	31·1	37·9		55·9	19		34	9
"	"	"	"	"	"	"	"	"	"	"	" OQ870 OQ810 T140	25·8	30·6	36·8		55·5	26		73	"
"	"	"	"	"	"	"	"	"	"	"	" SR870—380 SH810—220	21·3	24·4	29·2		49·5	34		94	"
"	"	"	"	"	"	"	"	"	"	"	" SR870—470 SH810—220	21·7	24·4	28·9		48·1	$32\frac{1}{2}$		95	"

SR = step refine. SH = step harden.

MECHANICAL PROPERTIES AT ROOM TEMPERATURE (Continued)

Process of Manufacture	Chemical Composition (%) C	Si	Mn	S	P	Ni	Cr	Mo	Section Size dia. (in.)	Heat Treatment °C	Tensile 0·1% P.S. (0·1Rp) tons/sq.in.	0·2% P.S. (0·2Rp) tons/sq.in.	0·5% P.S. (0·5Rp) tons/sq.in.	Y.P. (Rm) tons/sq.in.	T.S. (Rs) tons/sq.in.	El % (A)	R.A. % (Z)	Izod ft.lb.	Ref.
B.E.A.	0·15	0·16	0·72	0·012	0·012	0·54	0·65	0·13	0·169	BC900 OQ870 OQ810 T140	30·8				63·0	15	50		6
"	"	"	"	"	"	"	"	"	0·236	" " " "	43·5				65·7	15	50		"
"	"	"	"	"	"	"	"	"	0·367	" " " "	40·8				67·2	$15\frac{1}{2}$	50		"
"	"	"	"	"	"	"	"	"	0·574	" " " "	34·9				57·2	19	62	21	"
"	"	"	"	"	"	"	"	"	$\frac{7}{8}$	" " " "	20·8				51·6	24	55		"
"	"	"	"	"	"	"	"	"	$1\frac{1}{4}$	" " " "	16·0				44·4	27	52	27	"
"	"	"	"	"	"	"	"	"	$1\frac{1}{8}$	OQ870 OQ800					58·0	22	40	36	?
A.O.H.	0·16	0·33	0·83	0·010	0·016	0·59	0·64	0·11	$1\frac{1}{8}$	BC900 OQ820					52·8	26	47	22	8
"	"	"	"	"	"	"	"	"	"	" OQ840					53·6	24	50	43	"
A.O.H.	0·16	0·34	0·86	0·013	0·022	0·62	0·67	0·08	$1\frac{1}{8}$	" OQ820					50·8	27	$52\frac{1}{2}$	37	"
"	"	"	"	"	"	"	"	"	"	" OQ840					49·6	27	55	37	"
	0·16		0·86			0·58	0·71	0·10	$1\frac{1}{8}$	OQ880 OQ780					53·8	20		37	5
	0·17	0·27	0·80			0·58	0·62	0·10	$1\frac{1}{8}$	BC900 OQ860 OQ800					45·0	30	57	70	2
	0·17	0·30	0·80			0·75	0·58	0·02	$1\frac{1}{8}$	OQ870 OQ780				25·0	48·0	23	$38\frac{1}{2}$	46	7
A.O.H.	0·17	0·31	0·67	0·037	0·032	0·56	0·34	0·10	$1\frac{1}{8}$	OQ810				39·6	50·0	23	53	24	3
	0·17		0·75			0·57	0·68	0·11	$1\frac{1}{8}$	OQ840					53·2	21		49	5

EFFECT OF TEMPERING ON CASE HARDNESS

(TYPICAL CURVE)

Carburized 900°C.
Oil refined 860°C.
Oil hardened 800°C.

(Ref. 3)

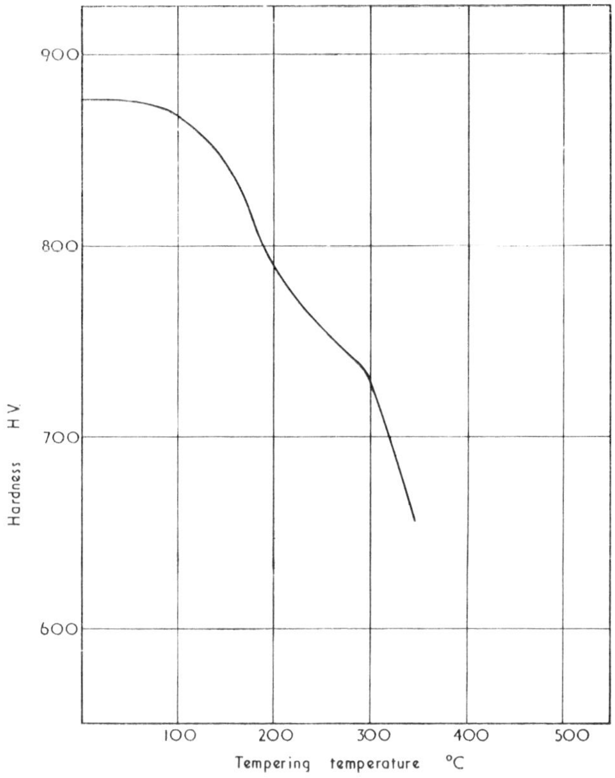

Fig. 361.4

EFFECT OF SECTION SIZE ON THE CORE PROPERTIES AT ROOM TEMPERATURE

(TYPICAL CURVES)

Blank carburized 900°C.
Oil refined 860°C.
Oil hardened 800°C.

(Ref. 3)

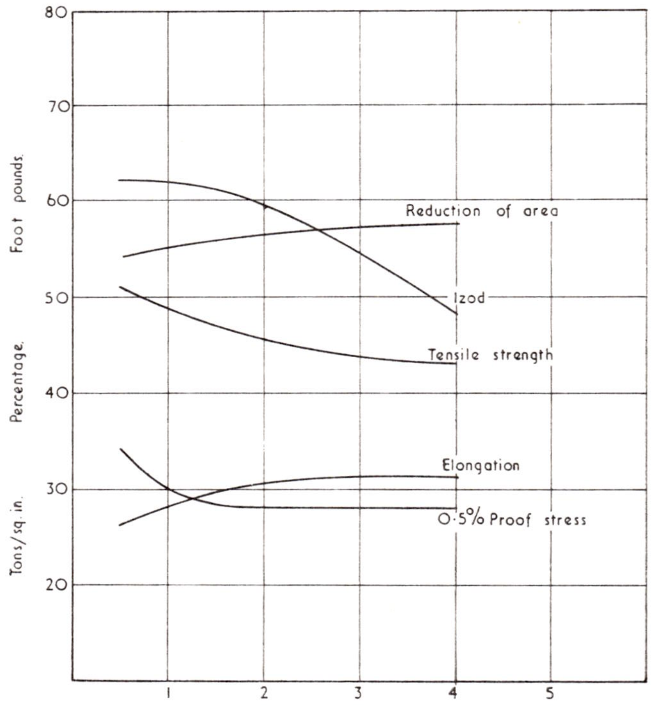

Fig. 361.5

Blank carburized 900°C.
Oil hardened 820°C.
(Single quench quality).

(Ref. 3)

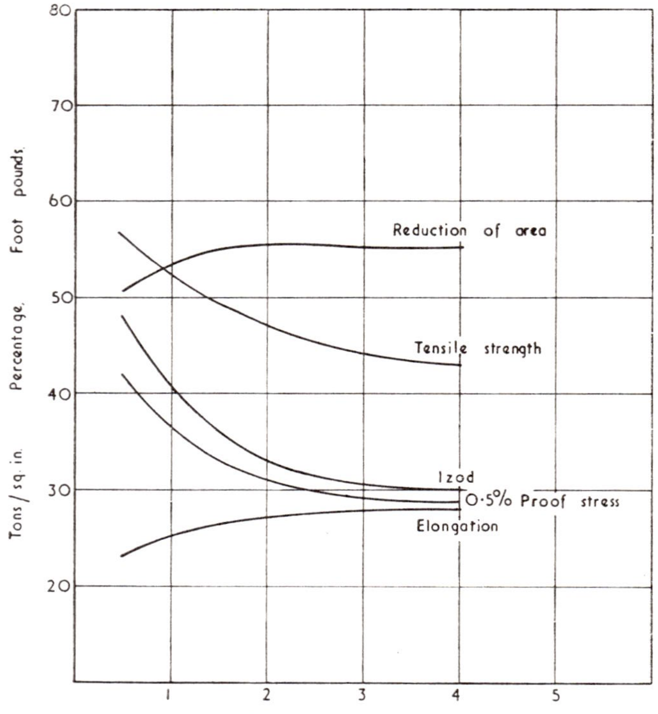

Fig. 361.6

LIST OF REFERENCES

1. Private correspondence with the Royal Aircraft Establishment, Farnborough.
2. Private correspondence with The English Steel Corporation Ltd.
3. Private correspondence with Firth Brown Ltd.
4. Private correspondence with The United Steel Companies Ltd.
5. Private correspondence with International Nickel Ltd.
6. Ministry of Supply, Alloy Steels Research Technical Committee Report.
7. Private correspondence with Kayser Ellison and Co. Ltd.
8. Private correspondence with Jessop–Saville Ltd.
9. Private correspondence with the De Havilland Aircraft Co. Ltd.

"20" CARBON LOW-ALLOY CASE-HARDENING STEEL

SPECIFICATION

CHEMICAL COMPOSITION (%)

C	Si	Mn	S	P	Ni	Cr	Mo
0·18—0·23	0·35 max.	0·70—1·00	0·050 max.	0·050 max.	0·40—0·70	0·55—0·80	0·08—0·15

MECHANICAL PROPERTIES

Tensile strength, tons/sq. in. min.(Rm)	55
Elongation, % min. (A)	15
Izod impact value, ft. lb., min.	15

The heat treatment given to test bars shall be blank carburize at a temperature between 880°C and 930°C.[1], [2] Refine at a temperature of 850°C to 880°C, cool in air, oil or water. Harden in oil from a temperature of 780°C to 820°C.

[1] By special arrangement with the purchaser the test may be carried out on samples from which the blank carburizing treatment has been omitted or modified to a short heating period.

[2] A single oil quench from 810°C to 830°C may be used in certain cases when the refining treatment after carburizing is omitted by special arrangement between the purchaser and the manufacturer.

RELATED SPECIFICATIONS

United Kingdom	America		France
En No.	SAE	AISI	AFNOR
362	8620	8620	20 NCD2

APPLICATIONS

A low-alloy case-hardening steel which may be carburized and hardened to produce a hard wear-resistant case and developing a core strength of the order of 55 tons/sq. in. The steel can be used as an alternative to En 35 or En 36A. Due to the high carbon content of the core, the core strength in small sections may be very high with consequent reduction in toughness. Typical applications are gears, pinions, gudgeon pins, layshafts, cams, camshafts, mining haulage, cage suspension and lifting gear.

WELDING

(See Introduction for key to symbols used in this table.)

	Welding Process					
	MA	G	IG	S	F	B
En 362	c	ax	g	e	a	h

Reproduced by kind permission of H.M.S.O.

MACHINABILITY

In the softened condition the machinability is approximately 80% of that for mild steel (En 3).

HOT WORKING AND HEAT TREATMENT TEMPERATURES

Forging, rolling and stamping	1150°C, finish above 900°C.
Softening	640°C—680°C, cool in air or furnace.
Carburizing	880°C—930°C.
Refining*	850°C—880°C, cool in air, oil or water.
Hardening*	780°C—820°C, oil quench.
Tempering (for stress relief)	150°C—200°C.

* A single oil quench from 810°C to 830°C may be adopted when steel is specifically requested for this treatment.

PHYSICAL PROPERTIES

SPECIFIC GRAVITY (d)

Chemical Composition (%)								Heat Treatment °C	Specific Gravity at 20°C	Ref.
C	Si	Mn	S	P	Ni	Cr	Mo			
Within specification								Softened	7·85	1
"	"							Softened	7·87	2
0·19	0·26	0·83	0·025	0·017	0·53	0·73	0·12	BC910 OQ810	7·84	3

SPECIFIC HEAT(c)

The mean specific heat between 20 and 100°C is 0·118 cal/g. deg. C (Ref. 1).

MEAN COEFFICIENT OF THERMAL EXPANSION (α)

Chemical Composition (%)								Heat Treatment °C	$10^6\times$ Mean Coefficient of Thermal Expansion per deg. C					Ref.
C	Si	Mn	S	P	Ni	Cr	Mo		20 to 100°C	20 to 200°C	20 to 300°C	20 to 500°C	20 to 700°C	
Within specification								Softened	12		13	14	14·5	1
" "								Softened	12·5		13·4	14·1	14·7	2
0·19	0·26	0·83	0·025	0·017	0·53	0·73	0·12	BC 910 OQ810		12·6				3

ELECTRICAL RESISTIVITY (ρ)

Chemical Composition (%)								Heat Treatment °C	Electrical Resistivity microhm-cm			Ref.
C	Si	Mn	S	P	Ni	Cr	Mo		20°C	100°C	200°C	
Within specification								Softened	25			1
0·19	0·26	0·83	0·025	0·017	0·53	0·73	0·12	BC910 OQ810	28·5	33·5	40·3	3

YOUNG'S MODULUS (E)

Chemical Composition (%)								Heat Treatment °C	Young's Modulus* tons/sq.in.			Ref.
C	Si	Mn	S	P	Ni	Cr	Mo		20°C	100°C	200°C	
0·19	0·26	0·83	0·025	0·017	0·53	0·73	0·12	BC910 OQ810	13,550	13,300	12,900	3

* Vibration method.

TRANSFORMATION CHARACTERISTICS

ISOTHERMAL TRANSFORMATION

CORE CHARACTERISTICS

Chemical Composition (%)					
C	Si	Mn	Ni	Cr	Mo
0·20	0·20	0·70	0·60	0·68	0·12

McQuaid Ehn Grain Size 8 (ASTM).

Austenitizing Temperature 820°C, 30 min.

(Ref. 2)

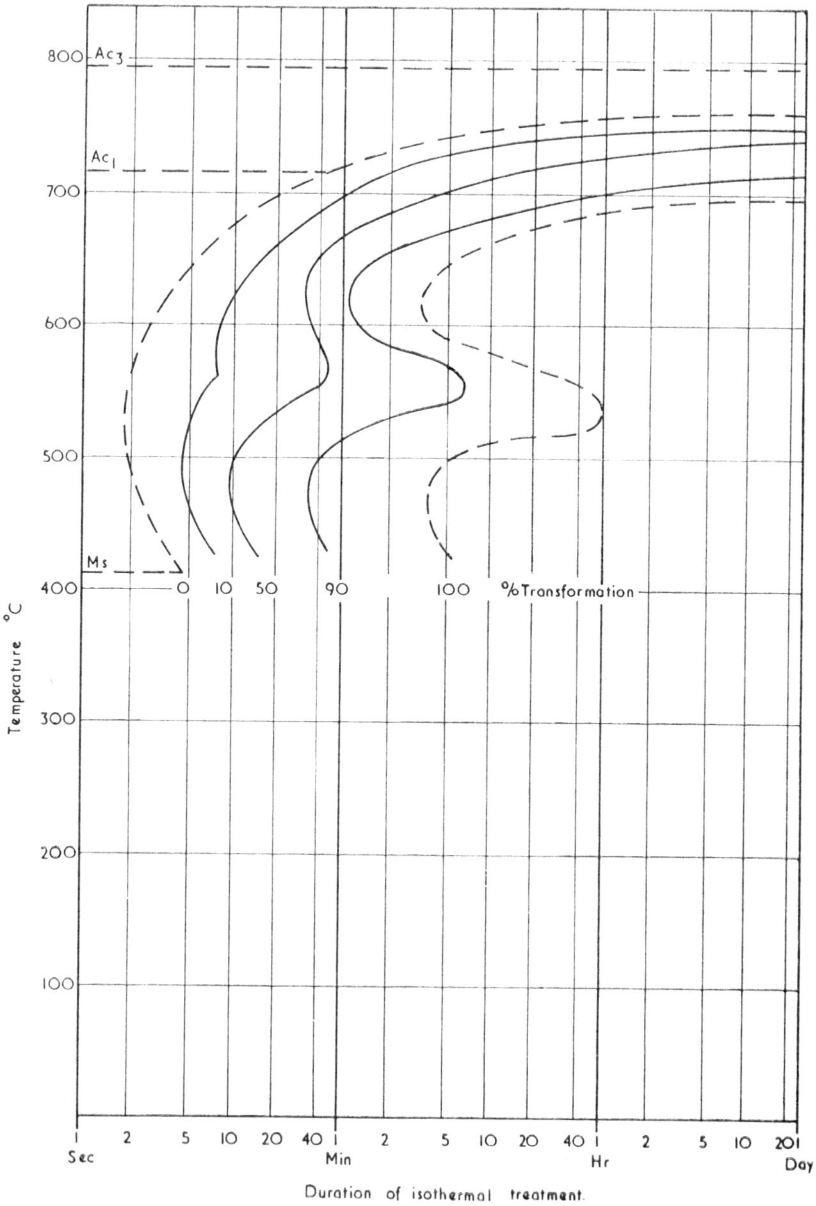

Fig. 362.1

ISOTHERMAL TRANSFORMATION (Continued)

CASE CHARACTERISTICS

Chemical Composition (%)					
C	Si	Mn	Cr	Mo	Ni
0·98	0·20	0·70	0·60	0·68	0·12

McQuaid Ehn Grain Size 8 (ASTM).

Austenitizing Temperature 820°C, 30 min.

(Ref. 2)

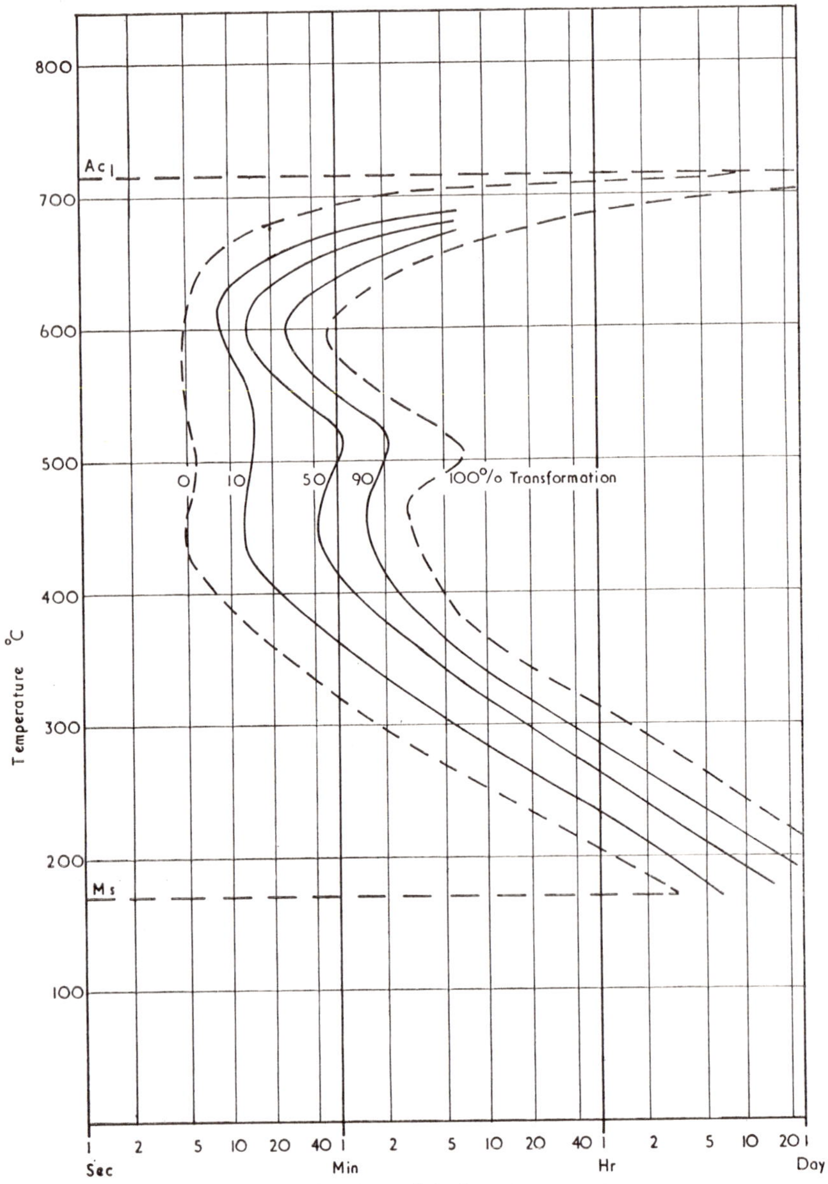

Fig. 362.2

HARDENABILITY

END QUENCH HARDENABILITY BAND

These curves have been constructed from the results obtained on several casts of En 362.

(Ref. 2)

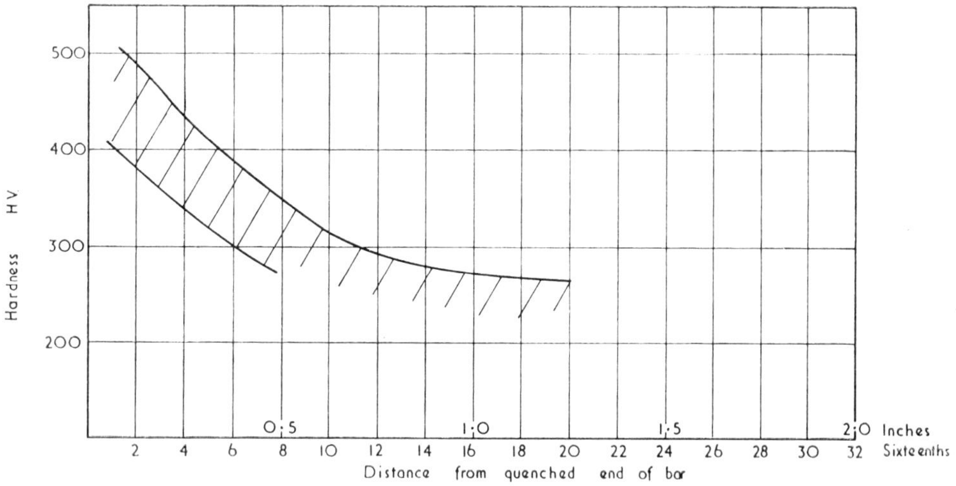

Fig. 362.3

MECHANICAL PROPERTIES AT ROOM TEMPERATURE

Process of Manufacture	ASTM Grain Size	Chemical Composition (%) C	Si	Mn	S	P	Ni	Cr	Mo	Section Size dia. (in.)	Heat Treatment °C	Tensile 0·1% P.S. (0·1Rp) tons/sq.in.	0·2% P.S. (0·2Rp)	0·5% P.S. (0·5Rp)	Y.P. (Rm)	T.S. (Rs)	El % (A)	R.A. % (Z)	Izod ft.lb.	Ref.
B.O.H.	6—8	0·18	0·23	0·83	0·033	0·045	0·54	0·66	0·12	1⅛	BC OQ800				50·0	60·3	17½	31	8	3
"	"	"	"	"	"	"	"	"	"	"	" OQ820				52·4	60·3	17½	39	10	"
"	"	"	"	"	"	"	"	"	"	"	" OQ840				52·7	60·9	21	42	19	"
"	"	"	"	"	"	"	"	"	"	"	" OQ860				51·6	60·8	21½	47	29	"
"	"	"	"	"	"	"	"	"	"	"	OQ860 OQ800				48·2	60·1	21½	44½	27	"
"	"	"	"	"	"	"	"	"	"	"	" OQ840				54·0	63·6	19	39	27	"
B.O.H.	5—6	0·18	0·25	0·87	0·038	0·027	0·56	0·74	0·13	1⅛	BC OQ800				51·8	59·2	18	39	11	"
"	"	"	"	"	"	"	"	"	"	"	" OQ820				51·2	59·2	19	42	14	"
"	"	"	"	"	"	"	"	"	"	"	" OQ840				49·0	57·9	21	49½	33	"
"	"	"	"	"	"	"	"	"	"	"	" OQ860				51·2	58·8	21½	55	32	"
"	"	"	"	"	"	"	"	"	"	"	OQ860 OQ800				51·6	60·0	22½	44½	28	"
"	"	"	"	"	"	"	"	"	"	"	" OQ840				52·4	62·3	20	55	42	"
H.F.		0·19	0·30	0·87	0·021	0·023	0·62	0·80	0·06	1⅛	BC OQ880 OQ780	34·0				62·4	15	31	18	4
"		"	"	"	"	"	"	"	"	"	" " OQ800	34·5				71·2	10	21½	15	"
H.F.		0·19	0·21	0·75	0·025	0·017	0·47	0·60	0·16	1⅛	OQ860 OQ800 T180				50·5	59·6	18	42	22	3
		0·20	0·20	0·70			0·60	0·68	0·12	1⅛	BC900 OQ860 OQ800					56·0	22	52	45	2
H.F.		0·20	0·19	0·73	0·025	0·016	0·47	0·61	0·08	1⅛	OQ860 OQ800 T180				47·0	55·4	19	39	20	3
B.O.H.	7	0·20	0·23	0·95	0·042	0·034	0·70	0·87	0·12	1⅛	BC OQ800				69·0	76·8	16	36½	9	"
"	"	"	"	"	"	"	"	"	"	"	" OQ820				67·6	76·8	16½	47	12	"
"	"	"	"	"	"	"	"	"	"	"	" OQ840				67·2	77·2	16	47	15	"
"	"	"	"	"	"	"	"	"	"	"	" OQ860				71·6	74·4	17	49½	15	"
"	"	"	"	"	"	"	"	"	"	"	OQ860 OQ800				69·2	81·4	13½	36½	15	"
"	"	"	"	"	"	"	"	"	"	"	" OQ840				76·8	81·5	14½	47	16	"
B.E.		0·21		0·90			0·65	0·60	0·17	1⅛	OQ860 OQ800				34·6	56·3	24½	46½	51	1
		0·21		0·87			0·64	0·67	0·13	¾	OQ820					65·1	19		39	5
		"		"			"	"	"	1⅛	"					58·5	22½		43	"
		"		"			"	"	"	2⅛	"					52·4	28		37	"
		"		"			"	"	"	3	"					50·2	25		45	"
		"		"			"	"	"	¾	OQ860 OQ800					65·5	19		34	5
		"		"			"	"	"	1⅛	" "					60·0	21		35	"
		"		"			"	"	"	2⅛	" "					53·1	25		37	"
		"		"			"	"	"	3	" "					51·4	25½		35	"
		"		"			"	"	"	4¼	" "					48·5	27½		35	"

MECHANICAL PROPERTIES AT ROOM TEMPERATURE (Continued)

Process of Manufacture	ASTM Grain Size	Chemical Composition (%) C	Si	Mn	S	P	Ni	Cr	Mo	Section Size dia. (in.)	Heat Treatment °C	Tensile 0·1% P.S. (0·1Rp) tons/sq.in.	0·2% P.S. (0·2Rp)	0·5% P.S. (0·5Rp)	Y.P. (Rm)	T.S. (Rs)	El % (A)	R.A. % (Z)	Izod ft.lb.	Ref.
		0·21	0·21	0·77			0·56	0·67	0·14	1⅛	BC900 OQ860 OQ800					59·0	22	51	41	2
		0·21	0·30	0·80			0·62	0·71	0·14	1⅛	" " "					62·0	21	52	40	"
		0·21	0·28	0·82			0·61	0·62	0·14	1⅛	" " "					64·5	21	44½	34	"
		0·21		0·72			0·47	0·71	0·13	½	OQ870 OQ810					65·3	18	46	37	1
		"		"			"	"	"	1⅛	" "					58·4	22	48	35	"
		"		"			"	"	"	2½	" "					52·4	26	45	35	"
		"		"			"	"	"	4	" "					47·4	30	45	35	"
H F.		0·22	0·21	0·97	0·025	0·019	0·67	0·77	0·15	1⅛	OQ860 OQ800 T180				78·4	85·1	13	33½	20	3
		0·22	0·23	0·82			0·57	0·62	0·11	1⅛	BC900 OQ860 OQ800					59·0	21	49½	49	2
		0·22	0·35	0·75			0·62	0·70	0·12	1⅛	" " "					59·5	26	47	41	"
		0·22	0·26	0·71			0·57	0·62	0·15	1⅛	" " "					60·5	20	44½	31	"
		0·22	0·32	0·83			0·60	0·65	0·12	1⅛	" " "					62·0	24	44½	35	"
		0·22	0·24	0·78			0·60	0·67	0·15	1⅛	" " "					62·0	20	49½	35	"
H.F.		0·23	0·20	0·91	0·024	0·019	0·67	0·69	0·08	1⅛	OQ860 OQ800 T180				68·4	74·6	13½	36½	15	3
		0·23		0·81			0·55	0·61	0·15	1⅛	OQ820					64·3	20		21	5
		"		"			"	"	"	"	OQ860 OQ820					70·0	16½		22	"
A.O.H.	8	0·23	0·32	0·79	0·012	0·022	0·66	0·63	0·17	1⅛	N900 OQ740	25·2	29·3	36·1		61·9	18½	33½	10	2
"	"	"	"	"	"	"	"	"	"	"	" OQ760	27·4	32·2	39·1		63·7	22½	39	15	"
"	"	"	"	"	"	"	"	"	"	"	" OQ780	29·2	34·2	42·0		65·4	17	31	11	"
"	"	"	"	"	"	"	"	"	"	"	" OQ800	31·8	37·5	45·2		62·9	21	46½	24	"
"	"	"	"	"	"	"	"	"	"	"	" OQ820	36·5	42·3	50·2		65·2	19	46	25	"
"	"	"	"	"	"	"	"	"	"	"	" OQ840	30·5	36·0	43·3		61·4	20	43½	26	"
"	"	"	"	"	"	"	"	"	"	0·424	N900 OQ820	48·3	57·4	69·3		82·6	13½	45½	27	2
"	"	"	"	"	"	"	"	"	"	¾	" "	34·6	44·0	49·2		65·8	18	45	27	"
"	"	"	"	"	"	"	"	"	"	1	" "	36·0	41·1	48·0		63·5	22	49	34	"
"	"	"	"	"	"	"	"	"	"	1⅛	" "	37·1	42·2	49·3		63·6	20	49	32	"
"	"	"	"	"	"	"	"	"	"	2⅛	" "	30·5	35·0	40·8		57·8	23	48½	38	"
"	"	"	"	"	"	"	"	"	"	3	" "	28·6	32·8	38·4		57·4	24½	44	37	"
"	"	"	"	"	"	"	"	"	"	4¼(O)	" "	29·8	33·4	38·4		55·1	27	50	40	"
"	"	"	"	"	"	"	"	"	"	4¼(C)	" "	27·3	31·1	36·6		54·8	24	42	31	"
"	"	"	"	"	"	"	"	"	"	0·424	N900 OQ860 OQ800	45·2	52·9	63·4		79·6	16	47	27	"
"	"	"	"	"	"	"	"	"	"	¾	" " "	35·6	42·6	51·7		68·5	18	46½	26	"
"	"	"	"	"	"	"	"	"	"	1	" " "	36·6	41·7	48·4		64·2	21	49½	34	"
"	"	"	"	"	"	"	"	"	"	1⅛	" " "	36·0	41·6	49·5		65·8	21	47	34	"
"	"	"	"	"	"	"	"	"	"	2⅛	" " "	31·2	35·6	41·6		59·5	22	45½	35	"
"	"	"	"	"	"	"	"	"	"	3	" " "	28·6	32·7	38·1		56·8	24	44½	35	"
"	"	"	"	"	"	"	"	"	"	4¼(O)	" " "	27·5	31·4	36·6		56·4	25½	42½	30	"
"	"	"	"	"	"	"	"	"	"	4¼(C)	" " "	23·6	27·4	32·9		53·2	24	43½	33	"

EFFECT OF TEMPERING TEMPERATURE ON CASE HARDNESS

Carburized 900°C.
Oil refined 860°C.
Oil hardened 800°C.

(Ref. 2)

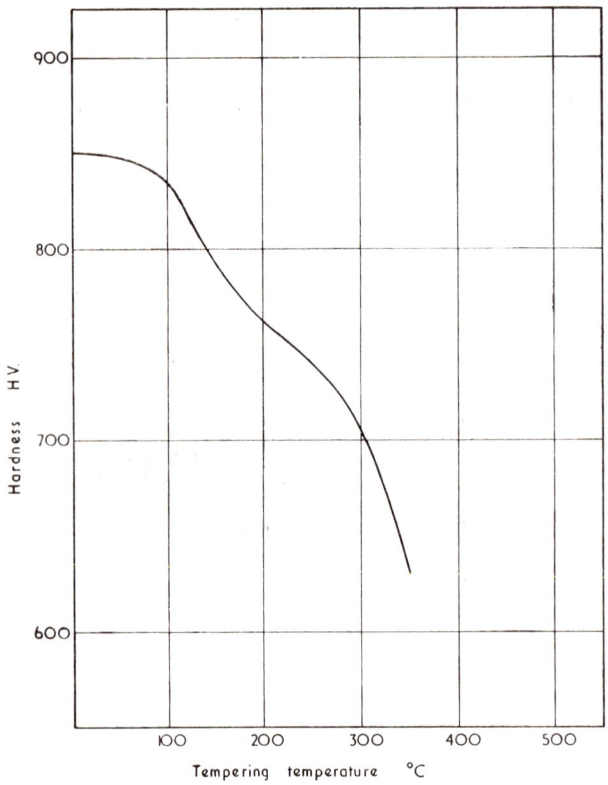

Fig. 362.4

EFFECT OF SECTION SIZE ON THE CORE PROPERTIES AT ROOM TEMPERATURE

Blank carburized 900°C.
Oil refined 860°C.
Oil hardened 800°C.

(Ref. 2)

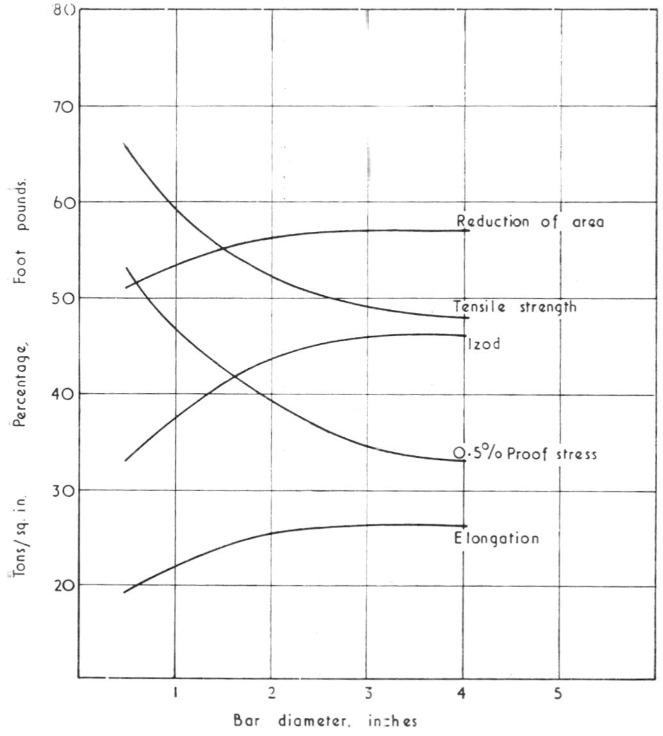

Fig. 362.5

Blank carburized 900°C.
Oil hardened 820°C.
(Single quenched quality.)

(Ref. 2)

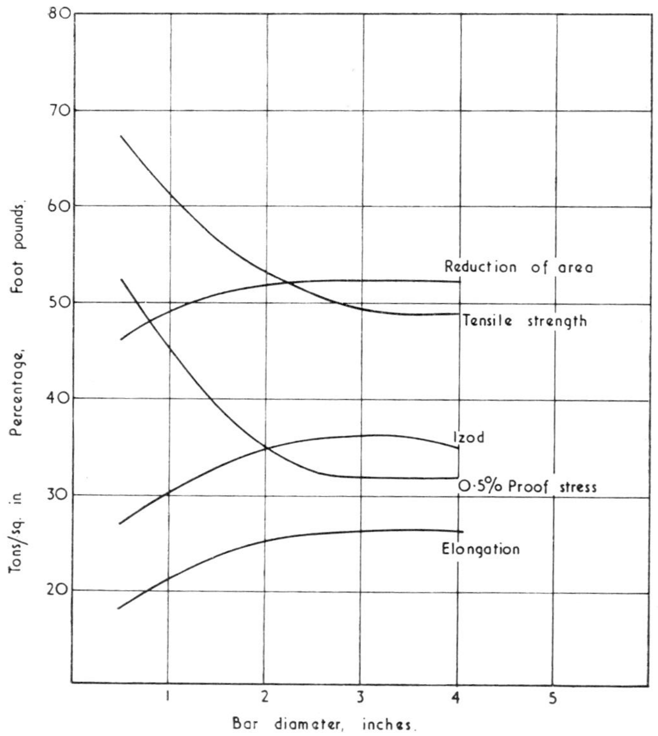

Fig. 362.6

En 362

IMPACT TRANSITION TEMPERATURE DATA

Chemical Composition (%)								Section Size dia. (in.)	Heat Treatment °C	Tensile				Izod fb.lb.	Ref.
C	Si	Mn	S	P	Ni	Cr	Mo			Y.P. (Rm) tons/sq.in.	T.S. (Rs) tons/sq.in.	El % (A)	R.A. % (Z)		
0·21	0·25	0·79	0·017	0·016	0·48	0·48	0·16	1	WQ870 T550 (½hr.)	57·3	61·6	20	61	71	6

Sub-standard test piece ⅜ in. dia., 1¾ in. overall length. Notch 45° included angle, 0·25 mm. root radius. Thickness under notch 0·220 in.

(Ref. 6)

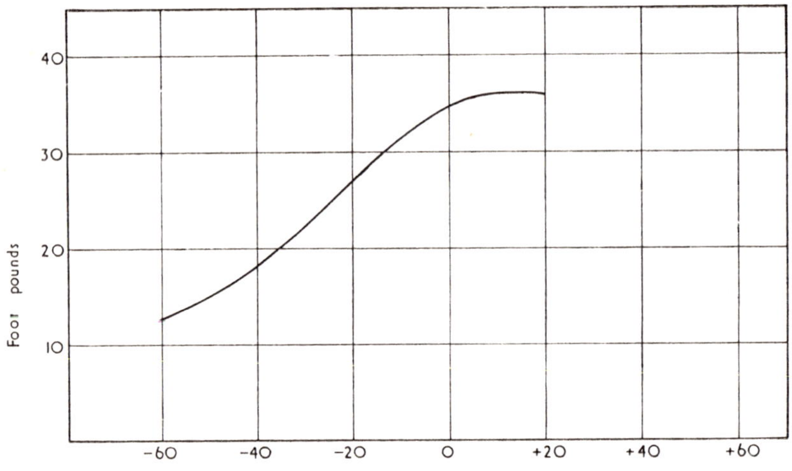

Fig. 362.7

LIST OF REFERENCES

1. Private correspondence with Firth Brown Ltd.
2. Private correspondence with The English Steel Corporation Ltd.
3. Private correspondence with The United Steel Companies Ltd.
4. Private correspondence with Jessop–Saville Ltd.
5. Private correspondence with International Nickel Ltd.
6. Private correspondence with the Ministry of Power, Safety in Mines Research Establishment.

"25" CARBON LOW-ALLOY STEEL CASE-HARDENING

SPECIFICATION

CHEMICAL COMPOSITION (%)

C	Si	Mn	S	P	Ni	Cr	Mo
0·22—0·26	0·35 max.	0·70—1·00	0·050 max.	0·050 max.	0·40—0·70	0·55—0·80	0·08—0·15

MECHANICAL PROPERTIES

Tensile strength tons/sq.in. min. (Rm)	65

The steel may be supplied to tensile strength only or, by special arrangement between the purchaser and manufacturer, on analysis only.

The heat treatment given to test bars shall be blank carburize at a temperature between 880°C and 930°C.[1],[2] Refine at a temperature of 850°C to 880°C, cool in air or oil, harden in oil from a temperature of 780°C to 820°C.

[1] By special arrangement with the purchaser the test may be carried out on samples from which the blank carburizing treatment has been omitted or modified to a short heating period.

[2] A single oil quench from 810°C to 830°C may be used in certain cases when the refining treatment after carburizing is omitted by special arrange- ment between the purchaser and the manufacturer.

RELATED SPECIFICATIONS

United Kingdom	America	
En No.	AISI	SAE
363	8622 8625	8622 8625

APPLICATIONS

A low-alloy case-hardening steel which may be carburized and hardened to produce a hard wear-resistant case and developing a core strength of the order of 65 tons/sq.in. Due to its high carbon content the steel, in small section sizes, will develop considerably greater core strengths with consequent reduction in toughness. En 363 is suitable in certain applications where toughness is not important and fairly uniform sections are used, e.g. gudgeon pins, camshafts, rocker arms, rocker shafts, levers and bushes.

WELDING

(See Introduction for key to symbols used in this table.)

	Welding Process					
	MA	G	IG	S	F	B
En 363	cx	ax	g	e	a	h

Reproduced by kind permission of H.M.S.O.

MACHINABILITY

In the softened condition the machinability is approximately 80% of that for mild steel (En 3).

HOT WORKING AND HEAT TREATMENT TEMPERATURES

Forging, rolling and stamping	1150°C, finish above 900°C.
Softening	640°C—680°C, cool in air or furnace.
Carburizing	880°C—930°C.
Refining*	850°C—880°C, cool in air or oil.
Hardening*	780°C—820°C, oil quench.
Tempering (for stress relief)	150°C—200°C.

* A single oil quench from 810°C to 830°C may be adopted when steel is specifically requested for this treatment.

PHYSICAL PROPERTIES

Few data are available for the physical properties of this steel. The following are typical values obtained for steels within the specified composition range.

SPECIFIC GRAVITY (d)

7·85 at 20°C (Ref. 1).
7·87 at 20°C (Ref. 2).

SPECIFIC HEAT (e)

0·118 cal/g. deg. C at 20°C to 100°C (Ref. 1).

MEAN COEFFICIENT OF THERMAL EXPANSION (α)

$10^6 \times$ Mean Coefficient of Thermal Expansion/deg.C				Ref.
20 to 100°C	20 to 300°C	20 to 500°C	20 to 700°C	
12	13	14	14·5	1
12·4	14·1	14·3	14·5	2

TRANSFORMATION CHARACTERISTICS

ISOTHERMAL TRANSFORMATION

CORE CHARACTERISTICS

Chemical Composition (%)					
C	Si	Mn	Ni	Cr	Mo
0·23	0·30	0·84	0·55	0·66	0·11

McQuaid Ehn Grain Size 8 (ASTM).
Austenitizing Temperature 820°C, 30 min. (Ref. 2)

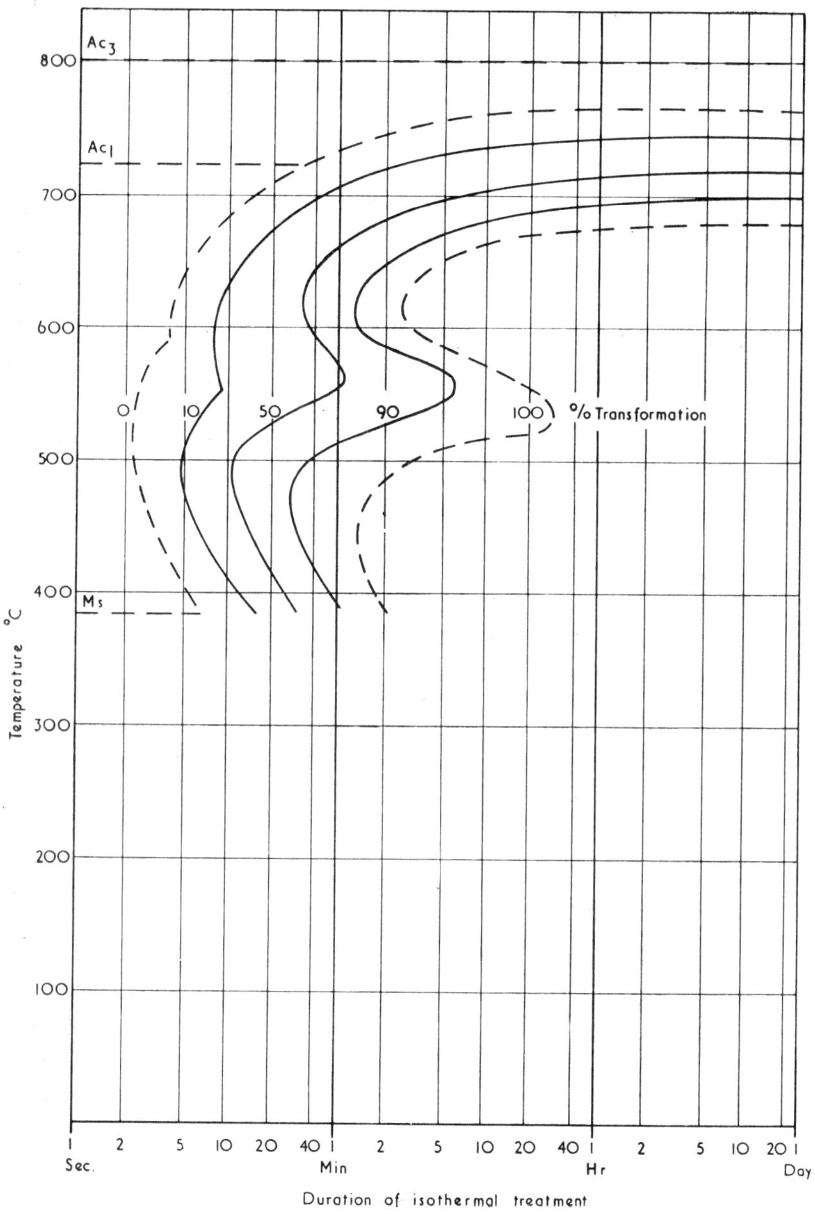

Fig. 363.1

En 363

ISOTHERMAL TRANSFORMATION (Continued)

CASE CHARACTERISTICS

Chemical Composition (%)					
C	Si	Mn	Ni	Cr	Mo
0·95	0·30	0·84	0·55	0·66	0·11

McQuaid Ehn Grain Size 8 (ASTM).
Austenitizing Temperature 820°C, 30 min.

(Ref. 2)

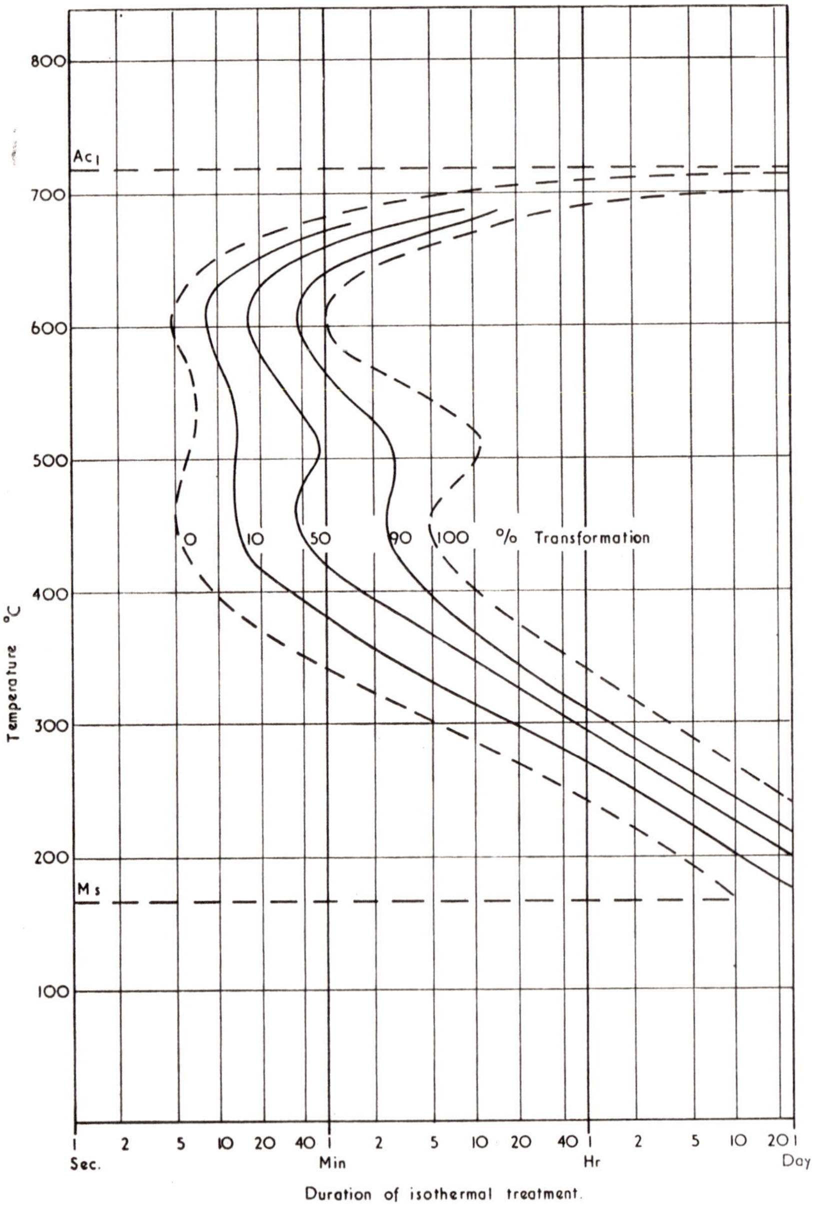

Fig 363.2

HARDENABILITY

END QUENCH HARDENABILITY BAND

These curves have been constructed from the results obtained on several casts of En 363.

(Ref. 2)

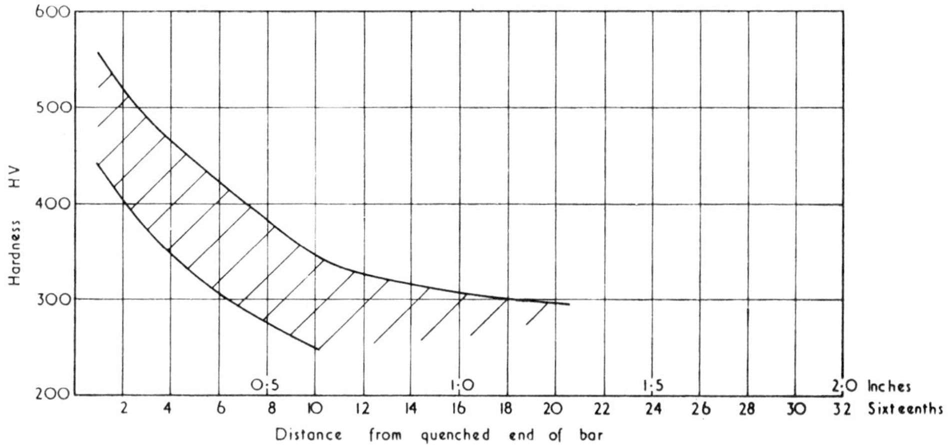

Fig. 363.3

MECHANICAL PROPERTIES AT ROOM TEMPERATURE

Process of Manufacture	ASTM Grain Size	Chemical Composition (%)								Section Size dia. (in.)	Heat Treatment °C	Tensile							Izod ft.lb.	Ref.
		C	Si	Mn	S	P	Ni	Cr	Mo			0·1% P.S. (0·1Rp)	0·2% P.S. (0·2Rp)	0·5% P.S. (0·5Rp)	Y.P. (Rm)	T.S. (Rs)	El % (A)	R.A. % (Z)		
												tons/sq.in.								
		0·23	0·21	0·80			0·62	0·79	0·11	1⅛	BC900 ƆQ860 OQ800					67·0	19	46	28	2
		0·23	0·30	0·84			0·55	0·66	0·11	1⅛	" " "					67·0	19	40½	32	"
		0·23	0·28	0·93			0·65	0·73	0·17	1⅛	" " "					69·0	17½	36½	20	"
		0·23		0·87			0·71	0·67	0·13	½	OQ̊870 OQ810				78·7	97·4	12½		14	1
		"		"			"	"	"	1⅛	" "				52·6	73·9	19		21	"
		"		"			"	"	"	2½	" "				43·3	66·0	20		15	"
		"					"	"	"	4	" "				32·4	54·4	24½		29	"
B.O.H.	7	0·23	0·24	0·95	0·036	0·040	0·60	0·72	0·11	1⅛	BC OQ800				59·9	74·9	16	31	9	3
"	"	"	"	"	"	"	"	"	"	"	" OQ820				60·2	74·5	16½	36½	10	"
"	"	"	"	"	"	"	"	"	"	"	" OQ840				64·8	72·9	16	42	18	"
"	"	"	"	"	"	"	"	"	"	"	" OQ860				68·0	73·8	18	42	24	"
B.O.H	5—8	0·24	0·31	0·84	0·040	0·050	0·50	0·77	0·13	⅜	BC OQ850 OQ790					83·2	14	33½		1
"	"	"	"	"	"	"	"	"	"	¾	" " "				46·7	75·0	18	36	14	"
"	"	"	"	"	"	"	"	"	"	1⅛	" " "				39·2	66·8	20	35	14	"
"	"	"	"	"	"	"	"	"	"	1¾	" " "				28·4	64·4	19	38	15	"
"	"	"	"	"	"	"	"	"	"	2½	" " "				30·8	61·2	21	38	16	"
		0·25	0·30	0·80			0·66	0·63	0·12	1⅛	BC900 OQ860 OQ800					67·0	21	43	31	2
		0·25	0·24	0·82			0·62	0·57	0·13	1⅛	" " "					68·8	15	31½	18	"
B.O.H.	7	0·25	0·24	0·83	0·043	0·031	0·57	0·70	0·11	1⅛	BC OQ800				59·2	72·8	14½	24½	9	3
"	"	"	"	"	"	"	"	"	"	"	" OQ820				60·0	69·5	18½	36½	20	"
"	"	"	"	"	"	"	"	"	"	"	" OQ840				59·6	67·8	18	39	21	"
"	"	"	"	"	"	"	"	"	"	"	" OQ860				62·2	71·9	17½	39	23	"
"	"	"	"	"	"	"	"	"	"	"	OQ860 OQ800				65·8	74·9	13	27½	10	"
"	"	"	"	"	"	"	"	"	"	"	" OQ820				62·4	72·8	17½	39	17	"
"	"	"	"	"	"	"	"	"	"	"	" OQ840				65·2	72·3	17½	42	22	"
"	"	"	"	"	"	"	"	"	"	"	" OQ860				66·8	73·2	17½	39	19	"
B.O.H.	7—8	0·25	0·23	0·83	0·041	0·026	0·53	0·59	0·13	1⅛	BC OQ800				55·0	69·9	15	31	8	"
"	"	"	"	"	"	"	"	"	"	"	" OQ820				56·5	68·4	20	39	23	"
"	"	"	"	"	"	"	"	"	"	"	" OQ840				53·2	68·0	19½	42	21	"
"	"	"	"	"	"	"	"	"	"	"	" OQ860				58·0	68·9	18½	42	22	"

MECHANICAL PROPERTIES AT ROOM TEMPERATURE (Continued)

Process of Manufacture	ASTM Grain Size	Chemical Composition (%) C	Si	Mn	S	P	Ni	Cr	Mo	Section Size dia. (in.)	Heat Treatment °C	Tensile 0·1% P.S. (0·1Rp) tons/sq.in.	0·2% P.S. (0·2Rp)	0.5% P.S. (0·5Rp)	Y.P. (Rm)	T.S. Rs)	El % (A)	R.A. % (Z)	Izod fb.lb.	Ref.
B.O.H.	7—8	0·25	0·23	0·83	0·041	0·026	0·53	0·59	0·13	$1\frac{1}{8}$	BC OQ860 OQ800				59·2	72·9	16	$18\frac{1}{2}$	9	3
"	"	"	"	"	"	"	"	"	"	"	" " OQ840				57·6	69·1	17	42	19	"
"	"	"	"	"	"	"	"	"	"	"	OQ860 OQ800				57·7	73·8	16	$21\frac{1}{2}$	11	"
"	"	"	"	"	"	"	"	"	"	"	" "				58·0	72·8	17	$36\frac{1}{2}$	19	"
A.O.H.	7	0·25	0·32	0·72	0·013	0·022	0·64	0·67	0·13	0·424	N900 OQ860 OQ800	41·0	49·0	63·9		80·9	13	$35\frac{1}{2}$	21	2
"	"	"	"	"	"	"	"	"	"	$\frac{3}{4}$	" " "	36·8	42·7	51·0		66·7	$18\frac{1}{2}$	$42\frac{1}{2}$	29	"
"	"	"	"	"	"	"	"	"	"	1	" " "	36·7	41·7	48·5		63·8	21	$45\frac{1}{2}$	31	"
"	"	"	"	"	"	"	"	"	"	$1\frac{1}{8}$	" " "	35·5	39·4	45·3		60·5	21	45	27	"
"	"	"	"	"	"	"	"	"	"	$2\frac{1}{8}$	" " "	25·8	30·2	36·3		56·4	$20\frac{1}{2}$	43	19	"
"	"	"	"	"	"	"	"	"	"	3	" " "	25·0	29·5	36·1		53·6	23	43	28	"
"	"	"	"	"	"	"	"	"	"	$4\frac{1}{4}$(O)	" " "	24·8	29·0	34·9		53·7	$23\frac{1}{2}$	$44\frac{1}{2}$	31	"
"	"	"	"	"	"	"	"	"	"	$4\frac{1}{4}$(C)	" " "	22·1	26·0	32·3		51·5	24	$39\frac{1}{2}$	26	"
"	"	"	"	"	"	"	"	"	"	0·424	N900 OQ820	48·0	53·4	64·7		76·8	15	$42\frac{1}{2}$	24	"
"	"	"	"	"	"	"	"	"	"	$\frac{3}{4}$	" "	37·9	43·0	57·0		65·7	20	46	27	"
"	"	"	"	"	"	"	"	"	"	1	" "	35·0	40·2	47·0		62·7	20	45	31	"
"	"	"	"	"	"	"	"	"	"	$1\frac{1}{8}$	" "	38·5	41·8	47·0		60·1	$20\frac{1}{2}$	50	31	"
"	"	"	"	"	"	"	"	"	"	$2\frac{1}{8}$	" "	28·4	32·6	39·1		54·5	$21\frac{1}{2}$	47	32	"
"	"	"	"	"	"	"	"	"	"	3	" "	25·4	29·7	35·4		52·8	$22\frac{1}{2}$	45	32	"
"	"	"	"	"	"	"	"	"	"	$4\frac{1}{4}$(O)	" "	27·8	31·4	36·7		52·5	24	$47\frac{1}{2}$	31	"
"	"	"	"	"	"	"	"	"	"	$4\frac{1}{4}$(C)	" "	22·1	25·9	31·5		49·6	25	43	25	"
"	"	"	"	"	"	"	"	"	"	0·424	N900	24·5	28·0	32·9		50·6	22	47	29	"
"	"	"	"	"	"	"	"	"	"	$\frac{3}{4}$	"	22·5	26·3	32·3		50·1	23	43	22	"
"	"	"	"	"	"	"	"	"	"	1	"	20·5	28·0	31·0		48·2	22	$43\frac{1}{2}$	20	"
"	"	"	"	"	"	"	"	"	"	$1\frac{1}{8}$	"	22·5	25·7	31·4		48·4	22	$44\frac{1}{2}$	20	"
"	"	"	"	"	"	"	"	"	"	$2\frac{1}{8}$	"	23·0	24·3	28·0		44·9	22	43	26	"
"	"	"	"	"	"	"	"	"	"	3	"	23·0	23·9	27·5		42·7	$25\frac{1}{2}$	$47\frac{1}{2}$	40	"
"	"	"	"	"	"	"	"	"	"	$4\frac{1}{4}$(O)	"			25·4		41·1	30	59	52	"
										$4\frac{1}{4}$(C)	"	22·3	22·4	25·3		41·1	22	$49\frac{1}{2}$	32	"
		0·26	0·30	0·81			0·57	0·67	0·11	$1\frac{1}{8}$	BC900 OQ860 OQ800					65·5	16	43	18	2
A.O.H.		0·26	0·27	0·81	0·029	0·025	0·69	0·63	0·09	$1\frac{1}{8}$	OQ800					70·2	$17\frac{1}{2}$	44	20	1
B.O.H.	7—8	0·26	0·25	0·81	0·038	0·029	0·57	0·60	0·13	$1\frac{1}{8}$	BC OQ800				66·0	74·0	13	$27\frac{1}{2}$	7	3
"	"	"	"	"	"	"	"	"	"	"	" OQ820				68·8	73·9	16	$36\frac{1}{2}$	11	"
"	"	"	"	"	"	"	"	"	"	"	" OQ840				68·5	74·5	16	$36\frac{1}{2}$	15	"
"	"	"	"	"	"	"	"	"	"	"	" OQ860				69·5	75·0	$14\frac{1}{2}$	$33\frac{1}{2}$	12	"
"	"	"	"	"	"	"	"	"	"	"	OQ860 OQ800				67·2	78·9	$11\frac{1}{2}$	$18\frac{1}{2}$	6	"
											" OQ840				79·2	84·2	15	$33\frac{1}{2}$	12	"

EFFECT OF TEMPERING TEMPERATURE ON CASE HARDNESS

(TYPICAL CURVE)

Carburized 900°C.
Oil refined 860°C.
Oil hardened 800°C.

(Ref. 2)

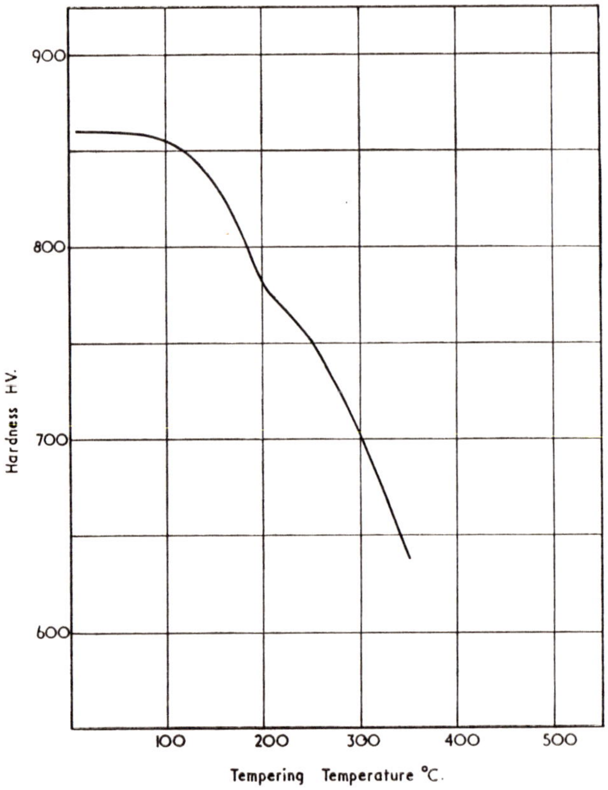

Fig. 363.4

EFFECT OF SECTION SIZE ON THE CORE PROPERTIES AT ROOM TEMPERATURE (TYPICAL CURVE)

Blank carburized 900°C.
Oil refined 860°C.
Oil hardened 800°C. (Ref. 2)

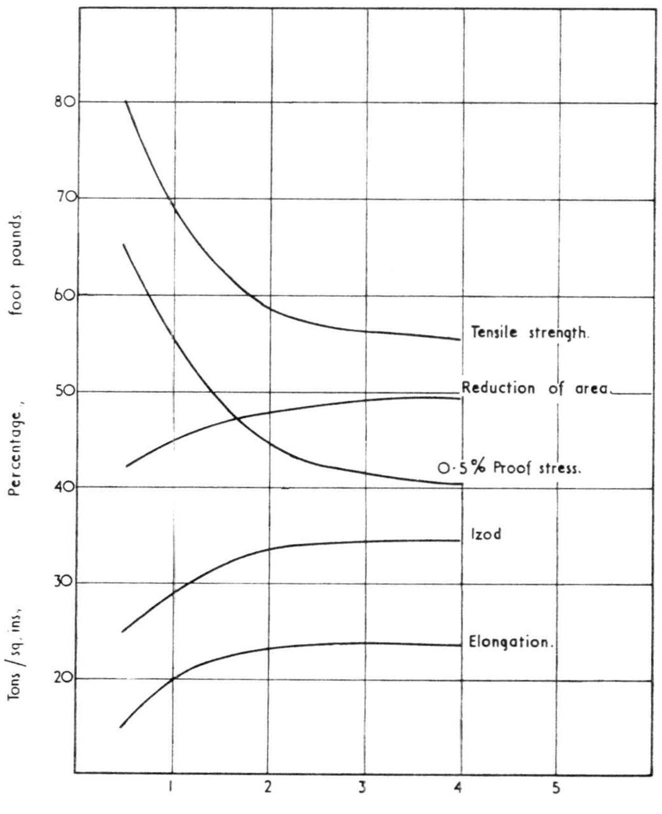

Fig. 363.5

Blank carburized 900°C.
Oil hardened 820°C.
(Single quench quality.) (Ref. 2)

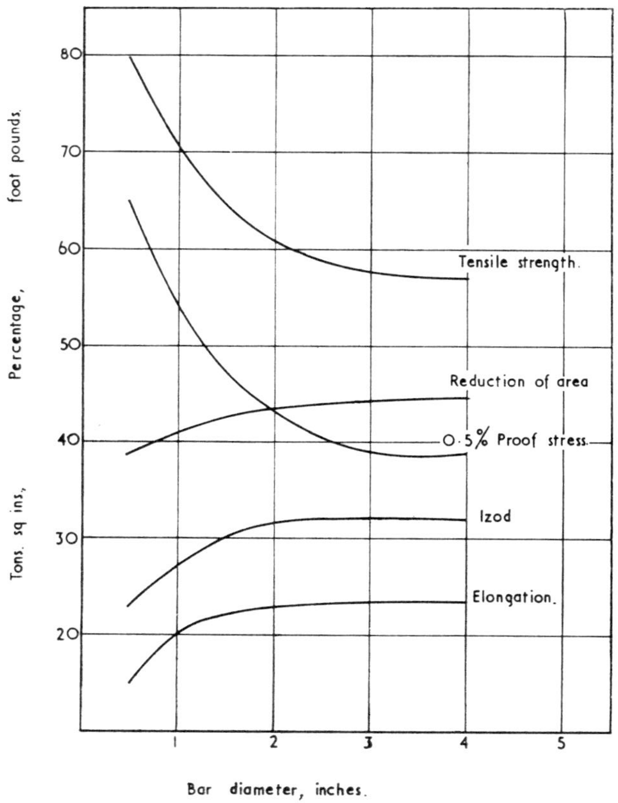

Fig. 363.6

LIST OF REFERENCES

1. Private correspondence with Firth Brown Ltd.
2. Private correspondence with The English Steel Corporation Ltd.
3. Private correspondence with The United Steel Companies Ltd.

Appendix I

ELECTRICAL RESISTIVITY (ϱ), THERMAL CONDUCTIVITY (k) AND LORENZ FUNCTION (L) OF 22 STEELS (N.P.L. DATA)

En No.	C	Si	Mn	Ni	Cr	Mo	W	Property	0°C	100°C	200°C	300°C	400°C	500°C	600°C	700°C	800°C	900C°	1000°C
2	0·06	0·01	0·38	0·05	0·02	0·03	–	ϱ	12·0	17·8	25·2	34·1	44·8	57·5	72·5	89·8	107·3	112·4	116·0
								k	0·156	0·144	0·133	0·122	0·111	0·098	0·090	0·081	0·072	0·065	0·066
Rimming								$10^3 \times L$	6·84	6·88	7·08	7·26	7·39	7·30	7·47	7·48	7·20	6·24	6·02
2	0·08	0·08	0·31	0·04	0·07	0·02	–	ϱ	13·2	19·0	26·3	35·2	45·8	58·4	73·4	90·5	108·1	113·0	116·5
BOH								k	0·142	0·138	0·128	0·118	0·107	0·096	0·086	0·076	0·068	0·064	0·066
								$10^3 \times L$	6·85	7·03	7·11	7·26	7·29	7·25	7·22	7·06	6·85	6·17	6·04
3	0·23	0·11	0·63	0·07	Trace	–	–	ϱ	15·9	21·9	29·2	38·1	48·7	60·1	75·8	92·5	109·4	113·68	116·7
BOH								k	0·124	0·122	0·116	0·106	0·102	0·094	0·085	0·076	0·062	0·063	0·055
								$10^3 \times L$	7·22	7·16	7·15	7·04	7·38	7·31	7·47	7·22	6·33	6·09	5·95
8	0·41	0·11	0·64	0·06	Trace	–	–	ϱ	16·0	22·1	29·6	38·7	49·3	61·9	76·6	93·2	111·1	114·9	117·9
BOH								k	0·124	0·121	0·115	0·109	0·100	0·091	0·031	0·072	0·059	0·061	0·064
								$10^3 \times L$	7·26	7·17	7 20	7·35	7·32	7·29	7·11	6·90	6·11	5·97	5·93
8	0·43	0·20	0·69	0·04	0·03	0·01	–	ϱ	17·7	23·8	31·2	39·9	50·4	62·6	77·0	93·6	111·3	114·9	118·0
AOH								k	0·115	0·115	0·111	0·105	0·098	0·092	0·086	0·075	0·064	0·062	0·064
								$10^3 \times L$	7·46	7·33	7·32	7·30	7·34	7·45	7·58	7·22	6·64	6·07	5·93
42G	0·80	0·13	0·32	0·13	Trace	–	–	ϱ	17·0	23·2	30·8	39·8	50·5	62·8	77·2	93·5	112·9	116·4	119·1
BEA								k	0·119	0·115	0·108	0·099	0·091	0·084	0·078	0·072	0·058	0·061	0·064
								$10^3 \times L$	7·40	7·15	7·03	6·89	6·83	6·82	6·90	6·93	6·10	6·05	5·99
42G	0·84	0·13	0·24	Trace	Trace	–	–	ϱ	16·5	22·8	30·3	39·2	49·8	62·2	76·7	92·7	112·9	116·6	119·7
BOH								k	0·122	0·117	0·110	0·100	0·092	0·085	0·079	0·073	0·058	0·060	0·064
								$10^3 \times L$	7·37	7·15	7·05	6·84	6·80	6·84	6·95	6·95	6·10	5·95	6·01
44E	1·22	0·16	0·35	0·13	0·11	0·01	–	ϱ	18·4	25·2	33·3	43·0	54·0	66·5	80·2	96·4	115·2	119·6	122·6
BOH								k	0·108	0·107	0·102	0·096	0·089	0·083	0·076	0·068	0·057	0·059	0·052
								$10^3 \times L$	7·28	7·23	7·18	7·20	7·14	7·15	6·98	6·73	6·12	6·02	5·96
14	0·23	0·12	1·51	0·04	0·05	0·025	–	ϱ	19·7	25·9	33·3	42·1	52·3	64·5	78·6	94·6	110·3	114·3	117·4
BOH								k	0·110	0·111	0·107	0·103	0·096	0·089	0·032	0·075	0·071	0·064	0·065
								$10^3 \times L$	7·93	7·70	7·53	7·56	7·45	7·43	7·38	7·29	7·29	6·24	5·99
18A	0·31	0·20	0·69	0·07	1·09	0·01	–	ϱ	20·0	25·9	33·0	41·7	51·7	63·6	77·8	93·4	110·6	114·5	117·7
AOH								k	0·116	0·111	0.106	0·101	0·092	0·085	0·076	0·069	0·062	0·064	0·067
								$10^3 \times L$	8·50	7·70	7·40	7·35	7·07	7·00	6·76	6·62	6·39	6·24	6·20
19	0·35	0·21	0·59	0·26	0·88	0·20	–	ϱ	21·1	27·1	34·2	42·8	52·9	64·6	78·6	94·4	110·3	113·8	117·1
AOH								k	0·103	0·102	0·100	0·097	0·093	0·087	0·081	0·074	0·063	0·065	0·057
								$10^3 \times L$	7·95	7·42	7·24	7·25	7·30	7·28	7·29	7·17	6·47	6·30	6·17
22	0·32	0·18	0·55	3·47	0·17	0·04	–	ϱ	25·9	32·0	39·0	47·0	56·7	67·9	81·4	99·2	112·2	114·9	118·0
AOH								k	0·037	0·090	0·092	0·091	0·087	0·082	0·076	0·068	0·060	0·063	0·066
								$10^3 \times L$	8·25	7·72	7·59	7·45	7·34	7·20	7·08	6·93	6·28	6·17	6·12
23	0·33	0·17	0·53	3·38	0·80	0·07	–	ϱ	25·6	31·7	38·7	46·9	56·7	68·1	81·7	98·1	111·5	114·8	117·8
AOH								k	0·082	0·036	0·088	0·088	0·087	0·084	0·080	0·070	0·062	0·053	0·066
								$10^3 \times L$	7·69	7·31	7·20	7·20	7·34	7·41	7·48	7·06	6·45	6·16	6·11

(continued)

En No.	C	Si	Mn	Ni	Cr	Mo	W	V	Property	0°C	100°C	200°C	300°C	400°C	500°C	600°C	700°C	800°C	900°C	1000°C
23 BEA	**0·32**	0·25	0·55	3·41	0·71	0·06	–		ϱ	26·5	32·5	39·5	47·7	57·2	68·6	82·0	99·0	111·7	115·0	118·0
									k	0·080	0·085	0·087	0·087	0·087	0·082	0·078	0·068	0·062	0·063	0·66
									$10^3 \times L$	7·76	7·40	7·26	7·25	7·40	7·29	7·15	6·93	6·46	6·18	6·12
27 BEA	**0·34**	0·27	0·55	3·53	**0·78**	**0·39**	–	–	ϱ	27·7	33·7	40·6	48·7	58·2	69·4	82·5	99·4	111·4	114·6	117·6
									k	0·079	0·081	0·084	0·085	0·085	0·080	0·073	0·67	0·064	0·066	0·068
									$10^3 \times L$	8·01	7·32	7·21	7·23	7·35	7·19	6·90	6·84	6·65	6·44	6·27
46 AOH	**0·48**	1·98	**0·90**	0·16	0·04	–	–	–	ϱ	41·9	47·0	52·9	60·1	68·5	78·8	91·1	105·7	117·3	119·7	122·3
									k	0·060	0·068	0·072	0·074	0·074	0·074	0·072	0·067	0·050	0·061	0·063
									$10^3 \times L$	9·21	8·57	8·06	7·77	7·53	7·55	7·52	7·28	6·57	6·23	6·05
56B BEA	**0·13**	**0·17**	**0·25**	0·14	12·95	–	–	–	ϱ	48·6	58·4	67·9	76·9	85·4	93·8	102·1	110·3	116·0	115·0	117·0
									k	0·064	0·066	0·066	0·067	0·066	0·065	0·063	0·061	0·060	0·064	0·066
									$10^3 \times L$	11·4	10·33	9·47	8·98	8·38	7·89	7·38	6·92	6·49	6·27	6·06
56D BEA	**0·27**	0·18	**0·28**	**0·20**	13·69	0·01	0·25		ϱ	50·3	59·5	68·4	76·9	85·3	93·5	101·5	109·9	115·7	114·6	117·9
									k	0·060	0·053	0·065	0·066	0·066	0·065	0·064	0·061	0·060	0·064	0·066
									$10^3 \times L$	11·1	10·05	9·39	8·85	8·37	7·87	7·44	6·88	6·47	6·24	6·11
58E BEA	**0·08**	**0·68**	**0·37**	8·14	19·11	–	0·60		ϱ	69·4	77·6	85·0	91·6	97·6	102·6	107·2	111·1	114·1	117·1	119·6
									k	0·038	0·039	0·041	0·044	0·048	0·052	0·057	0·061	0·064	0·064	0·067
									$10^3 \times L$	9·64	8·11	7·37	7·04	6·96	6·90	7·00	6·97	6·80	6·39	6·30
BEA	1·22	**0·22**	13·00	0·07	0·03	–	–		ϱ	66·5	75·7	84·7	93·1	100·4	105·9	110·0	115·6	120·4	124·5	127·5
									k	0·031	0·035	0·039	0·043	0·046	0·049	0·052	0·054	0·056	0·058	0·061
									$10^3 \times L$	7·55	7·09	6·98	6·93	6·87	6·72	6·54	6·42	6·28	6·15	6·11
BEA	0·28	0·15	0·89	29·37	Trace	–	–		ϱ	82·9	89·1	94 7	99·6	103·9	107·7	111.2	114·2	116·5	118·4	120·6
									k	0·030	0·035	0·039	0·042	0·045	0·049	0·053	0·056	0·060	0·063	0·066
									$10^3 \times L$	9·13	8·36	7·80	7·30	6·94	6·83	6·80	6·58	6·51	6·35	6·25
BHF	**0·71**	0·30	0·25	0·07	4·26	Trace	18·45	1·075	ϱ	40·6	47·2	54·4	62·7	71·8	81·5	92·2	103·7	115·2	117·3	120·9
									k	0·058	0·062	0·065	0·067	0·068	0·067	0·065	0·064	0·062	0·064	0·066
									$10^3 \times L$	8·63	7·84	7·48	7·34	7·25	7·06	6·87	6·83	6·65	6·40	6·28

Appendix II

COMPARISON OF ELONGATION ON $L=4\sqrt{A}$ AND $L=5{\cdot}65\sqrt{A}$

El % on $l=4\sqrt{A}$	Elongation Per Cent on $l=5{\cdot}65\sqrt{A}$ For Reduction of Area =																
	10	15	20	25	30	35	40	45	50	55	60	65	70	75	80	85	90
5	4·3	3·9	3·5	3·2	2·9												
6	5·3	4·9	4·5	4·2	3·9												
7	6·4	6·0	5·6	5·3	5·0												
8	7·5	7·1	6·7	6·4	6·1	5·9											
9	8·5	8·1	7·7	7·3	7·0	6·8											
10	9·5	9·1	8·7	8·3	8·0	7·7	7·4	7·0	6·7								
11		10·2	9·8	9·4	9·0	8·6	8·3	8·0	7·7								
12		11·3	10·8	10·4	10·0	9·6	9·3	9·0	8·7								
13		12·4	11·8	11·3	10·9	10·5	10·2	9·9	9·6								
14		13·6	12·9	12·2	11·8	11·5	11·2	10·2	10·6	10·3							
15		14·7	14·0	13·2	12·8	12·4	12·0	11·7	11·4	11·1							
16			15·0	14·3	13·7	13·2	12·8	12·5	12·2	11·9							
17			16·1	15·4	14·7	14·1	13·8	13·5	13·2	12·9							
18			17·1	16·4	15·7	15·1	14·8	14·5	14·3	14·0	13·8						
19			18·2	17·5	16·8	16·2	15·8	15·5	15·2	14·9	14·6	14·4					
20			19·3	18·6	17·9	17·3	16·8	16·4	16·1	15·8	15·5	15·2	15·0	14·8			
21				19·7	19·0	18·4	17·8	17·3	17·0	16·7	16·5	16·2	15·9	15·7			
22				20·8	20·0	19·4	18·8	18·3	17·9	17·6	17·4	17·1	16·9	16·7			
23				21·9	21·0	20·4	19·8	19·3	18·9	18·5	18·2	18·0	17·8	17·6			
24				23·0	22·0	21·4	20·8	20·3	19·9	19·5	19·2	19·0	18·8	18·6			
25				24·2	23·0	22·4	21·8	21·3	20·9	20·6	20·3	20·0	19·8	19·6	19·5	19·4	19·3
26				25·3	24·2	23·6	22·9	22·4	22·0	21·7	21·4	21·1	20·9	20·7	20·5	20·4	20·4
27				26·4	25·3	24·6	23·9	23·3	22·9	22·6	22·3	22·0	21·8	21·7	21·6	21·5	21·4
28				27·6	26·4	25·6	24·8	24·3	23·9	23·5	23·2	22·9	22·7	22·6	22·5	22·4	22·3
29				28·7	27·4	26·6	25·8	25·3	24·9	24·6	24·3	23·8	23·6	23·5	23·4	23·3	23·2
30				29·8	28·5	27·6	26·7	26·2	25·9	25·6	25·3	24·9	24·6	24·4	24·3	24·2	24·1
31					29·5	28·6	27·7	27·2	26·9	26·6	26·3	26·0	25·7	25·4	25·2	25·1	25·0
32					30·6	29·7	28·8	28·1	27·8	27·5	27·2	27·0	26·7	26·4	26·2	26·1	26.0
33					31·7	30·7	29·8	29·1	28·8	28·5	28·1	27·9	27·7	27·5	27·2	27·1	27.0
34					32·8	31·7	30·8	30·0	29·7	29·4	29·1	28·8	28·5	28·3	28·1	28·0	27.9
35					33·8	32·7	31·8	31·0	30·6	30·3	30·0	29·7	29·4	29·2	29.0	28·9	28.8
36					34·9	33·8	32·9	32·0	31·6	31·2	30·9	30·6	30·3	30·1	29.9	29·8	29.7
37					36·0	34·8	33·9	33·0	32·5	32·2	31·9	31·5	31·2	31·0	30·8	30·7	30.6
38					37·1	35·8	34·9	34·1	33·5	33·1	32·8	32·4	32·1	31·9	31·7	31·5	31.4
39					38·3	36·9	36·0	35·2	34·5	34·0	33·7	33·3	33·0	32·8	32·6	32·4	32.3
40					39·4	38·0	37·0	36·2	35·5	35·0	34·6	34·2	33·9	32·7	33·5	33·2	33·1
41						39·4	38·1	37·2	36·5	35·9	35·5	35·1	35·8	34·5	34·3	34·1	34·0
42						40·5	39·3	38·3	37·4	36·9	36·4	36·0	35·7	35·4	35·2	35·0	34·8
43						41·6	40·5	39·3	38·4	37·9	37·4	37·0	36·6	36·3	36·1	35·9	35·7
44						42·7	41·6	40·4	39·5	39·0	38·5	38·0	37·5	37·1	36·9	36·7	36·5
45						43·9	42·7	41·5	40·7	40·1	39·5	38·9	38·5	38·2	37·9	37·6	37·4
46							43·9	42·7	41·8	41·1	40·5	40·0	39·6	29·2	38·8	38·5	38.3
47							45·1	43·8	42·9	42·1	41·5	41·0	40·6	40·2	39·8	39·4	39.2
48							46·2	44·9	44·0	43·2	42·6	42·1	41·7	41·3	40·9	40·5	40.2
49							47·3	46·0	45·1	44·3	43·7	43·2	42·8	42·4	42·0	41·7	41.4
50							48·5	47·1	46·2	45·3	44·7	44·2	43·8	43·5	43·2	42·9	42·6

Note this table does not apply for Austenitic Steels

APPROXIMATE RELATIONSHIP BETWEEN VARIOUS HARDNESS NUMBERS

This table has been compiled from the average of a number of published conversions.

Vickers Diamond Hardness HV	Brinell Hardness HB	Rockwell C Hardness HRC	Rockwell B Hardness HRB	Tensile Strength		
				Tons/sq. in	1000 lbs/sq. in.	Kgm/sq. mm
1200	—	70½	—	—	—	—
1150	—	70	—	—	—	—
1100	—	69½	—	—	—	—
1050	—	68½	—	—	—	—
1000	—	68	—	—	—	—
980	—	67½	—	—	—	—
960	—	67	—	—	—	—
940	—	66½	—	—	—	—
920	—	66	—	—	—	—
900	—	65½	—	—	—	—
880	—	64½	—	—	—	—
860	—	64	—	—	—	—
840	—	63½	—	—	—	—
820	—	63	—	—	—	—
800	—	62	—	—	—	—
780	—	61½	—	—	—	—
760	—	60½	—	—	—	—
740	—	60	—	—	—	—
720	612	59	—	—	—	—
700	600	58½	—	131	293	206
680	587	57½	—	128	286	202
660	572	56½	—	125	280	197
640	558	55½	—	122	273	192
620	543	54½	—	119	266	187
600	530	53½	—	116	259	183
580	516	52½	—	113	253	178
560	502	51½	—	110	246	173
540	488	50½	—	107	239	169
520	475	49½	—	104	232	164
500	461	48	—	101	226	159
480	444	47	—	98	219	154
460	428	45½	—	95	212	149
440	412	44	—	92	206	144
420	397	42	—	90	201	141
400	380	40	111	88	197	138
380	363	38	110	84	188	132
360	345	36	109	79	176	124
340	328	34	108	76	170	119
320	312	32	107	72	161	113
300	294	30	106	68	152	107
280	275	27	104	64	143	100
260	255	24	102	59	132	93
240	235	20	99	51	114	80
220	215	16	96	46	103	72
200	195	11	92	42	94	65
180	175	6	87	38	85	60
160	155	—	82	34	76	53
140	135	—	76	30	67	74
120	115	—	67	26	58	41
100	96	—	56	22	49	35
80	77	—	35	18	40	28

Appendix IV

TONS PER SQUARE INCH TO KILOGRAMMES PER SQUARE MILLIMETRE

BASED ON 1 INCH = 25·4 MILLIMETRES; 1 POUND = 0·45359243 KILOGRAMMES

Tons/ sq. in.	0	0·1	0·2	0·3	0·4	0·5	0·6	0·7	0·8	0·9
0	—	0·157	0·315	0·472	0·630	0·787	0·945	1·10	1·26	1·42
1	1·57	1·73	1·89	2·05	2·20	2·36	2·52	2·68	2·83	2·99
2	3·15	3·31	3·46	3·62	3·78	3·94	4·09	4·25	4·41	4·57
3	4·72	4·88	5·04	5·20	5·35	5·51	5·67	5·83	5·98	6·14
4	6·30	6·46	6·61	6·77	6·93	7·09	7·24	7·40	7·56	7·72
5	7·87	8·03	8·19	8·35	8·50	8·66	8·82	8·98	9·13	9·29
6	9·45	9·61	9·76	9·92	10·1	10·2	10·4	10·6	10·7	10·9
7	11·0	11·2	11·3	11·5	11·7	11·8	12·0	12·1	12·3	12·4
8	12·6	12·8	12·9	13·1	13·2	13·4	13·5	13·7	13·9	14·0
9	14·2	14·3	14·5	14·6	14·8	15·0	15·1	15·3	15·4	15·6
10	15·7	15·9	16·1	16·2	16·4	16·5	16·7	16·9	17·0	17·2
11	17·3	17·5	17·6	17·8	18·0	18·1	18·3	18·4	18·6	18·7
12	18·9	19·1	19·2	19·4	19·5	19·7	19·8	20·0	20·2	20·3
13	20·5	20·6	20·8	20·9	21·1	21·3	21·4	21·6	21·7	21·9
14	22·0	22·2	22·4	22·5	22·7	22·8	23·0	23·2	23·3	23·5
15	23·6	23·8	23·9	24·1	24·3	24·4	24·6	24·7	24·9	25·0
16	25·2	25·4	25·5	25·7	25·8	26·0	26·1	26·3	26·5	26·6
17	26·8	26·9	27·1	27·2	27·4	27·6	27·7	27·9	28·0	28·2
18	28·3	28·5	28·7	28·8	29·0	29·1	29·3	29·5	29·6	29·8
19	29·9	30·1	30·2	30·4	30·6	30·7	30·9	31·0	31·2	31·3
20	31·5	31·7	31·8	32·0	32·1	32·3	32·4	32·6	32·8	32·9
21	33·1	33·2	33·4	33·5	33·7	33·9	34·0	34·2	34·3	34·5
22	34·6	34·8	35·0	35·1	35·3	35·4	35·6	35·7	35·9	36·1
23	36·2	36·4	36·5	36·7	36·9	37·0	37·2	37·3	37·5	37·6
24	37·8	38·0	38·1	38·3	38·4	38·6	38·7	38·9	39·1	39·2
25	39·4	39·5	39·7	39·8	40·0	40·2	40·3	40·5	40·6	40·8
26	40·9	41·1	41·3	41·4	41·6	41·7	41·9	42·0	42·2	42·4
27	42·5	42·7	42·8	43·0	43·2	43·3	43·5	43·6	43·8	43·9
28	44·1	44·3	44·4	44·6	44·7	44·9	45·0	45·2	45·4	45·5
29	45·7	45·8	46·0	46·1	46·3	46·5	46·6	46·8	46·9	47·1
30	47·2	47·4	47·6	47·7	47·9	48·0	48·2	48·3	48·5	48·7
31	48·8	49·0	49·1	49·3	49·5	49·6	49·8	49·9	50·1	50·2
32	50·4	50·6	50·7	50·9	51·0	51·2	51·3	51·5	51·7	51·8
33	52·0	52·1	52·3	52·4	52·6	52·8	52·9	53·1	53·2	53·4
34	53·5	53·7	53·9	54·0	54·2	54·3	54·5	54·6	54·8	55·0
35	55·1	55·3	55·4	55·6	55·8	55·9	56·1	56·2	56·4	56·5
36	56·7	56·9	57·0	57·2	57·3	57·5	57·6	57·8	58·0	58·1
37	58·3	58·4	58·6	58·7	58·9	59·1	59·2	59·4	59·5	59·7
38	59·8	60·0	60·2	60·3	60·5	60·6	60·8	60·9	61·1	61·3
39	61·4	61·6	61·7	61·9	62·1	62·2	62·4	62·5	62·7	62·8
40	63·0	63·2	63·3	63·5	63·6	63·8	63·9	64·1	64·3	64·4
41	64·6	64·7	64·9	65·0	65·2	65·4	65·5	65·7	65·8	66·0
42	66·1	66·3	66·5	66·6	66·8	66·9	67·1	67·2	67·4	67·6
43	67·7	67·9	68·0	68·2	68·4	68·5	68·7	68·8	69·0	69·1
44	69·3	69·5	69·6	69·8	69·9	70·1	70·2	70·4	70·6	70·7
45	70·9	71·0	71·2	71·3	71·5	71·7	71·8	72·0	72·1	72·3
46	72·4	72·6	72·8	72·9	73·1	73·2	73·4	73·5	73·7	73·9
47	74·0	74·2	74·3	74·5	74·6	74·8	75·0	75·1	75·3	75·4
48	75·6	75·8	75·9	76·1	76·2	76·4	76·5	76·7	76·9	77·0
49	77·2	77·3	77·5	77·6	77·8	78·0	78·1	78·3	78·4	78·6
50	78·7	78·9	79·1	79·2	79·4	79·5	79·7	79·8	80·0	80·2

TONS PER SQUARE INCH TO KILOGRAMMES PER SQUARE MILLIMETRE

(continued)

BASED ON 1 INCH = 25·4 MILLÍMETRES; 1 POUND = 0·45359243 KILOGRAMMES

Tons/ sq. in.	0	0·1	0·2	0·3	0·4	0·5	0·6	0·7	0·8	9·0
51	80·3	80·5	80·6	80·8	80·9	81·1	81·3	81·4	81·6	81·7
52	81·9	82·1	82·2	82·4	82·5	82·7	82·8	83·0	83·2	83·3
53	83·5	83·6	83·8	83·9	84·1	84·3	84·4	84·6	84·7	84·9
54	85·0	85·2	85·4	85·5	85·7	85·8	86·0	86·1	86·3	86·5
55	86·6	86·8	86·9	87·1	87·2	87·4	87·6	87·7	87·9	88·0
56	88·2	88·4	88·5	88·7	88·8	89·0	89·1	89·3	89·5	89·6
57	89·8	89·9	90·1	90·2	90·4	90·6	90·7	90·9	91·0	91·2
58	91·3	91·5	91·7	91·8	92·0	92·1	92·3	92·4	92·6	92·8
59	92·9	93·1	93·2	93·4	93·5	93·7	93·9	94·0	94·2	94·3
60	94·5	94·7	94·8	95·0	95·1	95·3	95·4	95·6	95·8	95·9
61	96·1	96·2	96·4	96 5	96·7	96·9	97·0	97·2	97·3	97·5
62	97·6	97·8	98·0	98·1	98·3	98·4	98·6	98·7	98·9	99·1
63	99·2	99·4	99·5	99·7	99·8	100·0	100·2	100·3	100·5	100·6
64	100·8	101·0	101·1	101·3	101·4	101·6	101·7	101·9	102·1	102·2
65	102·4	102·5	102·7	102·8	103·0	103·2	103·3	103·5	103·5	103·8
66	103·9	104·1	104·3	104·4	104·6	104·7	104·9	105·0	105·2	105·4
67	105·5	105·7	105·8	106·0	106·1	106·3	106·5	106·6	106·8	106·9
68	107·1	107·2	107·4	107·6	107·7	107·9	108·9	108·2	108·4	108·5
69	108·7	108·8	109·0	109·1	109·3	109·5	109·6	109·8	109·9	110·1
70	110·2	110·4	110·6	110·7	110·9	111·0	111·2	111·3	111·5	111·7
71	111·8	112·0	112·1	112·3	112·4	112·6	112·8	112·9	113·1	113·2
72	113·4	113·5	113·7	113·9	114·0	114·2	114·3	114·5	114·7	114·8
73	115·0	115·1	115·3	115·4	115·6	115·8	115·9	116·1	116·2	116·4
74	116·5	116·7	116·9	117·0	117·2	117·3	117·5	117·6	117·8	118·0
75	118·1	118·3	118·4	118·6	118·7	118·9	119·1	119·2	119·4	119·5
76	119·7	119·8	120·0	120·2	120·3	120·5	120·6	120·8	121·0	121·1
77	121·3	121·4	121·6	121·7	121·9	122·1	122·2	122·4	122·5	122·7
78	122·8	123·0	123·2	123·3	123·5	123·6	123·8	123·9	124·1	124·3
79	124·4	124·6	124·7	124·9	125·0	125·2	125·4	125·5	125·7	125·8
80	126·0	126·1	126·3	126·5	126·6	126·8	126·9	127·1	127·3	127·4
81	127·6	127·7	127·9	128·0	128·2	128·4	128·5	128·7	128·8	129·0
82	129·1	129·3	129·5	129·6	129·8	129·9	130·1	130·2	130·4	130·6
83	130·7	130·9	131·0	131·2	131·3	131·5	131·7	131·8	132·0	132·1
84	132·3	132·4	132·6	132·8	132·9	133·1	133·2	133·4	133·5	133·7
85	133·9	134·0	134·2	134·3	134·5	134·7	134·8	135·0	135·1	135·3
86	135·4	135·6	135·8	135·9	136·1	136·2	136·4	136·5	136·7	136·9
87	137·0	137·2	137·3	137·5	137·6	137·8	138·0	138·1	138·3	138·4
88	138·6	138·7	138·9	139·1	139·2	139·4	139·5	139·7	139·8	140·0
89	140·2	140·3	140·5	140·6	140·8	141·0	141·1	141·3	141·4	141·6
90	141·7	141·9	142·1	142·2	142·4	142·5	142·7	142·8	143·0	143·2
91	143·3	143·5	143·6	143·8	143·9	144·1	144·3	144·4	144·6	144·7
92	144·9	145·0	145·2	145·4	145·5	145·7	145·8	146·0	146·1	146·3
93	146·5	146·6	146·8	146·9	147·1	147·3	147·4	147·6	147·7	147·9
94	148·0	148·2	148·4	148·5	148·7	148·8	149·0	149·1	149·3	149·5
95	149·6	149·8	149·9	150·1	150·2	150·4	150·6	150·7	150·9	151·0
96	151·2	151·3	151·5	151·7	151·8	152·0	152·1	152·3	152·4	152·6
97	152·8	152·9	153·1	153·2	153·4	153·6	153·7	153·9	154·0	154·2
98	154·3	154·5	154·7	154·8	155·0	155·1	155·3	155·4	155·6	155·8
99	155·9	156·1	156·2	156·4	156·5	156·7	156·9	157·0	157·2	157·3
100	157·5	157·6	157·8	158·0	158·1	158·3	158·4	158·6	158·7	158·9

Appendix IV

TONS PER SQUARE INCH TO POUNDS PER SQUARE INCH

Tons	0	0·1	0·2	0·3	0·4	0·5	0·6	0·7	0·8	0·9
	lb.	lb.	lb.	lb.	lb.	lb.	lb.	lb.	lb.	lb.
—	—	224	448	672	896	1120	1344	1568	1792	2016
1	2240	2464	2688	2912	3136	3360	3584	3808	4032	4256
2	4480	4704	4928	5152	5376	5600	5824	6048	6272	6496
3	6720	6944	7168	7392	7616	7840	8064	8288	8512	8736
4	8960	9184	9408	9632	9856	10080	10304	10528	10752	10976
5	11200	11424	11648	11872	12096	12320	12544	12768	12992	13216
6	13440	13664	13888	14112	14336	14560	14784	15008	15232	15456
7	15680	15904	16128	16352	16576	16800	17024	17248	17472	17696
8	17920	18144	18368	18592	18816	19040	19264	19488	19712	19936
9	20160	20384	20608	20832	21056	21280	21504	21728	21952	22176
10	22400	22624	22848	23072	23296	23520	23744	23968	24192	24416
11	24640	24864	25088	25312	25536	25760	25984	26208	26432	26656
12	26880	27104	27328	27552	27776	28000	28224	28448	28672	28896
13	29120	29344	29568	29792	30016	30240	30464	30688	30912	31136
14	31360	31584	31808	32032	32256	32480	32704	32928	33152	33376
15	33600	33824	34048	34272	34496	34720	34944	35168	35392	35616
16	35840	36064	36288	36512	36736	36960	37184	37408	37632	37856
17	38080	38304	38528	38752	38976	39200	39424	39648	39872	40096
18	40320	40544	40768	40992	41216	41440	41664	41888	42112	42336
19	42560	42784	43008	43232	43456	43680	43904	44128	44352	44576
20	44800	45024	45248	45472	45696	45920	46144	46368	46592	46816
21	47040	47264	47488	47712	47936	48160	48384	48608	48832	49056
22	49280	49504	49728	49952	50176	50400	50624	50848	51072	51296
23	51520	51744	51968	52192	52416	52640	52864	53088	53312	53536
24	53760	53984	54208	54432	54656	54880	55104	55328	55552	55776
25	56000	56224	56448	56672	56896	57120	57344	57568	57792	58016
26	58240	58464	58688	58912	59136	59360	59584	59808	60032	60256
27	60480	60704	60928	61152	61376	61600	61824	62048	62272	62496
28	62720	62944	63168	63392	63616	63840	64064	64288	64512	64736
29	64960	65184	65408	65632	65856	66080	66304	66528	66752	66976
30	67200	67424	67648	67872	68096	68320	68544	68768	68992	69216
31	69440	69664	69888	70112	70336	70560	70784	71008	71232	71456
32	71680	71904	72128	72352	72576	72800	73024	73248	73472	73696
33	73920	74144	74368	74592	74816	75040	75264	75488	75712	75936
34	76160	76384	76608	76832	77056	77280	77504	77728	77952	78176
35	78400	78624	78848	79072	79296	79520	79744	79968	80192	80416
36	80640	80864	81088	81312	81536	81760	81984	82208	82432	82656
37	82880	83104	83328	83552	83776	84000	84224	84448	84672	84896
38	85120	85344	85568	85792	86016	86240	86464	86688	86912	87136
39	87360	87584	87808	88032	88256	88480	88704	88928	89152	89376
40	89600	89824	90048	90272	90496	90720	90944	91168	91392	91616
41	91840	92064	92288	92512	92736	92960	93184	93408	93632	93856
42	94080	94304	94528	94752	94976	95200	95424	95648	95872	96096
43	96320	96544	96768	96992	97216	97440	97664	97888	98112	98336
44	98560	98784	99008	99232	99456	99680	99904	100128	100352	100576
45	100800	101024	101248	101472	101696	101920	102144	102368	102592	102816
46	103040	103264	103488	103712	103936	104160	104384	104608	104832	105056
47	105280	105504	105728	105952	106176	106400	106624	106848	107072	107296
48	107520	107744	107968	108192	108416	108640	108864	109088	109312	109536
49	109760	109984	110208	110432	110656	110880	111104	111328	111552	111776
50	112000	112224	112448	112672	112896	113120	113344	113568	113792	114016

TONS PER SQUARE INCH TO POUNDS PER SQUARE INCH

(continued)

Tons	0	0·1	0·2	0·3	0·4	0·5	0·6	0·7	0·8	0·9
	lb.	lb.	lb.	lb.	lb.	lb.	lb.	lb.	lb.	lb.
51	114240	114464	114688	114912	115136	115360	115584	115808	116032	116256
52	116480	116704	116928	117152	117376	117600	117824	118048	118272	118496
53	118720	118944	119168	119392	119616	119840	120064	120288	120512	120736
54	120960	121184	121408	121632	121856	122080	122304	122528	122752	122976
55	123200	123424	123648	123872	124096	124320	124544	124768	124992	125216
56	125440	125664	125888	126112	126336	126560	126784	127008	127232	127456
57	127680	127904	128128	128352	128576	128800	129024	129248	129472	129696
58	129920	130144	130368	130592	130816	131040	131264	131488	131712	131936
59	132160	132384	132608	132832	133056	133280	133504	133728	133952	134176
60	134400	134642	134848	135072	135296	135520	135744	135968	136192	136416
61	136640	136864	137088	137312	137536	137760	137984	138208	138432	138656
62	138880	139104	139328	139552	139776	140000	140224	140448	140672	140896
63	141120	141344	141568	141792	142016	142240	142464	142688	142912	143136
64	143360	143584	143808	144032	144256	144480	144704	144928	145152	145376
65	145600	145824	146048	146272	146496	146720	146944	147168	147392	147616
66	147840	148064	148288	148512	148736	148960	149184	149408	149632	149856
67	150080	150304	150528	150752	150976	151200	151424	151648	151872	152096
68	152320	152544	152768	152992	153216	153440	153664	153888	154112	154336
69	154560	154784	155008	155232	155456	155680	155904	156128	156352	156576
70	156800	157024	157248	157472	157696	157920	158144	158368	158592	158816
71	159040	159264	159488	159712	159936	160160	160384	160608	160832	161056
72	161280	161504	161728	161952	162176	162400	162624	162848	163072	163296
73	163520	163744	163968	164192	164416	164640	164864	165088	165312	165536
74	165760	165984	166208	166432	166656	166880	167104	167328	167552	167776
75	168000	168224	168448	168672	168896	169120	169344	169568	169792	170016
76	170240	170464	170688	170912	171136	171360	171584	171808	172032	172256
77	172480	172704	172928	173152	173376	173600	173824	174048	174272	174496
78	174720	174944	175168	175392	175616	175840	176064	176288	176512	176736
79	176960	177184	177408	177632	177856	178080	178304	178528	178752	178976
80	179200	179424	179648	179872	180096	180320	180544	180768	180992	181216
81	181440	181664	181888	182112	182336	182560	182784	183008	183232	183456
82	183680	183904	184128	184352	184576	184800	185024	185248	185472	185696
83	185920	186144	186368	186592	186816	187040	187264	187488	187712	187936
84	188160	188384	188608	188832	189056	189280	189504	189728	189952	190176
85	190400	190624	190848	191072	191296	191520	191744	191968	192192	192416
86	192640	192864	193088	193312	193536	193760	193984	194208	194432	194656
87	194880	195104	195328	195552	195776	196000	196224	196448	196672	196896
88	197120	197344	197568	197792	198016	198240	198464	198688	198912	199136
89	199360	199584	199808	200032	200256	200480	200704	200928	201152	201376
90	201600	201824	202048	202272	202496	202720	202944	203168	203392	203616
91	203840	204064	204288	204512	204736	204960	205184	205408	205632	205856
92	206080	206304	206528	206752	206976	207200	207424	207648	207872	208096
93	208320	208544	208768	208992	209216	209440	209664	209888	210112	210336
94	210560	210784	211008	211232	211456	211680	211904	212128	212352	212576
95	212800	213024	213248	213472	213696	213920	214144	214368	214592	214816
96	215040	215264	215488	215712	215936	216160	216384	216603	216832	217056
97	217280	217504	217728	217952	218176	218400	218624	218848	219072	219296
98	219520	219744	219968	220192	220416	220640	220864	221088	221312	221536
99	221760	221984	222208	222432	222656	222880	223104	223328	223552	223776
100	224000									

Appendix V

TEMPERATURE CONVERSION TABLE

459·4 to 0			0 to 99						100 to 1000					
C	C / F	F	C	C / F	F	C	C / F	F	C	C / F	F	C	C / F	F
−273	−459·4		−17·8	*0*	32	10·0	*50*	122·0	38	*100*	212	260	*500*	932
−268	−450		−17·2	*1*	33·8	10·6	*51*	123·8	43	*110*	230	266	*510*	950
−262	−440		−16·7	*2*	35·6	11·1	*52*	125·6	49	*120*	248	271	*520*	968
−257	−430		−16·1	*3*	37·4	11·7	*53*	127·4	54	*130*	266	277	*530*	986
−251	−420		−15·6	*4*	39·2	12·2	*54*	129·2	60	*140*	284	282	*540*	1004
−246	−410		−15·0	*5*	41·0	12·8	*55*	131·0	66	*150*	302	288	*550*	1022
−240	−400		−14·4	*6*	42·8	13·3	*56*	132·8	71	*160*	320	293	*560*	1040
−234	−390		−13·9	*7*	44·6	13·9	*57*	134·6	77	*170*	338	299	*570*	1058
−229	−380		−13·3	*8*	46·4	14·4	*58*	136·4	82	*180*	356	304	*580*	1076
−223	−370		−12·8	*9*	48·2	15·0	*59*	138·2	88	*190*	374	310	*590*	1094
−218	−360		−12·2	*10*	50·0	15·6	*60*	140·0	93	*200*	392	316	*600*	1112
−212	−350		−11·7	*11*	51·8	16·1	*61*	141·8	99	*210*	410	321	*610*	1130
−207	−340		−11·1	*12*	53·6	16·7	*62*	143·6	100	*212*	413·6	327	*620*	1148
−201	−330		−10·6	*13*	55·4	17·2	*63*	145·4	104	*220*	428	332	*630*	1166
−196	−320		−10·0	*14*	57·2	17·8	*64*	147·2	110	*230*	446	338	*640*	1184
−190	−310		− 9·4	*15*	59·0	18·3	*65*	149·0	116	*240*	464	343	*650*	1202
−184	−300		− 8·9	*16*	60·8	18·9	*66*	150·8	121	*250*	482	349	*660*	1220
−179	−290		− 8·3	*17*	62·6	19·4	*67*	152·6	127	*260*	500	354	*670*	1238
−173	−280		− 7·8	*18*	64·4	20·0	*68*	154·4	132	*270*	518	360	*680*	1256
−169	−273	−459·4	− 7·2	*19*	66·2	20·6	*69*	156·2	138	*280*	536	366	*690*	1274
−168	−270	−454	− 6·7	*20*	68·0	21·1	*70*	158·0	143	*290*	554	371	*700*	1292
−162	−260	−436	− 6·1	*21*	69·8	21·7	*71*	159·8	149	*300*	572	377	*710*	1310
−157	−250	−418	− 5·6	*22*	71·6	22·2	*72*	161·6	154	*310*	590	382	*720*	1328
−151	−240	−400	− 5·0	*23*	73·4	22·8	*73*	163·4	160	*320*	608	388	*730*	1346
−146	−230	−382	− 4·4	*24*	75·2	23·3	*74*	165·2	166	*330*	626	393	*740*	1364
−140	−220	−364	− 3·9	*25*	77·0	23·9	*75*	167·0	171	*340*	644	399	*750*	1382
−134	−210	−346	− 3·3	*26*	78·8	24·4	*76*	168·8	177	*350*	662	404	*760*	1400
−129	−200	−328	− 2·8	*27*	80·6	25·0	*77*	170·6	182	*360*	680	410	*770*	1418
−123	−190	−310	− 2·2	*28*	82·4	25·6	*78*	172·4	188	*370*	698	416	*780*	1436
−118	−180	−292	− 1·7	*29*	84·2	26·1	*79*	174·2	193	*380*	716	421	*790*	1454
−112	−170	−274	− 1·1	*30*	86·0	26·7	*80*	176·0	199	*390*	734	427	*800*	1472
−107	−160	−256	− 0·6	*31*	87·8	27·2	*81*	177·8	204	*400*	752	432	*810*	1490
−101	−150	−238	0	*32*	89·6	27·8	*82*	179·6	210	*410*	770	438	*820*	1508
− 96	−140	−220	0·6	*33*	91·4	28·3	*83*	181·4	216	*420*	788	443	*830*	1526
− 90	−130	−202	1·1	*34*	93·2	28·9	*84*	183·2	221	*430*	806	449	*840*	1544
− 84	−120	−184	1·7	*35*	95·0	29·4	*85*	185·0	227	*440*	824	454	*850*	1562
− 79	−110	−166	2·2	*36*	96·8	30·0	*86*	186·8	232	*450*	842	460	*860*	1580
− 73	−100	−148	2·8	*37*	98·6	30·6	*87*	188·6	238	*460*	860	466	*870*	1598
− 68	− 90	−130	2·2	*38*	100·4	31·1	*88*	190·4	243	*470*	878	471	*880*	1616
− 62	− 80	−112	3·9	*39*	102·2	31·7	*89*	192·2	249	*480*	896	477	*890*	1634
− 57	− 70	− 94	4·4	*40*	104·0	32·2	*90*	194·0	254	*490*	914	482	*900*	1652
− 51	− 60	− 76	5·0	*41*	105·8	32·8	*91*	195·8				488	*910*	1670
− 46	− 50	− 58	5·6	*42*	107·6	33·3	*92*	197·6				493	*920*	1688
− 40	− 40	− 40	6·1	*43*	109·4	33·9	*93*	199·4				499	*930*	1706
− 34	− 30	− 22	6·7	*44*	111·2	34·4	*94*	201·2				504	*940*	1724
− 29	− 20	− 4	7·2	*45*	113·0	35·0	*95*	203·0				510	*950*	1742
− 23	− 10	14	7·8	*46*	114·8	35·6	*96*	204·8				516	*960*	1760
− 17·8	0	32	8·3	*47*	116·6	36·1	*97*	205·6				521	*970*	1778
			8·9	*48*	118·4	36·7	*98*	208·4				527	*980*	1796
			9·4	*49*	120·2	37·2	*99*	210·2				532	*990*	1814

Look up reading in middle column. If in degrees Centigrade, read Fahrenheit equivalent in right hand column; if in Fahrenheit degrees, read Centigrade equivalent in left hand column.

TEMPERATURE CONVERSION TABLE

(continued)

1000 to 1990						2000 to 3000					
C	C / F	F	C	C / F	F	C	C / F	F	C	C / F	F
538	1000	1832	816	1500	2732	1093	2000	3632	1371	2500	4532
543	1010	1850	821	1510	2750	1099	2010	3650	1377	2510	4550
549	1020	1868	827	1520	2768	1104	2020	3668	1382	2520	4568
554	1030	1886	832	1530	2786	1110	2030	3686	1388	2530	4586
560	1040	1904	838	1540	2804	1116	2040	3704	1393	2540	4604
566	1050	1922	843	1550	2822	1121	2050	3722	1399	2550	4622
571	1060	1940	849	1560	2840	1127	2060	3740	1404	2560	4640
577	1070	1958	854	1570	2858	1132	2070	3758	1410	2570	4658
582	1080	1976	860	1580	2876	1138	2080	3776	1416	2580	4676
588	1090	1994	866	1590	2894	1143	2090	3794	1421	2590	4694
593	1100	2012	871	1600	2912	1149	2100	3812	1427	2600	4712
599	1110	2030	877	1610	2930	1154	2110	3830	1432	2610	4730
604	1120	2048	882	1620	2948	1160	2120	3848	1438	2620	4748
610	1130	2066	888	1630	2966	1166	2130	3866	1443	2630	4766
616	1140	2084	893	1640	2984	1171	2140	3884	1449	2640	4784
621	1150	2102	899	1650	3002	1177	2150	3902	1454	2650	4802
627	1160	2120	904	1660	3020	1182	2160	3920	1460	2660	4820
632	1170	2138	910	1670	3038	1188	2170	3938	1466	2670	4838
638	1180	2156	916	1680	3056	1193	2180	3956	1471	2680	4856
643	1190	2174	921	1690	3074	1199	2190	3974	1477	2690	4874
649	1200	2192	927	1700	3092	1204	2200	3992	1482	2700	4892
654	1210	2210	932	1710	3110	1210	2210	4010	1488	2710	4910
660	1220	2228	938	1720	3128	1216	2220	4028	1493	2720	4928
666	1230	2246	943	1730	3146	1221	2230	4046	1499	2730	4946
671	1240	2264	949	1740	3164	1227	2240	4064	1504	2740	4964
677	1250	2282	954	1750	3182	1232	2250	4082	1510	2750	4982
682	1260	2300	960	1760	3200	1238	2260	4100	1516	2760	5000
688	1270	2318	966	1770	3218	1243	2270	4118	1521	2770	5018
693	1280	2336	971	1780	3236	1249	2280	4136	1527	2780	5036
699	1290	2354	977	1790	3254	1254	2290	4154	1532	2790	5054
704	1300	2372	982	1800	3272	1260	2300	4172	1538	2800	5072
710	1310	2390	988	1810	3290	1266	2310	4190	1543	2810	5090
716	1320	2408	993	1820	3308	1271	2320	4208	1549	2820	5108
721	1330	2426	999	1830	3326	1277	2330	4226	1554	2830	5126
727	1340	2444	1004	1840	3344	1282	2340	4244	1560	2840	5144
732	1350	2462	1010	1850	3362	1288	2350	4262	1566	2850	5162
738	1360	2480	1016	1860	3380	1293	2360	4280	1571	2860	5180
743	1370	2498	1021	1870	3398	1299	2370	4298	1577	2870	5198
749	1380	2516	1027	1880	3416	1304	2380	4316	1582	2880	5216
754	1390	2534	1032	1890	3434	1310	2390	4334	1588	2890	5234
760	1400	2552	1038	1900	3452	1316	2400	4352	1593	2900	5252
766	1410	2570	1043	1910	3470	1321	2410	4370	1599	2910	5270
771	1420	2588	1049	1920	3488	1327	2420	4388	1604	2920	5288
777	1430	2606	1054	1930	3506	1332	2430	4406	1610	2930	5306
782	1440	2624	1060	1940	3524	1338	2440	4424	1616	2940	5324
788	1450	2642	1066	1950	3542	1343	2450	4442	1621	2950	5342
793	1460	2660	1071	1960	3560	1349	2460	4460	1627	2960	5360
799	1470	2678	1077	1970	3578	1354	2470	4478	1632	2970	5378
804	1480	2696	1082	1980	3596	1360	2480	4496	1638	2980	5396
810	1490	2714	1088	1990	3614	1366	2490	4514	1643	2990	5414
									1649	3000	5432

Appendix VI

COMPARISON OF IMPACT VALUES OBTAINED WITH THE BRITISH STANDARD IZOD TEST AND VARIOUS ALTERNATIVE METHODS OF TESTING

These graphs have been derived from a large number of results, each show two bands. The inner dotted band represents the area within which 50 per cent of the results may be expected to fall, whilst the wider full band covers approximately 80 per cent of results.

These curves should not be used for comparing or compiling specifications.

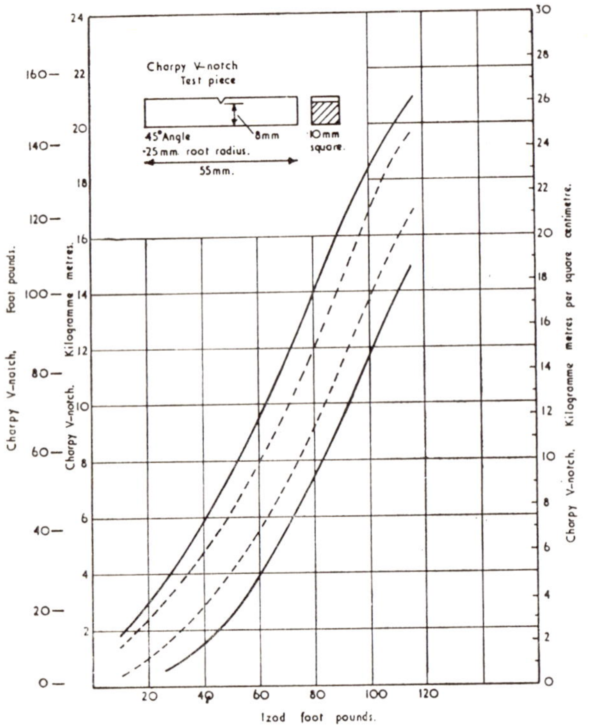

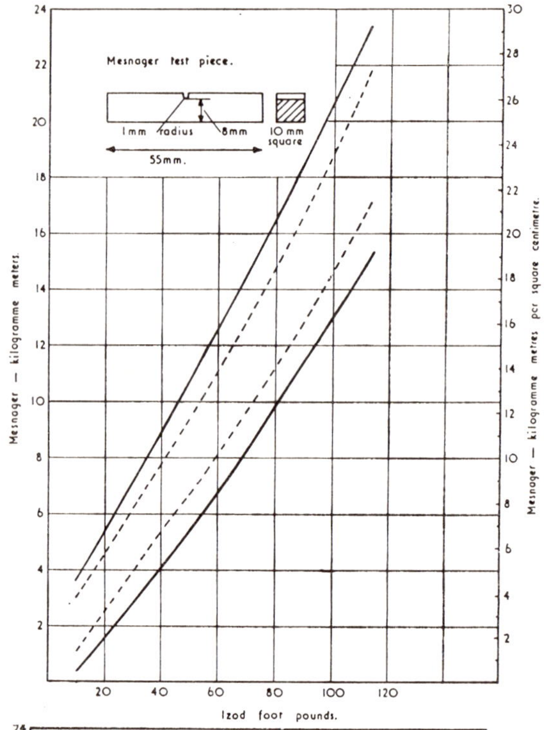

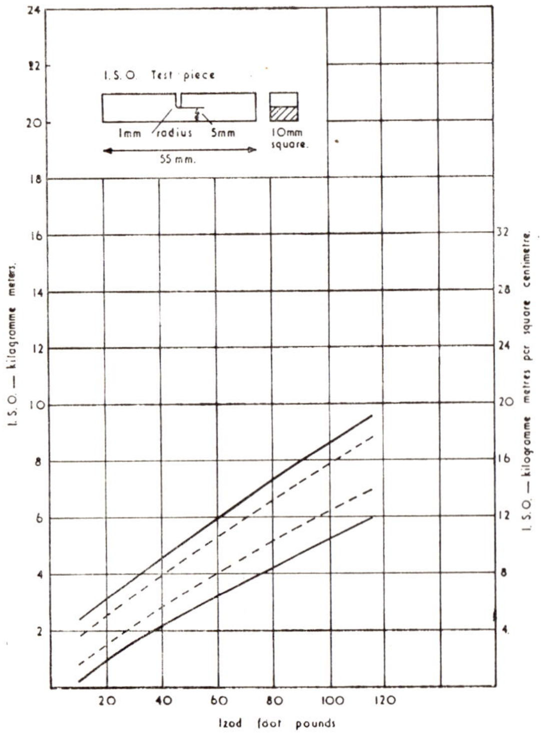

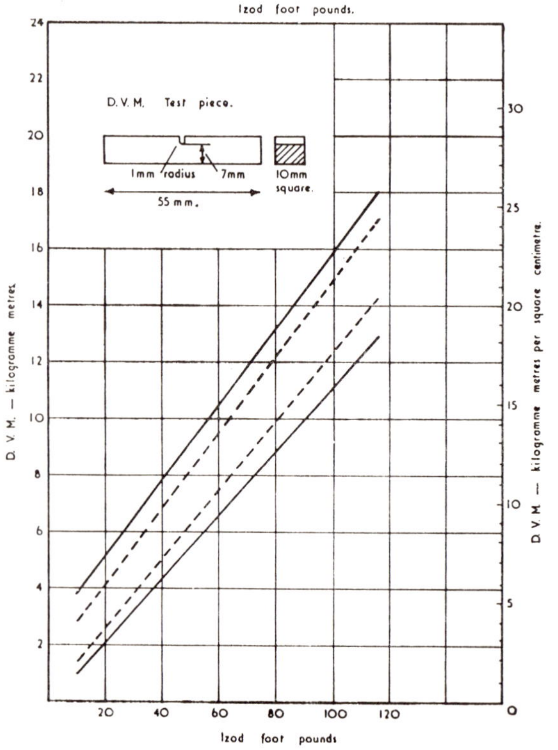

(Reproduced by kind permission of the English Steel Corqoration Ltd.)